THOMAS A. STELLMAN
G. V. KRISHNAN

HARNESSING

AutoCAD® 2000

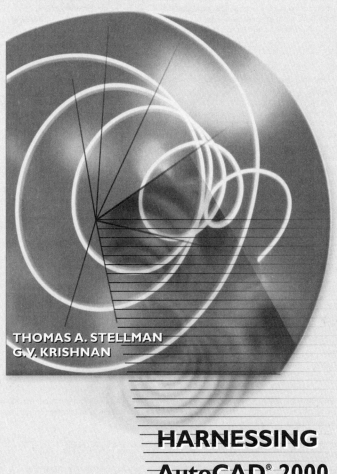

THOMAS A. STELLMAN
G. V. KRISHNAN

HARNESSING
AutoCAD® 2000

Autodesk.
Press

Thomson Learning™

Africa • Australia • Canada • Denmark • Japan • Mexico • New Zealand
Philippines • Puerto Rico • Singapore • Spain • United Kingdom • United States

NOTICE TO THE READER

Publisher does not warrant or guarantee any of the products described herein or perform any independent analysis in connection with any of the product information contained herein. Publisher does not assume, and expressly disclaims, any obligation to obtain and include information other than that provided to it by the manufacturer.

The reader is expressly warned to consider and adopt all safety precautions that might be indicated by the activities herein and to avoid all potential hazards. By following the instructions contained herein, the reader willingly assumes all risks in connection with such instructions.

The publisher makes no representation or warranties of any kind, including but not limited to, the warranties of fitness for particular purpose or merchantability, nor are any such representations implied with respect to the material set forth herein, and the publisher takes no responsibility with respect to such material. The publisher shall not be liable for any special, consequential, or exemplary damages resulting, in whole or part, from the readers' use of, or reliance upon, this material. Autodesk does not guarantee the performance of the software and Autodesk assumes no responsibility or liability for the performance of the software or for errors in this manual.

Trademarks

AutoCAD® and the AutoCAD logo are registered trademarks of Autodesk, Inc. Windows is a trademark of the Microsoft Corporation. All other product names are acknowledged as trademarks of their respective owners.

Autodesk Press Staff
Executive Director: Alar Elken
Executive Editor: Sandy Clark
Developmental Editor: John Fisher
Editorial Assistant: Allyson Powell
Executive Marketing Manager: Maura Theriault
Executive Production Manager: Mary Ellen Black
Production Coordinator: Jennifer Gaines
Art and Design Coordinator: Mary Beth Vought
Marketing Coordinator: Paula Collins
Technology Project Manager: Tom Smith

Cover design by Scott Keidong's Image Enterprises. AutoCAD® image © 1999, Autodesk, Inc.

COPYRIGHT © 2000 Thomson Learning™

Printed in Canada
    5 6 7 8 9 10 XXX 04 03 02 01 00

For more information, contact:
Autodesk Press
3 Columbia Circle, Box 15-015
Albany, New York USA 12212-15015;
or find us on the World Wide Web at http://www.autodeskpress.com

Library of Congress Cataloging-in-Publication Data
[To come]

CONTENTS

INTRODUCTION

HARNESSING THE POWER OF AUTOCAD 2000

AutoCAD 2000 is well named. It is positioning itself for entry into the new millennium. With the streamlining, ease of use, and internet connectivity, AutoCAD should remain the leader in the PC CADD realm.

Harnessing AutoCAD 2000 brings you comprehensive descriptions, explanations and examples of the basics of AutoCAD 2000 and its new innovations. It has been written to be used both in the classroom as a textbook as well as in industry by the professional CADD designer/drafter as a reference and learning tool. The chapter on dimensioning has been completely updated. The new AutoCAD Design Center is thoroughly covered along with in-depth explanations of the new AutoSnap and Tracking features.

Whether you're new to AutoCAD or a seasoned user upgrading your skills, Harnessing AutoCAD will show you how to rein in the power of AutoCAD to improve your professional skills and increase your productivity.

HIGHLIGHTS AND FEATURES OF THIS NEW EDITION

The improvements that AutoCAD2000 brings include the following:

- A new right-click Shortcut menu puts a list of common optional responses to commands within easy reach of your cursor.

- Multiple Design Environment (MDE) with Object Drag & Drop, Property Painter Support, Cut/Copy/Paste, and Concurrent Command Execution.

- AutoCAD Design Center (ADC) with Content Explorer.

- AutoSnap Enhancements with Extension & Parallel osnap modes, Status bar for toggling Polar and Align features on/off, Cursor tracking through incremental angles, Angle overrides similar to osnap overrides, Alignment paths are displayed during cursor movement, Tracking includes displacement points specifying. Developers can implement snap modifiers, and Right-click accessibility to Snap, Grid, Osnap & Alignment features.

- QDIM (Quick Dimensioning)

- PARTIAL OPEN and PARTIAL LOAD

- 3DORBIT
- In-place Reference Editing
- Object Property Manager (OPM), with Edit by type, No selection set, Tabbed Listing, Dynamic Updating, Quick Select Access.
- Quick Select
- Expanded Reach with Direct Browser Access, Object Hyperlinks, DXF Enhancements.
- Layouts (MULTIPLE PAPERSPACES), with Layout Creation, Plot Setup, Named Plot Setups, and WYSIWYG Layouts.
- Non-rectangular Viewports
- Integrated Visual Lisp

HOW TO USE THIS BOOK

OVERVIEW

The first chapter of this text provides an overview of the AutoCAD program, its interface, the commands, special features and warnings, and AutoCAD 2000 enhancements. Specific commands are described in detail throughout the book, along with lessons on how to use them.

FUNDAMENTALS

Harnessing AutoCAD contains five chapters devoted to teaching the fundamentals of AutoCAD 2000. Fundamentals I introduces some of the basic commands and concepts, and Fundamentals II through V continue to build logically on that foundation until the student has a reasonable competency in the most basic functions of AutoCAD.

INTERMEDIATE

After mastering the fundamentals, you move on to the intermediate topics, which include dimensioning, plotting and printing, hatching and boundaries, blocks and attributes, external references, and drawing environments. Other chapters teach students to make the most of AutoCAD using utility commands, scripts and slides, 3D commands, rendering, and the digitizing tablet.

ADVANCED

For the advanced AutoCAD user, this book offers a chapter on customizing AutoCAD 2000 (including toolbar customization) and Visual LISP. These two chapters teach you to make AutoCAD 2000 more individualized and powerful as you tailor them to your special needs.

APPENDICES

There are eight appendices in the back of this book. Appendix A is an introduction to hardware and software requirements of AutoCAD. Appendix B is a quick reference of AutoCAD commands with a brief description of their basic functions, and Appendix C provides a visual reference of AutoCAD toolbars.

Appendix D lists system variables, including default setting, type, whether or not it is read-only, and an explanation of the system variables. To see hatch and fill patterns, fonts, and linetypes provided with the AutoCAD program, refer to Appendices E, F, and G respectively. Appendix H lists available AutoCAD Command Aliases.

STYLE CONVENTIONS

In order to make this text easier for you to use, we have adopted certain conventions that are used throughout the book:

Convention	Example
Command names are in small caps	The MOVE command
Menu names appear with the first letter capitalized	Draw pull down menu
Toolbar menu names appear with the first letter capitalized	Standard toolbar
Command sequences are indented. User inputs are indicated by boldface. Instructions are indicated by italics and are enclosed in parentheses	Command: **move** Enter variable name or [?]: **snapmode** Enter group name: *(enter group name)*

HOW TO INVOKE COMMANDS

Methods of invoking a command are summarized in a table

Standard toolbar	Choose Redo (see Figure 3–75)
Edit menu	Choose Redo
Command:	Redo (ENTER)

EXERCISE ICONS

Step-by-step Project Exercises are identified by the special icon shown in the following table. Exercises that give you practice with types of drawings that are often found in a particular discipline are identified by the icons shown in the following table. Exercises that involve skills applicable to most or all disciplines are designated with the General icon.

Type of Exercise	Icon	Type of Exercise	Icon
Project Exercises		Civil	
General		Electrical	
Mechanical		Piping	
Architectural			

ONLINE COMPANION™

The Online Companion™ is your link to AutoCAD on the Internet. We've compiled supporting resources with links to a variety of sites. Not only can you find out about training and education, industry sites, and the online community, we also point to valuable archives compiled for AutoCAD users from various Web sites. In addition, there are pages specifically for users of Harnessing AutoCAD. These include an owner's page with updates, a swap bank where you can share your drawings with other AutoCAD students, and a page where you can send us your comments. You can find the Online Companion at:

http://www.autodeskpress.com/onlinecompanion.html

When you reach the Online Companion page, click on the title Harnessing AutoCAD.

E-RESOURCE

E- resource is an educational resource that creates a truly electronic Classroom. It is a CD-ROM containing tools and instructional resources that enrich your classroom and make your preparation time shorter. The elements of e-resource link directly to the text and tie together to provide a unified instructional system. Spend your time teaching, not preparing to teach.

Features contained in e-resource include:

Syllabus: Lesson plans created by chapter that list goals and discussion topics. You have the option of using these lesson plans with your own course information.

Chapter Hints: Objectives and teaching hints that provide direction on how to present the material and coordinate the subject matter with student projects.

Answers to Review Questions: These solutions enable you to grade and evaluate end of chapter tests.

PowerPoint® Presentation: These slides provide the basis for a lecture outline that helps you to present concepts and material. Key points and concepts can be graphically highlighted for student retention.

Computerized Test Bank: Over 800 questions of varying levels of difficulty are provided in true/false and multiple-choice formats. Exams can be generated to assess student comprehension or questions can be made available to the student for self-evaluation.

CADD Drawing Files: An assortment of completed drawing exercises, project exercises and examples from the textbook are provided.

World Class Manager: This allows instructors to automatically extract and track data from on-line tests, practice tests and tutorials created by World Class Test. Instructors can also create reports such as individual student reports, missing assignment reports, a summary of grade distribution, class lists and more. All statistics are updated automatically as data is entered or changed.

Animations: These .AVI files graphically depict the execution of key concepts and commands in drafting, design, and AutoCAD and let you bring multi-media presentations into the classroom.

Spend your time teaching, not preparing to teach!
ISBN: 0-7668-1233-2

WE WANT TO HEAR FROM YOU!

Many of the changes to the look and feel of this new edition were made by way of requests from users of our previous editions. We'd like to hear from you as well! If you have any questions or comments, please contact

The CADD Team
c/o Autodesk Press
3 Columbia Circle
P.O. Box 15015
Albany, NY 12212-5015

ABOUT THE AUTHORS

Thomas A. Stellman received a B.A. degree in architecture from Rice University and has over 20 years of experience in the architecture, engineering, and construction industry. He has taught at the college level for over ten years and has been teaching courses in AutoCAD since the introduction of version 1.4 in 1984. He conducts seminars covering both introductory and advanced AutoLISP. In addition, he develops and markets third-party software for AutoCAD. He currently is a CADD consultant, AutoLISP programmer, and project coordinator for Testengeer, Inc., in Port Lavaca, Texas.

G.V. Krishnan is director of the Applied Business and Technology Center, University of Houston—Downtown, a Premier Autodesk Training Center. He has used AutoCAD since the introduction of version 1.4 and writes about AutoCAD from the standpoint of a user, instructor, and general CADD consultant to area industries. Since 1985 he has taught courses ranging from basic to advanced levels of AutoCAD, including customizing, 3D AutoCAD, solid modeling, and AutoLISP programming.

ACKNOWLEDGMENTS

The authors would like to thank Dr. Abu Reza Wazil for contributing portions of the Visual Lisp chapter, Mr. Mike Flores for providing drawing exercises, and Mr. Roy Parisher for chapter review questions.

We would like to thank and acknowledge the many professionals who reviewed the manuscript to help us publish this Harnessing AutoCAD 2000. The technical edit and development of exam questions was performed by Max Johnston and Argie Nichols, instructors at Westark College. They were assisted by graduates of the Westark College CADD Program: Cynthia Cone, Marty Mason, and Vilayvanh Sayakoumane.

A special acknowledgment is due the following instructors, who reviewed the chapters in detail:

Tom Bledsaw, ITT Educational Services, Indianapolis, IN

Tony Cook, Gwinnett Tech, Lawrenceville, GA

Allen Frankley, Cerritos College, Norwalk, CA

Neil Jacobson, St. Paul Technical College, St. Paul, MN

Matt Mathan, Lincoln Technical Institute, Elizabeth, NJ

Gregory Nelson, University of Advancing Computer Technology, Tempe, AZ

Kevin Standiford, ITT Technical Institute, Little Rock, AR

Jesse Vaughn, Alamance Community College, Graham, NC

The authors would like to acknowledge and thank the following staff members of Delmar Publishers:

Publisher: Alar Elken
Acquisitions Editor: Sandy Clark
Developmental Editor: John Fisher
Production Coordinator: Jennifer Gaines
Art & Design Coordinator: Mary Beth Vought
Editorial Assistant: Allyson Powell

The authors also would like to acknowledge and thank the following people:

Copyediting: Rachel Pearce Anderson, Archer Editorial Services
Composition: John Shanley, Phoenix Creative Graphics

CHAPTER 1

Getting Started

INTRODUCTION

The key phrase for AutoCAD 2000 is Heads-Up. Heads-Up design means that your design/drafting work can proceed smoothly with a minimum of looking away from the screen. And new features like AutoCAD DesignCenter (ADC), Multiple Design Environment (MDE), AutoSnap Enhancements, QDIM (Quick Dimensioning), and PARTIAL OPEN and PARTIAL LOAD are just a few of the improvements to an already powerful and easy-to-use CADD program.

The improvements that AutoCAD 2000 brings include the following:

- A new right-click Shortcut menu to put a list of common optional responses to commands within easy reach of your cursor.
- Multiple Design Environment (MDE) with Object Drag & Drop, Property Painter Support, Cut/Copy/Paste, and Concurrent Command Execution.
- AutoCAD Design Center (ADC) with Content Explorer.
- AutoSnap Enhancements with Extension & Parallel osnap modes, status bar for toggling Polar and Align features on/off, cursor tracking through incremental angles, angle overrides similar to osnap overrides, alignment paths are displayed during cursor movement, tracking includes displacement points specifying, developers can implement snap modifiers, and right-click accessibility to Snap, Grid, Osnap & Alignment features.
- Quick Dimensioning.
- Partial Open and Partial Load.
- 3D Orbit.
- In-place Reference Editing.
- Object Property Manager (OPM), with Edit by type, Tabbed Listing, Dynamic Updating, Quick Select Access.
- Quick Select.
- Database Connectivity – dbConnect.
- Expanded reach with Direct Browser Access, Object Hyperlinks, and DXF Enhancements.

- Layouts (Multiple Paperspaces), with Layout Creation, Page Setup, Named Plot Setups, and WYSIWYG Layouts.
- Non-rectangular floating Viewports.
- Integrated Visual Lisp.

AUTOCAD COMMANDS

As much as possible, AutoCAD divides commands into related categories. For example, Draw is not a command, but a category of commands used for creating primary objects such as lines, circles, arcs, text (lettering), and other useful objects that are visible on the screen. Categories include Modify, View, and another group listed under the Options section of the Tools menu for controlling the electronic drawing environment. The commands under Format are also referred to as drawing aids and utility commands throughout the book. Learning the program can progress at a better pace if the concepts and commands are mentally grouped into their proper categories. This not only helps you find them when you need them, but also helps you grasp the fundamentals of computer-aided drafting more quickly.

STARTING AUTOCAD

Design/drafting is what AutoCAD (and this book) is all about. So how do you get into AutoCAD? Choose the Start button (Windows 95/98 and Windows NT 4.0 operating system), select the AutoCAD 2000 program group, and then select the AutoCAD 2000 program. AutoCAD displays the Startup dialog box, as shown in Figure 1–1.

Figure 1–1 *Startup dialog box*

Whenever you begin a new drawing, whether by means of one of the two available wizards or one of the available templates or by starting from scratch, AutoCAD creates a new drawing called *DRAWING.DWG*. You can begin working immediately and save the drawing to a file name later, using the SAVE or SAVEAS command.

STARTING A NEW DRAWING WITH THE WIZARDS

If from the Startup dialog box you select the **Use a Wizard** option, as shown in Figure 1–1, AutoCAD leads you through the basic steps of setting up a drawing, using either the Quick Setup or the Advanced Setup. The initial drawing settings correspond to those in either the template ACAD.DWT (English units) or the template ACADISO.DWT (metric units), based on the current setting of the MEASUREINIT system variable in the registry. When MEASUREINIT is set to 0, the drawing settings are based on the template ACAD.DWT; when it is set to 1, the drawing settings are based on the template ACADISO.DWT. Depending on which wizard you select, you can then set the values of such variables as limits, units, and angle direction.

Quick Setup

The Quick Setup dialog box has two pages: the Units page (see Figure 1–2) and the Area page.

Figure 1–2 *Quick Setup dialog box: Units page*

The **Units** page (Figure 1–2) applies to how the linear units of the drawing are entered. They also determine how the linear units are reported in the status bar. When you select one of the radio buttons, an example of how linear units will be written is displayed to the right of the radio buttons. AutoCAD allows you to choose from several formats for the display and entry of the coordinates and distances. For example, you can choose feet and fractional inches for architectural drafting. Other options include scientific notation and engineering formats. Selection of Decimal units allows you to draw in inches, feet, millimeters, or whatever units you require. This enables you to draw with real-world values and eliminates the possibility of scaling errors. Once the drawing is complete, you can plot it at whatever scale you like. As mentioned earlier, drawing to real-world size is an advantage of AutoCAD that some overlook. You can plot a drawing at several different scales, thereby eliminating the need for separate drawings at different scales.

After specifying the units, choose the **Next >** button. AutoCAD displays the Area page.

The **Area** page (see Figure 1–3) allows you to set your drawing area's width (left-to-right dimension) and length (bottom-to-top dimension), also known as the *limits*. You may set the limits to accommodate your drawing. For example, if you are drawing a printed circuit board that is 8 inches wide by 6 inches long, you can choose a decimal drawing unit and set the width to 8 and length to 6. If the drawing exceeds your original plans or the drawing limits become too restrictive, you can change the drawing limits. A detailed description of how to set up limits appears later in the chapter.

Figure 1–3 *Quick Setup dialog box: Area page*

Choose the **Finish** button to close the Quick Setup dialog box. AutoCAD automatically adjusts the scale factors for dimension settings and text height. The adjusted settings are based on the full-scale size of the objects you draw. The dimension variables that are adjusted include DIMASZ, DIMCEN, DIMDLI, DIMEXE, DIMEXO, DIMGAP, and DIMTXT. Refer to Chapter 7, "Dimensioning," for a detailed discussion of dimension variable settings. In addition, AutoCAD adjusts the linetype scale, and hatch pattern scale.

Advanced Setup

The Advanced Setup dialog box has five pages: Units, Angle, Angle Measure, Angle Direction, and Area.

> The **Units** page and **Area** page settings are the same as for the Quick Setup wizard. In the **Units** page, the **Precision:** text box lets you set the number of decimal places or the fractional precision of the denominator to which coordinates and linear distances are reported.
>
> In the **Angle** page (see Figure 1–4) you can set the type of units in which angular input and reporting is given. When you select one of the radio buttons, an example of how the angular units will be written is displayed to the right of the radio buttons. You can select the format used for the display and entry of angles. Degrees in decimal form are a common choice. However, you might also select gradient, radians, degrees/minutes/seconds, or surveyor's units. The **Precision:** text box lets you set the number of decimal places or the degrees, minutes, seconds, or decimal precision of seconds to which angles are reported. After specifying the angle, choose the **Next >** button. AutoCAD displays the **Angle Measure** page.

Figure 1–4 *Advanced Setup dialog box: Angle page*

The **Angle Measure** page (see Figure 1–5) lets you set zero degrees for your drawing relative to the universally accepted map compass, where North is up on the drawing and East is 90 degrees clockwise from North. You can select from the **East, North, West,** or **South** radio buttons, or the **Other** button in which you can enter the direction in the text box. After specifying the angle measure, choose the **Next >** button. AutoCAD displays the **Angle Direction** page.

Figure 1–5 *Advanced Setup dialog box: Angle Measure page*

The **Angle Direction** page (see Figure 1–6) lets you set the direction (clockwise or counterclockwise) in which angle values increase. After specifying the angle direction, choose the **Next >** button. AutoCAD displays the Area page.

Choose the **Finish** button after specifying the area to close the Advanced Setup dialog box. As with the Quick Setup wizard, AutoCAD automatically adjusts the scale factors for dimension settings and text height. The adjusted settings are based on the full-scale size of the objects you draw.

 Note: The Wizards let you set up the drawing parameters that most commonly vary from one drawing to the next. It is convenient to have those variables accessible in a single place, and lessens the chance of forgetting one. You should note, however, that the Wizards' Area page does not permit you to use a point other than 0,0 as a lower left corner of the limits.

Any settings you changed in the wizard can be changed again later by invoking the UNITS and LIMITS commands. Detailed discussions of the UNITS and LIMITS commands are provided later in the chapter.

Figure 1–6 *Advanced Setup dialog box: Angle Direction page*

STARTING A NEW DRAWING WITH A TEMPLATE

If from the Startup dialog box you select the **Use a Template** option, as shown in Figure 1–7, AutoCAD lists the available templates. An AutoCAD template is a drawing file with a file extension of .DWT instead of .DWG.

Figure 1–7 *Startup dialog box: Use a Template option*

The preview graphics box on the right of the Use a Template dialog box lets you see a small view of the template that is highlighted in the **Select a Template:** list box. If a description has been saved with the selected template, it will be displayed in the **Template Description** section. To begin your drawing in the identical setup as the selected template, either choose **OK** or double-click the highlighted name of the template in the **Select a Template:** list box. You can also choose the **Browse** button to invoke the Select A Template File dialog box to select a template file from a different folder.

When you use a template to create a new drawing, AutoCAD copies all the information from the template drawing to the new drawing. The most common use of drawing templates is to enable you to start with a border and title block already drawn, layers and styles already created, and the system variables set to values that suit the drawing you intend to make. For example, a template for a house floor plan might have the limits set for drawing the floor plan to draw at a full scale and to plot at a scale of 1/4" = 1'-0" on a 24" x 18" sheet, the units set to architectural, and the border and title block drawn to full scale. Dimension variables might be saved in the desired style, and separate layers might be set for drawing walls, doors, windows, cabinetry, plumbing, dimensions, text, and any other object you wish to have created on its own layer. You can even use an existing .DWG drawing file as a template to create a new drawing.

STARTING A NEW DRAWING FROM SCRATCH

From the Startup dialog box, if you select the **Start from Scratch** option, as shown in Figure 1–8, AutoCAD provides you two suboptions: English and Metric.

Figure 1–8 *Startup dialog box: Start from Scratch option*

The **English** suboption sets the units to feet and inches settings; the **Metric** suboption sets the units to metric settings. By default the limits are set to 0,0 for the lower left corner and 12,9 for the upper right corner. If necessary, you can change the settings at the beginning of the drawing session or at any time during the drawing session.

OPENING AN EXISTING DRAWING FROM THE STARTUP DIALOG BOX

If from the Startup dialog box you select the **Open a Drawing** option, as shown in Figure 1–9, AutoCAD lists the drawing files available from the current folder. If you need to open a drawing from a different folder, then choose **Browse** button. AutoCAD displays a Select File dialog box. Select the appropriate folder and drawing to open.

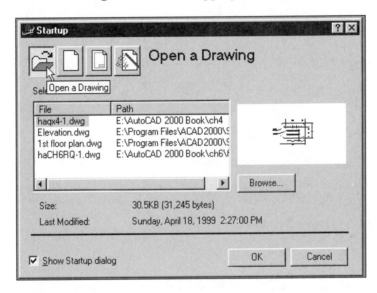

Figure 1–9 *Startup dialog box: Open a Drawing option*

After making any necessary changes, choose the **OK** button. AutoCAD closes the Startup dialog box and displays the AutoCAD screen.

 Note: The Open a Drawing option is available only in the Startup dialog box. If you need to open an existing drawing after starting a drawing, then you have to invoke the OPEN command to open an existing drawing.

AutoCAD SCREEN

The AutoCAD screen (see Figure 1–10) consists of the following elements: the graphics window, status bar, title bar, toolbars, menu bar, model tab/layout tabs, and the command window.

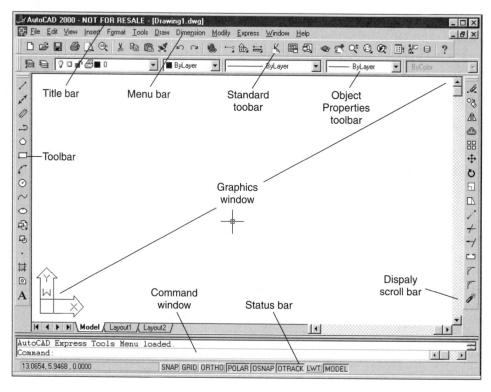

Figure 1–10 *The AutoCAD screen*

GRAPHICS WINDOW

The graphics window is where you can view the objects you create. In this window, AutoCAD displays the cursor, indicating your current working point. As you move your pointing device (usually a mouse or puck) around on a digitizing tablet, mouse pad, or other suitable surface, the cursor mimics your movements on the screen. When AutoCAD prompts you to select a point, the cursor is in the form of crosshairs. It changes to a small pick box when you are required to select an object on the screen. AutoCAD uses combinations of crosshairs, boxes, dashed rectangles, and arrows in various situations so you can quickly see what type of selection or pick mode to use.

 Note: It is possible to enter coordinates outside of the viewing area for AutoCAD to use for creating objects. As you become more adept at AutoCAD, you may find a need to do this. Until then, working within the viewing area is recommended.

STATUS BAR

The status bar at the bottom of the screen displays the cursor's coordinates and important information on the status of various modes.

TITLE BAR

The title bar displays the current drawing name for the AutoCAD application window.

TOOLBARS

The toolbars contain tools, represented by icons, from which you can invoke commands. Click a toolbar button to invoke a command, and then select options from a dialog box or respond to the prompts on the command line. If you position your pointer over a toolbar button and wait a moment, the name of the tool is displayed, as shown in Figure 1–11. This is called the ToolTip. In addition to the ToolTip, AutoCAD displays a very brief explanation of the function of the command on the status bar.

Figure 1–11 *Toolbar with a ToolTip displayed*

Some of the toolbar buttons have a small triangular symbol in the lower right corner of the button indicating that there are *flyout* buttons underneath that contain subcommands. Figure 1–12 shows the ZOOM command flyout located in the Standard Toolbar. When you pick a flyout option, it remains on top to become the default option.

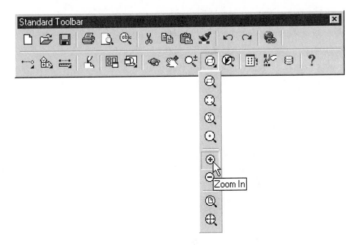

Figure 1–12 *Display of the Zoom flyout located in the Standard Toolbar*

You can display multiple toolbars on screen at once, change their contents, resize them, and dock or float them. A *docked* toolbar attaches to any edge of the graphics window. A *floating* toolbar can lie anywhere on the screen and can be resized.

Docking and Undocking a Toolbar

To dock a toolbar, position the cursor on the caption, and press the pick button on the pointing device. Drag the toolbar to a dock location to the top, bottom, or either side of the graphics window. When the outline of the toolbar appears in the docking area, release the pick button. To undock a toolbar, position the cursor anywhere on the border of the toolbar and drag-and-drop it outside the docking regions. To place a toolbar in a docking region without docking it, hold down CTRL as you drag. By default, the Standard toolbar and the Object Properties toolbar are docked at the top of the graphics window (see Figure 1–10). Figure 1–13 shows the Standard toolbar and the Object Properties toolbar docked at the top of the graphics window, the Draw toolbar docked on the left side of the graphics window, and the Modify toolbar docked on the right side of the graphics window.

Figure 1–13 *Docking of toolbars in the graphics window*

Resizing a Floating Toolbar

If necessary, you can resize a floating toolbar. To resize a floating toolbar, position the cursor anywhere on the border of the toolbar, and drag it in the direction you want to resize. Figure 1–14 shows different combinations of resizing the Draw toolbar.

Figure 1–14 *Draw toolbar in different resizing positions*

Closing a Toolbar

To close a toolbar, position the cursor on the X located in the upper right corner of the toolbar, as shown in Figure 1–15, and press the pick button on your pointing device. The toolbar will disappear from the graphics window.

Figure 1–15 *Positioning the cursor to close a toolbar*

Opening a Toolbar

AutoCAD 2000 comes with 24 toolbars. To open any of the available toolbars, invoke the TOOLBAR command:

View menu	Choose Toolbars
Command: prompt	**toolbar** (ENTER)

AutoCAD displays the Toolbars dialog box, as shown in Figure 1–16.

Figure 1–16 *Toolbars dialog box*

To open a toolbar, first select the checkbox of the toolbar from the **Toolbars:** list box. AutoCAD displays the selected toolbar. The **Show ToolTips** checkbox controls whether or not to display the ToolTips. The **Large Buttons** checkbox controls the size (large vs. regular) at which buttons are displayed. The default display size is 16 x 15 pixels. Setting the Large Buttons checkbox to ON displays the icons at 24 x 22 pixels.

Figure 1–17 shows the command icons available in the Standard Toolbar; Figure 1–18 shows the commands available in the Object Properties toolbar. Appendix C lists all the toolbars available in AutoCAD.

Figure 1–17 *Standard Toolbar*

Figure 1–18 *Object Properties toolbar*

You can also open toolbars from the shortcut menu. Place your cursor anywhere on a toolbar that is displayed in the graphic window, click your right-button on your pointing device, and a shortcut menu appears listing all the available toolbars (see Figure 1–19). Select the toolbar you want to open. You can also close a toolbar by removing the check mark next to the name of the toolbar.

Figure 1–19 *Shortcut menu with listing of toolbars*

MENU BARS

The menus are available from the menu bar at the top of the screen. To select any of the available commands, move the crosshairs cursor into the menu bar area and press the pick button on your pointing device, which pops that menu bar onto the screen (see Figure 1–20). Selecting from the list is a simple matter of moving the cursor until the desired item is highlighted and then pressing the designated pick button on the pointing device. If a menu item has an arrow to the right, it has a cascading submenu. To display the submenu, move the pointer over the item and press the pick button. Menu items that include ellipses (...) display dialog boxes. To select one of these, just pick that menu item.

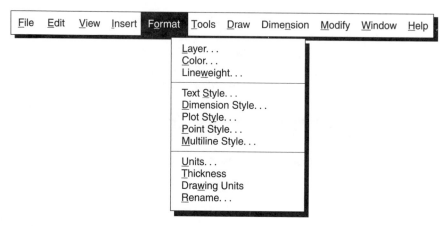

Figure 1–20 *Example of a menu bar*

MODEL TAB/LAYOUT TABS

AutoCAD allows you to switch your drawing from model (drawing) space and paper (layout) space. You generally create your designs in model space, and then create layouts to plot your drawing in paper space. Refer to Chapter 8 for a detailed explanation on plotting from layouts.

COMMAND WINDOW

The command window is a window in which you enter commands and in which AutoCAD displays prompts and messages. The command window can be a floating window with a caption and frame. You can move the floating command window anywhere on the screen and resize its width and height with the pointing size.

There are two components to the command window: the single command line where AutoCAD prompts for input and you see your input echoed back, as shown in Figure 1–21, and the command history area, which shows what has transpired in the current drawing session. One display of the single Command Line remains at the bottom of the screen.

Figure 1–21 *Command Line window*

The command history area can be enlarged like other windows by picking the top edge and dragging it to a new size. You can also scroll inside the enlarged area to see previous command activity by using the scroll bars (see Figure 1–22).

```
AutoCAD · Command Line · Drawing1.dwg
Command: L
LINE Specify first point: 3,4
Specify next point or [Undo]: 3,5
Specify next point or [Undo]: 6,0
Specify next point or [Close/Undo]: @3,4
Specify next point or [Close/Undo]: C
Command: C
CIRCLE Specify center point for circle or [3P/2P/Ttr (tan tan radius)
Specify radius of circle or [Diameter] <0.7293>: 1
Command: A
ARC Specify start point of arc or [CEnter]:
Specify second point of arc or [CEnter/ENd]:
Specify end point of arc:
Command: RECTANGLE
Specify first corner point or [Chamfer/Elevation/Fillet/Thickness/Wic
Point or option keyword required.
Specify first corner point or [Chamfer/Elevation/Fillet/Thickness/Wic
Specify other corner point:
Command:
```

Figure 1–22 *Command history*

By default, F2 causes the command history text box to switch between being displayed and being hidden. When the text box is displayed, you can scroll through the command history.

When you see "Command:" displayed in the Command Window, it signals that AutoCAD is ready to accept a command. After you enter a command name and press ENTER select a command from one of the menus or toolbars, the prompt area continues to inform you of the type of response(s) that you must furnish, until the command is either completed or terminated. For example, when you pick the LINE command, the prompt displays "Specify first point:"; after selecting a starting point by appropriate means, you will see "Specify next point or [Undo]:" asking for the endpoint of the line.

Each command has its own series of prompts. The prompts that appear when a particular command is used in one situation may differ from the prompts or sequence of prompts when invoked in another situation. You will become familiar with the prompts as you learn to use each command.

When you enter the command name or give any other response by typing from the keyboard, make sure to press ENTER or SPACEBAR. Pressing ENTER sends the input to the program for processing. For example, after you enter LINE, you must press ENTER or SPACEBAR in order for AutoCAD to start the part of the program that lets you draw lines. If you enter **LIN** and press ENTER or SPACEBAR, you will get an error message, unless someone has customized the program and created a command alias or command named "LIN." Likewise, entering **LINEZ** and pressing ENTER or SPACEBAR is not a standard AutoCAD command.

Pressing SPACEBAR has the same function as ENTER except when you are entering in strings of words, letters, or numbers in response to the TEXT and MTEXT commands.

To repeat the previous command, you can press ENTER or SPACEBAR at the "Command:" prompt. A few commands skip some of their normal prompts and assume default settings when repeated in this manner.

Terminating a Command

There are three ways to terminate a command:

- Complete the command sequence and return to the "Command:" prompt.
- Press ESC to terminate the command before it is completed.
- Invoke another command from one of the menus, which automatically cancels any command in progress.

INPUT METHODS

There are several ways to input an AutoCAD command: from the keyboard, toolbars, menu bars, the side screen menu, dialog boxes, the shortcut menu, or the digitizing tablet.

KEYBOARD

To invoke a command from the keyboard, simply enter the command name at the "Command:" prompt and then press ENTER or SPACEBAR (ENTER and SPACEBAR are interchangeable except when entering a space in a text string).

If at the "Command:" prompt you want to repeat a command you have just used, press ENTER, SPACEBAR, or right-click on your pointing device. Right-clicking causes a shortcut menu to appear on the screen from which you can choose the Repeat *<last command>* option. You can also repeat a command by using the up-arrow and down-arrow keys to display the commands you previously entered from the keyboard. Use the up-arrow key to display the previous line in the command history; use the down-arrow key to display the next line in the command history. Depending on the buffer size, AutoCAD stores all the information you entered from the keyboard in the current session.

AutoCAD also allows you to use certain commands transparently, which means they can be entered on the command line while you are using another command. Transparent commands frequently are commands that change drawing settings or drawing tools, such as GRID, SNAP, and ZOOM. To invoke a command transparently, enter an apostrophe (') before the command name while you are using another command. After you complete the transparent command, the original command resumes.

TOOLBARS

The toolbars contain tools that represent commands. Click a toolbar button to invoke the command, then select options from a dialog box or follow the prompts on the command line.

MENU BAR

The menus are available from the menu bar at the top of the screen. You can invoke almost all of the available commands from the menu bar. You can choose menu options in one of the following ways:

- First select the menu name to display a list of available commands, and then select the appropriate command.

- Press ALT and then enter the underlined letter in the menu name. For example, to invoke the LINE command, first hold down ALT while pressing **D** (that is, press ALT + D) to open the Draw menu, and then press **L**.

The default menu file is ACAD.MNU. You can load a different menu file by invoking the MENU command.

SIDE SCREEN MENU

The side screen menu provides another, traditional way to enter AutoCAD commands. By default, the side screen menu is turned off in AutoCAD 2000. While this book does not refer to the side screen menu, traditional DOS users of AutoCAD may be more comfortable using it. To display the side screen menu, enter **options** at the "Command:" prompt and press ENTER or SPACEBAR. AutoCAD displays the Options dialog box, as shown in Figure 1–23.

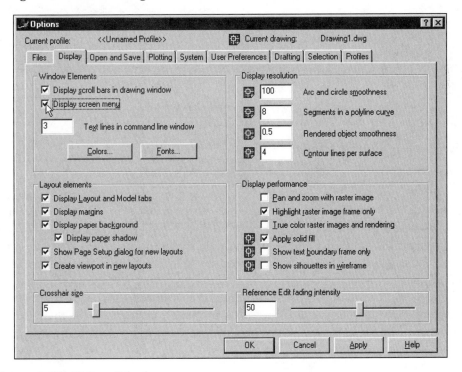

Figure 1–23 *Options dialog box*

Select the Display tab, and set the checkbox for **Display Screen Menu** to ON located in the Windows Elements section of the dialog box. Choose the **OK** button to close the dialog box and save the settings. AutoCAD displays the side screen menu, as shown in Figure 1–24.

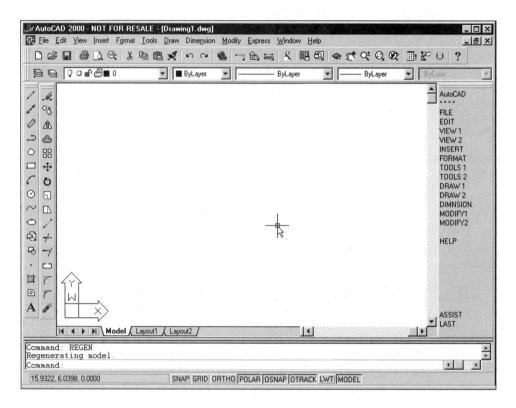

Figure 1–24 *AutoCAD screen window with the side screen menu*

Moving the pointing device to the right will cause the cursor to move into the screen menu area. Moving the cursor up and down in the menu area will cause selectable items to be highlighted. When the desired item is highlighted, you can choose that item by pressing the designated pick button on the pointing device. If the item is a command, either it will be put into action or the menu area will be changed to a list of actions that are options of that command. The screen menu is made up of menus and submenus. At the top of every screen menu is the word *AutoCAD*. When selected, it will return you to what is called the *root menu*. The root menu is the menu that is displayed when you first enter AutoCAD. It lists the primary classifications of commands or functions available.

DIALOG BOXES
(THE RULE IN AUTOCAD 2000 INSTEAD OF THE EXCEPTION)

Many commands, when invoked, cause a dialog box to appear. This is a change from AutoCAD Release 14, where in order to invoke the dialog box you had to prefix the command with the letters DD, such as in DDINSERT. Now the standard procedure for this and many other commands is to display a dialog box unless you prefix the command with a hyphen. In AutoCAD 2000, entering **insert** causes the dialog box to be displayed as shown in Figure 1–25, and entering **-insert** causes responses to be displayed in the Command: prompt area. Dialog boxes display the lists and descriptions of options, long rectangles for receiving your input data, and, in general, are the more convenient and user-friendly method of communicating with the AutoCAD program for that particular command.

AutoCAD 2000 has added more commands and responses that can be input through dialog boxes and refined the existing dialog boxes. The commands that are listed in the menu bar that include ellipses (...), such as PLOT... and HATCH..., display dialog boxes. For a detailed discussion of different dialog box components, see the section on "Using Dialog Boxes."

Figure 1–25 *Dialog box invoked from the* INSERT *command*

CURSOR MENU

The AutoCAD cursor menu (see Figure 1–26) appears at the location of the cursor by pressing the middle button on a three-or-more-button mouse. On a two-button mouse you can invoke this feature by pressing SHIFT + ENTER. On a two-button mouse, the ENTER button is usually the right button. The menu includes the handy Object Snap mode options along with the X,Y,Z/filters. The reason that the Object Snap modes and Tracking are in such ready access will become evident when you learn the significance of these functions.

Figure 1–26 *Cursor menu*

SHORTCUT MENU

The AutoCAD Shortcut menu is a new innovation for AutoCAD 2000. It appears at the location of the cursor when you press the right button (right-click) of the pointing device. The contents of the Shortcut menu depend on the situation at hand.

If there are no commands in effect, and right-click in the drawing window then the Shortcut menu will include: selection to repeat the last command, a section for editing objects such as cut and copy, a section with Undo, Pan, and Zoom, and a section with Quick Select, Find, and Options, as shown in Figure 1–27. Selections that cannot be invoked under the current situation will appear in lighter text than those that can be invoked.

Figure 1–27 *Shortcut menu when no command is in effect*

If you select one or more objects (system variable PICKFIRST and/or GRIPS set to ON), and no commands are in effect, then right-click and the Shortcut menu will include some of the editing commands, as shown in Figure 1–28.

Figure 1–28 *Shortcut menu with one or more objects selected when no command is in effect*

Whenever you have entered a command and do not wish to proceed with the default option, you can invoke the Shortcut menu and select the desired option with the mouse. For example, if instead of the default center-radius method of drawing a circle, you wished to use the TTR (tangent-tangent-radius), 2P (two-point), or 3P (three-point) option, you can select one of them from the Shortcut menu, as shown in Figure 1–29. You can also select the PAN and ZOOM commands (transparently) or cancel the command. If pressing ENTER is required, that is also available.

Figure 1–29 *Shortcut menu when the CIRCLE command is in effect*

Right-click anywhere in the Command window and the shortcut menu provides an access to the six most recently issued command options (see Figure 1–30) that are useful when working on the command line.

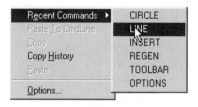

Figure 1–30 *Shortcut menu in the Command window*

Right-click on any of the buttons in the status bar and the shortcut menu provides toggle options for drawing tools and modify their settings.

Right-click on the Model tab or Layout tabs in the lower left corner of the drawing area and the shortcut menu provides display plotting, page setup, and various layout options.

Right-click on any of the open AutoCAD dialog boxes and windows and the shortcut menu provides context-specific options. Figure 1–31 shows an example for the Layer Properties Manager dialog box with a shortcut menu.

Figure 1–31 *Shortcut menu Layer Properties Manager dialog box.*

DIGITIZING TABLET

The most common input device after the mouse is the digitizing tablet. It combines the screen cursor control of a mouse with its own printed menu areas for selecting items. However, with the new Heads-up features and customizability in AutoCAD 2000, the tablet overlay with command entries is becoming obsolete. One powerful feature of the tablet (not related to entering commands) is that it allows you to lay a map or other picture on the tablet and trace over it with the puck (the specific point- ing device for a digitizing tablet), thereby transferring the objects to the AutoCAD drawing. The new interfaces with other platforms that allow you to insert a picture of an aerial photograph, for example, make tablet digitizing less of a demand.

USING DIALOG BOXES

When a dialog box appears, the crosshairs cursor changes to an arrow, pointing up and to the left. You can use the arrow keys on your keyboard to make selections in the dialog box, but it is much easier to use your pointing device. Another way to make selections in a dialog box is to use keyboard equivalents. You can move the cursor from one field to another by pressing TAB when the cursor is not in the text box.

TEXT BOX

An text box is an area that accepts one line of text entry. It is normally used to specify a name, such as a layer name or even a file name, including the drive and/or directory path. Text boxes often function as an alternative to selecting from a list of names when the desired name is not in the list box. Once the correct text is keyed in, enter it by pressing ENTER.

Moving the pointer into the text box causes the text cursor to appear in a manner similar to the cursor in a word processor. The text cursor, in combination with special editing keys, can be used to facilitate changes to the text. You can see both the text cursor and the pointer at the same time, making it possible to click the pointer at a character in the text box and relocate the text cursor to that character.

Right and Left Arrows < > keys
These move the cursor right or left (respectively) across text without having any effect on the text.

Backspace key
This deletes the character to the left of the cursor and moves the cursor to the space previously occupied by the deleted character.

Delete key
This deletes the character at the location of the cursor, causing any text to the right to move one space to the left.

BUTTONS

Actions are immediately initiated when you select one of the dialog buttons.

Default Buttons

If a heavy line surrounds a button (like the **OK** button in most cases), then it is the default button and pressing ENTER is the same as clicking that button.

Buttons with Ellipses (...)

Buttons with ellipses display a second dialog box, called a subdialog or child dialog box.

 Note: When a subdialog box appears, you must respond to the options in the subdialog box before the underlying dialog box can continue.

Screen Action Buttons

Buttons that are followed by an arrow (<) require a graphic response, such as selecting an object on the screen or picking/specifying coordinates.

Disabled Buttons

Buttons with actions that are not currently acceptable will be disabled. They appear grayed out.

Character Equivalents

A button with a label that has an underlined character in it can be activated by holding down ALT + pressing the underlined character.

Radio Buttons

Radio buttons are options when only one of two or more selections can be active at a time (see Figure 1–32). Selecting one radio button will deactivate any other in the group, like selecting a station button on the radio.

Figure 1–32 *Radio buttons*

CHECK BOXES

A check box acts as a toggle. When selected, the check box switches the named setting between ON and OFF, as shown in Figure 1–33. A check or X in the check box means the option is set to ON; no check (an empty box) means the option is set to OFF.

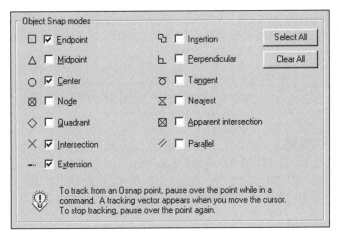

Figure 1–33 *Check boxes*

LIST BOXES AND SCROLL BARS

List boxes make it easy to view, select, and enter a name from a list of existing items, such as file names and fonts. Move the pointer to highlight the desired selection. When you click on the item, it appears in the text box. You accept this item by selecting **OK** or by double-clicking on the item. For example, Figure 1–34 shows the list box from the Standard File dialog box.

Figure 1–34 *List box from the Standard File dialog box*

List boxes are accompanied by scroll bars to facilitate moving long lists left to right and right to left in the list box.

Some boxes that have multiple options available are really just unexpanded list boxes. They will have a down arrow on the right side. Selecting the down arrow displays the expanded list.

 Note: Many dialog boxes have a Help button. If you are not sure how to use the feature in the dialog box, choose the Help button for a brief explanation of the dialog box.

GETTING HELP

When you are in the graphics window, AutoCAD provides a context-sensitive help facility to list its commands and what they do. The HELP command provides online assistance within AutoCAD. When an invalid command is entered, AutoCAD prints a message to remind you of the availability of the help facility.

Whenever you need AutoCAD help, invoke the HELP command:

Standard Toolbar	Choose the HELP command (see Figure 1–35)
Help menu	Choose AutoCAD Help Topics
Command: prompt	? or HELP (ENTER)

Figure 1–35 *Invoking the HELP command from the Standard Toolbar*

AutoCAD displays the **Help Topics: AutoCAD Help** dialog box, as shown in Figure 1–36.

The HELP command can be invoked while you are in the middle of another command. This is referred to as a *transparent command*. To invoke a command transparently (if it is one of those that can be used that way), simply prefix the command name with an apostrophe. For example, to use HELP transparently, enter **'help** or **'?** in response to any prompt that is not asking for a text string. AutoCAD displays help for the current command. Often the help is general in nature, but sometimes it is specific to the command's current prompt.

As an alternative, press the function key F1 to invoke the HELP command. When you ask for help in the middle of a command, AutoCAD displays context-sensitive help.

For example, if you press F1 in the middle of the LINE command, AutoCAD automatically selects the help information that describes how to use the LINE command.

Figure 1–36 *Help Topics: AutoCAD Help dialog box*

Help flips to an independent window, so you will need to switch to the AutoCAD 2000 program to continue.

The **Contents** tab of the **Help Topics: AutoCAD Help** dialog box (see Figure 1–36) has a list of subjects prefixed by a closed-book icon. You may scroll through the list, pressing ENTER when the item on which you need help is highlighted, or you may double-click that item in the text box. This causes the subject's icon to become an open book, and below the selected subject, a sublist will be displayed, indented, containing items that may be prefixed by a closed book or a ? button, as shown in Figure 1–37. Items with closed-book icons lead to sublists, and ? buttons display a window containing the help available on the highlighted item.

Figure 1–37 *Contents tab of the Help Topics: AutoCAD Help dialog box*

The **Index** tab provides help on a particular item. First begin typing the desired topic in the top text box, as shown in Figure 1–38. These items might be a command name; a concept, such as *color*; or maybe just a symbol, such as an asterisk or a backslash. The initial character(s) typed in will highlight the topic in the lower text box whose initial characters match the string of characters. You may scroll through the list, choosing the **Display** button when the item on which you need help is highlighted, or you may double-click that item in the text box. AutoCAD then displays a window containing the help available on the highlighted item.

The **Find** tab in the AutoCAD Help dialog box creates a list of indexed words from the *User's Guide* that can be searched. You may scroll through the list, pressing ENTER when the word on which you need help is highlighted, or you may double-click that item in the text box, as shown in Figure 1–39. A list of topics about the highlighted word is then displayed in the text box. You may scroll through this sublist, pressing ENTER when the topic on which you need help is highlighted, or you may double-click that item in the text box.

Figure 1–38 *Index tab of the Help Topics: AutoCAD Help dialog box*

Figure 1–39 *Find tab of the Help Topics: AutoCAD Help dialog box*

Learning Assistance

You can also call up an interactive presentation of "Learning Assistance" of AutoCAD 2000 functionality by selecting Learning Assistance from the Help menu. Subjects include Tutorials, Concepts, Fast Track To AutoCAD 2000, AutoCAD Migration Assistance, AutoCAD Support Assistance, AutoCAD Support Online, and AutoCAD Help.

 Note: To access Learning Assistance Program, it must be installed in your hard drive. This program requires about 600 MB of hard disk memory.

In addition, you can view an interactive presentation of "What's New" in AutoCAD 2000 by selecting What's New from the Help menu. A menu of five topics is provided including Heads-UP Design Environment, Improved Access And Usability, Expanded Reach, Streamlined Output, and Customizing And Extending AutoCAD.

BEGINNING A NEW DRAWING WITH THE NEW COMMAND

The NEW command allows you to begin a new drawing.

Invoke the NEW command:

Standard Toolbar	Chose the NEW command (see Figure 1–40)
File menu	Chose New
Command: prompt	**new** (ENTER)

Figure 1–40 *Invoking the NEW command from the Standard Toolbar*

AutoCAD displays the Create New Drawing dialog box, as shown in Figure 1–41.

The options provided in the Create New Drawing dialog box are the same as in the Startup dialog box, explained earlier is this chapter. The only option missing in the Startup dialog box is the Open a Drawing option. For a detailed description of all of the options available in the Startup dialog box, refer to the "Starting AutoCAD" section earlier in this chapter.

Figure 1–41 *Create New Drawing dialog box*

OPENING AN EXISTING DRAWING WITH THE OPEN COMMAND

The OPEN command allows you to open an existing drawing.

Invoke the OPEN command:

Standard Toolbar	Choose OPEN (see Figure 1–42)
File menu	Choose Open
Command: prompt	**open** (ENTER)

Figure 1–42 *Invoking the OPEN command from the Standard Toolbar*

AutoCAD displays the Select File dialog box, as shown in Figure 1–43. The dialog box is almost identical to the Standard File dialog box, except that it includes checkboxes for selecting an initial view and for setting read-only mode. In addition, when you click on the file name, AutoCAD displays a bitmap image in the Preview box.

Figure 1–43 *Select File dialog box*

The **Select Initial View** list box permits you to specify a view name in the named drawing to be the startup view. If there are named views in the drawing, an M or P beside their name will tell if the view is model or paper space, respectively.

You may open a drawing in the read-only mode, which permits you to view the drawing but not save it with its current name. However, you can edit and save it under a different name by invoking the SAVEAS command.

Choosing the **Find File...** button causes the **Browse/Search** dialog box to be displayed. Various drives and directories are searched using search criteria. The Browse/Search dialog box combines the usual Windows file/path search of files by name (and normally with the .DWG extension) with small pictures of the drawings on the specified path to enable you to distinguish visually and select the drawing you wish to open.

Browse

The Browse tab, as shown in Figure 1–44, displays bitmap images of the drawings on the specified path.

> **File Name:** permits you to read the names of the available drawings and change the currently selected drawing.
>
> **Directories:** permits you to read the names of the available directories and change the currently selected directory.

Drives: permits you to read the names of the available drives and change the currently selected drive.

List Files of Type: permits you to read the available file types and change the current file type.

Size: permits you to change the size of the bitmap images.

Network causes the Map Network Drive dialog box to be displayed, which is used for connecting to networked drives.

Select the appropriate drawing file to open from the Browse section.

Figure 1–44 *Browse tab of the Browse/Search dialog box*

Search

The Search tab, shown in Figure 1–45, permits you to search a specific path or paths and then displays bitmap images of the drawings that meet the search criteria and are on the specified path(s). It also displays each drawing's name and path.

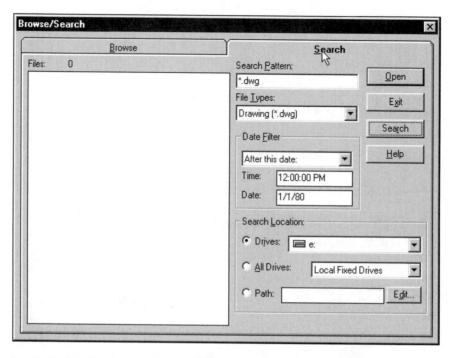

Figure 1–45 *Search tab of the Browse/Search dialog box*

Files: permits you to read a list of files that meet the search criteria.

Search Pattern: permits you to specify criteria by which to search for files with the specified file type.

File Types: permits you to specify the file type.

Date Filter: permits you to tell AutoCAD whether to search forward or backward from a specified date and time.

The **Time:** text field permits you to specify the time from which to search forward or backward.

The **Date:** text field permits you to specify the date from which to search forward or backward.

Search Location: permits you to specify the drive/path on which AutoCAD will search.

All **Drives:** causes AutoCAD to search all available drives.

Path: permits you to specify drive/directory paths for searching.

Edit: permits you to edit the path.

Select the appropriate drawing file to open from the Search section.

 Note: You can open and edit an AutoCAD Release 12, 13, or 14 drawing. If necessary, you can save the drawing in AutoCAD Release 12, 13, or 14 format by using the SAVEAS command. Certain limitations apply when doing this, which are explained in the SAVEAS section later in this chapter.

WORKING WITH MULTIPLE DRAWINGS

AutoCAD 2000 allows you to work in more than one drawing in a single AutoCAD session (Multiple Document). When multiple drawings are open, you can switch between the drawings, by selecting the appropriate name of the drawing from the Window menu. If the drawing is arranged in Tile or Cascade mode, then simply click anywhere in the drawing to make it active. You can also use CTRL + F6 or CTRL + TAB to switch between open drawings. However, you cannot switch between drawings during certain long operations such as regenerating the drawing.

You can copy and paste objects between drawings, and use Property Painter to transfer properties from objects in one drawing to objects in another drawing. You can also use AutoCAD object snaps, the Copy with Base point command, and the Paste to Original Coordinates command to ensure accurate placement, especially when copying objects from one drawing to another.

If you want to turn off the Multiple Document mode, open the Options dialog box:

Tools menu	Choose Options
Command: prompt	**options** (ENTER)

AutoCAD displays the Options dialog box. Select System tab, and under General options set the **Single -drawing compatibility mode** to ON (see Figure 1–46). Choose the **OK** button to save the changes. AutoCAD will allow you to open only one drawing at a time (similar to AutoCAD Release 14).

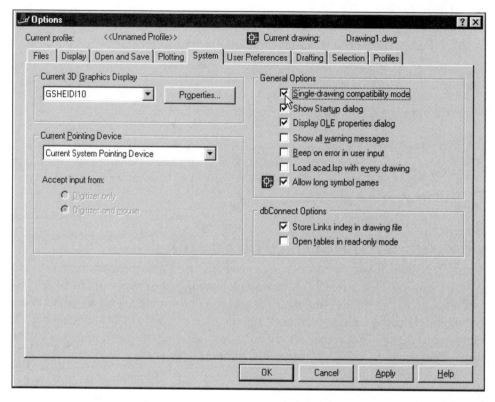

Figure 1–46 *Options dialog box – System tab*

CHANGING UNITS

The UNITS command lets you change the linear and angular units by means of the **Drawing Units** dialog box. In addition, it lets you set the display format measurement and precision of your drawing units. You can change any or all of the following:

Unit display format	Angle display precision
Unit display precision	Angle base
Angle display format	Angle direction

Invoke the UNITS command:

Format menu	Choose UNITS
Command: prompt	**units** (ENTER)

AutoCAD displays the Drawing Units dialog box, as shown in Figure 1–47.

Figure 1–47 *The Drawing Units dialog box*

Length

The Length section of the **Drawing Units** dialog box allows you to change the units of linear measurement. From the **Type:** list box select one of the five types of report formats you prefer. For the selected report format, choose precision from the **Precision:** list box.

The engineering and architectural report formats produce feet-and-inches displays. These formats assume each drawing unit represents 1 inch. The other formats (scientific, decimal, and fractional) make no such assumptions, and they can represent whatever real-world units you like.

Drawing a 150-ft-long object might, however, differ depending on the units chosen. For example, if you use decimal units and decide that 1 unit = 1 foot, then the 150-ft-long object will be 150 units long. If you decide that 1 unit = 1 inch, then the 150-ft-long object will be drawn 150 x 12 = 1,800 units long. In architectural and engineering units modes, the unit automatically equals 1 inch. You may then give the length of the 150-ft-long object as 150' or 1,800" or simply 1,800.

Angles

The Angles section of the Drawing Units dialog box allows you to set the drawing's angle measurement. From the **Type:** list box select one of the five types of report formats you prefer. For the selected format, choose precision from the **Precision:** list.

Select the direction in which the angles are measured, clockwise or counterclockwise. If the **Clockwise** checkbox is set to ON, then the angles will increase in value in the clockwise direction. If it is set to OFF, the angles will increase in value in the counter-clockwise direction. (see Figure 1–48).

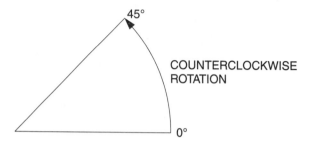

Figure 1–48 *The default, counterclockwise-angle direction of angle measurement*

Drawing Units For DesignCenter Blocks

The units that you select from the **When Inserting Blocks Into This Drawing Scale Them To:** list box determines the unit of measure used for block insertions from AutoCAD DesignCenter. If a block is created in units different than the units specified in the list box, they will be inserted and scaled in the specified units. If you select Unitless, the block will be inserted as is and the scale will not be adjusted to match the specified units.

 Note: Source content units and Target drawing units settings in the User Preferences tab of the Options dialog box are used when Insert Units are not defined.

Direction

To control the direction of angles, choose the **Direction** button; the Direction Control subdialog box appears, as shown in Figure 1–49.

Figure 1–49 *Direction Control subdialog box*

AutoCAD, by its default setting, assumes that 0 degrees is to the right (east, or 3 o'clock) (see Figure 1–50), and that angles increase in the counterclockwise direction.

Figure 1–50 *Default angle setting direction*

You can change measuring angles to start at any compass point by selecting one of the five radio buttons.

You can also show AutoCAD the direction you want for angle 0 by specifying two points. This can be done by selecting the radio button for **Other** and choosing the **Pick an angle** button. AutoCAD prompts you for two points and sets the direction for angle 0. Choose the **OK** button to close the Direction Control subdialog box.

Once you are satisfied with all of the settings in the **Drawing Units** dialog box, choose the **OK** button to set the appropriate settings to the current working drawing and close the dialog box.

Note: When AutoCAD prompts for a distance, displacement, spacing, or coordinates, you can always reply with numbers in integer, decimal, scientific, or fractional format. If engineering or architectural report format is in effect, you can also input feet, inches, or a combination of feet and inches. However, feet-and-inches input format differs slightly from the report format because it cannot contain a blank. For example, a distance of 75.5 inches can be entered in the feet/inches/fractions format as 6'3-1/2". Note the absence of spaces and the hyphen in the unconventional location between the inches and the fraction. Normally, it will be displayed in the status area as 6'-3 1/2.

If you wish, you can use the SETVAR command to set the UNITMODE system variable to 1 (default UNITMODE setting is 0) to display feet-and-inches output in the accepted format. For example, if you set UNITMODE to 1, AutoCAD displays the fractional value of 45 1/4 as you enter it: 45-1/4. The feet input should be followed by an apostrophe (') and inches with a trailing double quote (").

When engineering or architectural report format is in effect, the drawing unit equals 1 inch, so you can omit the trailing double quote (") if you like. When you enter feet-and-inches values combined, the inch values should immediately follow the apostrophe, without an intervening space. Distance input does not permit spaces because, except when entering text, pressing SPACEBAR functions the same as pressing ENTER.

CHANGING THE AREA

The LIMITS command allows you to place an imaginary rectangular drawing sheet in the CAD drawing space. But, unlike the limitations of the drawing sheet of the board drafter, you can move or enlarge the CAD electronic sheet (the limits) after you have started your drawing. The LIMITS command does not affect the current display on the screen. The defined area determined by the limits governs the portion of the drawing indicated by the visible grid (For more on the GRID command, see Chapter 3). Limits are also factors that determine how much of the drawing is displayed by the ZOOM ALL command (For more on the ZOOM ALL command, see Chapter 3).

The limits are expressed as a pair of *2D* points in the World Coordinate System, a lower left and an upper right limit. For example, to set limits for an A-size sheet, set lower left as 0,0 and upper right as 11,8.5 or 12,9; for a B-size sheet set lower left as 0,0 and upper right as 17,11 or 18,12. Most architectural floor plans are drawn at a scale of 1/4" = 1'-0". To set limits to plot on a C-size (22" x 17") paper at 1/4" = 1'-0", the limits are set lower left as 0,0 and upper right as 88',68' (4 x 22, 4 x 17).

Invoke the LIMITS command:

Format menu	Choose Drawing Limits
Command: prompt	**limits** (ENTER)

AutoCAD prompts:

Command: limits (ENTER)
Specify lower left corner or [ON/OFF] <current>: *(press ENTER to accept the current setting, specify the lower left corner, or right-click for shortcut menu and select one of the available options)*
Specify upper right corner <current>: *(press ENTER to accept the current setting, or specify the upper right corner)*

The response you give for the upper right corner gives the location of the upper right corner of the imaginary rectangular drawing sheet.

There are two additional options available for the LIMITS command. When AutoCAD prompts for the lower left corner, you may respond to the ON or OFF options. The

ON/OFF options determine whether or not you can specify a point outside the limits when prompted to do so.

When you select the ON option, limits checking is turned on and you cannot start or end an object outside the limits, nor can you specify displacement points required by the MOVE or COPY command outside the limits. You can, however, specify two points (center and point on circle) that draw a circle, part of which might be outside the limits. The limits check is simply an aid to help you avoid drawing off the imaginary rectangular drawing sheet. Leaving the limits checking ON is a sort of safety net to keep you from inadvertently specifying a point outside the limits. On the other hand, limits checking is a hindrance if you need to specify such a point.

When you select the OFF option (default), AutoCAD disables limits checking, allowing you to draw the objects and specify points outside the limits.

Whenever you change the limits, you will not see any change on the screen unless you use the All option of the ZOOM command. ZOOM ALL lets you see entire newly set limits on the screen. For example, if your current limits are 12 by 9 (lower left corner 0,0 and upper right corner 12,9) and you change the limits to 42 by 36 (lower left corner 0,0 and upper right corner 42,36), you still see the 12 by 9 area. You can draw the objects anywhere on the limits 42 by 36 area, but you will see on the screen the objects that are drawn only in the 12 by 9 area. To see the entire limits, invoke the ZOOM command using the All option.

Invoke the ZOOM ALL command:

View menu	Choose Zoom > All
Command: prompt	**zoom** (ENTER)

AutoCAD prompts:

Command: **zoom** (ENTER)
Specify corner of window, enter a scale factor (nX or nXP), or [All/
   Center/Dynamic/Extents/Previous/Scale/Window] <real time>: **all**
   (ENTER)

You see the entire limits or current extents (whichever is greater) on the screen. If objects are drawn outside the limits, ZOOM ALL displays all objects. (For a detailed explanation of the ZOOM command, see Chapter 3).

Whenever you change the limits, you should always invoke ZOOM ALL to see the entire limits or current extents on the screen.

For example, the following command sequence shows steps to change limits for an existing drawing (see Figures 1–51 and 1–52).

44

Figure 1–51 *The limits of an existing drawing before being changed by the* LIMITS *command*

Figure 1–52 *The new limits of the drawing, after being changed by the* LIMITS *command*

Command: **limits** (ENTER)
ON/OFF/<lower left corner><0.00,0.00>: (ENTER)
Upper right corner <22.00,17.00>: **42,36** (ENTER)

Command: **zoom** (ENTER)
Specify corner of window, enter a scale factor (nX or nXP), or [All/
    Center/Dynamic/Extents/Previous/Scale/Window] <real time>: **all**
    (ENTER)

SAVING A DRAWING

While working in AutoCAD, you should save your drawing once every 10 to 15 minutes without exiting AutoCAD. By saving your work periodically, you are protecting your work from possible power failures, editing errors, and other disasters. This can be done automatically by setting the SAVETIME system variable to a specific interval (in minutes). In addition, you can also manually save by using the SAVE, SAVEAS, and QSAVE commands.

The SAVE command saves an unnamed drawing with a file name. If the drawing is already named, then it works like the SAVEAS command, when the SAVE is invoked from the Comman: prompt.

The SAVEAS command saves an unnamed drawing with a file name or renames the current drawing. If the current drawing is already named, then AutoCAD saves the drawing to the current drawing name, or prompts for a new file name, and sets the current drawing to the new file name you specified. If the current drawing is already named and you accept the current default file name, AutoCAD saves the current drawing and continues to work on the updated drawing. If you specify a file name that already exists in the current folder, AutoCAD displays a message warning you that you are about to overwrite another drawing file. If you do not want to overwrite it, specify a different file name. The SAVEAS command also allows you to save in various formats, including Release 12,13, and 14 formats, and as a drawing template.

The QSAVE command saves an unnamed drawing with a file name. If the drawing is named, AutoCAD saves the drawing without requesting a file name.

Invoke the SAVE command:

Command: prompt	**save** (ENTER)

AutoCAD displays the Save Drawing As dialog box, as shown in Figure 1–53.

Figure 1–53 *Save Drawing As dialog box*

Select the appropriate folder in which to save the file, and type the name of the file in the **File name:** text field. The file name can contain up to 255 characters (Windows 95, Windows 98 and Windows NT 4.0 operating systems), including embedded spaces and punctuation. Following are the examples of valid filenames:

> this is my first drawing
>
> first house
>
> machine part one

AutoCAD automatically appends DWG as a file extension. If you save it as a template file, then AutoCAD appends DWT as a file extension.

Invoke the SAVEAS command:

File menu	Choose SAVEAS
Command: prompt	**saveas** (ENTER)

AutoCAD displays the Save Drawing As dialog box. Select the appropriate folder in which to save the file, and type the name of the file in the **File name:** text field. The file name can contain up to 255 characters (Windows 95, Windows 98 and Windows NT 4.0 operating systems), including embedded spaces and punctuation. Select the format type you want to save to from the **Save as type:** option menu.

If you want to save the current drawing to the given file name, then select the **Save** button without changing the file name. AutoCAD displays a message to warn you that you are about to overwrite another drawing file. Select **Yes** to save to the current file name and **No** to not replace the older file with the current file name.

Invoke the QSAVE command:

Standard toolbar	Select the SAVE command (see Figure 1–54)
File menu	Choose Save
Command: prompt	**qsave** (ENTER)

Figure 1–54 *Invoking the* SAVE *command from the Standard Toolbar*

If the drawing is unnamed, then AutoCAD displays the Save Drawing As dialog box. Select the appropriate folder in which to save the file, and type the name of the file in the **File name:** text field. The file name can contain up to 255 characters (Windows 95, Windows 98 and Windows NT 4.0 operating systems), including embedded spaces and punctuation. If the drawing is named, AutoCAD saves the drawing without requesting a file name.

If you are working in multiple drawings, the CLOSE command closes the active drawing. Invoke the CLOSE command:

File menu	Choose the CLOSE command
Command: prompt	**close** (ENTER)

AutoCAD closes the active drawing. If you have not saved the drawing, since the last change, AutoCAD displays the **AutoCAD Alert Box – Save changes to Filename.DWG.** If you select "No" then AutoCAD closes the drawing. If you select "Yes", then AutoCAD displays the Save Drawing As dialog box. Select the appropriate folder in which to save the file, and type the name of the file in the **File name:** text field.

 Note: The CLOSE command is not available when AutoCAD is in Single Document mode.

EXITING AUTOCAD

The EXIT or QUIT command allows you to exit AutoCAD. The EXIT or QUIT command exits the current drawing if there have been no changes since the drawing was last saved. If the drawing has been modified, AutoCAD displays the Drawing Modification dialog box to prompt you to save or discard the changes before quitting.

Invoke the EXIT command:

File menu	Choose the EXIT command
Command: prompt	**exit** (ENTER)

If the drawing has been modified, AutoCAD displays the Drawing Modification dialog box to prompt you to save or discard the changes before exiting.

Invoke the QUIT command:

Command: prompt	**quit** (ENTER)

If the drawing has been modified, AutoCAD displays the Drawing Modification dialog box to prompt you to save or discard the changes before quitting.

REVIEW QUESTIONS

1. If you executed the following commands in order: LINE, CIRCLE, ARC, and ERASE, what would you need to do to re-execute the CIRCLE command?

   a. press PGUP, PGUP, PGUP, ENTER

   b. press DOWNARROW, DOWNARROW, DOWNARROW, ENTER

   c. press UPARROW, UPARROW, UPARROW, ENTER

   d. press PGDN, PGDN, PGDN, ENTER

   e. press LEFTARROW, LEFTARROW, LEFTARROW, ENTER

2. In a dialog box, when there are a set of mutually exclusive options (a list of several from which you must select exactly one), these are called:

   a. text box

   b. check boxes

   c. radio buttons

   d. scroll bars

   e. list boxes

3. If you lost all pull down menus, what command could you use to load the standard menu?

   a. LOAD   d. MENU

   b. OPEN   e. PULL

   c. NEW

4. What is the extension used by AutoCAD for template drawing files used by the setup wizard?

   a. .DWG   d. .TEM

   b. .DWT   e. .WIZ

   c. .DWK

5. If you performed a ZOOM ALL and your drawing shrunk to a small portion of the screen, one possible problem might be:

   a. out of computer memory

   b. misplaced drawing object

   c. Grid and Snap set incorrectly

   d. this should never happen in AutoCAD

   e. Limits are set much larger than current drawing objects.

6. What is an external tablet used to input absolute coordinate addresses to AutoCAD by means of a puck or stylus called?

   a. digitizer

   b. input pad

   c. coordinate tablet

   d. touch screen

7. The menu which can be made to appear at the location of the cross hairs is called:

   a. mouse menu

   b. cross-hair menu

   c. cursor menu

   d. none of the above, there is no such menu

8. In order to save basic setup parameters (such as snap, grid, etc.) for future drawings, you should:

   a. create an AutoCAD Macro

   b. create a prototype file

   c. create a new configuration file

   d. modify the ACAD.INI file

9. To cancel an AutoCAD command, press:

   a. CTRL + A

   b. CTRL + X

   c. ALT + A

   d. ESC

   e. CTRL + ENTER

10. The SAVE command:

    a. saves your work

    b. does not exit you out of AutoCAD

    c. is a valuable feature for periodically storing information to disk

    d. all of the above

CHAPTER 2

Fundamentals I

INTRODUCTION

This chapter introduces some of the basic commands and concepts in AutoCAD that can be used to complete a simple drawing. The project drawing used in this chapter is relatively uncomplicated, but for the newcomer to AutoCAD it presents ample challenge. It has fundamental problems that provide useful material for lessons in drawing setup, and in creating and editing objects. When you learn how to access and use the commands, how to find your way around the screen, and how AutoCAD makes use of coordinate geometry, you can apply these skills to the chapters containing more advanced drawings and projects.

After completing this chapter, you will be able to do the following:

- Construct geometric figures with LINE, RECTANGLE, CIRCLE, and ARC commands
- Use coordinate systems
- Use various object selection methods
- Use the ERASE command

CONSTRUCTING GEOMETRIC FIGURES

AutoCAD gives you an ample variety of drawing elements, called objects. It also provides you with many ways to generate each object in your drawing. You will learn about the properties of these objects as you progress in this text. It is important to keep in mind that the examples in this text of how to generate the various lines, circles, arcs, and other objects are not always the only methods available. You are invited, even challenged, to find other more expedient methods to perform tasks demonstrated in the lessons. You will progress at a better rate if you make an effort to learn as much as possible as soon as possible about the descriptive properties of the individual objects. When you become familiar with how the CAD program creates, manipulates, and stores the data that describes the objects, you are then able to create drawings more effectively.

DRAWING LINES

The primary drawing object is the line. A series of connected straight line segments can be drawn by invoking the LINE command and then selecting the proper sequence of endpoints. AutoCAD connects the points with a series of lines. The LINE command is one of the few AutoCAD commands that automatically repeats in this fashion. It uses the ending point of one line as the starting point of the next, continuing to prompt you for each subsequent ending point. To terminate this continuing feature you must give a null response (press ENTER or right-click and choose Enter from the shortcut menu). Even though a series of lines is drawn using a single LINE command, each line is a separate object, as though it had been drawn with a separate LINE command.

You can specify the endpoints using either two dimensional (x,y) or three dimensional (x,y,z) coordinates, or a combination of the two. If you enter 2D coordinates, AutoCAD uses the current elevation as the Z element of the point (zero is the default). This chapter is concerned only with 2D points whose elevation is zero. (3D concepts and nonzero elevations are covered in later chapters.)

Invoke the LINE Command:

Draw toolbar	Choose the LINE command (see Figure 2–1)
Draw menu	Choose Line
Command: prompt	**line** (ENTER)

Figure 2–1 *Invoking the LINE command from the Draw toolbar*

AutoCAD prompts:

> Command: **line** (ENTER)
> Specify first point:

Where to Start

The first point of the first object in a drawing normally establishes where all of the points of other objects must be placed. It is like the cornerstone of a building. Careful thought should go into locating the first point.

You can specify the starting point of the line by entering absolute coordinates (see the section on Coordinate Systems for a detailed explanation) or by using your pointing device (mouse or puck). After specifying the first point, AutoCAD prompts:

> Specify next point or [Undo]:

Where to From Here?

In addition to the first point's being the cornerstone, the direction of the first object is also critical to where all other points of other objects are located with respect to one another.

You can specify the end of the line by means of absolute coordinates, relative coordinates, or by using your pointing device to specify the end of the line on the screen. Again, AutoCAD repeats the prompt:

Specify next point or [Undo]:

You can enter a series of connected lines. To save time, the LINE command remains active and prompts "Specify next point" after each point you specify. When you have finished entering a connected series of lines, give a null reply (press ENTER) to terminate the LINE command.

If you are placing points with a cursor instead of specifying coordinates, a rubber-band preview line is displayed between the starting point and the crosshairs. This helps you see where the resulting line will go. In Figure 2–2 the dotted lines represent previous cursor positions.

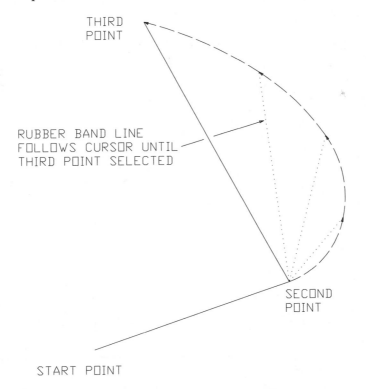

Figure 2–2 *Placing points with the cursor rather than with keyboard coordinates input*

Most of the AutoCAD commands have a variety of options. For the LINE command, three options are available: Continue, Close, and Undo. The available options can be selected from the shortcut menu that appears when you right-click on your pointing device after you invoke the LINE command.

Continue Option

When you invoke the LINE command and respond to the "Specify first point:" prompt by pressing ENTER, AutoCAD automatically sets the start of the line to the end of the most recently drawn line or arc. This provides a simple method for constructing a tangentially connected line in an arc-line continuation.

The subsequent prompt sequence depends on whether a line or arc was more recently drawn. If the line is more recent, the starting point of the new line will be set as the ending point of that most recent line, and the "Specify next point:" prompt appears as usual. If an arc is more recent, its end defines the starting point and the direction of the new line. AutoCAD prompts for:

Length of line:

Specify the length of the line to be drawn, and then AutoCAD continues with the normal "Specify next point:" prompt.

The following command sequence shows an example using the Continue option (see Figure 2–3).

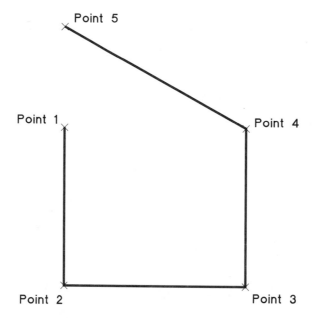

Figure 2–3 *Using the* LINE *command's Continue option*

Command: **line** (ENTER)
Specify first point: *(specify Point 1)*
Specify next point or [Undo]: *(specify Point 2)*
Specify next point or [Undo]: *(specify Point 3)*
Specify next point or [Close/Undo]: (ENTER)
Command: *(right-click and choose the* LINE *command again)*
Specify first point: *(to continue the line from Point 3, press* ENTER *or the* SPACEBAR)
Specify next point or [Undo]: *(specify Point 4)*
Specify next point or [Undo]: *(specify Point 5)*
Specify next point or [Close/Undo]: (ENTER)

Close Option

If you are drawing a sequence of lines to form a polygon, then you can use the Close option to join the last and first points automatically. AutoCAD draws the closing line segment if you respond to the "Specify next point:" prompt by right-clicking and choosing Close from the shortcut menu. AutoCAD performs two steps when you choose the Close option. The first step closes the polygon, and the second step terminates the LINE command (equivalent to a null response) and returns you to the "Command:" prompt.

The following command sequence shows an example of using the Close option (see Figure 2–4).

Figure 2–4 *Using the* LINE *command's Close option*

Command: **line** (ENTER)
Specify first point: *(specify Point 1)*
Specify next point or [Undo]: *(specify Point 2)*
Specify next point or [Undo]: *(specify Point 3)*
Specify next point or [Close/Undo]: *(specify Point 4)*
Specify next point or [Close/Undo]: *(specify Point 5)*
Specify next point or [Close/Undo]: *(right-click and choose Close option
    from the shortcut menu)*

Undo Option

While drawing a series of connected lines, you may wish to erase the most recent line segment and continue from the end of the previous line segment. You can do so and remain in the LINE command without exiting by using the Undo option. Whenever you wish to erase the most recent line segment, at the "Specify next point:" prompt, choose the Undo option from the shortcut menu. If necessary, you can select multiple Undos; this will erase the most recent line segment one at a time. Once you are out of the LINE command, it is too late to use the Undo option of the LINE command to erase the most recent line segment.

The following command sequence shows an example using the Undo option (see Figure 2–5).

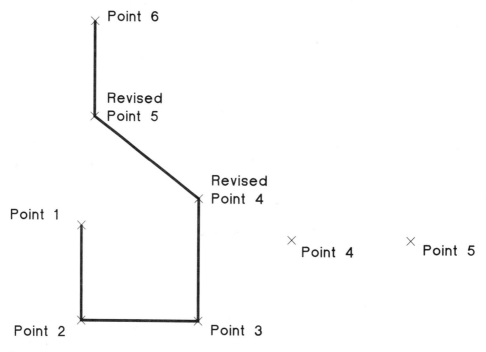

Figure 2–5 *Using the* LINE *command's Undo option*

Command: **line** (ENTER)
Specify first point: *(specify Point 1)*
Specify next point or [Undo]: *(specify Point 2)*
Specify next point or [Undo]: *(specify Point 3)*
Specify next point or [Close/Undo]: *(specify Point 4)*
Specify next point or [Close/Undo]: *(specify Point 5)*
Specify next point or [Close/Undo]: *(right-click and choose Undo option from the shortcut menu)*
Specify next point or [Close/Undo]: *(right-click and choose Undo option from the shortcut menu)*
Specify next point or [Close/Undo]: *(specify Revised Point 4)*
Specify next point or [Close/Undo]: *(specify Revised Point 5)*
Specify next point or [Close/Undo]: *(specify Point 6)*
Specify next point or [Close/Undo]: (ENTER)

DRAWING RECTANGLES

When it is necessary to create a rectangular box, you can use the RECTANGLE command.

Invoke the RECTANGLE command:

Draw toolbar	Choose the RECTANGLE command (see Figure 2–6)
Draw menu	Choose Rectangle
Command: prompt	**rectangle** (ENTER)

Figure 2–6 *Invoking the* RECTANGLE *command from the Draw toolbar*

AutoCAD prompts:

Command: **rectangle** (ENTER)
Specify first corner point or [Chamfer/Elevation/Fillet/Thickness/Width]: *(specify first corner point to define the start of the rectangle or right-click for shortcut menu and choose one of the avilable options)*
Specify other corner point: *(specify a point to define the opposite corner of the rectangle)*

Chamfer Option

The Chamfer option sets the chamfer distance for the rectangle to be drawn. Refer to Chapter 4 for a detailed explanation on the usage of the CHAMFER command and its available settings.

Elevation Option

The Elevation option specifies the elevation of the rectangle to be drawn. Refer to Chapter 15 for a detailed explanation on the usage of the Elevation setting.

Fillet Option

The Fillet option sets the fillet radius for the rectangle to be drawn. Refer to Chapter 4 for a detailed explanation on the usage of the FILLET command and its available settings.

Thickness Option

The Thickness option specifies the thickness of the rectangle to be drawn. Refer to Chapter 15 for a detailed explanation on the Thickness setting.

Width Option

The Width option selection allows you set the line width for the rectangle to be drawn. The default width is set to 0.0.

DRAWING WIDE LINES

When it is necessary to draw thick lines, the TRACE command may be used instead of the LINE command. Traces are entered just like lines except that the line width is set first. To specify the width, you can specify a distance or select two points and let AutoCAD use the measured distance between them. When you draw using the TRACE command, the previous trace segment is not drawn until the next endpoint is specified.

Invoke the TRACE command:

Command: prompt	**trace** (ENTER)

AutoCAD prompts:

Command: **trace** (ENTER)
Specify trace width <current>: (specify the trace width and press ENTER)
Specify start point: (specify the starting point of the trace)

Specify a series of connected trace lines. When you have finished specifying a connected series of lines, give a null reply (press ENTER) to terminate the TRACE command.

For example, the following command sequence shows the placement of connected lines using the TRACE command (see Figure 2–7).

Command: **trace** (ENTER)
Specify trace width <current>: .05 (ENTER)
Specify start point: (specify Point 1)
Specify next point: (specify Point 2)
Specify next point: (specify Point 3)
Specify next point: (specify Point 4)
Specify next point: (ENTER)

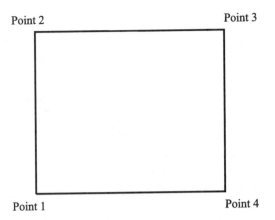

Figure 2–7 *Placing wide connected lines using the* TRACE *command*

COORDINATE SYSTEMS

In accordance with the conventions of the Cartesian coordinate system, horizontal distances increase in the positive X direction, toward the right, and vertical distances increase in the positive Y direction, upward. Distances perpendicular to the XY plane that you are viewing increase toward you in the positive Z direction. This set of axes defines the *World Coordinate System*, abbreviated as WCS.

The significance of the WCS is that it is always in your drawing; it cannot be altered. An infinite number of other coordinate systems can be established relative to it. These others are called *User Coordinate Systems* (UCSs) and can be created with the UCS command. Even though the WCS is fixed, you can view it from any angle, side, or rotation without changing to another coordinate system.

AutoCAD provides what is called a *coordinate system icon* to help you keep your bearings among different coordinate systems in a drawing. The icon will show you the orientation of your current UCS by indicating the positive directions of the X and Y axes. Figure 2–8 shows some examples of coordinate system icons.

Figure 2–8 *Examples of the UCS icons*

Computer-aided drafting permits you always to draw an object at its true size and then make the border, title block, and other nonobject associated features fit the object. The completed combination is reduced (or increased) to fit the plotted sheet size you require when you plot.

A more complicated situation is when you wish to draw objects at different scales on the same drawing. This can be handled easily by one of several methods with the more advanced commands provided in AutoCAD.

Drawing a schematic that is not to scale is one situation where the graphics and computing power are hardly used to their potential. But even though the symbols and the distances between them have no relationship to any real-life dimensions, the sheet size, text size, line widths, and other visible characteristics of the drawing must be considered in order to give your schematic the readability you desire. Some planning, including sizing, needs to be applied to all drawings.

When AutoCAD prompts for the location of a point, you can use one of several available point entry techniques, including absolute rectangular coordinates, relative rectangular coordinates, relative polar coordinates, spherical coordinates, and cylindrical coordinates.

ABSOLUTE RECTANGULAR COORDINATES

The rectangular coordinates method is based on specifying the location of a point by providing its distances from two intersecting perpendicular axes in a 2D plane or from three intersecting perpendicular planes for 3D space. Each point's distance is measured along the X axis (horizontal), Y axis (vertical), and Z axis (toward or away from the viewer). The intersection of the axes, called the origin ($X, Y, Z = 0,0,0$) divides the coordinates into four quadrants for 2D or eight sections for 3D (see Figure 2–9).

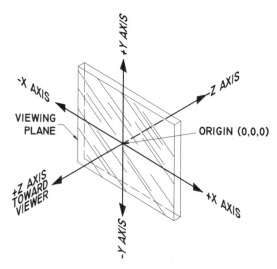

Figure 2–9 *Specifying rectangular coordinates using the intersections of the X, Y, and Z axes*

Points are located by absolute rectangular coordinates in relation to the origin. You specify the reference to the WCS origin or UCS origin. In AutoCAD, by default the origin (0,0) is located at the lower left corner of the drawing, as shown in Figure 2–10.

Figure 2–10 Default location of the AutoCAD origin

As mentioned earlier, the horizontal distance increases in the positive X direction from the origin, and the vertical distance increases in the positive Y direction from the origin. You specify a point by entering its X, Y, Z coordinates in decimal, fractional, or scientific notation separated by commas.

For example, the following command sequence shows placement of connected lines as shown in Figure 2–11 by absolute coordinates (see Figure 2–12):

> Command: **line** (ENTER)
> Specify first point: **2,2** (ENTER)
> Specify next point or [Undo]: **2,4** (ENTER)
> Specify next point or [Undo]: **3,5** (ENTER)
> Specify next point or [Close/Undo]: **5,5** (ENTER)
> Specify next point or [Close/Undo]: **5,7** (ENTER)
> Specify next point or [Close/Undo]: **7,9** (ENTER)
> Specify next point or [Close/Undo]: **10,9** (ENTER)

Specify next point or [Close/Undo]: **13,4** (ENTER)
Specify next point or [Close/Undo]: **13,2** (ENTER)
Specify next point or [Close/Undo]: **2,2** (ENTER)
Specify next point or [Close/Undo]: (ENTER)
Command:

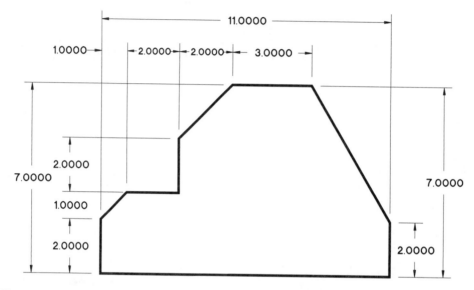

Figure 2–11 *Placing connected lines*

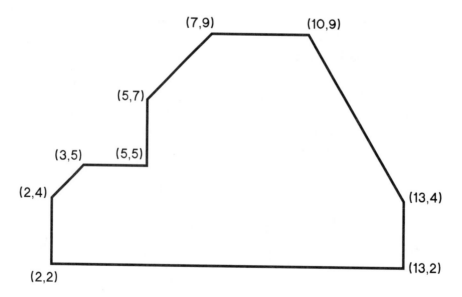

Figure 2–12 *Placing connected lines using absolute coordinates*

RELATIVE RECTANGULAR COORDINATES

Points are located by relative rectangular coordinates in relation to the last specified position or point, rather than the origin. This is like specifying a point as an offset from the last point you entered. In AutoCAD, whenever you specify relative coordinates, the @ ("at" symbol) must precede your entry. This symbol is selected by holding the SHIFT key and simultaneously pressing the key for the number 2 at the top of the keyboard.

The following command sequence shows placement of connected lines as shown in Figure 2–11 by relative rectangular coordinates (see Figure 2–13):

```
Command: line (ENTER)
Specify first point: 2,2 (ENTER)
Specify next point or [Undo]: @0,2 (ENTER)
Specify next point or [Undo]: @1,1 (ENTER)
Specify next point or [Close/Undo]: @2,0 (ENTER)
Specify next point or [Close/Undo]: @0,2 (ENTER)
Specify next point or [Close/Undo]: @2,2 (ENTER)
Specify next point or [Close/Undo]: @3,0 (ENTER)
Specify next point or [Close/Undo]: @3,–5 (ENTER)
Specify next point or [Close/Undo]: @0,–2 (ENTER)
Specify next point or [Close/Undo]: 2,2 (ENTER)
Specify next point or [Close/Undo]: (ENTER)
Command:
```

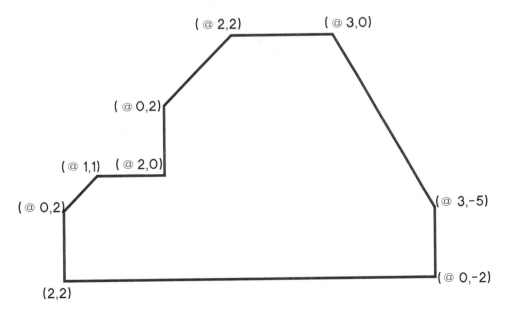

Figure 2–13 *Placing connected lines using relative rectangular coordinates*

RELATIVE POLAR COORDINATES

Polar coordinates are based on a distance from a fixed point at a given angle. In AutoCAD, a polar coordinate point is determined by the distance from a previous point and angle measured from the zero degree, rad, or gradient. In AutoCAD, by default the angle is measured in the counterclockwise direction. It is important to remember that for points located using polar coordinates they are to be positioned relative to the previous point and not the origin (0,0). You can specify a point by entering its distance from the previous point and its angle in the *XY* plane, separated by < (not a comma). This symbol is selected by holding the SHIFT key and simultaneously pressing the comma (",") key at the bottom of the keyboard. Failure to use the @ symbol will cause the point to be located relative to the origin (0,0).

The following command sequence shows the placement of the connected lines shown in Figure 2–11 by using a combination of polar and rectangular coordinates (see Figure 2–14).

```
Command: line (ENTER)
Specify first point: 2,2 (ENTER)
Specify next point or [Undo]: @2<90 (ENTER)
Specify next point or [Undo]: @1,1 (ENTER)
Specify next point or [Close/Undo]: @2<0 (ENTER)
Specify next point or [Close/Undo]: @2<90 (ENTER)
Specify next point or [Close/Undo]: @2,2 (ENTER)
Specify next point or [Close/Undo]: @3,0 (ENTER)
Specify next point or [Close/Undo]: @3,–5 (ENTER)
Specify next point or [Close/Undo]: @2<270 (ENTER)
Specify next point or [Close/Undo]: 2,2 (ENTER)
Specify next point or [Close/Undo]: (ENTER)
Command:
```

If you are working in a UCS and would like to enter points with reference to the WCS, enter coordinates preceded by an asterisk (*). For example, to specify the point with an *X* coordinate of 3.5 and a *Y* coordinate of 2.57 with reference to the WCS, regardless of the current UCS, enter:

*3.5,2.57

In the case of relative coordinates, the asterisk will be preceded by the @ symbol. For example:

@*4,5

This represents an offset of 4,5 from the previous point with reference to the WCS.

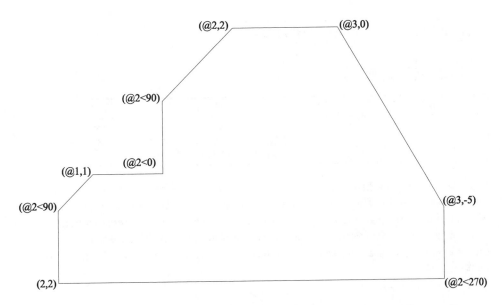

Figure 2–14 *Placing connected lines using a combination of polar and rectangular coordinates*

COORDINATE DISPLAY

The Coordinates Display is a report in the status bar of the cursor coordinates at the bottom of the screen. It has three settings. On most systems the F6 function toggles between the three settings. The three settings are as follows:

1. This setting causes the display to report the location of the cursor when the prompt is in the "Command:" status or when you are being prompted to select the first point selection of a command. It then changes to a relative polar mode when you are prompted for a second point that could be specified relative to the previous point. In this case the report is in the form of the direction/distance. The direction is given in terms of the current angular units setting and the distance in terms of the current linear units setting.

2. This setting is similar to the previous one, except that the display for the second location is given in terms of its absolute coordinates, rather than relative to the previous point.

3. This setting is used to save either the location in the display at the time you toggle to this setting or the last point entered. It does not change dynamically with the movement of the cursor.

EXERCISES 2–1 THROUGH 2–4

EXERCISE 2–1

Type of shape	Input Methods
Single Line Sequence	• Absolute rectangular coordinates • Relative rectangular coordinates • Relative polar coordinates

Any one of the three listed methods can be used to enter the points that determine the shape in this exercise. This is not always the case, as you will see in the next few exercises. Also, all of the lines that determine the shape are sequentially connected; that is, the ending point of each line can be used as the starting point of another line. That means that this figure can be drawn without exiting the LINE command once it is invoked. Again, this is not the case with most shapes.

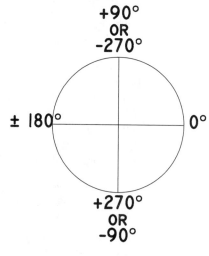

Points of Interest

- With rectangular coordinates, the distances are entered as horizontal (*X*) and vertical (*Y*) displacements.

- Absolute rectangular coordinates are *X* and *Y* displacements from the origin (0,0).

- Relative rectangular coordinates are *X* and *Y* displacements from the last point entered.

- @ Represents the last point entered.

- Relative polar coordinates are a combination of a direction and a distance from the last point entered.

- Substituting rectangular coordinate input with polar coordinate input (or vice versa) is not always done easily without the application of trigonometric functions.

Drawing Setup	Use Quick Setup wizard, set Units to Decimal, set Area to 9 x 13. Invoke the LINE command and then respond to the points as shown for each method.

Point by Point

Absolute	Relative Rectangular	Relative Polar
1. First point: **2,1** (ENTER)	1. First point: **2,1** (ENTER)	1. First point: **2,1** (ENTER)
2. Next point: **2,4** (ENTER)	2. Next point: **@0,3** (ENTER)	2. Next point: **@3<90** (ENTER)
3. Next point: **3,4** (ENTER)	3. Next point: **@1,0** (ENTER)	3. Next point: **@1<0** (ENTER)
4. Next point: **3,8** (ENTER)	4. Next point: **@0,4** (ENTER)	4. Next point: **@4<90** (ENTER)
5. Next point: **2,8** (ENTER)	5. Next point: **@–1,0** (ENTER)	5. Next point: **@1<180** (ENTER)
6. Next point: **2,10** (ENTER)	6. Next point: **@0,2** (ENTER)	6. Next point: **@2<90** (ENTER)
7. Next point: **3,10** (ENTER)	7. Next point: **@1,0** (ENTER)	7. Next point: **@1<0** (ENTER)
8. Next point: **3,11** (ENTER)	8. Next point: **@0,1** (ENTER)	8. Next point: **@1<90** (ENTER)
9. Next point: **5,11** (ENTER)	9. Next point: **@2,0** (ENTER)	9. Next point: **@2<0** (ENTER)
10. Next point: **5,4** (ENTER)	10. Next point: **@0,–7** (ENTER)	10. Next point: **@7<270** (ENTER)
11. Next point: **6,4** (ENTER)	11. Next point: **@1,0** (ENTER)	11. Next point: **@1<0** (ENTER)
12. Next point: **6,1** (ENTER)	12. Next point: **@0,–3** (ENTER)	12. Next point: **@3<270** (ENTER)
13. Next point: **2,1** (ENTER)	13. Next point: **@–4,0** (ENTER)	13. Next point: **@4<180** (ENTER)
14. Command: (ENTER)	14. Command: (ENTER)	14. Command: (ENTER)

EXERCISE 2-2

Type of shape	Input Methods
Multiple line sequences	• Absolute rectangular coordinates • Relative rectangular coordinates • Relative polar coordinates

As in Exercise 2–1, the points that determine the shapes in this exercise can be entered by any one of the three listed methods here. But because these are three separate shapes, all lines are not sequentially connected. The LINE command must be invoked at least three times.

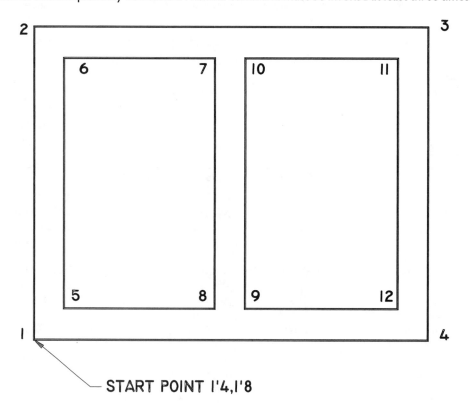

START POINT 1'4,1'8

Points of Interest

- Steps 5, 11, and 17 in the following tables illustrate how you can use the Close option of the LINE command to draw the last segment of a closed sequence of lines and exit the LINE command at the same time.

Drawing Setup	Use Quick Setup Wizard, set Units to Architectural, set Area to 8' x 6'.

Invoke the LINE command and then respond to the prompts as shown for each method. Steps 6 and 12 use automatic reinvoking of the previously used command by pressing (ENTER). In this case the LINE command is the previously used command.

Absolute	Relative Rectangular	Relative Polar
1. Start point: **1'4,1'8** (ENTER)	1. Start point: **1'4,1'8** (ENTER)	1. Start point: **1'4,1'8** (ENTER)
2. Next point: **1'4,5'** (ENTER)	2. Next point: **@0,3'4** (ENTER)	2. Next point: **@3'4<90** (ENTER)
3. Next point: **5'8,5'** (ENTER)	3. Next point: **@4'4,0** (ENTER)	3. Next point: **@4'4<0** (ENTER)
4. Next point: **5'8,1'8** (ENTER)	4. Next point: **@0,-3'4** (ENTER)	4. Next point: **@3'4<-90** (ENTER)
5. Next point: **c** (ENTER)	5. Next point: **c** (ENTER)	5. Next point:**c** (ENTER)
6. Command: (ENTER)	6. Command: (ENTER)	6. Command: (ENTER)
7. From point: **1'8,2'** (ENTER)	7. From point: **1'8,2'** (ENTER)	7. From point: **1'8,2'** (ENTER)
8. Next point: **1'8,4'8** (ENTER)	8. Next point: **@0,2'8** (ENTER)	8. Next point: **@2'8<90** (ENTER)
9. Next point: **3'4,4'8** (ENTER)	9. Next point: **@1'8,0** (ENTER)	9. Next point: **@1'8<0** (ENTER)
10. Next point: **3'4,2'** (ENTER)	10. Next point: **@0,-2'8** (ENTER)	10. Next point: **@2'8<270** (ENTER)
11. Next point: **c** (ENTER)	11. Next point: **c** (ENTER)	11. Next point: **c** (ENTER)
12. Command: (ENTER)	12. Command: (ENTER)	12. Command: (ENTER)
13. From point: **3'8,2'** (ENTER)	13. From point: **3'8,2'** (ENTER)	13. From point: **3'8,2'** (ENTER)
14. Next point: **3'8,4'8** (ENTER)	14. Next point: **@0,2'8** (ENTER)	14. Next point: **@2'8<90** (ENTER)
15. Next point: **5'4,4'8** (ENTER)	15. Next point: **@1'8,0** (ENTER)	15. Next point: **@1'8<0** (ENTER)
16. Next point: **5'4,2'** (ENTER)	16. Next point: **@0,-2'8** (ENTER)	16. Next point: **@2'8<270** (ENTER)
17. Next point: **c** (ENTER)	17. Next point: **c** (ENTER)	17. Next point: **c** (ENTER)

general EXERCISE

EXERCISE 2–3

Type of shape	Input Methods
Multiple line sequences	• Absolute rectangular coordinates for the first line segment of the first sequence
	• Relative polar coordinates for the remaining segments of the first sequence
	• Relative polar coordinates for entering the starting point of the second and third separate single line sections of the shape

Unlike the cases of Exercises 2–1 and 2–2, all of the points that determine the figure in this exercise cannot be entered with complete accuracy by all three of the methods listed (see Important Point below). Except for points A1, A2, A5, and A8, the points in the shape in this exercise can be accurately specified only by polar coordinates. For example, the coordinates of point A3 (2 units at 240° from point A2) are 5, 1.267949192431. This is derived from the X coordinate of $6 - 2 \cos 240°$ and the Y coordinate of $5 - 2 \sin 240°$. Even though the X coordinate happens to be a round number in this case, it is an exception rather than the rule. Keying in numbers like 1.267949192431 takes time and is subject to error. Any fewer digits might cause problems when endpoints of sequences are expected to end exactly at some other point. You cannot see the difference on a plot, but some of the computer functions that depend on this type of accuracy might not perform as expected. So not only is it easier to enter the polar coordinates @2<240, it is also quicker, more accurate, and less subject to error.

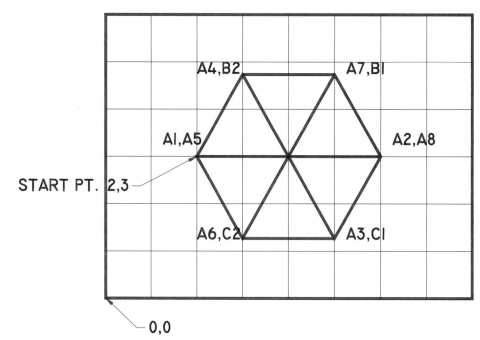

The LINE command must be invoked at least three times. And in order to enter the starting point for the second separate part of the shape (the line from B1 to B2), the last segment of the first sequence needs to end where shown, at point A8.

Important Point

AutoCAD has more advanced commands and features to make drawing these early exercises more easily than used here.

Points of Interest

- Steps 11 and 15 in the following Point by Point instructions show how you can specify starting points of lines (in this case points B1 and C1) by using the @ symbol and polar coordinate "distance and direction" data. Remember, the @ symbol means "last point."

- This method is possible because we ended the previous lines in predetermined points (A8 and B2) whose direction and distance from B1 and C1 we knew. If we had wanted to start segment B1 at B2, it would have been difficult to establish it from A8 by means of polar coordinates.

- Only the second point can be entered easily and accurately by means of any of the three methods (absolute, relative rectangular, or relative polar coordinates).

Drawing Setup	Use Quick Setup wizard, set Units to Decimal, set Area to 10 X 7.

Invoke the LINE command and then respond to the points prompt as shown. Steps 10 and 14 reinvoke the LINE command for separate sequences.

Point by Point

1. Specify first point: **2,3** (ENTER) *(starting point of line A1–A2)*
2. Specify next point or [Undo]: **@4<0** (ENTER) *(endpoint of line A1–A2)*
3. Specify next point or [Undo]: **@2<240** (ENTER) *(endpoint of line A2–A3)*
4. Specify next point or [Close/Undo]: **@4<120** (ENTER) *(endpoint of line A3–A4)*
5. Specify next point or [Close/Undo]: **@2<240** (ENTER) *(endpoint of line A4–A5)*
6. Specify next point or [Close/Undo]: **@2<300** (ENTER) *(endpoint of line A5–A6)*
7. Specify next point or [Close/Undo]: **@4<60** (ENTER) *(endpoint of line A6–A7)*
8. Specify next point or [Close/Undo]: **@2<300** (ENTER) *(endpoint of line A7–A8)*
9. Specify next point or [Close/Undo]: (ENTER) *(exits the LINE command)*
10. Command: (ENTER) *(automatic recall of LINE command for B1 to B2)*
11. Specify first point: **@2<120** (ENTER) *(starting point of line B1-B2)*
12. Specify next point or [Undo]: **@2<180** (ENTER) *(endpoint of line B1–B2)*
13. Specify next point or [Undo]: (ENTER) *(exits the LINE command)*
14. Command: (ENTER) *(automatic recall of LINE command for C1 to C2)*
15. Specify first point: **@4<300** (ENTER) *(starting point of line C1 to C2)*
16. Specify next point or [Undo]: **@2<180** (ENTER) *(endpoint of line C1 to C2)*
17. Specify next point or [Undo]: (ENTER) *(exits the LINE command)*

EXERCISE 2–4

Type of shape	Input Methods
Single line sequence	• Absolute rectangular coordinates for the first line segment of the sequence • Relative polar coordinates for the remaining segments Important Point

AutoCAD has more advanced commands and features to make drawing these early exercises easier than we've shown here. These methods are pabulum. Solid food will come later.

Points of Interest

Step 9 in the following Point by Point instructions can be substituted by simply selecting the Close option.

general EXERCISE

Drawing Setup	Use Start from Scratch option, then set Units to Architectural and Limits to -4'4",-1'8" for lower left corner and 10'0",11'4" for upper right corner.
Useful Point	Don't forget to ZOOM ALL after setting the limits.

Invoke the LINE command and respond to the "Specify first point:" as shown below.

Point by Point

1. Specify first point: **0,0** (ENTER) *(starting point of line A–B)*
2. Specify next point or [Undo]: **@4'<0** (ENTER) *(endpoint of line A–B)*
3. Specify next point or [Undo]: **@4'<45** (ENTER) *(endpoint of line B–C)*
4. Specify next point or [Close/Undo]: **@4'<90** (ENTER) *(endpoint of line C–D)*
5. Specify next point or [Close/Undo]: **@4'<135** (ENTER) *(endpoint of line D–E)*
6. Specify next point or [Close/Undo]: **@4'<180** (ENTER) *(endpoint of line E–F)*
7. Specify next point or [Close/Undo]: **@4'<225** (ENTER) *(endpoint of line F–G)*
8. Specify next point or [Close/Undo]: **@4'<270** (ENTER) *(endpoint of line G–H)*
9. Specify next point or [Close/Undo]: **@4'<315** (ENTER) *(endpoint of line H–A)*
10. Specify next point or [Close/Undo]: (ENTER) *(exits the LINE command)*

general EXERCISE

DRAWING CIRCLES

The CIRCLE command offers five different options for drawing circles: Center-Radius (default), Center-Diameter, 2 Point, 3 Point, and Tangent, Tangent, Radius (TTR).

Center-Radius Option

Draws a circle based on a center point and a radius. Invoke the CIRCLE command:

Draw toolbar	Choose the CIRCLE command (see Figure 2–15)
Draw menu	Choose Circle > Center, Radius
Command: prompt	**circle** (ENTER)

Figure 2–15 *Invoking the CIRCLE command from the Draw toolbar*

AutoCAD prompts:

    Command: **circle** (ENTER)
    Specify center point for circle or [3P/2P/Ttr (tan tan radius)]: *(specify
        center point to define the center of the circle or right-click for shortcut menu
        to select one of the available options)*
    Specify radius of circle or [Diameter]: *(specify the radius of the circle or right-
        click for shortcut menu to select one of the available options)*

The following command sequence shows an example (see Figure 2–16):

    Command: **circle** (ENTER)
    Specify center point for circle or [3P/2P/Ttr (tan tan radius)]: **2,2** (ENTER)
    Specify radius of circle or [Diameter]: **1** (ENTER)

The same circle can be drawn as follows (see Figure 2–17):

    Command: **circle** (ENTER)
    Specify center point for circle or [3P/2P/Ttr (tan tan radius)]: **2,2** (ENTER)
    Specify radius of circle or [Diameter]: **3,2** (ENTER)

In the last example, AutoCAD used the distance between the center point and the second point given as the value for the radius of the circle.

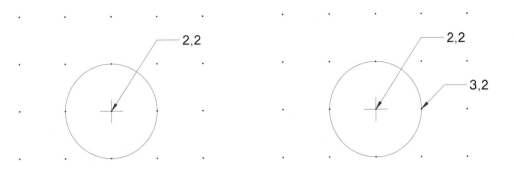

Figure 2–16 *A circle drawn with the* CIRCLE *command's default options: Center, Radius*

Figure 2–17 *A circle drawn with the Center-Radius option by specifying the distance*

Center-Diameter Option

Draws a circle based on a center point and a diameter. Invoke the Center-Diameter option:

Draw menu	Choose Circle > Center, Diameter
Command: prompt	**circle** (ENTER)

AutoCAD prompts:

> Command: **circle** (ENTER)
> Specify center point for circle or [3P/2P/Ttr (tan tan radius)]: *(specify center point to locate the center of the circle)*
> Specify radius of circle or [Diameter]: **d** (ENTER)
> Specify diameter of circle: *(specify the diameter of the circle)*

The following command sequence shows an example:

> Command: **circle** (ENTER)
> Specify center point for circle or [3P/2P/Ttr (tan tan radius)]: **2,2** (ENTER)
> Specify radius of circle or [Diameter]: **d** (ENTER)
> Specify diameter of circle: **2** (ENTER)

The same circle can be generated as follows (see Figure 2–18):

> Command: **circle** (ENTER)
> Specify center point for circle or [3P/2P/Ttr (tan tan radius)]: **2,2** (ENTER)
> Specify radius of circle or [Diameter]: **d** (ENTER)
> Specify diameter of circle: **4,2** (ENTER)

Note: Specifying a point causes AutoCAD to use the distance to the point specified from the previously selected center as the value for the diameter of the circle to be drawn.

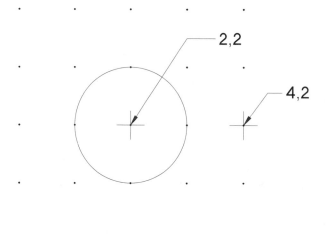

Figure 2–18 *A circle drawn using the Center-Diameter option*

Three-Point Circle Option

The Three-Point circle option draws a circle based on three points on the circumference.

Invoke the Three-Point circle option:

Draw menu	Choose Circle > 3 Points
Command: prompt	**circle** (ENTER)

AutoCAD prompts:

Command: **circle** (ENTER)
Specify center point for circle or [3P/2P/Ttr (tan tan radius)]: **3p** (ENTER)
Specify first point on circle: *(specify a point or a coordinate)*
Specify second point on circle: *(specify a point or a coordinate)*
Specify third point on circle: *(specify a point or a coordinate)*

The following command sequence shows an example (see Figure 2–19).

Command: **circle** (ENTER)
Specify center point for circle or [3P/2P/Ttr (tan tan radius)]: **3P** (ENTER)
Specify first point on circle: **2,1** (ENTER)
Specify second point on circle: **3,2** (ENTER)
Specify third point on circle: **2,3** (ENTER)

Note: The 3P response allows you to override the Center Point default. The 3 Point option can also be selected from the shortcut menu.

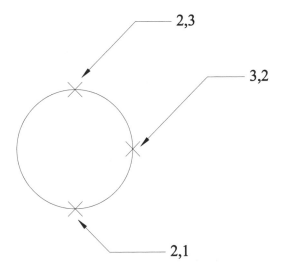

Figure 2–19 *A circle drawn with the Three-Point option*

Two-Point Circle Option

The Two-Point Circle option draws a circle based on two endpoints of the diameter.

Invoke the Two-Point circle option:

Draw menu	Choose Circle > 2 Points
Command: prompt	**circle** (ENTER)

AutoCAD prompts:

> Command: **circle** (ENTER)
> Specify center point for circle or [3P/2P/Ttr (tan tan radius)]: **2p** (ENTER)
> Specify first end point of circle's diameter: *(specify a point or a coordinate)*
> Specify second end point of circle's diameter: *(specify a point or a coordinate)*

The following command sequence shows an example (see Figure 2–20).

> Command: **circle** (ENTER)
> Specify center point for circle or [3P/2P/Ttr (tan tan radius)]: **2P** (ENTER)
> Specify first end point of circle's diameter: **1,2** (ENTER)
> Specify second end point of circle's diameter: **3,2** (ENTER)

78

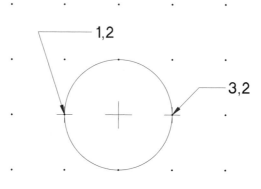

Figure 2–20 *A circle drawn with the Two-Point option*

Tangent, Tangent, Radius (TTR) Option

This option draws a circle tangent to two objects (either lines, arcs, or circles) with a specified radius.

Invoke the Tangent, Tangent, Radius option:

Draw menu	Choose Circle > Tan, Tan, Radius
Command: prompt	**circle** (ENTER)

AutoCAD prompts:

Command: **circle** (ENTER)
Specify center point for circle or [3P/2P/Ttr (tan tan radius)]: **ttr** (ENTER)
Specify point on object for first tangent of circle: *(specify a point on an object for first tangent of circle)*
Specify point on object for second tangent of circle: *(specify a point on an object for second tangent of circle)*
Specify radius of circle: *(specify radius)*

For specifying the "tangent-to" objects, it normally does not matter where on the objects you make your selection. However, if more than one circle can be drawn to the specifications given, AutoCAD will draw the one whose tangent point is nearest to the selection made.

Note: Until it is changed, the radius/diameter you specify in any one of the options becomes the default setting for subsequent circles to be drawn.

DRAWING ARCS

The ARC command makes an arc and offers eleven combinations to draw an arc:

1. Three-point (3 points)
2. Start, center, end (S,C,E)
3. Start, center, included angle (S,C,A)
4. Start, center, length of the chord (S,C,L)
5. Start, end, included angle (S,E,A)
6. Start, end, direction (S,E,D)
7. Start, end, radius (S,E,R)
8. Center, start, end (C,S,E)
9. Center, start, included angle (C,S,A)
10. Center, start, length of the chord (C,S,L)
11. Continuation from line or arc (LinCont or ArcCont)

Methods 8, 9, and 10 are just rearrangements of methods 2, 3, and 4, respectively.

Three-Point Arc Option

This option draws an arc using three specified points on the arc's circumference. The first point specifies the start point, the second point specifies a point on the circumference of the arc, and the third point is the arc endpoint. You can specify a three-point arc either clockwise or counterclockwise.

Invoke the Three-Point arc option:

Draw toolbar	Choose the ARC command (see Figure 2–21)
Draw menu	Choose Arc > 3 Points
Command: prompt	**arc** (ENTER)

Figure 2–21 *Invoking the ARC command from the Draw toolbar*

AutoCAD prompts:

> Command: **arc** (ENTER)
> Specify start point of arc or [Center]: *(specify second point of arc or right-click for shortcut menu to select one of the available options)*
> Specify second point of arc or [Center/End]: *(specify start point of arc or right-click for shortcut menu to select one of the available options)*
> Specify end point of arc: *(specify end point of arc)*

The following command sequence shows an example (see Figure 2–22).

> Command: **arc** (ENTER)
> Specify start point of arc or [Center]: **1,2** (ENTER)
> Specify second point of arc or [Center/End]: **2,1** (ENTER)
> Specify end point of arc: **3,2** (ENTER)

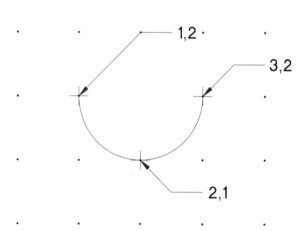

Figure 2–22 *An arc drawn with the* ARC *command's default option: 3 Points*

Start, Center, End Option (or S,C,E for short)

This option draws an arc using three specified points. The first point specifies the start point, the second point specifies the center point of the arc to be drawn, and the third point is the arc endpoint.

Invoke the Start, Center, End option:

Draw menu	Choose Arc > Start, Center, End
Command: prompt	**arc** (ENTER)

AutoCAD prompts:

> Command: **arc** (ENTER)
> Specify start point of arc or [Center]: *(specify start point of arc or right-click*
>   *for shortcut menu to select one of the available options)*
> Specify second point of arc or [Center/End]: **c** (ENTER)
> Specify center point of arc: (specify the center point of arc)
> Specify end point of arc or [Angle/chord Length]: *(specify end point of arc or*
>   *right-click for shortcut menu to select one of the available options)*

The following command sequence shows an example (see Figure 2–23).

> Command: **arc** (ENTER)
> Specify start point of arc or [Center]: **1,2** (ENTER)
> Specify second point of arc or [Center/End]: **c** (ENTER)
> Specify center point of arc: **2,2** (ENTER)
> Specify end point of arc or [Angle/chord Length]: **2,3** (ENTER)

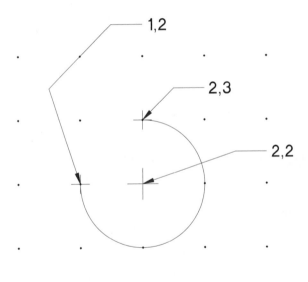

Figure 2–23 *An arc drawn with the Start, Center, End (S,C,E) option*

Arcs drawn by this method are always drawn counterclockwise from the starting point. The distance between the center point and the starting point determines the radius. Therefore, the point specified in response to "end point" needs only to be on the same radial line of the desired endpoint. For example, specifying the point 2,2.5 or 2,4 draws the same arc.

An alternative to this method is to specify the center point first, as follows (see Figure 2–24):

```
Command: arc (ENTER)
Specify start point of arc or [Center]: c (ENTER)
Specify center point of arc: 2,2 (ENTER)
Specify start point of arc: 1,2 (ENTER)
Specify end point of arc or [Angle/chord Length]: 2,3 (ENTER)
```

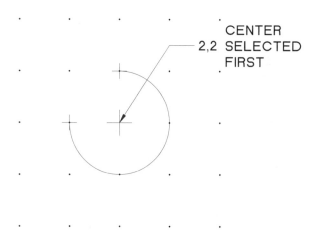

CENTER
2,2 SELECTED
FIRST

Figure 2–24 *An arc drawn by specifying the center point first*

Start, Center, Angle Option (or S,C,A for short)

This option draws an arc similar to the Start, Center, End option method, but it places the endpoint on a radial line at the specified angle from the line between the center point and the start point. If you specify a positive angle as the included angle, an arc is drawn counterclockwise; for a negative angle, the arc is drawn clockwise.

Invoke the Start, Center, Angle option:

Draw menu	Choose Arc > Start, Center, Angle
Command: prompt	arc (ENTER)

AutoCAD prompts:

```
Command: arc (ENTER)
Specify start point of arc or [CEnter]: (specify start point of arc or right-click
    for shortcut menu to select one of the available options)
Specify second point of arc or [CEnter/End]: c (ENTER)
Specify center point of arc: (specify the center point of arc)
Specify end point of arc or [Angle/chord Length]: a (ENTER)
Specify Included Angle: (specify the included angle of the arc to be drawn)
```

The following command sequence shows an example (see Figure 2–25).

> Command: **arc** (ENTER)
> Specify start point of arc or [CEnter]: **1,2** (ENTER)
> Specify second point of arc or [CEnter/End]: **c** (ENTER)
> Specify center point of arc: **2,2** (ENTER)
> Specify end point of arc or [Angle/chord Length]: **a** (ENTER)
> Specify Included Angle: **270** (ENTER)

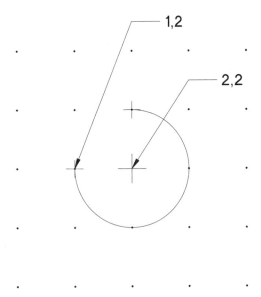

Figure 2–25 *An arc drawn with the Start, Center, Angle (S,C,A) option*

Note: If a point directly below the specified center were selected (in the previous example) in response to the "Included angle:" prompt, AutoCAD would read the angle of the line (270 degrees from zero) as the included angle for the arc. In Figure 2–25, the point selected in response to the "Included angle:" prompt causes AutoCAD to read the angle between the line it establishes from the center and the zero direction (east in the default coordinate system). It does not measure the angle between the line the point establishes from the center and the line established from the center to the start point.

Start, Center, Length of Chord Option (or S,C,L for short)

This method uses the specified chord length as the straight-line distance from the start point to the endpoint. With any chord length (equal to or less than the diameter length) there are four possible arcs that can be drawn: a major arc in either direction and a minor arc in either direction. Therefore, all arcs drawn by this method are counterclockwise from the start point. A positive value for the length of chord will cause AutoCAD to draw the minor arc; a negative value will result in the major arc.

84

Invoke the Start, Center, Length of chord Option:

Draw menu	Choose Arc > Start, Center, Length
Command: prompt	**arc** (ENTER)

AutoCAD prompts:

> Command: **arc** (ENTER)
> Specify start point of arc or [CEnter]: *(specify start point of arc or right-click*
>   *for shortcut menu to select one of the available options)*
> Specify second point of arc or [CEnter/End]: **c** (ENTER)
> Specify center point of arc: *(specify the center point of arc)*
> Specify end point of arc or [Angle/chord Length]: **l** (ENTER)
> Specify length of chord: *(specify the length of the chord of the arc to be drawn)*

The following command sequence shows an example of drawing a minor arc (shown in Figure 2–26).

> Command: **arc** (ENTER)
> Specify start point of arc or [CEnter]: **1,2** (ENTER)
> Specify second point of arc or [CEnter/End]: **c** (ENTER)
> Specify center point of arc: **2,2** (ENTER)
> Specify end point of arc or [Angle/chord Length]: **l** (ENTER)
> Specify length of chord: **1.414** (ENTER)

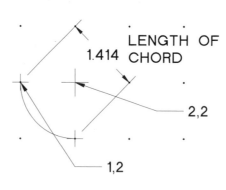

Figure 2–26 *A minor arc drawn with the Start, Center, Length of chord (S,C,L) option.*

The following command sequence shows an example of drawing a major arc (shown in Figure 2–27).

Command: **arc** (ENTER)
Specify start point of arc or [CEnter]: **1,2** (ENTER)
Specify second point of arc or [CEnter/End]: **c** (ENTER)
Specify center point of arc: **2,2** (ENTER)
Specify end point of arc or [Angle/chord Length]: **l** (ENTER)
Specify length of chord: **–1.414** (ENTER)

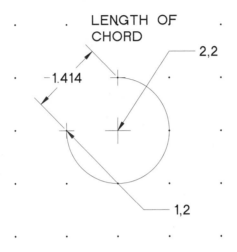

Figure 2–27 *A major arc drawn with the Start, Center, Length of chord (S,C,L) option*

Start, End, Angle Option (or S,E,A for short)

This option draws an arc similar to the Start, Center, Angle option method, and places the endpoint on a radial line at the specified angle from the line between the center point and the start point. If you specify a positive angle for the included angle, an arc is drawn counterclockwise; for a negative angle the arc is drawn clockwise.

Invoke the Start, End, Angle option:

Draw menu	Choose Arc > Start, Center, Angle
Command: prompt	**arc** (ENTER)

AutoCAD prompts:

Command: **arc** (ENTER)
Specify start point of arc or [CEnter]: *(specify start point of arc or right-click
    for shortcut menu to select one of the available options)*
Specify second point of arc or [CEnter/End]: **e** (ENTER)
Specify end point of arc: *(specify the end point of arc)*
Specify center point of arc or [Angle/Direction/Radius]: **a** (ENTER)
Specify included angle: *(specify the included angle of the arc to be drawn)*

The arc shown in Figure 2–28 is drawn using the following sequence:

Command: **arc** (ENTER)
Specify start point of arc or [CEnter]: **3,2** (ENTER)
Specify second point of arc or [CEnter/End]: **e** (ENTER)
Specify end point of arc: **2,3** (ENTER)
Specify center point of arc or [Angle/Direction/Radius]: **a** (ENTER)
Specify included angle: **90** (ENTER)

The arc shown in Figure 2–29 is drawn with a negative angle using the following sequence:

Command: **arc** (ENTER)
Specify start point of arc or [CEnter]: **3,2** (ENTER)
Specify second point of arc or [CEnter/End]: **e** (ENTER)
Specify end point of arc: **2,3** (ENTER)
Specify center point of arc or [Angle/Direction/Radius]: **a** (ENTER)
Specify included angle: **–270** (ENTER)

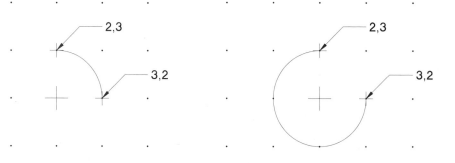

Figure 2–28 *An arc drawn counter-clockwise with the Start, End, Included Angle (S,E,A) option*

Figure 2–29 *An arc drawn clockwise with the Start, End, Included Angle (S,E,A) option*

Start, End, Direction Option (or S,E,D for short)

This method allows you to draw an arc between selected points by specifying a direction in which the arc will start from the selected start point. Either the direction can be keyed in or you can select a point on the screen with your pointing device.

If you select a point on the screen, AutoCAD uses the angle from the start point to the selected point as the starting direction.

Invoke the Start, End, Direction option:

Draw menu	Choose Arc > Start, End, Direction
Command: prompt	**arc** (ENTER)

AutoCAD prompts:

> Command: **arc** (ENTER)
> Specify start point of arc or [Center]: *(specify start point of arc or right-click for shortcut menu to select one of the available options)*
> Specify second point of arc or [CEnter/End]: **e** (ENTER)
> Specify end point of arc: *(specify end point of arc)*
> Specify center point of arc or [Angle/Direction/Radius]: **d** (ENTER)
> Direction from start point: *(specify the direction from the start point of the arc to be drawn)*

The arc shown in Figure 2–30 is drawn using the following sequence:

> Command: **arc** (ENTER)
> Specify start point of arc or [Center]: **3,2** (ENTER)
> Specify second point of arc or [CEnter/End]: **e** (ENTER)
> Specify end point of arc: **2,3** (ENTER)
> Specify center point of arc or [Angle/Direction/Radius]: **d** (ENTER)
> Direction from start point: **90** (ENTER)

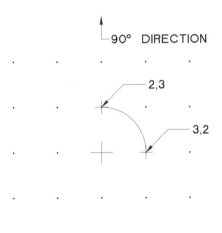

Figure 2–30 *An arc drawn with the Start, End, Direction (S,E,D) option*

Start, End, Radius Option (or S,E,R for short)

This method allows you to specify a radius after selecting the two endpoints of the arc. As with the Chord Length method, there are four possible arcs that can be drawn: a major arc in either direction and a minor arc in either direction. Therefore, all arcs

88

drawn by this method are counterclockwise from the start point. A positive value for the radius causes AutoCAD to draw the minor arc; a negative value results in the major arc.

Invoke the Start, End, Radius option:

Draw menu	Choose Arc > Start, End, Radius
Command: prompt	**arc** (ENTER)

AutoCAD prompts:

Command: **arc** (ENTER)
Specify start point of arc or [CEnter]: *(specify start point of arc or right-click for shortcut menu to select one of the available options)*
Specify second point of arc or [CEnter/End]: **e** (ENTER)
Specify end point of arc: *(specify the end point of arc)*
Specify center point of arc or [Angle/Direction/Radius]: **r** (ENTER)
Specify radius of arc: *(specify the radius of the arc to be drawn)*

The following command sequence shows an example of drawing a minor arc, as shown in Figure 2–31.

Command: **arc** (ENTER)
Specify start point of arc or [CEnter]: **1,2** (ENTER)
Specify second point of arc or [CEnter/End]: **e** (ENTER)
Specify end point of arc: **2,3** (ENTER)
Specify center point of arc or [Angle/Direction/Radius]: **r** (ENTER)
Specify radius of arc: **-1** (ENTER)

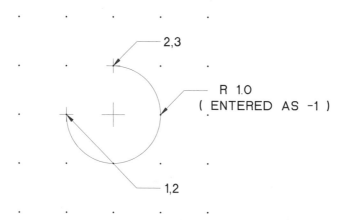

Figure 2–31 *A major arc drawn with the Start, End, Radius (S,E,R) option*

The following command sequence shows an example of drawing a major arc, as shown in Figure 2–32.

Command: **arc** (ENTER)
Specify start point of arc or [CEnter]: **2,3** (ENTER)
Specify second point of arc or [CEnter/End]: **e** (ENTER)
Specify end point of arc: **1,2** (ENTER)
Specify center point of arc or [Angle/Direction/Radius]: **r** (ENTER)
Specify radius of arc: **1** (ENTER)

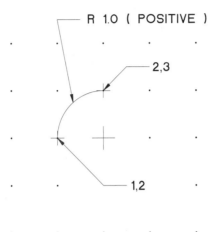

R 1.0 (POSITIVE)

2,3

1,2

Figure 2–32 *A minor arc drawn with the Start, End, Radius (S,E,R) option*

Center, Start, End Option (or C,S,E for short)

This method is similar to the Start, Center, End (S,C,E) method, except in this option the beginning point is the center point of the arc rather than the start point.

Center, Start, Angle Option (or C,S,A for short)

This method is similar to the Start, Center, Angle (S,C,A) method, except in this option the beginning point is the center point of the arc rather than the start point.

Center, Start, Length Option (or C,S,L for short)

This method is similar to the Start, Center, Length (S,C,L) method, except that the beginning point is the center point of the arc rather than the start point.

Line-Arc and Arc-Arc Continuation Option

You can use an automatic start point, endpoint, starting direction method to draw an arc by pressing ENTER as a response to the first prompt of the ARC command. After pressing ENTER, the only other input is to select or specify the endpoint of the arc you wish to draw. AutoCAD uses the endpoint of the previous line or arc (whichever was drawn last) as the start point of the new arc. AutoCAD then uses the ending direction of that last drawn object as the starting direction of the arc. Examples are shown in the following sequences and figures.

The start point of the existing arc is 2,1 and the endpoint is 3,2 with a radius of 1. This makes the ending direction of the existing arc 90 degrees, as shown in Figure 2–33.

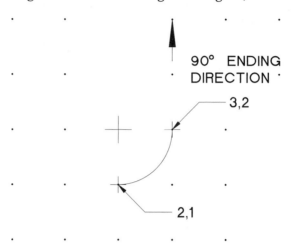

Figure 2–33 *An arc drawn with a start point (2,1), an endpoint (3,2), and a radius of 1.0*

The following command sequence continues drawing an arc, from the last-drawn arc, as shown in Figure 2–34 (arc-arc continuation).

Command: **arc** (ENTER)
Specify start point of arc or [Center]: (ENTER)
Specify end point of arc: **2,3** (ENTER)

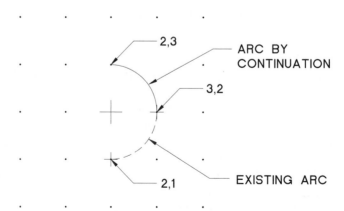

Figure 2–34 *An arc drawn by means of the Arc-Arc Continuation method*

The arc, as shown in Figure 2–33, is drawn clockwise instead, with its start point at 3,2 to an endpoint of 2,1 (see Figure 2–35).

The following command sequence will draw the automatic start point, endpoint, starting direction arc, as shown in Figure 2–36.

> Command: **arc** (ENTER)
> Specify start point of arc or [Center]: (ENTER)
> Specify end point of arc: **2,3** (ENTER)

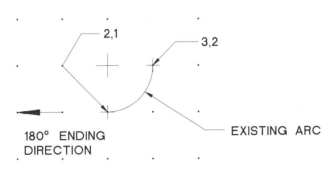

Figure 2–35 *An arc drawn clockwise with start point (3,2) and endpoint (2,1)*

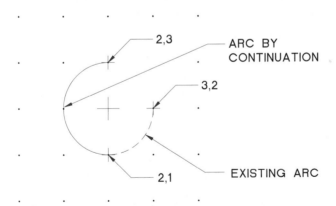

Figure 2–36 *An arc drawn by means of the automatic Start point–Endpoint-Starting direction method*

In the last case, the direction used is 180 degrees. The same arc would have been drawn if the last "line-or-arc" drawn was a line starting at 4,1 and ending at 2,1.

 Note: This method uses the last drawn arc or a line. If you draw an arc, then draw a line, then draw a circle, and then use this continuation method, AutoCAD will use the line as the basis for the start point and direction. This is because the line was the last of the "line-or-arc" objects drawn.

OBJECT SELECTION

Many AutoCAD modify and construct commands prompt you to select one or more objects for manipulation. When you select one or more objects, AutoCAD highlights them by displaying them with dashed lines. The group of objects selected for the manipulation is called the selection set. There are several different ways of selecting the objects for manipulation. The selection options include: Window, Window Polygon (WP), Crossing, Crossing Polygon (CP), Box, Fence, All, Last, Previous, Group Add, Remove, Single, Multiple, and Undo.

All modify and construct commands require a selection set, for which AutoCAD prompts:

> Select objects:

AutoCAD replaces the screen crosshairs with a small box called the object selection target. With the target cursor, select individual objects for manipulation. Using your pointing device (or the keyboard's cursor keys), position the target box so it touches only the desired object or a visible portion of it. The object selection target helps you select the object without having to be very precise. Every time you select an object, the "Select objects:" prompt reappears. To indicate your acceptance of the selection set, give a Null reply (press ENTER) at the "Select objects:" prompt.

Sometimes, it is difficult to select objects that are close together or lie directly on top of one another. You can use the pick button to cycle through these objects, one after the other, until you reach the one you want.

To cycle through objects for selection, at the "Select Objects" prompt, hold down the CTRL button. Select a point as near as possible to the object. Press the pick button on your pointing device repeatedly until the object you want is highlighted, and press ENTER to select the object.

WINDOW OPTION

The Window option for selecting objects allows you to select all the objects contained completely within a rectangular area or dynamically manipulated window.

The window can be defined by specifying a point at the appropriate location to the "Select Objects:" prompt and moving the cursor toward the right of the first data point. AutoCAD prompts:

> Specify opposite corner: (specify opposite corner)

You can also define a window for selection of objects, by entering w in response to the "Select Objects:" prompt, and AutoCAD prompts:

> Specify first corner: (specify first corner)
> Specify opposite corner: (specify opposite corner)

If there is an object that is partially inside the rectangular area, then that object is not included in the selection set. You can select only objects currently visible on the screen. To select a partially visible object, you must include all its visible parts within the window. See Figure 2–37, in which only the lines will be included, not the circles, because a portion of each of the circles is outside the rectangular area.

Figure 2–37 *Selecting objects by means of the Window option*

Crossing Option

The Crossing option for selecting objects allows you to select all the objects contained completely in a rectangular area as well as the objects that are crossing the window.

The crossing rectangle can be placed by specifying a point at the appropriate location to the "Select Objects:" prompt and moving the cursor toward the left of the first data point. AutoCAD prompts:

> Specify opposite corner: *(specify opposite corner)*

You can also define a crossing window for the selection of objects, by entering c to the "Select Objects:" prompt, and AutoCAD prompts:

> Specify first corner: (specify first corner)
> Specify opposite corner: (specify opposite corner)

See Figure 2–38, in which all the lines and circles are included, though parts of the circles are outside the rectangle.

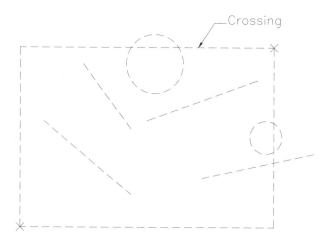

Figure 2–38 *Selecting objects by means of the Crossing option*

Previous Option

The Previous option enables you to perform several operations on the same object or group of objects. AutoCAD remembers the most recent selection set and allows you to reselect it with the Previous option. For example, if you moved several objects and now wish to copy them elsewhere, you can invoke the COPY command and respond to the "Select objects:" prompt by entering **p** to select the same objects again. (There is a command called SELECT that does nothing but create a selection set; you can then use the Previous option to refer to this set in subsequent commands.)

Last Option

The Last option is an easy way to select the most recently created object currently visible. Only one object is designated, no matter how often you use the Last option, when constructing a particular selection set. The Last option is invoked by entering **l** at the "Select Objects:" prompt.

The wpolygon, cpolygon, fence, all, group, box, auto, undo, single, multiple, add, and remove options are explained in Chapter 5.

MODIFY OBJECTS

AutoCAD not only allows you to draw objects easily, but also allows you to modify the objects you have drawn. Of the many modifying commands available, the ERASE command will probably be the one you use most often. Everyone makes mistakes, but in AutoCAD it is easier to erase them. Or, if you are through with an object that you have created for aid in constructing other objects, you may wish to erase it.

ERASING OBJECTS

To erase objects from a drawing, invoke the ERASE command:

Modify toolbar	Choose the ERASE command (see Figure 2–39)
Modify menu	Choose Erase
Command: prompt	**erase** (ENTER)

Figure 2–39 *Invoking the* ERASE *command from the Modify toolbar*

> Command: **erase** (ENTER)
> Select objects: (select objects to be erased, and then press the SPACEBAR
>   or ENTER)

You can use one or more available object selection methods. After selecting the object(s), press ENTER (null response) in response to the next "Select objects:" prompt to complete the ERASE command. All the objects that were selected will disappear.

The following command sequence shows an example of erasing individual objects (shown Figure 2–40).

> Command: **erase** (ENTER)
> Select objects: *(select Line 2, the line is highlighted)* I selected, I found
> Select objects: *(select Line 4, the line is highlighted)* I selected, I found
> Select objects: (ENTER)

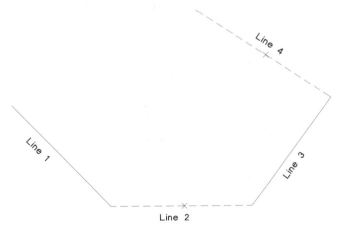

Figure 2–40 *Selection of individual objects to erase*

The following command sequence shows an example of erasing a selection of objects by the Window option (shown in Figure 2–41):

Command: **erase** (ENTER)
Select objects: (specify a point)
Specify opposite corner: *(Specify a point for the diagonally opposite corner)*
Select objects: (ENTER)

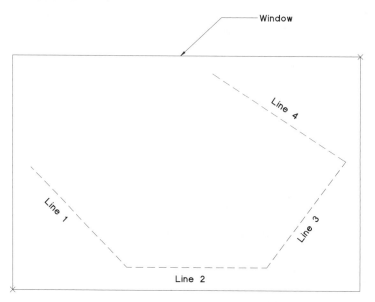

Figure 2–41 *Erasing a group of objects*

GETTING IT BACK

The OOPS command restores objects that have been unintentionally erased. Whenever the ERASE command is used, the last group of objects erased is stored in memory. The OOPS command will restore the objects; it can be used at any time. It only restores the objects erased by the most recent erase command. See Chapter 3 on the UNDO command, if you need to step back further than one ERASE command.

To restore objects erased by the last ERASE command, invoke the OOPS command:

Command: prompt	oops (ENTER)

AutoCAD restores the objects erased by the last erase command.

The following example shows the command sequence for using the OOPS command in conjunction with the erase command:

Command: erase (ENTER)
Select objects: *(specify a point)*
Other corner: *(specify a point for the diagonally opposite corner)*
Select objects: (ENTER)
Command: oops (ENTER) *(will restore the erased objects)*

PROJECT EXERCISE

This project exercise provides point-by-point instructions for setting up the drawing, laying out the border, and then creating the objects shown in Figure P2–1. In this exercise you will apply the skills acquired in Chapters 1 and 2 to drawing basic architectural, mechanical, electronic, and piping objects.

Figure P2–1 *Completed project drawing*

In this project you will learn to do the following:

- Apply the "Use a Wizard" option of the Create a New Drawing dialog box.
- Set the Units used for the drawing.
- Set the Limits for the drawing Area
- Use the LINE, RECTANGLE, CIRCLE, and ARC commands.

 Note: As you complete each step in the project procedure, place a check mark by the step to help you keep up with where you are in the project.

SET UP THE DRAWING AND DRAW A BORDER

This procedure describes the steps required to set up the drawing and to draw the border.

Step 1: Invoke the AutoCAD program.

Example: Under Microsoft Windows 98, find the AutoCAD 2000 program in the Start > Programs menu and select it.

Step 2: To create a new drawing, invoke the NEW command from the Standard toolbar, as shown in Figure P2–2, or select NEW from the File menu.

Figure P2–2 *Invoking the* NEW *command from the Standard toolbar*

Step 3: AutoCAD displays the Create New Drawing dialog box, as shown in Figure P2–3. Click the "Use a Wizard" button, and then select the Quick Setup and choose the **OK** button to close the dialog box.

Figure P2–3 *Create New Drawing dialog box*

Step 4 AutoCAD displays the Quick Setup dialog box with the listing of options available for unit of measurement, as shown in Figure P2–4. Select the "Decimal" radio button, and choose the **Next >** button.

Figure P2–4 *Quick Setup dialog box with the selection of the unit of measurement*

Step 5   AutoCAD displays the Quick Setup dialog box with the display for Area setting, as shown in Figure P2–5. Under the **Width:** text box enter 11, and under the **Length:** text box enter 8.5. Choose the **Finish** button to close the Quick Step dialog box.

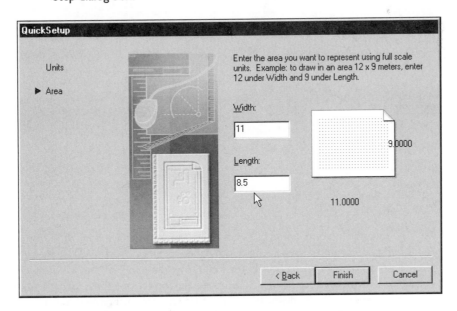

Figure P2–5 *Quick Setup dialog box with the Area setting*

Step 6   To draw the border, invoke the RECTANGLE command from the Draw toolbar, as shown in Figure P2–6, or enter **rectangle** at the "Command:" prompt and press ENTER.

Figure P2–6 *Invoking the* RECTANGLE *command from the Draw toolbar*

AutoCAD prompts:

> Command: **rectangle** (ENTER)
> Specify first corner point or [Chamfer/Elevation/Fillet/Thickness/
>   Width]: **0.25,0.25** (ENTER)
> Specify other corner point: **10.75,8.25** (ENTER)

AutoCAD draws a rectangle border as shown in Figure P2–7.

Figure P2–7 *Completed border outline*

Step 7   To save the current status of the drawing, invoke the SAVE command from the File menu or enter save at the "Command:" prompt. AutoCAD displays the Save Drawing As dialog box. Enter **PROJ2** as the name of the drawing in the **File name:** text box, and click the **Save** button to save the drawing and close the Save Drawing As dialog box. Make sure you are saving the drawing in the appropriate folder.

DRAW THE DOOR JAMB

This procedure describes the steps required to draw the door jamb shown in Figure P2–8. The door jamb is drawn with the LINE command using rectangular coordinates and polar coordinates.

DOOR JAMB

Figure P2–8 *Door Jamb*

Step I   To draw the door jamb, invoke the LINE command from the Draw Toolbar, as shown in Figure P2–9, or enter **line** and press ENTER at the "Command:" prompt.

Figure P2–9 *Invoking the* LINE *command from the Draw toolbar*

AutoCAD prompts:

    Command: **line** (ENTER)
    Specify first point: **2,1.5** (ENTER)
    Specify next point or [Undo]: **2,2.5** (ENTER)
    Specify next point or [Undo]: **1.625,2.5** (ENTER)
    Specify next point or [Close/Undo]: **1.625,3.5** (ENTER)
    Specify next point or [Close/Undo]: **2,3.5** (ENTER)
    Specify next point or [Close/Undo]: **2,4.25** (ENTER)
    Specify next point or [Close/Undo]: **3.125,4.25** (ENTER)

Specify next point or [Close/Undo]: **3.125,4** (ENTER)
Specify next point or [Close/Undo]: **3,4** (ENTER)
Specify next point or [Close/Undo]: **@.125<90** (ENTER)
Specify next point or [Close/Undo]: **@.875<180** (ENTER)
Specify next point or [Close/Undo]: **@.75<270** (ENTER)
Specify next point or [Close/Undo]: **@.375<180** (ENTER)
Specify next point or [Close/Undo]: **@0,–.75** (ENTER)
Specify next point or [Close/Undo]: **@.375,0** (ENTER)
Specify next point or [Close/Undo]: **@0,–1** (ENTER)
Specify next point or [Close/Undo]: **@.875,0** (ENTER)
Specify next point or [Close/Undo]: **@0,.125** (ENTER)
Specify next point or [Close/Undo]: **@0.125,0** (ENTER)
Specify next point or [Close/Undo]: **@0,–.25** (ENTER)
Specify next point or [Close/Undo]: **@–1.125,0** (ENTER)
Specify next point or [Close/Undo]: *(press the* ENTER *to terminate the command sequence)*

Step 2   To save the current status of the drawing, invoke the SAVE command from the File menu or enter **qsave** at the "Command:" prompt. AutoCAD saves the current status of the drawing.

DRAW THE INSTRUMENT PANEL

This procedure describes the steps required to draw the instrument panel shown in Figure P2–10. The instrument panel is drawn with the LINE, RECTANGLE, CIRCLE, and ARC commands.

DOOR JAMB   INSTRUMENT PANEL

Figure P2–10 *Door jamb and instrument panel*

Step 1 To draw the main outline of the instrument panel and the central rectangle, invoke the RECTANGLE command from the Draw Toolbar or enter **rectangle** and press ENTER at the "Command:" prompt.

AutoCAD prompts:

> Command: **rectangle** (ENTER)
> Specify first corner point or [Chamfer/Elevation/Fillet/Thickness/
>   Width]: **5.75,1.25** (ENTER)
> Specify other corner point: **9.875,4** (ENTER)
>
> Command: *(press ENTER to repeat the RECTANGLE command)*
> Specify first corner point or [Chamfer/Elevation/Fillet/Thickness/
>   Width]: **7.375,2.5** (ENTER)
> Specify other corner point: **8.375,3.125** (ENTER)

Step 2 To draw the trapezoidal shapes located in the top of the instrument panel, invoke the LINE command from the Draw Toolbar or enter **line** and press ENTER at the "Command:" prompt.

AutoCAD prompts:

> Command: **line** (ENTER)
> Specify first point: **7.375,3.75** (ENTER)
> Specify next point or [Undo]: **@.375<0** (ENTER)
> Specify next point or [Undo]: **7.625,3.5** (ENTER)
> Specify next point or [Close/Undo]: **@.125<180** (ENTER)
> Specify next point or [Close/Undo]: **c** (ENTER)
>
> Command: *(press ENTER to repeat the LINE command)*
> Specify first point: **8,3.75** (ENTER)
> Specify next point or [Undo]: **@.375<0** (ENTER)
> Specify next point or [Undo]: **8.25,3.5** (ENTER)
> Specify next point or [Close/Undo]: **@.125<180** (ENTER)
> Specify next point or [Close/Undo]: **c** (ENTER)

Step 3 To draw the circular gauges with the arrow pointers located in the top of the instrument panel, first invoke the CIRCLE command from the Draw Toolbar as shown in Figure P2–11, or enter **circle** and press ENTER at the "Command:" prompt.

Figure P2–11 *Invoking the* CIRCLE *command from the Draw toolbar*

AutoCAD prompts:

> Command: **circle** (ENTER)
> Specify center point for circle or [3P/2P/Ttr (tan tan radius)]:
>   **6.625,3.25** (ENTER)
> Specify radius of circle or [Diameter]: **@.5<0** (ENTER)
>
> Command: (press enter to repeat the CIRCLE command)
> Specify center point for circle or [3P/2P/Ttr (tan tan radius)]:
>   **9.125,3.25** (ENTER)
> Specify radius of circle or [Diameter]: **@.5<0** (ENTER)

To draw the arrow pointers, invoke the LINE command from the Draw Toolbar or enter **line** and press ENTER at the "Command:" prompt.

AutoCAD prompts:

> Command: **line** (ENTER)
> Specify first point: **6.5,3.5** (ENTER)
> Specify next point or [Undo]: **6.625,3.75** (ENTER)
> Specify next point or [Undo]: **6.75,3.5** (ENTER)
> Specify next point or [Close/Undo]: (press ENTER to terminate the
>   command sequence)
>
> Command: (press ENTER to repeat the LINE command)
> Specify first point: **6.625,3.25** (ENTER)
> Specify next point or [Undo]: **6.625,3.75** (ENTER)
> Specify next point or [Undo]: (press ENTER to terminate the command
>   sequence)
>
> Command: (press ENTER to repeat the LINE command)
> Specify first point: **9.375,3.375** (ENTER)
> Specify next point or [Undo]: **9.625,3.25** (ENTER)
> Specify next point or [Undo]: **9.375,3.125** (ENTER)
> Specify next point or [Close/Undo]: (press ENTER to terminate the
>   command sequence)
>
> Command: (press ENTER to repeat the LINE command)
> Specify first point: **9.125,3.25** (ENTER)
> Specify next point or [Undo]: **9.625,3.25** (ENTER)
> Specify next point or [Undo]: (press ENTER to terminate the command
>   sequence)

Step 4   To draw the two circles located in the bottom of the instrument panel, first invoke the CIRCLE command from the Draw Toolbar or enter **circle** and press ENTER at the "Command:" prompt.

AutoCAD prompts:

Command: **circle** (ENTER)
Specify center point for circle or [3P/2P/Ttr (tan tan radius)]: **7.5,1.875**
   (ENTER)
Specify radius of circle or [Diameter]: **.25** (ENTER)

Command: *(press* enter *to repeat the* CIRCLE *command)*
Specify center point for circle or [3P/2P/Ttr (tan tan radius)]:
   8.25,1.875 (ENTER)
Specify radius of circle or [Diameter]: **.25** (ENTER)

Step 5   To draw the switches located in the bottom of the instrument panel, first invoke the RECTANGLE command from the Draw toolbar, or enter **rectangle** at the "Command:" prompt.

AutoCAD prompts:

Command: **rectangle** (ENTER)
Specify first corner point or [Chamfer/Elevation/Fillet/Thickness/
   Width]: **6.4375,1.5** (ENTER)
Specify other corner point: **6.5625,2** (ENTER)

Command: *(press* ENTER *to repeat the* RECTANGLE *command)*
Specify first corner point or [Chamfer/Elevation/Fillet/Thickness/
   Width]: **9.1875,1.5** (ENTER)
Specify other corner point: **9.3125,2** (ENTER)

Invoke the ARC command from the Draw Toolbar, as shown in Figure P2–12, or enter **arc** at the "Command:" prompt.

Figure P2–12 *Invoking the* ARC *command from the Draw toolbar*

AutoCAD prompts:

Command: **arc** (ENTER)
Specify start point of arc or [CEnter]: **6.5625,1.875** (ENTER)
Specify second point of arc or [CEnter/ENd]: **c** (ENTER)
Specify center point of arc: **6.5,2** (ENTER)
Specify end point of arc or [Angle/chord Length]: **6.4375,1.875** (ENTER)

Command: *(press* ENTER *to repeat the* ARC *command)*
Specify start point of arc or [CEnter]: **9.3125,1.875** (ENTER)
Specify second point of arc or [CEnter/ENd]: **c** (ENTER)
Specify center point of arc: **9.25,2** (ENTER)
Specify end point of arc or [Angle/chord Length]: **9.1875,1.875** (ENTER)

Step 6 To save the current status of the drawing, invoke the SAVE command from the File menu or enter **qsave** at the "Command:" prompt. AutoCAD saves the current status of the drawing.

DRAW THE 90° ELL

This procedure describes the steps required to draw the 90-degree ell shown in Figure P2–13. The 90-degree ell is drawn with the LINE and ARC commands.

Figure P2–13 *Door jamb, instrument panel, and 90-degree ell*

Step 1 To draw the lower lines of the 90-degree ell, invoke the LINE command from the Draw toolbar or enter **line** and press ENTER at the "Command:" prompt.

AutoCAD prompts:

    Command: **line** (ENTER)
    Specify first point: **8.25,5.875** (ENTER)
    Specify next point or [Undo]: **@.125<0** (ENTER)
    Specify next point or [Undo]: **@.625<-90** (ENTER)
    Specify next point or [Close/Undo]: **@1.625<180** (ENTER)
    Specify next point or [Close/Undo]: **@.625<90** (ENTER)
    Specify next point or [Close/Undo]: **@.125<0** (ENTER)
    Specify next point or [Close/Undo]: (press ENTER to terminate the
        command sequence)

Step 2   To draw the large outer arc, invoke the ARC command from the Draw toolbar or enter **arc** and press ENTER at the "Command:" prompt.

AutoCAD prompts:

> Command: **arc** (ENTER)
> Specify start point of arc or [CEnter]: **6.875,5.875** (ENTER)
> Specify second point of arc or [CEnter/ENd]: **c** (ENTER)
> Specify center point of arc: **@1.875<0** (ENTER)
> Specify end point of arc or [Angle/chord Length]: **a** (ENTER)
> Specify included angle: **−90** (ENTER)

Step 3   To draw the upper lines of the 90-degree ell, invoke the LINE command from the Draw toolbar or enter **line** and press ENTER at the "Command:" prompt.

AutoCAD prompts:

> Command: **line** (ENTER)
> Specify first point: **8.75,7.75** (ENTER)
> Specify next point or [Undo]: **@.125<90** (ENTER)
> Specify next point or [Undo]: **@.625<0** (ENTER)
> Specify next point or [Close/Undo]: **@1.625<-90** (ENTER)
> Specify next point or [Close/Undo]: **@.625<180** (ENTER)
> Specify next point or [Close/Undo]: **@.125<90** (ENTER)
> Specify next point or [Close/Undo]: *(press ENTER to terminate the command sequence)*

Step 4   To draw the smaller outer arc, invoke the ARC command from the Draw toolbar or enter **arc** and press ENTER at the "Command:" prompt.

AutoCAD prompts:

> Command: **arc** (ENTER)
> Specify start point of arc or [CEnter]: *(enter @ and press enter to select the last point as the start point)*
> Specify second point of arc or [CEnter/ENd]: **c** (ENTER)
> Specify center point of arc: **@.5<-90** (ENTER)
> Specify end point of arc or [Angle/chord Length]: **@.5<180** (ENTER)

Step 5   To save the current status of the drawing, invoke the SAVE command from the File menu or enter **qsave** at the "Command:" prompt. AutoCAD saves the current status of the drawing.

DRAW THE COIL

This procedure describes the steps required to draw the coil shown in Figure P2–14. The coil is drawn with the ARC command.

Figure P2–14 *Door jamb, instrument panel, 90-degree ell, and coil*

Step 1 To draw the coil, invoke the ARC command from the Draw toolbar or enter **arc** and press ENTER at the "Command:" prompt.

AutoCAD prompts:

Command: **arc** (ENTER)
Specify start point of arc or [CEnter]: **1,7.25** (ENTER)
Specify second point of arc or [CEnter/ENd]: **1.625,7.625** (ENTER)
Specify end point of arc: **2.25,7.25** (ENTER)

Command: *(press ENTER to repeat the ARC command)*
Specify start point of arc or [CEnter]: *(press ENTER to continue from the previous point)*
Specify end point of arc: **2.25,5.25** (ENTER)

Command: *(press ENTER to repeat the ARC command)*
Specify start point of arc or [CEnter]: *(press ENTER to continue from the previous point)*
Specify end point of arc: **2,5.25** (ENTER)

Command: *(press* ENTER *to repeat the* ARC *command)*
Specify start point of arc or [CEnter]: *(press* ENTER *to continue from the previous point)*
Specify end point of arc: **2,7.25** (ENTER)

Command: *(press* ENTER *to repeat the* ARC *command)*
Specify start point of arc or [CEnter]: *(press* ENTER *to continue from the previous point)*
Specify end point of arc: **3.25,7.25** (ENTER)

(Take a breath. This completes the first loop of the coil. Even though the Three-Point option was used for the leading arc of the first loop, the leading arc of the second loop, which is the trailing arc of the first loop, was drawn with the arc-arc continuation method.)

Command: *(press* ENTER *to repeat the* ARC *command)*
Specify start point of arc or [CEnter]: *(press* ENTER *to continue from the previous point)*
Specify end point of arc: **3.25,5.25** (ENTER)

Command: *(press* ENTER *to repeat the* ARC *command)*
Specify start point of arc or [CEnter]: *(press* ENTER *to continue from the previous point)*
Specify end point of arc: **3,5.25** (ENTER)

Command: *(press* ENTER *to repeat the* ARC *command)*
Specify start point of arc or [CEnter]: *(press* ENTER *to continue from the previous point)*
Specify end point of arc: **3,7.25** (ENTER)

Command: *(press* ENTER *to repeat the* ARC *command)*
Specify start point of arc or [CEnter]: *(press* ENTER *to continue from the previous point)*
Specify end point of arc: **4.25,7.25** (ENTER)

(Take another breath. This completes the second loop and starts the last loop of the coil.)

Command: *(press* ENTER *to repeat the* ARC *command)*
Specify start point of arc or [CEnter]: *(press* ENTER *to continue from the previous point)*
Specify end point of arc: **4.25,5.25** (ENTER)

Command: (press ENTER to repeat the ARC command)
Specify start point of arc or [CEnter]: *(press* ENTER *to continue from the previous point)*
Specify end point of arc: **4,5.25** (ENTER)

Command: *(press* ENTER *to repeat the* ARC *command)*
Specify start point of arc or [CEnter]: *(press* ENTER *to continue from the previous point)*
Specify end point of arc: **4,7.25** (ENTER)

Command: *(press* ENTER *to repeat the* ARC *command)*
Specify start point of arc or [CEnter]: *(press* ENTER *to continue from the previous point)*
Specify end point of arc: **5.25,7.25** (ENTER)

Step 2   To save the current status of the drawing, invoke the SAVE command from the File menu or enter **qsave** at the command: prompt. AutoCAD saves the current status of the drawing.

Congratulations! You have just successfully applied several AutoCAD concepts in creating the drawing.

EXERCISES 2–5 THROUGH 2–8

EXERCISE 2–5

Type of shape	Input Methods
Multiple circles	• Absolute rectangular coordinates • Center-Radius option for circles

Points of Interest

- The main octagon in this figure was drawn precisely by using the relative polar coordinate point entry method. However, only the coordinates of the bottom line are nice, whole units. The other endpoints around the polygon cannot easily be entered exactly from the keyboard.

- The most significant point of this figure is probably its center. The relative distance/direction from the center to the endpoints of the bottom line are calculated by trigonometry. Therefore, at this level we will simply enter "very close" coordinates for the center. We will do the same for the "bolt circle" (circle of circles).

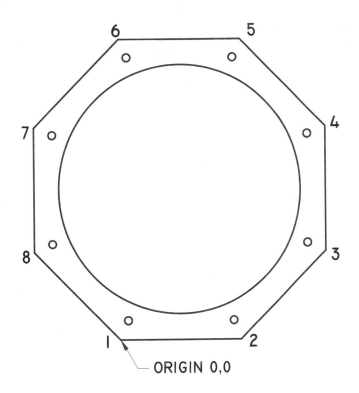

ORIGIN 0,0

Important Points

- Advanced features like the ARRAY command may be used to make the drawing exact if necessary.

- Note that arcs can be drawn both clockwise and counterclockwise with the three-point option. This is not the case with all options of the ARC command.

Drawing Setup	Open the drawing made in Exercise 2–4.
Power Point	Each center of the smaller circles can be established relative to the previous one drawn. But in order to start the first small circle, its center must be established by its absolute coordinates or by its relation to another point. We know that it is 2" distance at 68.5° from the left end of the bottom line (the origin 0,0). So, in order to make the origin the "last point," we will just start a line there and then not complete it. This will establish it as the point used by the "@" symbol. The next chapter will eliminate this trickery with the Tracking feature.

Point by Point

Invoke the CIRCLE command, and AutoCAD prompts:

> Command: **circle** (ENTER)
> Specify center point for circle or [3P/2P/Ttr (tan tan radius)]: **2'0,4'10**
>   (ENTER) *(center point of the large circle)*
> Specify radius of circle or [Diameter]: **48** (ENTER) *(radius of the large circle)*

Invoke the LINE command, and AutoCAD prompts:

> Command: **line** (ENTER)
> Specify start point: **0,0** (ENTER) *(establish the last point at 0,0)*
> Specify next point or [Undo]: *(press ENTER terminate the command sequence)*

Invoke the CIRCLE command, and AutoCAD prompts:

> Command: **circle**
> Specify center point for circle or [3P/2P/Ttr (tan tan radius)]: **@8<67.5**
>   (ENTER) *(center point of the first small circle)*
> Specify radius of circle or [Diameter]: **1.5** (ENTER) *(radius of the first small circle)*
>
> Command: *(press ENTER to invoke the CIRCLE command again)*
> Specify center point for circle or [3P/2P/Ttr (tan tan radius)]: **@42<0**
>   (ENTER) *(center point of the second small circle)*
> Specify radius of circle or [Diameter]: **1.5** (ENTER) *(radius of the second small circle)*

mechanical EXERCISE

Command: *(press ENTER to invoke the CIRCLE command again)*
Specify center point for circle or [3P/2P/Ttr (tan tan radius)]: **@42<45**
  (ENTER) *(center point of the third small circle)*
Specify radius of circle or [Diameter]: **1.5** (ENTER) *(radius of the second third circle)*

Command: *(press ENTER to invoke the CIRCLE command again)*
Specify center point for circle or [3P/2P/Ttr (tan tan radius)]: **@42<90**
  (ENTER) *(center point of the fourth small circle)*
Specify radius of circle or [Diameter]: **1.5** (ENTER) *(radius of the fourth small circle)*

(Draw the fifth through eighth small circles by continuing with the method explained above in the previous circles but adding 45 degrees to the angle each time)

mechanical EXERCISE

EXERCISE 2–6

Type of shapes	Input Methods
Multiple line/arc sequences	• Absolute rectangular coordinates • Relative rectangular coordinates • Relative polar coordinates

Some—but not all—of the points that determine the shape in this exercise can be entered by any one of the three listed methods. This exercise is primarily to emphasize when the line-arc continuation and the arc-line continuation options can be applied and when they cannot.

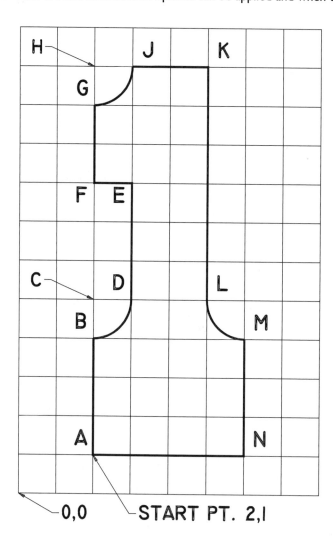

0,0   START PT. 2,1

Points of Interest

- The default line-line continuation of the LINE command does not require that lines drawn subsequent to the previous line segment start in the same direction that the previous line ends.

- In line-line continuation you may proceed in any desired direction with the continuing line.

- In line-arc and arc-line continuations, the connecting line/arc must start in the same direction that the previous arc/line ends.

Drawing Setup	Use Quick Setup Wizard, set Units to Decimal, set Area to 9 x 13.

Point by Point

Invoke the LINE command, and AutoCAD prompts:

> Command: **line** (ENTER)
> Specify first point: **2,1** (ENTER) *(starting point of line A–B)*
> Specify next point or [Undo]: **2,4** (ENTER) *(end point of the line A–B)*
> Specify next point or [Close]: (ENTER) *(exits the LINE command)*

Invoke the ARC command, and AutoCAD prompts:

> Command: **arc** (ENTER)
> Specify start point of arc or [CEnter]: **2,4** (ENTER) *(starting point of the arc B–D)*
> Specify second point of arc or [CEnter/ENd]: **c** (ENTER)
> Specify center point of arc: **2,5** (ENTER) *(arc center for the arc B–D)*
> Specify end point of arc or [Angle/chord Length]: **3,5** (ENTER) *(end point of the arc B–D)*

Invoke the LINE command, and AutoCAD prompts:

> Command: **line** (ENTER)
> Specify first point: (ENTER) *(invoke the continuation option)*
> Length of line: **3** (ENTER) (length of the line D–E)
> Specify next point or [Undo]: **2,8** (ENTER) (end point of the line E–F)
> Specify next point or [Undo]: **2,10** (ENTER) (end point of the line F–-G)
> Specify next point or [Close/Undo]: (ENTER) (exits the LINE command)

Invoke the ARC command, and AutoCAD prompts:

> Command: **arc** (ENTER)
> Specify start point of arc or [CEnter]: **2,10** (ENTER) *(starting point of the arc G–J)*
> Specify second point of arc or [CEnter/ENd]: **c** (ENTER)

Specify center point of arc: **2,11** (ENTER) *(arc center for the arc G–J)*
Specify end point of arc or [Angle/chord Length]: **3,11** (ENTER) *(end point of the arc G–J)*

Invoke the LINE command, and AutoCAD prompts:

Command: **line** (ENTER)
Specify first point: **3,11** (ENTER) *(start point of the line J–K)*
Specify next point or [Undo]: **5,11** (ENTER) *(end point of the line J–K)*
Specify next point or [Undo]: **5,5** (ENTER) (end point of the line K–L)
Specify next point or [Close/Undo]: (ENTER) (exits the LINE command)

Invoke the ARC command, and AutoCAD prompts:

Command: **arc** (ENTER)
Specify start point of arc or [CEnter]: (ENTER) *(invoke the continuation option to start the arc at L)*
Specify end point of arc: **6,4** (enter) *(endpoint of the arc L–M)*

Invoke the LINE command, and AutoCAD prompts:

Command: **line** (ENTER)
Specify first point: **6,4** (ENTER) *(start point of the line M–N)*
Specify next point or [Undo]: **6,1** (ENTER) *(end point of the line M–N)*
Specify next point or [Undo]: **2,1** (ENTER) *(end point of the line N–A)*
Specify next point or [Close/Undo]: (enter) *(exits the line command)*

EXERCISE 2–7

Type of shapes	Input Methods
Single line sequence Multiple arcs	• Absolute rectangular coordinates • Three-point option for arcs

Points of Interest

- The top view of this tent-shaped object gives some clues to its true shape, but not enough. A side view is required.

- In the front view the two outside arcs are represented in their true 2D appearance. However, the two inner arcs represent arcs that are rotated 60°. Their true representation would be ellipses. Since we have not covered ellipses, we will use arcs to give close approximations.

Important Points

- Advanced features like the ELLIPSE command and 3D drawing may used to make the drawing exact if necessary.

- Note that arcs can be drawn both clockwise and counterclockwise with the three point option. This is not the case with all options of the ARC command.

Drawing Setup	Open the drawing created in Exercise 2–3

Point by Point

Invoke the LINE command, and AutoCAD prompts:

Command: **line** (ENTER)
Specify first point: **4,0.5** (ENTER) *(starting point of line 1)*
Specify next point or [Undo]: **8,0.5** (ENTER) *(end point of the line 1)*
Specify next point or [Undo]: (ENTER) (exits the LINE command)

Invoke the ARC command, and AutoCAD prompts:

Command: **arc** (ENTER)
Specify start point of arc or [CEnter]: **4,0.5** (ENTER) *(starting point of the arc 1)*
Specify second point of arc or [CEnter/ENd]: **5,2** (ENTER) *(second point of the arc 1)*
Specify end point of arc or [Angle/chord Length]: **6,2.5** (enter) *(end point of the arc 1)*

Command: *(press ENTER to invoke the arc command again)*
Specify start point of arc or [CEnter]: **5,0.5** (ENTER) *(starting point of the arc 2)*
Specify second point of arc or [CEnter/ENd]: **5.5,1.875** (ENTER) *(second point of the arc 2)*
Specify end point of arc or [Angle/chord Length]: **6,2.5** (enter) *(end point of the arc 2)*

Command: (press ENTER to invoke the arc command again)
Specify start point of arc or [CEnter]: **7,0.5** (ENTER) *(starting point of the arc 3)*
Specify second point of arc or [CEnter/ENd]: **6.5,1.875** (ENTER) *(second point of the arc 3)*
Specify end point of arc or [Angle/chord Length]: **6,2.5** (enter) *(end point of the arc 3)*

Command: *(press (ENTER) to invoke the arc command again)*
Specify start point of arc or [CEnter]: **8,0.5** (ENTER) *(starting point of the arc 4)*
Specify second point of arc or [CEnter/ENd]: **7,2** (ENTER) *(second point of the arc 4)*
Specify end point of arc or [Angle/chord Length]: **6,2.5** (ENTER) *(end point of the arc 4)*

EXERCISE 2-8

Type of shape	Input methods
Multiple line/arc sequences	• Absolute rectangular coordinates • Relative rectangular coordinates • Relative polar coordinates

- This exercise is to be done without line-arc or arc-line continuation. Later, you will learn to use the command that draws Polylines; its options facilitate polyline-polyarc-polyline continuation.

- Some architectural illusionism is introduced.

architectural EXERCISE

Drawing Setup	Open the drawing created in Exercise 2–2.

Point by Point

Invoke the LINE command, and AutoCAD prompts:

> Command: **line** (ENTER)
> Specify first point: **2'5,3'3** (ENTER) *(starting point of line 1)*
> Specify next point or [Undo]: **1'10,3'3** (ENTER) *(end point of the line 1)*
> Specify next point or [Undo]: **1'10,2'10** (ENTER) (end point of the line 2)
> Specify next point or [Close/Undo]: (ENTER) (exits the LINE command)

Invoke the ARC command, and AutoCAD prompts:

> Command: **arc** (ENTER)

Specify start point of arc or [CEnter]: **1'10,2'10** (ENTER) *(starting point of the arc 1)*
Specify second point of arc or [CEnter/ENd]: **c** (ENTER)
Specify Center point of arc: **2'6,2'10** (ENTER) *(arc center for the arc 1)*
Specify end point of arc or [Angle/chord Length]: **a** (ENTER)
Specify Included angle: **45** (ENTER) *(arc included angle for the arc 1)*

Invoke the LINE command, and AutoCAD prompts:

Command: **line** (ENTER)
Specify first point: **1'10,3'5** (ENTER) *(starting point of line 3)*
Specify next point or [Undo]: **1'10,3'10** (ENTER) *(end point of the line 3)*
Specify next point or [Close/Undo]: (ENTER) *(exits the LINE command)*

Invoke the ARC command, and AutoCAD prompts:

Command: **arc** (ENTER)
Specify start point of arc or [CEnter]: **1'10,3'10** (ENTER) *(starting point of the arc 2)*
Specify second point of arc or [CEnter/ENd]: **c** (ENTER)
Specify Center point of arc: **2'6,3'10** (ENTER) *(arc center for the arc 2)*
Specify end point of arc or [Angle/chord Length]: **a** (ENTER)
Specify Included angle: **–82.82** (ENTER) *(arc included angle for the arc 2)*

Invoke the LINE command, and AutoCAD prompts:

Command: **line** (ENTER)
Specify first point: **2'7,3'5** (ENTER) *(starting point of line 4)*
Specify next point or [Undo]: **2'7,4'6** (ENTER) *(end point of the line 4)*
Specify next point or [Undo]: (enter) *(exits the line command)*

Invoke the ARC command, and AutoCAD prompts:

Command: **arc** (ENTER)
Specify start point of arc or [CEnter]: **2'7,4'6** (ENTER) *(starting point of the arc 3)*
Specify second point of arc or [CEnter/ENd]: **c** (ENTER)
Specify Center point of arc: **2'6,3'10** (ENTER) *(arc center for the arc 21)*
Specify end point of arc or [Angle/chord Length]: **a** (ENTER)
Specify Included angle: **–45** (ENTER) *(arc included angle for the arc 2)*

Invoke the LINE command, and AutoCAD prompts:

Command: **line** (ENTER)
Specify first point: **2'7,2'2** (ENTER) *(starting point of line 5)*
Specify next point or [Undo]: **2'7,3'3** (ENTER) *(end point of the line 5)*
Specify next point or [Undo]: **3'2,3'3** (ENTER) *(end point of the line 6)*
Specify next point or [Close/Undo]: (ENTER) *(exits the LINE command)*

(Second window: repeat the preceding points, adding 2'–0" to each X coordinate; slashed (glass) lines can be drawn freehand with the LINE command.)

EXERCISES 2–9 THROUGH 2–24

Create the drawings (orthographic projections) according to the settings given in the following table:

Settings	Value
1. Units	Decimal
2. Limits	Lower left corner 0,0
	Upper right corner 12,9

 Note: Grid lines in the drawings are spaced 0.25 units apart.

EXERCISE 2–9

EXERCISE 2–10

EXERCISE 2–11

EXERCISE 2–12

EXERCISE 2–13

EXERCISE 2–14

mechanical EXERCISE

EXERCISE 2–15

EXERCISE 2–16

EXERCISE 2–17

EXERCISE 2–18

EXERCISE 2–19

EXERCISE 2–20

EXERCISE 2–21

EXERCISE 2–22

EXERCISE 2–23

EXERCISE 2–24

REVIEW QUESTIONS

1. The RECTANGLE command requests what information when drawing a rectangle?

   a. an initial corner, the width, and the height

   b. the coordinates of the four corners of the rectangle

   c. the coordinates of diagonally opposite corners of the rectangle

   d. the coordinates of three adjacent corners of the rectangle

2. When drawing a trace line, after selecting the second point:

   a. nothing appears on the screen

   b. you are prompted for the trace width

   c. the segment is drawn and the command terminates

   d. the first segment is drawn and you are prompted for the next point

3. To draw multiple connected line segments, you must invoke the LINE command multiple times?

   a. True

   b. False

4. The file extension .BAK stands for:

   a. backup drawing file

   b. binary file

   c. binary attribute file

   d. drawing file

   e. both b and c

5. The HELP command cannot be used:

   a. while in the LINE command

   b. while in the CIRCLE command option TTR

   c. to list commands

   d. for a text string

6. Points are located by relative rectangular coordinates in relation to:

   a. the last specified point or position

   b. the global origin

   c. the lower left corner of the screen

   d. all of the above

7. Polar coordinates are based on a distance from:

    a. the global origin

    b. the last specified position at a given angle

    c. the center of the display

    d. all of the above

8. To enter a command from the keyboard, simply enter the command name at the "Command:" prompt:

    a. in lower-case letters

    b. in upper-case letters

    c. either a or b or mixed case

    d. commands cannot be entered via the keyboard

9. The "C" option used in the LINE command at the "From Point:" prompt will:

    a. Continue the line from the last line or arc that was drawn

    b. Close the previous set of line segments

    c. Displays an error message

10. A rectangle generated by the RECTANGLE command will always have horizontal and vertical sides.

    a. True

    b. False

11. Which of the following coordinates will define a point at the screen default origin?

    a. 000

    b. 00

    c. 0.0,0.0

    d. 112

    e. @0,0

12. Switching between graphics and text screen can be accomplished by:

    a. pressing (ENTER) twice

    b. entering (CTRL) and (ENTER) at the same time

    c. pressing the (ESC) key

    d. pressing the (F2) function key

    e. both b and c

13. To draw a line a length of eight feet, four and five-eights inches in the 12 o'clock direction from the last point selected, enter:

    a. @8'4-5/8<90

    b. 8'-4-5/8<90

    c. @8-45/8<90

    d. 8'-45/8<90

    e. None of the above

14. By default, what direction does a positive number indicate when specifying angles in degrees?

    a. Clockwise

    b. Counterclockwise

    c. Has no impact when specifying angles in degrees

    d. None of the above

15. When erasing objects, if you select a point which is not on any object, AutoCAD will:

    a. Terminate the ERASE command

    b. delete the selected objects and continue with the ERASE command

    c. allow you to drag a window to select multiple objects within the area

    d. ignore the selection and continue with the ERASE command

16. Regarding the ARC options, what does "S,C,E" mean?

    a. Start, Center, End

    b. Second, Continue, Extents

    c. Second, Center, End

    d. Start, Continue, End

17. The number of different methods by which a circle can be drawn is:

    a. 1

    b. 3

    c. 4

    d. 7

    e. None of the above

18. When using the ERASE command, AutoCAD deletes each object from the drawing as you select it.

    a. True

    b. False

19. Once an object is erased from a drawing, which of the following commands could restore it to the drawing?

    a. OOPS

    b. RESTORE

    c. REPLACE

    d. CANCEL

20. When drawing a circle with the two-point potion, the distance between the two points is equal to:

    a. the circumference

    b. the perimeter

    c. the shortest chord

    d. the radius

    e. the diameter

21. A circle may be created by any of the following options, except:

    a. 2P

    b. 3P

    c. 4P

    d. Cen,Rad

    e. TTR

CHAPTER 3

Fundamentals II

INTRODUCTION

AutoCAD provides various tools to make your drafting and design work easier. The drawing tools described in this chapter will assist you in creating drawings rapidly, while ensuring the highest degree of precision.

After completing this chapter, you will be able to do the following:

- Use and control drafting settings (e.g., GRID, SNAP, ORTHO, POLAR TRACKING, and OBJECT SNAP)
- Use Tracking and Direct Distance
- Use display control commands (e.g., ZOOM, PAN, REDRAW, and REGEN)
- Create tiled viewports
- Use layering techniques
- Use the UNDO and REDO commands

DRAFTING SETTINGS

The GRID, SNAP, ORTHO, POLAR TRACKING, and OBJECT SNAP commands do not create objects. However, they make it possible to create and modify them more easily and accurately. Each of these Drafting Settings commands can be readily toggled on when needed and off when not. These commands, when turned on, operate according to settings which can also be changed easily. When used appropriately, these commands provide the power, speed, and accuracy associated with Computer Aided Design/Drafting.

SNAP COMMAND

The SNAP command provides an invisible reference grid in the drawing area. When set to ON, the Snap feature forces the cursor to lock onto the nearest point on the specified Snap Grid. Using the SNAP command, you can specify points quickly, letting AutoCAD ensure that they are placed precisely. You can always override the snap spacing by entering absolute or relative coordinate points by means of the keyboard, or by simply turning the Snap mode OFF. When the Snap mode is turned OFF, it has no effect on the cursor. When it is turned ON, you cannot place the cursor on a point that is not on one of the specified snap locations.

Turning Snap ON and OFF

In AutoCAD, five primary methods to turn the SNAP ON and OFF are available. Of the five explained in this section, the first method (selecting SNAP from the status bar) is the easiest and most commonly used with the pointing device. The second method (pressing the function key F9) is the most commonly used from the keyboard. The others, while not quite as convenient as the first two, might be convenient in certain situations.

1. In the status bar at the bottom of the screen, when the snap button appears to be pressed in, then the Snap is ON. When the snap button appears to be out, then the Snap is OFF. To change its setting, simply place the cursor over the snap button and press the pointing device pick button.

2. To change the Snap setting to ON when it is OFF or to OFF when it is ON, simply press the function key F9.

3. Right-click the Snap button in the status bar at the bottom of the screen and from the shortcut menu, choose ON to turn the Snap ON or choose OFF to turn the Snap OFF.

4. From the Tools menu choose Drafting Settings. Then select the Snap and Grid tab if it is not already selected. In the Snap and Grid section, the Snap setting is ON when the **Snap On (F9)** check box is checked, and is OFF when it is not checked. To change its setting, simply place the cursor over the check box as shown in Figure 3–1 and press the pointing device pick button. You can also enter **ddrmodes** at the Command: prompt to display the Drafting Settings dialog box, from which the desired tab can be selected if it is not already selected.

5. At the Command: prompt:

Command: **snap** (ENTER)
Specify snap spacing or [ON/OFF/Aspect/Rotate/Style/Type] <current>:
*(enter **on** to turn the Snap ON and **off** to turn the Snap OFF or right-click and choose ON or OFF from the shortcut menu)*

Figure 3–1 *Choosing the Snap On (F9) check box from the Snap and Grid section of the Drafting Settings menu*

Changing Snap Spacing

In AutoCAD there are three primary methods to change the Snap spacing:

1. Right-click the Snap button in the status bar at the bottom of the screen and choose Settings from the shortcut menu. AutoCAD will display the Drafting Settings dialog box with the Snap and Grid tab selected. From the Snap and Grid section, you can change the settings of the X and Y spacings by entering the desired values in the **Snap X Spacing** and **Snap Y Spacing** text boxes, respectively, as shown in Figure 3–2. You can also enter **ddrmodes** at the Command: prompt to display the Drafting Settings dialog box, from which the desired tab can be selected if it is not already selected.

2. From the Tools menu, choose Drafting Settings. Then select the Snap and Grid tab. See the explanation in Method 1 for changing settings.

3. At the Command: prompt:

Command: **snap** (ENTER)
Specify snap spacing or [ON/OFF/Aspect/Rotate/Style/Type] <current>:
  (specify distance to be used for both X and Y Snap Spacings)

Figure 3–2 *Changing the SNAP settings from the Snap and Grid section of the Drafting Settings dialog box*

Aspect

The Aspect option allows you to set the Y Snap Spacing different from the X Snap Spacing. When you select the Aspect option, AutoCAD prompts for the X and Y values independently. This is handy if the X and Y modular dimensions of your design are of unequal multiples. There are three primary methods to changing the Snap Aspect setting.

1. From the Snap and Grid section of the Drafting Settings dialog box, you can set the Y spacing different from the X spacing. This is done by entering a value in the **Snap Y Spacing** text box that is different from the value in the **Snap X Spacing** text box. However, any change to the X spacing after that will cause the Y spacing to default to the new X spacing. You can also enter **ddrmodes** at the Command: prompt to display the Drafting Settings dialog box, from which the desired tab can be selected if it is not already selected.

2. From the Tools menu, choose Drafting Settings. Then select the Snap and Grid tab. See the explanation in Method 1 for setting the Y spacing different from the X spacing.

3. At the Command: prompt:

Command: **snap** (ENTER)
Specify snap spacing or [ON/OFF/Aspect/Rotate/Style/Type] <1.0000>:
  a (ENTER)

Specify horizontal spacing <1.0000>: *(specify X Snap spacing)*
Specify vertical spacing <1.0000>: *(specify Y Snap spacing)*

For example, the following sequence shows how you can set a horizontal snap spacing of 0.5 and vertical snap spacing of 0.25.

Command: **snap** (ENTER)
Specify snap spacing or [ON/OFF/Aspect/Rotate/Style/Type] <1.0000>:
  a (ENTER)
Specify horizontal spacing <1.0000>: **0.5** (ENTER)
Specify vertical spacing <1.0000>: **0.25** (ENTER)

Rotate

The Rotate option allows you to specify an angle to rotate both the visible Grid and the invisible Snap grid. It is a simple version of the more complicated User Coordinate System. It permits you to set up a Snap grid with an origin (*X* coordinate, *Y* coordinate of 0,0) and an angle of rotation specified with respect to the default origin and Zero-East system of direction. In conjunction with the X and Y spacing of the Snap grid, the Rotate option can make it easier to draw certain shapes.

The plot plan in Figure 3–3 is an example of where the Rotate option of the SNAP command can be applied. The property lines are drawn using surveyor's units of angular display. In this example, the decimal linear units and the surveyor's angular units are selected. The limits are set up with the lower left corner at –20',–10', and the upper right corner at 124',86'. The sequence for drawing the property lines is as follows:

Figure 3–3 *Example drawing where the Rotate option of the SNAP command can be used*

Command: **snap** (ENTER)
Specify snap spacing or [ON/OFF/Aspect/Rotate/Style/Type] <current>:
  r (ENTER)
Base point <0'-0.0",0'-0.0">: *(press ENTER to accept the default value)*
Rotation angle <0r>: **4d45'08"** (ENTER)

Command: **line** (ENTER) *(invoking the LINE command)*
Specify first point: **0,0** (ENTER) *(selecting the first point)*
Specify next point or [Undo]: **85'** (ENTER) *(point selected with the cursor
  north of the first point)*
Specify next point or [Undo]: **120'** (ENTER) *(point selected with the cursor
  east of the previous point)*
Specify next point or [Undo]: **85'** (ENTER) *(point selected with the cursor
  south of the previous point)*
Specify next point or [Undo]: **c** *(completes the property boundary by closing
  the rectangle and exiting the LINE command)*

Figure 3–4 shows the property boundary drawn.

Figure 3–4 *Layout of the property boundary lines*

Style

The Style option permits you to select one of two available formats, Standard or
Isometric. The Standard option refers to the normal rectangular grid (default), and

the Isometric option refers to a Grid and Snap designed for Isometric drafting purposes (see Figure 3–5).

AutoCAD prompts:

> Command: **snap** (ENTER)
> Specify snap spacing or [ON/OFF/Aspect/Rotate/Style/Type] <current>: **s**
>   *(invoke Style option)*
> Enter snap grid style [Standard/Isometric] <S>: **i** (ENTER) *(change Style to Isometric)*
> Specify vertical spacing <1.0000>: **0.5** (ENTER) *(specify Snap spacing)*

Figure 3–5 *Setting the snap for isometric drafting*

You can switch the Isoplanes between Left (90- and 150-degree angles), Top (30- and 150-degree angles), and Right (30- and 90-degree angles) by CTRL + E (the combination key strokes of holding down CTRL and then pressing E) or by simply press the function key F5.

Type

The Type option permits you to select one of two available Snap types, Polar or Grid. Choosing the Polar type sets the snap to Polar Tracking angles. See the explanations on Polar Tracking and ORTHO, later in this chapter. Choosing the Grid type sets the snap to grid.

AutoCAD prompts:

> Command: **snap** (ENTER)
> Specify snap spacing or [ON/OFF/Aspect/Rotate/Style/Type] <current>: **t**
> Enter snap type [Polar/Grid] <Grid>: **p**

GRID COMMAND

The GRID command is used to display a visible array of dots with row and column spacings that you specify. AutoCAD creates a grid that is similar to a sheet of graph paper. You can turn the grid display ON and OFF, and can change the dot spacing. The grid is a drawing tool and is not part of the drawing; it is for visual reference and is never plotted. In the World Coordinate System, the grid fills the area defined by the limits.

The grid has several uses within AutoCAD. First, it shows the extent of the drawing limits. For example, if you set the limits to 42 by 36 units and grid spacing is set to 0.5 units, then each row will have 85 dots and each column will have 73 dots. This will give you a better sense of the drawing's size relative to the limits than if it were on a blank background.

Second, using the GRID command with the SNAP command is helpful when you create a design in terms of evenly spaced units. For example, if your design is in multiples of 0.5 units, then you can set grid spacing as 0.5 to facilitate point entry. You could check your drawing visually by comparing the locations of the grid dots and the crosshairs. Figure 3–6 shows a drawing with a grid spacing of 0.5 units, with limits set to 0,0 and 17,11.

Note: While the GRID, SNAP, ORTHO, POLAR TRACKING, and OBJECT SNAP commands do not create objects, they make it possible to create them more easily and accurately. Each of these drafting settings commands, when toggled ON, operate according to the value(s) to which you have set it and, when toggled OFF, has no effect. You are advised to identify and master the two skills involved in using these utilities; one is to learn how to change the settings of the utility commands, and the other is to learn how and when to best turn them ON and OFF. Changing the settings of these utilities is normally done more easily by means of their associated dialog box(es). You can also turn them on and off from a dialog box. But, because these features are normally switched on and off so frequently during a drawing session, special buttons are provided in the status bar at the bottom of the screen for this purpose.

Figure 3–6 *A grid spacing of 0.5 units, with limits set to (0,0) and (17,11)*

Turning the Grid ON and OFF

In AutoCAD, five primary methods to turn the GRID ON and OFF are available. Of the five explained in this section, the first method (selecting GRID from the status bar) is the easiest and most commonly used with the pointing device. The second method (pressing the function key F7) is the most commonly used from the keyboard. The others, while not quite as convenient as the first two, might be convenient in certain situations.

1. In the status bar at the bottom of the screen, when the GRID button appears to be pressed in, then the Grid is ON. When the GRID button appears to be out, then the Grid is OFF. To change its setting, simply place the cursor over the GRID button and press the pointing device pick button.

2. To change the Grid setting to ON when it is OFF or to OFF when it is ON, simply press the function key F7.

3. Right-click the Grid button in the status bar at the bottom of the screen and from the shortcut menu, choose ON to turn the Grid ON or choose OFF to turn the Grid OFF.

4. From the Tools menu, choose Drafting Settings. Then select the Snap and Grid tab if it is not already selected. In the Snap and Grid section, the Grid setting is ON when the **Grid On (F7)** check box is checked, and is OFF when it is not checked. To change its setting, simply place the cursor over the check box, as shown in Figure 3–7, and press the pointing device pick button. You can also enter **ddrmodes** at the Command: prompt to display the Drafting Settings dialog box, from which the desired tab can be selected if it is not already selected.

5. At the Command: prompt:

Command: **grid** (ENTER)
Specify grid spacing(X) or [ON/OFF/Snap/Aspect] <current >: *(enter **on** to turn the Grid ON and **off** to turn the Grid OFF, or right-click and choose ON or OFF from the shortcut menu)*

Figure 3–7 *Choosing the Grid On (F7) check box from the Snap and Grid section of the Drafting Settings dialog box*

Changing Grid Spacing

In AutoCAD there are three primary methods to change the Grid spacing:

1. From the status bar at the bottom of the screen, right-click on the GRID button and choose Settings. AutoCAD will display the Drafting Settings dialog box

with the Snap and Grid tab selected. From the Snap and Grid section, you can change the settings of the X and Y spacings by entering the desired values in the **Grid X Spacing** and **Grid Y Spacing** text boxes, respectively, as shown in Figure 3–8.

2. From the Tools menu, choose Drafting Settings. Then select the Snap and Grid tab. See the explanation in Method 1 for changing settings.

3. At the Command: prompt:

Command: **grid** (ENTER)
Specify grid spacing(X) or [ON/OFF/Snap/Aspect] <current>: *(specify distance to be used for both X and Y Grid Spacings)*

Figure 3–8 *Changing the GRID settings from the Snap and Grid section of the Drafting Settings dialog box*

Aspect

The Aspect option allows you to set the Y Grid Spacing different from the X Grid Spacing. When you select the Aspect option, AutoCAD prompts for the *X* and *Y* values independently. This is handy if the *X* and *Y* modular dimensions of your design are of unequal multiples.

1. From the Snap and Grid section of the Drafting Settings dialog box, you can set the Y spacing different from the X spacing. This is done by entering a value in the **Grid Y Spacing** text box that is different from the value in the **Grid X Spacing** text box. However, any change to the X spacing after that will cause the Y spacing to default to the X spacing. You can also enter **ddrmodes** at the Command: prompt to display the Drafting Settings dialog box, from which the desired tab can be selected if it is not already selected.

2. From the Tools menu, choose Drafting Settings. Then select the Snap and Grid tab. See the explanation in Method 1 for setting the Y spacing different from the X spacing.

3. At the Command: prompt:

Command: **grid** (ENTER)
Specify grid spacing(X) or [ON/OFF/Snap/Aspect] <A>: **a** (ENTER)
Specify the horizontal spacing(X) <1.0000>: *(specify X Grid spacing)*
Specify the vertical spacing(Y) <1.0000>: *(specify Y Grid spacing)*

For example, the following sequence of commands shows how you can set a grid spacing of 0.5 and vertical grid spacing of 0.25.

Command: **grid** (ENTER)
Specify grid spacing(X) or [ON/OFF/Snap/Aspect] <A>: **a** (ENTER)
Specify the horizontal spacing(X) <1.0000>: **0.5** (ENTER)
Specify the vertical spacing(Y) <1.0000>: **0.25** (ENTER)

Applying the Aspect option in this example provides the Grid dot spacing shown in Figure 3–9.

Relationship to Snap Setting

It is often useful to set the grid spacing equal to the snap resolution, or make it a multiple of it.

To specify the grid spacing the same as the snap value, enter s in response to the Specify grid spacing(X) or [ON/OFF/Snap/Aspect] <A>: prompt:

Command: **grid** (ENTER)
Specify grid spacing(X) or [ON/OFF/Snap/Aspect] <current>: **s** (ENTER)

To specify the grid spacing as a multiple of the snap value, enter x after the value. For example, to set up the grid value as three times the current snap value (snap = 0.5 units), enter 3x for the prompt, which is the same as setting it to 1.5 units in response to the Specify grid spacing(X) or [ON/OFF/Snap/Aspect] <A>: prompt:

Command: **grid** (ENTER)
Specify grid spacing(X) or [ON/OFF/Snap/Aspect] <current>: **3x** (ENTER)

Figure 3–9 *Display after setting the grid aspect to 0.5 for horizontal and 0.25 for vertical spacing*

If the spacing of the visible grid is set too small, AutoCAD displays the following message and does not show the dots on the screen:

Grid too dense to display

To display the grid, invoke another GRID command and specify a larger spacing.

Note: The relationship between the Grid setting and the Snap setting, when established as described in the previous section, is based on the current Snap setting. If the Snap setting is subsequently changed, the Grid setting does not change accordingly. For example, if the Snap setting is 1.00 and you enter **s** in response to the Specify grid spacing(X) or [ON/OFF/Snap/Aspect] <current>: prompt, the Grid setting becomes 1.00 and remains 1.00 even if the Snap setting is later set to something else. Likewise, if you set the Grid setting to **3x**, it becomes 3.00 and will not change with a subsequent change in the Snap setting.

ORTHO COMMAND

The ORTHO command lets you draw lines and specify point displacements that are parallel to either the *X* or *Y* axis. Lines drawn with the Ortho mode ON are therefore either parallel or perpendicular to each other. This mode is helpful when you need to draw lines that are exactly horizontal or vertical. Also, when the Snap Style is set to Isometric, it forces lines to be parallel to one of the three isometric axes.

Turning Ortho ON and OFF

In AutoCAD, four primary methods to turn the ORTHO command ON and OFF are available. Of the four explained in this section, the first method (selecting ORTHO from the status bar) is the easiest and most commonly used with the pointing device. The second method (pressing the function key F8) is the most commonly used from the keyboard. The others, while not quite as convenient as the first two, might be convenient in certain situations.

1. In the status bar at the bottom of the screen, when the ORTHO button appears to be pressed in, then the ORTHO is ON. When the ORTHO button appears to be out, then the ORTHO is OFF. To change its setting, simply place the cursor over the ORTHO button and press the pointing device pick button.

2. To change the ORTHO setting to ON when it is OFF or to OFF when it is ON, simply press the function key F8.

3. Right-click the ORTHO button in the status bar at the bottom of the screen and from the shortcut menu, choose ON to turn the ORTHO ON or choose OFF to turn the ORTHO OFF.

4. At the Command: prompt:

Command: **ortho** (ENTER)
Enter mode [ON/OFF] <OFF>: *(enter* **on** *to turn* ORTHO *ON and* **off** *to turn* ORTHO *OFF, or right-click and choose ON or OFF from the shortcut menu)*

 Note: The ORTHO and POLAR TRACKING modes (explained later in this chapter) cannot both be turned ON at the same time. They can both be turned OFF, or either one can be turned ON.

When the Ortho mode is active, you can draw lines and specify displacements only in the horizontal or vertical directions, regardless of the cursor's on-screen position. The direction in which you draw is determined by the change in the X value of the cursor movement compared to the change in the cursor's distance to the Y axis. AutoCAD allows you to draw horizontally if the distance in the X direction is greater than the distance in the Y direction; conversely, if the change in the Y direction is greater than the change in the X direction, then it forces you to draw vertically. The ORTHO mode does not affect keyboard entry of points.

OBJECT SNAP

The Object Snap (or Osnap, for short) feature lets you specify points on existing objects in the drawing. For example, if you need to draw a line from an endpoint of an existing line, you can apply the Object Snap mode, called ENDpoint. In response to the "Specify first point:" prompt, place the cursor so that it touches the line nearer the desired endpoint. AutoCAD will lock onto the endpoint of the existing line when you press the pick button of your pointing device. The endpoint becomes the starting point of the new line. This feature is similar to the basic SNAP command, which locks to invisible reference grid points.

You can invoke an Object Snap mode whenever AutoCAD prompts for a point.

Object snap modes can be invoked while executing an AutoCAD command that prompts for a point, such as the LINE, CIRCLE, MOVE, and COPY commands.

Applying Object Snap Modes

Object Snap modes can be applied in either of two ways:

1. From the Osnap Toolbar, choose Object Snap Settings (see Figure 3–10), or from the status bar at the bottom of the screen, right-click on the OSNAP button and choose Settings. The Drafting Settings dialog box will be displayed with the Object Snap tab selected, as shown in Figure 3–11. An Object Snap mode has been chosen if there is a check in its associated check box. Choose the desired Object Snap mode(s) by placing the cursor over its check box and pressing the pointing device pick button. This makes it possible to use the checked mode(s) any time you are prompted to specify a point. This is referred to as the *Running Osnap* method. You can also enter **ddrmodes** at the Command: prompt to display the Drafting Settings dialog box, from which the Object Snap tab can be selected if it is not already selected. Running Osnap can be turned ON and OFF in the same manner as the GRID command. The most common method is to use the toggle button from the status bar at the bottom of the screen. When the OSNAP button appears to be pressed in, then the Object Snap is ON. When the OSNAP button appears to be out, then the Object Snap is OFF. To change its setting, simply place the cursor over the OSNAP button and press the pointing device pick button. It is just as easy to simply press the function key F3, to change the OSNAP setting to ON when it is OFF or to OFF when it is ON.

Figure 3–10 *Choosing Object Snap Settings from the Object Snap toolbar*

2. When prompted to specify a point, enter the first three letters of the name of the desired Object Snap mode, or select it from the Object Snap Toolbar before specifying the point. This is a *one-time-only* Osnap method. This will override any other Running Osnap mode for this one point selection only.

Figure 3–11 *Drafting Settings dialog box displaying the Object Snap tab*

For example, to draw a line from an end point of an existing line, invoke the LINE command, and AutoCAD prompts:

Command: **line** (ENTER)
Specify first point: **end** *(enter the first three letters of the Osnap mode endpoint)*
of *(move the cursor near the desired end of a line or arc and press the pick button)*

The first point of the new line is drawn from the end point of the selected object.

Osnap Markers and Tooltips

Whenever one or more Object Snap modes are activated and you move the cursor target box over a snap point, AutoCAD displays a geometric shape (Marker) and Tooltip. By displaying a Marker on the snap points with a Tooltip, you can see the point that will be selected and the Object Snap mode in effect. AutoCAD displays the Marker depending on the Object Snap mode selected. In the Object Snap section of the Drafting Settings dialog box, each Marker is displayed next to the name of its associated Object Snap mode.

From the Object Snap section of the Drafting Settings dialog box, choose the **Options** button. AutoCAD displays the Options dialog box with the Drafting tab selected, where the settings for displaying the Markers and Tooltips can be changed. They are check boxes in the AutoSnap Settings section (see Figure 3–12).

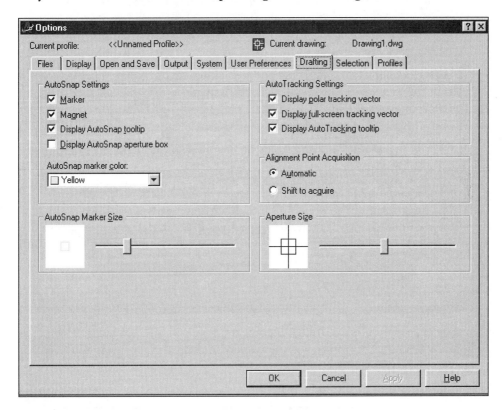

Figure 3–12 *Options dialog box with the Drafting tab selected*

In the AutoSnap Settings section:

The Marker setting is ON when the **Marker** check box is checked, and is OFF when it is not checked. The Marker is the geometric shape that is displayed when the cursor moves over a snap point.

The Magnet setting is ON when the **Magnet** check box is checked, and is OFF when it is not checked. The Magnet locks the cursor onto the snap point.

The Tooltip setting is ON when the **Display AutoSnap Tooltip** check box is checked, and is OFF when it is not checked. The Tooltip is the flag that displays the name of the Object Snap mode.

The Aperture setting is ON when the **Display AutoSnap Aperture** box check box is checked, and is OFF when it is not checked. The Aperture is the square target box that is displayed when AutoCAD is prompting to specify a point.

To change the ON/OFF setting of the Marker, Magnet, Tooltip, or Aperture, simply place the cursor over its check box and press the pointing device pick button. To change the color of the Marker, select the desired color from the **AutoSnap Marker Color** list box.

To change the size of the Marker, hold the pick button of the pointing device over the slide bar in the **Autosnap Marker Size** section and move it to the right to make the Marker larger and to the left to make it smaller. The Image tile shows the current size of the Marker.

To change the size of the Aperture, hold the pick button of the pointing device over the slide bar in the **Aperture Size** section and move it to the right to make the Aperture larger and to the left to make it smaller. The Image tile shows the current size of the Aperture.

After making the necessary changes to the AutoSnap settings, choose the **OK** button to close the Options subdialog box and then choose **OK** again to close the Drafting Settings dialog box.

Osnap Modes

The following section explains the available Object Snap modes, with examples:

Object Snap—ENDpoint The ENDpoint mode allows you to snap to the closest endpoint of a line, arc, elliptical arc, mline, polyline segment, or ray, or to the closest corner of a trace, solid, or 3Dface.

During an appropriate AutoCAD command, invoke the Object Snap—ENDpoint mode:

Object Snap toolbar	Select the ENDpoint mode (see Figure 3–13)
Command: prompt	Enter **end** and press ENTER whenever an AutoCAD command: prompt requests a point to be specified

Figure 3–13 *Invoke the Object Snap ENDpoint mode from the Object Snap toolbar*

For example, to connect a line to the endpoint of an existing line, as shown in Figure 3–14, the following command sequence is used:

Command: **line** (ENTER)
Specify first point: **end** *(enter the first three letters of the Osnap mode Endpoint)*
of *(move the aperture cursor near the end of Line A and specify it)*
Specify next point or [Undo]: *(specify other end of line)*
Specify next point or [Undo]: (ENTER)

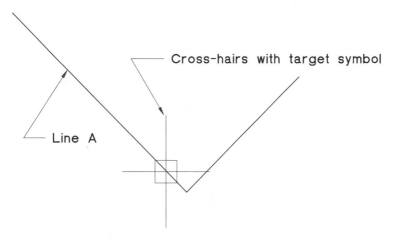

Cross-hairs with target symbol

Line A

Figure 3–14 *Connecting a line to the endpoint of another line using the ENDpoint Object Snap mode*

Object Snap—MIDpoint The MIDpoint mode allows you to snap to the midpoint of a line, arc, elliptical arc, mline, polyline segment, xline, solid, or spline.

During an appropriate AutoCAD command, invoke the Object Snap—MIDpoint mode:

Object Snap toolbar	Select the MIDpoint mode (see Figure 3–15)
Command: prompt	Enter **mid** and press ENTER whenever an AutoCAD command: prompt requests a point to be specified

Figure 3–15 *Invoking the Object Snap—MIDpoint mode from the Object Snap toolbar*

For example, to connect a line to the midpoint of an existing line, as shown in Figure 3–16, the following command sequence is used:

Command: **line** (ENTER)
Specify first point: **mid** *(enter the first three letters of the Osnap mode Midpoint)*
of *(move the aperture cursor to anywhere on Line A and specify it)*
Specify next point or [Undo]: *(specify other end of line)*
Specify next point or [Undo]: *(ENTER)*

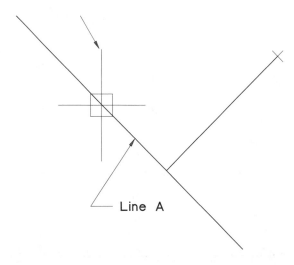

Line A

Figure 3–16 *Connecting a line to another line's midpoint using the MIDpoint Object Snap mode*

Object Snap—CENter The CENter mode allows you to snap to the center of an arc, circle, ellipse, or elliptical arc.

During an appropriate AutoCAD command, invoke the Object Snap—CENter mode:

Object Snap toolbar	Select the CENter mode (see Figure 3–17)
Command: prompt	Enter **cen** and press ENTER whenever an AutoCAD command: prompt requests a point to be specified

Figure 3–17 *Invoking the Object Snap—CENter mode from the Object Snap toolbar*

For example, to draw a line to the center of a circle, as shown in Figure 3–18, the following command sequence is used:

Command: **line** (ENTER)
Specify first point: *(specify point P)*
Specify next point or [Undo]: **cen** *(enter the first three letters of the Osnap mode Center)*
of *(specify point P1 on circle)*
Specify next point or [Undo]: (ENTER)

P1

P

Figure 3–18 *Selecting a point with the CENter Object Snap mode*

Object Snap—NODe The NODe mode allows you to snap to a point object.

During an appropriate AutoCAD command, invoke the Object Snap—NODe mode:

Object Snap toolbar	Select the NODe mode (see Figure 3–19)
Command: prompt	Enter **nod** and press ENTER whenever an AutoCAD command: prompt requests a point to be specified

Figure 3–19 *Invoking the Object Snap NODe mode from the Object Snap toolbar*

Object Snap—QUAdrant The QUAdrant mode allows you to snap to one of the quadrant points of a circle, arc, ellipse, or elliptical arc. The quadrant points are located at 0 degrees, 90 degrees, 180 degrees, and 270 degrees from the center of the circle or arc, as shown in Figure 3–20. The quadrant points are determined by the zero degree direction of the current coordinate system.

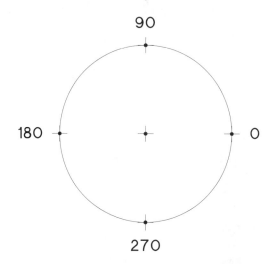

Figure 3–20 *The quadrant points recognized by the QUAdrant Object Snap mode*

During an appropriate AutoCAD command, invoke the Object Snap—QUAdrant mode:

Object Snap toolbar	Select the QUAdrant mode (see Figure 3–21)
Command: prompt	Enter **qua** and press ENTER whenever an AutoCAD command: prompt requests a point to be specified

Figure 3–21 *Invoking the Object Snap—QUAdrant mode from the Object Snap toolbar*

In Figure 3–22 a line is drawn to a quadrant as follows:

Command: **line** (ENTER)
Specify first point: *(specify point A)*
Specify next point or [Undo]: **qua** *(enter the first three letters of the Osnap mode Quadrant)*
of *(specify point A1 on circle)*
Specify next point or [Undo]: (ENTER)

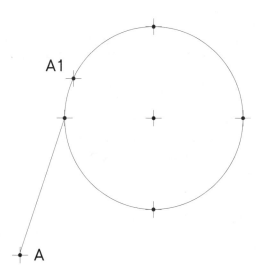

Figure 3–22 *Drawing a line to a circle's quadrant point (Method 1)*

A line can also be drawn to a quadrant as follows (see Figure 3–23):

Command: **line** (ENTER)
Specify first point: *(specify point A)*
Specify next point or [Undo]: **qua** *(enter the first three letters of the Osnap mode Quadrant)*
of *(specify point A2 on circle)*
Specify next point or [Undo]: (ENTER)

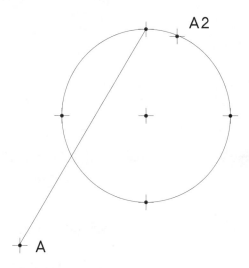

Figure 3–23 *Drawing a line to a circle's quadrant (Method 2)*

Note: Special precautions should be taken when attempting to select circles or arcs in blocks or ellipses that are rotated at an angle that is not a multiple of 90 degrees. When a circle or an arc in a block is rotated, the point of that QUAdrant Object Snap mode is also rotated. But when a circle/arc not in a block is rotated, the QUAdrant Object Snap points stay at the 0, 90, 180, or 270 degree points.

Object Snap—INTersection The INTersection mode allows you to snap to the intersection of two objects that can include arcs, circles, ellipses, elliptical arcs, lines, mlines, polylines, rays, splines, or xlines. An example with valid intersection points is shown in Figure 3–24.

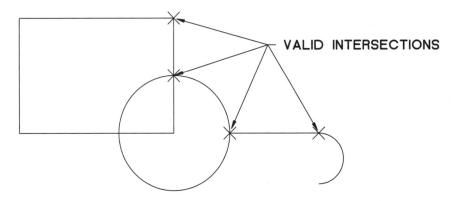

Figure 3–24 *Valid intersections that can be selected using object snap—INTersection mode*

During an appropriate AutoCAD command, invoke the Object Snap—INTersection mode:

Object Snap toolbar	Select the INTersection mode (see Figure 3–25)
Command: prompt	Enter **int** and press ENTER whenever an AutoCAD command: prompt requests a point to be specified

Figure 3–25 *Invoking the Object Snap—INTersection mode from the Object Snap toolbar*

AutoCAD prompts you to select two objects to establish the intersection point.

Object Snap—EXTension   The EXTension mode allows you to snap to a point that is the extension of a line or an arc.

During an appropriate AutoCAD command, invoke the Object Snap - EXTension mode:

Object Snap toolbar	Select the EXTension mode (see Figure 3–26)
Command: prompt	Enter **ext** and press ENTER whenever an AutoCAD command: prompt requests a point to be specified

This example is to show how to apply the EXTension Object Snap mode to draw LINE A, as shown in Figures 3–27.

Figure 3–26 *Invoking the Object Snap EXTension mode from the Object Snap toolbar*

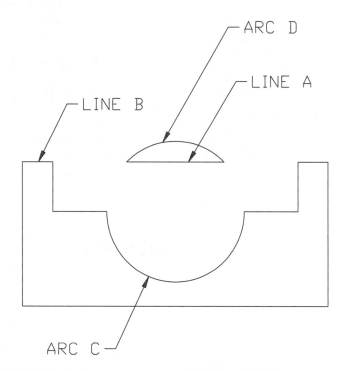

Figure 3–27 *LINE A is drawn using extensions of LINE B and ARC C*

Figure 3–28 *LINE B and ARC C extensions shown with hidden lines*

LINE A is an extension of LINE B between points P1 and P2 (see Figure 3-28), where LINE B intersects the circle of which ARC C is a part, as shown in Figure 3–27. To draw LINE A, invoke the LINE command, and AutoCAD prompts:

Command: **line** (ENTER)
Specify first point: **ext** *(enter the first three letters of the Osnap mode Extension)*
of *(move the aperture cursor to the right end of LINE B and without pressing the pick button move it off of the line to the right)*

As you move the cursor away from LINE B, a dashed line follows as long as the cursor is near the extension of LINE B, as shown in Figure 3–29.

Figure 3–29 *A dashed line follows the cursor indicating the extension of LINE B*

Next, move the cursor to the left end of ARC C and then upward away from it. Another dashed arc follows until the cursor is near the extension of LINE B, as shown in Figure 3–30. At this time you can press the pick button on the pointing device and point P1 is established as the start point of LINE A.

Figure 3–30 *A dashed arc follows the cursor indicating the extension of ARC C to the dashed extension of LINE B, at which point P1 is specified*

After pressing the pick button to specify point P1, move the cursor to LINE B again and then move the cursor to the right end of ARC C and then upward away from it. Another dashed arc follows until the cursor is near the extension of LINE B, as shown in Figure 3–31.

Figure 3–31 *A dashed arc follows the cursor indicating the extension of ARC C to the dashed extension of LINE B, at which point P2 is specified*

At this time, press the pick button on the pointing device and point P2, the end point of line A, is established, as shown in Figure 3–32.

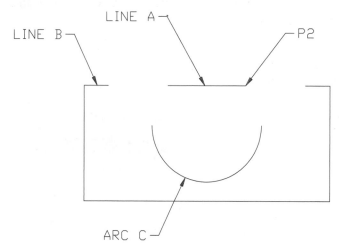

Figure 3–32 *LINE A is drawn*

After exiting the LINE command, ARC D is drawn with the Start, Center, End method (SCE). The start point P2 and the end point P3 of the arc is specified by using the Endpoint Object Snap mode and selecting the endpoints of LINE A. The center of the arc is specified by using the Center Object Snap mode and selecting ARC C.

Object Snap—INSertion The INSertion mode allows you to snap to the insertion point of a block, text string, attribute, or shape.

During an appropriate AutoCAD command, invoke the Object Snap—INSertion mode:

Object Snap toolbar	Select the INSert mode (see Figure 3–33)
Command: prompt	Enter **ins** and press ENTER whenever an AutoCAD command: prompt requests a point to be specified

Figure 3–33 *Invoking the Object Snap INSert mode from the Object Snap toolbar*

Object Snap—PERpendicular The PERpendicular mode allows you to snap to a point perpendicular to a line, arc, circle, elliptical arc, mline, polyline, ray, solid, spline, or xline.

During an appropriate AutoCAD command, invoke the Object Snap—PERpendicular mode:

Object Snap toolbar	Select the PERpendicular mode (see Figure 3–34)
Command: prompt	Enter **per** and press ENTER whenever an AutoCAD command: prompt requests a point to be specified

Figure 3–34 *Invoking the Object Snap—PERpendicular mode from the Object Snap toolbar*

The following command sequence demonstrates how you can draw lines using the PERpendicular object snap mode.

From inside the circle:

Command: **line** (ENTER)
Specify first point: *(specify point A)*
Specify next point or [Undo]: **per** *(enter the first three letters of the Osnap mode PERpendicular)*
to *(specify near-side point AN, as shown in Figure 3–35)*
Specify next point or [Undo]: (ENTER)

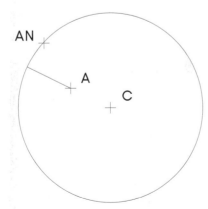

Figure 3–35 *Specifying a line's endpoint perpendicular to a circle (from inside of the circle to the near side of the circle) with the PERpendicular object snap mode*

Command: **line** (ENTER)
Specify first point: *(specify point A)*
Specify next point or [Undo]: **per** *(enter the first three letters of the Osnap mode PERpendicular)*
to *(specify far-side point AF, as shown in Figure 3–36)*

Specify next point or [Undo]: (ENTER)

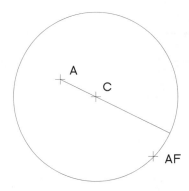

Figure 3–36 *Specifying a line's endpoint perpendicular to a circle (from inside of the circle to the far side of the circle) with the PERpendicular Object Snap mode*

From outside the circle:

Command: **line** (ENTER)
Specify first point: *(specify point B)*
Specify next point or [Undo]: **per** *(enter the first three letters of the Osnap mode PERpendicular)*
to *(specify near-side point BN, as shown in Figure 3–37)*

Specify next point or [Undo]: (ENTER)

Figure 3–37 *Specifying a line's endpoint perpendicular to a circle (from outside of the circle to the near side of the circle) with the PERpendicular object snap mode*

Command: **line** (ENTER)
Specify first point: *(specify point B)*
Specify next point or [Undo]: **per** *(enter the first three letters of the Osnap mode PERpendicular)*
to *(specify far-side point BF, as shown in Figure 3–38)*
Specify next point or [Undo]: (ENTER)

Figure 3–38 *Specifying a line's endpoint perpendicular to a circle (from outside of the circle to the far side of the circle) with the PERpendicular object snap mode*

When drawing a line perpendicular to another line, the point that AutoCAD establishes can be off the line selected (in response to the "perpendicular to" prompt) and the new line will still be drawn to that point. In Figure 3–39 lines from both A and B can be drawn perpendicular to line L.

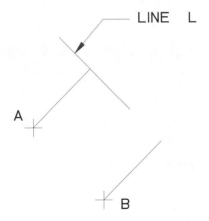

Figure 3–39 *Using the PERpendicular osnap mode, a line is drawn perpendicular to another line*

Applying the PERpendicular osnap mode to an arc works in a manner similar to a circle. Unlike drawing a perpendicular to a line, the point established must be on the arc.

In Figure 3–40 a line is drawn perpendicular to an arc from A but not from B.

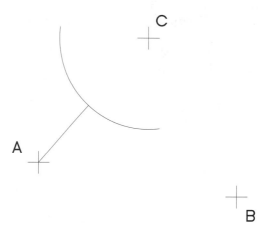

Figure 3–40 *Using the PERpendicular osnap mode, a line is drawn perpendicular to an arc*

Object Snap–TANgent The TANgent mode allows you to snap to the tangent of an arc, circle, ellipse, or elliptical arc.

During an appropriate AutoCAD command, invoke the Object Snap—TANgent mode:

Object Snap toolbar	Select the TANgent mode (see Figure 3–41)
Command: prompt	Enter **tan** and press ENTER whenever an AutoCAD command: prompt requests a point to be specified

Figure 3–41 *Invoking the Object Snap—TANgent mode from the Object Snap toolbar*

The following command sequence draws a line from point A tangent to a point on the circle, as shown in Figure 3–42.

> Command: **line** (ENTER)
> Specify first point: *(specify point A)*
> Specify next point or [Undo]: **tan** *(enter the first three letters of the Osnap mode Tangent)*
> to *(specify point AL toward left semicircle)*
> Specify next point or [Undo]: (ENTER)

Figure 3–42 *Using the TANgent osnap mode, a line is drawn from a point outside a circle tangent to a point on the left side of the circle*

A line can also be drawn from point A tangent to a point on the circle, as shown in Figure 3–43.

> Command: **line** (ENTER)
> Specify first point: *(specify point A)*
> Specify next point or [Undo]: **tan** *(enter the first three letters of the Osnap mode Tangent)*
> to *(specify point AR toward right semicircle)*
> Specify next point or [Undo]: (ENTER)

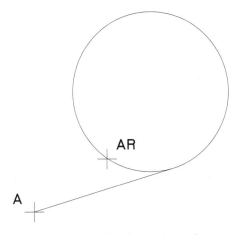

Figure 3–43 *Using the TANgent osnap mode, a line is drawn from a point outside a circle tangent to a point on the right side of the circle*

With the TANgent osnap mode you can also select an arc. Like the PERpendicular osnap mode, the tangent point must be on the arc selected (see Figure 3–44).

Figure 3–44 *Using the TANgent osnap mode, a line is drawn tangent to a point on an arc*

Object Snap—NEArest The NEArest mode lets you select any object (except text and shape) in response to a prompt for a point, and AutoCAD snaps to the point on that object nearest the cursor.

During an appropriate AutoCAD command, invoke the Object Snap—NEArest mode:

Object Snap toolbar	Select the NEArest mode (see Figure 3–45)
Command: prompt	Enter **nea** and press ENTER whenever an AutoCAD command: prompt requests a point to be specified

Figure 3–45 *Invoking the Object Snap NEArest mode from the Object Snap toolbar*

Object Snap—APParent intersection The APParent intersection mode allows you to snap to the apparent intersection of an arc, circle, ellipse, elliptical arc, line, mline, polyline, ray, spline, or xline, which may or may not actually intersect in 3D space.

During an appropriate AutoCAD command, invoke the Object Snap—APParent intersection mode:

Object Snap toolbar	Select the APParent intersection mode (see Figure 3–46)
Command: prompt	Enter **app** and press ENTER whenever an AutoCAD command: prompt requests a point to be specified

Figure 3–46 *Invoking the Object Snap—APParent intersection mode from the Object Snap toolbar*

AutoCAD prompts you to select two objects to establish the apparent intersection point.

 Note: The INTersection and APParent Intersection modes should not be in effect at the same time.

Object Snap—PARallel The PARallel mode allows you to draw a line that is parallel to another object. Once the first point of the line has been specified (and the Parallel Object Snap mode is selected), move the cursor over the object to which you wish to make the new line parallel. Then move the cursor near a line from the first point that is parallel to the object selected and a construction line will appear. While the construction line is visible, specify a point and the new line will be parallel to the selected object.

During an appropriate AutoCAD command, invoke the Object Snap - PARallel mode:

Object Snap toolbar	Select the PARallel mode (see Figure 3–47)
Command: prompt	Enter **par** and press ENTER whenever an AutoCAD command: prompt requests a point to be specified

Figure 3–47 *Invoking the Object Snap PARallel mode from the Object Snap toolbar*

TRACKING OPTION

TRACKING, or moving through nonselected point(s) to a selected point, could be called a command "enhancer." It can be used whenever a command prompts for a point. If the desired point can best be specified relative to some known point(s), you can "make tracks" to the desired point by invoking the Tracking option and then specifying one or more points relative to previous point(s) "on the way to" the actual point that the command is prompting for. These intermediate tracking points are not necessarily associated with the object being created or modified by the command. The primary significance of tracking points is that they are used to establish a path to the point you wish to specify as the response to the command prompt. Some of the objects in the partial plan shown in Figure 3–48 can be drawn more easily by means of Tracking.

Figure 3–48 *Example of a partial plan to demonstrate the Tracking option*

The following example will use Tracking to draw lines A and B in the Partial Plan in Figure 3–49, leaving the 3'-0" door opening in the correct place. By means of Tracking, we can draw the lines with the given dimension information without having to calculate the missing information.

In Figure 3–50, line A from TK1/SP1 (tracking point 1 and starting point 1) to EP1 (ending point 1) and line B from SP2 (starting point 2) to TK2/EP2 (tracking point 2 and ending point 2) can be drawn by using the TRACKING command enhancer. Invoke the LINE command and AutoCAD prompts:

Figure 3–49 *Lines A and B to be drawn with the help of the Tracking option*

Figure 3–50 *Various points to be drawn with the help of the Tracking option*

Command: **line** (ENTER)
Specify first point: **0,12'** (ENTER) *(specify point TK1/SP1)*
Specify next point or [Undo]: **tk** (ENTER) *(invoke the Tracking feature)*
First tracking point: *(specify point SP1/TK1 again as the first tracking point)*
Next point (Press ENTER to end tracking): **@13'8,0** (ENTER) *(locates the second tracking point, TK2, as shown in Figure 3–50)*
Next point (Press ENTER to end tracking): **@2'2<180** (ENTER) *(locates the third tracking point, TK3, as shown in Figure 3–50)*
Next point (Press ENTER to end tracking): **@1'6<180** (ENTER) *(locates the fourth tracking point, EP1, as shown in Figure 3–50)*
Next point (Press ENTER to end tracking): *(press ENTER to exit Tracking; by this you are designating the point to which you have "made tracks" as the response to the prompt that was in effect when you entered Tracking)*
Specify next point or [Undo]: *(press ENTER to exit the LINE command)*

Once the Tracking option is invoked, you establish a path to EP1 by specifying the initial tracking point, TK1, and then each subsequent point relative to the previous point, that is, TK2 relative to TK1, TK3 relative to TK2, and EP1 relative to TK3. The first track in this case is specified by the Relative Rectangular method, and the next two are Relative Polar. Also, because the tracking points are all on one horizontal line, you could use the direct distance feature by placing the cursor in the correct direction (with ORTHO set to ON) and entering the distance from the keyboard.

Note: If you knew the coordinates of one of the intermediate tracking points, then it probably should be the initial tracking point. The idea behind Tracking is to establish a point by means of a path from and through other points. Thus, the shortest path is the best. If the coordinates of TK2 were known, or if you could specify it by some other method, it could become the initial tracking point. Keep this in mind as you learn to use Object Snap. You don't necessarily need to know the coordinates if you can use Object Snap to select a point from which a tracking path could be specified.

The line from SP2 to TK2/EP2 can be started and ended in a similar manner as the line from TK1/SP1 to EP1 was started, with some minor modifications. The sequence (which involves using Tracking twice) is as follows:

```
Command: line (ENTER) (invoke the LINE command)
Specify first point: tk (ENTER) (invoke the Tracking option)
First tracking point: 0,12' (ENTER) (specify point TK1/SP1 as the first tracking
    point)
Next point (Press ENTER to end tracking): @13'8,0 (ENTER) (locates the
    second tracking point, TK2, as shown in Figure 3–50)
Next point (Press ENTER to end tracking): @2'2<180 (ENTER) (locates the
    third tracking point, TK3, as shown in Figure 3–50)
Next point (Press ENTER to end tracking): @1'6<0 (ENTER) (locates the
    fourth tracking point, SP2, as shown in Figure 3–50)
Next point (Press ENTER to end tracking): (press ENTER to exit Tracking;
    AutoCAD establishes point SP2)
Specify next point or [Undo]: tk (ENTER) (invoke the Tracking option again)
First tracking point: (specify point TK1/SP1 as the first tracking point)
Next point (Press ENTER to end tracking): @13'8,0 (ENTER) (locates the
    second tracking point, TK2/EP2, as shown in Figure 3–50)
Next point (Press ENTER to end tracking): (press ENTER to exit Tracking,
    AutoCAD establishes point EP2)
Specify next point or [Undo]: (press ENTER to terminate the LINE command)
```

This example shows an application of the Tracking option in which the points were established in reference to some known points.

DIRECT DISTANCE OPTION

The Direct Distance option for specifying a point relative to another point can be used with a command like LINE to permit a variation of the Relative Coordinates mode. In the case of the Direct Distance option, the distance is keyed in and the direction is determined by the current location of the cursor. This option is very useful when you know the exact distance but specifying the exact angle is not as easy as placing the cursor on a point that is at the exact angle desired.

Figure 3–51 shows a shape that can be drawn more easily by using the Direct Distance option along with setting the Snap and Ortho modes to ON and OFF at the appropriate times. Points A, B, and C are on the Snap grid (*X,Y* coordinates 3,3 for A, 3,6 for B, and 9,6 for C). By setting the Snap mode to ON with the value set to 1 or perhaps 0.5, the cursor can be placed on the required points to draw lines A-B and B-C. After drawing lines A-B and B-C (with Ortho set to either ON or OFF), the cursor should be placed on point A (with Ortho set to OFF).

If you have exited the LINE command after drawing line B-C, you must invoke the LINE command and specify C as the first point before moving the cursor to A. The rubber-band line indicates that the next line is drawn from C to A, as shown in Figure 3–52. However, you wish to draw a line only 2 units long but in the same direction as a line from C to A. With the cursor placed on A, enter 2 and press ENTER.

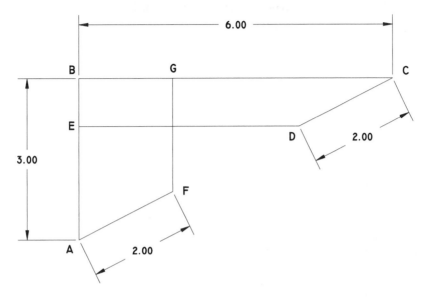

Figure 3–51 *Example of a direct distance application*

Figure 3–52 *Drawing a line from C by means of the Direct Distance option*

AutoCAD draws the line of 2 units, and Point D is established. To draw a line from D to E without exiting the LINE command, first set Ortho and Snap to ON, and then place the cursor on line A-B. The rubber-band line will indicate line D-E, as shown in Figure 3–53. In this case, you do not know the distance, but you do know that the line terminates on line A-B. Therefore, simply press the pick button on the cursor, and line D-E is drawn. You do not have to specify point D as the starting point of the line. Using the line-line continuation after drawing line C-D does that for you.

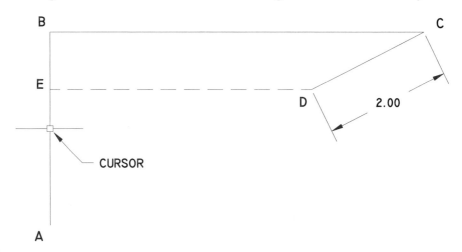

Figure 3–53 *Drawing a line from D to E*

Line A-F can be drawn in a similar manner as line C-D was drawn. Unlike line C-D, line A-F is not a continuation of another line. However, the starting point, A, can be selected with the Snap set to ON. After starting point A has been specified, place the cursor on point C and enter 2. From line A-F, line F-G can be drawn in a similar manner as line D-E was drawn. This example shows an application of the Direct Distance option in which the distance was entered in and the direction was controlled by the cursor.

POLAR TRACKING AND OBJECT SNAP TRACKING

The POLAR TRACKING feature lets you draw lines and specify point displacements in directions that are multiples of a specified increment angle. You can specify whether the increment angles are measured from the current coordinate system or from the angle of a previous object. You can also add up to 10 additional angles to which a line or displacement can be diverted from a base direction. The OBJECT SNAP TRACKING feature lets you track your cursor from a strategic (Osnap) point on an object in a specified direction, either orthogonal or preset polar angles.

Turning Polar Tracking ON and OFF

In AutoCAD, four primary methods are available to turn POLAR TRACKING ON and OFF:

1. In the status bar at the bottom of the screen, when the POLAR button appears to be pressed in, then Polar Tracking is ON. When the POLAR button appears to be out, then Polar Tracking is OFF. To change its setting, simply place the cursor over the POLAR button and press the pointing device pick button.

2. To change the Polar Tracking to ON when it is OFF or to OFF when it is ON, simply press the function key F10.

3. Right-click the POLAR button in the status bar at the bottom of the screen and from the shortcut menu, choose Settings. The Drafting Settings dialog box will be displayed with the Polar Tracking tab selected. Polar Tracking is ON when the **Polar Tracking On (F10)** check box is checked, and is OFF when it is not checked. To change its setting, simply place the cursor over the check box and press the pointing device pick button. You can also enter **ddrmodes** at the Command: prompt to display the Drafting Settings dialog box from which the Polar Tracking tab can be selected if it is not already selected.

4. Right-click the POLAR button in the status bar at the bottom of the screen and from the shortcut menu, and choose ON to turn Polar Tracking ON or choose OFF to turn Polar Tracking OFF.

When the Polar Tracking mode is ON, it allows you to specify a displacement or draw in a direction which is a multiple of a specified increment angle. But even with Polar Tracking ON, you can still specify a "next point" of a line or displacement with the cursor that is not at one of the multiples of the specified Polar Tracking increment angle. This is different from the ORTHO and SNAP features where you are forced (when specifying a point with the cursor) to accept a point orthogonally or on the Snap Grid when the respective mode is turned ON. You will notice, however, that when the Polar Tracking mode is ON and you move the cursor near one of the radials (multiple of increment angle), the cursor will snap to that radial and the radial will appear as a dotted construction line indicating the direction of the line that will be drawn if you specify that point by pressing the pick button on the pointing device.

Changing Polar Tracking Increment & Additional Angles

In AutoCAD there are two primary methods to change the Polar Tracking increment angle and add additional angles:

1. From the status bar at the bottom of the screen, right-click on the POLAR button and choose Settings. AutoCAD displays the Drafting Settings dialog box with the Polar Tracking tab selected. From the Polar Angle Settings section, enter the desired increment angle in the **Increment Angle:** text box. This is the base increment angle whose multiples are used by Polar Tracking. Up to 10 additional angles can be added in the text box under the **Additional Angles** check box. Simply select the **New** button and then enter the desired additional angle(s). If there is no check in the Additional Angles check box, the additional angles are not available for use when Polar Tracking

is ON. To make the additional angle(s) available when the Polar Tracking mode is ON, place the cursor over the **Additional Angles** check box and press the pointing device pick button (see Figure 3–54).

2. From the Tools menu, choose Drafting Settings. Then select the Polar Tracking tab. See the explanation in Method I for changing settings.

 Note: The POLAR TRACKING and the ORTHO modes cannot both be turned ON simultaneously. They can both be turned OFF, or either one can be turned ON.

Figure 3–54 *Selecting the* POLAR TRACKING ON *(F10) check box and setting the increment angle to 18° and one additional angle to 9°*

With the 18° increment angle and the Polar Angle Measurement set to **Absolute**, you can draw lines and specify displacements at 18°, 36°, 54°, 72°, 90°, 108°… …324°, 342°, 360° around the compass by just snapping to the construction lines when they appear. The accompanying AutoTrack Tooltip will appear when you have snapped to a multiple of the increment angle if the **Display AutoTrack Tooltip** check box is checked in the Drafting section of the Options dialog box. The Tooltip displays the cursor's distance and direction from the first specified point. You can, with the cursor snapped to one of the Polar Tracking angles, use the Direct Distance option and enter a distance from the keyboard and press ENTER. AutoCAD will draw the line or apply the displacement in accordance with the distance entered and the direction set by the cursor. For example, in Figure 3–55, the line from P1 to P2 was drawn 2 units long at 0°. This could have been done with Direct Distance and either ORTHO turned ON or with Polar Tracking. But with P2 to P3 being at 18° and from P3 to P4 being at 36°, it is easier to use Polar Tracking with the settings shown in the example in Figure 3–54. Figure 3–56 shows the three line segments drawn using Polar Tracking with the increment angle set to 18° and the Polar Angle Measurement set to Absolute. Figure 3–57 shows how you can use the Additional Angle setting of 9° to draw a 1-unit-long line at 9° from absolute zero.

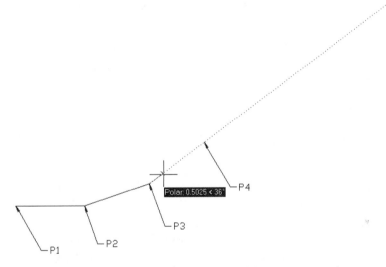

Figure 3–55 *Drawing line segments at multiples of the 18° increment angle*

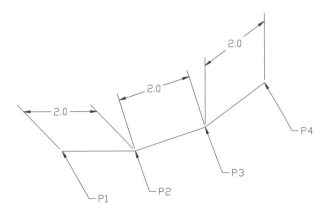

Figure 3–56 *Three line segments drawn using Polar Tracking with the increment angle set to*

18° and the Polar Angle Measurement set to Absolute

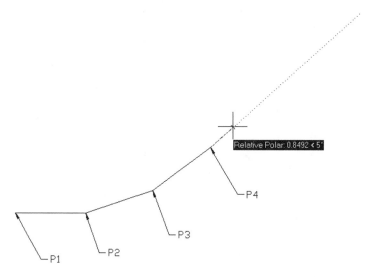

Figure 3–57 *Four line segments drawn using Polar Tracking with the increment angle set to 18° and at one additional angle at 9° from Polar Angle Measurement set to Absolute*

After completing the three segments drawn at multiples of 18°, instead of adding the segment at 9°, let's change the increment angle to 5° and the Polar Angle Measurement set to *Relative To Last Segment.* Continuing with three more segments, each 1 unit long, the first increment angle will be measured from the last segment which is at 36°. The Tooltip displays "Relative Polar (distance) < 5°" but the resulting angle is 41°. Figure 3–58 shows the three additional line segments drawn using Polar Tracking with the increment angle set to 5° and the Polar Angle Measurement set to *Relative To Last Segment.* The last two segments will actually be at 46° and 51°.

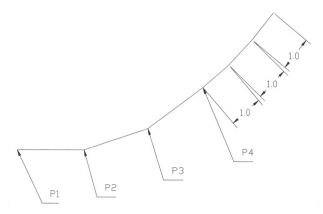

Figure 3–58 *Three line segments drawn using Polar Tracking with the increment angle set to 5° and the Polar Angle Measurement set to Relative To Last Segment*

Polar Snap

The Polar Snap feature can be used to make drawing the 2-unit-long and 1-unit-long segments in the previous examples even easier. From the Snap and Grid tab of the Drafting Settings dialog box, check the **Polar Snap** check box in the Snap Type & Style section. Then in the Polar distance text box of the Polar Spacing section, enter the desired distance. In the case of the above mentioned example, the 2-unit-long segments at 0°, 18°, and 36°, you can enter 2.0 for the distance. For the 41°, 46°, and 51° segment (additions of 5°, each, in the *Relative To Last Segment mode*), you can use a distance of 1.0.

Object Snap Tracking

Polar Tracking can be used, making one of an object's strategic points as the base point for tracking, by using the Object Snap Tracking feature, along with the Object Snap mode, that will allow acquiring the particular type of strategic point you wish such as endpoint, midpoint, or center. If you wish to start a line 1 unit and 18° from an object's endpoint, simply turn on the Object Snap Tracking by checking the **Object Snap Tracking ON (F11)** check box in the Object Snap tab of the Drafting Settings dialog box (or press the function key F11). Be sure the Polar Tracking angle and Polar Tracking Distance are set to 18° and 1.0, respectively. Then invoke the Object Snap mode Endpoint and acquire the desired endpoint by moving the cursor over the object's end temporarily before moving it to the second (end of tracking) point.

This next example is to show how to apply Polar Tracking and Object Snap Tracking to draw the vertical line B and the horizontal line C from existing line A and arc D, as shown in Figures 3–59 and 3–60.

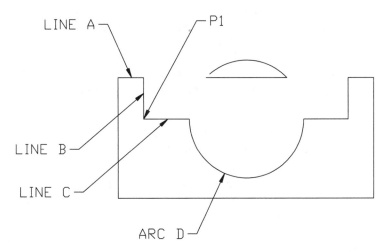

Figure 3–59 *Completed object including lines A, B, and C, and arc D*

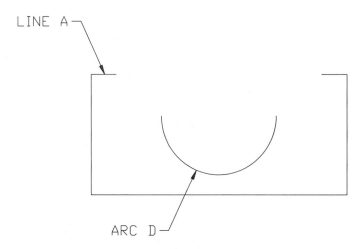

Figure 3–60 *Line A and arc D from which lines B and C will be drawn using Polar Tracking*

From the status bar at the bottom of the screen, right-click on the POLAR button and choose Settings. The Drafting Settings dialog box is displayed with the Polar Tracking tab selected. From the Object Snap Tracking Settings section choose the **Track Orthoganol Only** radio button. Check to be sure that Polar Tracking is ON. Invoke the LINE command:

> Command: **line** (ENTER)
> Specify first point: **end** *(enter the first three letters of the Osnap mode Endpoint and then select the right end of LINE A to establish the start point of LINE B)*
> Specify next point or [Undo]: *(move the aperture cursor to the right end of LINE A, keep it there until the Endpoint marker appears and without pressing a button move it downward from the line)*

As you move the cursor downward from line A, a dashed construction line will follow as long as the line between the end of line A and the cursor is vertical or horizontal, as shown in Figure 3–61.

Next, move the cursor to the left end of arc D, keep it there until the Endpoint marker appears and then move it away from it toward the left, again without pressing the pick button. A horizontal construction line will follow until the cursor is near the vertical construction line through the end of line A as shown in Figure 3–62. At this time, with both the vertical and horizontal construction lines being displayed, you can press the pick button on the pointing device and point P1, the end point of line B will be established.

After pressing the pick button to specify point P1 and complete line B, move the cursor to the left end of arc D, invoke the Endpoint Osnap mode, and select arc D. This will complete line C, as shown in Figure 3–63.

Figure 3–61 *A vertical construction line passes through the cursor and the end of LINE A*

Figure 3–62 *A dashed arc follows the cursor indicating the extension of ARC C to the dashed extension of LINE B, at which point P1 is specified*

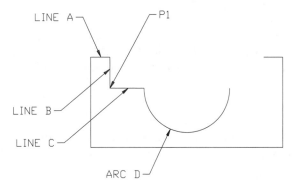

Figure 3–63 *Lines B and C are drawn with point P1 at their intersection*

DISPLAY CONTROL

There are many ways to view a drawing in AutoCAD. These viewing options vary from on-screen viewing to hard-copy plots. The hard-copy options are discussed in Chapter 8. Using the display commands, you can select the portion of the drawing to be displayed, establish 3D perspective views, and much more. By being able to view the drawing in different ways, AutoCAD provides a way to draw faster. The commands that are explained in this section are like utility commands. They make your job easier and help you to draw more accurately.

ZOOM COMMAND

The ZOOM command is like a zoom lens on a camera. You can increase or decrease the viewing area, although the actual size of objects remains constant. As you increase the visible size of objects, you view a smaller area of the drawing in greater detail as though you were closer. As you decrease the visible size of objects, you view a larger area as though you were farther away. This ability provides a close-up view for better accuracy and detail or a distant view to get the whole picture.

Invoke the ZOOM command:

Zoom toolbar	Choose one of the available options (see Figure 3–64)
View menu	Zoom > Choose one of the available options
Command: prompt	**zoom** (ENTER)

Figure 3–64 *Zoom toolbar*

AutoCAD prompts:

> Command: **zoom**
> Specify corner of window, enter a scale factor (nX or nXP), or [All/
>   Center/Dynamic/Extents/Previous/Scale/Window] <real time>: *(right-
>   click and choose one of the available options from the shortcut menu)*

Default Option

The default option of the ZOOM command is a ZOOM WINDOW. After the ZOOM command displays its list of options, simply specify two points on the screen that represent diagonally opposite corners of a rectangle. The view in the rectangle is enlarged to fill the drawing area.

Realtime Option

The ZOOM REALTIME option lets you zoom interactively to a logical extent. Once you invoke the command with the Realtime option, the cursor changes to a magnifying glass with a "±" symbol. To zoom in closer, hold the pick button and move the cursor vertically toward the top of the window. To zoom out further, hold the pick button and move the cursor vertically toward the bottom of the window. To discontinue the zooming, release the pick button. To exit the ZOOM REALTIME option, press ENTER, ESC, or from shortcut menu select ENTER.

The current drawing window is used to determine the zooming factor. If the cursor is moved by holding the pick button from the bottom of the window to the top of the window vertically, the zoom-in factor would be 200%. Conversely, when holding the pick button from the top of the window and moving vertically to the bottom of the window, the zoom-out factor would be 200%.

When you reach the zoom-out limit, the "-" symbol on the cursor disappears while attempting to zoom-out indicating that you can no longer zoom out. Similarly, when you reach the zoom-in limit, the "+" symbol on the cursor disappears while attempting to zoom in indicating that you can no longer zoom in.

Invoke the ZOOM REALTIME command:

Standard toolbar	Choose Zoom Realtime (see Figure 3–65)
View menu	Zoom > Realtime
Command: prompt	**zoom** (ENTER)

Figure 3–65 *Invoking the Zoom Realtime option from the Standard toolbar*

AutoCAD prompts:

Command: **zoom** (ENTER)
Specify corner of window, enter a scale factor (nX or nXP), or
[All/Center/Dynamic/Extents/Previous/Scale/Window] <real time>: (ENTER)
Press ESC or ENTER to exit, or right-click to activate the shortcut menu.

Control the display by moving the cursor appropriately; to exit the ZOOM REALTIME option, press the ESC or ENTER key. You can also exit by selecting Exit from the shortcut menu that is displayed when you right-click on your pointing device. In addition, you can perform other operations related to ZOOM and PAN by selecting appropriate commands from the shortcut menu.

Scale Option

The ZOOM SCALE command lets you enter a display scale (or magnification) factor. The scale factor, when entered as a number (it must be a numerical value and must not be expressed in units of measure), is applied to the area covered by the drawing limits. For example, if you enter a scale value of 3, each object appears three times as large as in the ZOOM ALL view. A scale factor of 1 displays the entire drawing (the full view), which is defined by the established limits. If you enter a value less than 1, AutoCAD decreases the magnification of the full view. For example, if you enter a scale of 0.5, each object appears half its size in the full view while the viewing area is twice the size in horizontal and vertical dimensions. When you use this option, the object in the center of the screen remains centered.

See Figures 3–66 and 3–67 for the difference between a full view and a 0.5 zoom.

Figure 3–66 *A drawing at ZOOM All (full view)*

Figure 3–67 *The same drawing after setting ZOOM to 0.5*

Invoke the ZOOM SCALE command:

Zoom toolbar	Choose Zoom Scale (see Figure 3–68)
View menu	Zoom > Scale
Command: prompt	**zoom** (ENTER)

Figure 3–68 *Invoking the ZOOM SCALE command from the Zoom toolbar*

AutoCAD prompts:

> Command: **zoom** (ENTER)
> Specify corner of window, enter a scale factor (nX or nXP), or
> [All/Center/Dynamic/Extents/Previous/Scale/Window] <real time>: **s**
> Enter a scale factor (nX or nXP): *(specify a scale factor)*

186

If you enter a number followed by X, the scale is determined relative to the current view. For instance, entering 2X causes each object to be displayed two times its current size on the screen.

The scale factor XP option is explained in Chapter 7, related to the layout of the drawing.

All Option

The ZOOM ALL command lets you see the entire drawing. In a plan view, it zooms to the drawing's limits or current extents, whichever is larger. If the drawing extends outside the drawing limits, the display shows all objects in the drawing.

Invoke the ZOOM ALL command:

Zoom toolbar	Choose Zoom All (see Figure 3–69)
View menu	Zoom > All
Command: prompt	**zoom** (ENTER)

Figure 3–69 *Invoking the* ZOOM ALL *command from the Zoom toolbar*

AutoCAD prompts:

Command: **zoom** (ENTER)
Specify corner of window, enter a scale factor (nX or nXP), or
[All/Center/Dynamic/Extents/Previous/Scale/Window] <real time>:
   a (ENTER)

Center

The ZOOM CENTER command lets you select a new view by specifying its center point and the magnification value or height of the view in current units. A smaller value for the height increases the magnification; a larger value decreases the magnification.

Invoke the ZOOM CENTER command:

Zoom toolbar	Choose Zoom Center (see Figure 3–70)
View menu	Zoom > Center
Command: prompt	**zoom** (ENTER)

Figure 3–70 *Invoking the* ZOOM CENTER *command from the Zoom toolbar*

AutoCAD prompts:

> Command: **zoom** (ENTER)
> Specify corner of window, enter a scale factor (nX or nXP), or
> [All/Center/Dynamic/Extents/Previous/Scale/Window] :
>   **c** (ENTER)
> Center point: *(specify the center point)*
> Magnification or Height <current height>: *(specify the magnification or height)*

The following command sequence produces the example of the ZOOM Center option shown in Figure 3–71.

> Command: **zoom** (ENTER)
> Specify corner of window, enter a scale factor (nX or nXP), or
> [All/Center/Dynamic/Extents/Previous/Scale/Window] :
>   **c** (ENTER)
> Specify center point: **8,6** (ENTER)
> Enter magnification or height <current height>: **4**

BEFORE   AFTER

Figure 3–71 *Using the* ZOOM CENTER *command*

In addition to providing coordinates for center point, you can also specify the center point by placing a point on the view window. The height can also be specified in terms of the current view height by specifying the magnification value followed by an X. A response of 3X will make the new view height three times as large as the current height.

Dynamic Option

The AutoCAD ZOOM DYNAMIC command provides a quick and easy method to move to another view of the drawing. With ZOOM DYNAMIC, you can see the entire drawing and then simply select the location and size of the next view by means of cursor manipulations. Using ZOOM DYNAMIC is one means by which you can visually select a new display area that is not entirely within the current display.

Invoke the ZOOM DYNAMIC command:

Zoom toolbar	Choose Zoom Dynamic (see Figure 3–72)
View menu	Zoom > Dynamic
Command: prompt	**zoom** (ENTER)

Figure 3–72 *Invoking the* ZOOM DYNAMIC *command from the Zoom toolbar*

AutoCAD prompts:

> Command: **zoom** (ENTER)
> Specify corner of window, enter a scale factor (nX or nXP), or
> [All/Center/Dynamic/Extents/Previous/Scale/Window] <real time>:
>   **d** (ENTER)

The current viewport is then transformed into a selecting view that displays the drawing extents, as shown in the example in Figure 3–73.

When the selecting view is displayed, you see the drawing extents marked by a white or black box, the current display marked by a blue or magenta dotted box. A new view box, the same size as the current display, appears. Its location is controlled by the movement of the pointing device. Its size is controlled by a combination of the pick button and cursor movement. When the new view box has an X in the center, the box pans around the drawing in response to cursor movement. After you press the pick button on the pointing device, the X disappears and an arrow appears at the right edge of the box. The new view box is now in Zoom mode. While the arrow is in the box, moving the cursor left decreases the box size; moving the cursor right increases the size.

When the desired size has been chosen, press the pick button again to pan, or press ENTER to accept the view defined by the location/size of the new view box. Pressing ESC cancels the ZOOM DYNAMIC and returns you to the current view.

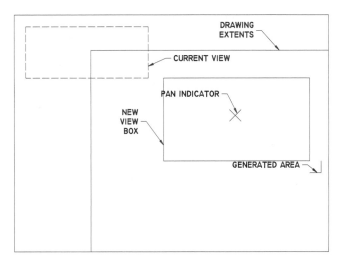

Figure 3–73 *Using the* ZOOM DYNAMIC *command to display the drawing extents*

Extents Option

The ZOOM EXTENTS command lets you see the entire drawing on screen. Unlike the All Option, the Extents Option uses only the drawing extents and not the drawing limits. See Figures 3–74 and 3–75, which illustrate the difference between the options All and Extents.

Figure 3–74 *A drawing after a ZOOM All*

Figure 3–75 *The same drawing after a ZOOM Extents*

Invoke the ZOOM EXTENTS command:

Zoom toolbar	Choose Zoom Extents (see Figure 3–76)
View menu	Zoom > Extents
Command: prompt	**zoom** (ENTER)

Figure 3–76 *Invoking the ZOOM EXTENTS command from the Zoom toolbar*

AutoCAD prompts:

Command: **zoom** (ENTER)
Specify corner of window, enter a scale factor (nX or nXP), or
[All/Center/Dynamic/Extents/Previous/Scale/Window] <real time>: **e** (ENTER)

Previous Option

The ZOOM PREVIOUS command displays the last displayed view. While editing or creating a drawing, you may wish to zoom into a small area, back out to view the larger area, and then zoom into another small area. To do this, AutoCAD saves the coordinates of the current view whenever it is being changed by any of the zoom options or other view commands, so you can return to the previous view by entering the Previous option, which can restore the previous 10 views.

Invoke the ZOOM PREVIOUS command:

View menu	Zoom > Previous
Command: prompt	**zoom** (ENTER)

AutoCAD prompts:

Command: **zoom** (ENTER)
Specify corner of window, enter a scale factor (nX or nXP), or
[All/Center/Dynamic/Extents/Previous/Scale/Window] <real time>: **p** (ENTER)

Window Option

The ZOOM WINDOW command lets you specify a smaller area of the part of the drawing being currently displayed and have that portion fill the drawing area. This is done by specifying two diagonally opposite corners of a rectangle. The center of the area selected becomes the new display center, and the area inside the window is enlarged to fill the drawing area as completely as possible.

You can enter two opposite corner points to specify an area by means of coordinates or the pointing device (see Figures 3–77 and 3–78).

Figure 3–77 *Specifying a ZOOM Window area*

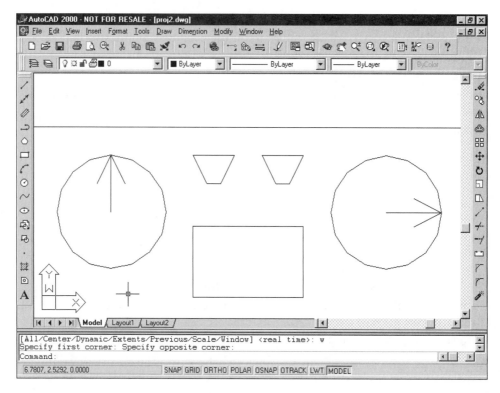

Figure 3–78 *After specifying a window area*

Invoke the ZOOM WINDOW command:

Zoom toolbar	Choose Zoom Window (see Figure 3–79)
View menu	Zoom > Window
Command: prompt	**zoom** (ENTER)

Figure 3–79 *Invoking the* ZOOM WINDOW *command from the Zoom toolbar*

AutoCAD prompts:

Command: **zoom** (ENTER)
Specify corner of window, enter a scale factor (nX or nXP), or

[All/Center/Dynamic/Extents/Previous/Scale/Window] <real time>:
   w (ENTER)
Specify first corner: *(specify a point to define the first corner of the window)*
Specify opposite corner: *(specify a point to define the diagonally opposite corner of the window)*

PAN COMMAND

AutoCAD lets you view a different portion of the drawing in the current view without changing the magnification. You can move your viewing area to see details that are currently off screen. Imagine that you are looking at your drawing through the display window and that you can slide the drawing left, right, up, and down without moving the window.

The PAN REALTIME command lets you pan interactively to the logical extent (edge of the drawing space). Once you invoke the command, the cursor changes to a hand cursor. To pan, hold the pick button on the pointing device to lock the cursor to its current location relative to the viewport coordinate system, and move the cursor in any direction. Graphics within the window are moved in the same direction as the cursor. To discontinue the panning, release the pick button.

When you reach a logical extent (edge of the drawing space), a line-bar is displayed on the hand cursor on the side. The line-bar is displayed at the top, bottom, or left or right side of the drawing, depending upon whether the logical extent is at the top, bottom, or side of the drawing.

Invoke the PAN REALTIME command:

Standard toolbar	Choose Pan Realtime (see Figure 3–80)
View menu	Pan > Real Time
Command: prompt	**pan** (ENTER)

Figure 3–80 *Invoking the* PAN REALTIME *command from the Standard toolbar*

AutoCAD prompts:

Command: **pan** (ENTER)
Press Esc or Enter to exit, or right-click to display shortcut menu.

To exit PAN REALTIME, press ESC or ENTER. You can also exit by selecting Exit from the shortcut menu that is displayed when you right-click on your pointing device. In addition, you can perform other operations related to zoom and pan by selecting the appropriate commands from the shortcut menu.

You can specify two points by entering the −pan command at the Command: prompt, in which case AutoCAD computes the displacement from the first point to the second.

For example, the following command sequence views a different portion of the drawing by placing two data points as shown in Figure 3–81.

```
Command: -pan (ENTER)
Specify base point or displacement: (Specify first point of displacement)
Specify second point: (Specify second point of displacement)
```

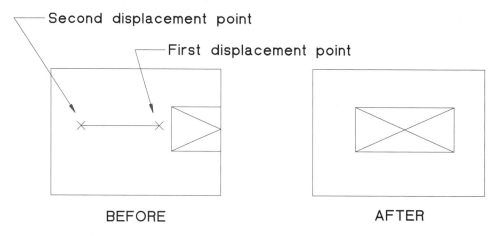

Figure 3–81 *Moving the view with the PAN command by specifying two displacement points*

AutoCAD calculates the distance and direction between the two points and pans the drawing accordingly. Sometimes it is useful to pan in exactly the horizontal or the vertical direction. In that case, set ORTHO mode to ON before invoking the PAN command to constrain cursor movement to the X or Y axis directions. You can enter a single coordinate pair indicating the relative displacement of the drawing with respect to the screen. If you give a null response to the "Second point" prompt, you are indicating that the coordinates provided represent the displacement of the drawing with respect to the origin. If you provide the coordinates for the second point instead of giving a null response, then AutoCAD computes the displacement from the first point to the second.

For example, the following command sequence views a different portion of the drawing 2 units to the left and 0.75 units up, as shown in Figure 3–82.

Command: **-pan** (ENTER)
Specify base point or displacement: **-2,.75** (ENTER)
Specify second point: (ENTER)

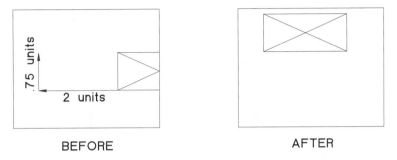

BEFORE   AFTER

Figure 3–82 *Moving the view with the* PAN *command by specifying a pair of coordinates*

Pan Left, Right, Up, or Down

By choosing one of the Left, Right, Up, or Down options from the View > Pan flyout menu, it causes AutoCAD to pan the view left, right, up, or down accordingly.

AERIAL VIEW

The DSVIEWER command is used to activate the Aerial View, which provides a quick method of visually panning and zooming. By default, AutoCAD displays the Aerial View window, with the entire drawing displayed in the window, as shown in Figure 3–83. You can select any portion of the drawing in the Aerial View window by visually panning and zooming; in turn, AutoCAD displays the selected portion in the view window (current viewport).

Figure 3–83 *The Aerial View dialog window*

Invoke the DSVIEWER command:

View menu	Aerial View
Command: prompt	**dsviewer** (ENTER)

AutoCAD displays the Aerial View window. Two option menus are provided to pan and zoom visually.

View Menu

The View menu in the Aerial View window has three options. The ZOOM IN option causes the view to appear closer, enlarging the details of objects, but covering a smaller area. The ZOOM OUT causes the view to appear farther away, decreasing the size of objects, but covering a larger area. The GLOBAL option causes the entire drawing to be viewable in the Aerial View window. You can also select the three options from the toolbar provided in the Aerial View window.

Options Menu

The options menu in the Aerial View window has three options. AUTO VIEWPORT causes the active viewport to be displayed in model space. DYNAMIC UPDATE toggles whether the view is updated or not in response to editing. The REALTIME ZOOM controls whether or not the AutoCAD window updates in real time when you zoom using the Aerial View.

CONTROLLING DISPLAY WITH INTELLIMOUSE

AutoCAD also allows you to control the display of the drawing with the small wheel provided with the IntelliMouse (two-button mouse). You can use the wheel to zoom and pan in your drawing any time without using any AutoCAD commands. By rotating the wheel forward you can zoom in and backwards you can zoom out. When double-clicking the wheel button, AutoCAD displays the drawing to the extent of the view window. To pan the display of the drawing, press the wheel button and drag the mouse. By default, each increment in the wheel rotation changes the zoom level by 10 percent. The ZOOMFACTOR system variable controls the incremental change, whether forward or backward. The higher the setting, the smaller the change.

AutoCAD also allows you to display the Object Snap shortcut menu when you click the wheel button. To do so, set the MBUTTONPAN system variable to 0. By default, the MBUTTONPAN is set to 1.

REDRAW COMMAND

The REDRAW command is used to refresh the on-screen image. You can use this command whenever you see an incomplete image of your drawing. If you draw two lines in the same place and erase one of the lines, it appears as if both the lines are erased. By invoking the REDRAW command, the second line will reappear. Also, you can use the REDRAW command to remove the blip marks on the screen. A redraw is considered a screen refresh as opposed to database regeneration.

Invoke the REDRAW command:

View menu	Choose Redraw
Command: prompt	**redraw** (ENTER)

AutoCAD prompts:

Command: **redraw** (ENTER)

AutoCAD does not provide any options for the REDRAW command.

REGEN COMMAND

The REGEN command is used to regenerate the drawing's data on the screen. In general, you should use the REGEN command if the image presented by REDRAW does not correctly reflect your drawing. REGEN goes through the drawing's entire database and projects the most up-to-date information on the screen; this command will give you the most accurate image possible. Because of the manner in which it functions, a REGEN takes significantly longer than a REDRAW.

There are certain AutoCAD commands for which REGEN takes place automatically unless REGENAUTO is set to OFF.

Invoke the REGEN command:

View menu	Choose Regen
Command: prompt	**regen** (ENTER)

AutoCAD prompts:

Command: **regen** (ENTER)

AutoCAD does not provide any options for the REGEN command.

SETTING MULTIPLE VIEWPORTS

The ability to divide the display into two or more separate viewports is one of the most useful features of AutoCAD. Multiple viewports divide your drawing screen into rectangles, permitting several different areas for drawing instead of just one. It is like having multiple zoom lens cameras, with each camera being used to look at a different portion of the drawing. You retain your menus and "Command:" prompt area.

Each viewport maintains a display of the current drawing independent of the display shown by other viewports. You can simultaneously display a viewport showing the entire drawing, and another viewport showing a closeup of part of the drawing in greater detail. A view in one viewport can be from a different point of view than those in other viewports. You can begin drawing (or modifying) an object in one viewport and complete it in another viewport. For example, three viewports could be used in a 2D drawing, two of them to zoom in on two separate parts of the drawing, showing two widely separated features in great detail on the screen simultaneously, and the third to show the entire drawing, as in Figure 3–84. In a 3D drawing, four viewports could be used to display simultaneously four views of a wireframe model; top, front, right side, and isometric, as in Figure 3–85.

Figure 3–84 *Multiple viewports show different parts of the same 2D drawing*

AutoCAD allows you to divide the graphics area of your display screen into multiple, non-overlapping (tiled) viewports, as in Figures 3–84 and 3–85, only when you are in Model Space (system variable TILEMODE set to 1). The maximum number of tiled viewports that you can have is 64. In addition you can also create multiple overlapping (floating) viewports when the system variable TILEMODE is set to 0. For a detailed explanation on to create floating viewports, refer to Chapter 8.

You can work in only one viewport at a time. It is considered the current viewport. A viewport is set to current by moving the cursor into it with your pointing device and then pressing the pick button. You can even switch viewports in midcommand (except during some of the display commands). For example, to draw a line using two viewports, you must start the line in the current viewport, make another viewport current by clicking in it, and then specify the endpoint of the line in the second viewport. When a viewport is current, its border will be thicker than the other viewport borders.

The cursor is active for specifying points or selecting objects only in the current viewport; when you move your pointing device outside the current viewport, the cursor appears as an arrow pointer.

Figure 3–85 *Using viewports to show four views simultaneously for a 3D wireframe model*

Display commands like ZOOM and PAN and drawing tools like GRID, SNAP, ORTHO, and UCS icon modes are set independently in each viewport. The most important thing to remember is that the images shown in multiple viewports are all of the same drawing. An object added to or modified in one viewport will affect its image in the other viewports. You are not making copies of your drawing, just viewing its image in different viewports.

When you are working in tiled viewports, visibility of the layers is controlled globally in all the viewports. If you turn off a layer, AutoCAD turns it off in all viewports.

CREATING TILED VIEWPORTS
AutoCAD allows you to display tiled viewports in various configurations. Display of the viewports depends on the number and size of the views you need to see. By default, whenever you start a new drawing, AutoCAD displays a single viewport that fills the entire drawing area.

To create multiple viewports, open the Viewports dialog box:

Layouts toolbar	Choose Display Viewports Dialog (see Figure 3–86)
View menu	Choose Viewports > New Viewports...
Command: prompt	**vports** (ENTER)

Figure 3–86 *Invoking the* VPORTS *command from the Layout toolbar*

AutoCAD displays Viewports dialog box similar to Figure 3–87.

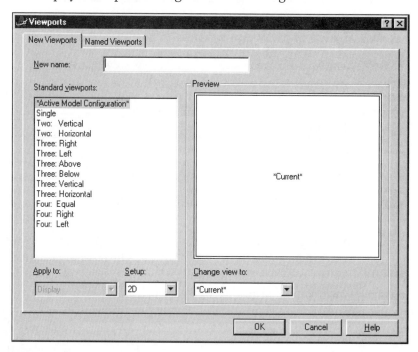

Figure 3–87 *Viewports dialog box*

Choose the name of the configuration you want to use from the **Standard viewports** list. AutoCAD displays the corresponding configuration in the preview window. If necessary, you can save the selected configuration by providing a name in the **New name:** text box. Select Display option from the **Apply to:** menu and select 2D from the **Setup:** menu for 2D viewport setup or select 3D for 3D viewport setup. Choose the **OK** button to create the selected viewport configuration. If you need additional viewports other than the standard configurations, you can subdivide a selected viewport. First select the viewport which you want to subdivide, then open the Viewports dialog box, select the appropriate configuration, and select Current Viewport from the **Apply to:** menu.

The Named Viewports tab lists all saved viewport configurations. At any time you can restore one of the saved viewport configurations.

You can also create tiled viewports from the command version of the VPORTS command:

Command: prompt	**-vports** (ENTER)

AutoCAD prompts:

> Command: **-vports** (ENTER)
> Enter an option [Save/Restore/Delete/Join/SIngle/2/<3>/4]: *(select one of the available options)*

Save Option

The Save option allows you to save the current viewport configuration. The configuration includes the number and placement of active viewports and their associated settings. You can save any number of configurations with the drawing, to be recalled at any time. When you select this option, AutoCAD prompts:

> Enter name for new viewport configuration or [?]:

You can use the same naming conventions for your configuration as for layer names. Instead of providing the name, you can respond with ? to request a list of saved viewport configurations.

Restore Option

The Restore option allows you to display a saved viewport configuration. When you select this option, AutoCAD prompts:

> Enter name of viewport configuration to restore or [?]: *(Specify the name of the viewport configuration to restore)*

Delete Option

The Delete option deletes a named viewport configuration. When you select this option, AutoCAD prompts:

> Enter name of viewport configuration to delete: *(Specify the name of the viewport configuration you want to delete)*

Join Option

The Join option combines two adjoining viewports into a single viewport. The view for the resulting viewport is inherited from the dominant viewport. When you select this option, AutoCAD prompts:

> Select dominant viewport <current viewport>:

You can give a null response to show the current viewport as the dominant viewport, or you can move the cursor to the desired viewport and press the pick button. Once you identify the dominant viewport, then AutoCAD prompts:

> Select viewport to join:

Move the cursor to the desired viewport to join, and press the pick button. If the two viewports selected are not adjacent or do not form a rectangle, AutoCAD displays an error message and reissues the prompts.

Single Option
The Single option allows you to make the current viewport the single viewport.

? Option
The ? option displays the identification numbers and screen positions of the active viewports. When you select this option, AutoCAD prompts:

> Enter name(s) of viewport configuration(s) to list <*>:

To list all saved configurations, give a null response. You also can use wild cards to list saved viewport names. All viewports are given an identification number by AutoCAD. This number is independent of any name you might give the viewport configuration. Each viewport is given a coordinate location, with respect to 0.0000,0.0000 as the lower left corner of the graphics area and 1.0000,1.0000 as the upper right corner.

2 Option
The 2 option splits the current viewport in half. When you select this option, AutoCAD prompts:

> Enter a configuration option [Horizontal/Vertical] <Vertical>:

You can select a horizontal or a vertical split, as shown in Figure 3–88. Vertical is the default.

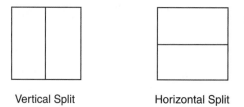

Vertical Split   Horizontal Split

Figure 3–88 *Using the 2 option to split a current viewport in half by vertical or horizontal division*

3 Option
The 3 option divides the current viewport into three viewports. This is the default option. When you select this option, AutoCAD prompts:

> Enter a configuration option [Horizontal/Vertical/Above/Below/Left/Right]
>   <Right>:

You can select the Horizontal or Vertical option to split the current viewport into thirds by horizontal or vertical division, as shown in Figure 3–89. The other options let you split into two small ones and one large one, specifying whether the large is to be placed above, below, or to the left or right (see Figure 3–89).

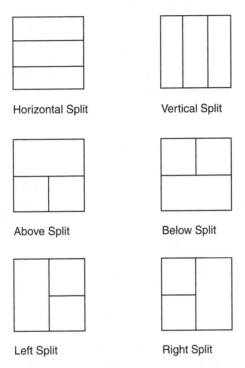

| Horizontal Split | Vertical Split |

| Above Split | Below Split |

| Left Split | Right Split |

Figure 3–89 *Using the 3 option to split a current viewport into various three-way divisions*

4 Option

The 4 option divides the current viewport into four viewports of equal size both horizontally and vertically, as shown in Figure 3–90.

Four Viewports

Figure 3–90 *Using the 4 option to split a current viewport into four viewports of equal size*

When you are working in multiple viewports, the REDRAW and REGEN commands will affect only the current viewport. To redraw or regenerate all the viewports simultaneously, use the REDRAWALL or REGENALL commands, respectively.

CREATING AND MODIFYING LAYER SYSTEM

AutoCAD offers a means of grouping objects in layers in a manner similar to the manual drafter's separating complex drawings into simpler ones on individual transparent sheets superimposed in a single stack. Under these conditions, the manual drafter would be able to draw on the top sheet only. Likewise, in AutoCAD you can draw only on the current layer. However, AutoCAD permits you to transfer selected objects from one layer to another (neither of which needs to be the current layer) with commands called CHANGE, CHPROP, and several others. (Let's see the manual drafter try that!)

A common application of the layer feature is to use one layer for construction (or layout) lines. You can create geometric constructions with objects, such as lines, circles, and arcs. These generate intersections, endpoints, centers, points of tangency, midpoints, and other useful data that might take the manual drafter considerable time to calculate with a calculator or to hand-measure on the board. From these you can create other objects using intersections or other data generated from the layout. Then the layout layer can be turned off (making it no longer visible) or set the layout not to plot. The layer is not lost, but can be recalled (turned on) for viewing later as required. The same drawing limits, coordinate system, and zoom factors apply to all layers in a drawing.

To draw an object on a particular layer, first make sure that that layer is set as the "current layer." There is one, and only one, current layer. Whatever you draw will be placed on the current layer. The current layer can be compared to the manual drafter's top sheet on the stack of transparencies. To draw an object on a particular layer, that layer must first have been created; if it is not the current layer, you must make it the current layer.

You can always move, copy, or rotate any object, whether it is on the current layer or not. When you copy an object that is not on the current layer, the copy is placed on the layer that the original object is on. This is also true with the mirror or an array of an object or group of objects.

A layer can be visible (ON) or invisible (OFF). Only visible layers are displayed or plotted. If necessary, AutoCAD allows you to set a visible layer not to plot. The layer(s) that are visible and set not to plot, will not be plotted. Invisible layers are still part of the drawing; they are just not displayed or plotted. You can turn layers on and off at will, in any combination. It is possible to turn off the current layer. If this happens and you draw an object, it will not appear on the screen; it will be placed on the current layer and will appear on the screen when that layer is turned ON (provided you are viewing the area in which the object was drawn). This is not a common occurrence, but it can cause concern to both the novice and the more experienced operator who has not faced the problem before. Do not turn off the current layer; the results can be very confusing. When the TILEMODE system variable is set to OFF, you can make specified layers visible only in certain viewports.

Each layer in a drawing has an associated name, color, lineweight, and linetype. The name of a layer may be up to 255 characters long. It may contain letters, digits, and the special characters dollar ($), hyphen (-), underscore (_), and spaces. Always give a descriptive name appropriate to your application, such as floor-plan or plumbing. The first several characters of the current layer's name are displayed in the layer list box located in the Object Properties toolbar (see Figure 3–91). You can change the name of a layer any time you wish, and you can delete unused layers except layer 0.

Color

AutoCAD allows you to assign a color to a layer in a drawing. You can assign to a layer any one of the 255 available colors. If necessary, you can assign the same color to more than one layer. If your graphics monitor is monochrome, all color numbers will produce the same visual effect. Even in this case, color numbers are useful, because they can be assigned to different pens on a pen plotter to plot in different line weights and colors. This works even for single-pen plotters; you can instruct AutoCAD to pause for pen changes.

Figure 3–91   *The current layer name is displayed in the list box located in the Object Properties toolbar*

Lineweight

Similar to assigning color, you can also assign a specific lineweight to a layer. Lineweights add width to your objects, both on screen and on paper. Using lineweights, you can create heavy and thin lines to show varying object thicknesses in details. For example, by assigning varying lineweights to different layers, you can differentiate between new, existing, and demolition construction.

AutoCAD allows you to assign a lineweight to a specific layer or an object in either inches or millimeters, with millimeters being the default. If necessary, you can change the default setting by invoking the LINEWEIGHT command.

A lineweight value of 0 is displayed as one pixel in model space and plots at the thinnest lineweight available on the specified plotting device. Appendix G lists all the available lineweights as well as associated industry standards.

Lineweights are displayed differently in Model Space than in a Paper Space layout (refer to Chapter 8 for a detailed explanation on layouts). In Model Space, lineweights are displayed in relation to pixels. In a Paper Space layout, lineweights are displayed in the exact plotting width. You can recognize that an object has a thick or thin lineweight in model space but the lineweight does not represent an object's real-world width. A lineweight of 0 will always be displayed on screen with the minimum display width of one pixel. All other lineweights are displayed using a pixel width in proportion to its real-world unit value. A lineweight displayed in model space does not change with the zoom factor. For example, a lineweight value that is represented by a width of four pixels is always displayed using four pixels regardless of how close you zoom in to your drawing.

If necessary, you can change the display scale of lineweights of the objects in the Model space to appear thicker or thinner. Changing the display scale does not affect the lineweight plotting value. However, AutoCAD regeneration time increases with lineweights that are represented by more than one pixel. By default, all lineweight values that are less than or equal to 0.01 in or 0.25 mm are displayed at one pixel and do not slow down performance in AutoCAD. If you want to optimize AutoCAD performance when working in the Model Space, set the lineweight display scale to the minimum value. The LINEWEIGHT command allows you to change the display scale of lineweights.

In Model Space, AutoCAD allows you to turn OFF the display of lineweights by toggling the LWT toggle button in the status bar. By setting the LWT toggle button to OFF, AutoCAD displays all the objects with a lineweight of 0 and reduces regeneration time.

When exporting drawings to other applications or cutting/copying objects to the Clipboard, objects retain lineweight information. The system variables LWUNITS and LWDEFAULT set the units in which lineweight is displayed and applied to newly created objects and layers, and set the default value for lineweights, respectively.

Linetype

A linetype is a repeating pattern of dashes, dots, and blank spaces. AutoCAD adds the capability of including the repeated objects in the custom linetypes. The assigned linetype is normally used to draw all objects on the layer unless you set the current linetype to some other linetype.

The following are some of the linetypes that are provided in AutoCAD in a library file called ACAD.LIN:

Border	Dashdot	Dot
Center	Dashed	Hidden
Continuous	Divide	Phantom

See Appendix G for examples of each of these linetypes. Linetypes are another means of conveying visual information. You can assign the same linetype to any number of layers. In some drafting disciplines, conventions have been established giving specific meanings to particular dash-dot patterns. If a line is too short to hold even one dash-dot sequence, AutoCAD draws a continuous line between the endpoints. When you are working on large drawings, you may not see the gap between dash-dot patterns in a linetype, unless the scaling for the linetype is set for a large value. This can be done by means of the LTSCALE command. This command is discussed in more detail later in this chapter.

Every drawing will have a layer named 0 (zero). By default, layer 0 is set to ON and assigned the color white, the default lineweight, and the linetype continuous. Layer 0 cannot be renamed or deleted.

If you need additional layers, you must create them. By default, each new layer is assigned the properties of the layer directly above it in the table. If necessary, you can always reassign the color, lineweight, and linetype of the newly created layer.

CREATING AND MANAGING LAYERS WITH THE LAYER PROPERTIES MANAGER DIALOG BOX

The Layer Properties Manager dialog box or command line version of the layer can be used to set up and control layers.

To open the Layer Properties Manager dialog box, invoke the LAYER command:

Object Properties toolbar	Choose Layers (see Figure 3–92)
Format menu	Choose Layer...
Command: prompt	**layer** (ENTER)

Figure 3–92 *Invoking the LAYER command from the Object Properties toolbar*

AutoCAD displays the Layer Properties Manager dialog box, as shown in Figure 3–93.

AutoCAD lists the available layer names in the Layer Properties Manager dialog box. By default, AutoCAD provides one layer called 0 that is set to ON and is assigned the color white, the lineweight set to default, and the linetype continuous.

Figure 3–93 *The Layer Properties Manager dialog box*

Creating New Layers

To create new layer(s), choose the **New** button. AutoCAD then creates a new layer by assigning the name **Layer1** and setting it to ON, with the properties of the layer directly above it in the table. When first listed in the layer list box, the name Layer1 is highlighted and ready to be edited. Just enter the desired name for the new layer. If you accept the name Layer1 you can still change it later. To rename the layer, click anywhere on the line where the layer is listed and the whole line is highlighted. Then double-click on the layer name, enter a new name, and press ENTER.

Note: The layer name cannot contain wild-card characters (such as * and ?). You cannot duplicate existing names.

Making a Layer Current

To make a layer current, first choose the layer name from the layer list box in the Layer Properties Manager dialog box, and then choose the **Current** button located on the top right side of the dialog box. You can also make a layer current by double clicking on the name. AutoCAD displays the name of the current layer just above the listing of layers.

Filters for Showing/Not Showing Layers

In the **Named Layer Filters** section of the Layer Properties Manager dialog box, the list box at the left shows the sets of filters by which layers are displayed in the layer list box. Choosing the down-arrow button displays the available options. The default options include *Show all layers, Show all used layers,* and *Show all Xref dependent layers.* The Show all layers selection lists all the layers created in the current drawing, the Show all used layers selection lists all the layers that have objects drawn, and the Show all Xref dependent layers selection lists all the layers that belong to external reference drawings (see Chapter 11 for a detailed explanation of external references). You can create and name other sets of filters to add to this list. Choosing the ellipsis button (...) causes the Named Layer Filters dialog box to be displayed, as shown in Figure 3–94. From this dialog box you can create and name a set of filters that cause AutoCAD to filter layers with respect to their name, color, lineweight, linetype, plot style, and whether they are ON/OFF, frozen/thawed, locked/unlocked (including viewports), and plotted/not plotted.

Figure 3–94 *Named Layer Filters dialog box*

When the **Invert Filter** check box has no check in it, then the layers that are shown in the list are those that have all the matching characteristic(s) of the specified filter. For example, if you were to create and name a set of filters that specified the color yellow and the linetype dashed, then the list would include only layers whose color was yellow and linetype was dashed. If you put a check in the Invert Filter check box, then all layers would be listed *except* those with both the color yellow and the linetype dashed.

When the **Apply to Object Properties Toolbar** check box has no check in it, then the layers that are filtered (or Invert Filtered) are shown in the list box of the Layer Properties Manager dialog box only. When there is a check in the **Apply to Object Properties Toolbar** check box, then the list in the text box for layers in the Object Properties Toolbar will also be filtered (or Invert Filtered), depending on the status of the Invert Filter check box.

Show Details

The Details section of the Layer Properties Manager dialog box is displayed by choosing the **Show Details** button, as shown in Figure 3–95.

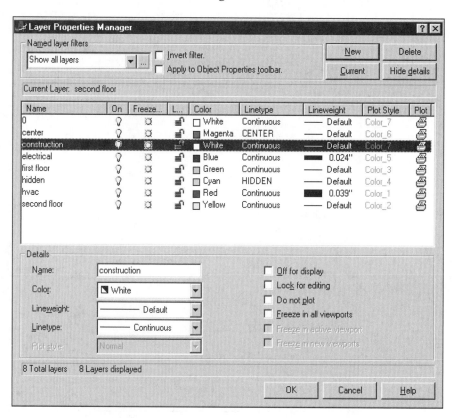

Figure 3–95 *The Layer Properties Manager dialog box with the Details section showing*

When the Details section is displayed, the Show Details button becomes the **Hide Details** button which, when chosen, causes the Details section to be hidden. The Details section has text boxes and check boxes showing the status of the layer that is highlighted in the list box above it. The text boxes include the Name, Color, Lineweight, Linetype, and Plotstyle of the highlighted layer. The check boxes include Off for Display, Lock for Editing, Do Not Plot, Freeze in All Viewports, Freeze in Active Viewport, and Freeze in New Viewport. The Details section provides a means of changing the properties and status of layers quickly and easily.

 Note: You cannot delete layer 0 or layers that contain objects. To select more than one layer, hold down CTRL and select the layer(s) from the list box.

Visibility of Layers

When you turn a layer OFF, the objects on that layer are not displayed in the drawing area and they are not plotted. The objects are still in the drawing, but they are not visible on the screen. And they are still calculated during regeneration of the drawing, even though they are invisible.

To change the setting for the visibility of layer(s), select the icon corresponding to the layer name located under the On column (second column from the left) in the Layer Properties Manager dialog box. The icon is a toggle for ON/OFF of layers.

Freezing and Thawing Layers

In addition to turning the layers OFF, you can freeze layers. The layers that are frozen will not be visible in the view window, nor will be they be plotted. In this respect, frozen layers are similar to layers that are OFF. However, layers that are simply turned OFF still go through a screen regeneration each time the system regenerates your drawing, whereas the layers that are frozen are not considered during a screen regeneration. If you want to see the frozen layer later, you simply thaw it, and automatic regeneration of the drawing area takes place.

To change the setting for the visibility of layer(s) by freezing/thawing, select the icon corresponding to the layer name located under the Freeze in All Viewports column (third column from the left). The icon is a toggle for FREEZE/THAW of layers.

Assigning Plot Style to Layers

Plot Styles are a collection of property settings (such as color, linetype, and lineweight) that can be assigned to a layer or to individual objects. These property settings are contained in a named plot style table (for a detailed explanation on creating Plot Styles, refer to Chapter 8). When applied, the plot style can affect the appearance of the plotted drawing. By default, AutoCAD assigns the Plot Style to Normal to a newly created layer. To change the assigned plot style, choose the plot style name corresponding to the layer name located under the Linetype column (eighth column from the left). AutoCAD displays the **Select Plot Style** dialog box, which allows you to change the Plot Style of the selected layer(s). Select the appropriate plot style from the list box, and choose the OK button to accept the plot style selection. To cancel the selection, choose the **Cancel** button. A layer that is assigned a Normal plot style assumes the properties that have already been assigned to that layer. You can create new plot styles, name them, and assign them to individual layers.

Plotting Layers

AutoCAD allows you to turn plotting ON or OFF for visible layers. For example, if a layer contains construction lines that need not be plotted, you can specify that the layer is not plotted. If you turn OFF plotting for a layer, the layer is displayed but is not plotted. At plot time, you do not have to turn OFF the layer before you plot the drawing.

To change the setting for the plotting of layers, select the icon corresponding to the layer name located under the Plot (ninth column from the left). The icon is a toggle for plotting or not plotting layers.

Locking and Unlocking Layers

Objects on locked layers are visible in the view window but cannot be modified by means of the modifying commands. However, it is still possible to draw on a locked layer by making it the current layer, changing the linetypes, lineweights, and colors, freezing them, and using any of the inquiry commands and Object Snap modes on them.

To change the setting for the Lock/Unlock of layer(s), select the icon corresponding to the layer name located under the Lock column (fourth column from the left). The icon is a toggle for LOCK/UNLOCK of layers.

Changing the Color of Layers

By default, AutoCAD assigns the color of the layer directly above it to a newly created layer. To change the assigned color, choose the corresponding icon to the layer name located under the Color column (fifth column from the left). AutoCAD displays the Select Color dialog box, which allows you to change the color of the selected layer(s). You can select one of the 256 colors. Use the cursor to select the color you want, or enter its name or number in the **Color:** text box. Choose the **OK** button to accept the color selection. To cancel the selection, choose the **Cancel** button.

Changing the Linetype of Layers

By default, AutoCAD assigns the linetype of the layer directly above it to a newly created layer. To change the assigned linetype, choose the linetype name corresponding to the layer name located under the Linetype column (sixth column from the left). AutoCAD displays the Select Linetype dialog box, which allows you to change the linetype of the selected layer(s). Select the appropriate linetype from the list box, and choose the **OK** button to accept the linetype selection. To cancel the selection, choose the **Cancel** button.

Note: AutoCAD lists only the linetypes that are loaded in the current drawing in the Select Linetype dialog box. To load additional linetypes in the current drawing, choose the **Load...** button, and AutoCAD displays the Load or Reload Linetypes dialog box. AutoCAD lists the available linetypes from the default linetype file ACAD.LIN. Select all the linetypes that need to be loaded, and select the **OK** button to load the linetypes into the current drawing. If necessary, you can change the default linetype file ACAD.LIN to another file by choosing the **File...** button and selecting the appropriate linetype file. You can also load the linetypes into the current drawing from the Linetype tab in the Layer and Linetype Properties dialog box.

Changing the Lineweight of Layers

By default, AutoCAD assigns the lineweight of the layer directly above it to a newly created layer. To change the assigned lineweight, choose the lineweight name corresponding to the layer name located under the Lineweight column (seventh column from the left). AutoCAD displays the Lineweight dialog box, which allows you to change the lineweight of the selected layer(s). Select the appropriate lineweight from the list box, and choose the **OK** button to accept the lineweight selection. To cancel the selection, choose the **Cancel** button.

If necessary, you can drag the widths of the column headings to see additional characters of the layer name, full legend for each symbol and color name, or number in the list box. If you want to sort the order in which layers are displayed in the list box, choose the column headings. The first selection lists the layers in descending order (Z to A, then numbers), and a second selection lists the layers in ascending order (numbers, A to Z). Choosing the status column headers lists the layers by the property in the list.

You can also create New Layers, Select All, Clear All, Select all but current, Invert selection, Invert layer filter, and select one of the available options from Layer filters from the shortcut menu (see Figure 3–96) that is displayed by right-clicking on your pointing device in the Layer Properties Manager dialog box.

After making the necessary changes, choose the **OK** button to keep the changes and close the dialog box. To discard the changes, choose the **Cancel** button and close the dialog box.

Figure 3–96 *The shortcut menu displayed in the Layer Properties Manager dialog box*

CHANGING LAYER PROPERTIES FROM THE OBJECT PROPERTIES TOOLBOX

You can also toggle on/off, freeze/thaw, lock/unlock, or plot/no plot, in addition to making a layer current from the layer list box provided in the Object Properties toolbar. Select the appropriate icon next to the layer name you wish to toggle, as shown in Figure 3–97.

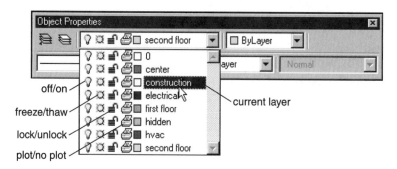

Figure 3–97 *Layer list box in the Object Properties toolbar*

MAKING AN OBJECT'S LAYER CURRENT

AutoCAD allows you to select an object in the drawing to make its layer the current layer. To do so, invoke the command:

Object Properties toolbar	Choose Make Object's Layer Current (See Figure 3–98)

Figure 3–98 *Invoking the Make Object's Layer Current from the Object Properties toolbar*

AutoCAD prompts:

> Select object whose layer will become current: *(select the object in the drawing to make its layer current)*

AutoCAD makes the selected object's layer current.

SETTING UP AND CONTROLLING LAYERS BY MEANS OF THE COMMAND VERSION OF THE LAYER COMMAND

Invoke the command version of the LAYER command:

Command: prompt	**-layer** (ENTER)

AutoCAD prompts:

> Command: **-layer** (ENTER)
> Enter an option
> [?/Make/Set/New/ON/OFF/Color/Ltype/LWeight/Plot/Freeze/Thaw/LOck/
>   Unlock]: *(select one of the available options)*

? Option

The ? option lists the names of layers defined in the current drawing, showing their on/off, freeze/thaw, lock/unlock, lweight, plot status, color, and linetypes. When you select this option, AutoCAD prompts:

> Enter layer name(s) to list <*>:

You can enter a list of layer names, using wild cards if you wish, or press the SPACEBAR or ENTER to accept the default to list all layer names.

Make

The Make option effectively does three things. When this option is invoked, AutoCAD prompts you for a layer name. Once you enter a name, Make does the following:

1. Searches for the layer to determine whether it exists.

2. Creates it if the layer does not exist, and assigns the color white and the linetype continuous.

3. Sets the newly made layer to be the current layer.

When you choose this option, AutoCAD prompts:

Enter name for new layer (becomes the current layer) <0>:

If the layer you selected exists and is presently turned OFF, AutoCAD turns it ON automatically, using the color and linetype previously assigned to that layer, and makes it the current layer.

Set

The Set option tells AutoCAD which layer you want to draw on (making it the current layer). When you choose this option, AutoCAD prompts:

Enter layer name to make current <0>:

Enter the layer's name (it must be an existing layer) and press the SPACEBAR or ENTER. This layer becomes the current layer. The first several characters of the layer name are displayed in the Object Properties layer list box.

New

The New option allows you to create new layers. When you choose this option, AutoCAD prompts:

Enter name list for new layer(s):

You can enter more than one name at a time by separating the names with a comma. If you need to separate characters in the name, use the underscore (_) instead of pressing the SPACEBAR (in AutoCAD, pressing the SPACEBAR is the same as pressing ENTER). Each layer thus created is automatically turned ON and assigned the color white and linetype continuous.

OFF

The OFF option allows you to turn OFF selected layers. When you choose this option, AutoCAD prompts:

Enter name list of layer(s) to turn off:

The list should contain only existing layer names. The names may include wild-card characters. When you turn OFF a layer, its associated objects will not be displayed in the new window and they are not plotted. The objects still exist in the drawing; they are just invisible. And they are still calculated during the regeneration of the drawing, even though they are not visible.

ON

The ON option allows you to turn ON layers that have been turned OFF. When you choose this option, AutoCAD prompts:

> Enter name list of layer(s) to turn on:

This list should contain only existing layer names. The name may include wild-card characters. Each designated layer is turned on using the color and linetype previously associated with it. Turning a layer ON does not cause it to be the current layer.

Color

The Color option allows you to change the color associated with that specific layer. When you choose this option, AutoCAD first prompts:

> Enter color name or number (1-255):

Reply with one of the standard color names or with a legal color number from 1 to 255 (for example, 2 for yellow). After you specify the color, AutoCAD prompts:

> Enter name list of layer(s) for color 2 (yellow) <0>:

Reply with the names of existing layers, separated by commas. The names may include wild-card characters. The specified layers are given the color you designated and are then automatically turned ON if they are OFF. If you would prefer to assign the color but to turn the layers OFF, precede the color with a minus sign (–).

Linetype

The Linetype option allows you to change the linetype associated with a specific layer. When you choose this option, AutoCAD prompts:

> Enter loaded linetype name or [?] <Continuous>:

Reply with the name of an existing defined linetype. AutoCAD then asks for a list of layer names to which the linetype should be applied. For example, if you had replied to the first prompt with the linetype named HIDDEN, the next prompt would be:

> Enter name list of layer(s) for linetype "hidden" <0>:

Reply with the names of existing layers, separated by commas. The names may include wild-card characters.

Lineweight

The Lweight option allows you to change the lineweight associated with a specific layer. When you choose this option, AutoCAD prompts:

> Enter lineweight (0.0mm - 2.11mm):

Enter a numeric value for the lineweight. AutoCAD then asks for a list of layer names to which the lineweight should be applied. For example, if you had replied to the first prompt with the lineweight of 0.1, the next prompt would be:

Lineweight rounded to nearest valid value of 0.09 mm

Enter name list of layers(s) for lineweight 0.09mm <0>:

Reply with the names of existing layers, separated by commas.

Freeze and Thaw

The layers that are frozen will not be visible on the display, nor will they be plotted. In this respect, Freeze is similar to OFF. However, layers that are simply turned OFF still go through a screen regeneration each time the system regenerates your drawing. If you want to see the frozen layer later, you simply Thaw it, and automatic regeneration of the screen takes place. The layers that are frozen will not be regenerated.

When you choose the Freeze option, AutoCAD prompts:

Enter name list of layer(s) to freeze:

You can enter more than one name at a time by separating the names with a comma.

When you choose the Thaw option, AutoCAD prompts:

Enter name list of layer(s) to thaw:

You can enter more than one name at a time by separating the names with a comma.

Lock and Unlock

Objects on locked layers are visible on the display but cannot be modified with the modify commands. When you choose the Lock option, AutoCAD prompts:

Enter name list of layer(s) to lock:

You can enter more than one name at a time by separating the names with a comma.

When you choose the Unlock option, AutoCAD prompts:

Enter name list of layer(s) to unlock:

You can enter more than one name at a time by separating the names with a comma.

Plot and No Plot

Objects on layers whose Plot Status is Plot will be plotted. Objects on layers whose Plot Status is No Plot will not be plotted. When you choose the Plot option, AutoCAD prompts:

Enter a plotting preference [Plot/No plot] <Plot>: p

Enter layer name(s) for this plot preference <0>:

You can enter more than one name at a time by separating the names with a comma.

When you choose the No Plot option, AutoCAD prompts:

> Enter a plotting preference [Plot/No plot] <Plot>: **n**

> Enter layer name(s) for this plot preference <0>:

You can enter more than one name at a time by separating the names with a comma.

Whenever you give a null response to the LAYER command, it returns you to the "Command:" prompt.

SETTING THE LINETYPE SCALE FACTOR

The linetype scale factor allows you to change the relative lengths of dashes and spaces between dashes and dots linetypes per drawing unit. The definition of the linetype instructs AutoCAD on how many units long to make dashes and the spaces between dashes and dots. As long as the linetype scale is set to 1.0, the displayed length of dashes and spaces coincides with the definition of the linetype. The LTSCALE command allows you to set the linetype scale factor.

Invoke the LTSCALE command:

Command: prompt	**ltscale** (ENTER)

AutoCAD prompts:

> Command: **ltscale** (ENTER)
> Enter new linetype scale factor <1.0000>: *(specify the scale factor)*

Changing the linetype scale affects all linetypes in the drawing. If you want dashes that have been defined as 0.5 units long in the dashed linetype to be displayed as 10 units long, you set the linetype scale factor to 20. This also makes the dashes that were defined as 1.25 units long in the center linetype display as 25 units long, and the short dashes (defined as 0.25 units long) display as 5 units long. Note that the 1.25-unit-long dash in the center linetype is 2.5 times longer than the 0.5-unit-long dash in the dashed linetype. This ratio will always remain the same, no matter what the setting of LTSCALE. So if you wish to have some other ratio of dash and space lengths between different linetypes, you will have to change the definition of one of the linetypes in the ACAD.LIN file.

Remember that linetypes are for visual effect. The actual lengths of dashes and spaces are bound more to how they should look on the final plotted sheet than to distances or sizes of any objects on the drawing. An object plotted full size can probably use an LTSCALE setting of 1.0. A 50'-long object plotted on an 18" x 24" sheet might be plotted at a 1/4" = 1'-0" scale factor. This would equate to 1 = 48. An LTSCALE setting of 48 would make dashes and spaces plot to the same lengths as the full-size plot with a setting of 1.0. Changing the linetype scale factor causes the drawing to regenerate.

WILD CARDS AND NAMED OBJECTS

AutoCAD lets you use a variety of wild cards for specifying selected groups of named objects when responding to prompts during commands. By placing one or more of these wild cards in the string (your response), you can specify a group that includes (or excludes) all of the objects with certain combinations or patterns of characters.

The types of objects associated with a drawing that are referred to by name include blocks, layers, linetypes, text styles, dimension styles, named User Coordinate Systems, named views, shapes, plot styles and named viewport configurations.

The wild card characters include the two commonly used in DOS (* and ?), as well as eight more that come from the UNIX operating system. Here is a table listing the wild cards and their uses.

Wild Card	Use
# (pound)	Matches any numeric digit
@ (at)	Matches any alphanumeric character
. (period)	Matches any character except alphanumeric
* (asterisk)	Matches any string. It can be used anywhere in the search pattern: the beginning, middle, or end of the string.
? (question mark)	Matches any single character
~ (tilde)	Matches anything but the pattern
[...]	Matches any one of the characters enclosed
[~...]	Matches any character not enclosed
[-] (hyphen)	Specifies single-character range
' (reverse quote)	Reads characters literally

The following table shows some examples of wild card patterns.

Pattern	Will match or include . . .	But not . . .
ABC	Only ABC	
~ABC	Anything but ABC	
?BC	ABC through ZBC	AB, BC, ABCD, XXBC
A?C	AAC through AZC	AC, ABCD, AXXC, ABCX
AB?	ABA through ABZ	AB, ABCE, XAB

Pattern	Will match or include . . .	But not . . .
A*	Anything starting with A	XAAA
A*C	Anything starting with A and ending with C	XA, ABCDE
*AB	Anything ending with AB	ABCX, ABX
AB	AB anywhere in string	AXXXB
~*AB*	All strings without AB	AB, ABX, XAB, XABX
[AB]C	AC or BC	ABC, XAC
[A-K]D	AD, BD, through KD	ABC, AKC, KD

U, UNDO, AND REDO COMMANDS

The UNDO command undoes the effects of the previous command or group of commands, depending on the option employed. The U command reverses the most recent operation, and the REDO command is a one-time reversal of the effects of the previous U and UNDO commands.

U COMMAND

The U command undoes the effects of the previous command by displaying the name of that command. Pressing (ENTER) after using the U command undoes the next-previous command, and continues stepping back with each repetition until it reaches the state of the drawing at the beginning of the current editing session.

When an operation cannot be undone, AutoCAD displays the command name but performs no action. An operation external to the current drawing, such as plotting or writing to a file, cannot be undone.

Invoke the U command:

Standard toolbar	Choose Undo (see Figure 3–99)
Pull-down menu	Edit > Undo
Command: prompt	u (ENTER)

Figure 3–99 *Invoking the* U *command from the Standard toolbar*

AutoCAD reverses the most recent operation.

For example, if the previous command sequences drew a circle and then copied it, two u commands would undo the two previous commands in sequence, as follows:

Command: **u (**ENTER)
COPY
Command: (ENTER)
CIRCLE

Using the u command after commands that involve transparent commands or subcommands causes the entire sequence to be undone. For example, when you set a dimension variable and then perform a dimension command, a subsequent u command nullifies the dimension drawn and the change in the setting of the dimension variable.

UNDO COMMAND

The UNDO command permits you to select a specified number or marked group of prior commands for undoing.

Invoke the UNDO command:

Command: prompt	**undo** (ENTER)

AutoCAD prompts:

Command: **undo** (ENTER)
Enter the number of operations to undo or [Auto/Control/BEgin/End/
    Mark/Back]: *(specify number of undo operations to undo or select one of the
    available options)*

Control Option

The Control option controls the number of available options. By limiting the number of options, you can free up the memory and disk space that is otherwise being used to save undoing operation information. AutoCAD prompts as follows when the Control option is selected:

Command: **undo** (ENTER)
Enter the number of operations to undo or [Auto/Control/BEgin/End/
    Mark/Back]: **c** (ENTER)
Enter an UNDO control option [All/None/One]<All>: *(select one of the three
    available options)*

Selection of the **All** option (the default setting of the UNDO command) enables all Undo options. The **None** option disables the u and UNDO commands but not the Control option of the UNDO command that re-enables various options. And the **One** option sets the UNDO command to work like the u command.

Mark and Back Options

If you are at a point in the editing session at which you would like to experiment but would still like the option of undoing the experiment, you can mark that point. An example of the use of the Mark and Back options is as follows:

> Command: **line** (ENTER) *(draw a line)*
> Command: **circle** (ENTER) *(draw a circle)*
> Command: **undo** (ENTER)
> Auto/Control/Begin/End/Mark/Back/<Number>: **m** (ENTER)
> Command: **text** (ENTER) *(enter text)*
> Command: **arc** (ENTER) *(draw an arc)*
> Command: **undo** (ENTER)
> Auto/Control/Begin/End/Mark/Back/<Number>: **b** (ENTER)

The Back option returns you to the state of the drawing that has the line and the circle. Following this UNDO Back with a U removes the circle. Another U removes the line. Another U displays the following prompt:

> Everything has been undone

Using the Back option when no Mark has been established will prompt:

> This will undo everything. OK? <Y>

Responding Y undoes everything done since the current editing session was begun or since the last SAVE command.

 Note: The default is Y; think twice before pressing ENTER in response to the "This will undo everything" prompt.

Begin and End Options

AutoCAD's U and UNDO commands treat the operations between an UNDO Begin and an UNDO End as one command. A Begin option entered after another Begin option (before an UNDO End) will automatically invoke an UNDO End option, thereby grouping the operations since that prior Begin option. If the UNDO Control has been set to None or One, the Begin option will not work. Using the U command is permissible after Begin and before a U End, to undo operations, but only back to the UNDO Begin.

The Begin and End options are normally intended for use in strings of menu commands where a menu pick involves several operations.

Auto Option

The Auto option causes multiple operations invoked by a single menu pick to be treated as one command by the U or UNDO command. UNDO Begin should be placed at the beginning of a menu string, with UNDO End at the end of the string. It has no effect if the UNDO Control has been set to None or One, however.

The effects of the following commands cannot be undone:

AREA, ATTEXT, DBLIST, DELAY, DIST, DXFOUT, END, FILES, FILMROLL, GRAPHSCR, HELP, HIDE, ID, IGESOUT, LIST, MSLIDE, PLOT, PRPLOT, QUIT, REDRAW, REDRAWALL, REGENALL, RESUME, SAVE, SHADE, SHELL, STATUS, and TEXTSCR.

REDO COMMAND

The REDO command permits one reversal of a prior U or UNDO command. This will reverse the undo. In order to function, the REDO command must be used following the U or UNDO command and prior to any other action.

Invoke the REDO command:

Standard toolbar	Choose Redo (see Figure 3–100)
Edit menu	Choose Redo
Command: prompt	**redo** (ENTER)

Figure 3–100 *Invoking the REDO command from the Standard toolbar*

AutoCAD reverses the prior U or UNDO command.

The REDO command does not have any options.

PROJECT EXERCISE

This project exercise provides point-by-point instructions for setting up the drawing with layers and then creating the objects shown in Figure P3–1. In this exercise you will apply the skills acquired in Chapters 1, 2, and 3.

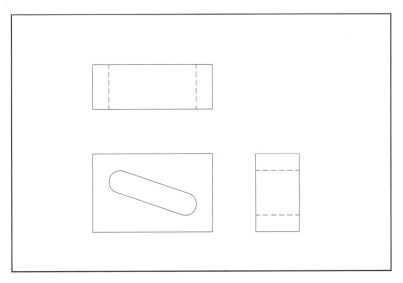

Figure P3–1 *Completed project drawing*

In this project you will learn to do the following:

- Set up the drawing, including setting up the layers.
- Use the LINE and RECTANGLE commands with Object Snap modes

SET UP THE DRAWING AND DRAW A BORDER

Step 1:   Start the AutoCAD program.

Step 2:   To create a new drawing, invoke the NEW command from the Standard toolbar or choose New from the File menu.

AutoCAD displays the Create New Drawing dialog box. Choose the "Use a Wizard" button, and then select the Quick Setup and choose the OK button to close the dialog box.

AutoCAD displays the Quick Setup dialog box with the listing of the options available for unit of measurement. Select the "Decimal" radio button, and choose the **Next>** button.

AutoCAD displays the Quick Setup dialog box with the display for Area setting. Under the **Width:** text box enter **18**, and under the **Length:** text box enter **12**. Choose the **Finish** button to close the Quick Step dialog box.

Step 3: Invoke the LAYER command from the Object Properties toolbar, or choose Layer from the Format menu. AutoCAD displays the Layer Properties Manager dialog box.

Create four layers, rename them as shown in the table, and assign appropriate colors and linetypes.

Layer Name	Color	Linetype	Lineweight
Border	Cyan	Continuous	Default
Object	Green	Continuous	Default
Hidden	Blue	Hidden	Default
Const	Red	Continuous	Default

Set Border as the current layer, and close the Layer Properties Manager dialog box.

Step 4: To draw the border, invoke the RECTANGLE command from the Draw toolbar or enter **rectangle** at the Command: prompt and press (ENTER). AutoCAD prompts:

Command: **rectangle** (ENTER)
Specify first corner point or [Chamfer/Elevation/Fillet/Thickness/ Width]: **0.25,0.25** (ENTER)
Specify other corner point: **17.75,11.75** (ENTER)

AutoCAD draws a rectangle border.

DRAW THE OBJECTS

Step 1: From the Object Properties toolbar, choose the Layer Control down-arrow icon to display the layer list box. Select the Object layer as the current layer.

Step 2: Invoke the RECTANGLE command from the Draw toolbar to draw a rectangle.

Command: **rectangle**
Specify first corner point or [Chamfer/Elevation/Fillet/Thickness/ Width]: **2,7.5**
Specify other corner point: **7.5,9.5**

Invoke the RECTANGLE command again from the Draw toolbar to draw another rectangle.

Command: **rectangle**
Specify first corner point or [Chamfer/Elevation/Fillet/Thickness/ Width]: **9.5,2**
Specify other corner point: **11.5,5.5**

Step 3: Invoke the RECTANGLE command again from the Draw toolbar to draw a rectangle by invoking appropriate object snaps.

Command: **rectangle**
Specify first corner point or [Chamfer/Elevation/Fillet/Thickness/ Width]: *(invoke the APPint object snap, and select line 1 and line 4, as shown in Figure P3–2)*

Specify other corner point: *(invoke the App Int Object Snap mode, and select line 2 and line 3, as shown in Figure P3–2)*

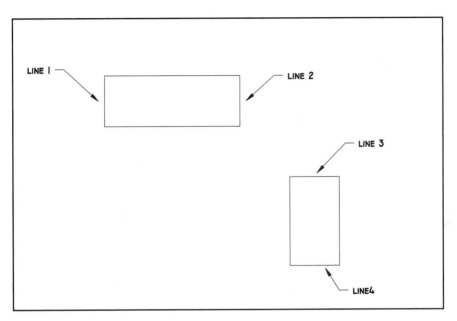

Figure P3–2 *Identifying the lines to draw a rectangle*

After drawing the rectangle, the drawing should look like Figure P3–3.

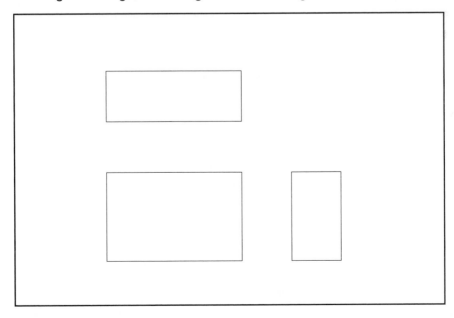

Figure P3–3 *Completed drawing of rectangles*

Step 4: From the Object Properties toolbar, choose the Layer Control down-arrow icon to display the layer list box. Select the Const layer as the current layer.

Step 5: Invoke the CIRCLE command from the Draw toolbar and draw two circles.

Respond to the prompts in the command area as follows:

Command: **circle**
Specify center point for circle or [3P/2P/Ttr (tan tan radius)]: **3.25,4.25** (ENTER)
Specify radius of circle or [Diameter]: **.5** (ENTER)

Command: (ENTER)
Specify center point for circle or [3P/2P/Ttr (tan tan radius)]: **6.25,3.25** (ENTER)
Specify radius of circle or [Diameter]: **.5** (ENTER)

Step 6: Invoke the LINE command from the Draw toolbar and draw series of lines by invoking appropriate object snaps.

Command: **line**
Specify first point: *(invoke the QUAdrant object snap, and select a point QUA 1, as shown in Figure P3–4)*
Specify next point or [Undo]: *(invoke the PERpendicular object snap, and select line 5, as shown in Figure P3–4)*
Specify next point or [Undo]: (ENTER)

Command: (ENTER)
Specify first point: *(invoke the QUAdrant object snap, and select a point QUA 2, as shown in Figure P3–4)*
Specify next point or [Undo]: *(invoke the PERpendicular object snap, and select line 5, as shown in Figure P3–4)*
Specify next point or [Undo]:(ENTER)

Command: (ENTER)
Specify first point: *(invoke the QUAdrant object snap, and select a point QUA 3, as shown in Figure P3–4)*
Specify next point or [Undo]: *(invoke the PERpendicular object snap, and select line 6, as shown in Figure P3–4)*
Specify next point or [Undo]: (ENTER)

Command: (ENTER)
Specify first point: *(invoke the QUAdrant object snap, and select a point QUA 4, as shown in Figure P3–4)*
Specify next point or [Undo]: *(invoke the PERpendicular object snap, and select line 6, as shown in Figure P3–4)*
Specify next point or [Undo]: (ENTER)

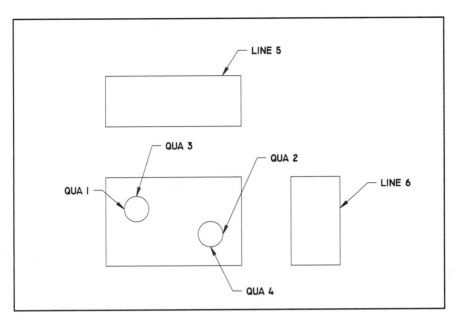

Figure P3–4 *Identifying the points to draw additional lines*

The drawing should look like Figure P3–5.

Figure P3–5 *Completed drawing*

Step 7: From the Object Properties toolbar, choose the Layer Control down-arrow icon to display the layer list box. Select the Hidden layer to make it the current layer.

Step 8: Invoke the LINE command from the Draw toolbar. Draw a series of lines by invoking appropriate object snaps.

> Command: **line**
> Specify first point: *(invoke the INTersection object snap, and select point 1, as shown in Figure P3–6)*
> Specify next point or [Undo]: *(invoke the INTersection object snap, and select point 2, as shown in Figure P3–6)*
> Specify next point or [Undo]: (ENTER)

Draw three additional lines:

From:	To:
point 3	point 4
point 5	point 6
point 7	point 8

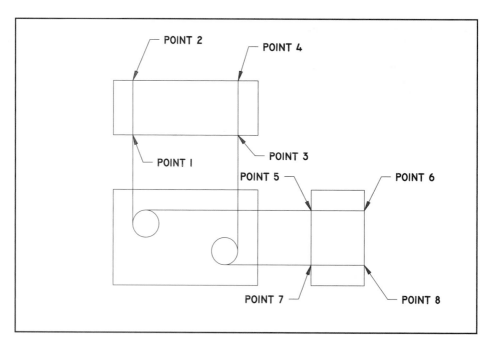

Figure P3–6 *Identifying the points to draw additional lines*

Step 9: Invoke the ERASE command from the Modify toolbar, and erase line 7, line 8, line 9, and line 10, as shown in Figure P3–7.

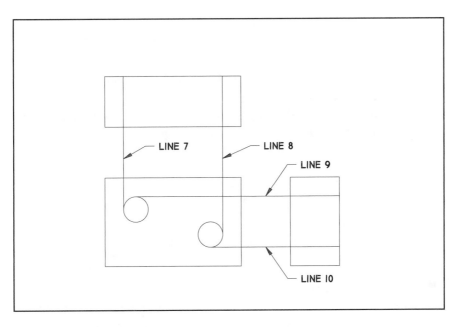

Figure P3–7 *Identifying the lines to erase*

The completed drawing should look like Figure P3–8.

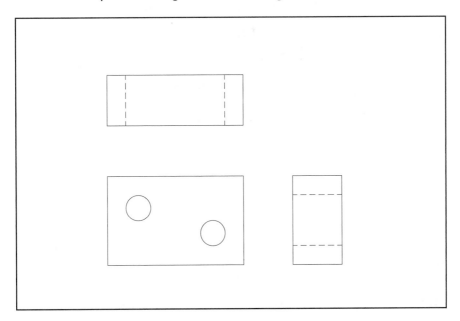

Figure P3–8 *Completed drawing*

Step 10: From the Object Properties toolbar, choose the Layer Control down-arrow icon to display the layer list box. Select the Object layer to make it the current layer.

Step 11: Invoke the LINE command from the Draw toolbar and draw a series of lines.

> Command: **line**
> Specify first point: *(invoke the TANgent object snap, and select a point in the upper half of circle 1, as shown in Figure P3–9)*
> Specify next point or [Undo]: *(invoke the TANgent object snap, and select a point in the upper half of circle 2, as shown in Figure P3–9)*
> Specify next point or [Undo]: (ENTER)
>
> Command: (ENTER)
> Specify first point: *(invoke the TANgent object snap, and select a point in the lower half of circle 1, as shown in Figure P3–9)*
> Specify next point or [Undo]: *(invoke the TANgent object snap, and select a point in the lower half of circle 2, as shown in Figure P3–9)*
> Specify next point or [Undo]: (ENTER)

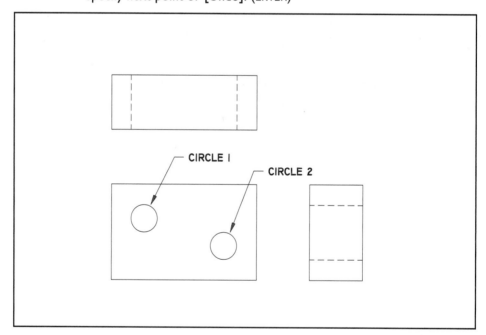

Figure P3–9 *Identifying the object snap points to draw additional lines*

Step 12: Invoke the ARC 3-Points command from the Draw toolbar.

Command: **arc**

Specify start point of arc or [CEnter]: *(invoke the ENDpoint object snap and select endpoint 1. as shown in Figure P3–10)*

Specify second point of arc or [CEnter/ENd]: *(invoke the NEArest object snap and select a point on circle 1, where indicated in Figure P3–10)*

Specify end point of arc: *(invoke the ENDpoint object snap and select endpoint 2, as shown in Figure P3–10)*

Command: (ENTER)

Specify start point of arc or [CEnter]: *(invoke the ENDpoint object snap and select endpoint 3, as shown in Figure P3–10)*

Specify second point of arc or [CEnter/ENd]: *(invoke the NEArest object snap and select a point on circle 2, where indicated in Figure P3–10)*

Specify end point of arc: *(invoke the ENDpoint object snap and select endpoint 4, as shown in Figure P3–10)*

Step 13: From the Object Properties toolbar, choose the Layer Control down-arrow icon to display the layer list box. Set the Const layer to OFF.

The completed drawing should look like Figure P3–11.

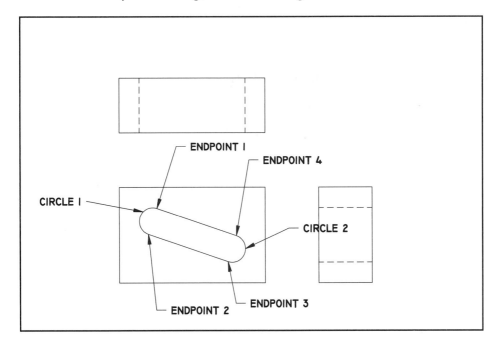

Figure P3–10 *Identifying the Object Snap points to draw arcs*

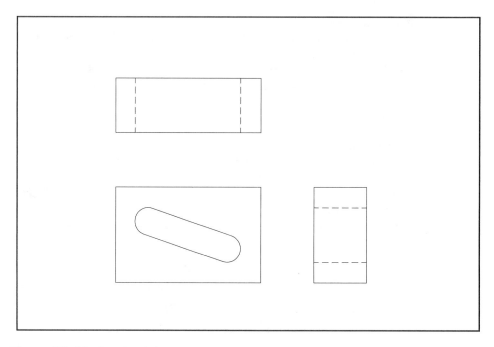

Figure P3–11 *Completed drawing*

Step 14: Invoke the SAVE command from the File menu to save the current drawing as PROJ3.DWG.

EXERCISES 3–1 THROUGH 3–6

EXERCISE 3–1

Create the drawing according to the settings given in the following table:

Settings	Value
1. Units	Decimal with two decimal places
2. Limits	Lower left corner: 0,0
	Upper right corner: 18,12
3. Grid	0.25
4. Snap	0.25
5. Layers	*NAME* *COLOR* *LINETYPE*
	Solid Red Continuous
	Hidden Blue Hidden

 Hint: To draw the border, use the RECTANGLE command with diagonally opposite corners at coordinates 0.25,0.25 and 16.75,11.75

mechanical E X E R C I S E

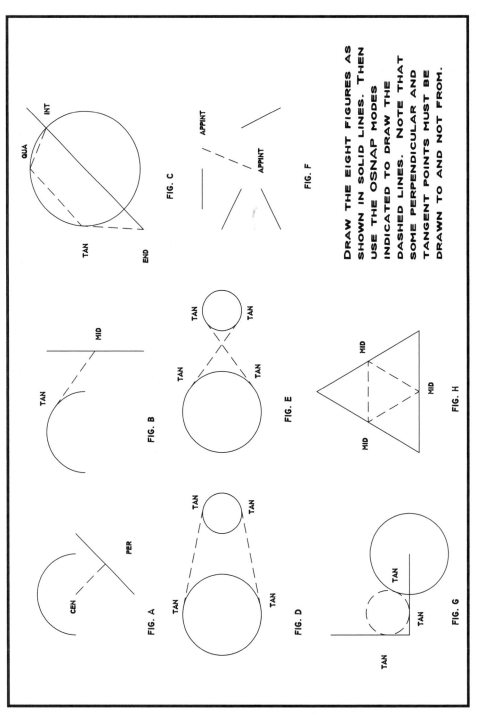

DRAW THE EIGHT FIGURES AS SHOWN IN SOLID LINES. THEN USE THE OSNAP MODES INDICATED TO DRAW THE DASHED LINES. NOTE THAT SOME PERPENDICULAR AND TANGENT POINTS MUST BE DRAWN TO AND NOT FROM.

Figure Ex3–1

EXERCISE 3–2

Create the drawing according to the settings given in the following table:

Settings	Value		
1. Units	Decimal with two decimal places		
2. Limits	Lower left corner: 0,0		
	Upper right corner: 17,11		
3. Grid	0.50		
4. Snap	0.25		
5. Layers	*NAME*	*COLOR*	*LINETYPE*
	Border	Red	Continuous
	Object	Green	Continuous
	Center	Blue	Center

Hints: Use the LINE and CIRCLE commands to complete the drawing. Make sure to draw the objects in the appropriate layers. Do not dimension the drawing.

A good place to start is the center of the three circles. With the radius of 4 units, the center could be at coordinates 5,5.

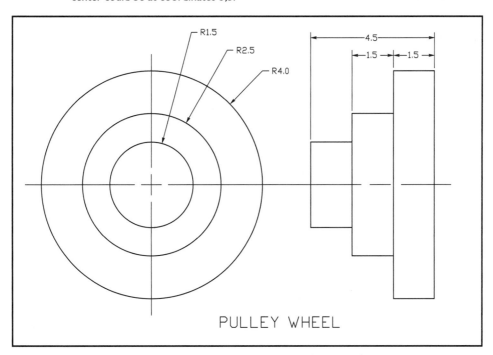

PULLEY WHEEL

Figure Ex3–2

EXERCISE 3-3

Create the drawing according to the settings given in the following table:

Settings	Value
1. Units	Decimal with two decimal places
2. Limits	Lower left corner: 0,0
	Upper right corner: 22,17
3. Grid	0.50
4. Snap	0.25
5. Layers	*NAME* *COLOR* *LINETYPE*
	Border Red Continuous
	Object Green Continuous
	Center Blue Center

Hints: Use the LINE and CIRCLE commands to complete the drawing. Make sure to draw the objects in the appropriate layers. Do not dimension the drawing.

A good place to start is the center of the three circles. With the radius of 5.5 units, the center could be at coordinates 7,7.

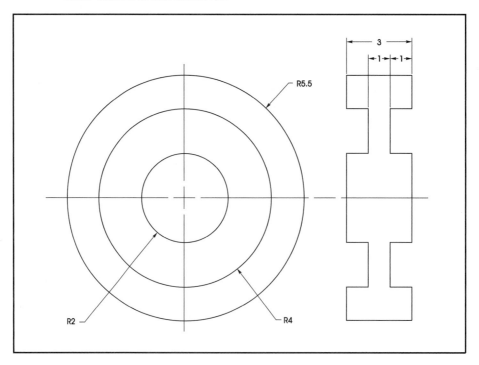

Figure Ex3-3

EXERCISE 3–4

Create the drawing according to the settings given in the following table:

Settings	Value		
1. Units	Decimal with two decimal places		
2. Limits	Lower left corner: 0,0		
	Upper right corner: 24,18		
3. Grid	1.00		
4. Snap	0.50		
5. Layers	_NAME_	_COLOR_	_LINETYPE_
	Border	Red	Continuous
	Object	Green	Continuous
	Center	Blue	Center

Hints: Use the LINE and CIRCLE and ARC commands to complete the drawing. Make sure to draw the objects in the appropriate layers. Do not dimension the drawing.

A good place to start is the center of the large circle. The large circle is 4.5 units from the bottom of the plate and 8.5 units from the left side, so its center could be at coordinates 10,6.

CRANE PULLEY PLATE

Figure Ex3–4

EXERCISE 3–5

In the following exercise the outline of the house shown in Figure Ex3–5a will be drawn. Create the drawing according to the settings given in the following table:

Settings	Value
1. Units	Linear UNITS set to Architectural Angular UNITS set to Surveyor's
2. Limits	Lower left corner: –20',–10' Upper right corner: 124',86'
3. Grid	5'
4. Snap	Set to Rotated snap at an angle of 4d45'08".

Hints: To draw the property lines, you can use the steps in the example following the text section on the Direct Distance Option regarding the Rotate option of the SNAP command. Do not draw the text; that is for reference only. After invoking the LINE command, begin tracking at TK1 (0,0) (refer to Figure Ex3–5b). Turn Ortho ON and track through point TK2 to TK3/A by forcing the cursor in the right direction and entering the proper distance (35.3' with the cursor to the East for TK2, and 7.1' with the cursor to the North for TK3/A). Draw the first line, A-B, by forcing the cursor to the north and entering the distance of 16.4'. Continue the lines around the outline of the house without exiting the LINE command. Each line's length can be entered when the cursor is forced in the right direction, thus using the DIRECT DISTANCE option for the endpoint of each line.

POINT OF INTEREST

- It is important to realize that using the Rotate option of the SNAP command to establish a new snap origin and snap angle does not change the X and Y coordinates of the points on the screen. Nor does it change input angles. These can be done with the more complicated UCS command. This means that if you use Absolute Coordinate entry, AutoCAD will use the coordinates based on the origin and angle of rotation of the coordinate system in effect before changing the snap origin and/or angle of rotation. Also, Relative Rectangular or Polar Coordinate input is based on the unrotated system of coordinates. For example, if the snap origin and angle were changed to 4,4 (X,Y coordinates) and 45 degrees, respectively, and you were specifying a point relative to a point at the new snap origin as @4,0 or @4<0, the specified point would be at the X,Y coordinates of 8,4. They would not be 4 units distance at 45 degrees from 4,4 as you might expect. A rotated snap grid with a relocated origin is for use primarily in conjunction with the Snap and/or Ortho set to ON. Otherwise, SNAP Rotate is not that functional, as you will learn with experience.

Figure Ex3–5a *Outline of a house*

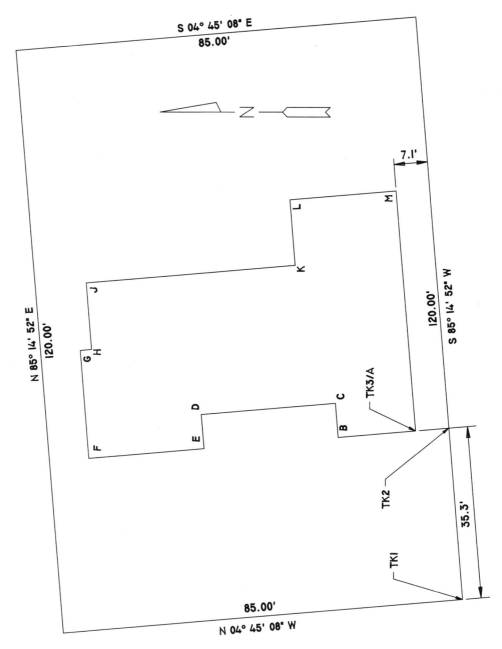

Figure Ex3–5b *Outline of a house with dimensions*

EXERCISE 3–6

Create the drawing according to the settings given in the following table:

Settings	Value
1. Units	Decimal
2. Limits	Lower left corner: –2,–2
	Upper right corner: 10,7
3. Grid	1.0
4. Snap	1.0

Hints: Draw a sequence of lines from A (7,1) through B, C, and D, to E (3,4) with the Snap set to 1 (or 0.50, 0.25, or 0.125) and set to ON. Using the SNAP Rotate option, select point E as the new origin and point A to designate the rotation angle. Using the Continue option of the LINE command, draw the lines from E through F and G and back to point A. Points F and G can be specified only with the Snap rotated, as explained, and the Snap set to ON.

POINT OF INTEREST

- This exercise seems simple enough. However, it is not too often that two systems of coordinates share the same common points like A and E in this exercise. This is because E-A is a 5-unit hypotenuse of a right triangle whose base and altitude are 4 and 3 units.

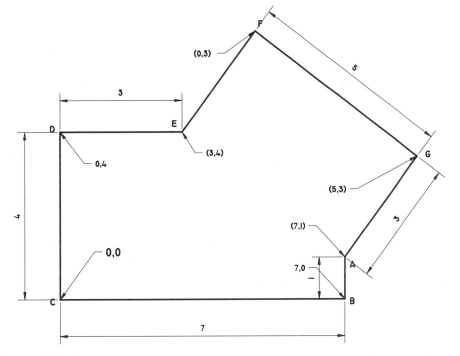

Figure Ex3–6

REVIEW QUESTIONS

1. The invisible grid, which the crosshairs lock onto, is called:

   a. SNAP

   c. ORTHO

   b. GRID

   d. Cursor Lock

2. To globally change the sizes of the dashes for all dashed lines, you should adjust:

   a. Line Scale

   c. SCALE

   b. LTSCALE

   d. Layer Scale

3. To reverse the effect of the last 11 commands, you could:

   a. Use the UNDO command

   b. Use the U command multiple times

   c. Either A or B

   d. It is not possible, AutoCAD only retains the last 10 commands

4. The following are AutoCAD tools available, except:

   a. GRID

   d. TSNAP

   b. SNAP

   e. OSNAP

   c. ORTHO

5. If you just used the U command, what command would restore the drawing to the state before the U command?

   a. RESTORE

   c. REDO

   b. U

   d. OOPS

6. Which of the following can not be modified in the Drawing Settings dialog box?

   a. Snap

   d. Limits

   b. Grid

   e. All of the above

   c. Ortho

7. If the spacing of the visible grid is set too small, AutoCAD responds as follows:

   a. does not accept the command

   b. produces a "Grid too dense to display" messages

   c. produces a display that is distorted

   d. automatically adjusts the size of the grid so it will display

   e. displays the grid anyway

8. The smallest number that can be displayed in the denominator when setting units to architectural units is:

   a. 8

   b. 16

   c. 64

   d. 128

   e. none of the above

9. All of the following are considered valid options of the -LAYER command, except:

   a. On

   b. Use

   c. Make

   d. Lock

   e. Set

10. Which of the following is not a valid option of the Layer Properties dialog box ?

    a. Close

    b. Lock

    c. On

    d. Freeze

    e. Color

11. After having drawn a 3-point circle, you want to begin a line at the exact center of the circle. What tool in AutoCAD would you use?

    a. Snap

    b. Object Snap

    c. Entity Snap

    d. Geometric Calculator

12. How many previous Zooms are available with the previous option of the ZOOM command?

    a. 4

    b. 6

    c. 8

    d. 10

13. The assignment of a specific color to a specific layer is permitted by the -LAYER command option:

    a. Set

    b. New

    c. Color

    d. Make

    e. Both C and D

14. In general, a REDRAW is quicker than a REGEN

    a. True

    b. False

15. When a layer is ON and THAWed:

    a. the objects on that layer are visible on the monitor

    b. the objects on that layer are not visible on the monitor

    c. the objects on that layer are ignored by a REGEN

    d. the drawing REDRAW time is reduced

    e. the objects on that layer cannot be selected

16. To ensure the entire limits of the drawing are visible on the display, you should perform a ZOOM-

    a. All

    b. Previous

    c. Extents

    d. Limits

17. A layer where objects may not be edited or deleted, but are still visible on the screen and may be OSNAPed to is considered:

    a. Frozen

    b. Locked

    c. On

    d. unSet

    e. fiXed

18. The -LAYER command will allow you to:

    a. assign colors

    b. assign linetypes

    c. list previously created layers

    d. selectively turn ON and OFF the layers

    e. all of the above

19. Which OSNAP option allows you to select the closest endpoint of a line, arc, or polyline segment?

    a. ENDpoint

    b. MIDpoint

    c. CENter

    d. INSertion point

    e. PERpendicular

20. Which OSNAP option allows you to select the point in the exact center of a line?

    a. ENDpoint

    b. MIDpoint

    c. CENter

    d. INSertion point

    e. PERpendicular

21. Which OSNAP option allows you to select the location where blocked objects have been located?

    a. ENDpoint

    b. MIDpoint

    c. CENter

    d. INSertion point

    e. PERpendicular

22. When drawing a line which OSNAP option allows you to select the point on a line or polyline segment where the angle formed with the line is a 90 degree angle?

    a. ENDpoint

    b. MIDpoint

    c. CENter

    d. INSertion point

    e. PERpendicular

23. Which OSNAP option allows you to select the location where two lines, arcs, or polyline segments cross each other?

    a. NODe

    b. QUADrant

    c. TANgent

    d. NEArest

    e. INTersection

24. Which OSNAP option allows you to select the location where a "point" has been established?

    a. NODe   d. NEArest

    b. QUADrant   e. INTersection

    c. TANgent

25. Which OSNAP option allows you to select a point on a circle that is 0, 90, 180, or 270 degrees from the circle's center?

   a. NODe

   b. QUADrant

   c. TANgent

   d. NEArest

   e. INTersection

26. Which OSNAP option allows you to select a point on any object, except text, that is closest to the cursor's position?

   a. NODe

   b. QUADrant

   c. TANgent

   d. NEArest

   e. INTersection

27. Which OSNAP option allows you to select the location where a "point" has been established?

   a. NODe

   b. QUADrant

   c. APParent intersection

   d. NEArest

   e. INTersection

28. Which OSNAP option allows you to select the location where a two lines or arcs may or may not cross each other in 3D space?

   a. QUADrant

   b. APParent intersection

   c. NEArest

   d. INTersection

CHAPTER 4

Fundamentals III

INTRODUCTION

After completing this chapter, you will be able to do the following:

- Draw construction lines using the XLINE and RAY commands
- Construct geometric figures with polygons, ellipses, and polylines
- Create single and multi-text using appropriate styles and sizes, to annotate drawings
- Use the construct commands: COPY, ARRAY, OFFSET, MIRROR, FILLET, and CHAMFER.
- Use the modify commands MOVE, TRIM, BREAK, and EXTEND.

DRAWING CONSTRUCTION LINES

AutoCAD provides a powerful tool for drawing lines (XLINE and RAY commands) that extend infinitely in one or both directions. These lines have no effect, however, on the ZOOM EXTENTS command. They can be moved, copied, and rotated like any other objects. If necessary, you can trim the lines, break them anywhere with the BREAK command, draw an arc between two non-parallel construction lines with the FILLET command, and draw a chamfer between two non-parallel construction lines with the CHAMFER command.

The construction lines can be used as reference lines for creating other objects. To keep the construction lines from being plotted, you can draw them on a layer that can be turned off or set not to plot at the same time the objects are visible on the drawing.

XLINE COMMAND

The XLINE command allows you to draw lines that extend infinitely in both directions from the point selected when being created.

Invoke the XLINE command:

Draw toolbar	Choose the CONSTRUCTION LINE command (see Figure 4–1)
Draw menu	Choose Construction Line
Command: prompt	**xline** (ENTER)

Figure 4–1 *Invoking the* XLINE *command from the Draw toolbar*

AutoCAD prompts:

> Command: **xline** (ENTER)
> Specify a point or [Hor/Ver/Ang/Bisect/Offset]: *(specify a point or right-click for shortcut menu and choose one of the available options))*

When you specify a point to define the root of the construction line, this point becomes the conceptual midpoint of the construction line. AutoCAD prompts:

> Specify through point: *(specify a point through which the construction line should pass)*

AutoCAD draws a line that passes through two points and extends infinitely. AutoCAD continues to prompt for additional points to draw construction lines. To terminate the command sequence, press ENTER or the SPACEBAR.

Horizontal Option

The Horizontal option allows you to draw a construction line through a point that you specify and is parallel to the *X* axis of the current UCS.

Vertical Option

The Vertical option allows you to draw a construction line through a point that you specify and is parallel to the *Y* axis of the current UCS.

Angle Option

The Angle option allows you to draw a construction line at a specified angle. AutoCAD prompts:

Enter angle of xline or [Reference]>: *(specify an angle at which to place the construction line)*

Specify through point: *(specify a point through which the construction line should pass)*

AutoCAD draws the construction line through the specified point, using the specified angle.

The **Reference** option allows you to draw a construction line at a specific angle for a selected reference line. The angle is measured counterclockwise from the reference line.

Bisect Option

The Bisect option allows you to draw a construction line through the first point bisecting the angle determined by the second and third points, with the first point being the vertex. AutoCAD prompts:

Specify angle vertex point: *(specify a point for the vertex of an angle to be bisected and through which the construction line will be drawn)*

Specify angle start point: *(specify a point to determine one boundary line of an angle)*

Specify angle endpoint: *(specify a point to determine second boundary line of angle)*

The construction line lies in the plane determined by the three points.

Offset Option

The Offset option allows you to draw a construction line parallel to and at the specified distance from the line object selected and on the side selected. AutoCAD prompts:

Specify offset distance or [Through]: *(specify an offset distance or right-click for shortcut menu and choose one of the available options)*

Select a line object: *(select a line, pline, ray, or xline)*

Specify side to offset? *(specify a point to draw a construction line parallel to the selected object)*

The **Through** option allows you to specify a point through which a construction line is drawn to the line object selected.

RAY COMMAND

The RAY command allows you to draw lines that extend infinitely in one direction from the point selected when the line is being created.

Invoke the RAY command:

Draw menu	Choose RAY
Command: prompt	**ray** (ENTER)

AutoCAD prompts:

> Command: **ray** (ENTER)
> Specify start point: *(specify the start point to draw the ray)*
> Specify through point: *(specify a point through which you want the ray to pass)*
> Specify through point: *(specify a point to draw additional rays or press ENTER to terminate the command sequence)*

The ray is drawn starting at the first point and extending infinitely in one direction through the second point. AutoCAD continues to prompt for through points until you provide a null response to terminate the command sequence.

DRAWING POLYGONS

AutoCAD allows you to draw 2D polygons (edges with equal length) with the POLYGON command. The number of sides can be anywhere from 3 (which forms an equilateral triangle) to 1024. AutoCAD draws a polygon as a polyline with zero width and no tangent information. If necessary, with the help of the PEDIT command you can modify the polygon, such as changing its width.

Invoke the POLYGON command:

Draw toolbar	Choose the POLYGON command (see Figure 4–2)
Draw menu	Choose Polygon
Command: prompt	**polygon** (ENTER)

Figure 4–2 *Invoking the POLYGON command from the Draw toolbar*

AutoCAD prompts:

> Command: **polygon** (ENTER)
> Enter number of sides: *(specify the number of sides for the polygon to be drawn)*
> Specify center of polygon or [Edge]: *(specify center point for polygon or right-click for shortcut menu and choose Edge option)*

After specifying the center point of polygon, AutoCAD provides two options to draw the polygon: Inscribed in circle and Circumscribed about circle.

The **Inscribed in circle** option selection draws the polygon of equal length for all sides inscribed inside an imaginary circle (see Figure 4–3) having the same diameter as the distance across opposite polygon corners (for an even number of sides).

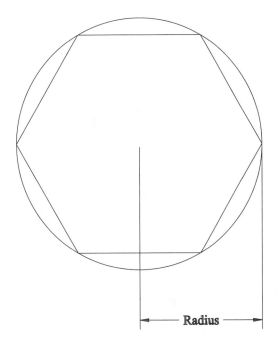

Figure 4–3 *Example of displaying a polygon inscribed inside an imaginary circle*

For example, the following command sequence shows steps in drawing a six-sided polygon (see Figure 4–4).

```
Command: polygon (ENTER)
Enter number of sides: 6 (ENTER)
Specify center of polygon or [Edge]: 3,3 (ENTER)
Enter an option [Inscribed in circle/Circumscribed about circle]: I (ENTER)
Specify radius of circle: 2 (ENTER)
```

AutoCAD draws a polygon with six sides, centered at 3,3, whose edge vertices are 2 units from the center of the polygon.

Specifying the radius with a specific value draws the bottom edge of the polygon at the current snap rotation angle. If instead, you specify the radius with your pointing device or by means of coordinates, AutoCAD determines the rotation and size of the polygon.

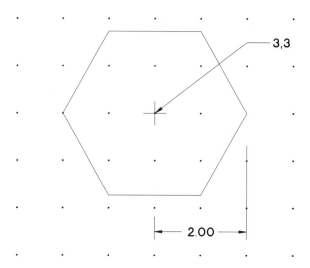

Figure 4–4 *Polygon drawn with six sides by selecting the Inscribed in circle option*

The following command sequence is an example of drawing a polygon by specifying the radius with relative coordinates.

Command: **polygon** (ENTER)
Enter number of sides: **6** (ENTER)
Specify center of polygon or [Edge]: **3,3** (ENTER)
Enter an option [Inscribed in circle/Circumscribed about circle]: **I** (ENTER)
Specify radius of circle: **@2<90** (ENTER)

The **Circumscribed about circle** option selection draws a polygon circumscribed around the outside of an imaginary circle having the same diameter as the distance across the opposite polygon sides (for an even number of sides) (see Figure 4–5).

For example, the following command sequence shows steps in drawing an eight-sided polygon with the selection of **Circumscribed about circle** option (see Figure 4–6).

Command: **polygon** (ENTER)
Enter number of sides: **8** (ENTER)
Specify center of polygon or [Edge]: **3,3** (ENTER)
Enter an option [Inscribed in circle/Circumscribed about circle]: **c** (ENTER)
Specify radius of circle: **2** (ENTER)

AutoCAD draws a polygon with eight sides, centered at 3,3, whose edge vertices are 2 units from the center of the polygon.

Specifying the radius with a value draws the bottom edge of the polygon at the current snap rotation angle. If instead, you specify the radius with your pointing device or by means of coordinates, AutoCAD determines the rotation and size of the polygon.

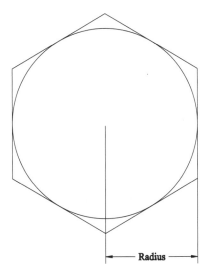

Figure 4–5 *Example of displaying a polygon circumscribed around the outside an imaginary circle*

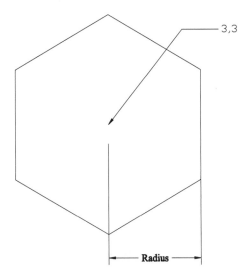

Figure 4–6 *Polygon drawn with eight sides by selecting the Circumscribed about circle option*

Edge Option

The Edge option allows you to draw a polygon by specifying the endpoints of the first edge.

For example, the following command sequence shows steps in drawing a seven-sided polygon (see Figure 4–7).

Command: **polygon** (ENTER)
Enter number of sides: **7** (ENTER)
Specify center of polygon or [Edge]: **e** (ENTER)
Specify first endpoint of edge: **1,1** (ENTER)
Specify second endpoint of edge: **3,1** (ENTER)

Figure 4–7 *Polygon drawn with seven sides by selecting the Edge option*

DRAWING ELLIPSES

AutoCAD allows you to draw an ellipse or an elliptical arc with the ELLIPSE command. Invoke the ELLIPSE command:

Draw toolbar	Choose the ELLIPSE command (see Figure 4–8)
Draw menu	Choose Ellipse
Command: prompt	**ellipse** (ENTER)

Figure 4–8 *Invoking the ELLIPSE command from the Draw toolbar*

AutoCAD prompts:

> Command: **ellipse** (ENTER)
> Specify axis endpoint of ellipse or [Arc/Center]: *(specify axis endpoint of the ellipse to be drawn or right-click for shortcut menu and choose one of the available options)*

Drawing an Ellipse by Defining Axis Endpoints

This option allows you to draw an ellipse by specifying the endpoints of the axes. AutoCAD prompts for two endpoints of the first axis. The first axis can define either the major or the minor axis of the ellipse. Then AutoCAD prompts for an endpoint of the second axis as the distance from the midpoint of the first axis to the specified point.

For example, the following command sequence shows steps in drawing an ellipse by defining axis endpoints (see Figure 4–9).

> Command: **ellipse** (ENTER)
> Specify axis endpoint of ellipse or [Arc/Center]: **1,1** (ENTER)
> Specify other endpoint of axis: **5,1** (ENTER)
> Specify distance to other axis or [Rotation]: **3,2** (ENTER)

AutoCAD draws an ellipse whose major axis is 4.0 units long in a horizontal direction and whose minor axis is 2.0 units long in a vertical direction, as shown in Figure 4–9.

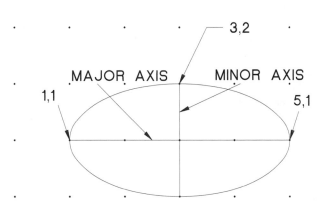

Figure 4–9 *An ellipse drawn by specifying the major and minor axes*

Drawing an Ellipse by Defining the Center of the Ellipse

This option allows you to draw an ellipse by defining the center point and axis endpoints. First, AutoCAD prompts for the ellipse center point. Then AutoCAD prompts for an endpoint of an axis as the distance from the center of the ellipse to the specified point. The first axis can define either the major or the minor axis of the ellipse. Then AutoCAD prompts for an endpoint of the second axis as the distance from the center of the ellipse to the specified point.

For example, the following command sequence shows steps in drawing an ellipse by defining the ellipse center point.

```
Command: ellipse (ENTER)
Specify axis endpoint of ellipse or [Arc/Center]: c (ENTER)
Specify center of ellipse: 3,1 (ENTER)
Specify endpoint of axis: 1,1 (ENTER)
Specify distance to other axis or [Rotation]: 3,2 (ENTER)
```

AutoCAD draws an ellipse similar to the previous example, with a major axis 4.0 units long in a horizontal direction and a minor axis 2.0 units long in a vertical direction.

Drawing an Ellipse by Specifying the Rotation Angle

AutoCAD allows you to draw an ellipse by specifying a rotation angle after defining two endpoints of one of the two axes. The rotation angle defines the major-axis-to-minor-axis ratio of the ellipse by rotating a circle about the first axis. The greater the rotation angle value, the greater the ratio of major to minor axes. AutoCAD draws a circle if you set the rotation angle to 0 degrees.

For example, the following command sequence shows steps in drawing an ellipse by specifying the rotation angle.

```
Command: ellipse (ENTER)
Specify axis endpoint of ellipse or [Arc/Center]: 3,-1 (ENTER)
Specify other endpoint of axis: 3,3 (ENTER)
Specify distance to other axis or [Rotation]: r (ENTER)
Specify rotation around major axis: 30 (ENTER)
```

See Figure 4–10 for examples of ellipses with various rotation angles.

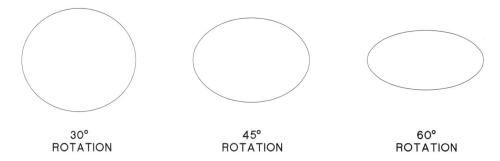

| 30°
ROTATION | 45°
ROTATION | 60°
ROTATION |

Figure 4–10 *Ellipses drawn with different rotation angles*

Drawing an Elliptical Arc with the Arc option

The Arc option allows you to draw an elliptical arc. After you specify the major and minor axis endpoints, AutoCAD prompts for the start and end angle points for the elliptical arc to be drawn. Instead of specifying the start angle or the end angle, you can toggle to the Parameter option, which prompts for the Start parameter and End parameter point locations. AutoCAD creates the elliptical arc using the following parametric vector equation:

$$p(u) = c + a^x + \cos(u) + b^x \sin(u)$$

where c is the center of the ellipse, and a and b are its major and minor axes, respectively. Instead of specifying the end angle, you can also specify the included angle of the elliptical arc to be drawn.

For example, the following command sequence shows steps in drawing an elliptical arc.

```
Command: ellipse (ENTER)
Specify axis endpoint of ellipse or [Arc/Center]: a (ENTER)
Specify axis endpoint of elliptical arc or [Center]: 1,1 (ENTER)
Specify other endpoint of axis: 5,1 (ENTER)
Specify distance to other axis or [Rotation]: 3,2 (ENTER)
Specify start angle or [Parameter]: 3,2 (ENTER)
Specify end angle or [Parameter/Included angle]:
Arc/Center/<Axis endpoint 1>: 1,1 (ENTER)
```

AutoCAD draws an elliptical arc with the start angle at 3,2 and the ending angle at 1,1.

Isometric Circles (or Isocircles) Option

By definition, Isometric Planes (*iso* meaning "same" and *metric* meaning "measure") are all being viewed at the same angle of rotation (see Figure 4–11). The angle is approximately set to 54.7356 degrees. AutoCAD uses this angle of rotation automatically when you wish to represent circles in one of the isoplanes by drawing ellipses with the Isocircle option.

Normally, a circle 1 unit in diameter being viewed in one of the isoplanes will project a short axis dimension of 0.577350 units. One of its diameters parallel to an isoaxis will project a dimension of 0.816497 units. A line drawn in isometric that is parallel to one of the three main axes will also project a dimension of 0.816497 units. We would like these lines and circle diameters to project a dimension of exactly 1.0 unit. Therefore, you automatically increase the entire projection by a fudge factor of 1.22474 (the reciprocal of 0.816497) in order to be able to use true dimensioning parallel to one of the isometric axes.

This means that circles 1 unit in diameter will be measured along one of their isometric diameters rather than along their long axis. This facilitates using true lengths as the lengths of distances projected from lines parallel to one of the isometric axes. So a 1-unit-diameter isocircle will project a long axis that is 1.224744871 units and a short axis that is 0.707107 (0.577350 x 1.22474) units. These "fudge" factors are built into AutoCAD isocircles.

The Isometric Circle method is available as one of the options of the ELLIPSE command when you are in the isometric Snap mode.

> Command: **ellipse** (ENTER)
> Specify axis endpoint of ellipse or [Arc/Center/Isocircle]: **i** (ENTER)
> Specify center of isocircle: *(select the center of the isometric circle)*
> Specify radius of isocircle or [Diameter]: *(specify the radius or right-click for the shortcut menu and choose one of the available options)*

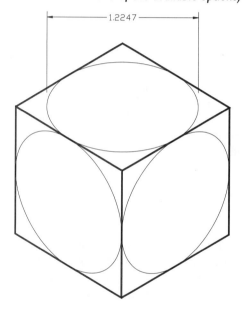

Figure 4–11 *Ellipses drawn using the Isocircle option of the* ELLIPSE *command*

If you override the last prompt default by typing d, the following prompt will appear:

> Circle diameter: *(specify the desired diameter)*

 Note: The Iso and Diameter options will work only when you are in the isometric Snap mode.

DRAWING POLYLINES

The poly in polyline refers to a single object with multiple connected straight-line and/or arc segments. The polyline is drawn by invoking the PLINE command and then selecting a series of points. In this respect, PLINE functions much like the LINE command. However, when completed, the segments act like a single object when operated on by modify commands. You specify the endpoints using only 2D (*X,Y*) coordinates.

The versatile PLINE command also draws lines and arcs of different widths, linetypes, tapered lines, and a filled circle. The area and perimeter of a 2D Polyline can be calculated.

By default, polylines are drawn as optimized polylines. The optimized polyline provides most of the functionality of 2D polylines but with much improved performance and reduced drawing file size. The vertices are stored as an array of information on one object. When you use the PEDIT command to edit the polyline to spline fitting or curve fitting, the polyline loses its optimization feature and vertices are stored as separate entities, but it still behaves as a single object when operated on by modify commands.

 Note: If you open an AutoCAD Release 13 drawing containing polylines, they are transparently converted to optimized polylines, unless the 2D polyline is a curve-fit or splined polyline.

Invoke the PLINE command:

Draw toolbar	Choose the Polyline command (see Figure 4–12)
Draw menu	Choose Polyline
Command: prompt	**pline** (ENTER)

Figure 4–12 *Invoking the* PLINE *command from the Draw toolbar*

AutoCAD prompts:

> Command: **pline** (ENTER)
> Specify start point: *(specify the start point of the polyline)*
> Specify next point or [Arc/Close/Halfwidth/Length/Undo/Width]: *(specify next point or right-click for shortcut menu and choose one of the available options)*

You can specify the end of the line by means of absolute coordinates, relative coordinates, or by using your pointing device to specify the end of the line on the screen. After you do, AutoCAD repeats the prompt:

> Specify next point or [Arc/Close/Halfwidth/Length/Undo/Width]:

Having drawn a connected series of lines, you can give a null reply (press ENTER) to terminate the PLINE command. The resulting figure is recognized by AutoCAD modify and construct commands as a single object.

The following command sequence presents an example of connected lines, as shown in Figure 4–13, drawn by means of the PLINE command.

Figure 4–13 *Example of connected line segments drawn by means of the* PLINE *command*

```
Command: pline (ENTER)
Specify start point: 2,2 (ENTER)
Current line-width is 0.0000
Specify next point or [Arc/Close/Halfwidth/Length/Undo/Width]: 4,2
    (ENTER)
Specify next point or [Arc/Close/Halfwidth/Length/Undo/Width]: 5,1
    (ENTER)
Specify next point or [Arc/Close/Halfwidth/Length/Undo/Width]: 7,1
    (ENTER)
Specify next point or [Arc/Close/Halfwidth/Length/Undo/Width]: 8,2
    (ENTER)
Specify next point or [Arc/Close/Halfwidth/Length/Undo/Width]: 10,2
    (ENTER)
Specify next point or [Arc/Close/Halfwidth/Length/Undo/Width]: 10,4
    (ENTER)
Specify next point or [Arc/Close/Halfwidth/Length/Undo/Width]: 9,5
    (ENTER)
Specify next point or [Arc/Close/Halfwidth/Length/Undo/Width]: 8,5
    (ENTER)
Specify next point or [Arc/Close/Halfwidth/Length/Undo/Width]: 7,4
    (ENTER)
Specify next point or [Arc/Close/Halfwidth/Length/Undo/Width]: 5,4
    (ENTER)
```

    Specify next point or [Arc/Close/Halfwidth/Length/Undo/Width]: **4,5**
      (ENTER)
    Specify next point or [Arc/Close/Halfwidth/Length/Undo/Width]: **3,5**
      (ENTER)
    Specify next point or [Arc/Close/Halfwidth/Length/Undo/Width]: **2,4**
      (ENTER)
    Specify next point or [Arc/Close/Halfwidth/Length/Undo/Width]: **c**
      (ENTER)

Close and Undo Options

The Close and Undo options work similarly to the corresponding options in the LINE command.

Width Option

After selecting a starting point, you may enter **w** or right-click for the shortcut menu and choose the width option to specify a starting and an ending width for a wide segment. When you select this option, AutoCAD prompts:

    Specify starting width <default>: *(specify starting width)*
    Specify ending width <default>: *(specify ending width)*

You can specify a width by entering a value at the prompt or by selecting a width determining points on the screen. When you specify points on the screen, AutoCAD uses the distance from the starting point of the polyline to the point selected as the starting width. You can accept the default value for the starting width by providing a null response (press ENTER), or enter in a new value. The starting width you enter becomes the default for the ending width. If necessary, you can change the ending width to another width, which results in a tapered segment or an arrow. The ending width, in turn, becomes the uniform width for all subsequent segments until you change the width again.

The following command sequence presents an example of connected lines with tapered width, as shown in Figure 4–14, drawn by means of the PLINE command.

    Command: **pline** (ENTER)
    Specify start point: **2,2** (ENTER)
    Current line-width is 0.0000
    Specify next point or [Arc/Close/Halfwidth/Length/Undo/Width]: **w**
      (ENTER)
    Specify starting width <default>: **0** (ENTER)
    Specify ending width <default>: **.25** (ENTER)
    Specify next point or [Arc/Close/Halfwidth/Length/Undo/Width]: **2,2.5**
      (ENTER)
    Specify next point or [Arc/Close/Halfwidth/Length/Undo/Width]: **2,3**
      (ENTER)

Specify next point or [Arc/Close/Halfwidth/Length/Undo/Width]: **w**
(ENTER)
Specify starting width <0.2500>: (ENTER)
Specify ending width <0.2500>: **0** (ENTER)
Specify next point or [Arc/Close/Halfwidth/Length/Undo/Width]: **2,3.5**
(ENTER)
Specify next point or [Arc/Close/Halfwidth/Length/Undo/Width]: **w**
(ENTER)
Specify starting width <0.0000>: (ENTER)
Specify ending width <0.0000>: **.25** (ENTER)
Specify next point or [Arc/Close/Halfwidth/Length/Undo/Width]: **2.5,3.5**
(ENTER)
Specify next point or [Arc/Close/Halfwidth/Length/Undo/Width]: **3,3.5**
(ENTER)
Specify next point or [Arc/Close/Halfwidth/Length/Undo/Width]: **w**
(ENTER)
Specify starting width <0.2500>: (ENTER)
Specify ending width <0.2500>: **0** (ENTER)
Specify next point or [Arc/Close/Halfwidth/Length/Undo/Width]: **3.5,3.5**
(ENTER)
Specify next point or [Arc/Close/Halfwidth/Length/Undo/Width]: **w**
(ENTER)
Specify starting width <0.0000>: (ENTER)
Specify ending width <0.0000>: **.25** (ENTER)
Specify next point or [Arc/Close/Halfwidth/Length/Undo/Width]: **3.5,3**
(ENTER)
Specify next point or [Arc/Close/Halfwidth/Length/Undo/Width]: **3.5,2.5**
(ENTER)
Specify next point or [Arc/Close/Halfwidth/Length/Undo/Width]: **w**
(ENTER)
Specify starting width <0.2500>: (ENTER)
Specify ending width <0.2500>: **0** (ENTER)
Specify next point or [Arc/Close/Halfwidth/Length/Undo/Width]: **3.5,2**
(ENTER)
Specify next point or [Arc/Close/Halfwidth/Length/Undo/Width]: **w**
(ENTER)
Specify starting width <0.0000>: (ENTER)
Specify ending width <0.0000>: **.25** (ENTER)
Specify next point or [Arc/Close/Halfwidth/Length/Undo/Width]: **3,2**
(ENTER)
Specify next point or [Arc/Close/Halfwidth/Length/Undo/Width]: **2.5,2**
(ENTER)
Specify next point or [Arc/Close/Halfwidth/Length/Undo/Width]: **w**
(ENTER)

Specify starting width <0.2500>: (ENTER)
Specify ending width <0.2500>: **0** (ENTER)
Specify next point or [Arc/Close/Halfwidth/Length/Undo/Width]: **c**
  (ENTER)

Figure 4–14 *Example of connected line segments with tapered width drawn by means of the*
  PLINE *command*

Halfwidth Option

The Halfwidth option is similar to the width option, including the prompts, except it
lets you specify the width from the center of a wide polyline to one of its edges. In
other words, you specify half of the total width. For example, it is easier to input
1.021756 as the halfwidth than to figure out the total width by doubling. You can
specify a halfwidth by selecting points on the screen in the same manner used to
specify the full width.

Arc Option

The Arc option allows you to draw a polyline arc. When you select the **Arc** option,
AutoCAD displays another submenu:

Specify endpoint of arc or [Angle/CEnter/CLose/Direction/Halfwidth/Line/
  Radius/Second pt/Undo/Width]: *(specify end point of arc or right-click for
  shortcut menu and choose one of the available options)*

If you respond with a point, it is interpreted as the endpoint of the arc. The endpoint
of the previous segment is the starting point of the arc, and the starting direction of

the new arc will be the ending direction of the previous segment (whether the previous segment is a line or an arc). This resembles the ARC command's Start, End, Direction (S,E,D) option, but requires only the endpoints to be specified or selected on the screen.

The **CLose, Width, Halfwidth,** and **Undo** options are similar to the corresponding options for the straight-line segments described earlier.

The **Angle** option lets you specify the included angle by prompting:

> Specify included angle: *(specify an angle)*

The arc is drawn counterclockwise if the value is positive, clockwise if it is negative. After the angle is specified, AutoCAD prompts for the endpoint of the arc.

The **CEnter** option lets you override with the location of the center of the arc and AutoCAD prompts:

> Specify center point: *(specify center point)*

When you provide the center point of the arc, AutoCAD prompts for additional information:

> Specify endpoint of arc or [Angle/Length]: *(specify end point of arc or right-click for shortcut menu and choose one of the available options)*

If you respond with a point, it is interpreted as the endpoint of the arc. Selecting Angle or Length allows you to specify the arc's included angle or chord length.

The **Direction** option lets you override the direction of the last segment, and AutoCAD prompts:

> Specify the tangent direction for the start point of arc: *(specify the direction)*

If you respond with a point, it is interpreted as the starting point of the direction and AutoCAD prompts for the endpoint for the arc.

The **Line** option reverts to drawing straight-line segments.

The **Radius** option allows you to specify the radius by prompting:

> Specify radius of arc: *(specify the radius of the arc)*

After the radius is specified, you are prompted for the endpoint of the arc.

The **Second point** option causes AutoCAD to use the three-point method of drawing an arc by prompting:

> Specify second point on arc: *(specify second point)*

If you respond with a point, it is interpreted as the second point and then you are prompted for the endpoint of the arc. This resembles the ARC command's Three-point option.

Length Option

The Length option continues the polyline in the same direction as the last segment for a specified distance.

DRAWING TEXT

You have learned how to draw some of the geometric shapes that make up your design. Now it is time to learn how to annotate your design. When you draw on paper, adding descriptions of the design components and the necessary shop and fabrication notes is a tedious, time-consuming process. AutoCAD provides several text commands and tools (including a spell checker) that greatly reduce the tedium of text placement and the time it takes.

Text is used to label the various components of your drawing and to create the necessary shop or field notes needed for fabrication and construction of your design. AutoCAD includes a large number of text fonts. Text can be stretched, compressed, obliqued, mirrored, or drawn in a vertical column by applying a style. Each text string can be sized, rotated, and justified to meet your drawing needs. You should be aware that AutoCAD considers a text string (all the characters that comprise the line of text) as one object.

 Note: If you do not know how to type, you can place text quickly and easily after a period of learning the keyboard and developing typing skills. If you create drawings that require a lot of text entry, it may be worth your time to learn to type with all ten fingers ("touch-typing"). There are several computer programs that can help you teach yourself to type, and almost all colleges offer typing classes. If you do not have time to learn touch-typing, there is no need to worry, many "two-finger" typists productively place text in their drawings.

CREATING A SINGLE LINE OF TEXT

The TEXT command allows you to create several lines of text in the current style. If necessary, you can change the current style. To modify a style or to create a new style, refer to the section Creating and Modifying Text Styles, later in this chapter.

Invoke the TEXT command:

Draw menu	Choose Text > Single Line Text
Command: prompt	**text** (ENTER)

AutoCAD prompts:

Command: **text** (ENTER)
Specify start point of text or [Justify/Style]: *(specify start point of text or right-click for shortcut menu and choose one of the available options)*

The start point indicates the lower left corner of the text. If necessary, you can change the location of the justification point. You can specify the starting point in absolute coordinates or by using your pointing device. After you specify the starting point, AutoCAD prompts:

Specify height <default>: *(specify the text height)*

This allows you to select the text height. You can accept the default text height by giving a null response, by using your pointing device, or you can enter the appropriate text height. Next, AutoCAD prompts:

Specify rotation angle of text: *(specify the rotation angle)*

This allows you to place the text at any angle in reference to 0 degrees (default is 3 o'clock, or east, measured in the counterclockwise direction). The default value of the rotation angle is 0 degrees, and the text is placed horizontally at the specified start point. The last prompt is:

Enter text: *(type the desired text and press ENTER)*

A cursor appears on the screen at the starting point you have selected. After you enter the first line of text and press ENTER, you will notice the cursor drop down to the next line, anticipating that you wish to enter more text. If this is the case, enter the next line of text string; when you are through with typing text strings, press ENTER at the "Enter Text:" prompt to terminate the command sequence.

If you are in the TEXT command and notice a mistake (or simply want to change a value or word), backspace to the text you want to change. This, however, deletes all of the text you backspaced over to get to the point you want to change. If this involves erasing several lines of text, it may be faster to use the CHANGE or DDEDIT commands to make changes to the text string.

One feature of TEXT that will speed up text entry on your drawing is the ability to move the crosshairs cursor to a new point on the drawing while staying in the TEXT command. As you move the crosshairs cursor and specify a new point, you will notice the cursor moves to this new point, allowing you to enter a new string of text. However, you must remember to give a null response (press ENTER) to the "Enter text:" prompt to terminate the command.

For example, the following command sequence shows placement of left-justified text by providing the starting point of the text, as shown in Figure 4–15.

> Command: **text** (ENTER)
> Specify start point of text or [Justify/Style]: *(specify start point of text as shown in Figure 4–15)*
> Specify height <default>: **.25** (ENTER)
> Specify rotation angle of text: (ENTER)
> Enter text: **Sample Text Left Justified** (ENTER)
> Enter text: (ENTER)

Sample Text Left Justified

─ Start Point

Figure 4–15 *Using the* TEXT *command to place left-justified text by specifying a start point*

Justify Option

The Justify option allows you to place text in one of the 14 available justification points. When you select this option, AutoCAD prompts:

> Enter an option [Align/Fit/Center/Middle/Right/TL/TC/TR/ML/MC/MR/BL/BC/BR]: *(select one of the available options or right-click for shortcut menu and choose one of the available options)*

The **Center** option allows you to select the center point for the baseline of the text. Baseline refers to the line along which the bases of the capital letters lie. Letters with descenders, such as g, q, and y, dip below the baseline. After providing the center point, enter the text height and rotation angle. The justification does not occur on the screen until ENTER key is pressed to exit the TEXT command.

For example, the following command sequence shows placement of center-justified text, by providing the center point of the text, as shown in Figure 4–16.

> Command: **text** (ENTER)
> Specify start point of text or [Justify/Style]: **j** (ENTER)
> Enter an option [Align/Fit/Center/Middle/Right/TL/TC/TR/ML/MC/MR/BL/BC/BR]: **c** (ENTER)
> Specify height <default>: **.25** (ENTER)
> Specify rotation angle of text: **0** (ENTER)
> Enter text: **Sample Text Center Justified** (ENTER)

Sample Text Center Justified

Center Point

Sample Text Middle Justified

Middle Point

Sample Text Right Justified

End Point

Figure 4–16 *Using the* TEXT *command to place text by specifying a center point (center justified), a middle point (middle justified), or an endpoint (right justified)*

The **Middle** option allows you to center the text both horizontally and vertically at a given point. After providing the middle point, enter the text height and rotation angle.

For example, the following command sequence shows placement of middle-justified text by providing the middle point of the text, as shown in Figure 4–16.

> Command: **text** (ENTER)
> Specify start point of text or [Justify/Style]: **j** (ENTER)
> Enter an option [Align/Fit/Center/Middle/Right/TL/TC/TR/ML/MC/MR/BL/
>   BC/BR]: **m** (ENTER)
> Specify height <default>: **.25** (ENTER)
> Specify rotation angle of text: **0** (ENTER)
> Enter text: **Sample Text Middle Justified** (ENTER)

The **Right** option allows you to place the text in reference to its lower right corner (right justified). Here, you provide the point where the text will end. After providing the right point, enter the text height and rotation angle.

For example, the following command sequence shows placement of right-justified text, as shown in Figure 4–16.

> Command: **text** (ENTER)
> Specify start point of text or [Justify/Style]: **j** (ENTER)
> Enter an option [Align/Fit/Center/Middle/Right/TL/TC/TR/ML/MC/MR/BL/
>   BC/BR]: **r** (ENTER)
> Specify height <default>: **.25** (ENTER)
> Specify rotation angle of text: 0 (ENTER)
> Enter text: **Sample Text Right Justified** (ENTER)

Other options are combinations of the previously mentioned options:

TL   top left

TC   top center

TR   top right

ML   middle left

MC   middle center

MR   middle right

BL   bottom left

BC   bottom center

BR   bottom right

The **Align** option allows you to place the text by designating the endpoints of the baseline. AutoCAD computes the text height and orientation so that the text just fits proportionately between two points. The overall character size adjusts in proportion to the height. The height and width of the character will be the same.

For example, the following command sequence shows placement of text using the Align option as shown in Figure 4–17.

```
Command: text (ENTER)
Specify start point of text or [Justify/Style]: j (ENTER)
Enter an option [Align/Fit/Center/Middle/Right/TL/TC/TR/ML/MC/MR/BL/
    BC/BR]: a (ENTER)
Specify first endpoint of text baseline: (specify the first point)
Specify second endpoint of text baseline: (specify the second point)
Enter text: Sample Text Aligned (ENTER)
```

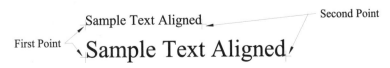

Figure 4–17 *Using the Align and Fit option of the* TEXT *command to place text*

The **Fit** option is similar to the Align option, but in the case of the Fit option AutoCAD uses the current text height and adjusts only the text's width, expanding or contracting it to fit between the points you specify.

For example, the following command sequence shows placement of text using the Fit option as shown in Figure 4–17.

```
Command: text (ENTER)
Specify start point of text or [Justify/Style]: j (ENTER)
Enter an option [Align/Fit/Center/Middle/Right/TL/TC/TR/ML/MC/MR/BL/
    BC/BR]: f (ENTER)
Specify first endpoint of text baseline: (specify the first point)
Specify second endpoint of text baseline: (specify the second point)
Specify height <default>: 0.25 (ENTER)
Enter text: Sample Text Fit (ENTER)
```

Style Option

The **Style** option allows you to select one of the available styles in the current drawing. To modify a style or to create a new style, refer to the section Creating and Modifying Text Styles, later in this chapter.

CREATING MULTILINE TEXT

AutoCAD offers the MTEXT command for drawing text by "processing" the words in paragraph form; the width of the paragraph is determined by the user-specified rectangular boundary. It is an easy way to have your text automatically formatted as a multiline group, with left, right, or center justification as a group. Each multiline text object is a single object, regardless of the number of lines it contains. The text boundary remains part of the object's framework, although it is not plotted or printed.

Invoke the MTEXT command:

Draw toolbar	Choose the MTEXT command (see Figure 4–18)
Draw menu	Choose Text > Multiline Text...
Command: prompt	**mtext** (ENTER)

Figure 4–18 *Invoking the* MULTILINE TEXT *command from the Draw toolbar*

AutoCAD prompts:

> Command: **mtext** (ENTER)
> Specify first corner: *(specify the first corner of the rectangular boundary)*
> Specify opposite corner or [Height/Justify/Line spacing/Rotation/Style/
>   Width]: *(specify the opposite corner of the rectangular boundary, or right-
>   click for shortcut menu and choose one of the available options)*

When you drag the cursor after specifying the first corner of the rectangular bound-ary, AutoCAD displays an arrow within the rectangle to indicate the direction of the paragraph's text flow. After you specify the opposite corner of the rectangular bound-ary, AutoCAD displays the Multiline Text Editor dialog box, shown in Figure 4–19.

Figure 4–19 *Multiline Text Editor dialog box*

Enter the text in the text editor dialog box. If you right-click in the Multiline Text Editor dialog box, AutoCAD displays a shortcut menu with options that include Undo, Cut, Copy, Paste, Select All, Change Case, Remove Formatting, and Com-bine Paragraphs.

The Multiline Text Editor dialog box contains various control options, provided via four tabs: Character, Properties, Line Spacing, and Find/Replace. The **Character** tab provides options to control character formatting for text entered in the Multiline Text Editor dialog box or imported into the text editor. The **Properties** tab provides the options to control the properties that apply to the multiline text object. The **Line Spacing** tab provides the options to set the line spacing. And the **Find/Replace** tab provides the options to search for specified text strings and replace them with new text strings.

The text can be highlighted in the following three ways: holding down the pick but-ton while dragging across the selected text, double-clicking to select the entire word, or triple-clicking to select the entire line of text.

Character Options

The options provided in the Character tab (see Figure 4–19) include character formatting: Font, Height, Bold, Italic, Underline, Undo, Stack, Color, and Symbol.

The **Font** option allows you to specify a font for new text or changes the font of selected text. All of the available TrueType fonts and SHX fonts are listed in the drop-down list box.

The **Height** option sets the character height in drawing units. The default value for the height is based on the current style. (A detailed discussion is provided later in the chapter on creating or modifying a text style.) If the current style is set to 0, then the value of the height is based on the value stored in the TEXTSIZE system variable. Each multiline text object can contain a text string of varying text size. When you highlight the text string in the dialog box, AutoCAD displays the selected text height in the list box. If necessary, you can specify a new height in addition to those listed.

You can also set the text height by selecting the Height option when AutoCAD prompts for the opposite corner of the rectangular boundary. Once the text height is specified, AutoCAD returns to the previous prompt until the opposite corner of the rectangular boundary is specified.

The **Bold** option allows you to turn ON and OFF bold formatting for new text or selected text. The Bold option is available only to the characters that belong to the TrueType font.

The **Italic** option allows you to turn ON and OFF italic formatting for new text or selected text. The Italic option is available only to the characters that belong to the TrueType font.

The **Underline** option allows you to turn ON and OFF underlining for new text or selected text.

The **Undo** option undoes the last edit action in the Multiline Text Editor dialog box that includes changes in the content of the text string or formatting.

The **Stack** option is used to place one part of a selected group of text over the remaining part. Before using the Stack option, the selected text must contain a forward slash (/) to separate the top part (to the left of the /) from the bottom part (to the right of the /). The slash will cause the horizontal bar to be drawn between the upper and lower parts, necessary for fractions with center justification. Instead of a slash (/), you can use the caret (^) symbol. In this case, AutoCAD will not draw a horizontal bar between the upper and lower parts with left justification, which is useful for placing tolerance values.

The **Color** option sets the color for new text or changes it for the selected text. You can assign the color by BYLAYER, BYBLOCK, or one of the available colors.

The **Symbol** option allows you to insert a listed symbol or nonbreaking space at the cursor position in the dialog box. In addition, you can also insert the symbols manually by means of the control characters.

Properties Options

The options provided in the Properties tab (see Figure 4–20) control properties of the multiline text object such as Style, Justification, Width, and Rotation.

Figure 4–20 *Multiline Text Editor dialog box displaying the Properties tab*

The **Style** option allows you to apply an existing style to new text or selected text. If you apply a new style to an existing text object, AutoCAD overrides character formatting such as font, height, bold, and italic attributes. Styles that have backwards or upside-down effects are not applied.

You can also select the text style by means of the Style option when AutoCAD prompts for the opposite corner of the rectangular boundary. Once the text style is set, AutoCAD returns to the previous prompt until the opposite corner of the rectangular boundary is specified.

The **Justification** option sets the justification and alignment for new or selected text. Text is center, left, or right justified with respect to the left and right text boundaries, and aligned from the middle, top, or bottom of the paragraph with respect to the top and bottom text boundaries.

You can also set the justification by selecting the Justify option when AutoCAD prompts for the opposite corner of the rectangular boundary. The available justification options are the same as in the TEXT command. Once justification is specified, AutoCAD returns to the previous prompt until the opposite corner of the rectangular boundary is specified.

The **Width** option sets the paragraph width for new or selected text. If it is set to the No Wrap option, the resulting multiline text object appears on a single line.

You can also set the width by selecting the Width option when AutoCAD prompts for the opposite corner of the rectangular boundary. Once the width is specified, AutoCAD returns to the previous prompt until the opposite corner of the rectangular boundary is specified.

The **Rotation** option sets the rotation angle for new or selected text, in the current unit of angle measurement.

You can also set the rotation by selecting the Rotation option when AutoCAD prompts for the opposite corner of the rectangular boundary. Once the rotation is specified, AutoCAD returns to the previous prompt until the opposite corner of the rectangular boundary is specified.

Line Spacing Options

The options that are provided specifies line spacing for the text objects. Two options are provided for line spacing: At Least and Exactly.

The **At Least** selection adjusts lines of text automatically based on the height of the largest character in the line. When At Least is selected, lines of text with taller characters have added space between lines.

The **Exactly** selection forces the line spacing to be the same for all lines of text in the mtext object. Spacing is based on the text height of the object or text style.

The **Line Spacing** text box sets the vertical distance between the bottom (or baseline) of one line of text and the bottom of the next line of text. You can set the spacing increment to a multiple of single-line spacing, or to an absolute value.

Single spacing is 1.66 times the height of the text characters. You can enter the spacing increment as a number followed by x to indicate a multiple of single spacing.

You can also set the line spacing by selecting the Line Spacing option when AutoCAD prompts for the opposite corner of the rectangular boundary. Once the line spacing is specified, AutoCAD returns to the previous prompt until the opposite corner of the rectangular boundary is specified.

 Note: Exact spacing is recommended when you use mtext to create a table.

Find/Replace Options

The options that are provided in the Find/Replace tab (see Figure 4–21) include Find, Replace with, Match Case, and Whole Word searches for specified text strings and replaces them with new text.

Enter the text string to be searched for in the **Find** text box, then choose the **Find** button to start the search. AutoCAD highlights the appropriate text string in the dialog box. To continue the search, choose the **Find** button again.

Figure 4–21 *Multiline Text Editor dialog box displaying the Find/Replace tab*

In the **Replace with** text box type the text string that you want as replacement for the text string in the **Find** text box. Then choose the **Replace with** button to replace the highlighted text with the text in the Replace with text box.

When the **Match Case** toggle button is set to ON, AutoCAD finds text only if the case of all characters in the text object is matched to that of the text characters in the Find text box. When it is set to OFF, AutoCAD finds a match for the specified text string regardless of the case of the characters.

When the **Whole Word** toggle button is set to ON, AutoCAD finds text only if the text string is a single word. If the text is part of another text string, it is ignored. When it is set to OFF, AutoCAD finds a match for the specified text string whether it is a single word or part of another word.

To import text into the Multiline Text Editor dialog box, choose the **Import Text...** button; AutoCAD displays the Open dialog box. Select the appropriate file (ASCII or RTF format) from the dialog box. Inserted text retains its original character formatting and style properties. If necessary, you can replace either selected text or all text of the current selection within the text boundary.

 Note: Imported text is limited to 16 KB file size.

To keep the settings and draw the paragraph text, click the **OK** button. AutoCAD closes the Multiline Text Editor dialog box and draws the paragraph text at the appropriate location.

SPECIAL TEXT CHARACTER AND SYMBOL OPTIONS

In addition to the options provided in the Multiline Text Editor dialog box for drawing special characters, you can draw them by means of the control characters. The control characters for a symbol begin with a double percent sign (%%). The next character you enter represents the symbol. The control sequences defined by AutoCAD are presented in Table 4–1.

Table 4–1 Control Character Sequences for Drawing Special Characters and Symbols

Special Character or Symbol	Control Character Sequence	Example	
		Text String	**Control Character Sequence**
° (degree symbol)	%%d	104.5°F	104.5%%dF
± (plus/minus tolerance symbol)	%%p	34.5±3	34.5%%p3
Æ (diameter symbol)	%%c	56.06Æ	56.06%%c
% (single percent sign; necessary only when it must precede another control sequence)	%%%	34.67%±1.5	34.67%%%%%%P1.5
Special coded symbols (where nnn stands for a three-digit code)	%%nnn	@	%%064

EDITING TEXT

The DDEDIT command allows you to edit text and attributes. An attribute is informational text associated with a block. See Chapter 10 for a detailed discussion of blocks and attributes.

Invoke the DDEDIT command:

Modify II toolbar	Choose the EDIT TEXT command (see Figure 4–22)
Modify menu	Chose Text...
Command: prompt	**ddedit** (ENTER)

Figure 4–22 *Invoking the DDEDIT command from the Modify II toolbar*

AutoCAD prompts:

Command: **ddedit** (ENTER)
Select an annotation object or [Undo]: *(select the text or attribute definition, or right-click for shortcut menu and choose one of the available options)*

If you select a text string created by means of a TEXT command, AutoCAD displays the Edit Text dialog box, as shown in Figure 4–23. Make the necessary changes in the text string and choose the **OK** button to keep the changes.

If instead you select text created by means of the MTEXT command, AutoCAD displays the Multiline Text Editor dialog box shown in Figure 4–24. Make the necessary changes in the text string, and click the **OK** button to keep the changes.

Figure 4–23 *Edit Text dialog box*

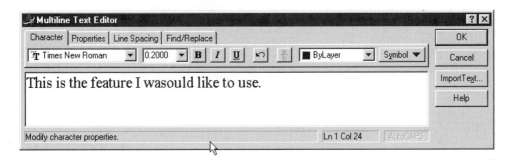

Figure 4–24 *Multiline Text Editor dialog box*

AutoCAD continues to prompt you to select a new text string to edit, or you can enter U to undo the last change made to the text. To terminate the command sequence, give a null response (press ENTER).

SPELL-CHECKING
The SPELL command is used to correct the spelling of text objects created with the TEXT or MTEXT command.

Invoke the SPELL command:

Tools menu	Spelling
Command: prompt	**spell** (ENTER)

AutoCAD prompts:

Command: **spell** (ENTER)
Select objects: *(select one or more text strings, and press* ENTER *to terminate object selection)*

AutoCAD displays the Check Spelling dialog box, similar to Figure 4–25, only when it finds a dubious word in the selected text objects.

Figure 4–25 *Check Spelling dialog box*

AutoCAD displays the name of the current dictionary in the top of the Check Spelling dialog box. If necessary, you can change to a different dictionary by clicking the **Change Dictionaries...** button and selecting the appropriate dictionary from the Change Dictionaries dialog box.

AutoCAD displays each misspelled word in the **Current Word** section and lists the suggested alternate spellings in the **Suggestions:** list box. Click the **Change** button to replace the current word with the selected suggested work, or click the **Change All** button to replace all instances of the current word. Alternatively, click the **Ignore** button to skip the current word, or click the **Ignore All** button to ignore all subsequent entries of the current word.

The **Add** button allows you to include the current word (up to 63 characters) in the current or custom dictionary. The **Lookup** button allows you to check the spelling of the word in the suggestions box.

After completion of the spelling check, AutoCAD displays an AutoCAD message informing you that the spelling check is complete.

CONTROLLING THE DISPLAY OF TEXT

The QTEXT command is a utility command for TEXT and MTEXT that is designed to reduce the redraw and regeneration time of a drawing. Regeneration time becomes a significant factor if the drawing contains a great amount of text and attribute information and/or if a fancy text font is used. Using QTEXT, the text is replaced with rectangular boxes of a height corresponding to the text height. These boxes are regenerated in a fraction of the time required for the actual text.

If a drawing contains many text and attribute items, it is advisable to set QTEXT to ON. However, before plotting the final drawing, or inspection of text details, the QTEXT command is set to OFF and is followed by the REGEN command.

Invoke the QTEXT command:

Command: prompt	**qtext** (ENTER)

AutoCAD prompts:

> Command: **qtext** (ENTER)
> ON/OFF <current>: *(right-click and select one of the available options from the shortcut menu and choose one of the available options)*

CREATING AND MODIFYING TEXT STYLES

The Style option of the TEXT and MTEXT commands (in conjunction with the STYLE command) lets you determine how text characters and symbols appear, other than adjusting the usual height, slant, and angle of rotation. To specify a text style from the Style option of the TEXT and MTEXT commands, it must have been defined by using the STYLE command. In other words, the STYLE command creates a new style or modifies an existing style. The Style option under the TEXT or MTEXT command allows you to choose a specific style from the styles available.

There are three things to consider when using the STYLE command.

First, you must name the newly defined style. Style names may contain up to 255 characters, numbers, and special characters ($, –, and _). Names like "title block," "notes," and "bill of materials" can remind you of the purpose for which the particular style was designed.

Second, you may apply a particular font to a style. The font that AutoCAD uses as a default is called TXT. It has blocky looking characters, which are economical to store in memory. But the TXT.SHX font, made up entirely of straight-line (noncurved) segments, is not considered as attractive or readable. Other fonts offer many variations in characters, including those for foreign languages. All fonts are stored for use in files of their font name with an extension of .SHX. The most effective way to get a distinctive appearance in text strings is to use a specially designed font. You can also use TrueType fonts and Type 2 postscript. See Appendix F for a list of fonts that come with AutoCAD. If necessary, you can buy additional fonts from third-party vendors. AutoCAD can also read hundreds of PostScript fonts available in the marketplace.

The third consideration of the STYLE command is in how AutoCAD treats general physical properties of the characters, regardless of the font that is selected. These properties are the height, width-to-height ratio, obliquing angle, backwards, upside-down, and orientation (horizontal/vertical) options.

Invoke the STYLE command:

Format menu	Text Style
Command: prompt	**style** (ENTER)

AutoCAD displays the Text Style dialog box, similar to Figure 4–26.

Figure 4–26 *Text Style dialog box*

Choose the **New...** button to create a new style. AutoCAD displays the New Text Style dialog box shown in Figure 4–27. Enter the appropriate name for the text style and choose the OK button to create the new style.

Figure 4–27 *New Text Style dialog box*

To rename an existing style, first select the style from the **Style Name** list box in the New Text Style dialog box, and then click the **Rename...** button. AutoCAD displays the Rename Text Style dialog box. Make the necessary changes in the name of the style, and choose the **OK** button to rename the text style.

To delete an existing style, first select the style from the **Style Name** list box in the Text Style dialog box, and then click the **Delete** button. AutoCAD displays the AutoCAD Alert dialog box to confirm the deletion of the selected style. Click the **Yes** button to confirm the deletion or the **No** button to cancel the deletion of the selected style.

To assign a font to the selected text style, select the appropriate font from the **Font Name:** list box. Similarly, select a font style from the **Font Style:** list box. The font style specifies font character formatting, such as italic, bold, or regular.

The **Height:** text box sets the text height to the value you enter. If you set the height to 0 (zero), then when you use this style in the TEXT or MTEXT command, you are given an opportunity to change the text height with each occurrence of the command. If you set it to any other value, then that value will be used for this style and you will not be allowed to change the text height.

The **Upside down** and **Backwards** toggle buttons control whether the text is drawn right to left (with the characters backward) or upside down (left to right), respectively. See Figure 4–28.

Figure 4–28 *Examples of backward and upside-down text*

The **Vertical** toggle button controls the display of the characters aligned vertically. The Vertical option is available only if the selected font supports dual orientation. See Figure 4–29 for an example of vertically oriented text.

Figure 4–29 *Example of vertically oriented text*

The **Width Factor:** text box sets the character width relative to text height. If it is set to more than 1.0, the text widens; if set to less than 1.0, it narrows.

The **Oblique Angle:** text box sets the obliquing angle of the text. If it is set to 0 (zero) degrees, the text is drawn upright (or in AutoCAD, 90 degrees). A positive value slants the top of the characters toward the right, or in the clockwise direction. A negative value slants the characters in the counterclockwise direction. See Figure 4–30 for examples of oblique angle settings applied to a text string.

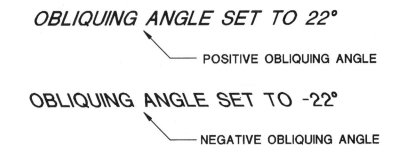

Figure 4–30 *Example of oblique angle settings applied to a text string*

The **Preview** section of the Text Style dialog box displays sample text that changes dynamically as you change fonts and modify the effects. To change the sample text, enter characters in the box below the larger preview image.

After making the necessary changes in the Text Style dialog box, choose the **Apply** button to apply the changes. Click the **Close** button to close the Text Style dialog box.

CREATING OBJECTS FROM EXISTING OBJECTS

AutoCAD not only allows you to draw objects easily, but also allows you to create additional objects from existing objects. This section discusses six important commands that will make your job easier: COPY, ARRAY, OFFSET, MIRROR, FILLET, and CHAMFER.

COPYING OBJECTS

The COPY command places copies of the selected objects at the specified displacement, leaving the original objects intact. The copies are oriented and scaled the same as the original. If necessary, you can make multiple copies of selected objects. Each resulting copy is completely independent of the original and can be edited and manipulated like any other object.

Invoke the COPY command:

Modify toolbar	Choose the COPY command (see Figure 4–31)
Modify menu	Choose Copy
Command: prompt	**copy** (ENTER)

Figure 4–31 *Invoking the* COPY *command from the Modify toolbar*

AutoCAD prompts:

> Command: **copy** (ENTER)
> Select objects: *(select the objects and press* ENTER *to complete the selection)*
> Specify base point or displacement, or [Multiple]: *(specify base point or right-click for shortcut menu and choose one of the available options)*
> Specify second point of displacement or <use first point as displacement>: *(specify a point for displacement, or press* ENTER *to use first point as displacement)*

You can use one or more object selection methods to select the objects. If you specify two data points, AutoCAD computes the displacement and places a copy accordingly. If you provide a null response to the second point of displacement, AutoCAD considers the point provided as the second point of a displacement vector with the origin (0,0,0) as the first point, indicating how far to copy the objects and in what direction.

The following command sequence shows an example of copying a group of objects selected by means of the Window option, as shown in Figure 4–32, by placing two data points:

Command: **copy** (ENTER)
Select objects: *(specify a point to place one corner of a window)*
Other corner: *(specify a point to place the opposite corner of the window)*
Select objects: (ENTER)
Specify base point or displacement, or [Multiple]: *(specify the base point as shown in Figure 4–32)*
Specify second point of displacement or <use first point as displacement>: *(specify the second point as shown in Figure 4–32)*

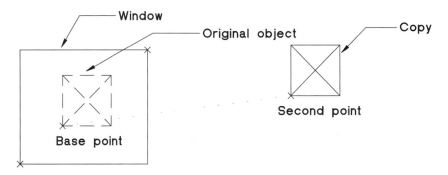

Figure 4–32 *Using the COPY Window option to copy a group of objects by specifying two data points*

Creating Multiple Copies

To make multiple copies, first invoke the COPY command, and respond to the "Specify second point of displacement or <use first point as displacement>:" prompt by entering **m**, for multiple, or right-click and choose Multiple from the shortcut menu. The "Specify base point" prompt then reappears, followed by repeated "Specify second point of displacement or <use first point as displacement>:" prompts, and a copy of the selected objects is made at a location determined by each displacement you enter. Each displacement is relative to the original base point. When you have made all the copies you need, give a null response to the "Specify second point of displacement or <use first point as displacement>:" prompt to terminate the command sequence.

The following command sequence shows an example of placing multiple copies of a group of objects selected by means of the Window option, as shown in Figure 4–33.

Command: **copy** (ENTER)
Select objects: *(specify a point to place one corner of the window)*
Other corner: *(specify a point to place the opposite corner of the window)*
Select objects: (ENTER)

Specify base point or displacement, or [Multiple]: (right-click and choose
    Multiple from shortcut menu and choose one of the available options)
Specify base point: *(select base point as shown in Figure 4–33)*
Specify second point of displacement or <use first point as displacement>:
    (specify the second point for Copy 1 as shown in Figure 4–33)
Specify second point of displacement or <use first point as displacement>:
    (specify the second point for Copy 2 as shown in Figure 4–33)
Specify second point of displacement or <use first point as displacement>:
    (specify the second point for Copy 3 as shown in Figure 4–33)
Specify second point of displacement or <use first point as displacement>:
    (ENTER)

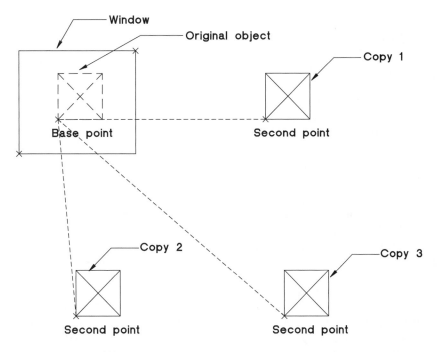

Figure 4–33 *Using the COPY Window option to make multiple copies of a group of objects*

CREATING A PATTERN OF COPIES

The ARRAY command is used to make multiple copies of selected objects in either rectangular or polar arrays (patterns). In the rectangular array, you can specify the number of rows, the number of columns, and the spacing between rows and columns (row and column spacing may differ). The whole rectangular array can be rotated at a selected angle. In the polar array, you can specify the angular intervals, the number of copies, the angle that the group covers, and whether or not the objects maintains their orientation as they are arrayed.

Rectangular Array

Invoke the ARRAY command:

Modify toolbar	Choose the ARRAY command (see Figure 4–34)
Modify menu	Choose Array
Command: prompt	**array** (ENTER)

Figure 4–34 *Invoking the ARRAY command from the Modify toolbar*

AutoCAD prompts:

> Command: **array** (ENTER)
> Select objects: *(select the objects and then give a null response to complete the selection)*
> Enter the type of array [Rectangular/Polar] <R>: *(choose **r** for rectangular array or right-click for shortcut menu and choose one of the available options)*
> Enter the number of rows (—) <1>: *(specify a nonzero integer for number of rows)*
> Enter the number of columns (||||) <1> *(specify a nonzero integer for number of columns)*
> Enter the distance between rows or specify unit cell (—):*(specify distance between rows, or specify two points to measure the distance between rows)*
> Specify the distance between columns (||||):*(specify a distance between columns, or specify two points to measure the distance between columns)*

Any combination of whole numbers of rows and columns may be entered (except both 1 row and 1 column, which would not create any copies). AutoCAD includes the original object in the number you enter. Row and column spaces can be different from each other. They can be entered separately when prompted, or you can select two points that specify the opposite corners of a rectangle called a unit cell. AutoCAD uses the width of the unit cell as the horizontal distance(s) between columns and the height as the vertical distance(s) between rows.

A positive number for the column and row spacing causes the elements to array toward the right and upward, respectively. Negative numbers for the column and row spacing cause the elements to array toward the left and downward, respectively.

AutoCAD creates rectangular arrays along a baseline defined by the current snap rotation. By default, the snap rotation is set to 0 degrees, so that rows and columns are orthogonal with respect to the *X* and *Y* drawing axes. The **Rotate** option of the SNAP command allows you to change the rotation angle and creates a rotated array.

The following command sequence shows an example of placing a rectangular array with 4 rows and 6 columns, as shown in Figure 4–35.

> Command: **array** (ENTER)
> Select objects: *(select objects)*
> Enter the type of array [Rectangular/Polar] <R>: **r** (ENTER)
> Enter the number of rows (—) <1>: **4** (ENTER)
> Enter the number of columns (||||) <1> **6** (ENTER)
> Enter the distance between rows or specify unit cell (—): **1** (ENTER)
> Specify the distance between columns (||||): **1.5** (ENTER)

Figure 4–35 *Using the* ARRAY *command to place a rectangular array*

Polar Array

Invoke the ARRAY command:

Modify toolbar	Choose the ARRAY command (see Figure 4–36)
Modify menu	Choose Array
Command: prompt	**array** (ENTER)

Figure 4–36 *Invoking the* ARRAY *command from the Modify toolbar*

AutoCAD prompts:

Command: **array** (ENTER)
Select objects: *(select the objects and then give a null response to complete the selection)*
Enter the type of array [Rectangular/Polar] <R>: *(choose **p** for polar array or right-click for shortcut menu and choose one of the available options)*
Specify center point of array: *(specify a point around which you want the array to form)*
Enter the number of items in the array: *(specify a positive integer for number of items in the array, include the original item or press ENTER)*
Specify the angle to fill (+=ccw, -=cw) <360>: *(specify an angle to fill or press ENTER)*
Specify Angle between items (+=CCW, -=CW): *(specify an angle between items; this prompt appears only if a null response is provided for Angle to fill)*
Rotate arrayed objects? [Yes/No] <Y>: *(choose **y**, for yes to Rotate as they are copied, or **n**, for no to Rotate as they are copied)*

If you specify the number of items for an array, you must specify either the angle to fill or the angle between items. If you provide a null response to the number of items, then you must specify the angle to fill and the angle between items.

The following command sequence presents an example of placing a rotated polar array as shown in Figure 4–37.

Command: **array** (ENTER)
Select objects: *(select objects)*
Enter the type of array [Rectangular/Polar] <R>: **p** (ENTER)
Specify center point of array: *(specify the center point as shown in Figure 4–37)*
Enter the number of items in the array: **8** (ENTER)
Specify the angle to fill (+=ccw, -=cw) <360>: **360** (ENTER)
Rotate arrayed objects? [Yes/No] <Y>: (ENTER)

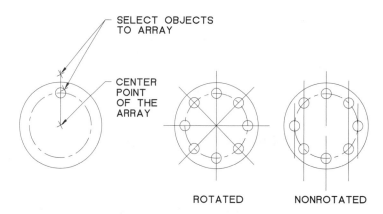

Figure 4–37 *Using the ARRAY command to place rotated and nonrotated polar arrays*

Figure 4–37 shows both non-rotated and rotated polar arrays.

CREATING PARALLEL LINES, PARALLEL CURVES, AND CONCENTRIC CIRCLES

The OFFSET command creates parallel lines, parallel curves, and concentric circles relative to existing objects, as shown in Figure 4–38. Special precautions must be taken when using the OFFSET command to prevent unpredictable results from occurring when using the command on arbitrary curve/line combinations in polylines.

Figure 4–38 *Examples created using the* OFFSET *command*

Invoke the OFFSET command:

Modify toolbar	Choose the OFFSET command (see Figure 4–39)
Modify menu	Choose Offset
Command: prompt	**offset** (ENTER)

Figure 4–39 *Invoking the* OFFSET *command from the Modify toolbar*

AutoCAD prompts:

> Command: **offset** (ENTER)
> Specify offset distance or [Through] <1.0000>: *(specify offset distance, or right-click for shortcut menu and choose one of the available options)*

Select object to offset or <exit>: *(select an object to offset)*

Specify point on side to offset: *(specify a point to one side of the object to offset)*

Select object to offset or <exit>: *(continue selecting additional objects for offset, and specify the side of the object to offset, or press* ENTER *to terminate the command sequence)*

Instead of specifying the offset distance, select the **Through** option, and AutoCAD prompts for a through point. Specify a point, and AutoCAD creates an object passing through the specified point.

Valid Objects to Offset

Valid objects include the line, spline curve, arc, circle, and 2D polyline. If you select another type of object, such as text, you will get the following error message:

Cannot offset that object.

The object selected for offsetting must be in a plane parallel to the current coordinate system. Otherwise you will get the following error message:

Object not parallel with UCS.

Offsetting Miters and Tangencies

The OFFSET command affects single objects in a manner different from a polyline made up of the same objects. Polylines whose arcs join lines and other arcs in a tangent manner are affected differently than polylines with nontangent connecting points. For example, in Figure 4–40 the seven lines are separate objects. When you specify side to offset as shown, there are gaps and overlaps at the ends of the newly created lines.

Figure 4–40 *Using the* OFFSET *command with single objects*

In Figure 4–41, the lines have been joined together (see PEDIT in Chapter 5) as a single polyline. See how the OFFSET command affects the corners where the new polyline segments join.

Figure 4–41 *Using the* OFFSET *command with polylines*

 Note: The results of offsetting polylines with arc segments that connect other arc segments and/or line segments in dissimilar (nontangent) directions might be unpredictable. Examples of offsetting such polylines are shown in Figure 4–42.

Figure 4–42 *Using the* OFFSET *command with nontangent arc and/or line segments*

If you are not satisfied with the resulting new polyline configuration, you can use the PEDIT command to edit it. Or, you can explode the polyline and edit the individual segments.

CREATING A MIRROR COPY OF OBJECTS

The MIRROR command creates a copy of selected objects in reverse, that is, mirrored about a specified line. Invoke the MIRROR command:

Modify toolbar	Choose the MIRROR command (see Figure 4–43)
Modify menu	Choose Mirror
Command: prompt	**mirror** (ENTER)

Figure 4–43 *Invoking the MIRROR command from the Modify toolbar*

AutoCAD prompts:

> Command: **mirror** (ENTER)
> Select objects: *(select the objects and then give a null response to complete the selection)*
> Specify first point of mirror line: *(specify a point to define the first point of the mirror line)*
> Specify second point of mirror line: *(specify a point to define the second point of the mirror line)*
> Delete source objects? [Yes/No] <N>: *(enter **y**, for yes to delete the original objects, or **n**, not to delete the original objects, that is, to retain them)*

The first and second points of the mirror line become the endpoints of an invisible line about which the selected objects will be mirrored.

The following command sequence shows an example of mirroring a group of selected objects by means of the Window option, as shown in Figure 4–44:

> Command: **mirror** (ENTER)
> Select objects: *(specify POINT 1 to place one corner of a window)*
> Second corner: *(specify POINT 2 to place the opposite corner of the window)*
> Select objects: (ENTER)
> Specify first point of mirror line: *(specify POINT 3, as shown in Figure 4–44)*
> Specify second point of mirror line: *(specify POINT 4, as shown in Figure 4–44)*
> Delete source objects? [Yes/No] <N>: (ENTER)

The text as mirrored is located relative to other objects within the selected group. But the text will or will not retain its original orientation, depending on the setting of the system variable called MIRRTEXT. If the value of MIRRTEXT is set to 1, then text items in the selected group will have their orientations and location mirrored. That is, if their characters were normal and they read left to right in the original group, in the mirrored copy they will read right to left and the characters will be backwards. If MIRRTEXT is set to 0 (zero), then the text strings in the group will have their locations mirrored, but the individual text strings will retain their normal, left-to-right, character appearance. The MIRRTEXT system variable, like other system variables, is changed by the SETVAR command or by typing MIRRTEXT at the "Command:" prompt, as follows:

Command: **mirrtext** (ENTER)
Enter New Value for MIRRTEXT <1>: **0** (ENTER)

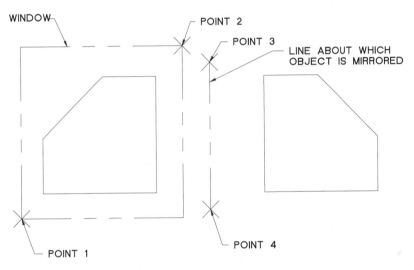

Figure 4–44 *Mirroring a group of objects selected by means of the Window option*

This setting causes mirrored text to retain its readability. Figures 4–45 and 4–46 show the result of the MIRROR command when the MIRRTEXT variable is set to 1 and 0, respectively.

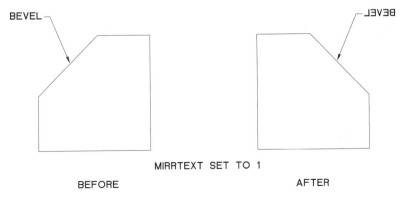

Figure 4–45 *The* MIRROR *command with the* MIRRTEXT *variable set to 1*

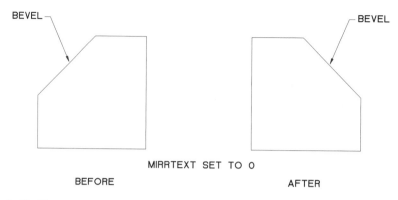

Figure 4–46 *The* MIRROR *command with the* MIRRTEXT *variable set to 0*

CREATING A FILLET BETWEEN TWO OBJECTS

The FILLET command fillets (rounds) the intersecting ends of two arcs, circles, lines, elliptical arcs, polylines, rays, xlines, or splines with an arc of a specified radius.

If the TRIMMODE system variable is set to 1 (default), then the FILLET command trims the intersecting lines to the endpoints of the fillet arc. And if TRIMMODE is set to 0 (zero), then the FILLET command leaves the intersecting lines at the endpoints of the fillet arc.

Invoke the FILLET command:

Modify toolbar	Choose the FILLET command (see Figure 4–47)
Modify menu	Choose Fillet
Command: prompt	**fillet** (ENTER)

Figure 4–47 *Invoking the* FILLET *command from the Modify toolbar*

AutoCAD prompts:

> Command: **fillet** (ENTER)
> Current settings: Mode = TRIM, Radius = <current>
> Select first object or [Polyline/Radius/Trim]: *(select one of the two objects to fillet, or right-click for the shortcut menu and choose one of the available options)*

By default, AutoCAD prompts you to select an object. If you select an object to fillet, then AutoCAD prompts:

> Select second object: *(select the second object to fillet)*

AutoCAD joins the two objects with an arc having the specified radius. If the objects selected to be filleted are on the same layer, AutoCAD creates the fillet arc on the same layer. If not, AutoCAD creates the fillet arc on the current layer.

AutoCAD allows you to draw a fillet between parallel lines, xlines, and rays. The first selected object must be a line or ray, but the second object can be a line, xline, or ray. The diameter of the fillet arc is always equal to the distance between the lines. The current fillet radius is ignored and remains unchanged.

Radius Option

The Radius option allows you to change the current fillet radius. The following command sequence sets the fillet radius to 0.5 and draws the fillet between two lines, as shown in Figure 4–48.

```
Command: fillet (ENTER)
Current settings: Mode = TRIM, Radius = <current>
Select first object or [Polyline/Radius/Trim]: r (ENTER)
Specify fillet radius <current>: 0.25 (ENTER)

Command: (ENTER) (to repeat the FILLET command)
Current settings: Mode = TRIM, Radius = 0.25
Select first object or [Polyline/Radius/Trim]: (select one of the lines, as shown
    in Figure 4–48)
Select second object: (select the other line to fillet, as shown in Figure 4–48)
```

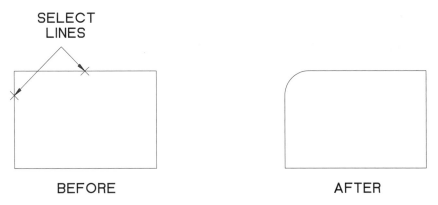

Figure 4–48 *Fillet drawn with a radius of 0.25*

If you select lines or arcs, AutoCAD extends these lines or arcs until they intersect or trims them at the intersection, keeping the selected segments if they cross. The following command sequence sets the fillet radius to 0 and draws the fillet between two lines as shown in Figure 4–49.

Command: **fillet** (ENTER)
Current settings: Mode = TRIM, Radius = <current>
Select first object or [Polyline/Radius/Trim]: **r** (ENTER)
Specify fillet radius <current>: 0 (ENTER)

Command: (ENTER) *(to repeat the* FILLET *command)*
Current settings: Mode = TRIM, Radius = 0.00
Select first object or [Polyline/Radius/Trim]: *(select the first object, as shown in Figure 4–49)*
Select second object: *(select the second object to fillet as shown in Figure 4–49)*

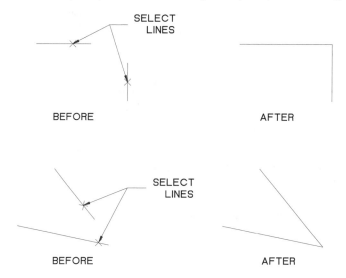

Figure 4–49 *Fillet drawn with a radius of 0*

Polyline Option

With the polyline option, AutoCAD draws fillet arcs at each vertex of a 2D polyline where two line segments meet. The following command sequence sets the fillet radius to 0.5 and draws the fillet at each vertex of a 2D polyline, as shown in Figure 4–50:

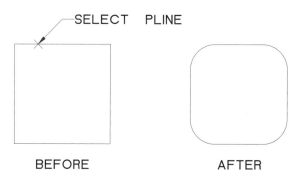

Figure 4–50 *Fillet with a radius of 0.5 drawn to a polyline*

Command: **fillet** (ENTER)
Current settings: Mode = TRIM, Radius = <current>
Select first object or [Polyline/Radius/Trim]: **r** (ENTER)
Specify fillet radius <current>: 0.5 (ENTER)

Command: (ENTER) *(to repeat the* FILLET *command)*
Current settings: Mode = TRIM, Radius = 0.50
Select first object or [Polyline/Radius/Trim]: **p** (ENTER)
Select 2D polyline: *(select the polyline as shown in Figure 4–50)*

Trim Option

The Trim option (Trim/No Trim) controls whether or not AutoCAD trims the selected edges to the fillet arc endpoints. This option is similar to setting the TRIMMODE system variable from 1 to 0 or 0 to 1.

CREATING A CHAMFER BETWEEN TWO OBJECTS

The CHAMFER command allows you to draw an angled corner between two lines. The size of the chamfer is determined by the settings of the first and the second chamfer distances. If it is to be a 45-degree chamfer for perpendicular lines, then the two distances are set to the same value.

If the TRIMMODE system variable is set to 1 (default), then the CHAMFER command trims the intersecting lines to the endpoints of the chamfer line. And if TRIMMODE is set to 0 (zero), then the CHAMFER command leaves the intersecting lines at the endpoints of the chamfer line.

Invoke the CHAMFER command:

Modify toolbar	Choose the CHAMFER command (see Figure 4–51)
Modify menu	Choose Chamfer
Command: prompt	**chamfer** (ENTER)

Figure 4–51 *Invoking the* CHAMFER *command from the Modify toolbar*

AutoCAD prompts:

Command: **chamfer** (ENTER)
(TRIM mode) Current chamfer Dist1 = <current>, Dist2 = <current>
Select first line or [Polyline/Distance/Angle/Trim/Method]: *(select one of the two lines to chamfer, or right-click for the shortcut menu and choose one of the available options)*

By default, AutoCAD prompts you to select the first line to chamfer. If you select a line to chamfer, then AutoCAD prompts:

Select second line: *(select the second line to chamfer)*

AutoCAD draws a chamfer to the selected lines. If the selected lines to be chamfered are on the same layer, AutoCAD creates the chamfer on the same layer. If not, AutoCAD creates the chamfer on the current layer.

Distance Option

The Distance option allows you to set the first and second chamfer distances. The following command sequence sets the first chamfer and second chamfer distance to 0.5 and 1.0, respectively, and draws the chamfer between two lines as shown in Figure 4–52.

```
Command: chamfer (ENTER)
(TRIM mode) Current chamfer Dist1 = <current>, Dist2 = <current>
Select first line or [Polyline/Distance/Angle/Trim/Method]: d (ENTER)
Specify first chamfer distance <0.5000>: 0.5 (ENTER)
Specify second chamfer distance <0.5000>: 1.0 (ENTER)

Command: (ENTER) (to repeat the CHAMFER command)
(TRIM mode) Current chamfer Dist1 = 0.5000, Dist2 = 1.0000
Select first line or [Polyline/Distance/Angle/Trim/Method]: (select the first
    line, as shown in Figure 4–52)
Select second line: (select the second line, as shown in Figure 4–52)
```

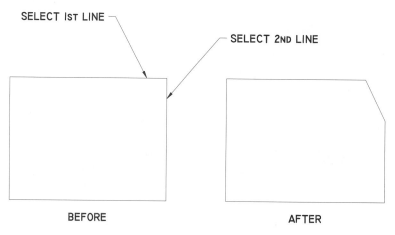

SELECT 1ST LINE

SELECT 2ND LINE

BEFORE   AFTER

Figure 4–52 *Chamfer drawn with distances of 0.5 and 1.0*

Polyline Option

With the Polyline option, AutoCAD draws chamfers at each vertex of a 2D polyline where two line segments meet. The following command sequence sets the chamfer distances to 0.5 and draws the chamfer at each vertex of a 2D polyline, as shown in Figure 4–53.

```
Command: chamfer (ENTER)
(TRIM mode) Current chamfer Dist1 = <current>, Dist2 = <current>
Select first line or [Polyline/Distance/Angle/Trim/Method]: d (ENTER)
Specify first chamfer distance <0.5000>: 0.5 (ENTER)
Specify second chamfer distance <0.5000>: 0.5 (ENTER)

Command: (ENTER) (to repeat the CHAMFER command)
(TRIM mode) Current chamfer Dist1 = 0.5000, Dist2 = 0.5000
Select first line or [Polyline/Distance/Angle/Trim/Method]: p (ENTER)
Select 2D polyline: (select the polyline, as shown in Figure 4–53)
```

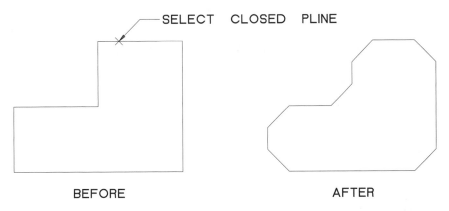

Figure 4–53 *Chamfer with distances of 0.5 drawn on a polyline*

Angle Option

The Angle option is similar to the Distance option, but instead of prompting for the first and second chamfer distances, AutoCAD prompts for the first chamfer distance and an angle from the first line. This is another method by which to create the chamfer line.

Method Option

The Method option controls whether AutoCAD uses two distances or a distance and an angle to create the chamfer line.

Trim Option

The Trim option (Trim/No Trim) controls whether or not AutoCAD trims the selected edges to the chamfer line endpoints. This option is similar to setting the TRIMMODE system variable from 1 to 0 or from 0 to 1.

 Note: The CHAMFER command set to zero distance operates the same way the FILLET command operates set to zero radius.

MODIFYING OBJECTS

In this section, four additional MODIFY commands are explained: Move, Trim, Break, and Extend. (The ERASE command was explained in Chapter 2.)

MOVING OBJECTS

The MOVE command lets you move one or more objects from their present location to a new one without changing orientation or size.

Invoke the MOVE command:

Modify toolbar	Choose the MOVE command (see Figure 4–54)
Modify menu	Choose Move
Command: prompt	**move** (ENTER)

Figure 4–54 *Invoking the* MOVE *command from the Modify toolbar*

AutoCAD prompts:

> Command: **move** (ENTER)
> Select objects: *(select the objects)*
> Specify base point or displacement: *(specify a point)*
> Specify second point of displacement: *(specify a point for displacement, or press* ENTER*)*

You can use one or more object selection methods to select the objects. If you specify two data points, AutoCAD computes the displacement and moves the selected objects accordingly. If you specify the points on the screen, AutoCAD assists you in visualizing the displacement by drawing a rubber-band line from the first point as you move the crosshairs to the second point. If you provide a null response by pressing ENTER to the second point of displacement, then AutoCAD interprets the base point as relative X,Y,Z displacement.

The following command sequence shows an example of moving a group of objects, selected by means of the Window option, by relative displacement, as shown in Figure 4–55.

> Command: **move** (ENTER)
> Select objects: *(specify Point 1 to place one corner of a window)*
> Other corner: *(specify Point 2 to place the opposite corner of the window)*

Select objects: (ENTER)
Specify base point or displacement: **2,3** (ENTER)
Specify second point of displacement: (ENTER)

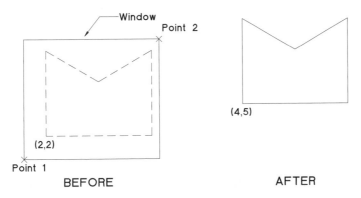

Figure 4–55 *Using the Window option of the* MOVE *command to move a group of objects by means of vector displacement*

The following command sequence shows an example of moving a group of objects selected by the Window option and moving the objects by specifying two data points, as shown in Figure 4–56.

Command: **move** (ENTER)
Select objects: *(pick Point 1 to place one corner for a window)*
Other corner: *(pick Point 2 to place the opposite corner of the window)*
Select objects: (ENTER)
Specify base point or displacement: *(specify base point)*
Specify second point of displacement: *(specify second point)*

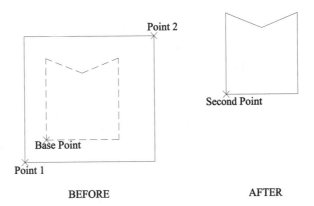

Figure 4–56 *Using the Window option of the* MOVE *command to move a group of objects by specifying two data points*

TRIMMING OBJECTS

The TRIM command is used to trim the portion of the object(s) that is drawn past a cutting edge or from an implied intersection defined by other objects. Objects that can be trimmed include lines, arcs, elliptical arcs, circles, 2D and 3D polylines, xlines, rays, and splines. Valid cutting edge objects include lines, arcs, circles, ellipses, 2D and 3D polylines, floating viewports, xlines, rays, regions, splines, and text.

Invoke the TRIM command:

Modify toolbar	Choose the TRIM command (see Figure 4–57)
Modify menu	Choose Trim
Command: prompt	**trim** (ENTER)

Figure 4–57 *Invoking the* TRIM *command from the Modify toolbar*

AutoCAD prompts:

> Command: **trim** (ENTER)
> Select cutting edge(s)...
> Select objects: *(select the objects and press* ENTER *to terminate the selection of the cutting edges)*
> Select object to trim or [Project/Edge/Undo]: *(select object(s) to trim or right-click for shortcut menu and choose one of the available options)*

The TRIM command initially prompts you to "Select cutting edge(s):". After selecting one or more cutting edges to trim, press ENTER. You are then prompted to "Select object to trim:". Select one or more objects to trim and then press ENTER to terminate the command. If you press ENTER in response to "Select cutting edge(s)" prompt without selecting any objects, by default AutoCAD selects all the objects.

 Note: Don't forget to press ENTER after selecting the cutting edge(s). Otherwise, the program will not respond as expected. In fact, trim continues expecting more cutting edges until you terminate the cutting edge selecting mode.

Edge Option

The Edge option determines whether objects that extend past a selected cutting edge or to an implied intersection are trimmed. AutoCAD prompts as follows when the Edge option is selected:

> Enter an implied edge extension mode [Extend/No extend] <current>:
> *(right-click for the shortcut menu and select one of the two available options)*

The **Extend** selection extends the cutting edge along its natural path to intersect an object in 3D (implied intersection).

The **No Extend** selection specifies that the object is to be trimmed only at a cutting edge that intersects it in 3D space.

Undo Option

The Undo option reverses the most recent change made by TRIM command.

Project Option

The Project option specifies the projection mode and AutoCAD uses when trimming objects. By default, it is set to the current UCS.

Figure 4–58 shows examples of using the TRIM command.

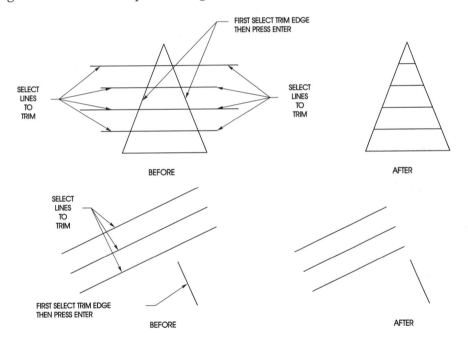

Figure 4–58 *Examples of using the* TRIM *command*

ERASING PARTS OF OBJECTS

The BREAK command is used to remove parts of objects or to split an object in two parts, and it can be used on lines, xlines, rays, arcs, circles, ellipses, splines, donuts, traces, and 2D and 3D polylines.

Invoke the BREAK command:

Modify toolbar	Choose the BREAK command (see Figure 4–59)
Modify menu	Choose Break
Command: prompt	**break** (ENTER)

Figure 4–59 *Invoking the* BREAK *command from the Modify toolbar*

AutoCAD prompts:

> Command: **break** (ENTER)
> Select object: *(select an object)*
> Specify second break point or [First point]: *(specify the second break point, or right-click for shortcut menu and choose one of the available options)*

See Figure 4–60 for examples of applications of the BREAK command.

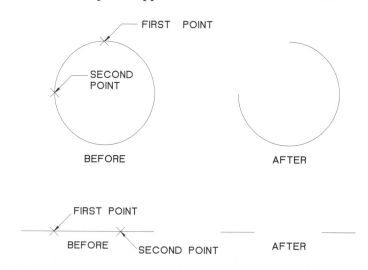

Figure 4–60 *Examples of applications of the* BREAK *command*

AutoCAD erases the portion of the object between the first point (the point where the object was selected) and second point. If the second point is not on the object, then AutoCAD selects the nearest point on the object. If you need to erase an object to one end of a line, arc, or polyline, then specify the second point beyond the end to be removed.

If instead of specifying the second point, you choose First point from the shortcut menu, AutoCAD prompts for the first point and then for the second point.

An object can be split into two parts without removing any portion of the object by selecting the same point as the first and second points. You can do so by entering @ to specify the second point.

If you select a circle, then AutoCAD converts it to an arc by erasing a piece, moving counterclockwise from the first point to the second point. For a closed polyline, the part is removed between two selected points, moving in direction from the first to the last vertex. And in the case of 2D polylines and traces with width, the BREAK command will produce square ends at the break points.

EXTENDING OBJECTS TO MEET ANOTHER OBJECT

The EXTEND command is used to change one or both endpoints of selected lines, arcs, elliptical arcs, open 2D and 3D polylines, and rays to extend to lines, arcs, elliptical arcs, circles, ellipses, 2D and 3D polylines, rays, xlines, regions, splines, text string, or floating viewports.

Invoke the EXTEND command:

Modify toolbar	Choose the EXTEND command (see Figure 4–61)
Modify menu	Choose Extend
Command: prompt	**extend** (ENTER)

Figure 4–61 *Invoking the* EXTEND *command from the Modify toolbar*

AutoCAD prompts:

> Command: **extend** (ENTER)
> Select boundary edges...
> Select objects: *(select the objects and then press* ENTER *to complete the selection)*
> Select object to extend or [Project/Edge/Undo]: *(select the object(s) to extend and press* ENTER *to terminate the selection process or right-click for shortcut menu and choose one of the available options)*

The EXTEND command initially prompts you to "Select boundary edges:". After selecting one or more boundary edges, press ENTER to terminate the selection process. Then AutoCAD prompts you to "Select object to extend:" (default option). Select one or more objects to extend to the selected boundary edges. After selecting the required objects to extend, press ENTER to complete the selection process.

The EXTEND and TRIM commands are very similar in this method of selecting. With EXTEND you are prompted to select the boundary edge to extend to; with TRIM you are prompted to select a cutting edge. If you press ENTER key in response to "Select boundary edge(s)" prompt with out selecting any objects, by default AutoCAD selects all the objects.

Edge Option

The Edge option determines whether objects are extended past a selected boundary or to an implied edge. AutoCAD prompts as follows when the Edge option is selected:

Enter an implied edge extension mode [Extend/No extend] <current>:
(right-click for the shortcut menu and select one of the two available options)

The **Extend** selection extends the boundary object along its natural path to intersect another object in 3D space (implied edge).

The **No Extend** selection specifies that the object is to extend only to a boundary object that actually intersects it in 3D space.

Undo Option

The Undo option reverses the most recent change made by the EXTEND command.

Project Option

The Project option specifies the projection and AutoCAD uses when trimming objects. By default, it is set to the current UCS.

Figure 4–62 shows examples of the use of the EXTEND command.

Figure 4–62 *Examples of applications of the* EXTEND *command*

PROJECT EXERCISE

This project exercise provides point-by-point instructions for setting up the drawing with layers and then creating the objects shown in Figure P4–1.

Figure P4–1 *Completed project drawing*

In this project you will learn how to do the following:

- Set up the drawing, including Units, Limits and Layers
- Use the LINE, RECTANGLE, POLYGON, and CIRCLE commands to create objects
- Use the Osnap modes of int, tan, ttr, and per
- Use the TRIM, OFFSET, BREAK, MIRROR, and ERASE commands to modify objects

SET UP THE DRAWING AND DRAW A BORDER

Step 1: Start the AutoCAD program.

Step 2: To create a new drawing, invoke the NEW command from the Standard toolbar or select New from the File menu.

AutoCAD displays the Create New Drawing dialog box. Click the "Use a Wizard" button, select the Quick Setup, and choose the OK button to close the dialog box.

AutoCAD displays the Quick Setup dialog box with the display of the Units section. Select the "Decimal" radio button and choose the **Next>** button.

AutoCAD displays the Quick Setup dialog box with the display of the Area section. Under the **Width:** text box enter 18, and under the **Length** text box enter 12. Choose the **Finish** button to close the Quick Setup dialog box.

Step 3: Invoke the LAYER command from the Object Properties toolbar, or select Layer... from the Format menu. AutoCAD displays the Layer Properties Manager dialog box.

Create three layers, and rename them as shown in the following table, assigning appropriate color and linetype.

Layer Name	Color	Linetype
Border	Red	Continuous
Centerline	Green	Center
Object	Blue	Continuous

Set Border as the current layer, and close the Layer Properties Manager dialog box.

Step 4: Open the Drafting Settings dialog box from the Tools menu, and set Grid to **0.5** and Snap to **0.5**, and then set the grid to ON.

Step 5: Invoke the RECTANGLE command to draw the border (17" by 11"), as shown in Figure P4–2.

Command: **rectangle** (ENTER)
Specify first corner point or [Chamfer/Elevation/Fillet/Thickness/
Width]: **.5,.5** (ENTER)
Specify other corner point: **@17,11** (ENTER)

.5,.5

Figure P4–2 *Border for a mechanical drawing*

Step 6: Begin the layout of the drawing by drawing the centerlines as shown in Figure P4–3. Set the CENTERLINE layer as the current layer. Invoke the LINE command from the Draw toolbar and draw lines 1, 2, 3, and 4.

> Command: **line** (ENTER)
> Specify first point: **4,8** (ENTER)
> Specify next point or [Undo]: **@8.5<0** (ENTER)
> Specify next point or [Undo]: (ENTER)
> Command: (ENTER)
>
> Specify first point: **7,5.5** (ENTER)
> Specify next point or [Undo]: **@5<90** (ENTER)
> Specify next point or [Undo]: (ENTER)
> Command: (ENTER)
>
> Specify first point: **7,8** (ENTER)
> Specify next point or [Undo]: **@3.5<225** (ENTER)
> Specify next point or [Undo]: (ENTER)
>
> Command: (ENTER)
> Specify first point: **7,8** (ENTER)
> Specify next point or [Undo]: **@6.5<315** (ENTER)
> Specify next point or [Undo]: (ENTER)

Figure P4–3 *Placement of centerlines*

Step 7: Use the OFFSET command to construct the additional centerlines necessary for completion of the project. Invoke the OFFSET command from the Modify toolbar, and construct centerlines 2A, 3A, 3B, and 4A, as shown in Figure P4–4.

project EXERCISE

Figure P4–4 *Placement of additional centerlines*

Command: **offset** (ENTER)
Specify offset distance or [Through] <1.0000>: **4**
Select object to offset or <exit>: *(select LINE 2)*
Specify point on side to offset: *(specify to the right of LINE 2 to construct LINE 2A, as shown in Figure P4–4)*
Select object to offset or <exit>: (ENTER)

Command: (ENTER)
OFFSET
Specify offset distance or [Through] <4.0000>: **3** (ENTER)
Select object to offset or <exit>: *(select LINE 3)*
Specify point on side to offset: *(specify a point to the right of line 3 to construct LINE 3A, as shown in Figure P4–4)*
Select object to offset or <exit>: (ENTER)

Command: (ENTER)
OFFSET
Specify offset distance or [Through] <3.0000>: **2** (ENTER)
Select object to offset or <exit>: *(select line 3A)*
Specify point on side to offset: *(specify a point to the right of line 3A to construct LINE 3B, as shown in Figure P4–4)*
Select object to offset or <exit>: *(select LINE 4)*
Specify point on side to offset: *(specify a point below LINE 4 to construct line 4A, as shown in Figure P4–4)*
Select object to offset or <exit>: (ENTER)

Step 8: Erase lines 3 and 4, as shown in Figure P4–4, by invoking the ERASE command from the Modify toolbar.

> Command: **erase** (ENTER)
> Select objects: *(select lines 3 and 4)*

Your drawing will appear as shown in Figure P4–5.

Figure P4–5 *Drawing after lines 3 and 4 have been erased*

Step 9: The intersection of the centerlines you laid out in the previous steps will provide the center points for the circles. Set the OBJECT layer as the current layer. Invoke the CIRCLE command from the Draw toolbar and draw the three circles shown in Figure P4–6.

Figure P4–6 *Placement of circles*

Command: **circle** (ENTER)
CIRCLE Specify center point for circle or [3P/2P/Ttr (tan tan radius)]:
   7,8 (ENTER)
Specify radius of circle or [Diameter]: **2** (ENTER)

Command: (ENTER)
CIRCLE Specify center point for circle or [3P/2P/Ttr (tan tan radius)]:
   11,8 (ENTER)
Specify radius of circle or [Diameter] <2.0000>: **1** (ENTER)

Command: (ENTER)
CIRCLE Specify center point for circle or [3P/2P/Ttr (tan tan radius)]:
   (use the Object snap tool "intersection" to snap to the intersection of lines
   3B and 4A to identify the center of circle 3)
Specify radius of circle or [Diameter] <1.0000>: (ENTER)

Step 10: Invoke the LINE command from the Draw toolbar, and draw LINE 5 tangent to
two circles (use the Osnap tool "tangent"), as shown in Figure P4–7.

Command: **line** (ENTER)
Specify first point: *(invoke the tan object snap and select the upper part of*
   CIRCLE 1)
Specify next point or [Undo]: *(invoke the tan object snap and select the*
   upper part of CIRCLE 2)
Specify next point or [Undo]: *(press ENTER to terminate the command*
   sequence)

Figure P4–7 *Lines 5 and 6 drawn tangent to two circles*

Step 11:   Invoke the MIRROR command to create line 6 from line 5 as shown in Figure P4–7.

> Command: **mirror** (ENTER)
> Select objects: *(select LINE 5)*
> Select objects: *(press ENTER to terminate the selection of objects)*
> Specify first point of mirror line: **7,8** (ENTER)
> Specify second point of mirror line: **11,8** (ENTER)
> Delete source objects? [Yes/No] <N>: (ENTER)

Step 12:   Invoke the LINE command and draw LINE 1A using the Object snap modes of intersection and perpendicular as shown in Figure P4–8.

> Command: **line** (ENTER)
> Specify first point: *(invoke the intersection object snap, and select the intersection of LINE 3B and CIRCLE 3)*
> Specify next point or [Undo]: *(invoke the perpendicular object snap, and select LINE 3A)*
> Specify next point or [Undo]: *(press* (ENTER) *to terminate the command sequence)*

Your drawing will appear as shown in Figure P4–8.

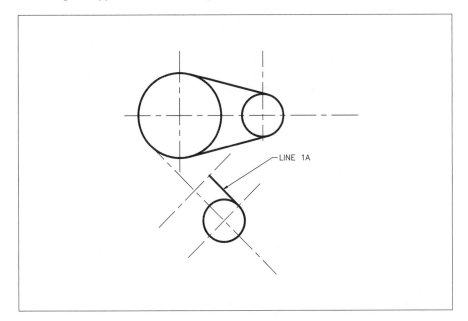

Figure P4–8 *Drawing after drawing LINE 1A*

Step 13:   Draw two circles, CIRCLE 4 and CIRCLE 5, as shown in Figure P4–9, by invoking the CIRCLE command's Tan Tan Radius option.

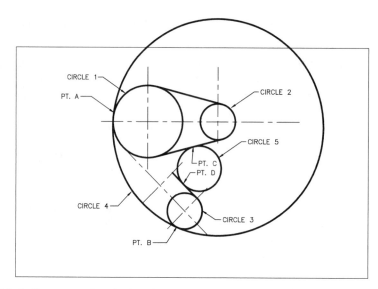

Figure P4–9 *Placement of circle 4 and circle 5 tangent to two objects*

> Command: **circle** (ENTER)
> Specify center point for circle or [3P/2P/Ttr (tan tan radius)]: **t** (ENTER)
> Specify point on object for first tangent of circle: *(select CIRCLE 1 at point A, as shown in Figure P4–9)*
> Specify point on object for second tangent of circle: *(select CIRCLE 3 at point B, as shown in Figure P4–9)*
> Specify radius of circle: **6** (ENTER)

Invoke the CIRCLE command again:

> Command: **circle** (ENTER)
> Specify center point for circle or [3P/2P/Ttr (tan tan radius)]: **t** (ENTER)
> Specify point on object for first tangent of circle: *(select the line at point C, as shown in Figure P4–9)*
> Specify point on object for second tangent of circle: *(select the line at point D, as shown in Figure P4–9)*
> Specify radius of circle <6.0000>: **1.25** (ENTER)

Your drawing will appear as shown in Figure P4–9.

Step 14: Invoke the TRIM command from the Modify toolbar to modify the two circles previously drawn in Step 13.

> Command: **trim** (ENTER)
> Current settings: Projection = UCS Edge = None
> Select cutting edges ...
> Select objects: *(select CIRCLE 1, CIRCLE 2, LINE 1, and LINE 2, as in Figure P4–10, as the cutting edges, and press ENTER)*

Select object to trim or [Project/Edge/Undo]: *(select CIRCLE 3 at point A, as shown in Figure P4–10)*

Select object to trim or [Project/Edge/Undo]: *(select CIRCLE 4 at point B, as shown in Figure P4–10)*

Select object to trim or [Project/Edge/Undo]: *(press* ENTER *to complete the command sequence)*

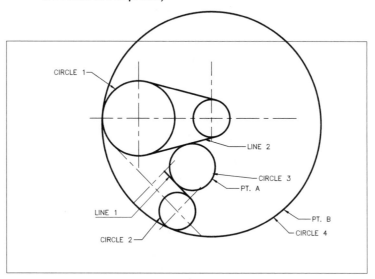

Figure P4–10 *Drawing indicating the objects as the cutting edges*

Your drawing will appear as shown in Figure P4–11.

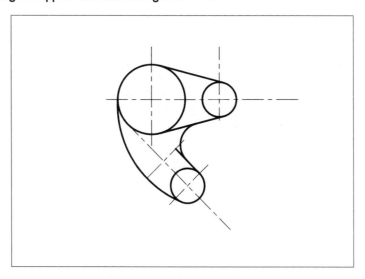

Figure P4–11 *Circles modified with the* TRIM *command*

Step 15: Use the TRIM command to modify the circles and lines to achieve the layout shown in Figure P4–12. Invoke the TRIM command and select lines 1, 2, 3, and 4 as cutting edges, as shown in Figure P4–13. Select circles 1, 2, and 3, as shown in Figure P4–13, as objects to trim.

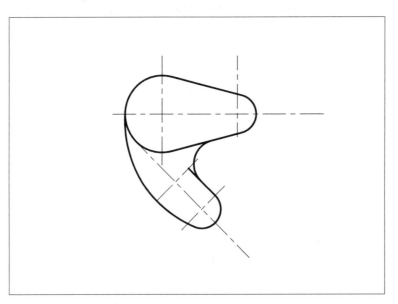

Figure P4–12 *Drawing as it will look after modifying the circles and lines*

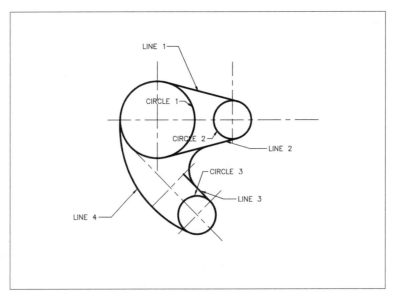

Figure P4–13 *Drawing showing the selection of the objects as cutting and trim edges*

Step 16:   Continue to use the TRIM command to modify the figure to achieve the layout shown in Figure P4–14. Invoke the TRIM command again and select arcs 1 and 2 as cutting edges, as shown in Figure P4–15. Select lines 1, 2, and arc 3 to trim, as shown in Figure P4–15. Your drawing will appear as shown in Figure P4–14.

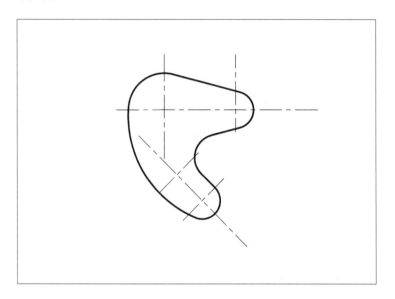

Figure P4–14 *Drawing as it will look after modifying the arcs and lines*

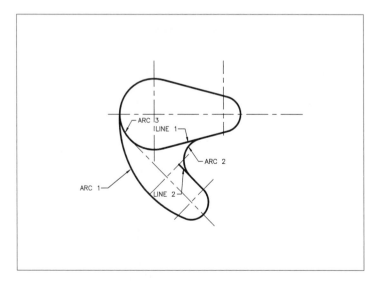

Figure P4–15 *Drawing showing the selection of the objects as cutting and trim edges*

Step 17: Draw the circles necessary to complete the design. Invoke the CIRCLE command and draw circles with points 1, 2, and 3 as center points and with a radius of 0.5. Draw another circle with center point at point 4 and with a radius of 1.5, as shown in Figure P4–16.

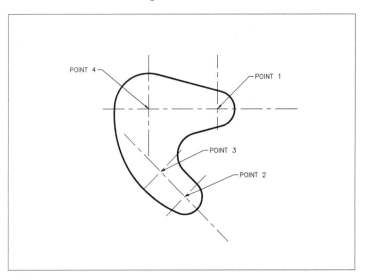

Figure P4–16 *Drawing showing the center points to draw the circles*

 Note: Make sure to use the object snap tool intersection when selecting the center points.

After drawing the circles, your drawing will appear as shown in Figure P4–17.

Figure P4–17 *Design with the addition of circles*

Step 18: Draw lines A and B, needed to form the slot, by invoking the LINE command, as shown in Figure P4–18.

Figure P4–18 *Addition of lines drawn to form the slot*

 Note: Use the object snap tool intersection to snap to the intersection of the centerlines and the circle, as shown in Figure P4–18.

Step 19: Invoke the TRIM command and select lines A and B as cutting edges. Next, select the small circles as objects to trim at points 1 and 2, as shown in Figure P4–19.

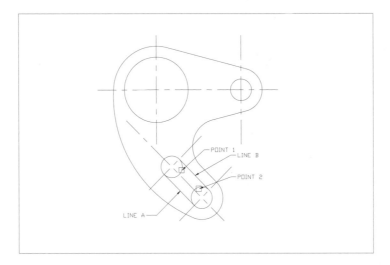

Figure P4–19 *Trim edges and circles to trim*

Your drawing will appear as shown in Figure P4–20.

Figure P4–20 *Drawing after the circles have been trimmed*

Step 20: Invoke the POLYGON command from the Draw toolbar to draw the polygon shown in Figure P4–21.

Figure P4–21 *Design with a six-sided polygon*

Command: **polygon** (ENTER)
Enter number of sides <4>: **6** (ENTER)
Specify center of polygon or [Edge]: *(select center point as in Figure P4–21)*
Enter an option [Inscribed in circle/Circumscribed about circle] <I>: **c** (ENTER)
Specify radius of circle: **@0.875<67.5** (ENTER)

Your drawing will appear as shown in Figure P4–21.

Step 21: Invoke the BREAK command from the Modify toolbar and select points 1 and 2, as shown in Figure P4–22, to break the centerline.

POINT 2

POINT 1

Figure P4–22 *Showing the points where the* BREAK *command will remove part of the centerline*

Your drawing will appear as shown in Figure P4–23 after removing the specified portion of the centerline.

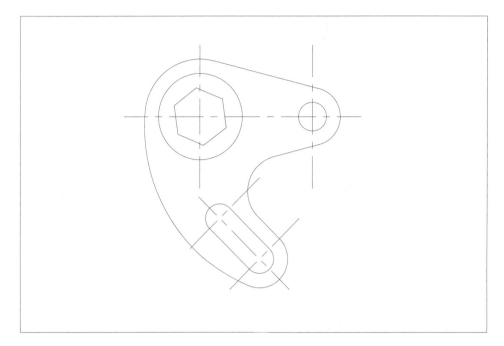

Figure P4–23 *Drawing with the specified portion of the centerline removed*

Step 22: Save the drawing as PROJ4.DWG.

Congratulations! You have just successfully applied several AutoCAD concepts in creating a rather complex mechanical drawing.

EXERCISES 4–1 THROUGH 4–7

EXERCISE 4–1

Create the accompanying drawing of the finishing unit vent piping according to the settings given in the following table.

Settings	Value
1. Units	Decimal with 2 decimal places
2. Limits	Lower left corner: 0,0
	Upper right corner: 12,9
3. Grid	0.125
4. Snap	0.0625
5. Layers	*NAME* *COLOR* *LINETYPE*
	Construction Cyan Continuous
	Border Red Continuous
	Pipes Green Continuous
	Sources Magenta Continuous
	Text Blue Continuous

Hints: Plan ahead before drawing the layout.

The border can be drawn from coordinates .125,.125 to 11.875,8.875 using the RECT-ANGLE command.

Set Construction as the current layer. Draw a horizontal construction line (xline) through a point whose Y coordinate is 1.0.

Use the ARRAY command with the Rectangular option to make eight copies of the construction line, at a spacing of 0.875 units. Don't forget to array 9 objects (the original is included in the count). The drawing should look like Figure Ex4–1a.

Set Sources as the current layer. Draw a circle (0.5 dia.) for one of the sources and a rectangle (0.5 x 0.5) for a pipe ID, and then copy them using the Multiple option of the COPY command.

The alphanumeric characters can be drawn in their respective circles and boxes using the TEXT or MTEXT command. Select the M (middle) or MC (middle center) mode of justification, and then make the center of the circle/box the insertion point of the character.

After you have finished drawing, either the Construction layer can be turned OFF or the construction lines can be erased. Your completed drawing should look like the drawing of the finishing unit vent piping.

Figure Ex4–1a *Completed drawing*

Figure Ex4–1b *Construction lines*

EXERCISE 4–2

Create the structural steel framing plan shown, according to the settings given in the following table.

Settings	Value		
1. Units	Architectural		
2. Limits	Lower left corner: –10'–0",–10'–0"		
	Upper right corner: 50'–0",35'–0"		
3. Grid	12"		
4. Snap	6"		
5. Layers	*NAME*	*COLOR*	*LINETYPE*
	Construction	Cyan	Continuous
	Border	Red	Continuous
	Columns	Green	Continuous
	Beams	Magenta	Continuous
	Text	Blue	Continuous

Hints: Plan ahead before drawing the layout. The columns and beams in the structural steel framing plan can be drawn by using the PLINE command, and the text objects can be drawn with the TEXT or MTEXT commands. Construction lines can be drawn by using the XLINE and RAY commands combined with the ARRAY command.

Draw the border from coordinates –9'–6",–9'–6" to 49'–6",34'–6" using the RECTANGLE command.

Draw a vertical construction line by means of the XLINE command through a point whose X coordinate is 0'–0".

Copy the vertical construction line using the Multiple option of the COPY command.

Draw two horizontal construction lines by means of the RAY command from the point whose coordinates are 24'–0",0'–0". One of the lines is drawn to the left and the other to the right. Use the ARRAY command with the Rectangular option set to four rows for the left construction ray, at a spacing of 8'–0", and five rows of the right construction ray, at a spacing of 6'–0". The drawing should look like Figure Ex4–2a.

Using the PLINE command, draw the columns with a width of 2" and the beams with a width of 1". Draw the beams from intersection to intersection of the construction lines, and then, using the BREAK command, remove the sections of the polylines either 6" or 12" (make sure Snap is set to ON) from the intersections, depending on which direction the column is oriented.

After you have finished drawing, either the Construction layer can be turned OFF or the construction lines can be erased. Your completed drawing should look like the finished framing plan. Do *not* dimension the drawing.

Figure Ex4–2a *The completed drawing*

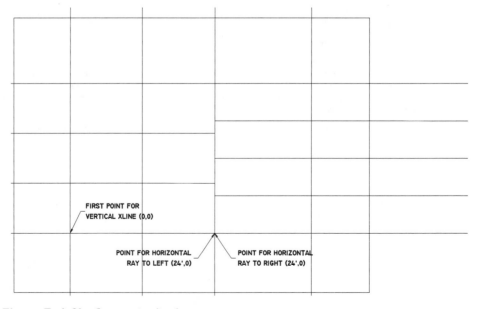

Figure Ex4–2b *Construction line layout*

EXERCISE 4–3

Create the drawing shown of the sleeve block according to the settings given in the following tables.

Settings	Value
1. Units	Decimal
2. Limits	Lower left corner: 0,0
	Upper right corner: 12,9
3. Grid	0.50
4. Snap	0.25
Settings	Value

Settings	Value		
5. Layers	*NAME*	*COLOR*	*LINETYPE*
	Object	Green	Continuous
	Center	Red	Center
	Hidden	Magenta	Hidden
	Text	Blue	Continuous

Hints: The object lines for the sleeve block can be drawn by using the LINE, POLYGON, and CIRCLE commands, and the text objects can be drawn with the TEXT or MTEXT commands. The arcs can be created using the FILLET command. Do not dimension this drawing. Use the TEXT command to label the views only.

Use two circles and two lines as the basis of the slot. Then use the TRIM command to remove the inner halves of the circles, leaving the outer arcs that, with the two lines, comprise the slot.

Other arcs (.5 and .125 radii) can be created by using the FILLET command with the correctly specified radius.

You will probably have to object snap to the "intersections" of the apexes of the hexagon in the top view and draw lines perpendicular to the base of the front view in order to locate them correctly. Then use the TRIM command to remove the unwanted portions of the lines.

Figure Ex4–3 *Sleeve block*

EXERCISE 4–4

Create the drawing shown of the input-output card according to the settings given in the following tables.

Settings	Value
1. Units	Decimal
2. Limits	Lower left corner: 0,0
	Upper right corner: 12,9
3. Grid	0.25
4. Snap	0.25

Settings	Value		
5. Layers	*NAME*	*COLOR*	*LINETYPE*
	Object	Green	Continuous
	Center	Red	Center
	Hidden	Magenta	Hidden
	Text	Blue	Continuous

Hints: The input-output card can be created using the PLINE, LINE, and CIRCLE commands, and the text objects can be drawn with the TEXT or MTEXT commands. Do not dimension the drawing.

Create a circle with radius 0.01 in the lower left corner of the array of dots, and then, using the ARRAY command, create an array of 15 rows and 25 columns.

The internal circuits can be drawn using the PLINE command, with the narrower segments set at 0.04 and the wider, "connection" end set at 0.10. This change should be done without exiting the PLINE command.

The wide-line circles at the ends of the circuits can be created by using the PLINE command with the Arc option, drawing two half-circles with a line width of 0.02 and a radius of 0.0625. The snap resolution will have to be set to 0.03125.

INPUT/OUTPUT CARD

Figure Ex4–4 *Completed drawing*

EXERCISES 4–5 & 4–6

Create the drawings shown of the brackets and clamps according to the settings given in the following table.

Settings	Value
1. Units	Architectural
2. Limits	Lower left corner: 0,0
	Upper right corner: 12'-0",9'-0"
3. Grid	4"
4. Snap	0.5"
5. Text Height	0.375"
6. Layers	

NAME	COLOR	LINETYPE
Object	Green	Continuous
Center	Red	Center
Hidden	Magenta	Hidden
Text	Blue	Continuous

Hint: To lay out the views, you need to estimate the dimensions of the spaces between the objects. Then add up the total of the objects and estimated distances between to determine the spaces left over around the whole grouping.

Figure Ex4–5 *Brackets*

mechanical **EXERCISE**

DETAIL "B"
PART# 3317

DETAIL "A"
PART# 3315

Figure Ex4–6 *Clamps*

EXERCISE 4–7

Create the survey plot plan drawing shown to the settings given in the following table. (refer to Exhibit A for description)

Settings	Value
1. Units/Angle	Decimal/Surveyor
2. Limits	Lower left corner: 0,0
	Upper right corner: 680,440
3. Dimscale	40
4. Ltscale	20
5. Layers	_NAME_ _COLOR_ _LINETYPE_
	Property Green Continuous
	Text Red Center

EXHIBIT A: METES & BOUNDS

Tracts 1

That certain 0.3443 acres of land, said tract being in the Ruben White Survey Abstract 24, same being out Lot 5, Blk. 5 Section 2 of DreamLand Place Subdivision, Harris County, Texas, Vol.   Pg.   Harris County Map Records, and being more particularly described by metes and bounds as follows;

Commencing at the intersection of F.M. 1942 (Gulf Pump Road) variable right of way, ie 60' at Lot 5, and the southwest corner of Blanchard Drive, a 60' right of way.

THENCE South 89 degrees 30' minutes 00" seconds West along the North line of F.M. 1942, a distance of 120.00' to a point for a corner, a SET #4 rebar and the _PLACE OF BEGINNING THE TRACT HEREIN DESCRIBED;_

THENCE continuing South 89 degrees 30' minutes 00" seconds West along the North line of F.M. 1942, and a South line of lot 5, a distance of 75.00' to a found #4 rebar for a corner in the West line of this Tract 1 and the East line of Tract 2, same being the Westerly portion of Lot 5, a point for a corner;

THENCE North 00 degrees 05' minutes 00" seconds West along the east line of Tract 2 and the West line of this tract a distance of 200.01 to a set #4 rebar in the center line of a drainage easement and the North line of Lot 5 a set #4 rebar for a corner;

THENCE North 89 degrees 30' minutes 00" seconds East along the North line of this tract and the South line of lot 7, a distance 75.00 to a set #4 rebar in the West line of Lots 3&4, on Blanchard Street, a point for a corner;

THENCE South 00 degrees 05' minutes 00" seconds East along the West line of Lots 3, 2 &1, on Blanchard Street, a distance of 200.00' to a set #4 rebar in the North right of way line of F.M. 1942 for a corner and _THE PLACE OF BEGINNING AND CONTAINING 0.3443 acres OF LAND OR 15,000.00 square feet of land._

Figure Ex4–7 *Survey Plot Plan*

REVIEW QUESTIONS

1. In order to draw two rays with different starting points, you must use the RAY command twice.

   a. True   b. False

2. When you place xlines on a drawing, they:

   a. may effect the limits of the drawing

   b. may effect the extents of the drawing

   c. always appear as construction lines on layer 0

   d. can be constructed as offsets to an existing line

   e. none of the above

3. Filleting two non-parallel, non-intersecting line segments with a zero radius will:

   a. return an error message

   b. have no effect

   c. create a sharp corner

   d. convert the lines to rays

4. Objects can be trimmed at the points where they intersect existing objects.

   a. True   b. False

5. The default justification for text is:

   a. TL   d. BR

   b. BL   e. None of the above

   c. MC

6. Which command allows you to change the location of the objects and allows a duplicate to remain intact?

   a. CHANGE

   b. MOVE

   c. COPY

   d. MIRROR

7. The maximum number of sides accepted by the POLYGON command is:

   a. 8   d. 1024

   b. 32   e. infinite (limited by computer memory, but VERY large)

   c. 128

8. Portions of objects can be erased or removed by using the command:

   a. ERASE   d. EDIT

   b. REMOVE   e. PARERASE

   c. BREAK

9. Ellipses are drawn by specifying:

   a. The major and minor axes

   b. The major axis and a rotation angle

   c. Any three points on the ellipse

   d. Any of the above

   e. Both a and b

10. Which of the following commands can be used to place text on a drawing?

    a. TEXT   d. All of the above

    b. DTEXT   e. None of the above

    c. MTEXT

11. Two lines are drawn. Then the first line is erased and the second line is moved. Executing the OOPS command at the "Command:" prompt will:

    a. restore the erased line

    b. execute the LINE command automatically

    c. replace the second line at its original position

    d. restore both lines to their original positions

    e. none of the above

12. To efficiently move multiple objects, which option would be more efficient?

    a. Objects   d. Add

    b. Last   e. Undo

    c. Window

13. While using the TEXT command, AutoCAD will display the text you are typing:

    a. in the command prompt area

    b. in the drawing screen area

    c. both a and b

    d. neither a or b

14. While using the DTEXT command, AutoCAD will display the text you are typing:

    a. in the command prompt area

    b. in the drawing screen area

    c. both a and b

    d. neither a or b

15. The MOVE command allows you to:

    a. move objects to new locations on the screen

    b. dynamically drag objects on the screen

    c. move only the objects that are on the current layer

    d. move an object from one layer to another

    e. both a and b

16. To create a rectangular array of objects, you must specify:

    a. the number of items and the distance between them

    b. the number of rows, the number of items, and the unit cell size

    c. the number of rows, the number of columns, and the unit cell size

    d. none of the above

17. A polyline:

    a. can have width

    b. can be exploded

    c. is one object

    d. all of the above

18. Polylines are:

    a. made up of line and arc segments, each of which is treated as an individual object

    b. are connected sequences of lines and arcs

    c. both a and b

    d. none of the above

19. To create an arc which is concentric with an existing arc, you could use what command?

    a. ARRAY

    b. COPY

    c. OFFSET

    d. MIRROR

20. The following are all options of the PLINE command **except:**

   a. Undo   d. Ltype

   b. Halfwidth   e. Width

   c. Arc

Directions: Answer the following questions based on Figure QX4–1.

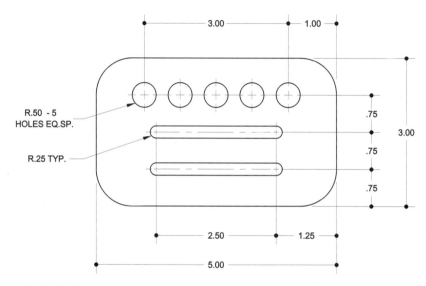

Figure QX4–1 *Mechanical part*

21. If the RECTANGLE command were used to create the object outline with the first corner at 0,0 what "*specify other corner point:*" coordinate location input would be required to complete the rectangle shape?

   a.   @3.0,3.0

   b.   5.0,3.0

   c.   3.0,5.0

   d.   @3.0,5.0

22. Which command would be the easiest and quickest to create rounded corners on the object?

   a.   ARC

   b.   ROUND

   c.   FILLET

   d.   CHAMFER

23. Creating the five holes equally spaced can be best achieved with which command?

   a.   MIRROR

   b.   OFFSET

   c.   COPY

   d.   ARRAY

24. Which ARRAY type would the five equally spaced holes are created with?

   a.   POLAR

   b.   RECTANGULAR

   c.   COPY

   d.   ARRAY

25. If the first corner of the object outline is at 0,0 what coordinate location input is required to position the center of the circle on the left?

   a.   @1.25,2.5

   b.   -2.25,1.25

   c.   @1.00,2.25

   d.   1.25,2.25

26. What is the *distance between cells* (holes) value required when using the ARRAY command to locate the five holes?

   a. 3.00

   b. 0.75

   c. 0.50

   d. 0.25

27. If the top horizontal line of the long slot is drawn with the LINE command, which command can easily establish the location of the remaining horizontal lines for the slots?

   a. XLINE

   b. MIRROR

   c. OFFSET

   d. MOVE

Fundamentals IV

INTRODUCTION

After completing this chapter, you will be able to do the following:

- Construct geometric figures with the DONUT, SOLID, and POINT commands
- Create freehand line segments with the SKETCH command
- Use advanced object selection methods and modes to modify objects
- Use the modify commands: LENGTHEN, STRETCH, ROTATE, SCALE, PEDIT (edit Polyline), and MATCHPROP

CONSTRUCTING GEOMETRIC FIGURES

DRAWING SOLID-FILLED CIRCLES

The DOUGHNUT (or DONUT) command lets you draw solid-filled circles and rings by specifying outer and inner diameters of the filled area. The fill display depends on the setting of the FILLMODE system variable.

Invoke the DONUT command:

Draw menu	Choose Donut
Command: prompt	**donut** (ENTER)

AutoCAD prompts:

> Command: **donut** (ENTER)
> Specify inside diameter of donut <current>: *(specify a distance, or press ENTER to accept the current setting)*
> Specify outside diameter of donut <current>: *(specify a distance, or press ENTER to accept the current setting)*
> Specify center of donut or <exit>: *(specify a point to draw the donut)*

You may specify the inside and outside diameters of the donut to be drawn by specifying two points at the appropriate distance apart on the screen for either or both diameters, and AutoCAD will use the measured distance for the diameter(s). Or you may enter the distances from the keyboard.

You can select the center point by entering its coordinates or by selecting it with your pointing device. After you specify the center point, AutoCAD prompts for the center of the next donut and continues prompting for subsequent center points. To terminate the command, enter a null response.

 Note: Be sure the FILLMODE system variable is set to ON (a value of 1). Check at the Command: prompt by entering Fill and pressing ENTER. If FILLMODE is set to OFF (value of 0), then the PLINE, TRACE, DONUT, and SOLID commands display only the outline of the shapes. With FILLMODE set to ON, the shapes you create with these commands appear solid. If FILLMODE is reset to ON after it has been set to OFF, you must use the REGEN command in order for the screen to display as filled any unfilled shapes created by these commands. Switching between ON and OFF affects only the appearance of shapes created with the PLINE, TRACE, DONUT, and SOLID commands. Solids can be selected or identified by specifying the outlines only. The interior of a solid area is not recognized as an object when specified with the pointing device.

For example, the following command sequence shows placement of a solid-filled circle, as shown in Figure 5–1, by use of the DONUT command.

Command: **donut** (ENTER)
Specify inside diameter of donut <0.5000>: **0** (ENTER)
Specify outside diameter of donut <1.0000>:**1** (ENTER)
Specify center of donut or <exit>: **3,2** (ENTER)
Specify center of donut or <exit>: (ENTER)

INSIDE DIA. O   INSIDE DIA..5

Figure 5–1 *Using the* DONUT *command to place a solid-filled circle and a filled circular shape*

The following command sequence shows placement of a filled circular shape, as shown in Figure 5–1, by use of the DONUT command.

Command: **donut** (ENTER)
Specify inside diameter of donut <0.0000>: **0.5** (ENTER)

Specify outside diameter of donut <1.0000>: **1** (ENTER)
Specify center of donut or <exit>: **6,2** (ENTER)
Specify center of donut or <exit>: (ENTER)

DRAWING SOLID-FILLED POLYGONS

The SOLID command creates a solid-filled straight-sided area whose outline is determined by points you specify on the screen. Two important factors should be kept in mind when using the SOLID command: (1) the points must be selected in a specified order or else the four corners generate a bowtie instead of a rectangle; and (2) the polygon generated has straight sides. (Closer study reveals that even filled donuts and PLINE-generated curved areas are actually straight-sided, just as arcs and circles generate as straight-line segments of small enough length to appear smooth.)

Invoke the SOLID command:

Surfaces toolbar	Choose the SOLID command (see Figure 5–2)
Draw menu	Choose > Surfaces > 2D Solid
Command: prompt	**solid** (ENTER)

Figure 5–2 *Invoking the* SOLID *command from the Surfaces toolbar*

AutoCAD prompts:

Command: **solid** (ENTER)
First point: *(specify a first point)*
Second point: *(specify a second point)*
Third point: *(specify a third point diagonally opposite the second point)*
Fourth point: *(specify a fourth point, or press* ENTER*)*

When you specify a fourth point, AutoCAD draws a quadrilateral area. If instead you press ENTER, AutoCAD creates a filled triangle.

For example, the following command sequence shows how to draw a quadrilateral area, as shown in Figure 5–3.

Command: **solid** (ENTER)
Specify first point: *(pick point 1)*
Specify second point: *(pick point 2)*
Specify third point: *(pick point 3)*
Specify fourth point or <exit>: *(pick point 4)*

Figure 5–3 *The pick order for creating a quadrilateral area with the SOLID command*

To create the solid shape, the odd-numbered picks must be on one side and the even-numbered picks on the other side. If not, you get an effect such as what is shown in Figure 5–4.

You can use the SOLID command to create an arrowhead or triangle, such as the one shown in Figure 5–5. Polygon shapes can be created with the SOLID command by keeping the odd picks along one side and the even picks along the other side of the object, as shown in Figure 5–6.

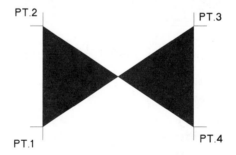

Figure 5–4 *Results of using the SOLID command when odd/even points are not specified correctly*

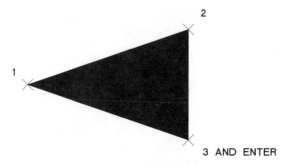

Figure 5–5 *Using the SOLID command to create a solid triangular shape*

Figure 5–6 *Using the* SOLID *command to create a polygonal shape*

DRAWING POINT OBJECTS

The POINT command draws points on the drawing, and these points are drawn on the plotted drawing sheet with a single "pen down." You can enter such points to be used as reference points for object snapping when necessary. When the drawing is finished, simply erase them from the drawing or freeze their layer. Points are entered by specifying coordinates or with the pointing device. You can specify a three-dimensional location for a point and object snap to a point by using the Node option of the Object Snap (Osnap) feature.

Invoke the POINT command:

Draw toolbar	Choose the POINT command (see Figure 5–7)
Draw menu	Choose Point
Command: prompt	**point** (ENTER)

AutoCAD prompts:

> Command: **point** (ENTER)
> Point: *(specify a point)*

Figure 5–7 *Invoking the* POINT *command from the Draw toolbar*

Point Modes

When you draw the point, it appears on the display as a blip (+) if the BLIPMODE system variable is set to ON (default is ON). After a REDRAW command, it appears as a dot (.).

You can make the point appear as a +, x, 0, or | by changing the PDMODE system variable. This can be done by entering PDMODE at the "Command:" prompt and entering the appropriate value. You can also change the PDMODE value by using the icon menu, as shown in Figure 5–8, invoked by typing DDPTYPE at the Command: prompt and pressing ENTER. The default value of PDMODE is zero, which means the point appears as a dot. If PDMODE is changed, all previous points remain the same until you regenerate the drawing. After a screen regeneration, all points appear as the last PDMODE value entered.

Figure 5–8 *The Point Style icon menu lets you select the shape and size of the point object*

Point Size

The size that the point appears on the screen depends on the value to which the PDSIZE system variable is set. If necessary, you can change the size via the PDSIZE command. The default for PDSIZE is zero (one pixel in size). Any positive value larger than this will increase the size of the point accordingly.

DRAWING SKETCH LINE SEGMENTS

The SKETCH command creates a series of freehand line segments. It is useful for freehand drawings, contour mapping, and signatures. The sketched lines are not added to the drawing until they are recorded.

Invoke the SKETCH command:

| Command: prompt | **sketch** (ENTER) |

AutoCAD prompts:

> Command: **sketch** (ENTER)
> Record increment <current>: *(specify a distance, or press* ENTER *to accept the current value)*

You may also respond by specifying two points, either keyed in or specified on the screen, causing AutoCAD to use the distance between the points as the record increment distance. Once a record increment is specified, AutoCAD displays the following list of options:

> Sketch   Pen   eXit   Quit   Record   Erase   Connect

You can use any of the available options while you are in the SKETCH command. They are accessible either as single-key entries or as a mouse/puck button, provided your mouse/puck has the number of buttons corresponding to the option. The following table shows the optional subcommands, and their key, button number, and function. Normal button functions are not available while in the SKETCH mode.

Command Character	Pointer Button	Function
P	Pick	Raise/lower pen
.(period)	1	Line to point
R	2	Record lines
X, Spacebar, or ENTER	3	Record lines and exit
Q, or ESC	4	Discard lines and exit
E	5	Erase
C	6	Connect

P (Pen Up and Down)

An imaginary pen follows the cursor movement. When the pen is down, AutoCAD sketches a connected segment whenever the cursor moves the specified increment distance from the previously sketched segment. When the pen is up, the pen follows the cursor movement without drawing.

The pen is raised (up) and lowered (down) either by pressing the pick button on the mouse/puck or by pressing P on the keyboard. When you invoke a PEN UP, the current location of the pen will be the endpoint of the last segment drawn, which will be shorter than a standard increment length.

A PEN UP does not take you out of the SKETCH mode. Nor does a PEN UP permanently record the lines drawn during the current SKETCH session.

. (Period; Line-to-Point)

While the pen is up, you cause AutoCAD to draw a straight line from the last segment to the current cursor location and return to the PEN UP status by entering. (period) from the keyboard. This is convenient for long, straight lines that might occur in the middle of irregular shapes.

R (Record)

Lines being displayed while the cursor is moved (with the pen down) are temporary. They will appear green (or red if the current color for that layer or object is green) on color monitors until they are permanently recorded. These temporary segments are subject to being modified with special Sketch options until you press R to record the latest lines. These may include several groups of connected lines drawn during PEN DOWN sequences separated by PEN UPs. When the Record option is invoked by entering either R ENTER or the third mouse/puck button, the total number of recorded segments is reported as follows:

    nnn lines recorded

E (Erase)

Prior to any group(s) of connected lines being recorded with the Record option, you may use the E (Erase) option to remove any or all of the lines from the last segment back to the first. The sequence of prompts is as follows:

    Erase:
    Select end of delete

The pen is automatically set to UP, and you may then use the cursor to remove segments, starting from the last segment. When you are satisfied with the lines remaining, press P or the pick button to accept the erasure. To abort the erasure and return to the SKETCH mode, press E again (or any other option) and the following will be displayed:

    Erase aborted

C (Connect)

Whenever a disconnect occurs (PEN UP or ERASE), you can reconnect and continue sketching from the point of the last disconnect as long as you have not exited the SKETCH command. The sequence is as follows:

    Connect:
    Move to endpoint of line.

At this prompt you can move the cursor near the end of the last segment. When you are within a specified increment length, sketching begins, connected to that last endpoint. This option is meaningless if invoked during PEN DOWN. A message also tells you:

No last point known

if no last point exists. The Connect option can be canceled by pressing C a second time.

X (Record and Exit)

The X option exits SKETCH mode after recording all temporary lines. This can also be accomplished by pressing either ENTER or the SPACEBAR.

Q (Quit)

The Q option exits SKETCH mode without recording any temporary lines. It is the same as pressing ESC.

OBJECT SELECTION

As mentioned earlier, all the modify commands initially prompt you to select objects. In most modify commands, the prompt allows you to select any number of objects. In some of the modify commands, however, AutoCAD limits your selection to only one object, for instance, the BREAK, PEDIT, DIVIDE, and MEASURE commands. In the case of the FILLET and CHAMFER commands, AutoCAD requires you to select two objects. And, whereas in the DIST and ID commands AutoCAD requires you to select a point, in the AREA command AutoCAD permits selection of either a series of points or an object.

Compared to the basic object selection option, the options covered in this section give you more flexibility and greater ease of use when you are prompted to select objects for use by the modify commands. The options that are explained in this section include WPolygon (WP), CPolygon (CP), Fence, All, Multiple, Box, Auto, Undo, Add, Remove, and Single.

WPOLYGON (WP) OPTION

The WPolygon option is similar to the Window option, but it allows you to define a polygon-shaped window rather than a rectangular area. You define the selection area as you specify the points about the objects you want to select. The polygon is formed as you select the points. The polygon can be of any shape but may not intersect itself. The polygon is formed as you specify points and includes rubber-band lines to the graphics cursor. When the selected points define the desired polygon, press ENTER. Only those objects that are totally inside the polygon shape are selected. To select the WPolygon option, enter WP and press ENTER at the "Select Objects:" prompt. The UNDO option lets you undo the most recent polygon pick point.

CPOLYGON (CP) OPTION

The CPolygon option is similar to the WPolygon option, but it selects all objects within or crossing the polygon boundary. If there is an object that is partially inside the polygon area, then the whole object is included in the selection set. To select the CPolygon option, enter CP and press ENTER at the "Select Objects" prompt.

FENCE (F) OPTION

The Fence option is similar to the CPolygon option, except you do not close the last vector of the polygon shape. The selection fence selects only those objects it crosses or intersects. Unlike the WPolygons and CPolygon, the fence can cross over and intersect itself. To select the Fence option enter F and press ENTER at the Select Objects: prompt.

ALL OPTION

The All option selects all the objects in the drawing, including objects on frozen or locked layers. After selecting all the objects, you may use the REMOVE (R) option to remove some of the objects from the selection set. The All option must be spelled out in full (All) and not applied as an abbreviation as you may do with the other options.

MULTIPLE OPTION

The Multiple option helps you overcome the limitations of the Point, Window, and Crossing options. The Point option is time-consuming for use in selecting many objects. AutoCAD does a complete scan of the screen each time a point is picked. By using the Multiple modifier option, you can pick many points without delay and when you press ENTER, AutoCAD applies all of the points during one scan.

Selecting one or more objects from a crowded group of objects is sometimes difficult with the Point option. It is often impossible with the Window option. For example, if two objects are very close together and you wish to point to select them both, AutoCAD normally selects only one no matter how many times you select a point that touches them both. By using the Multiple option, AutoCAD excludes an object from being selected once it has been included in the selection set. As an alternative, use the Crossing option to cover both objects. If this is not feasible, then the Multiple modifier may be the best choice.

BOX OPTION

The Box option is usually employed in a menu macro to give the user a double option of window and Crossing, depending on how and where the picks are made on the screen. It must be spelled out in full (Box) and not applied as an abbreviation as you may do with the others, for example, W for Window. Because of this, you normally would not go to the trouble to use the Box option from the keyboard when a single key (W or C) provides a decidedly faster selection.

When the Box modifier is invoked as a response to a prompt to select an object, the options are applied as follows:

> If the picks are made left to right (the first point is to the left of the second), then the two points become diagonally opposite corners of a rectangle that is used as a Window option. That is, all visible objects totally within the rectangle are part of that selection.

If the picks are right to left, then the selection rectangle becomes the Crossing option. That is, all visible objects that are within or partially within the rectangle are part of that selection.

AUTO OPTION

The Auto option is actually a triple option. It includes the Point option with the two Box options. If the target box touches an object, then that object is selected as you would in using the Point option. If the target box does not touch an object, then the selection becomes either a Window or Crossing option, depending on where the second point is picked in relation to the first.

 Note: The Auto option is the default option when you are prompted to Select Objects:.

UNDO OPTION

The Undo option allows you to remove the last item(s) selected from the selection set without aborting the Select Objects: prompt and then to continue adding to the selection set. It should be noted that if the last option to the selection process includes more than one object, the Undo option will remove all the objects from the selection set that were selected by that last option.

ADD OPTION

The Add option lets you switch back from the Remove mode in order to continue adding objects to the selection set by however many options you wish to use.

REMOVE OPTION

The Remove option lets you remove objects from the selection set. The Select Objects: prompt always starts in the Add mode. The Remove mode is a switch from the Add mode, not a standard option. Once invoked, the objects selected by whatever and however many options you use will be removed from the selection set. It will be in effect until reversed by the Add option.

SINGLE OPTION

The Single option causes the object selection to terminate and the command in progress to proceed after you use only one object selection option. It does not matter if one object is selected or a group is selected with that option. If no object is selected and the point selected cannot be the first point of a Window or Crossing rectangle, AutoCAD will not abort the command in progress; however, once there is a successful selection, the command proceeds.

OBJECT SELECTION MODES

AutoCAD provides six selection modes that will enhance object selection. You can toggle on/off one or more object selection modes from the Selection tab of the Options dialog box. By having the appropriate selection mode set to ON, you have various methods that give you more flexibility and greater ease of use in the selection of the objects.

To open the Selection Modes settings box, invoke the DDSELECT command:

Tools menu	Choose Options *(then select the Selection tab)*
Command: prompt	**ddselect** (ENTER)

AutoCAD displays the Options dialog box, as shown in Figure 5–9.

Figure 5–9 *Options with Selection tab selection dialog box*

You can toggle any one or more combinations of the settings provided under the selection modes. Noun/Verb Selection, Implied Windowing, and Object Grouping are the defaults.

NOUN/VERB SELECTION

The NOUN/VERB Selection feature allows the traditional verb-noun command syntax to be reversed for most modifying commands. When the NOUN/VERB Selection is set to ON, you can select the objects first at the Command: prompt and then invoke the appropriate modifying command you want to use on the selection set. For example, instead of invoking the COPY command followed by selecting the objects to be copied, with NOUN/VERB Selection set to ON, you can select the objects first and then invoke the COPY command, and AutoCAD skips the object selection prompt.

When NOUN/VERB Selection is set to ON, the cursor at the Command: prompt changes to resemble a Running-Osnap cursor. Whenever you want to use the NOUN/VERB selection feature, first create a selection set at the Command: prompt. Subsequent modify commands you invoke execute using the objects in the current selection set without prompting for object selection. To clear the current selection set, press ESC key at the Command: prompt. This clears the selection set, so any subsequent editing command will once again prompt for object selection.

Note: Another way to turn ON/OFF Noun/Verb Selection is to use the PICKFIRST system variable. TRIM, EXTEND, BREAK, CHAMFER, and FILLET are the commands not supported by the Noun/Verb selection feature.

USE SHIFT TO ADD

The Use Shift to Add feature controls how you add objects to an existing selection set. When **Use Shift to Add** is set to ON, it activates an additive selection mode in which SHIFT must be held down while adding more objects to the selection set. For example, if you first pick an object, it is highlighted. If you pick another object, it is highlighted and the first object is no longer highlighted. The only way you can add objects to the selection set is to select objects by holding down the SHIFT key. Similarly, the way to remove objects from the selection set is to select the objects by holding down the SHIFT key.

When **Use Shift to Add** is set to OFF (default), objects are added to the selection set by just picking them individually or by using one of the selection options; AutoCAD adds the objects to the selection set.

Note: Another way to select objects with this method is to set the PICKADD system variable appropriately.

PRESS AND DRAG

The Press and Drag feature controls the manner by which you draw the selection window with your pointing device. When Press and Drag is set to ON, you can create a selection window by holding down the pick button and dragging the cursor diagonally while you create the window.

When **Press and Drag** is set to OFF (default), you need to use two separate picks of the pointing device to create the selection window. In other words, you need to pick once to define one corner of the selection window, and pick a second time to define its diagonal corner.

Note: Another way to control how selection windows are drawn is to set the PICKDRAG system variable appropriately.

IMPLIED WINDOWING

The Implied Windowing feature allows you to create a selection window automatically when the "Select objects:" prompt appears. When **Implied Windowing** is set to

ON (default), it works like the box option, explained earlier. If Implied Windowing is set to OFF, you can create a selection window by using the Window or Crossing selection set methods.

 Note: Another way to control the Implied Windowing option is to set the PICKAUTO system variable appropriately.

OBJECT GROUPING

The Object Grouping feature controls the automatic group selection. If **Object Grouping** is set to ON, then selecting an object that is a member of a group selects the whole group. Refer to Chapter 6 for a detailed description of how to create groups.

ASSOCIATIVE HATCH

The Associative Hatch feature controls which objects will be selected when you select an associative hatch. If Associative Hatch is set to ON, then selecting an associative hatch also selects the boundary objects. Refer to Chapter 9 for a detailed description of hatching.

ADJUSTING PICKBOX

AutoCAD allows you to adjust the size of the pickbox using the slider bar provided in the dialog box. You can also set the sorting of objects for processing in the order in which they were created in the drawing.

MODIFYING OBJECTS

In this section, seven modify commands are described: LENGTHEN, STRETCH, MODIFY, ROTATE, SCALE, PEDIT, and MATCH PROPERTIES. Additional modify commands were explained in Chapters 2 and 3.

LENGTHENING OBJECTS

The LENGTHEN command is used to increase or decrease the length of line objects or the included angle of an arc.

Invoke the LENGTHEN command:

Modify toolbar	Choose the LENGTHEN command (see Figure 5–10)
Modify menu	Choose Lengthen
Command: prompt	**lengthen** (ENTER)

Figure 5–10 *Invoking the* LENGTHEN *command from the Modify toolbar*

AutoCAD prompts:

> Command: **lengthen** (ENTER)
> Select an object or [DElta/Percent/Total/DYnamic]: *(select an object or*
> *right-click for shortcut menu and choose one of the avilable options)*

When you select an object, AutoCAD displays its length in the current units display and, where applicable, the included angle of the selected object.

Delta

The Delta option changes the length or, where applicable, the included angle from the endpoint of the selected object closest to the pick point. A positive value results in an increase in extension; a negative value results in a trim. When you select the **Delta** option, AutoCAD prompts:

> Select an object or [DElta/Percent/Total/DYnamic]: **de** (ENTER)
> Enter delta length or [Angle] (current)>: *(specify positive or negative value)*
> Select an object to change or [Undo]: *(select an object, and its length is*
> *changed on the end nearest the selection point)*
> Select an object to change or [Undo]: *(select additional objects; when done,*
> *press* ENTER *to exit the command sequence)*

Instead of specifying the delta length, if you select the Angle option, AutoCAD prompts:

> Enter delta length or [Angle] <current>: **a** (ENTER)
> Enter delta angle <current>: *(specify positive or negative angle)*
> Select an object to change or [Undo]: *(select an object, and its included angle*
> *is changed on the end nearest the selection point)*
> Select an object to change or [Undo]: *(select additional objects; when done,*
> *press* ENTER *to exit the command sequence)*

The **undo** option reverses the most recent change made by the LENGTHEN command.

Percent

The Percent option sets the length of an object by a specified percentage of its total length. It will increase the length/angle for values greater than 100 and decrease them for values less than 100. For example, a 12-unit-long line will be changed to 15 units by using a value of 125. A 12-unit-long line will be changed to 9 units by using a value of 75. When you select the **Percent** option, AutoCAD prompts:

> Enter delta length or [Angle] <current>: **p** (ENTER)
> Enter percent length <current>: (specify positive nonzero value and
> press ENTER)
> Select an object to change or [Undo]: *(select an object, and its length is*
> *changed on the end nearest the selection point)*
> Select an object to change or [Undo]: *(select additional objects; when done,*
> *press* ENTER *to exit the command sequence)*

Total

The Total option changes the length/angle of an object to the value specified. When you select the **Total** option, AutoCAD prompts:

> Select an object or [DElta/Percent/Total/DYnamic]: **t** (ENTER)
> Specify total length or [Angle] <current>: (specify distance, or enter **a**, for angle, and then specify an angle for change)
> Select an object to change or [Undo]: (select an object, and its length is changed to the specified distance on the end nearest the selection point)
> Select an object to change or [Undo]: (select additional objects; when done, press ENTER to exit the command sequence)

Dynamic

The Dynamic option changes the length/angle of an object in response to the cursor's final location, relative to the endpoint nearest to where the object is selected. When you select the **Dynamic** option, AutoCAD prompts:

> Select an object or [DElta/Percent/Total/DYnamic]: **dy** (ENTER)
> Select an object to change or [Undo]: (select an object to change the endpoint)
> Select an object to change or [Undo]:(reposition the object if necessary, or press ENTER to exit the command sequence).

The options and prompts are similar to those for the Delta option.

STRETCHING OBJECTS

The STRETCH command allows you to stretch the shape of an object without affecting other crucial parts that remain unchanged. A common example is to stretch a square into a rectangle. The length is changed while the width remains the same.

AutoCAD stretches lines, polyline segments, rays, arcs, elliptical arcs, and splines that cross the selection window. The STRETCH command moves the endpoints that lie inside the window, leaving those outside the window unchanged. If the entire object is inside the window, then the STRETCH command operates like the MOVE command.

Invoke the STRETCH command:

Modify toolbar	Choose the STRETCH command (see Figure 5–11)
Modify menu	Choose Stretch
Command: prompt	**stretch** (ENTER)

Figure 5–11 *Invoking the* STRETCH *command from the Modify toolbar*

AutoCAD prompts:

> Command: **stretch** (ENTER)
> Select objects to stretch by crossing-window or crossing-polygon...
> Select objects: *(select the first corner of a crossing window or polygon)*
> Specify opposite corner: *(select the opposite corner of the crossing window or polygon)*
> Specify base point or displacement: *(specify a base point or press ENTER)*
> Specify second point of displacement: *(specify the second point of displacement or press ENTER)*

If you provide the base point and second point of displacement, AutoCAD stretches the selected objects the vector distance from the base point to the second point. If you press ENTER at the prompt for the second point of displacement, then AutoCAD considers the first point as the X,Y displacement value.

Figure 5–12 shows some examples of using the STRETCH command.

Figure 5–12 *Examples of using the* STRETCH *command*

ROTATING OBJECTS

The ROTATE command changes the orientation of existing objects by rotating them about a specified point, labeled as the base point. Design changes often require that an object, feature, or view be rotated. By default, a positive angle rotates the object in the counterclockwise direction, and a negative angle rotates in the clockwise direction.

Invoke the ROTATE command:

Modify toolbar	Choose the ROTATE command (see Figure 5–13)
Modify menu	Choose Rotate
Command: prompt	**rotate** (ENTER)

Figure 5–13 *Invoking the* ROTATE *command from the Modify toolbar*

AutoCAD prompts:

> Command: **rotate** (ENTER)
> Select objects: (*select the objects to rotate and press* ENTER *to complete the selection*)
> Specify base point: (*specify a base point about which selected objects are to be rotated*)
> Specify rotation angle or [Reference]: (*specify a positive or negative rotation angle, or enter **r** to select the Reference option*)

The base point can be anywhere in the drawing. If the base point selected is on the selected object itself, then the selected base point becomes an anchor point for rotation. The Rotation Angle determines how far an object rotates around the base point.

The following command sequence shows an example of rotating a group of objects selected by the Window option, as shown in Figure 5–14.

> Command: **rotate** (ENTER)
> Select objects: (*select the objects to rotate as shown in Figure 5–14 and press* ENTER *to complete the selection*)
> Specify base point: (*pick the base point as shown in Figure 5–14*)
> Specify rotation angle or [Reference]: **45** (ENTER)

AutoCAD rotates the selected objects by 45 degrees, as shown in Figure 5–14.

Reference Angle

If an object has to be rotated in reference to the current orientation, you can use the **Reference** option to do that. Specify the current orientation as referenced by the angle, or show AutoCAD the angle by pointing to the two endpoints of a line to be rotated, and specify the desired new rotation. AutoCAD automatically calculates the rotation angle and rotates the object appropriately. This method of rotation is very useful when you want to straighten an object or align it with other features in a drawing.

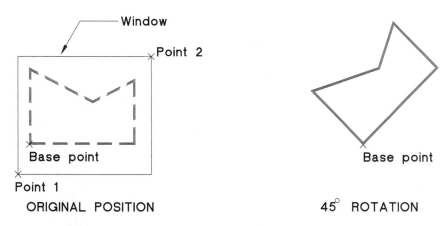

Figure 5–14 *Rotating a group of objects by means of the* ROTATE *command*

The following command sequence shows an example of rotating a group of objects selected by the Window option in reference to the current orientation, as shown in Figure 5–15.

> Command: **rotate** (ENTER)
> Select objects: *(select the objects to rotate, as shown in Figure 5–15)*
> Specify base point: *(pick the base point, as shown in Figure 5–15)*
> Specify rotation angle or [Reference]: **r** (ENTER)
> Specify the reference angle <0>: **60** (ENTER)
> Specify the new angle: **30** (ENTER)

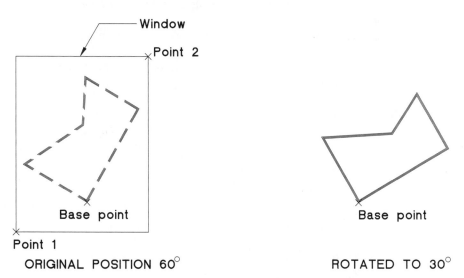

Figure 5–15 *Rotating a group of objects by means of the* ROTATE *command in reference to the current orientation*

SCALING OBJECTS

The SCALE command lets you change the size of selected objects or the complete drawing. Objects are made larger or smaller; the same scale factor is applied to the X, Y, and Z directions. To enlarge an object, specify a scale factor greater than 1. For example, a scale factor of 3 makes the selected objects 3 times larger. To reduce the size of an object, use a scale factor between 0 and 1. Do not specify a negative scale factor. For example, a scale factor of 0.75 would reduce the selected objects to three-quarters of their current size.

Invoke the SCALE command:

Modify toolbar	Choose the SCALE command (see Figure 5–16)
Modify menu	Choose Scale
Command: prompt	**scale** (ENTER)

Scale

Figure 5–16 *Invoking the SCALE command from the Modify toolbar*

AutoCAD prompts:

Command: **scale** (ENTER)
Select objects: *(select the objects to scale and press ENTER to complete the selection)*
Specify base point: *(specify a base point about which selected objects are to be scaled)*
Specify scale factor or [Reference]: *(specify a scale factor, or enter **r** to select the Reference option)*

The base point can be anywhere in the drawing. If the base point selected is on the selected object itself, then the selected base point becomes an anchor point for scaling. The scale factor multiplies the dimensions of the selected objects by the specified scale.

The following command sequence shows an example of enlarging a group of objects by a scale factor of 3, as shown in Figure 5–17.

Command: **scale** (ENTER)
Select objects: *(select the objects to enlarge, as shown in Figure 5–17, and press ENTER to complete the selection)*
Specify base point: *(pick the base point, as shown in Figure 5–17)*
Specify scale factor or [Reference]: **3** (ENTER)

AutoCAD enlarges the selected objects by a factor of 3, as shown in Figure 5–17.

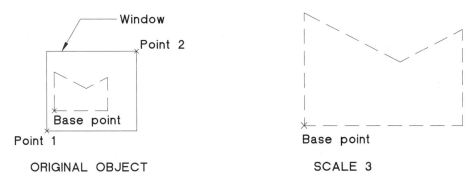

Figure 5–17 *Enlarging a group of objects by means of the* SCALE *command*

Reference Scale

The Reference option is used to scale objects relative to the current dimension. Specify the current dimension as a reference length, or select two endpoints of a line to be scaled, and specify the desired new length. AutoCAD will automatically calculate the scale factor and enlarge or shrink the object appropriately.

The following command sequence shows an example of using the SCALE command to enlarge a group of objects selected by means of the Window option in reference to the current dimension, as shown in Figure 5–18.

```
Command: scale (ENTER)
Select objects: (select the objects to enlarge, as shown in Figure 5–18)
Specify base point:(pick the base point, as shown in Figure 5–18)
Specify scale factor or [Reference]: r (ENTER)
Specify reference length <1>:3.8 (ENTER)
Specify new length: 4.8 (ENTER)
```

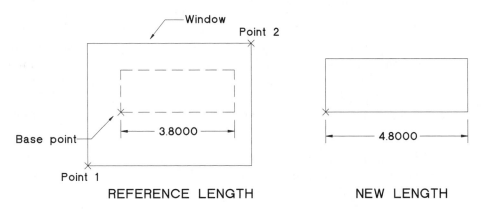

Figure 5–18 *Enlarging a group of objects by means of the* SCALE *command in reference to the current dimension*

MODIFYING POLYLINES

The PEDIT command allows you to modify polylines. In addition to using such modify commands as MOVE, COPY, BREAK, TRIM, and EXTEND, you can use the PEDIT command to modify polylines. The PEDIT command has special editing features for dealing with the unique properties of polylines and is perhaps the most complex AutoCAD command, with several multioption submenus totaling some 70 command options.

Invoke the PEDIT command:

Modify II toolbar	Choose the PEDIT command (see Figure 5–19)
Modify menu	Choose Object > Polyline
Command: prompt	**pedit** (ENTER)

Figure 5–19 *Invoking the* EDIT POLYLINE *command from the Modify II toolbar*

AutoCAD prompts:

Command: **pedit** (ENTER)
Select polyline: *(select polyline, line, or arc)*

If you select a line or an arc instead of a polyline, you are prompted as follows:

Object selected is not a polyline.
Do you want it to turn into one? <Y>

Responding Y or pressing ENTER turns the selected line or arc into a single-segment polyline that can then be edited. Normally this is done in order to use the **Join** option to add other connected segments that, if not polylines, will also be transformed into polylines. It should be emphasized at this time that in order to join segments together into a polyline, their endpoints must coincide. This occurs during line-line, line-arc, arc-line, and arc-arc continuation operations, or by using the endpoint Object Snap mode.

The second prompt does not appear if the first segment selected is already a polyline. It may even be a multisegment polyline. After the object selection process, you will be returned to the multioption prompt as follows:

Enter an option [Close/Join/Width/Edit vertex/Fit/Spline/Decurve/Ltype gen/Undo]: *(select one of the available options)*

Close

The Close option performs in a manner similar to the **Close** option of the LINE command. If, however, the last segment was a polyline arc, then the next segment will be similar to the arc-arc continuation, using the direction of the last polyarc as the starting direction and drawing another polyarc, with the first point of the first segment as the ending point of the closing polyarc.

Figures 5–20 and 5–21 show some examples of the application of the **Close** option.

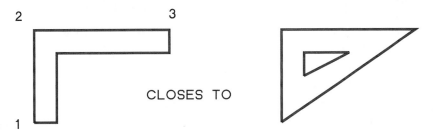

Figure 5–20 *Using the* PEDIT *command Close option with polylines*

Figure 5–21 *Using the* PEDIT *command Close option with polyacrs*

Open

The Open option deletes the segment that was drawn with the **Close** option. If the polyline had been closed by drawing the last segment to the first point of the first segment without using the **Close** option, then the **Open** option will not have a visible effect.

Join

The Join option takes selected lines, arcs, or polylines and combines them with a previously selected polyline into a single polyline if all segments are connected at sequential and coincidental endpoints.

Width

The Width option permits uniform or varying widths to be specified for polyline segments.

Edit Vertex

A vertex is the point where two segments join. When you select the **Edit Vertex** option, the visible vertices are marked with an X to indicate which one is to be modified. You can modify vertices of polylines in several ways. When you select the **Edit Vertex** option, AutoCAD prompts you with additional suboptions:

```
Command: pedit (ENTER)
Select polyline: (select a polyline)
Enter an option [Close/Join/Width/Edit vertex/Fit/Spline/Decurve/Ltype
   gen/Undo]: e (ENTER)
[Next/Previous/Break/Insert/Move/Regen/Straighten/Tangent/Width/
   eXit]<N>:
```

Next and **Previous** Whether or not you have modified the marked vertex, when you wish to move the mark to the next or previous vertex, you can use the N (Next) or P (Previous) option.

Break The Break option establishes the marked vertex as one vertex for the Break option and then prompts:

```
Enter an option [Next/Previous/Go/eXit] <N>:
```

The choices of the Break option permit you to step to another vertex for the second break point, or to initialize the break, or to exit the option. If two vertices are selected, you may use the Go option to have the segment(s) between the vertices removed. If you select the endpoints of a polyline, this option will not work. If you select the Go option immediately after the Break option, the polyline will be divided into two separate polylines. Or, if it is a closed polyline, it will be opened at that point.

Insert The Insert option allows you to specify a point and have the segment between the marked vertex and the next vertex become two segments meeting at the specified point. The selected point does not have to be on the polyline segment.

For example, the following command sequence shows the application of the Insert option as shown in Figure 5–22.

```
Command: pedit (ENTER)
Select polyline: (select the polyline as shown in Figure 5–22)
Enter an option [Close/Join/Width/Edit vertex/Fit/Spline/
Decurve/Ltype gen/Undo]:e (ENTER)
[Next/Previous/Break/Insert/Move/Regen/Straighten/Tangent/
   Width/eXit]<N>: i (ENTER)
Specify location for new vertex: (specify a new vertex)
```

BEFORE   AFTER

Figure 5–22 *Using the* PEDIT *command Insert option*

Move The Move option allows you to specify a point and have the marked vertex be relocated to the selected point.

For example, the following command sequence shows the application of the Move option, as shown in Figure 5–23.

```
Command: pedit (ENTER)
Select polyline: (select the polyline as shown in Figure 5–23)
Enter an option [Close/Join/Width/Edit vertex/Fit/Spline/
    Decurve/Ltype gen/Undo]: e (ENTER)
Enter a vertex editing option
[Next/Previous/Break/Insert/Move/Regen/Straighten/Tangent/
    Width/eXit]<N>: m (ENTER)
Specify new location for marked vertex: (specify the new location)
```

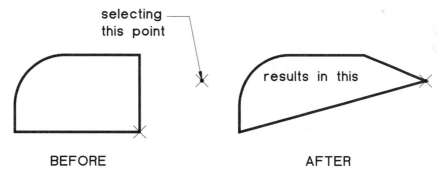

BEFORE   AFTER

Figure 5–23 *Using the* PEDIT *command Move option*

Regen The Regen option regenerates the polyline without having to cancel the PEDIT command to invoke the REGEN command at the Command: prompt.

Straighten The Straighten option establishes the marked vertex as one vertex for the Straighten option and then prompts:

```
Enter an option [Next/Previous/Go/eXit]<N>:
```

These choices of the Straighten option permit you either to step first to another vertex for the second point or to exit the option. When the two vertices are selected, you may use the Go option to have the segment(s) between the vertices replaced with a single straight-line segment.

For example, the following command sequence shows the application of the Straighten option, as shown in Figure 5–24.

```
Command: pedit (ENTER)
Select polyline: (select the polyline as shown in Figure 5–24)
Enter an option [Close/Join/Width/Edit vertex/Fit/Spline/
    Decurve/Ltype gen/Undo]: e (ENTER)
Enter a vertex editing option
[Next/Previous/Break/Insert/Move/Regen/Straighten/Tangent/
    Width/eXit]<N>: s (ENTER)
Enter an option [Next/Previous/Go/eXit]<N>: n (ENTER)
Enter an option [Next/Previous/Go/eXit]<N>: g (ENTER)
Enter an option [Next/Previous/Go/eXit]<N>: x (ENTER)
```

results in

BEFORE   AFTER

Figure 5–24 *Using the* PEDIT *command Straighten option*

Tangent The Tangent option permits you to assign to the marked vertex a tangent direction that can be used for the curve fitting option. The prompt is as follows:

Specify direction of vertex tangent:

You can either specify the direction with a point or type in the coordinates at the keyboard.

Width The Width option permits you to specify the starting and ending widths of the segment between the marked vertex and the next vertex. The prompt is as follows:

Specify new width for all segments:

For example, the following command sequence shows the application of the Width option as shown in Figure 5–25.

```
Command: pedit (ENTER)
Select polyline: (select a polyline)
Enter a vertex editing option
[Next/Previous/Break/Insert/Move/Regen/Straighten/Tangent/
    Width/eXit]<N>: e (ENTER)
Enter an option [Close/Join/Width/Edit vertex/Fit/Spline/
    Decurve/Ltype gen/Undo]: w (ENTER)
Specify new width for all segments: 0.25 (ENTER)
```

eXit The Exit option exits from the Vertex Editing option and returns to the PEDIT multioption prompt.

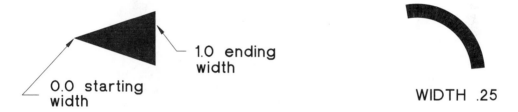

Figure 5–25 *Using the* PEDIT *command Width option*

Fit Curve

The Fit Curve option draws a smooth curve through the vertices, using any specified tangents.

Spline Curve

The Spline Curve option provides several ways to draw a curve based on the polyline being edited. These include Quadratic B-spline and Cubic B-spline curves.

Decurve

The Decurve option returns the polyline to the way it was drawn originally. See Figure 5–26 for differences between Fit Curve, Spline, and Decurve.

Undo

The Undo option reverses the latest PEDIT operation.

eXit

The Exit option exits the PEDIT command.

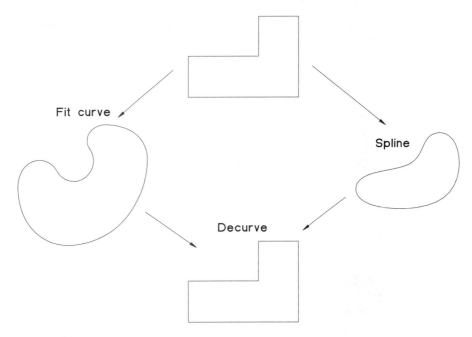

Figure 5–26 *Comparing* PEDIT *command Fit Curve, Spline Curve, and Decurve options*

MATCHING PROPERTIES

The MATCHPROP command allows you to copy selected properties from one object to one or more other objects located in the current drawing or any other drawing currently open. Properties that can be copied include color, layer, linetype, linetype scale, lineweight, thickness, plot style, and in some cases, dimension, text, and hatch.

Invoke the MATCHPROP command:

Standard toolbar	Choose the MATCH PROPERTIES **command** (see Figure 5–27)
Modify menu	Choose Match Properties
Command: prompt	**matchprop or painter** (ENTER)

Figure 5–27 *Invoking the* MATCH PROPERTIES *command from the Standard toolbar*

AutoCAD prompts:

> Command: **matchprop** (ENTER)
> Select Source Object: *(select the object whose properties you want to copy)*
> Select destination object(s) or [Settings]: *(select the destination objects and*
>   *press* ENTER *to terminate the selection, or enter* **s** *to select the Settings option*
>   *to display the Property Settings)*

Selection of the Settings option displays the Property Settings dialog box, similar to Figure 5–28. Use the Settings option to control which object properties are copied. By default, all object properties in the Property Settings dialog box are set to ON for copying. When the toggle button is set to ON, AutoCAD changes the settings of the destination object to that of the source object.

Figure 5–28 *Property Settings dialog box*

Choose the OK button to close the Property Settings dialog box. AutoCAD continues with the Select Destination Object(s): prompt. Press ENTER to complete the object selection.

PROJECT EXERCISE

This project exercise provides point-by-point instructions for setting up the drawing with layers and then creating the objects shown in the accompanying figure.

TYP.

3/8"

2 PL'S Ix7x2'-I"
(I-TOP & I-BOT.)

PL I/2xI2xI'-3"
(N.S. ONLY)

ALL BOLT HOLES
I5/I6" Ø FOR 7/8" Ø
A325 BOLTS

DETAIL - A

Figure P5–I *Completed project drawing (but with dimensions added for reference)*

In this project you will learn how to do the following:

- Set up the drawing, including Limits and Layers.
- Use the SKETCH, SOLID, and DONUT commands to create objects.
- Use the COPY command to create objects from other objects.
- Use the ROTATE and STRETCH commands to modify objects.
- Use the TEXT command to create text objects.

SET UP THE DRAWING AND DRAW A BORDER
Step I: Start the AutoCAD program.

Step 2: To create a new drawing, invoke the NEW command from the Standard toolbar or select New from the File menu.

AutoCAD displays the Create New Drawing dialog box. Choose the "Use a Wizard" button, then select the Quick Setup, and choose the **OK** button to close the dialog box.

AutoCAD displays the Quick Setup dialog box with the listing of the options available for unit of measurement. Select the "Architectural" radio button, and choose the **Next>** button.

AutoCAD displays the Quick Setup dialog box with the display for Area setting. Under the **Width:** text box enter **9'**, and under the **Length:** text box enter **7'**. Choose the **Finish** button to close the Quick Step dialog box.

Step 3: Invoke the LAYER command from the Object Properties toolbar, or select Layer... from the Format menu. AutoCAD displays the Layer Properties Manager dialog box.

Create six layers, and rename them as shown in the following table, assigning the appropriate color and linetype.

Layer Name	Color	Linetype	Lineweight
Border	Red	Continuous	Default
Object	White	Continuous	Default
Solid-Donut	White	Continuous	Default
Center	Red	Center	Default
Hidden	Magenta	Hidden	Default
Const	Blue	Continuous	Default

Set Border as the current layer, and close the Layer Properties Manager dialog box.

CREATING OBJECTS

Step 4: Draw the border using the PLINE command, with a width of 0.5" from 0'-2",0-2" to 0'-2",6'-10" to 8'-10",6'-10" to 8'-10",0'-2", and then close back to 0'-2",0-2".

Step 5: Make the Object layer current and draw the bottom plate using the RECTANGLE command from 3'-4",1'-6" to 5'-5",1'-7". Your drawing should look like Figure P5–2.

Figure P5–2 *Bottom plate*

Step 6: Copy the rectangle twice, with displacements of 0'-0",2'-1" and 0'-0",3'-1". See Figure P5–3.

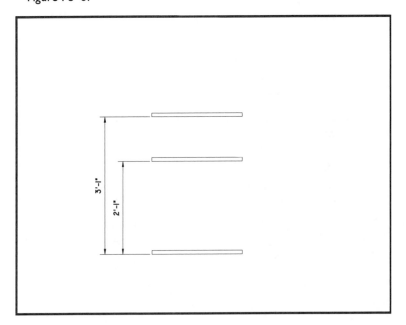

Figure P5–3 *Setting of the base*

Step 7:   Stretch the top rectangle with a displacement of 0",6" to change it from a 2'-1" x 1" rectangle to a 2'-1" x 7" rectangle. See Figures P5–4 and P5–5.

Figure P5–4 *Display with the stretch crossing*

Figure P5–5 *Result after stretching the top rectangle*

Step 8:   Draw a horizontal line through the top of the bottom rectangle, and then copy it five times up 1", up 1'-11", up 2'-0", up 2'-11", and up 3'-8". Your drawing should look like Figure P5–6.

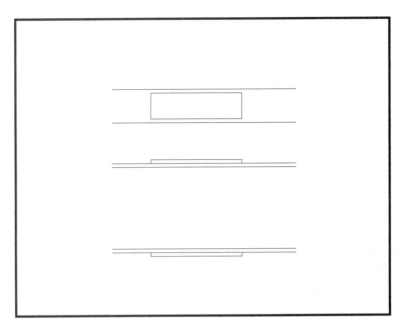

Figure P5–6 *Basic layout*

Step 9: Use the SKETCH command to create the end "breaks" of the top and side views of the beam. Your drawing should look like Figure P5–7.

Figure P5–7 *Drawing with sketch lines*

Step 10: While the Object layer is still current, draw the added lines shown in Figure P5–8. Refer to Figure P5–1 for the size of the objects, and make sure to draw in appropriate layers (Center and Hidden). When this step is completed, your drawing should look like Figure P5–9.

Figure P5–8 *Additional objects added to the layout*

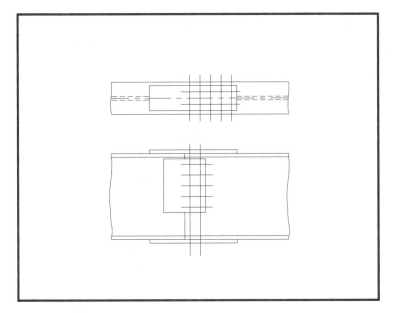

Figure P5–9 *Drawing after changing the linetypes*

Step 11: Set the Solid-Donut layer as the current layer. Create the solid-filled rectangles with the SOLID command. You can create each one individually, or you can create one and then COPY it to the required locations. Your drawing should then look like Figure P5–10.

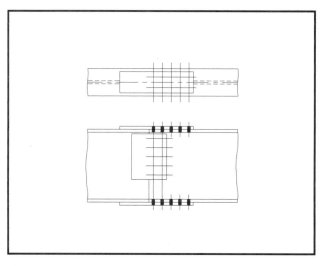

Figure P5–10 *Drawing with the solid objects drawn*

Step 12: Create the solid-filled circles with the DONUT command. The DONUT command permits you to place multiple created objects after you have specified the inside and outside diameters. These (the donuts) might be more easily created one by one simply by picking their individual insertion points without exiting the DONUT command. Your drawing should now look like Figure P5–11.

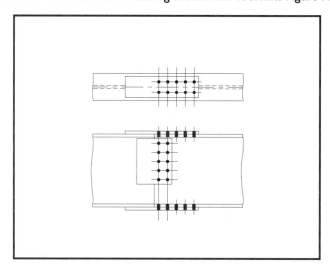

Figure P5–11 *Drawing with the donut objects drawn*

Step 13: Create the solid filled triangular arrowhead with the SOLID command. Draw the horizontal line and the circle as shown in Figure P5–12. Then use the ROTATE command to rotate the arrowhead, line, and circle by -45°, as shown in Figure P5–13.

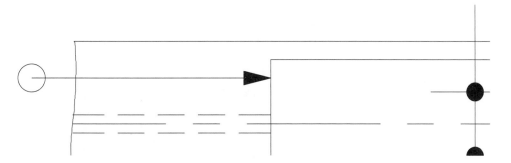

Figure P5–12 *Drawing with the solid-filled triangular arrowhead*

Figure P5–13 *Result after rotating the arrowhead -45 degrees*

Step 14: Copy and draw additional solid triangles, lines, circles, and open triangles as shown in Figure P5–14.

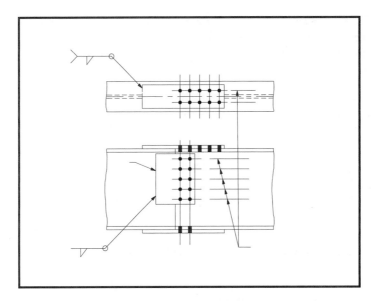

Figure P5–14 *Drawing after additional solid triangles have been created*

Step 15: Set the Text layer as the current layer. Using the TEXT command, draw the text objects shown in Figure P5-15. Save the drawing as Proj5.DWG.

Note: You'll have to use the BREAK command to create a gap in the "break" sketch line on the left end of the lower view. This is to leave a clear space for the text objects. FigureP5–15 is the completed drawing.

Figure P5–15 *Completed drawing*

EXERCISES 5–1 THROUGH 5–5

EXERCISE 5–1

Create the drawing shown according to the settings given in the following table:

Settings	Value
1. Units	Architectural
2. Limits	Lower left corner: 0',0'
	Upper right corner: 9',6'
3. Grid	4"
4. Snap	2"
5. Dimscale	8
6. Ltscale	4

Settings	Value		
7. Layers	*NAME*	*COLOR*	*LINETYPE*
	Center	Cyan	Center
	Border	Red	Continous
	Backboard	Green	Continuous
	Rim	Magenta	Hidden
	Text	Blue	Continuous

Hints: Make sure to make each layer current before beginning to draw any objects that are on it.

The border can be drawn from coordinates 3",3" to 8'-9",5'-6" using the RECTANGLE command.

Draw a vertical construction line (xline) through a point whose X coordinate is 4'-0".

Offset the vertical xline 2'-3" to each side of the line just created.

Draw a horizontal construction line (xline) through a point whose Y coordinate is 2'-0".

Offset the horizontal construction line 3" above and below the line just created. Offset the lower line 1'-2½" toward the bottom of the drawing.

The construction lines just created will guide you in drawing the circles and lines required for completing the outline of the object.

Using the intersections of the lines and/or circles, draw the sides and sloped bottom lines of the backboard on the Backboard layer.

Using the TRIM command, trim the lines and circles to complete the shape.

After the drawing is finished, either the Construction Layer can be turned OFF or the construction lines can be erased. Do not dimension.

Figure Ex5–1

EXERCISE 5–2

Create the drawing shown according to the settings given in the following table:

Settings	Value
1. Units	Architectural
2. Limits	Lower left corner: 0',0'
	Upper right corner: 12',9'
3. Grid	6"
4. Snap	3"

Settings	Value		
5. Layers	*NAME*	*COLOR*	*LINETYPE*
	Construction	Cyan	Continuous
	Border	Red	Continuous
	Object	Green	Continuous
	Center	Magenta	Center
	Text	Blue	Continuous
	Hidden	White	Hidden

Hints: Make sure to make each layer current before beginning to draw any objects that are on it.

The border can be drawn from coordinates 3",3" to 11'-9",8'-6" using the RECTANGLE command.

The front view of the wheel handle can be drawn using the PLINE command. The top of the handle is 8" above the top of the closure plate.

The circle part of the top view of the wheel handle can be drawn using either the DONUT command or the PLINE command with the arc option, creating two 180-degree polyarcs with a width of 1".

Figure Ex5–2

EXERCISE 5–3

Create the drawing shown according to the settings given in the following table:

Settings	Value
1. Units	Architectural
2. Limits	Lower left corner: 0',0'
	Upper right corner: 70',50'
3. Grid	12"
4. Snap	3"

Settings	Value		
5. Layers	*NAME*	*COLOR*	*LINETYPE*
	Construction	Cyan	Continuous
	Border	Red	Continuous
	Object	Green	Continuous
	Center	Magenta	Center
	Text	Blue	Continuous
	Hidden	White	Hidden

Hints: Open the drawing that you created in Exercise 5–2 and save is as EX5–3. Change the limits. Resize the border. Erase the front view and centerlines connecting the views. Be sure to make each layer current before beginning to draw any objects that are on it.

Now you can move the top view of the containment closure device to become the lower left object in the array, and use the ROTATE command to reorient the hinge from the right side to the top (+90°).

Before using the ARRAY command to create the 11 other closure devices, the hidden-line rectangle can be created and arrayed with the closure device.

Figure Ex5–3

EXERCISE 5-4

Create the drawing shown according to the settings given in the following table:

Settings	Value		
1. Units	Architectural		
2. Limits	Lower left corner: 0',0'		
	Upper right corner: 60',45'		
3. Grid	12"		
4. Snap	6"		
5. Layers	*NAME*	*COLOR*	*LINETYPE*
	Construction	Cyan	Continuous
	Border	Red	Continuous
	Object	Green	Continuous
	Text	Blue	Continuous

Hints: Be sure to make each layer current before beginning to draw any objects that are on it.

The border can be drawn from coordinates 2',2" to 58',43" using the RECTANGLE command.

The outline of the foundation wall can be drawn with the PLINE command. Then, by using the OFFSET command (set at 6"), create the additional four lines, each one from the previous.

The Detail of Stair Riser and Sections "A" and "B" can be drawn to true dimensions. Then you can use the SCALE command to enlarge the view of the Detail by a factor of 1.5 and to enlarge the view of the Sections by a factor of 2.0.

Figure Ex5–4

EXERCISE 5–5

Create the drawing shown according to the settings given in the following table:

Settings	Value		
1. Units	Architectural		
2. Limits	Lower left corner: 0',0'		
	Upper right corner: 60',45'		
3. Grid	12"		
4. Snap	6"		
5. Layers	*NAME*	*COLOR*	*LINETYPE*
	Construction	Cyan	Continuous
	Border	Red	Continuous
	Object	Green	Continuous
	Text	Blue	Continuous
	Hidden	White	Hidden

Hints: Open the drawing that you created in Exercise 5–4 and save it as EX5–5. Erase the Detail of Stair Riser and Sections "A" and "B."

Erase the internal double squares and the inside line of the foundation wall in the plan.

Change the foundation wall lines to be on the Hidden layer.

The roof lines can be drawn with polylines with a width of 1/2".

ROOFING PLAN
SCALE 1/2"=1'-0"

Figure Ex5–5

REVIEW QUESTIONS

1. The command that allows you to draw freehand lines is:

   a. SKETCH

   b. DRAW

   c. PLINE

   d. FREE

2. To create a six-sided area which would select all the objects completely within it, you should respond to the "Select Objects:" prompt with:

   a. WP

   b. CP

   c. W

   d. C

3. The CHPROP command does not allow you to modify an object's:

   a. linetype

   b. line width

   c. color

   d. layer

4. By default, most selection set prompts default to which option:

   a. All

   b. Auto

   c. Box

   d. Window

   e. Crossing

5. The multiple option selection sets allows you to:

   a. select multiple objects which lie on top of each other

   b. scan the database only once to find multiple objects

   c. use the window or crossing options

   d. both A and B

   e. both B and C

6. In regard to using the SOLID command, which of the following statements is true:

    a. The order of point selection is unimportant

    b. FILL must be turned in order to use the SOLID command

    c. The points must be selected on existing objects

    d. The points must be selected in a clockwise order

    e. None of the above

7. What command is commonly used to create a filled rectangle?

    a. FILL-ON

    b. PLINE

    c. RECTANGLE

    d. LINE

    e. SOLID

8. To turn a series of line segments into a polyline, one uses which option of the PEDIT command?

    a. Join

    b. Fit curve

    c. Spline

    d. Connect

    e. Line segments cannot be connected.

9. Which command allows you to change the size of an object, where the X and Y scale factors re changed equally?

    a. ROTATE

    b. SCALE

    c. SHRINK

    d. MODIFY

    e. MAGNIFY

10. The BOX option for creating selection sets is most useful for:

    a. creating custom functions

    b. creating rectangular selection areas

    c. extending a selection into 3D

    d. Nothing, it is not a valid option

11. Using the SCALE command, what number would you enter to enlarge an object by 50%?

    a. 0.5

    b. 50

    c. 3

    d. 1.5

12. To avoid changing the location of an object when using the SCALE command:

    a. the reference length should be less than the limits

    b. the scale factor should be less than one

    c. the base point should be on the object

    d. the base point should be at the origin

13. The ROTATE command is used to rotate objects around:

    a. any specified point

    b. the point -1,-1 only

    c. the origin

    d. is only usable in 3D drawings

    e. none of the above

14. The remove option for forming selection sets deletes objects from the drawing in much the same manner as the ERASE command.

    a. True

    b. False

15. Which of the following is not supported by the noun/verb feature?

    a. COPY

    b. MOVE

    c. TRIM

    d. ROTATE

    e. ERASE

16. When noun/verb selection is set to ON, you may add objects to the selection set by:

    a. picking the objects

    b. windowing the objects

    c. using SHIFT to add

    d. using CTRL to add

Directions: Use Figure CH5RQ–I to answer the following questions.

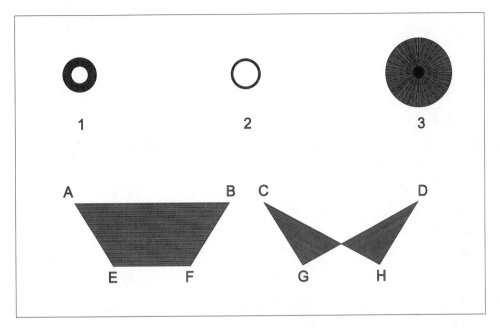

Figure QX5–1

17. Which of the figures were drawn with the DONUT command?

   a. I
   b. 2
   c. 3
   d. I & 2
   e. I, 2 & 3

18. Which of the DONUT objects has the smallest inside diameter?

   a. I
   b. 2
   c. 3

19. DONUTs are positioned in a manner similar to which OSNAP option?

   a. QUADrant
   b. CENter
   c. TsANgent
   d. NODe

20. Which of the options below is the correct sequence to create the SOLID object on the left?

   a. A, B, F, E

   b. A, E, F, B

   c. A, B, E, F

   d. F, E, A, B

   e. B, F, A, E

21. Which of the options below is the correct sequence to create the SOLID object on the right?

   a. H, G, C, D

   b. G, H, D, C

   c. G, D, H, C

   d. D, C, H, G

   e. C, G, H, D

22. If the space between the two SOLID shapes were to be filled what sequence of points would completely fill this space?

   a. G, B, F, C

   b. B, G, F, C

   c. C, B, F, G

   d. F, G, C, B

   e. B, F, C, G

CHAPTER 6

Fundamentals V

INTRODUCTION

After completing this chapter, you will be able to do the following:

- Construct geometric figures by means of the MULTILINE (MLINE) and SPLINE commands
- Modify multilines and splines
- Create or modify multiline styles
- Use grips to modify objects, group objects, and filter certain types of objects by Quick Select for modification
- Use INQUIRY commands
- Change the settings of system variables

MULTILINES

AutoCAD allows you to draw multiple parallel line segment with the MLINE command. In addition, you can also modify the intersections of two or more multilines or cut gaps with the MLEDIT command. AutoCAD also allows you to create a new multiline style or edit an existing one comprised of up to 16 lines, called elements, with the MLSTYLE command.

DRAWING MULTIPLE PARALLEL LINES

The MLINE command allows you to draw multiple parallel line segments, similar to polyline segments that have been offset one or more times. Examples of applying the MLINE command are shown in Figure 6–1.

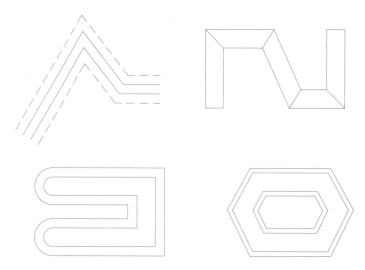

Figure 6–1 *Examples of multilines*

The properties of each element of a multiline are determined by the style that is current when the multiline is drawn. The properties of the multiline that can be determined by the style include whether to display the line at the joints (miters) and ends and whether to close the ends with a variety of half circles, connecting inner and/or outer elements. In addition, the style controls the element properties, such as color, linetype, and offset distance between two parallel lines.

Invoke the MLINE command:

Draw toolbar	Choose the MULTILINE command (see Figure 6–2)
Draw menu	Choose Multiline
Command: prompt	**mline** (ENTER)

Figure 6–2 *Invoking the MULTILINE command from the Draw toolbar*

AutoCAD prompts:

> Command: **mline** (ENTER)
> Specify start point or [Justification/Scale/STyle]: *(specify a style or right-click for a shortcut menu and select one of the avilable options)*

Start Point

The Start point option (default) lets you specify the starting point of the multiline, known as its origin. Once you specify the starting point for a multiline, AutoCAD prompts:

> Specify next point:

When you respond by selecting a point, the first multiline segment is drawn according to the current style. You are then prompted:

> Specify next point or [Undo]:

If you specify a point, the next segment is drawn, along with segments of all other elements specified by the current style. After two segments have been drawn the prompt will include the Close option:

> Specify next point or [Close/Undo]:

Choosing the Close option causes the next segment to join the starting point of the multiline, fillets all elements, and exits the command.

Selecting Undo (u) after any segment is drawn (and the MLINE command has not been terminated) causes the last segment to be erased, and you are prompted again for a point.

Justification

The Justification option determines the relationship between the elements of the multiline and the line you specify by way of the placement of the points. The justification is set by selecting one of the three available suboptions.

> Enter justification type [Top/Zero/Bottom] <top>:

The **Top** option causes the element with the greatest offset value to be drawn on the line of selected points. All other elements will be to the right of the line of points as viewed from the starting point to the ending point of each segment. In other words, if the line is drawn left to right, the line of points (and the element with the greatest offset value) will be on top of (above) all other elements.

The **Zero** Option causes the baseline to coincide with the line of selected points. Elements with positive offsets will be to the right of, and those with negative offsets to the left of, the line of selected points, as viewed from the starting point to the ending point of the each segment.

The **Bottom** Option causes the element with the least offset value to be drawn on the line of selected points. All other elements will be to the left of the line of points as viewed from the starting point to the ending point of each segment. In other words, if the line is drawn left to right, the line of points (and the element with the least offset value) will be on the bottom of (below) all other elements.

Figure 6–3 shows the location for various justifications.

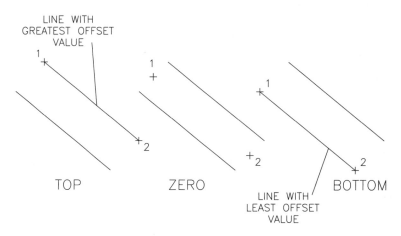

Figure 6–3 *Location of various justifications*

Scale

The Scale option determines the value used for offsetting elements when drawing them relative to the values assigned to them in the style. For instance, if the scale is changed to 3.0, elements that are assigned 0.5 and −1.5 will be drawn with offsets of 1.5 and −4.5, respectively. If a negative value is given for the scale, then the signs of the values assigned to them in the style will be changed (positive to negative and negative to positive). The value can be entered in decimal form or as a fraction. A 0 (zero) scale value produces a single line.

Style

The Style option sets the current multiline style from the available styles. AutoCAD prompts:

> Specify start point or [Justification/Scale/STyle]: **st** (ENTER)
> Enter mline style name or [?]: *(specify the name of an existing style)*

Detailed explanation is provided later in this chapter for creating or modifying multiline styles.

EDITING MULTIPLE PARALLEL LINES

The MLEDIT command helps you modify the intersections of two or more multilines or cut gaps in the lines of one multiline. The tools are available for the type of intersection operated on (cross, tee, or vertex) and if one or more elements need to be cut or welded.

Invoke the MLEDIT command:

Modify II toolbar	Choose the EDIT MULTILINE command (see Figure 6–4)
Modify menu	Choose > Object > Multiline
Command: prompt	**mledit** (ENTER)

Figure 6–4 *Invoking the* EDIT MULTILINE *command from the Modify II toolbar*

AutoCAD displays the Multiline Edit Tools image dialog box, as shown in Figure 6–5.

Figure 6–5 *Multiline Edit Tools dialog box*

To choose one of the available options, select one of the image tiles. AutoCAD then prompts for appropriate information.

The first column in the Multiline Edit Tools image dialog box works on multilines that cross, the second works on multilines that form a tee, the third works on corner joints and vertices, and the fourth works on multilines to be cut or welded.

The **Closed Cross** option cuts all lines that make up the second multiline you select at the point where it crosses the first multiline, as shown in Figure 6–6. Choose the Closed Cross image tile, as shown in Figure 6–7, to invoke the Closed Cross option. AutoCAD then prompts:

Select first mline: *(select the first multiline, as shown in Figure 6–6)*
Select second mline: *(select the second multiline, as shown in Figure 6–6)*

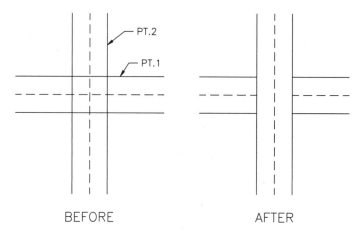

BEFORE   AFTER

Figure 6–6 *An example of a closed cross*

Figure 6–7 *Invoking the Closed Cross option from the Multiline Edit Tools image dialog box*

After the closed cross intersection is created, AutoCAD prompts:

Select first mline (or Undo): *(select another multiline, enter **u**, or press ENTER)*

Selecting another multiline repeats the prompt for the second mline. Entering **u** undoes the closed cross just created.

The **Open Cross** option cuts all lines that make up the first multiline you select and cuts only the outside line of the second multiline, as shown in Figure 6–8. Choose the Open Cross image tile, as shown in Figure 6–9, to invoke the Open Cross option; AutoCAD prompts:

Select first mline: *(select a multiline, as shown in Figure 6–8)*
Select second mline: *(select a multiline that intersects the first multiline, as shown in Figure 6–8)*

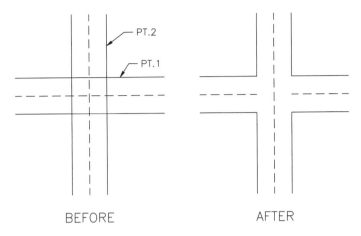

BEFORE   AFTER

Figure 6–8 *An example of an open cross*

Figure 6–9 *Invoking the Open Cross option from the Multiline Edit Tools image dialog box*

After the open cross intersection is created, AutoCAD prompts:

Select first mline (or Undo): *(select another multiline, enter **u**, or press* ENTER*)*

Selecting another multiline repeats the prompt for the second mline. Entering **u** undoes the open cross just created.

The **Merged Cross** option cuts all lines that make up the intersecting multiline you select except the centerlines, as shown in Figure 6–10. Choose the Merged Cross image tile, as shown in Figure 6–11, to invoke the Merged Cross option; AutoCAD prompts:

Select first mline: *(select a multiline, as shown in Figure 6–10)*
Select second mline: *(select a multiline that intersects the first multiline, as shown in Figure 6–10)*

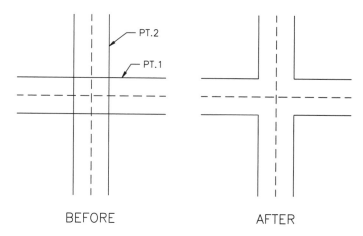

BEFORE   AFTER

Figure 6–10 *An example of a merged cross*

Figure 6–11 *Invoking the Merged Cross option from the Multiline Edit Tools image dialog box*

After the merged cross intersection is created, AutoCAD prompts:

Select first mline (or Undo): *(select another multiline, enter **u**, or press* ENTER*)*

Selecting another multiline repeats the prompt for the second mline. Entering u undoes the merged cross just created. The order in which the multilines are selected is not important.

The **Closed Tee** option extends or shortens the first multiline you identify to its intersection with the second multiline, as shown in Figure 6–12. Choose the Closed Tee image tile, as shown in Figure 6–13, to invoke the Closed Tee option; AutoCAD prompts:

Select first mline: *(select the multiline to trim or extend, as shown in Figure 6–12)*
Select second mline: *(select the intersecting multiline, as shown in Figure 6–12)*

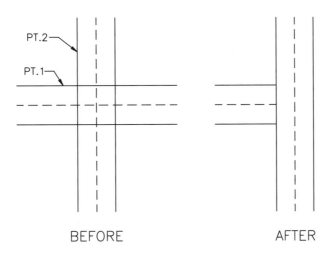

Figure 6–12 *An example of a closed tee*

Figure 6–13 *Invoking the Closed Tee option from the Multiline Edit Tools image dialog box*

After the closed tee intersection is created, AutoCAD prompts:

> Select first mline (or Undo): *(select another multiline, enter **u**, or press* ENTER)

Selecting another multiline repeats the prompt for the second mline. Entering **u** undoes the closed tee just created.

The **Open Tee** option is similar to the Closed Tee option, except it leaves an open end at the intersecting multiline, as shown in Figure 6–14. Choose the Open Tee image tile, as shown in Figure 6–15, to invoke the Open Tee option; AutoCAD prompts:

> Select first mline: *(select the multiline to trim or extend, as shown in Figure 6–14)*
> Select second mline: *(select the intersecting multiline, as shown in Figure 6–14)*

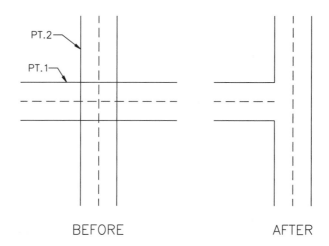

Figure 6–14 *An example of an open tee*

Figure 6–15 *Invoking the Open Tee option from the Multiline Edit Tools image dialog box*

After the open tee intersection is created, AutoCAD prompts:

> Select first mline (or Undo): *(select another multiline, enter **u**, or press* ENTER*)*

Selecting another multiline repeats the prompt for the second mline. Entering **u** undoes the open tee just created.

The **Merged Tee** option is similar to the Open Tee option, except the centerline of the first multiline is extended to the center of the intersecting multiline, as shown in Figure 6–16. Choose the Merged Tee image tile, as shown in Figure 6–17, to invoke the Merged Tee option; AutoCAD prompts:

> Select first mline: *(select the multiline to trim or extend, as shown in Figure 6–16)*
> Select second mline: *(select the intersecting multiline, as shown in Figure 6–16)*

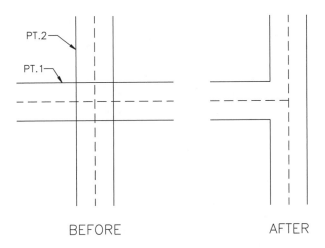

BEFORE   AFTER

Figure 6–16 *An example of a merged tee*

Figure 6–17 *Invoking the Merged Tee option from the Multiline Edit Tools image dialog box*

After the merged tee intersection is created, AutoCAD prompts:

> Select first mline (or Undo): *(select another multiline, enter **u**, or press* ENTER*)*

Selecting another multiline repeats the prompt for the second mline. Entering **u** undoes the merged tee just created.

The **Corner Joint** option lengthens or shortens each of the two multilines you select as necessary to create a clean intersection, as shown in Figure 6–18. Choose the Corner Joint image tile, as shown in Figure 6–19, to invoke the Corner Joint option; AutoCAD prompts:

> Select first mline: *(select the multiline to trim or extend, as shown in Figure 6–18)*
> Select second mline: *(select the intersecting multiline, as shown in Figure 6–18)*

BEFORE   AFTER

Figure 6–18 *An example of a corner joint*

Figure 6–19 *Invoking the Corner Joint option from the Multiline Edit Tools image dialog box*

After the corner joint intersection is created, AutoCAD prompts:

Select first mline (or Undo): *(select another multiline, enter **u**, or press* ENTER*)*

Selecting another multiline repeats the prompt for the second mline. Entering **u** undoes the corner joint just created.

The **Add Vertex** option adds a vertex to a multiline, as shown in Figure 6–20. Choose the Add Vertex image tile, as shown in Figure 6–21, to invoke the Add Vertex option; AutoCAD prompts:

Select mline: *(select a multiline to add a vertex, as shown in Figure 6–20)*

Figure 6–20 *An example of adding a vertex*

Figure 6–21 *Invoking the Add Vertex option from the Multiline Edit Tools image dialog box*

AutoCAD adds the vertex at the selected point and prompts:

Select mline (or Undo): *(select another multiline, enter **u**, or press ENTER)*

Selecting another multiline allows you to add another vertex. Entering **u** undoes the vertex just created.

The **Delete Vertex** option deletes a vertex from a multiline, as shown in Figure 6–22. Choose the Delete Vertex image tile, as shown in Figure 6–23, to invoke the Delete Vertex option; AutoCAD prompts:

Select mline: *(select a multiline to delete a vertex, as shown in Figure 6–22)*

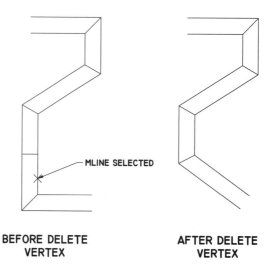

BEFORE DELETE
VERTEX

AFTER DELETE
VERTEX

Figure 6–22 *An example of deleting a vertex*

Figure 6–23 *Invoking the Delete Vertex option from the Multiline Edit Tools image dialog box*

AutoCAD deletes the vertex at the selected point and prompts:

> Select mline (or Undo): *(select another multiline, enter **u**, or press* ENTER)

Selecting another multiline allows you to delete another vertex. Entering **u** undoes the operation and displays the "Select mline:" prompt.

The **Cut Single** option cuts a selected element of a multiline between two cut points, as shown in Figure 6–24. Choose the Cut Single image tile, as shown in Figure 6–25, to invoke the Cut Single option; AutoCAD prompts:

> Select mline: *(select a multiline and the selected point becomes the first cut point, as shown in Figure 6–24)*

Select second point: *(select the second cut point on the multiline, as shown in Figure 6–24)*

BEFORE   AFTER

Figure 6–24 *An example of removing a selected element of a multiline between two cut points*

Figure 6–25 *Invoking the Cut Single option from the Multiline Edit Tools image dialog box*

AutoCAD cuts the multiline and prompts:

Select mline (or Undo): *(select another multiline, enter **u**, or press* ENTER*)*

Select another multiline to continue, or enter **u** to undo the operation and display the Select mline: prompt.

The **Cut All** option removes a portion of the multiline you select between two cut points, as shown in Figure 6–26. Choose the Cut All image tile, as shown in Figure 6–27, to invoke the Cut All option; AutoCAD prompts:

Select mline: *(select a multiline and the selected point becomes first cut point, as shown in Figure 6–26)*
Select second point: *(select the second cut point on the multiline, as shown in Figure 6–26)*

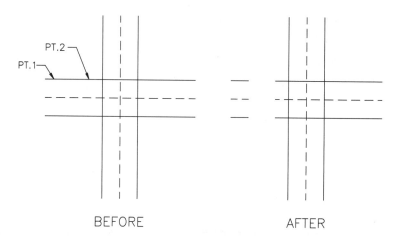

Figure 6-26 *An example of removing a portion of a multiline between two cut points*

Figure 6-27 *Invoking the Cut All option from the Multiline Edit Tools image dialog box*

AutoCAD cuts all the elements of the multiline and prompts:

> Select mline (or Undo): *(select another multiline, enter **u**, or press ENTER)*

Select another multiline to continue, or enter **u** to undo the operation and display the Select mline: prompt.

The **Weld All** option rejoins multiline segments that have been cut, as shown in Figure 6-28. Choose the Weld All image tile, as shown in Figure 6-29, to invoke the Weld All option; AutoCAD prompts:

> Select mline: *(select the multiline, as shown in Figure 6-28)*
> Select second point: *(select the endpoint on the multiline to be joined, as shown in Figure 6-28)*

BEFORE   AFTER

Figure 6–28 *An example of rejoining multiline segments that have been cut*

Figure 6–29 *Invoking the Weld All option from the Multiline Edit Tools image dialog box*

AutoCAD joins the multiline and prompts:

> Select mline (or Undo): *(select another multiline, enter **u**, or press* ENTER*)*

Select another multiline to continue, or enter **u** to undo the operation and display the Select mline: prompt.

CREATING AND MODIFYING MULTILINE STYLES

The MLSTYLE command is used to create a new Multiline Style or edit an existing one. You can define a multiline style comprised of up to 16 lines, called elements. The style controls the number of elements and the properties of each element. In addition you can specify the background color and the end caps of each multiline.

Invoke the MLSTYLE command:

Format menu	Choose Multiline Style
Command: prompt	**mlstyle** (ENTER)

AutoCAD displays the Multiline Styles dialog box, as shown in Figure 6–30.

Figure 6–30 *Multiline Styles dialog box*

The **Current:** list box allows you to choose from the available multiline styles loaded in the current drawing. To make a specific multiline style current, first choose the appropriate multiline style from the list box and then choose **OK** to close the dialog box. If necessary, you can load additional multiline styles into the current drawing by means of the Load option.

The **Name:** text box allows you to specify a name for a new Multiline Style or rename an existing one.

To create a new Multiline Style, first define the element and multiline properties, then enter a name for the newly created Multiline Style in the **Name:** text box. Choose the **Save...** button, and AutoCAD displays the Save Multiline Style dialog box shown in Figure 6–31. By default, AutoCAD saves the Multiline Style definition in the ACAD.MLN library file. If necessary, you can select another file or provide a new file name with .MLN as the file extension. Choose the **Save** button to save the multiline .MLN library file.

To make the newly created Multiline Style current, choose the **Add** button. AutoCAD then adds the newly created multiline style to the **Current:** list box and makes it the current Multiline Style.

The **Description:** text box allows you to add a description of up to 255 characters, including spaces.

Figure 6–31 *Save Multiline Style dialog box*

The **Load** button allows you to load a Multiline Style from a multiline library file into the current drawing. To load a Multiline Style into the current drawing, choose the **Load** button. AutoCAD then displays the Load Multiline Styles dialog box, as shown in Figure 6–32. Select one of the available Multiline Styles from the list box. If you need to load a Multiline Style from a different library file, then choose the **File** button. AutoCAD lists the available multiline library files. Select the appropriate library file, and AutoCAD lists the Multiline Styles available from the library file selected. After Choosing the Multiline Style, choose the **OK** button to close the dialog box.

Figure 6–32 *Load Multiline Styles dialog box*

The **Add** button, as mentioned earlier, allows you to add the newly created Multiline Style to the Current: list box.

ELEMENT PROPERTIES
The **Element Properties** button displays an Element Properties dialog box similar to the one shown in Figure 6–33, with options to set or change the offset, number, line type, or color of the multiline elements.

Figure 6–33 *Element Properties dialog box*

The **Elements:** list box displays existing elements along with the color and linetype of each.

Choose the **Add** or **Delete** button to add a line element to the multiline style or delete a line element from the multiline style, respectively.

The **Offset** text box sets the distance from the baseline for the elements selected from the **Elements:** list box. AutoCAD will use this distance to offset the element when the multiline is drawn and the scale factor is set to 1.0. Otherwise, the offset will be a ratio determined by the scale value.

The **Color** button displays the Select Color dialog box for setting the color for line elements (selected from Elements: list box) in the multiline style.

The **Linetype** button displays the Select Linetype dialog box for setting the linetype for line elements (selected from the Elements: list box) in the multiline style.

Multiline Properties

The **Multiline Properties** button displays a Multiline Properties dialog box similar to the one shown in Figure 6–34, with options to change the type of joints, end caps (and their angles), and background color for the multiline.

The **Display Joints** toggle button controls the display of the joints (miter) at the vertices of each multiline segment, as shown in Figure 6–35.

The **Caps** option has four suboptions to specify the appearance of multiline start and end caps. The Line toggle button controls the display of the start and end caps by a straight line, as shown in Figure 6–36.

The **Outer Arc** toggle button controls the display of the start and end caps by connecting the ends of the outermost elements with a semicircular arc, as shown in Figure 6–37.

Figure 6–34 *Multiline Properties dialog box*

DISPLAY JOINTS
OFF

DISPLAY JOINTS
ON

Figure 6–35 *Display of the joints*

BEFORE LINE

AFTER LINE

Figure 6–36 *Display of the line for start and end caps*

BEFORE OUTER
ARC

AFTER OUTER
ARC

Figure 6–37 *Display of the outer arc for start end caps*

The **Inner Arcs** toggle button controls the display of the start and end caps by connecting the ends of the innermost elements with a semicircular arc, as shown in Figure 6–38. For a multiline with an odd number of elements, the center element is not connected. For an even number of elements, connected elements are paired with elements that are the same number from each edge. For example, the second element from the left outer will be connected to the second element from the right outer, the third to the third, and so forth.

BEFORE INNER
ARC

AFTER INNER
ARC

Figure 6–38 *Display of the inner arc for start and end caps*

The **Angle** edit field sets the angle of endcaps, as shown in Figure 6–39.

The **Fill** toggle button controls the background fill of the multiline.

The **Color** button displays the Select Color dialog box and sets the color of the background fill.

BEFORE ANGLE   AFTER ANGLE

Figure 6–39 *Display of the end caps with an angular cap*

Once you have set the appropriate element properties and multiline properties for a new multiline style, enter the name and description in the **Name:** and **Description:** edit fields, respectively. Choose the **Save** button to save the newly created multiline style.

SPLINE CURVES

DRAWING SPLINE CURVES

The SPLINE command is used to draw a curve through or near points in a series. The type of curve is a nonuniform rational B-spline (NURBS). This type is used for drawing curves with irregularly varying radii, such as topographical contour lines.

The spline curve is drawn through a series of two or more points, with options either to specify end tangents or to use Close to join the last segment to the first. Another option lets you specify a tolerance, which determines how close to the selected points the curve is drawn.

Invoke the SPLINE command:

Draw toolbar	Choose the SPLINE command (see Figure 6–40)
Draw menu	Choose Spline
Command: prompt	**spline** (ENTER)

Figure 6–40 *Invoking the SPLINE command from the Draw toolbar*

AutoCAD prompts:

> Command: **spline** (ENTER)
> Specify first point or [Object]: *(specify a point or enter **o** for the Object option)*

The default option lets you specify the point from which the spline starts and to which it can be closed. After entering the first point, a preview line appears. You will then be prompted:

> Specify next point:

When you respond by specifying a point, the spline segments are displayed as a preview spline, curving from the first point, through the second point, and ending at the cursor. You are then prompted:

> Specify next point or [Close/Fit tolerance] <start tangent>:

If you specify a point, the next segment is added to the spline. This will occur with each additional point specified until you use the **Close** option or enter a null response by pressing ENTER.

Choosing a null response terminates the specifying of segment determining points. You are then prompted for a start tangent determining point as follows:

> Specify start tangent:

If you specify a point (for tangency), its direction from the start point determines the start tangent. If you press ENTER, the direction from the first point to the second point determines the tangency. After the start tangency is established, you are prompted:

> Specify end tangent:

If you specify a point (for tangency), its direction from the endpoint determines the end tangent. If you press ENTER, the direction from the last point to the previous point determines the tangency.

The **Close** option uses the original starting point of the first spline segment as the endpoint of the last segment and terminates segment placing. You are then prompted:

> Specify tangent:

You can specify a point to determine the tangency at the connection of the first and last segments. If you press ENTER, AutoCAD calculates the tangency and draws the spline accordingly. You can also use the **Perp** or **Tan** options to cause the tangency of the spline to be perpendicular or tangent to a selected object.

The **Fit Tolerance** option lets you vary how the spline is drawn relative to the selected points. You are then prompted:

> Specify fit tolerance <current>:

Entering 0 (zero) causes the spline to pass through the specified points. A positive value causes the spline to pass within the specified value of the points.

Object option

The **Object** option is used to change spline-fit polylines into splines. This can be used for 2D or 3D polylines, which will be deleted depending on the setting of the DELOBJ system variable.

EDITING SPLINE CURVES

Splines created by means of the SPLINE command have numerous characteristics that can be changed with the SPLINEDIT command. These include quantity and location of fit points, end characteristics such as open/close and tangencies, and tolerance of the spline (how near the spline is drawn to fit points).

SPLINEDIT operations on control points (which are different than fit points) of the selected spline include adding control points (with the ADD or the ORDER option) and changing the weight of individual control points, which determines how close the spline is drawn to individual control points.

Invoke the SPLINEDIT command:

Modify II toolbar	Choose the EDIT SPLINE command (see Figure 6–41)
Modify menu	Choose Object > Spline
Command: prompt	**splinedit** (ENTER)

Edit Spline

Figure 6–41 *Invoking the* EDIT SPLINE *command from the Modify II toolbar*

AutoCAD prompts:

> Command: **splinedit** (ENTER)
> Select Spline: *(specify a spline curve)*
> Enter an option [Fit data/Open/Move vertex/Refine/rEverse/Undo]: *(right-click to choose one of the available options from the shortcut menu)*

Control points appear in the grip color, and, if the spline has fit data, fit points also appear in the grip color. If you select a spline whose fit data is deleted, then the Fit Data option is not available. A spline can lose its fit data if you use the purge option while editing fit data, refine the spline, move its control vertices, fit the spline to a tolerance, or open or close the spline.

The Open option will replace Close if you select a closed spline, and vice versa.

Fit Data

The **Fit Data** option allows you to edit the spline by providing the following suboptions:

> Enter a fit data option [Add/Open/Delete/Move/Purge/Tangents/
> toLerance/eXit]<eXit>:

The **Add** suboption allows you to add fit points to the selected spline. AutoCAD prompts:

> Specify control point <exit>: *(select a fit point)*

After you select one of the fit points, AutoCAD highlights it and you are prompted for the next point:

> Specify new point <exit>: *(specify a point)*
> Specify new point <exit>: *(specify another point or press* ENTER*)*

Selecting a point places a new fit point between the highlighted ones.

The **Close** suboption closes an open spline smoothly with a segment or smoothes a spline with coincidental starting and ending points.

The **Open** suboption opens a closed spline, disconnecting it and changing the starting and ending points.

The **Delete** suboption deletes a selected fit point.

The **Move** suboption moves fit options to a new location by prompting:

> Specify new location or [Next/Previous/Select point/eXit]<N>:

The **Next** option steps forward through fit points.

The **Previous** option steps backwards through fit points.

The **Select** option permits you to select a fit point.

The **eXit** option exits this set of suboptions.

The **Enter New Location** option moves the highlighted point to a specified point.

The **Purge** suboption deletes fit data for the selected spline.

The **Tangents** suboption edits the start and end tangents of a spline by prompting:

> Specify start tangent or [System default]: *(specify a point, enter an option, or press* ENTER*)*
> Specify end tangent or [System default]: *(specify a point, enter an option, or press* ENTER*)*

For a closed spline, the prompt is:

> Specify tangent or [System default]:

If you choose the System Default, AutoCAD calculates the tangents. You can choose Tan or Perp and select an object for the spline tangent to be tangent to or perpendicular to the object selected.

The **toLerance** suboption refits the spline to the existing points with new tolerance values by prompting:

> Enter fit tolerance <current>: *(enter a value or press* ENTER*)*

The value you enter determines how near the spline will be fit to the points.

The **eXit** suboption exits the Fit Data options and returns to the main prompt.

Close

The **Close** option of the SPLINEDIT command causes the spline to be joined smoothly at its start point.

Open

The **Open** option opens a closed spline. Previously open splines with coincidental starting and ending points will lose their tangency. Others will be restored to a previous state.

Move Vertex

The **Move Vertex** option relocates a spline's control vertices by providing the following suboptions:

> Specify new location or [Next/Previous/Select point/eXit]<N>:

The **Specify New Location** suboption moves the highlighted point to a specified point.

The **Next** suboption steps forward through fit points.

The **Previous** suboption steps backward through fit points.

The **Select Point** suboption permits you to select a fit point.

The **eXit** suboption exits this set of suboptions.

Refine

The **Refine** option of the SPLINEDIT command allows you to fine-tune a spline definition by providing the following suboptions:

> Enter a refine option [Add control point/Elevate order/Weight/ eXit]<eXit>:

The **Add Control Point** suboption increases the number of control points that control a portion of a spline.

The **Elevate Order** suboption increases the order of the spline. You can increase the current order of a spline up to 26 (the default is 4), causing an increase in the number of control points.

The **Weight** suboption changes the weight at various spline control points by providing the following sub-options:

> Enter new weight (current = 1.0000) or [Next/Previous/Select point/eXit]<N>:

The **Next** suboption steps forward through fit points.

The **Previous** suboption steps backward through fit points.

The **Select Point** suboption permits you to select a fit point.

The **eXit** suboption exits this set of suboptions and returns you to the Refine prompt.

The default weight value for a control point is 1.0. Increasing it causes the spline to be drawn near the selected point. A negative or zero value is not valid.

From the Refine prompt the **eXit** suboption returns you to the main prompt.

rEverse

The **Reverse** option of the SPLINEDIT command reverses the direction of the spline. Reversing the spline does not delete the Fit Data.

Undo

The **Undo** option undoes the effects of the last subcommand.

eXit

The **eXit** option terminates the SPLINEDIT command.

EDITING WITH GRIPS

The grips feature allows you to edit AutoCAD drawings in an entirely different way than using the traditional AutoCAD modify commands. With grips you can move, stretch, rotate, copy, scale, and mirror selected objects without invoking one of the regular AutoCAD modify commands. When you select an object with grips enabled, small squares appear at strategic points on the object that enable you to edit the selected objects.

To enable the grips feature, first open the Options dialog box:

Tools menu	Choose Options
Command: prompt	**options** (ENTER)

AutoCAD displays the Options dialog box, as shown in Figure 6–42.

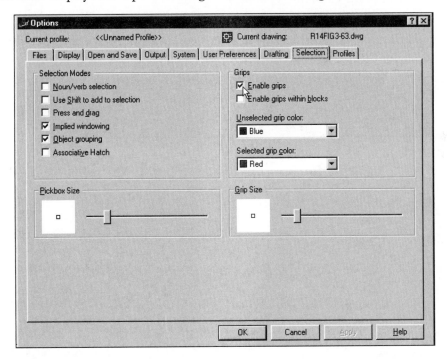

Figure 6–42 *Options dialog box*

Choose the Selection tab, and AutoCAD displays various options related to Grips, as shown in Figure 6–42.

The **Enable Grips** toggle button controls the display of the grips. If it is set to ON, the grips display is enabled; if it is set OFF, the grips display is disabled.

You can also enable the grips feature by setting the system variable GRIPS to 1.

The **Enable Grips Within Blocks** toggle button controls the display of grips on objects within blocks. If it is set to ON, the grips are displayed on all objects within the block; if it is set to OFF, a grip is displayed only on the insertion point of the block.

The **Unselected Grip Color** list box allows you to change the color to the unselected grips.

The **Selected Grip Color** list box allows you to change the color to the selected grips.

The **Grip Size** slider bar allows you to change the size of the grips. To adjust the size of grips, move the slider box left or right. As you move the slider, the size is illustrated to the right of the slider.

After making the necessary settings in the Grips dialog box, choose the **OK** button to keep the changes and close the Options dialog box.

AutoCAD gives you a visual cue when grips are enabled by displaying a pick box at the intersection of the crosshairs, even when you are at the Command: prompt, as shown in Figure 6–43.

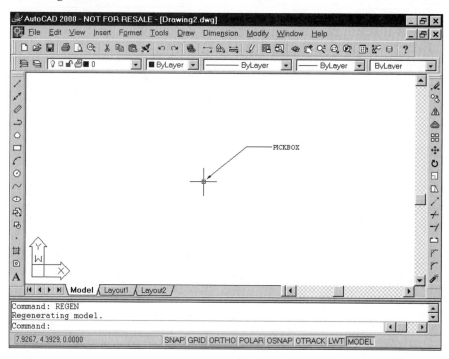

Figure 6–43 *The pick box displayed at the intersection of the crosshairs at the Command: prompt*

Note: The pick box is also displayed on the crosshairs when the PICKFIRST (Noun/Verb selection) system variable is set to ON.

To place grips directly from the Command: prompt, select one or more objects you wish to manipulate.

 Note: To place grips, you can either select objects individually or select multiple objects by placing two points to specify the diagonally opposite corners of a rectangle.

Grips appear on the endpoints and midpoint of lines and arcs, on the vertices and endpoints of polylines, on quadrants and the center of circles, on dimensions, text, solids, 3dfaces, 3dmeshes, and viewports, and on the insertion point of a block. Figure 6–44 shows location of the grips on some of the commonly used objects.

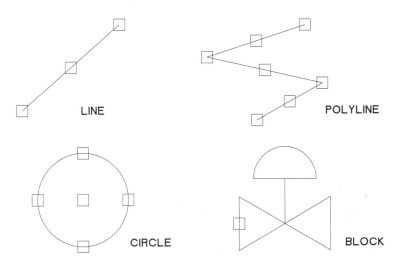

Figure 6–44 *Locations of grips on commonly used objects*

USING GRIPS

This section explains how to utilize grips in modifying your drawing. Learning to use grips speeds up the editing of your drawing while at the same time it maintains the accuracy of your work.

Snapping to Grips

When you move your cursor over a grip, it automatically snaps to the grip point. This allows you to specify exact locations in the drawing without having to use grid, snap, ortho, object snap, or coordinate entry tools.

Condition of Grips

Grips are categorized as being hot, warm, or cold, depending on their use.

A grip becomes hot when it is selected with your cursor. It has a solid-filled color and is the base point unless another grip is selected. You can make more than one grip hot. Hold down the SHIFT key while selecting the grips.

A grip is said to be warm if it is on an object in the current selection set that you haven't selected with the cursor. The object(s) are highlighted to indicate that they are in the selection set.

A grip is said to be cold if it is on an object that is not in the current selection set. Although the objects with cold grips are not highlighted, they will look identical to those with warm grips, and you can still use the grips to snap to. Figure 6–45 shows examples of hot, warm, and cold grips.

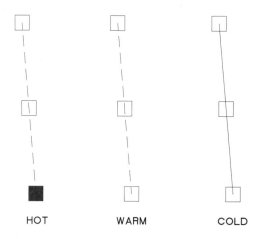

HOT   WARM   COLD

Figure 6–45 *Examples of hot, warm, and cold grips*

Clearing Grips

To clear grips from a selection set, press the ESC key twice. The first time all the warm grips will turn to cold grips. The second time all the grips will clear. When you invoke a non-modifying AutoCAD command, such as LINE or CIRCLE, AutoCAD clears the grips from a selection set.

To use grips to edit the selected objects, specify a grip to act as the base point for the editing operation at the "Command: prompt. Specifying a grip starts the Grip modes, which includes STRETCH, MOVE, ROTATE, SCALE, and MIRROR. You can cycle through the Grip modes by pressing the SPACEBAR, ENTER, entering a keyboard shortcut, or selecting from the right-click shortcut menu. To cancel Grip mode, enter x (for the mode's eXit option); AutoCAD returns to the Command: prompt. You can also use a combination of the current Grip mode and a multiple copy operation on the selection set.

As mentioned earlier, you can also select the available Grip modes from the shortcut menu when you specify a grip at the Command: prompt. Right-click after selecting a grip, and AutoCAD displays the shortcut menu, listing of the Grip modes, as shown in Figure 6–46.

Figure 6–46 *Shortcut menu displaying the Grip modes*

Stretch Mode

The Stretch mode works similar to the STRETCH command. It allows you to stretch the shape of an object without affecting other crucial parts that remain unchanged. When you are in the Stretch Mode, the following prompt appears:

STRETCH
Specify stretch point or [Base point/Copy/Undo/eXit]:

The **Specify stretch point** refers to the stretch displacement point. As you move the cursor, you see that the shape of the object is stretched dynamically from the base point. You can specify the new point with the cursor or by entering coordinates. The displacement is applied to all selected hot grips.

If necessary, you can change the base point to be other than the base grip, by selecting Base point from the shortcut menu. Then specify the point with the cursor, or enter the coordinates.

To make multiple copies while stretching objects, choose **copy** from the shortcut menu. Then specify destination copy points with the cursor, or enter their coordinates.

Move Mode

The Move mode works similar to the MOVE command. It allows you to move one or more objects from their present location to a new one without changing their orientation or size. In addition, you can make copies of the selected objects at the specified displacement, leaving the original objects intact. To invoke the Move mode, cycle through the modes by entering a null response until it takes you to the Move mode, or choose **Move** from the shortcut menu. When you are in the Move mode, the following prompt appears:

MOVE
Specify move point or [Base point/Copy/Undo/eXit]:

The **Specify move point** refers to the move displacement point. As you move the cursor, AutoCAD moves all the objects in the current selection set to a new point relative to the base point. You can specify the new point with the cursor or by entering the coordinates.

If necessary, you can change the base point to be other than the base grip by selecting Base Point from the shortcut menu. Then specify the point with the cursor or enter the coordinates.

To make multiple copies while moving objects, choose **Copy** from the shortcut menu and then specify the destination copy points with the cursor, or enter their coordinates.

Rotate Mode

The Rotate mode works similar to the ROTATE command. It allows you to change the orientation of objects by rotating them about a specified base point. In addition, you can make copies of the selected objects and at the same time rotate them about a specified base point.

To invoke the Rotate mode, cycle through the modes by entering a null response until it takes you to the Rotate mode, or choose **Rotate** from the shortcut menu. When you are in the Rotate Mode, the following prompt appears:

ROTATE
Specify rotation angle or [Base point/Copy/Undo/Reference/eXit]:

The **Specify rotation angle** refers to the rotation angle to which objects are rotated. As you move the cursor, AutoCAD allows you to drag the rotation angle, to position all the objects in the current selection set at the desired orientation. You can specify the new orientation with the cursor or from the keyboard. If you specify an angle by entering a value from the keyboard, this is taken as the angle that the objects should be rotated from their current orientation, around the base point. A positive angle rotates counterclockwise, and a negative angle rotates clockwise. Similar to the RO-TATE command, you can use the **Reference** option to specify the current rotation and the desired new rotation.

If necessary, you can change the base point to be other than the base grip by choosing Base Point from the shortcut menu. Then specify the point with the cursor, or enter the coordinates.

To make multiple copies while rotating objects, choose **Copy** from the shortcut menu. Then specify destination copy points with the cursor, or enter the coordinates.

Scale Mode

The Scale mode works similar to the SCALE command. It allows you to change the size of objects about a specified base point. In addition, you can make copies of the selected objects and at the same time change the size about a specified base point. To invoke the Scale mode, cycle through the modes by entering a null response until it takes you to the Scale mode, or choose **Scale** from the shortcut menu. When you are in the Scale mode, the following prompt appears:

> **SCALE***
> Specify scale factor or [Base point/Copy/Undo/Reference/eXit]:

The **Specify scale factor** refers to the scale factor to which objects are made larger or smaller. As you move the cursor, AutoCAD allows you to drag the scale factor, to change all the objects in the current selection set to the desired size. You can specify the new scale factor with the cursor or from the keyboard. If you specify the scale factor by entering a value from the keyboard, this is taken as a relative scale factor by which all dimensions of the objects in the current selection set are to be multiplied. To enlarge an object, enter a scale factor greater than 1; to shrink an object, use a scale factor between 0 and 1. Similar to the SCALE command, you can use the **Reference** option to specify the current length and the desired new length.

If necessary, you can change the base point to be other than the base grip, by choosing Base Point from the shortcut menu. Then specify the point with the cursor, or enter the coordinates.

To make multiple copies while scaling objects, choose **Copy** from the shortcut menu, and then specify the destination copy points with the cursor, or enter their coordinates.

Mirror Mode

The Mirror mode works similar to the MIRROR command. It allows you to make mirror images of existing objects. To invoke the Mirror mode, cycle through the modes by entering a null response until it takes you to the Mirror mode, or choose **Mirror** from the shortcut menu. When you are in the Mirror Mode, the following prompt appears:

> **MIRROR**
> Specify second point or [Base point/Copy/Undo/eXit]:

Two points are required in AutoCAD to define a line about which the selected objects are mirrored. AutoCAD considers the base grip point as the first point; the second point is the one you specify or enter in response to the Second point.

If necessary, you can change the base point to be other than the base grip, by choosing Base Point from the shortcut menu. Then specify the point with the cursor, or enter the coordinates.

To make multiple copies while retaining original objects, choose **Copy** from the shortcut menu. Then specify mirror points by specifying the point(s) with the cursor, or enter the coordinates.

Changing Object Properties

In AutoCAD, when you select an object at the Command: prompt with grips enabled, AutoCAD displays the selected object's layer, color, linetype, and lineweight in the Object Properties toolbar, replacing the current settings. If you select more than one object having multiple properties, then the corresponding Object Properties drop-down list is displayed blank. For example, if you select objects from the Object and Text layers with different settings for Color and Linetype, all three drop-down lists are displayed blank.

From the Layer drop-down list, you can change the on/off, unlocked/locked, and thawed/frozen status, as well as the layer itself, for the selected object(s). It is like using the CHANGE or PROPERTIES commands. Once you remove the display of the grips from the selected objects, AutoCAD returns the settings of the object properties, such as Layer, Color, and Linetype, to the current settings.

SELECTING OBJECTS BY QUICK SELECT

The QSELECT command is used to create selection sets based on objects that have or do not have similar characteristics or properties as determined by filters that you specify. For example, you can create a selection set of all lines that are equal to or less than 2.5 units long. Or you can create a selection set of all objects that are not text objects on one certain layer. The combinations of possible filters are almost limitless. You can cause the created selection set to replace the current selection set or be appended to it.

To create a filtered selection set, invoke the QSELECT command:

Tools menu	Choose Quick Select
Command: prompt	**qselect** (ENTER)

You can also invoke QSELECT command from shortcut menu any time when you are at the Command: prompt. AutoCAD displays the Quick Select dialog box, similar to shown in Figure 6–47.

The **Apply to:** text box lets you select whether to apply the specified filters to the Current Selection or to the Entire Drawing. If there is a current selection, then Current Selection is the default, otherwise, Entire Drawing is the default. If the **Append to Current Selection Set** check box is checked, then Current Selection is not an option. If you wish to create a selection set, choose the **Select Objects** button (located next to the **Apply to:** text box). The **Select Objects** button is available only when **Append to Current Selection Set** is not checked.

Figure 6–47 *Quick Select dialog box*

The **Select Objects** button returns you to the drawing screen so you can select objects. After selecting the objects to be included in the selection set, press ENTER.

The **Object Type:** text box lets you select whether to include or exclude only certain types of objects or multiple objects.

The **Properties** section list the properties that can be used for filters for the type of object(s) specified in the Object Type text box.

The **Operator** text box lets you apply logical operators to values. These include Greater Than, Less Than, Equal To, Not Equal To, and Wildcard Match. Greater Than and Less Than apply primarily to numeric values and Wildcard Match applies to text strings.

The **Value** text box lets you specify a value to which the Operator applies. If you specify the Greater Than Operator to the Value of 1.0 for the Length of lines, then the lines with lengths greater than 1.0 will be filtered to be either included in or excluded from the selection set, depending on which radio button in the **How To Apply** section has been chosen.

The **How to Apply** section lets you specify whether the filtered objects will be included in or excluded from the selection set.

The **Append to Current Selection Set** lets you specify whether the filtered selection set replaces the current selection or is appended to it.

Once you have all the required selection criteria set, choose the **OK** button to close the dialog box. AutoCAD displays grips on the selected objects. You can proceed with the appropriate modification required for the selected objects.

SELECTION SET BY FILTER TOOL

The FILTER command displays a dialog box that lets you create filter lists that you can apply to the selection set. This is another method of selecting objects. With the FILTER command, you can select objects based on object properties, such as location, object type, color, linetype, layer, block name, text style, and thickness. For example, you could use the FILTER command to select all the blue lines and arcs with a radius of 2.0 units. You can even name filter lists and save them to a file.

The new selection set that is created by the FILTER command can be used as the Previous option at the next Select object: prompt. If you use the FILTER command transparently, then AutoCAD passes the new selection set directly to the command in operation. This will save you a considerable amount of time.

Invoke the FILTER command:

> Command: **filter** (ENTER)

AutoCAD displays the Object Selection Filters dialog box shown in Figure 6–48.

Figure 6–48 *Object Selection Filters dialog box*

The list box displays the filters currently being used as a selection set. The first time you use the FILTER command in the current drawing, the list box is empty.

The **Select Filter** section lets you add filters to the list box based on object properties. Select the object or logical operator from the list box. You can use the grouping operators AND, OR, XOR, and NOT from the list box. The grouping operators must be paired and balanced correctly in the filter list. For example, each Begin OR operator must have a matching End OR operator. If you select more than one filter, AutoCAD by default uses an AND as a grouping operator between each filter.

The **Select...** button displays a dialog box that lists all items of the specified type within the drawing. From the list, you can select as many items as you want to filter. This process saves you from typing the specific filter parameters.

The **Add to List** button lets you add the current Select Filter selection to the filter list.

The **Add Selected Object** button allows you to select an object from the drawing and add it to the filter list.

The **Substitute** button replaces the selected filter with the current one in the Select Filter box.

The **Edit Item** button moves the selected filter into the Select Filter area for editing. First select the filter from the filter list box and then choose the **Edit Item** button. Make the necessary changes in the Select Filter section and choose the **Substitute** button. The edited filter replaces the selected filter.

The **Delete** button allows you to delete the selected filter in the filter list box.

The **Clear List** button allows you to clear the filter list from the filter list box.

To save the current filter list, enter the name for the filter list in the **Save As:** edit field, and choose the **Save As:** button to save the list in the given name.

The **Current:** list box displays the saved filter lists. Select a list to make it current.

The **Delete Current Filter List** button allows you to delete filter lists from the default filter file.

The **Apply** button lets you close the dialog box; AutoCAD displays the Select Objects: prompt, where you create a selection set. AutoCAD uses the filter list on the objects you select.

CHANGING PROPERTIES OF SELECTED OBJECTS

The PROPERTIES command is used to manage and change properties of selected objects by means of the **Properties Window**.

To manage or change the objects in a selection set, invoke the PROPERTIES command:

Tools menu	Choose Properties
Command: prompt	**properties** (ENTER)

AutoCAD displays the Properties Window, as shown in Figure 6–49.

Figure 6–49 *Properties Window*

Options available for displaying the Properties Window are accessible by right-clicking anywhere on the Properties Window except the title bar. Available Options include **Allow Docking, Hide, Description,** and **Undo.**

When **Allow Docking** is checked, the Properties Window can be moved to a docking location. Choosing **Hide** causes the Properties Window to be hidden. Checking **Description** causes a description section to be displayed at the bottom of the Properties Window describing a property when it has been selected. The **Undo** option undoes the last operation.

The Properties Window is called a window instead of a dialog box because you can draw and modify objects in the drawing area while it is being displayed. When the Properties Window is being displayed and one object, say a circle, is selected, then the Properties Window lists all the properties of that circle. Not only are properties like color, linetype and layer listed, but the center X, Y, and Z coordinates, radius, diameter, circumference, and area also listed. However, if two circles are selected, only the values of properties that are the same for both circles are listed. Other properties that are not the same but are common properties are left blank. If you enter a value in the blank area, the new value will be applied to both circles. For example, if the centers of the two circles are different and you enter X and Y coordinates in the Center X and Center Y edit fields respectively, both circles will be moved to have their centers coincide with the specified center. If different types of objects are in the selection set, then the properties listed in the Properties Window will include only those common to the selected objects such as color, layer and linetype. Whenever you change the properties of selected objects in the Properties Window, AutoCAD reflects the changes immediately in the drawing window.

You can also change properties of the selected objects by invoking the DDCHPROP or CHANGE commands.

The DDCHPROP command allows you to change properties of objects, such as color, layer, linetype, linetype scale, and thickness. Invoke the DDCHPROP command:

> Command: **ddchprop** (ENTER)
> Select objects: *(select objects to change properties and press* ENTER*)*

AutoCAD displays the Change Properties dialog box, as shown in Figure 6–50

Figure 6–50 *Change Properties dialog box*

Make the necessary changes in the dialog box by choosing the appropriate buttons for Color, Layer, and Linetype. If necessary, you can also change the Linetype Scale and Thickness. Choose the **OK** button to accept the changes and close the Change Properties dialog box.

The CHANGE command also allows you to change properties for selected objects such as their color, lineweight, linetype, lineweight, layer, elevation, thickness, and plotstyle. In addition, the CHANGE command lets you modify some of the characteristics of lines, circles, text, and blocks. Invoke the CHANGE command:

> Command: **change** (ENTER)
> Select objects: *(select objects and press ENTER to complete the selection)*
> Specify change point or [Properties]: *(specify change point for the selected objects, or choose properties to change one of the properties of the selected object)*

The **Change point** option allows you to modify some of the characteristics of lines, circles, text, and blocks. If you select one or more lines, the closest endpoint(s) are moved to the new change point. If you select a circle, change point allows you to change the radius of the circle. If you selected more than one circle, AutoCAD moves on to the next circle and repeats the prompt. If the selected object is a text string, then AutoCAD allows you to change one or more parameters such as text height, text style, text rotation angle and text string. And if the selected object is a block, then specifying a new location repositions the block.

Instead of changing point, you can change the properties of selected objects by choosing the Properties option. The properties that can be changed include color, lineweight, linetype, lineweight, layer, elevation, thickness, and plotstyle.

GROUPING OBJECTS

The GROUP command adds flexibility in modifying a group of objects. It allows you to name a selection set. Naming a selection set combines two powerful AutoCAD drawing features. One is being able to modify a group of unrelated objects as a group. It is similar to using Previous to select the last selection set when prompted to Select object: for a modify command. The advantage of using GROUP instead of Previous is that you are not restricted to only the last selection set. The other feature combined in the GROUP command is that of giving a name to a selected group of objects for recalling it later by the name of the group. This is similar to the BLOCK command. The advantage of using GROUP instead of BLOCK is that the GROUP command's "selectable" switch can be set to OFF for modifying an individual member without losing its "membership" in the group. Also, named groups, like blocks, are saved with the drawing.

A named group can be selected for modifying as a group only when its "selectable" switch is set to ON. Figure 6–51 shows the result of trimming an object with the selectable switch set to ON and to OFF. Modifying objects (such as with the MOVE or COPY command) that belong to a group can be selected by two methods. One is to select one of its members. The other method is to select the Group option by typing **g** at the Select Objects: prompt. AutoCAD prompts for the group's name. Enter the group name and press ENTER or the SPACEBAR. AutoCAD highlights the objects that belong to the selected group.

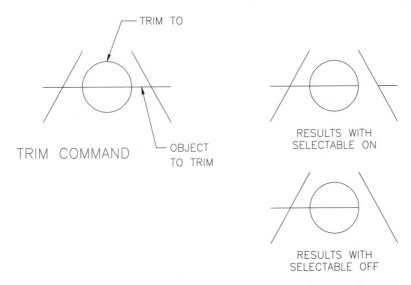

Figure 6–51 *Trimming an object with the selectable switch set to ON and to OFF*

To create a new group or edit an existing group, invoke the GROUP command:

Tools menu	Choose Object Group
Command: prompt	**group** (ENTER)

AutoCAD displays the Object Grouping dialog box, similar to Figure 6–52.

The dialog box is divided into four sections:

> Group Name list box
> Group Identification
> Create Group
> Change Group

Figure 6–52 *Object Grouping dialog box*

GROUP NAME LIST BOX

The Group Name list box lists the names of the existing groups defined in the current drawing. The Selectable column indicates whether a group is selectable. If it is listed as selectable, then selecting a single group member selects all the members except those on locked layers. If it is listed as unselectable, then selecting a single group member selects only that object.

GROUP IDENTIFICATION

The Group Identification section is where AutoCAD displays the group name and a description when a group is selected in the Group Name list.

The **Find Name** button lists the groups to which an object belongs. AutoCAD prompts for the selection of an object and displays the Group Member List dialog box, which lists the group or groups to which the selected object belongs.

The **Highlight** button lets you see the members of the selected group from the Group Name list box.

The **Include Unnamed** toggle box controls the listing of the unnamed groups in the Group Name list box.

CREATE GROUP

The **Create Group** section is used for creating a new group with or without a group name. In addition, you can set whether or not it is initially selectable.

To create a new group, enter the group name and description in the **Group Name:** and **Description:** text boxes. Group names can be up to 31 characters long and can include letters, numbers, and the special characters $ and _. To create an unnamed group, set the **Unnamed** toggle box to ON. AutoCAD assigns a default name, *An, to unnamed groups. The n represents a number that increases with each new group.

Set the **Selectable** box to ON or OFF, and then choose the **New>** button. AutoCAD prompts for the selection of objects. Select all the objects to be included in the new group, and press ENTER or the SPACEBAR to complete the selection.

CHANGE GROUP

The Change Group section is for making changes to individual members of a group or to the group itself. The buttons are disabled until a group name is selected in the **Group Name:** list box.

The **Remove** button lets you remove selected objects from the selected group. To remove objects from the selected group, choose the **Remove** button; AutoCAD prompts:

> Remove objects: *(select objects that are to be removed from the selected group and press* ENTER *or the* SPACEBAR*)*

AutoCAD redisplays the Object Grouping dialog box.

The **Add** button allows you to add the selected objects to the selected group. To add objects from the selected group, choose the **Add** button; AutoCAD prompts:

> Select objects: *(select objects that are to be added to the selected group and press* ENTER *or the* SPACEBAR*)*

AutoCAD redisplays the Object Grouping dialog box.

The **Rename** button allows you to change the name of the selected group to the name entered in the **Group Name:** text box in the Group Identification section.

The **Re-order** option allows you to change the numerical order of objects within the selected group. Initially, the objects are numbered in the order in which they were selected to form a group. Reordering is useful when creating tool paths. Choose the **Re-order** button; AutoCAD displays the Order Group subdialog box shown in Figure 6–53.

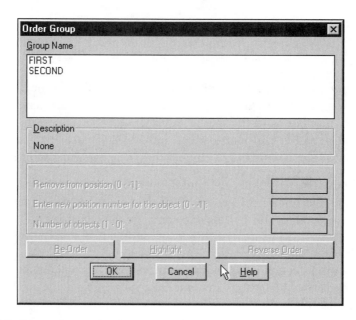

Figure 6–53 *Object Group subdialog box*

The **Group Name** list box gives the names of the groups defined in the current drawing. Members of a group are numbered sequentially starting with 0 (zero).

> **Remove from position (0 – n):** identifies the position number of an object.
>
> **Replace at position (0 – n):** identifies the new position number of the object.
>
> **Number of objects (1 – n):** identifies the number/range of objects to reorder.
>
> The **Re-Order** and **Reverse Order** buttons allow you to change the numerical order of objects as specified and reverses the order of all members, respectively.
>
> The **Highlight** button allows AutoCAD to display the members of the selected group in the graphics area.
>
> The **Description** button assigns an optional description up to 64 characters long.

The **Explode** button in the Change Group section of the Object Grouping dialog box deletes the selected group from the current drawing. Thus, the group no longer exists as a group. The members remain in the drawing and in any other group(s) of which they are members.

The **Description** button allows you to change the description of the selected group.

The **Selectable** button controls whether the group is selectable.

After making the necessary changes to the Object Grouping dialog box, choose the OK button to keep the changes and close the dialog box.

INFORMATION ABOUT OBJECTS

AutoCAD provides several commands for displaying useful information about the objects in the drawing. These commands do not create anything, nor do they modify or have any effect on the drawing or objects therein. The only effect on the AutoCAD editor is that on single-screen systems, the screen switches to the AutoCAD Text window (not to be confused with the TEXT command) and the information requested by the particular INQUIRY command is then displayed on the screen. If you are new to AutoCAD, it is helpful to know the FLIP SCREEN feature that returns you to the graphics screen so you can continue with your drawing. On most systems this is accomplished with the F2 function key. You can also change back and forth between graphic and text screens with the GRAPHSCR and TEXTSCR commands, entered at the Command: prompt, respectively. The Inquiry commands include LIST, AREA, ID, DBLIST, and DIST.

LIST COMMAND

The LIST command displays information about individual objects stored by AutoCAD in the drawing database. The information includes:

> The location, layer, object type, and space (model or paper) of the selected object as well as the color, lineweight, and linetype if not set to BYLAYER or BYBLOCK.
>
> The distance in the main axes between the endpoints of a line, that is, the delta-*X*, delta-*Y*, and delta-*Z*.
>
> The area and circumference of a circle or the area of a closed polyline.
>
> The insertion point, height, angle of rotation, style, font, obliquing angle, width factor, and actual character string of a selected text object.
>
> The object handle, reported in hexadecimal.

Invoke the LIST command:

Inquiry toolbar	Choose the LIST command (see Figure 6–54)
Tools menu	Choose Inquiry > List
Command: prompt	**list** (ENTER)

Figure 6–54 *Invoking the* LIST *command from the Inquiry toolbar*

AutoCAD prompts:

> Command: **list** (ENTER)
> Select objects: *(select the objects and press* ENTER *to terminate object selection)*

AutoCAD lists the information about the selected objects.

DBLIST COMMAND

The DBLIST command lists the data about all of the objects in the drawing. It can take a long time to scroll through all the data in a large drawing. DBLIST can, like other commands, be terminated by canceling by pressing ESC key.

Invoke the DBLIST command:

Command: prompt	**dblist** (ENTER)

AutoCAD lists the information about all the objects in the drawing.

AREA COMMAND

The AREA command is used to report the area (in square units) and perimeter of a selected closed geometric figure on the screen, such as a circle, polygon, or closed polyline. You may also specify a series of points that AutoCAD considers a closed polygon, compute, and report the area.

Invoke the AREA command:

Inquiry toolbar	Choose the AREA command (see Figure 6–55)
Tools menu	Choose Inquiry > Area
Command: prompt	**area** (ENTER)

Figure 6–55 *Invoking the* AREA *command from the Inquiry toolbar*

AutoCAD prompts:

> Command: **area** (ENTER)
> Specify first corner point or [Object/Add/Subtract]: *(specify a point or select one of the available options from the short-cut menu)*

The **default** option calculates the area when you select the vertices of the objects. If you want to know the area of a specific object such as a circle, polygon, or closed polyline, select the Object option.

The following command sequence is an example of finding the area of a polygon using the Object option, as shown in Figure 6–56.

Command: **area** (ENTER)
Specify first corner point or [Object/Add/Subtract]: **o** (ENTER)
Select objects: *(select an object, as shown in Figure 6–56)*
Area = 12.21 Perimeter = 13.79

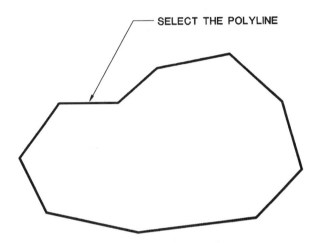

SELECT THE POLYLINE

Figure 6–56 *Finding the area of a polygon using the* AREA *command Object option*

The **Add** option allows you to add selected objects to form a total area; then you can use the **Subtract** option to remove selected objects from the running total.

The following example demonstrates the application of the Add and Subtract options. In this example, the area is determined for the closed shape after subtracting the area of the four circles, as shown in Figure 6–57.

Command: **area** (ENTER)
Specify first corner point or [Object/Add/Subtract]: **a** (ENTER)
Specify first corner point or [Object/Subtract]: **o** (ENTER)
(ADD mode) Select objects: *(select the polyline, as shown in Figure 6–57)*
Area = 12.9096, Perimeter = 15.1486
Total area = 12.9096
(ADD mode) Select objects: (ENTER)
Specify first corner point or [Object/Add/Subtract]: **s** (ENTER)
Specify first corner point or [Object/Add]: **o** (ENTER)
(SUBTRACT mode) Select objects: *(select circle A, as shown in Figure 6–57)*
Area = 0.7125, Circumference = 2.9992
Total area = 12.1971
(SUBTRACT mode) Select objects: *(select circle B, as shown in Figure 6–57)*
Area = 0.5452, Circumference = 2.6175

Total area = 11.6179
(SUBTRACT mode) Select objects: *(select circle C, as shown in Figure 6–57)*
Area = 0.7125, Circumference = 2.9922
Total area = 10.9394
(SUBTRACT mode) Select objects:*(select circle D, as shown in Figure 6–57)*
Area = 0.5452, Circumference = 2.6175
Total area = 10.3942
(SUBTRACT mode) Select objects: (ENTER)
Specify first corner point or [Object/Add]: (ENTER)

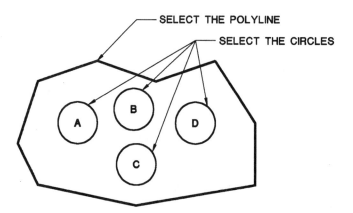

Figure 6–57 *Using the Add and Subtract options of the* AREA *command*

ID COMMAND

The ID command is used to obtain the coordinates of a selected point. If you do not use an Object Snap mode to select a point that is not in the current construction plane, AutoCAD assigns the current elevation as the Z coordinate of the point selected.

Invoke the ID command:

Inquiry toolbar	Choose the LOCATE POINT command (see Figure 6–58)
Tools menu	Choose Inquiry > ID Point
Command: prompt	**id** (ENTER)

Figure 6–58 *Invoking the* LOCATE POINT *command from the Inquiry toolbar*

AutoCAD prompts:

> Command: **id** (ENTER)
> Specify point: *(select a point)*

AutoCAD displays the information about the selected point.

If the BLIPMODE system variable is set to ON (the default), a blip appears on the screen at the specified point, provided it is in the viewing area.

DIST COMMAND

The DIST command prints out the distance, in the current units, between two points, either selected on the screen or keyed in from the keyboard. Included in the report are the horizontal and vertical distances (delta-X and delta-Y, respectively) between the points and the angles in and from the *XY* plane.

Invoke the DIST command:

Inquiry toolbar	Choose the DISTANCE command (see Figure 6–59)
Tools menu	Choose Inquiry > Distance
Command: prompt	**dist** (ENTER)

Figure 6–59 *Invoking the* DIST *command from the Inquiry toolbar*

AutoCAD prompts:

> Command: **dist** (ENTER)
> Specify first point: *(specify the first point to measure from)*
> Specify second point: *(specify the endpoint to measure to)*

AutoCAD displays the distance between two selected points.

SYSTEM VARIABLES

AutoCAD stores the settings (or values) for its operating environment and some of its commands in system variables. Each system variable has an associated type: integer (for switching), integer (for numerical value), real, point, or text string. Unless they are read-only, you can examine and change these variables at the Command: prompt by typing the name of the system variable, or you can change them by means of the SETVAR command.

Integers (for Switching) System variables that have limited nonnumerical settings can be switched by setting them to the appropriate integer value. For example, the snap can be either ON or OFF. The purpose of the SNAPMODE system variable is to turn the snap ON or OFF by using the AutoCAD SETVAR command or the AutoLISP (setvar) function.

Turning the snap ON or OFF is demonstrated in the following example by changing the value of its SNAPMODE system variable. First, its current value is set to "0", which is OFF.

```
Command: setvar (ENTER)
Enter variable name or [?]: snapmode (ENTER)
Enter new value for SNAPMODE <0>: I (ENTER)
```

This sequence may seem unnecessary, because the snap mode is so easily toggled with a press of a function key. Changing the snap with the SETVAR command is inconvenient, but doing so does allow you to view the results immediately.

For any system variable whose status is associated with an integer, the method of changing the status is just like the preceding example. In the case of SNAPMODE, "0" turns it OFF and "1" turns it ON. In a similar manner, you can use SNAPISOPAIR to switch one isoplane to another by setting the system variable to one of three integers: 0 is the left isoplane, 1 is the top, and 2 is the right isoplane.

It should be noted that the settings for the Osnap system variable named OSMODE are members of the binomial sequence. The integers are 1, 2, 4, . . . , 512, 1024, 2048, 4096, 8192. See Table 6–1 for the meaning of OSMODE values. While the settings are switches, they are more than just ON and OFF. There may be several Object Snap modes active at one time. It is important to note that the value of an integer (switching) has nothing to do with its numerical value.

Table 6–1 Values for the OSMODE System Variable

NONe	0
ENDpoint	1
MIDpoint	2
CENter	4
NODe	8
QUAdrant	16
INTersection	32
INSert	64
PERpendicular	128
TANgent	256

Table 6–1 Values for the OSMODE System Variable (continued)

NEArest	512
QUIck	1024
APP int	2048
EXTension	4096
PARallel	8192

Integers (for Numerical Value) System variables such as APERTURE and AUPREC are changed by using an integer whose value is applied numerically in some way to the setting, rather than just as a switch. For instance, the size of the aperture (the target box that appears for selecting Osnap points) is set in pixels (picture elements) according to the integer value entered in the SETVAR command. For example, setting the value of APERTURE to 9 should render a target box that is three times larger than setting it to 3.

AUPREC is the variable that sets the precision of the angular units in decimal places. The value of the setting is the number of decimal places; therefore, it is considered a numerical integer setting.

Real System variables that have a real number for a setting, such as VIEWSIZE, are called real.

Point (X Coordinate, Y Coordinate) LIMMIN, LIMMAX, and VIEWCTR are examples of system variables whose settings are points in the form of the *X* coordinate and *Y* coordinate.

Point (Distance, Distance) Some system variables, whose type is point, are primarily for setting spaces rather than a particular point in the coordinate system. For instance, the SNAPUNIT system variable, though called a point type, uses its X and Y distances from (0,0) to establish the snap X and Y resolution, respectively.

String These variables have names like CLAYER, for the current layer name, and DWGNAME, for the drawing name.

PROJECT EXERCISE

This project exercise provides point-by-point instructions to draw the objects shown in Figure P6–1. In this exercise you will apply the skills acquired in Chapters 1 through 6.

Figure P6–1 *Completed project drawing (with dimensions added for reference)*

In this project you will learn how to do the following:

- Set up the drawing, including limits and layers.
- Use the XLINE, PLINE, and MLINE commands to create objects.
- Use the MLEDIT command for special joining of multilines.
- Use the TRIM and FILLET commands to modify objects.
- Use the TEXT command to create text objects.

SET UP THE DRAWING AND DRAW A BORDER

Step 1: Start the AutoCAD program.

Step 2 : To create a new drawing, invoke the NEW command from the Standard toolbar or select New from the File menu.

AutoCAD displays the Create New Drawing dialog box. Choose the Start from Scratch button (with default settings set to English), and then set UNITS, LIMITS, SNAP, and GRID to the values shown in the nearby SETTINGS/VALUE table.

Step 3: Invoke the LAYER command from the Object Properties toolbar, or select **Layer...** from the pull-down Format menu. AutoCAD displays the Layer & Properties Manager dialog box. Create five layers, and rename them as shown in the table, assigning appropriate color and linetype.

SETTINGS	VALUE		
UNITS	Architectural		
LIMITS	Lower left corner: –5'-0",–6'-0"		
	Upper right corner: 24'-4",15'-4"		
GRID	4		
SNAP	4		
LAYERS	*NAME*	*COLOR*	*LINETYPE*
	Construction	Cyan	Continuous
	Border	Red	Continuous
	Object	White	Continuous
	Hidden	Magenta	Hidden
	Text	Blue	Continuous

ESTABLISHING CONSTRUCTION LINES

Step 4: Set Construction as the current layer, and set ORTHO, GRID, and SNAP all to ON. Invoke the ZOOM ALL command to display the screen to the limits.

Step 5: Invoke the XLINE command from the Draw toolbar. AutoCAD prompts:

Command: **xline** (ENTER)
Specify a point or [Hor/Ver/Ang/Bisect/Offset]: **v** (ENTER)
Specify through point: *(place the cursor so that the X coordinate shown in the status bar is at—4'-8" and press the pick button)*
Specify through point: *(continue placing six additional vertical construction lines whose X coordinates are 0'-0", 5'-0", 10'-0", 15'-0", 19'-4", and **24'-0"**, and press ENTER to terminate the command sequence)*
Command: *(press ENTER to repeat the XLINE command)*

Command: XLINE Specify a point or [Hor/Ver/Ang/Bisect/Offset]: **h** (ENTER)
Specify through point: *(place the cursor so that the Y coordinate shown in the status bar is at –5'-8" and press the pick button)*
Specify through point: *(continue placing four additional horizontal construction lines whose Y coordinates are **-4'-0", 0'-0", 12'-0",** and **15'-0"**, and press ENTER to terminate the command sequence)*

The display should appear as shown in Figure P6–2.

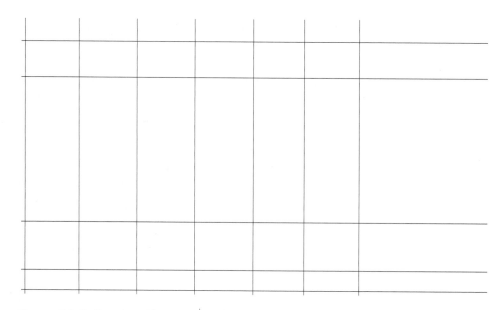

Figure P6–2 *Drawing with construction lines*

Step 6: Set Border as the current layer. Specify Intersection as the Running Object mode.

Step 7: Invoke the PLINE command from the Draw toolbar. AutoCAD prompts:

> Command: **pline** (ENTER)
> Specify start point: *(specify point 1 as shown in Figure P6–3)*
> Current line-width is 0'-0"
> Specify next point or [Arc/Close/Halfwidth/Length/Undo/Width]:
>   **w** (ENTER)
> Specify starting width <0'-0">: **2** (ENTER)
> Specify ending width <0'-2">: (ENTER)
> Specify next point or [Arc/Close/Halfwidth/Length/Undo/Width]:
>   *(specify point 2 as shown in Figure P6–3)*
> Specify next point or [Arc/Close/Halfwidth/Length/Undo/Width]:
>   *(specify point 3 as shown in Figure P6–3)*
> Specify next point or [Arc/Close/Halfwidth/Length/Undo/Width]:
>   *(specify point 4 as shown in Figure P6–3)*
> Specify next point or [Arc/Close/Halfwidth/Length/Undo/Width]:
>   **c** (ENTER)

Figure P6–3 *Points (intersections) for drawing the border*

Command: *(press* ENTER *to repeat the* PLINE *command)*
Specify start point: *(specify point 5 as shown in Figure P6–3)*
Current line-width is 0'-2"
Specify next point or [Arc/Close/Halfwidth/Length/Undo/Width]:
  (specify point 6 as shown in Figure P6–3)
Specify next point or [Arc/Close/Halfwidth/Length/Undo/Width]: *(press*
  ENTER *to terminate the command sequence)*

Command: *(press* ENTER *to repeat the* PLINE *command)*
Specify start point: *(specify point 7 as shown in Figure P6–3)*
Current line-width is 0'-2"
Specify next point or [Arc/Close/Halfwidth/Length/Undo/Width]:
  (specify point 8 as shown in Figure P6–3)
Specify next point or [Arc/Close/Halfwidth/Length/Undo/Width]: *(press*
  ENTER *to terminate the command sequence)*

Command: *(press* ENTER *to repeat the* PLINE *command)*
Specify start point: *(specify point 9 as shown in Figure P6–3)*
Current line-width is 0'-2"
Specify next point or [Arc/Close/Halfwidth/Length/Undo/Width]:
  (specify point 10 as shown in Figure P6–3)
Specify next point or [Arc/Close/Halfwidth/Length/Undo/Width]: *(press*
  ENTER *to terminate the command sequence)*

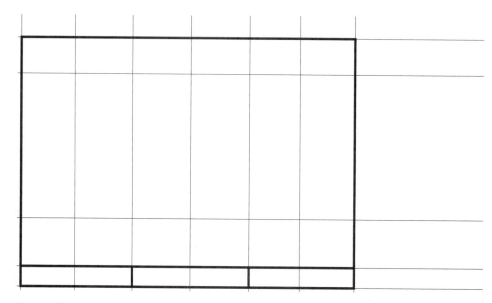

Figure P6–4 *Drawing with the border*

The display should appear as shown in Figure P6–4.

DRAWING TEXT IN THE TITLE BLOCK

Step 8: Set Construction as the current layer.

Step 9: To create additional construction lines for drawing text, invoke the LINE command from the Draw toolbar. AutoCAD prompts:

> Command: **line** (ENTER)
> Specify first point: *(specify a point on line A, as shown in Figure P6–5, at coordinates –4'8",–4'8")*
> Specify next point or [Undo]: *(invoke the Object Snap mode Perpendicular, and select line B, as shown in Figure P6–5)*
> Specify next point or [Undo]: *(press ENTER to terminate the command sequence)*
>
> Command: *(press ENTER to repeat the LINE command)*
> LINE Specify first point: *(specify a point on line A, as shown in Figure P6–5, at coordinates –4'8",–5',4")*
> Specify next point or [Undo]: *(invoke the Object Snap mode Perpendicular, and select line B, as shown in Figure P6–5)*
> Specify next point or [Undo]: *(press ENTER to terminate the command sequence)*
>
> Command: *(press ENTER to repeat the LINE command)*
> LINE Specify first point: *(invoke the Object Snap mode Midpoint, and select line C, as shown in Figure P6–5)*

Specify next point or [Undo]: *(use the Object Snap mode Perpendicular, and select line D, as shown in Figure P6–5)*

Specify next point or [Undo]: *(press* ENTER *to terminate the command sequence)*

Figure P6–5 *Points for creating lines to draw text objects*

Step 10: Set Text as the current layer.

Step 11: Invoke the TEXT command from the Draw toolbar to draw the text. AutoCAD prompts:

> Command: **text** (ENTER)
> Specify start point of text or [Justify/Style]: **j** (ENTER)
> Enter an option [Align/Fit/Center/Middle/Right/TL/TC/TR/ML/MC/MR/ BL /BC/BR]: **c** (ENTER)
> Specify center point of text: *(invoke the Object Snap mode Midpoint, and select the first construction line drawn in the previous step to establish point 1, as shown in Figure P6–6)*
> Specify height <current>: **4** (ENTER)
> Specify rotation angle of text <current>: **0** (ENTER)
> Enter text: *(enter* **SCHOOL NAME:***)*
>
> Enter text: (ENTER)

Figure P6–6 *Points to draw text for the title block*

Command: *(press* ENTER *to repeat the* TEXT *command)*
Specify start point of text or [Justify/Style]: **j** (ENTER)
Enter an option [Align/Fit/Center/Middle/Right/TL/TC/TR/ML/MC/MR/
    BL /BC/BR]: **c** (ENTER)
Specify center point of text: *(invoke the Object Snap mode Midpoint, and
    select the second construction line drawn in the previous step to establish
    point 2 as shown in Figure P6–6)*
Specify height <0'-4">: (ENTER)
Specify rotation angle of text <0>: (ENTER)
Enter text: *(enter the name of your school)*
Enter text: (ENTER)

Command: *(press* ENTER *to repeat the* TEXT *command)*
Specify start point of text or [Justify/Style]: *(specify point 3, as shown in
    Figure P6–6, at coordinates 5'-4",–4'-8")*
Specify height <0'-4">: (ENTER)
Specify rotation angle of text <0>: (ENTER)
Enter text: **NAME:** *(enter* **NAME:** *followed by your name)*
Enter text: (ENTER)

Command: *(press* ENTER *to repeat the* TEXT *command)*
Specify start point of text or [Justify/Style]: *(specify point 4, as shown in
    Figure P6–6 at coordinates 5'-4", -5'-4")*

Specify height <0'-4">: (ENTER)
Specify rotation angle of text <0>: (ENTER)
Enter text: **DATE:** *(enter **DATE:** followed by today's date)*
Enter text: (ENTER)

Command: *(press* ENTER *to repeat the* TEXT *command)*
Specify start point of text or [Justify/Style]: **j** (ENTER)
Enter an option [Align/Fit/Center/Middle/Right/TL/TC/TR/ML/MC/MR/
    BL /BC/BR]: **m** (ENTER)
Specify middle point of text: *(invoke the Object Snap mode Midpoint, and
    select the first construction line drawn in the previous step to establish
    point 5, as shown in Figure P6–6)*
Specify height <0'-4">: **6** (ENTER)
Specify rotation angle of text <0>: (ENTER)
Enter text: *(ENTER **LAUNDRY & STORAGE**)*

Enter text: (ENTER)

After drawing the text as indicated, the display should appear as shown in
Figure P6–7.

Figure P6–7 *Drawing with the border and title block (Construction layer ON)*

With the Construction layer set to OFF (do *not* turn it OFF at this time), the
display would appear as shown in Figure P6–8.

Figure P6–8 *Drawing with the border and title block (Construction layer OFF)*

Step 12: From the Standard toolbar, invoke the ZOOM SCALE command. AutoCAD prompts:

> Command: **zoom** (ENTER)
> Specify corner of window, enter a scale factor (nX or nXP), or
> [All/Center/Dynamic/Extents/Previous/Scale/Window] <real time>:
> **.8** (ENTER)

Step 13: Erase the construction lines that are not needed (see Figure P6–9) by invoking the ERASE command from the Modify toolbar. AutoCAD prompts:

> Command: **erase** (ENTER)
> Select objects: *(select all of the Construction xlines with the designated symbol, plus the Construction xlines created for the title block text, as shown in Figure P6–9).*

Figure P6–9 *Drawing with the lines to be erased indicated*

Invoke the ZOOM ALL command to display the whole drawing.

DRAWING THE WALLS

Step 14: Set Object as the current layer. To draw the walls, invoke the MLINE command from the Draw toolbar. Set the Scale factor to 4, and draw the closed rectangle by *specifying* the points indicated in Figure P6–10. AutoCAD prompts:

> Command: **mline** (ENTER)
> Current settings: Justification = Top, Scale = 1.00, Style = STANDARD
> Specify start point or [Justification/Scale/STyle]: **s** (ENTER)
> Enter mline scale <1.00>: **4** (ENTER)
> Current settings: Justification = Top, Scale = 4.00, Style = STANDARD
> Specify start point or [Justification/Scale/STyle]: *(specify point 1 as shown in Figure P6–10)*
> Specify next point: *(specify point 2 as shown in Figure P6–10)*
> Specify next point or [Undo]: *(specify point 3 as shown in Figure P6–10)*
> Specify next point or [Close/Undo]: *(specify point 4 as shown in Figure P6–10)*
> Specify next point or [Close/Undo]: **c** (ENTER)

Figure P6–10 *Points for drawing the walls*

Step 15: To draw the interior wall, again invoke the MLINE command from the Draw toolbar. Set the Scale to 8 and the Justification to 0, and draw the interior wall by *specifying* the points indicated in Figure P6–11. AutoCAD prompts:

> Command: **mline** (ENTER)
> Current settings: Justification = Top, Scale = 4.00, Style = STANDARD
> Specify start point or [Justification/Scale/STyle]: **s** (ENTER)

Enter mline scale <4.00>: **8** (ENTER)
Current settings: Justification = Top, Scale = 8.00, Style = STANDARD
Specify start point or [Justification/Scale/STyle]: **j** (ENTER)
Enter justification type [Top/Zero/Bottom] <top>: **z** (ENTER)
Current settings: Justification = Zero, Scale = 8.00, Style = STANDARD
Specify start point or [Justification/Scale/STyle]: *(specify point 1 as shown in Figure P6–11)*
Specify next point: *(specify point 2 as shown in Figure P6–11)*
Specify next point or [Undo]: *(press ENTER to terminate the command sequence)*

Figure P6–11 *Points for drawing the interior walls*

Step 16: Set the Running Object Snap mode to OFF.

Step 17: To edit the multiline as shown in Figure P6–12, invoke the MLEDIT command from the Modify II toolbar. AutoCAD displays the Multiline Edit Tools dialog box. Choose the Open Tee option, and then choose the OK button to close the dialog box. AutoCAD prompts:

Command: **mledit** (ENTER)
Select first mline: *(select the interior wall at point A, as shown in Figure P6–12)*
Select second mline: *(select the exterior wall at point B, as shown in Figure P6–12)*
Select first mline or [Undo]: *(select the interior wall at point C, as shown in Figure P6–12)*
Select second mline: *(select the exterior wall at point D, as shown in Figure P6–12)*
Select first mline or [Undo]: *(press ENTER to terminate the command sequence)*

Figure P6–12 *Points for editing multilines*

After drawing the mline for the interior and exterior walls and "opening" the tees as indicated, the display should appear as shown in Figure P6–13.

Figure P6–13 *Drawing with interior and exterior walls*

Step 18: Invoke the MOVE command from the Modify toolbar. AutoCAD prompts:

> Command: **move** (ENTER)
> Select objects: *(select the middle vertical construction line)*
> Select objects: *(press ENTER to complete the selection of objects)*
> Specify base point or displacement: *(specify a point on the selected line)*

Specify second point of displacement or <use first point as displacement>: *(specify a point that is 2'-4" left of the line)*

Command: *(press* ENTER *to repeat the* MOVE *command)*
Select objects: *(select the right vertical construction line)*
Select objects: *(press* ENTER *to complete the selection of objects)*
Specify base point or displacement: *(specify a point on the selected line)*
Specify second point of displacement or <use first point as displacement>: *(specify a point that is 1'-4" left of the line)*

Command: *(press* ENTER *to repeat the* MOVE *command)*
Select objects: *(select the lower horizontal construction line)*
Select objects: *(press* ENTER *to complete the selection of the objects)*
Specify base point or displacement: *(specify a point on the selected line)*
Specify second point of displacement or <use first point as displacement>: *(specify a point that is 5'-4"above the line)*

DRAWING THE COUNTER FOR THE SINK

Step 19: Set Intersection as the Running Object Snap mode. Invoke the LINE command from the Draw toolbar. AutoCAD prompts:

Command: **line** (ENTER)
Specify first point: *(specify point 1 as shown in Figure P6–14)*
Specify next point or [Undo]: *(specify point 2 as shown in Figure P6–14)*
Specify next point or [Undo]: *(specify point 3 as shown in Figure P6–14)*
Specify next point or [Undo]: *(press* ENTER *to terminate the command sequence)*

Figure P6–14 *Points for drawing the counter for the sink*

Command: (*press* ENTER *to repeat the* LINE *command*)
LINE Specify first point: (*specify point 4 as shown in Figure P6–14*)
Specify next point or [Undo]: (*specify point 5 as shown in Figure P6-14*)
Specify next point or [Undo]: (*press* ENTER *to terminate the command sequence*)

Step 20: Invoke the ERASE command from the Modify toolbar to erase the remaining construction lines. Set the Running Object Snap mode to OFF.

OPENING THE WALLS FOR DOORS AND WINDOWS

Step 21: To create the opening of the walls, invoke the MLEDIT command from the Modify II toolbar. AutoCAD displays the Multiline Edit Tools dialog box. Choose the Cut All option, and then choose OK to close the dialog box. AutoCAD prompts:

Command: **mledit** (ENTER)
Select mline: (*specify the lower wall at point 1'-0",0'-0"*)
Select second point: (*specify the lower wall at point 3'-0",0'-0"*)
Select mline or [Undo]: (*specify the lower wall at point 3'-8",0'-0"*)
Select second point: (*specify the lower wall at point 6'-8",0'-0"*)
Select mline or [Undo]: (*specify the lower wall at point 7'-4",0'-0"*)
Select second point: (*specify the lower wall at point 9'-4",0'-0"*)
Select mline or [Undo]: (*specify the lower wall at point 12'-0",0'-0"*)
Select second point: (*specify the lower wall at point 16'-0",0'-0"*)
Select mline or [Undo]: (*specify the upper wall at point 3'-8",12'-0"*)
Select second point: (*specify the upper wall at point 6'-8",12'-0"*)
Select mline or [Undo]: (*press* ENTER *to terminate the command sequence*)

DRAWING THE WINDOWS AND THE WATER HEATER

Step 22: Invoke the ZOOM WINDOW command to zoom in as shown in Figure P6–15.

Figure P6–15 *Drawing with* ZOOM WINDOW *in the area indicated*

Step 23: Change the SNAP setting to 2. Invoke the MLINE command from the Draw toolbar to draw the windows. AutoCAD prompts:

> Command: **mline** (ENTER)
> Current settings: Justification = Zero, Scale = 8.00, Style = STANDARD
> Specify start point or [Justification/Scale/STyle]: **s** (ENTER)
> Enter mline scale <8.00>: **2** (ENTER)
> Current settings: Justification = Zero, Scale = 2.00, Style = STANDARD
> Specify start point or [Justification/Scale/STyle]: *(specify point A, as shown in Figure P6–16, at coordinates 7'-4",0'-2")*
> Specify next point: *(specify the endpoint at coordinates 9'-4",0'-2")*
> Specify next point or [Undo]: *(press ENTER to terminate the command sequence)*

point A

Figure P6–16 *Points indicating where the starting point of the window has to be drawn*

> Command: *(press ENTER to repeat the MLINE command)*
> Current settings: Justification = Zero, Scale = 2.00, Style = STANDARD
> Specify start point or [Justification/Scale/STyle]: **s** (ENTER)
> Enter mline scale <2.00>: **8** (ENTER)
> Current settings: Justification = Zero, Scale = 8.00, Style = STANDARD
> Specify start point or [Justification/Scale/STyle]: *(specify point A at coordinates 7'-4",0'-2")*
> Specify next point: *(specify the endpoint at coordinates 9'-4",0'-2")*
> Specify next point or [Undo]: *(press ENTER to terminate the command sequence)*

Step 24: Invoke the TRIM command from the Modify toolbar to trim the line for the counter even with the line for the window. AutoCAD prompts:

> Command: **trim** (ENTER)
> Select cutting edges: *(select the mline for the cutting edge as shown in Figure P6–17)*

Select objects: *(press* ENTER *to complete selection of the cutting edge)*
Select object to trim or [Project/Edge/Undo]: *(select the object to trim as shown in Figure P6–17)*
Select object to trim or [Project/Edge/Undo]: *(press* ENTER *to complete selection of the object to trim)*

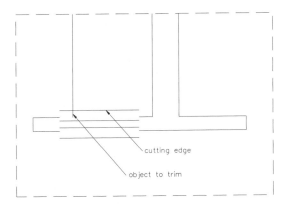

cutting edge

object to trim

Figure P6–17 *Indication of the lines to trim*

Step 25: Use the LINE command to close the end of the multilines for the window openings and door jambs.

After trimming the line for the cabinet and drawing the lines for closing the ends of the window, the display should appear as shown in Figure P6–18.

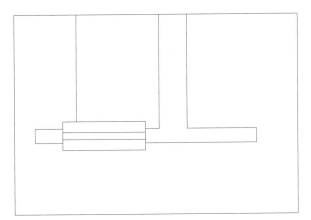

Figure P6–18 *Drawing showing the window opening*

Step 26: Draw two additional windows with the MLINE command with Scales 8 and 2 and Justification set to 0, and close the ends of the other multilines representing window and door jambs.

Step 27: Invoke the ZOOM ALL command to display the complete drawing. Draw the water heater with the CIRCLE command, with center at coordinates 11'-8",10'-4" and radius of 1'-0".

After drawing the remaining windows, closing the jambs, and drawing the water heater, the display should appear as shown in Figure P6–19.

Figure P6–19 *Drawing showing all the window openings and the water heater*

DRAWING THE DOORS

Step 28: Using the ZOOM WINDOW option, enlarge the area as shown in Figure P6–20 to begin drawing doors.

Figure P6–20 *Points indicating the area to be enlarged by means of the* ZOOM WINDOW *command*

Step 29   Invoke the MLINE command from the Draw toolbar. Set the Scale to 1, and draw the double line representing the door on the right by specifying the points indicated in Figure P6–21. AutoCAD prompts:

> Command: **mline** (ENTER)
> Current settings: Justification = Zero, Scale = 8.00, Style = STANDARD
> Specify start point or [Justification/Scale/STyle]: **j** (ENTER)
> Enter justification type [Top/Zero/Bottom] <top>: **t** (ENTER)
> Current settings: Justification = Top, Scale = 8.00, Style = STANDARD
> Specify start point or [Justification/Scale/STyle]: **s** (ENTER)
> Enter mline scale <2.00>: **1** (ENTER)
> Current settings: Justification = Top, Scale = 1.00, Style = STANDARD
> Specify start point or [Justification/Scale/STyle]: *(specify point A, as shown in Figure P6–21)*
> Specify next point: *(specify point B, as shown in Figure P6–21, at absolute coordinates 16'-0",-2'-0", or use the relative coordinates of @24<270)*
> Specify next point or [Undo]: *(press ENTER to terminate the command sequence)*

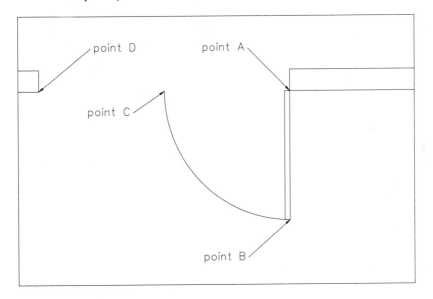

Figure P6–21 *Points indicating the location of the door*

Step 30:   Invoke the LINE command from the Draw toolbar to close the ends of the "door" just drawn.

Step 31:   Invoke the ARC command with Start/Center/Angle option to draw an arc to represent the swing of the door. AutoCAD prompts:

> Command: Specify start point of arc or [CEnter]: *(specify point B, as shown in Figure P6–21)*

Specify second point of arc or [CEnter/ENd]: **c** (right-click and choose Center)

Specify center point of arc: *(specify point A, as shown in Figure P6–21)*

Specify end point of arc or [Angle/chord Length]: *(right-click and select Angle from the shortcut menu)*

Specify included angle: **-90** (ENTER)

Step 32: Invoke the MIRROR command from the Modify toolbar. AutoCAD prompts:

Command: **mirror** (ENTER)

Select objects: *(select the multiline, end closing lines, and arc representing the door)*

Select objects: *(press ENTER to complete object selection)*

Specify first point of mirror line: *(specify point C, as shown in Figure P6–21)*

Specify second point of mirror line: *(specify a point on the left side of point C)*

Delete source objects? [Yes/No] <N>: (ENTER)

Step 33: Use the PAN command to move to the viewing area to include the opening made in the wall on the left, and repeat the preceding method to draw the multiline, end lines, and arc to represent the 3'-0" door on the left, as shown in Figure P6–22.

Figure P6–22 *Drawing showing the doors and windows*

DRAWING THE SINK

Step 34: Use the PAN and/or ZOOM command to move and resize the viewing area to include the counter and to draw a sink as shown in Figure P6–23.

Figure P6–23 *Drawing showing the closeup of the counter*

Step 35: Invoke the RECTANGLE command from the Draw toolbar to draw the sink. AutoCAD prompts:

> Command: **rectangle** (ENTER)
> Specify first corner point or [Chamfer/Elevation/Fillet/Thickness/
>   Width]: *(specify the lower left point at 7'-10",2'-0")*
> Specify other corner point: *(specify the upper right point at 9'-2",3'-8")*

Step 36: Invoke the FILLET command from the Modify toolbar. AutoCAD prompts:

> Command: **fillet** (ENTER)
> Current settings: Mode = TRIM, Radius = 0'-0 1/2"
> Select first object or [Polyline/Radius/Trim]: **r** (ENTER)
> Specify fillet radius <0'-0 1/2">: **2** (ENTER)
>
> Command: *(press ENTER to repeat the FILLET command)*
> Current settings: Mode = TRIM, Radius = 0'-2"
> Select first object or [Polyline/Radius/Trim]: **p** (ENTER)
> Select 2D polyline: *(select the rectangle drawn in the previous step)*

The results should appear as shown in Figure P6–24.

Figure P6–24 *Drawing with counter and sink*

DRAWING THE WASHERS AND DRYERS

Step 37: Invoke the ZOOM ALL command to display the complete drawing. Set Hidden as the current layer.

Step 38: Invoke the RECTANGLE command from the Draw toolbar. AutoCAD prompts:

Command: **rectangle** (ENTER)
Specify first corner point or [Chamfer/Elevation/Fillet/Thickness/ Width]: *(specify the lower left point at 0'-6",0'-8")*
Specify other corner point: *(specify the upper right point at 3'-0",3'-2")*

Command: *(press ENTER to repeat the RECTANGLE command)*
Specify first corner point or [Chamfer/Elevation/Fillet/Thickness/ Width]: *(specify the lower left point at 6'-8",5'-8")*
Specify other corner point: *(specify the lower left point at 9'-4",8'-4")*

Step 39: Invoke the ARRAY command from the Modify toolbar. AutoCAD prompts:

Command: **array** (ENTER)
Select objects: *(select the first rectangle drawn in step 38)*
Select objects: *(press ENTER to complete selection)*
Enter the type of array [Rectangular/Polar] <R>: **r** (ENTER)
Enter the number of rows (—) <1>: **4**
Enter the number of columns (|||) <1>: (ENTER)
Enter the distance between rows or specify unit cell (—): **2'8**

Step 40: Invoke the COPY command from the Modify toolbar. AutoCAD prompts:

Command: **copy** (ENTER)
Select objects: *(select the second rectangle drawn in step 38)*
Select objects: *(press ENTER to complete selection)*
Specify base point or displacement, or [Multiple]: *(specify a point)*

Specify second point of displacement or <use first point as displacement>: *(specify a relative displacement point at 3'-0" at 90 degrees)*

Figure P6–25 shows the result after drawing the washers and dryers.

Figure P6–25 *Drawing with washers and dryers*

COMPLETING THE SHELVES

Step 41: To draw the shelves, invoke the MLINE command from the Draw toolbar. AutoCAD prompts:

> Command: **mline** (ENTER)
> Current settings: Justification = Top, Scale = 1.00, Style = STANDARD
> Specify start point or [Justification/Scale/STyle]: **j** (ENTER)
> Enter justification type [Top/Zero/Bottom] <top>: **z** (ENTER)
> Current settings: Justification = Zero, Scale = 1.00, Style = STANDARD
> Specify start point or [Justification/Scale/STyle]: *(specify a point at 18'-0",8'-0")*
> Specify next point:*(specify a point at 19'-0",8'-0")*
> Specify next point or [Undo]: *(press ENTER to terminate the command sequence)*
>
> Command: *(press ENTER to repeat the MLINE command)*
> Current settings: Justification = Zero, Scale = 1.00, Style = STANDARD
> Specify start point or [Justification/Scale/STyle]: *(specify a point at 18'-0",4'-0")*
> Specify next point: *(specify a point at 19'-0",4'-0")*
> Specify next point or [Undo]: *(press ENTER to terminate the command sequence)*

Figure P6–26 shows the result after drawing the shelves.

Figure P6–26 *Drawing with shelves*

DRAWING TEXT OBJECTS

Step 42   Set Text as the current layer. Invoke the TEXT command, and draw the text at appropriate text size, as shown in Figure P6–27.

Figure P6–27 *Drawing with text objects*

Congratulations! You have just successfully applied several AutoCAD concepts in creating the drawing.

EXERCISES 6–1 THROUGH 6–6

EXERCISE 6–1

The drawing shown here is typical of informational graphics/text drawings whose objects are not dimensioned but parametric in nature. This cross-section of a reinforced concrete column is typical of those with varying dimensions and member of reinforcing bars. It is for reference only. The important information is given in the written text. Create the drawing shown according to the settings given in the following table.

Settings	Value		
1. Units	Decimal		
2. Limits	Lower left corner: 0",0"		
	Upper right corner: 12",9"		
3. Grid	.5		
4. Snap	.5		
5. Layers	*NAME*	*COLOR*	*LINETYPE*
	Center	Cyan	Continuous
	Border	Red	Continuous
	Object	Green	Continuous
	Text	Blue	Continuous
	Hidden	Cyan	Continuous
	Pattern	White	Continuous

Hints: Make sure you are in the appropriate layer before you draw the objects.

Use the DONUT command to draw two circles with large OD/ID and multiple circles with small OD and zero ID.

The patterning in the cross section can be generated by drawing one small triangle with the PLINE command and then copying and rotating it randomly throughout the area. The stipples that are conventional in representing concrete can also be randomly placed, by means of the POINT command. One of the conventions of patterning for concrete is that the stippling is usually more concentrated near the edges of the cross-section than in the center.

The MTEXT command is used to create blocks of text within prescribed rectangles. The justification determines where the block of text anchors and where the overage spills out of the rectangle.

SPIRAL REINFORCEMENT SHALL NOT HAVE DEFORMATIONS.

PILES FOR INDIVIDUAL COLUMN FOOTINGS SHALL BE OF 60 TON CAPACITY.

ALL REINFORCING BARS
ASTM A617 GRADE 60

PLACE ALL INTERIOR TOP
BEAM & SLAB BARS
SYMMETRICALLY ABOUT
COLUMN CENTER LINES
UNLESS OTHER WISE NOTED.

PLACE ALL BOTTOM BARS
SYMMETRICALLY UNLESS
OTHERWISE NOTED.

SECOND FLOOR
FRAMING PLAN

XYZ STEEL CO	HOUSTON, TX
DRAWN BY: I.B.WISE	CONTRACT 853
DATE: 2/2/2002	SHEET 5 OF 8

REINFORCED
CONCRETE COLUMN

SEQUENCE OF PLACING BAR SUPPORTS AND BARS

1. PLACE CONTINUOUS LINES OF SLAB BOLSTER IN E-W
   DIRECTION AT 4'-0" MAXIMUM O.C. BETWEEN COLUMNS.
2. LAY N-S BOTTOM BARS IN COLUMN & MIDDLE STRIPS.
3. LAY E-W BOTTOM BARS IN COLUMN & MIDDLE STRIPS.
4. PLACE 4 ROWS OF #4 SUPPORT BARS X 10'-0" AT 4'-0"
   MAXIMUM O.C. ON LHIGH CHAIRS AT 3'-0" MAXIMUM O.C.
   IN E-W DIRECTION AT COLUMN HEAD.

Figure Ex6–1 *Completed drawing*

EXERCISE 6–2

Create the drawing shown according to the settings given in the following table. (Do *not* dimension this drawing.)

Settings	Value
1. Units	Architectural
2. Limits	Lower left corner: 0'-0,0'-0
	Upper right corner: 65'-0,54'-0
3. Grid	12
4. Snap	6

Settings	Value		
5. Layers	*NAME*	*COLOR*	*LINETYPE*
	Center	Cyan	Continuous
	Border	Red	Continuous
	Object	Green	Continuous
	Text	Blue	Continuous
	Hidden	Cyan	Hidden

Hints: Make sure you are in the appropriate layer before you draw the objects.

Use the MLINE command to draw the beams in continuous linetype and the footings in hidden linetype.

Use the MLEDIT command to connect the beams in open tees.

The MTEXT command is used to create blocks of text within prescribed rectangles. The justification determines where the block of text anchors and where the overage spills out of the rectangle.

Figure Ex6–2 *Foundation plan*

EXERCISE 6–3

Create the drawing shown according to the settings given in the following table. (Do *not* dimension this drawing.)

Settings	Value
1. Units	Architectural
2. Limits	Lower left corner: 0'-0",0'-0"
	Upper right corner: 50'-0",40'-0"
3. Grid	12"
4. Snap	6"

Settings	Value		
5. Layers	*NAME*	*COLOR*	*LINETYPE*
	Center	Cyan	Continuous
	Border	Red	Continuous
	Object	Green	Continuous
	Text	Blue	Continuous
	Hidden	Cyan	Hidden

Hints: Make sure you are in the appropriate layer before you draw the objects.

Use the MLINE command to draw the beam edges in continuous linetype and the webs in hidden linetype.

Detail #1 can be drawn to true size and then the SCALE command used to increase its size by a factor of 1.5 (the scale ratio of 1/4" = 1'-0" to 3/8" = 1'-0").

Figure Ex6–3

EXERCISE 6–4

Create the drawing shown according to the settings given in the following table.

Settings	Value
1. Units	Architectural
2. Limits	Lower left corner: 10'-0",10'-0" Upper right corner: 132'-0",102'-0"
3. Grid	60"
4. Snap	12"
5. Added Layers	*NAME*　　　　*COLOR*　　　*LINETYPE*
	Spline　　　　Green　　　Continuous
	Text　　　　　Blue　　　　Continuous

 Hints: Use the SPLINE command to draw splines. They do not need to be drawn through exact points. Estimate their locations and reproduce the splines shown in the drawing. In the real world, the topographic point locations will be established by a surveying crew in the field and given to you to make your drawing.

See Figure Ex6–4b for points showing the approximate locations used to draw the splines.

Figure Ex6–4a *Topographic drawing*

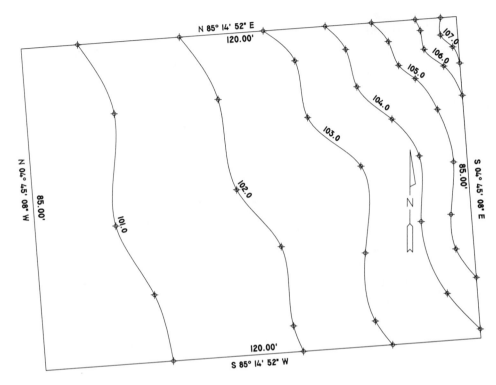

Figure Ex6–4b *Topographic drawing showing the approximate locations*

EXERCISE 6–5

Create the drawing shown according to the settings given in the following table. (Do *not* dimension this drawing.)

Settings	Value
1. Units	Architectural
2. Limits	Lower left corner: 0',0'
	Upper right corner: 12',9'
3. Grid	6"
4. Snap	3"

Settings	Value		
5. Layers	*NAME*	*COLOR*	*LINETYPE*
	Border	Red	Continuous
	Object	Green	Continuous
	Text	Blue	Continuous

Hints: Use the MLINE command to draw the door panel perimeters with a style that displays the joints. One of the top panels of the front door can be copied down. Then, using the STRETCH command, you can lengthen the copy and then copy the newly created longer panel. Repeat for the lower panels of all the doors.

Draw the door handles with a style that closes the ends of the mline.

The window inside the back door will have to be drawn using a style that does not display the joints or close the ends.

Use the MLEDIT command to connect the window mullions in open tees.

The MTEXT command is used to create text objects.

Figure Ex6–5 *Detailed drawing of a front door and a back door*

EXERCISE 6–6

Create the drawing shown according to the settings given in the following table. (Do *not* dimension this drawing.)

Settings	Value		
1. Units	Architectural		
2. Limits	Lower left corner: 0',0'		
	Upper right corner: 12',9'		
3. Grid	6"		
4. Snap	3"		
5. Layers	*NAME*	*COLOR*	*LINETYPE*
	Border	Red	Continuous
	Object	Green	Continuous
	Text	Blue	Continuous

Hints: Use the MLINE command to draw the drawer perimeters with a style that displays the joints.

Draw the drawer handles and desk top and bases with a style that closes the ends of the mline.

Draw Details "A" and "B" to true size, and then use the SCALE command to make them larger, as in the drawing shown.

Figure Ex6–6 *Detailed drawing of a desk*

REVIEW QUESTIONS

1. The elements of a multiline can have different colors.

   a. True

   b. False

2. A SPLINE object:

   a. does not actually pass through the control points

   b. is always shown as a continuous line

   c. requires you to specify tangent information for each control point

   d. requires you to specify tangent information for the first and last points only

3. When spell-checking a drawing, AutoCAD automatically checks the entire drawing.

   a. True

   b. False

4. To use incline lettering so that the vertical portions of the characters point to 2 o'clock for a horizontal line of text, the obliquing angle should be:

   a. –60

   b. –30

   c. 0

   d. 30

   e. 60

5. Once objects are grouped together, they can be ungrouped by:

   a. using the EXPLODE command

   b. using the UNGROUP command

   c. using the GROUPcommand

   d. they cannot be ungrouped

6. To insert the diameter symbol into a string of text, you should enter:

   a. %%d

   b. %%c

   c. %%phi

   d. %%dia

7. It is possible to force all text on a drawing to display as an open rectangle in order to speed up the redisplay of the drawing by using the:

a. TEXT command

b. RTEXT command

c. QTEXT command

d. DTEXT command

8. In order to select a different font for use in the TEXT command, a text style must be created.

a. True

b. False

9. The grips dialog box allows you to change all of the following, except:

a. grip size

b. grip color

c. toggle grips system variable ON/OFF

d. specify the location of the grips

10. Grips do not allow you to _____ an object.

a. trim

b. erase

c. mirror

d. move

e. stretch

11. To edit a word in the middle of a string of text, a reasonable command to use would be:

a. DDEDIT

b. CHPROP

c. DDCHPROP

d. MODIFY

e. CHANGE

12. Multiline styles can be saved to an external file, thus allowing their use in multiple drawings.

a. True

b. False

13. The ID command will:

    a. display the serial number of the AutoCAD program

    b. display the X, Y, and Z coordinates of a selected point

    c. allow you to password protect a drawing

    d. none of the above

14. A quick way to find the length of a line is to use what command?

    a. LIST   d. LENGTH

    b. DISTANCE   e. COORDINATES

    c. DIMENSION

15. To obtain a full listing of all the objects contained in the current drawing, you should use:

    a. LIST   d. DBDUMP

    b. DBLIST   e. DUMP

    c. LISTALL

16. When filleting multilines, you must specify:

    a. the radius of the innermost fillet

    b. the radius of the center line

    c. the radius of the outermost fillet

    d. multilines cannot be filleted

17. Which of the following are not valid options when creating a text style?

    a. width factor

    b. upside down

    c. vertical

    d. backwards

    e. none of the above (all answers are valid)

18. You can change the setting of a system variable by using the command:

    a. SYSVAR

    b. SETVAR

    c. VARSET

    d. VARSYS

    e. none of the above

19. Which of the following is not a valid type of system variable?

a. real

b. integer

c. point

d. string

e. double

20. To locate text such that it is centered exactly within a circle, a reasonable justification would be:

a. Center

b. Middle

c. Full

d. Both A and B

e. none, simply zoom in and approximate it

Directions: Use the following X and Y locations to construct figure Qx6–1. Your completed drawing will be used to answer the remaining questions.

Point	X value	Y value
A	1.5000	1.0000
B	2.1717	1.0000
C	2.7074	1.7500
D	4.5000	1.7500
E	4.8200	1.0000
F	5.5000	1.0000
G	4.1838	4.0711
H	3.5071	3.8100
J	3.7974	3.3861
K	3.6104	3.3139
L	2.8125	2.2500
M	4.2843	2.2500

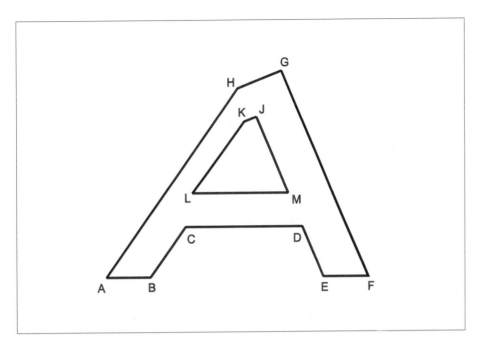

Figure Qx6–1

21. Using the appropriate AutoCAD command what is the length of Line AH?

   a. 3.5432

   b. 3.4523

   c. 3.4532

22. Using the appropriate AutoCAD command what is the angle of Line AH?

   a. 324 degrees

   b. 234 degrees

   c. 432 degrees

23. Using the appropriate AutoCAD command what is the length of Line CD?

   a. 1.7962

   b. 1.7926

   c. 1.7629

24. Using the appropriate AutoCAD command what is the length of Line FG?

 a. 3.3143

 b. 3.3314

 c. 3.3413

25. Using the appropriate AutoCAD command what is the angle of Line FG?

 a. 131 degrees

 b. 311 degrees

 c. 113 degrees

26. Using the appropriate AutoCAD command what is the length of Line GH?

 a. 0.7253

 b. 0.7532

 c. 0.7325

27. Using the appropriate AutoCAD command what is the angle of Line GH?

 a. 210 degrees

 b. 102 degrees

 c. 201 degrees

28. What is the AREA of shape JKLM?

 a. .8833

 b. .8383

 c. .8838

29. What is the perimeter of shape JKLM?

 a. 4.1281

 b. 4.1812

 c. 4.1182

30. What is the linear distance from point E to point J?

 a. 2.2589

 b. 2.5859

 c. 2.5958

CHAPTER 7

Dimensioning

INTRODUCTION

AutoCAD provides a full range of dimensioning commands and utilities to enable the drafter to comply with the conventions of most disciplines, including architecture and civil, electrical, and mechanical engineering, among many others.

After completing this chapter, you will be able to do the following:

- Draw linear dimensioning
- Draw aligned dimensioning
- Draw angular dimensioning
- Draw diameter and radius dimensioning
- Draw leaders with annotation and geometric tolerance
- Draw ordinate dimensioning
- Draw baseline and continue dimensioning
- Edit dimension text
- Create and modify dimension styles

AutoCAD makes drawing dimensions easy. For example, the width of the rectangle shown in Figure 7–1 can be dimensioned by specifying the two endpoints of the top corners and then specifying a point to determine the location of the Dimension Line. AutoCAD drags a preview image of the dimension to indicate how it will look while you move the cursor to specify the location of the Dimension Line.

Once Linear Dimensioning is mastered, other types of dimensions, such as Diameter, Radius, Angular, Baseline, and Ordinate Dimensioning, can be drawn quickly and accurately.

The dimension types available are Linear, Angular, Diameter, Radius, and Ordinate. There are primary and secondary commands available for each of these types. There are also other general utility, editing, and style-related commands and subcommands that assist you in drawing the correct dimensions quickly and with accuracy.

Figure 7–1 *An example of linear dimensioning*

The Linear Dimensioning command options include Horizontal, Vertical, Aligned, and Rotated. Angular dimensioning is covered by the ANGULAR command. Diameter, radius, and ordinate dimensioning are covered by the DIAMETER, RADIUS, and ORDINATE commands, respectively.

Approximately 60 Dimensioning System Variables are available specifically for dimensioning. Most of these have names that begin with "DIM." They are used for such purposes as determining the size of the gap between the extension line and the point specified on an object or whether one or both of the extension lines will be drawn or suppressed. It is combinations of these variable settings that can be named and saved as Dimension Styles and later recalled for applying when needed. See Appendix B for the listing of available Dimensioning System Variables.

Dimension variables change when their associated settings are changed in the Modify Dimension Style section of the Dimension Style Manager dialog box. For example, the value of the DIMEXO (extension line offset) dimensioning system variable is established by the number in the Offset From Origin text box in the Modify Dimension Style subdialog box. See the section later in this chapter on "Dimension styles."

Dimension utilities include Override, Center, Leader, Baseline, Continue, and Feature Control Frames for adding tolerancing information.

The dimension editing command options include Hometext, Newtext, Oblique, Tedit, and Trotate.

DIMENSION TERMINOLOGY

Following are the terms for the different parts that make up dimensions in AutoCAD.

Figure 7–2 shows the different components of a typical dimension.

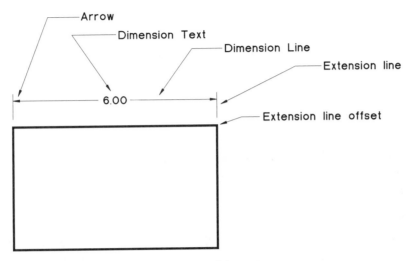

Figure 7–2 *The different components of a typical dimension*

DIMENSION LINE

The Dimension Line is offset from the measured feature. The Dimension Line (sometimes drawn into two segments outside the Extension Lines if a single line with its related text will not fit between the extension lines) indicates the direction and length of the measured distance, but is usually offset for clarity. If the dimension is measured between parallel lines of one or two objects, then it may not be offset, but is drawn on the object or between two objects. Dimension Lines are usually terminated with markers such as arrows or ticks (short slanted lines). Angular Dimension Lines become arcs whose centers are at the vertex of the angle.

ARROWHEAD

The arrowhead is a mark at the end of the Dimension Line to indicate its termination. Shapes other than arrows are used in some styles.

EXTENSION LINE

When the Dimension Line is offset from the measured feature, the Extension Lines (sometimes referred to as witness lines) indicate such offset. Unless you have invoked the Oblique option, the Extension Lines will be perpendicular to the direction of the measurement.

DIMENSION TEXT

The Dimension Text consists of numbers, words, characters, and symbols used to indicate the measured value and type of dimension. Unless the text has been altered in the standard Dimension Style, the number/symbol format is decimal. This conforms to the same linear and angular units as the default settings of the units. The text style usually conforms to that of the current text style.

LEADER

The leader is a radial line used to point from the Dimension Text to the circle or arc whose diameter or radius is being dimensioned. A leader can also be used for general annotation.

CENTER MARK

The center mark is made up of lines or a series of lines that cross in the center of a circle for the purpose of marking its center.

ASSOCIATIVE/NON-ASSOCIATIVE DIMENSIONS

Dimensions in AutoCAD can be drawn either as associative or non-associative, depending on the setting of the DIMASO dimensioning system variable.

ASSOCIATIVE

An associative dimension drawn with the DIMASO dimensioning system variable set to ON has all of its separate parts become members of a single associative dimension. Therefore, if any one of its members is selected for modifying, all members are highlighted and subject to being modified. This is similar to the manner in which member objects of a block reference are treated. In addition to the customary visible parts, AutoCAD draws point objects at the ends, where the measurement actually occurs on the object. If you have dimensioned the width of a rectangle with an associative dimension and then specified one end of the rectangle to stretch, the dimension will also be stretched, and the Dimension Text will be changed to correspond to the new measurement.

NON-ASSOCIATIVE

The non-associative dimensions are drawn while the DIMASO dimensioning system variable is set to OFF, and the members are drawn as separate objects. If one of the components of the dimension is selected for modifying, it will be the only one modified.

 Note: An associative dimension can be converted to a non-associative dimension with the EXPLODE command. Once the dimension is exploded, you cannot recombine the separate parts back into the associative dimension from which they were exploded (except by means of the UNDO command if feasible). Note that when you explode an associative dimension, the measurement-determining points (nodes) remain in the drawing as point objects.

DIMENSIONING COMMANDS

In AutoCAD three primary ways to invoke dimensioning commands (DIMALIGNED in the following examples) are available:

1. At the "Command:" prompt, enter **dimaligned** (or some other special dimensioning command).

2. From the Dimension toolbar (see Figure 7–3), choose one of the dimensioning command buttons.

3. From the Dimension menu (see Figure 7–4), choose one of the dimensioning commands.

Figure 7–3 *Choosing the* ALIGNED DIMENSION *command from the Dimension toolbar*

Figure 7–4 *Choosing the* ALIGNED *command from the Dimension menu*

LINEAR DIMENSIONING

Invoke the LINEAR DIMENSION command:

Dimension toolbar	Choose the LINEAR DIMENSION command (see Figure 7–5)
Dimension menu	Choose Linear
Command: prompt	**dimlinear** (ENTER)

Figure 7–5 *Invoking the* LINEAR DIMENSION *command from the Dimension toolbar*

AutoCAD prompts:

> Command: **dimlinear** (ENTER)
> Specify first extension line origin or <select object>: *(specify a point)*

AutoCAD uses it as the start point (origin) for the first extension line. This point can be the endpoint of a line, the intersection of objects, the center point of a circle, or even the insertion point of a text object. You can specify a point on the object itself. AutoCAD provides a gap between the object and the extension line that is equal to the value of the DIMEXO dimensioning system variable, which you can change at any time. After you specify the start point (origin), AutoCAD prompts:

> Specify second extension line origin: *(specify a point at which the second extension line should start)*

Dynamic Horizontal/Vertical Dimensioning

Dynamic horizontal/vertical dimensioning is an option after you have specified two points in response to the DIMLINEAR command. If you specify two points on the same horizontal line, moving the cursor above or below the line causes the preview image of the dimension to appear. AutoCAD assumes you wish to draw a horizontal dimension. Likewise, AutoCAD assumes a vertical dimension if the specified points are on the same vertical line.

Dynamic Drag (switching between horizontal and vertical) is more applicable when the two points specified are not on the same horizontal or vertical line. That is, they can be considered diagonally opposite corners of an imaginary rectangle with both width and height. After the two points are specified, you are prompted to specify the location of the dimension line. You will also be shown a preview image of where the dimension will be located, by where the cursor is located relative to the imaginary rectangle formed by the two points. If the cursor is above the top line or below the bottom line of the rectangle, then the dimension will be horizontal. If the cursor is to the right of the right side or to the left of the left side of the rectangle, then the dimension will be vertical. If the cursor is dragged to one of the outside quadrants or inside of the rectangle, it will maintain the type of dimension in effect before the cursor was moved.

After two points have been specified, AutoCAD prompts:

Specify dimension line location or [Mtext/Text/Angle/Horizontal/Vertical/ Rotated]: *(specify location for dimension line or right-click for the shortcut menu and select one of the available options)*

Drawing The Dimension Line

After specifying the two points (or using the right-click shortcut menu to change the dimension text or type), specify a point through which the dimension line is to be drawn. AutoCAD draws the line on which the dimension text is drawn. If there is enough room between the extension lines, the dimension text will be centered in line with or above this line. If the dimension text is to be in line with the dimension line, then the dimension line will be broken to allow room for the text. However, if the dimension line, arrows, and text do not fit between the extension lines, they are drawn outside. The text will be drawn near the second extension line.

Changing Dimension Text

The **Mtext** and **Text** options allow you to change the measured dimension text. The **Angle** option allows you to change the rotation angle of the dimension text. After responding appropriately for Text or Angle, AutoCAD repeats the prompts for the dimension line location.

Changing Dimension Type

The **Horizontal** option forces a horizontal dimension to be drawn (even when the dynamic drag switching calls for a vertical dimension). Likewise, the **Vertical** option forces a vertical dimension to be drawn (even when the dynamic drag switching calls for a horizontal dimension).

The **Rotated** option allows you to draw the dimension at a specified angle that is neither horizontal nor vertical. Nor is the desired angle at the angle determined by the first two points specified as it is in the case of aligned dimensioning (see the section on Aligned Dimensioning in this chapter). Figure 7–6 shows a situation where Rotated Dimension is applicable. Here is a case where the desired dimension is from point A to point B. However, the desired angle of dimensioning is the angle created by a line from point B to point C. The group of dimensions can be started by choosing the Rotated option, after specifying points A and B. Then points B and C are specified to determine the angle.

Right-click and choose Rotated from the shortcut menu, and then AutoCAD prompts:

Specify angle of dimension line <0>: *(Specify the dimension line angle)*

In lieu of specifying an angle, you can specify two points to determine the angle to rotate the dimension line. The points specified do not have to be parallel to the direction of the dimensioned distance.

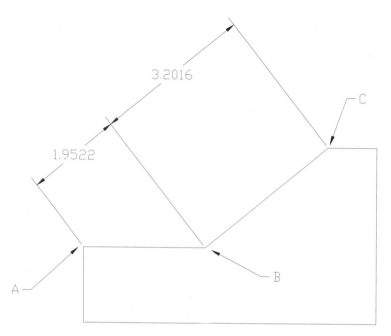

Figure 7–6 *Example of using Rotated Dimensioning*

The following command sequence is an example of drawing linear dimensioning for a horizontal line by providing two data points for the first and second line origins, respectively, as shown in Figure 7–7.

> Command: **dimlinear** (ENTER)
> Specify first extension line origin or <select object>: *(specify the origin of the first extension line)*
> Specify second extension line origin: *(specify the origin of the second extension line)*
> Specify dimension line location or [Mtext/Text/Angle/Horizontal/Vertical/Rotated]: *(specify the location for the dimension line)*

The following command sequence is an example of drawing a linear dimensioning for a vertical line by providing two data points for the first and second line origins, respectively, as shown in Figure 7–8.

> Command: **dimlinear** (ENTER)
> Specify first extension line origin or <select object>: *(specify the origin of the first extension line)*
> Specify second extension line origin: *(specify the origin of the second extension line)*
> Specify dimension line location or [Mtext/Text/Angle/Horizontal/Vertical/Rotated]: *(specify the location for the dimension line)*

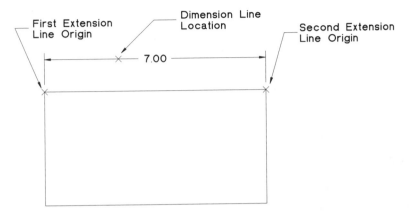

Figure 7–7 *Drawing a linear dimensioning for a horizontal line*

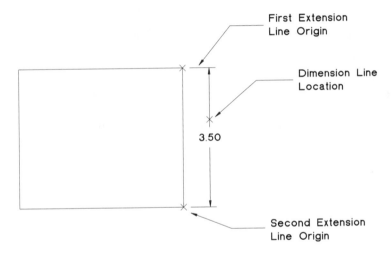

Figure 7–8 *Drawing linear dimensioning for a vertical line*

LINEAR DIMENSIONING BY SELECTING AN OBJECT

If the dimension you wish to draw is between endpoints of a line, arc, or circle diameter, AutoCAD allows you to bypass having to specify those endpoints separately. This speeds up drawing the dimension, especially if you have to invoke the endpoint object snap feature. When you have invoked the DIMLINEAR command and you press ENTER in response to the "Specify first extension line origin or <select object>:" prompt, AutoCAD prompts:

> Select object to dimension: *(select line, polyline segment, arc, or circle object)*
> Specify dimension line location or[Mtext/Text/Angle/Horizontal/Vertical/
>   Rotated]: *(specify location for dimension line or right-click for the
>   shortcut menu and select one of the available options)*

If you select a line object, AutoCAD automatically uses the line's endpoints as the first and second points to determine the distance to measure. You will be prompted to specify the location of the dimension line. If you right-click and choose the Horizontal mode, a horizontal dimension is drawn accordingly. In a similar manner, if you choose Vertical mode, a vertical dimension is drawn. If you do not choose the type of dimension to be drawn, then a horizontal dimension is drawn if the specified point for the dimension line is above or below the line object, and a vertical dimension is drawn if the specified point for the dimension line is to the right or left of the line object. If you right-click and choose the Aligned mode from the shortcut menu, AutoCAD uses the endpoints of the selected line object as the first and second points and its direction as the direction to measure. If you right-click and choose the Rotated mode from the shortcut menu, AutoCAD will use the endpoints of the line for the distance reference points and then use the next two points specified to determine the direction of the dimension and measure the distance between the two endpoints of the line object in that direction. The dimension line passes through the last point specified.

The following command sequence shows an example of drawing a linear dimension to a single object, as shown in Figure 7–9.

Command: **dimlinear** (ENTER)
Specify first extension line origin or <select object>: (ENTER)
Select object to dimension: *(select the line)*
Specify dimension line location or [Mtext/Text/Angle/Horizontal/Vertical/
   Rotated]: *(specify the location for the dimension line)*

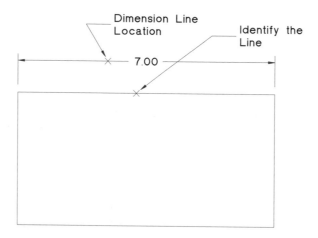

Figure 7–9 *Drawing a linear dimension for a single object*

If the object is a circle, AutoCAD automatically uses the diameter of the circle as the distance to measure and the point at which you selected the circle as one diameter endpoint for one end of the measured direction. If you are in the Horizontal mode, a horizontal dimension is drawn using the endpoints of the horizontal diameter. Likewise, with the Vertical mode the endpoints of the vertical diameter is used. If you are in the Rotated mode, AutoCAD uses the first two points specified to determine the direction of the dimension and to measure the distance between the two endpoints of a diameter in that direction. The dimension line passes through the last point specified.

If the object is an arc, AutoCAD automatically uses the arc's endpoints for the first and second points in determining the distance to measure. You will be prompted to specify the location of the dimension line. If you are in the Horizontal mode, a horizontal dimension is drawn accordingly. Likewise with the Vertical mode. If you have not yet determined the type of dimension to be drawn, then a horizontal dimension will be drawn if the specified point for the dimension line is above or below the arc, and a vertical dimension is drawn if the specified point for the dimension line is to the right or left of the arc. If you are in the Aligned mode, AutoCAD uses the endpoints of the selected arc as the first and second points and its direction as the direction to measure. If you are in the Rotated mode, AutoCAD uses the first two points specified to determine the direction of the dimension and to measure the distance between the two endpoints of the arc in that direction. The dimension line passes through the last point specified.

ALIGNED DIMENSIONING

When dimensioning a line drawn at an angle, it may be necessary to align the dimension line with an object line.

Invoke the ALIGNED DIMENSION command:

Dimension toolbar	Choose the ALIGNED DIMENSION command (see Figure 7–10)
Dimension menu	Choose Aligned
Command: prompt	**dimaligned** (ENTER)

Figure 7–10 *Invoking the* DIMALIGNED *command from the Dimension toolbar*

AutoCAD prompts:

> Command: **dimaligned** (ENTER)
> Specify first extension line origin or <select object>: *(specify the origin of the first extension line)*
> Specify second extension line origin: *(specify the origin of the second extension line)*

After two points have been specified, AutoCAD prompts:

> Specify dimension line location or [Mtext/Text/Angle]: *(specify the location for the dimension line or right-click for the shortcut menu and select one of the available options)*

Specify a point where the dimension line is to be drawn or choose one of the available options. The **Mtext, Text,** and **Angle** options are the same as in Linear dimensioning, explained earlier in this section.

The following command sequence shows an example of drawing aligned dimensioning for an angular line by providing two data points for the first and second line origins, respectively, as shown in Figure 7–11.

> Command: **dimaligned** (ENTER)
> Specify first extension line origin or <select object>: *(specify the origin of the first extension line)*
> Specify second extension line origin: *(specify the origin of the second extension line)*
> Specify dimension line location or [Mtext/Text/Angle]: *(specify the location of the dimension line)*

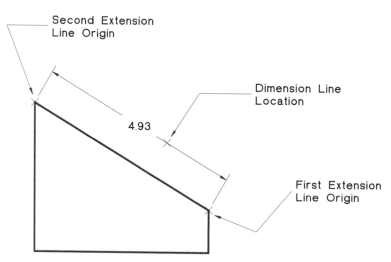

Figure 7–11 *Drawing aligned dimensioning for a line drawn at an angle*

ORDINATE DIMENSIONING

AutoCAD uses the mutually perpendicular X and Y axes of the World Coordinate System or current User Coordinate System as the reference lines from which to base the X or Y coordinate displayed in an ordinate dimension (sometimes referred to as a datum dimension). See Chapter 15 for a detailed discussion of creating a User Coordinate System.

Invoke the ORDINATE DIMENSION command:

Dimension toolbar	Choose the ORDINATE DIMENSION command (see Figure 7–12)
Dimension menu	Choose Ordinate
Command: prompt	**dimordinate** (ENTER)

Figure 7–12 *Invoking the* ORDINATE DIMENSION *command from the Dimension toolbar*

AutoCAD prompts:

Command: **dimordinate** (ENTER)
Specify feature location: *(specify a point)*

Although the default prompt is "Specify feature location:", AutoCAD is actually looking for a point that is significant in locating a feature point on an object, such as the endpoint/intersection where planes meet or the center of a circle representing a hole or shaft. Therefore, an Object Snap mode, such as endpoint, intersection, quadrant, or center, will normally need to be invoked when responding to the "Specify feature location:" prompt. Specifying a point determines the origin of a single orthogonal leader that will point to the feature when the dimension is drawn. AutoCAD prompts:

Specify leader endpoint or [Xdatum/Ydatum/Mtext/Text/Angle]: *(specify a point or right-click for the shortcut menu and select one of the available options)*

If the Ortho mode is set to ON, the leader will be a single horizontal line for a Ydatum ordinate dimension, as shown in Figure 7–13, or a single vertical line for an Xdatum ordinate dimension, as shown in Figure 7–14.

Figure 7–13 *Ydatum dimension*

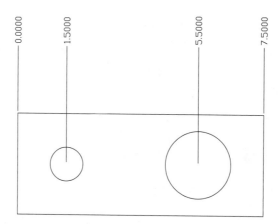

Figure 7–14 *Xdatum dimension*

If the Ortho mode is set to OFF, the leader will be a three part line consisting of orthogonal lines on each end joined by a diagonal line in the middle. It may be necessary to use the non-orthogonal leader if the text has to be offset to keep from interfering with other objects in the drawing. The type of dimension drawn (Ydatum or Xdatum) depends on which is greater of the horizontal and vertical distances between the specified "feature location" point and the "leader endpoint" point. A preview image of the dimension is displayed during specification of the "leader endpoint".

Right-click and choose Xdatum or Ydatum from the shortcut menu, and AutoCAD then draws an Xdatum dimension or Ydatum dimension, respectively, regardless of the location of the "leader endpoint" point relative to the "feature location" point.

RADIUS DIMENSIONING

The radius dimensioning feature provides commands to create radius dimensions, as shown in Figure 7–15 for a circle and an arc. The type of dimensions that AutoCAD utilizes depends on the Dimensioning System Variable settings (see the section on "Dimension styles," later in this chapter, for how to change the Dimensioning System Variable settings to draw appropriate radius dimensioning).

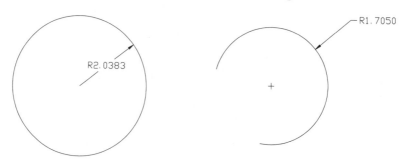

Figure 7–15 *Radius dimensioning of a circle and an arc*

Invoke the RADIUS DIMENSION command:

Dimension toolbar	Choose the RADIUS DIMENSION command (see Figure 7–16)
Dimension menu	Choose Radius
Command: prompt	**dimradius** (ENTER)

Radius Dimension

Figure 7–16 *Invoking the* RADIUS DIMENSION *command from the Dimension toolbar*

AutoCAD prompts:

Command: **dimradius** (ENTER)
Select arc or circle: *(select an arc or a circle to dimension)*
Specify dimension line location or [Mtext/Text/Angle]: *(specify the location for the dimension leader line or right-click for the shortcut menu and select one of the available options)*

The **Mtext, Text,** and **Angle** options are the same as in Linear dimensioning, explained earlier in this section. Dimension text for radius dimensioning is preceded by the letter R.

The following command sequence shows an example of drawing radius dimensioning for a circle, as shown in Figure 7–17.

Command: **dimradius** (ENTER)
Select arc or circle: *(select the circle object)*
Specify dimension line location or [Mtext/Text/Angle]: *(specify a point to draw the dimension leader line or right-click for the shortcut menu and select one of the available options)*

R2.0383

Figure 7–17 *Radius dimensioning of a circle*

DIAMETER DIMENSIONING

The diameter dimensioning feature provides commands to create diameter dimensions, as shown in Figure 7–18 for a circle and an arc. The type of dimensions that AutoCAD utilizes depends on the Dimensioning System Variable settings (see the section on "Dimension styles," later in this chapter, for how to change the Dimensioning System Variable settings to draw an appropriate diameter dimensioning).

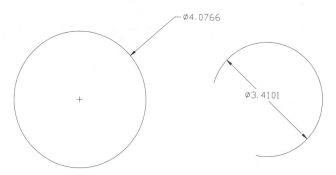

Figure 7–18 *Diameter dimensioning of a circle and an arc*

Invoke the DIAMETER DIMENSION command:

Dimension toolbar	Choose the DIAMETER DIMENSION command (see Figure 7–19)
Dimension menu	Choose Diameter
Command: prompt	**dimdiameter** (ENTER)

Figure 7–19 *Invoking the DIAMETER DIMENSION command from the Dimension toolbar*

AutoCAD prompts:

Command: **dimdiameter** (ENTER)
Select arc or circle: *(select an arc or a circle to dimension)*
Specify dimension line location or [Mtext/Text/Angle]: *(specify the location for the dimension line or right-click for the shortcut menu and select one of the available options)*

The **Mtext, Text,** and **Angle** options are the same as in Linear dimensioning, explained earlier in this section.

Specifying a point determines the location of the diameter dimension. Note that the dimension for an arc of less than 180 degrees cannot be forced to where neither end of the dimension is on the arc.

The following command sequence shows an example of drawing diameter dimensioning for a circle, as shown in Figure 7–20.

> Command: **dimdiameter** (ENTER)
> Select arc or circle: *(select an arc or a circle to dimension)*
> Specify dimension line location or [Mtext/Text/Angle]: *(specify a point to draw the dimension)*

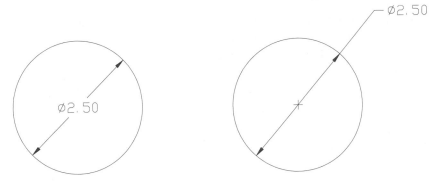

Figure 7–20 *Diameter dimensioning of a circle*

ANGULAR DIMENSIONING

The ANGULAR DIMENSION command allows you to draw angular dimensions using three points (vertex, point, point), between two nonparallel lines, on an arc (between the two endpoints of the arc, with the center as the vertex), and on a circle (between two points on the circle, with the center as the vertex).

Invoke the ANGULAR DIMENSION command:

Dimension toolbar	Choose the ANGULAR DIMENSION command (see Figure 7–21)
Dimension menu	Choose Angular
Command: prompt	**dimangular** (ENTER)

Figure 7–21 *Invoking the* ANGULAR DIMENSION *command from the Dimension toolbar*

AutoCAD prompts:

> Command: **dimangular** (ENTER)
> Select arc, circle, line, or <specify vertex>: *(select an object or press* ENTER *to specify a vertex)*

The default method of angular dimensioning is to select an object.

If the object selected is an arc, as shown in Figure 7–22, AutoCAD automatically uses the arc's center as the vertex and its endpoints for the first angle endpoint and second angle endpoint to determine the three points of a Vertex/Endpoint/Endpoint angular dimension. AutoCAD prompts:

> Specify dimension arc line location or [Mtext/Text/Angle]: *(specify the location for the dimension arc line or right-click for the shortcut menu and select one of the available options)*

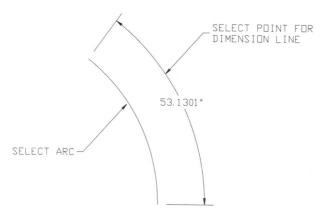

Figure 7–22 *Angular dimensioning of an arc*

The **Mtext, Text,** and **Angle** options are the same as in Linear dimensioning, explained earlier in this section. Specify a point for the location of the dimension. AutoCAD automatically draws radial extension lines.

If the object selected is a circle, as shown in Figure 7–23, AutoCAD automatically uses the circle's center as the vertex and the point at which you selected the circle as the endpoint for the first angle endpoint, and prompts:

> Specify second angle endpoint: *(specify a point to determine the second angle endpoint)*

Specify a point, and AutoCAD makes this point the second angle endpoint to use along with the previous two points as the three points of a Vertex/Endpoint/Endpoint angular dimension. Note that the last point does not have to be on the circle; however, it does determine the origin for the second extension line. Then, AutoCAD prompts:

Specify dimension arc line location or [Mtext/Text/Angle]: *(specify the location for the dimension arc line or right-click for the shortcut menu and select one of the available options)*

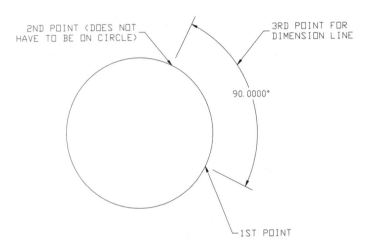

Figure 7–23 *Angular dimensioning of a circle*

The **Mtext, Text,** and **Angle** options are the same as in Linear dimensioning, explained earlier in this section.

If you specify a point for the location of the dimension arc line, AutoCAD automatically draws radial extension lines and draws either a minor or a major angular dimension, depending on whether the point used to specify the location of the dimension arc line is in the minor or major projected sector.

If the object selected is a line, as shown in Figure 7–24, then AutoCAD prompts:

Select second line: *(select a line object)*

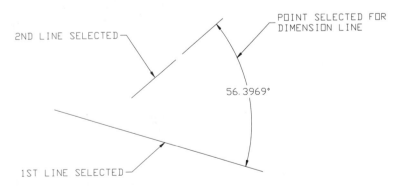

Figure 7–24 *Angular dimensioning of a line*

Select another line, and AutoCAD uses the apparent intersection of the two lines as the vertex for drawing a Vertex/Vector/Vector angular dimension. You are then prompted to specify the location of the dimension arc, which will always be less than 180 degrees.

If the dimension arc is beyond the end of either line, AutoCAD adds the necessary radial extension line(s). Then AutoCAD prompts:

> Specify dimension arc line location or [Mtext/Text/Angle]: *(specify the location for the dimension arc line or right-click for the shortcut menu and select one of the available options)*

The **Mtext, Text,** and **Angle** options are the same as in Linear dimensioning, explained earlier in this section.

If you specify a point for the location of the dimension arc, AutoCAD automatically draws extension lines and draws the dimension text.

If, instead of selecting an arc, a circle, or two lines for angular dimensioning, you press ENTER, AutoCAD allows you to do three-point angular dimensioning. The following command sequence shows an example of drawing angular dimensioning by providing three data points, as shown in Figure 7–25.

> Command: **dimangular** (ENTER)
> Select arc, circle, line, or ENTER: (ENTER)
> Specify angle vertex: *(specify a point)*
> Specify first angle endpoint: *(specify a point)*
> Specify second angle endpoint: *(specify a point)*
> Specify dimension arc line location or [Mtext/Text/Angle]: *(specify the dimension arc line location or right-click for the shortcut menu and select one of the available options)*

Figure 7–25 *Using three points for angular dimensioning*

BASELINE DIMENSIONING

Baseline dimensioning (sometimes referred to as parallel dimensioning) is used to draw dimensions to multiple points from a single datum baseline, as shown in Figure 7–26. The first extension line origin of the initial dimension (it can be a linear, angular, or ordinate dimension) establishes the base from which the baseline dimensions are drawn. That is, all of the dimensions in the series of baseline dimensions will share a common first extension line origin. AutoCAD automatically draws a dimension line/arc beyond the initial (or previous baseline) dimension line/arc. The location of the new dimension line/arc is an offset distance established by the DIMDLI (for dimension line increment) dimensioning variable.

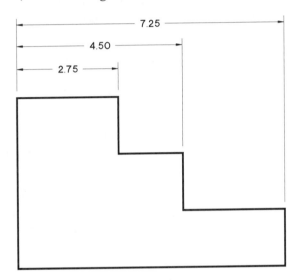

Figure 7–26 *Baseline dimensioning*

Invoke the BASELINE DIMENSION command:

Dimension toolbar	Choose the BASELINE DIMENSION command (see Figure 7–27)
Dimension menu	Choose Baseline
Command: prompt	**dimbaseline** (ENTER)

Figure 7–27 *Invoking the BASELINE DIMENSION command from the Dimension toolbar*

AutoCAD prompts:

Command: **dimbaseline** (ENTER)
Specify a second extension line origin or [Undo/Select>]: *(specify a point or right-click for the shortcut menu and select one of the available options)*

After you specify a point for the second extension line origin, AutoCAD will use the first extension line origin of the previous linear, angular, or ordinate dimension as the first extension line origin for the new dimension, and the prompt is repeated. To exit the command, right-click and choose ENTER from the shortcut menu.

For the BASELINE DIMENSION command to be valid, there must be an existing linear, angular, or ordinate dimension. If the previous dimension was not a linear, angular, or ordinate dimension, or if you press ENTER without providing the second extension line origin, AutoCAD prompts:

> Select base dimension: *(select a dimension object)*

You may select the base dimension, with the baseline being the extension line nearest to where you select the dimension.

The following command sequence shows an example of drawing baseline dimensioning for a circular object to an existing angular dimension, as shown in Figure 7–28.

> Command: **dimbaseline** (ENTER)
> Second extension line origin or [Undo<Select>]: *(specify a point)*
> Second extension line origin or [Undo<Select>]: *(specify a point)*
> Second extension line origin or [Undo<Select>]: *(press ESC or right-click and choose ENTER from the shortcut menu)*

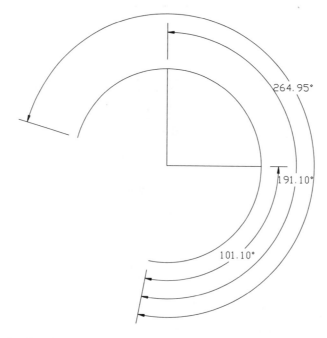

Figure 7–28 *Baseline dimensioning of a circular object*

CONTINUE DIMENSIONING

Continue dimensioning, shown in Figure 7–29, is used for drawing a string of dimensions, each of whose second extension line origin coincides with the next dimension's first extension line origin.

Figure 7–29 *Continue dimensioning*

Invoke the CONTINUE DIMENSION command:

Dimension toolbar	Choose the CONTINUE DIMENSION command (see Figure 7–30)
Dimension menu	Choose Continue
Command: prompt	**dimcontinue** (ENTER)

Figure 7–30 *Invoking the* CONTINUE DIMENSION *command from the Dimension toolbar*

AutoCAD prompts:

> Command: **dimcontinue** (ENTER)
> Specify a second extension line origin or [Undo<Select>]: *(specify a point or right-click for the shortcut menu and select one of the available options)*

After you specify a point for the second extension line origin, AutoCAD will use the second extension line origin of the previous linear, angular, or ordinate dimension as the first extension line origin for the new dimension, and the prompt is repeated. To exit the command, right-click and select ENTER from the shortcut menu.

For the CONTINUE DIMENSION command to be valid, there must be an existing linear, angular, or ordinate dimension. If the previous dimension was not a linear, angular, or ordinate dimension, or if you press ENTER without providing the second extension line origin, AutoCAD prompts:

Select continued dimension: *(select a dimension object)*

You may select the continued dimension, with the coincidental extension line origin being the one nearest to where the existing dimension is selected.

The following command sequence shows an example of drawing continue dimensioning for a linear object to an existing linear dimension, as shown in Figure 7–31.

Command: **dimcontinue** (ENTER)
Specify a second extension line origin or [Undo<Select>]: *(specify the right end of the 3.60-unit line)*
Specify a second extension line origin or [Undo<Select>]: *(specify the right end of the right 1.80-unit line)*
Specify a second extension line origin or [Undo<Select>]: *(press ESC or right-click and choose ENTER from the shortcut menu)*

Figure 7–31 *Continue dimensioning of a linear object*

QUICK DIMENSIONING

Quick Dimensioning or the QDIM command is used to draw a string of dimensions between all of the end and center points of the selected object(s).

Invoke the CONTINUE DIMENSION command:

Dimension toolbar	Choose the QDIM command (see Figure 7–32)
Dimension menu	Choose Quick Dimension
Command: prompt	**qdim** (ENTER)

Figure 7–32 *Invoking the* QUICK DIMENSION *command from the Dimension toolbar*

AutoCAD prompts:

> Command: **qdim** (ENTER)
> Select geometry to dimension: *(select one or more objects)*
> Specify dimension line position, or [Continuous/Staggered/Baseline/
>   Ordinate/Radius/Diameter/datumPoint/Edit] <Continuous>: *(specify
>   location for dimension line or right-click for the shortcut menu and select one
>   of the available options)*

If you specify a location for a dimension line, AutoCAD will draw continuous dimensioning between all end or center points of the objects selected horizontally or vertically, depending on where the dimension line location is specified (see Figure 7–33 for an example). You can right-click and select one of the options that will draw the type of dimension chosen.

Figure 7–33 *An example of Quick Dimensioning*

LEADER AND QLEADER

The LEADER command minimizes the steps required to draw the text with the line(s) and arrowhead pointing to an object (or feature on an object) for annotations and callouts used to describe them. The QLEADER command (a shortcut version of the LEADER command) assumes that you wish to enter multiline text that is top left justified.

Invoke the LEADER command:

Command: prompt	**leader** (ENTER)

Invoke the QLEADER command:

Dimension toolbar	Choose the LEADER command (see Figure 7–34)
Dimension menu	Choose Leader

Figure 7–34 *Invoking the* QLEADER *command from the Dimension toolbar*

When you invoke the LEADER command at the Command: prompt, AutoCAD prompts:

> Command: **leader** (ENTER)
> Specify leader start point: *(specify a point for the arrowhead end of the leader)*
> Specify next point: *(specify another point for the end of the first leader segment opposite the arrowhead)*
> Specify next point or [Annotation/Format/Undo] <Annotation>: *(specify a point or right-click for the shortcut menu)*

If you specify another point, AutoCAD connects another leader segment to the previous one and repeats the last prompt. If the leader radial direction is more than 15 degrees from horizontal, AutoCAD will add a horizontal segment pointing toward the annotation that is equal in length to an arrow, as determined by the DIMASZ dimensioning system variable.

Annotation

The Annotation option allows you to draw alphanumerics or symbols as text for the annotation at the end of the last leader segment opposite the arrowhead. AutoCAD prompts:

> Enter first line of annotation text or <options>: *(you can enter characters now or press* ENTER*)*

Enter the alphanumerics or symbols that will be used as text for the annotation at the end of the leader line opposite the arrowhead. If you press ENTER instead, AutoCAD prompts:

> Enter an annotation option [Tolerance/Copy/Block/None/Mtext] <Mtext>: *(right-click and choose one of the available options from the shortcut menu)*

Choosing the **Mtext** option displays the Mtext dialog box, shown in Figure 7–35, for creating or editing single-line or multiline text for annotation.

Figure 7–35 *Multiline Text Editor dialog box*

AutoCAD draws the primary dimension where it finds the < > (left arrow/right arrow) in a string of characters in the text box of the Mtext dialog box. If the measured dimension is 1/2" and **COPE** < > is displayed in the text box, the dimension would be drawn as **COPE 1/2"**. In a similar manner, [ ] (left bracket/right bracket) causes the secondary dimension to be drawn.

The **Tolerance** option allows you to draw a Feature Control Frame for use in describing standard tolerances by means of the Geometric Tolerance dialog box, as described in the section on "Tolerances" in this chapter.

The **Copy** option allows you to copy Text, Feature Control Frames, Mtext, or Block Reference.

The **Block** option allows you to insert a named Block Reference. (Refer to Chapter 10 for a discussion on Blocks.)

The **None** option allows you to draw the Leader without the annotation.

Format
The Format option is used to determine the appearance of the Leader and the arrowhead. AutoCAD prompts:

> Spline/STraight/Arrow/None/<Exit>: *(select one of the available options)*

The **Spline** option allows you to draw a leader in a manner similar to the SPLINE command.

The **STraight** option allows you to draw a leader with straight-line segments.

The **Arrow** option allows you to draw an arrowhead at the first specified point of the leader.

The **None** option allows you to draw a leader without an arrowhead.

The **Exit** option exits the Format option.

Undo

The Undo option undoes the last segment of the leader line.

Qleader

The LEADER command, when invoked from the Dimension toolbar or the Dimension menu, invokes QLEADER, and AutoCAD prompts:

> Command: _qleader (ENTER)
> Specify first leader point, or [Settings]: *(specify a point for the arrowhead end of the leader line or right-click for the shortcut menu)*
> Specify next point: *(specify another point for the end of the first leader segment opposite the arrowhead)*
> Specify text width <0.0000>: *(specify a point to determine the maximum width of the multiline text)*
> Enter first line of annotation text <Mtext>: *(enter the first line of annotation text or right-click for the Multiline Text Editor dialog box)*
> Enter next line of annotation text: *(enter the next line of annotation text or ENTER to complete the QLEADER command)*

Note that the text width that is specified causes the text to wrap to the next line when the text on the current line exceeds the specified width. If no width is specified, then the entire text is drawn on one line.

TOLERANCES

Tolerance symbology and text can be included with the dimensions that you draw in AutoCAD. AutoCAD provides a special set of subcommands for the two major methods of specifying tolerances. One set is for LATERAL TOLERANCES; the other is for GEOMETRIC TOLERANCES.

LATERAL TOLERANCES are the traditional tolerances. Though they are often easier to apply, they do not always satisfy tolerancing in all directions (circularly and cylindrically), and are more subject to misinterpretations, especially in the international community.

GEOMETRIC TOLERANCE values are the maximum allowable distances that a form or its position may vary from the stated dimension(s).

 Note: Lateral tolerance symbols and text are accessible from the Dimension Style dialog box. Geometric tolerance symbols and text are accessible from the Dimension toolbar, from the Dimension menu, and at the "Command:" prompt.

Lateral Tolerance

Lateral tolerancing draws the traditional symbols and text for Limit, Plus or Minus (unilateral and bilateral), Single Limit, and Angular tolerance dimensioning.

Lateral tolerance is the range from the smallest to the greatest that a dimension is allowed to deviate and still be acceptable. For example, if a dimension is called out as 2.50 ± 0.05, then the tolerance is 0.1 and the feature being dimensioned may be anywhere between 2.45 and 2.55. This is the symmetrical plus-or-minus convention. If the dimension is called out as 2.50 $^{+0.10}_{-0.00}$, then the feature may be 0.1 greater than 2.50, but may not be smaller than 2.50. This is referred to as unilateral. The Limits tolerance dimension may also be shown as $^{2.55}_{2.45}$.

To set the tolerance method, first open the Dimension Style Manager dialog box. Choose Modify, and AutoCAD displays the Modify Dimension Style dialog box. Select the Tolerance tab and select the tolerance method from the Method option menu. Figure 7–36 shows all the options available in the Tolerance Format section and the Alternate Unit Tolerance section when the Tolerances tab is chosen from the Modify Dimension Style dialog box. Selecting a method (other than NONE) from the **Method:** text box applies only to lateral tolerancing.

Figure 7–36 *Modify Dimension Style dialog box with the Tolerances tab selected.*

Choosing the **Method:** list box in the Tolerance Format section of the Modify Dimension Style dialog box Tolerances tab displays the following options: None, Symmetrical, Deviation, Limits, and Basic. See Table 7–1 for examples of the various options.

Table 7–1   Examples of Lateral Tolerance Methods

Tolerance Method	Description	Example
Symmetrical	Only the Upper Value box is usable. Only one value is required.	1.00 ± 0.05
Deviation	Both Upper Value and Lower Value boxes are active. A 0.00 value may be entered in either box, indicating that the variation is allowed in only one direction.	$1.00 \, ^{+0.07}_{-0.03}$
Limits	Both Upper and Lower Value boxes are active. In this case, there is no base dimension. The Justification box is not active, because the values are not tolerance values but are actual dimensions to be drawn in full height-text. Using an Upper Value of 0.0500 and a Lower Value of −0.0250 will cause annotation of a 1.000 basic dimension to be $^{0.0500}_{0.0250}$.	1.070 0.930
Basic	The dimension value is drawn in a box, indicating that it is the base value from which a general tolerance is allowed. The general tolerance is usually given in other notes or specifications on the drawing or other documents.	$\boxed{1.00}$

Note: To get a preview of how your tolerance will appear in relation to the dimension, the format is set in the Primary Units section of the Modify Dimension Style dialog box.

In the Tolerance Format and Alternate Unit Tolerance sections, the **Precision:** text box lets you determine how many decimals the text will be shown in decimal units. This value is recorded in DIMDEC. The **Upper Value:** and **Lower Value:** text boxes allow you to preset the values where they apply. The **Scaling for Height:** text box lets you set the height of the tolerance value text. The **Vertical Position:** selection box determines if the tolerance value that is drawn after the dimension will be at the top, middle, or bottom of the text space.

In the Tolerance Format and Alternate Unit Tolerance sections, the **Zero Suppression** subsection lets you specify whether or not zeros are displayed, as follows:

Selecting the **Leading** check box causes zeros ahead of the decimal point to be suppressed. *Example:* .700 is displayed instead of 0.700.

Selecting the **Trailing** check box causes zeros behind the decimal point to be suppressed. *Example:* 7 or 7.25 is displayed instead of 7.000 or 7.250, respectively.

Selecting the **0 Feet** check box causes zeros representing feet to be suppressed if the dimension text represents inches and/or fractions only. *Example:* 7" or 7 1/4" is displayed instead of 0'-7" or 0'-7 1/4".

Selecting the **0 Inches** check box causes zeros representing inches to be suppressed if the dimension text represents feet only. *Example:* 7' is displayed instead of 7'-0".

Geometric Tolerance

Geometric tolerancing draws a Feature Control Frame for use in describing standard tolerances according to the geometric tolerance conventions. Geometric tolerancing is applied to forms, profiles, orientations, locations, and runouts. Forms include squares, polygons, planes, cylinders, and cones.

Invoke the Geometric TOLERANCE command:

Dimension toolbar	Choose the TOLERANCE command (see Figure 7–37)
Dimension menu	Choose Tolerance. . .
Command: prompt	**tolerance** (ENTER)

Figure 7–37 *Invoking the Geometric* TOLERANCE *command from the Dimension toolbar*

AutoCAD displays the Geometric Tolerance dialog box, shown in Figure 7–38.

Figure 7–38 *Geometric Tolerance dialog box*

The conventional method of expressing a geometric tolerance for a single dimensioned feature is in a Feature Control Frame, which includes all necessary tolerance information for that particular dimension. A Feature Control Frame has the Geometric Characteristic Symbol box and the Tolerance Value box. Datum reference/ material condition datum boxes may be added where needed. The Feature Control Frame is shown in Figure 7–39. An explanation of the Characteristic Symbols is given in Figure 7–40. The supplementary material conditions of datum symbols are shown in Figure 7–41.

Once the symbol button located in the Symbol column in the Geometric Tolerance dialog box is chosen, AutoCAD displays the Symbol dialog box, as shown in Figure 7–42. Select one of the available symbols and specify appropriate tolerance values in the Tolerance field.

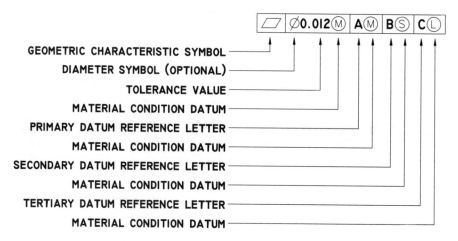

Figure 7–39 *Feature control frame*

Maximum Materials Condition (MMC) means a feature contains the maximum material permitted within the tolerance dimension; that is, the minimum hole size, maximum shaft size.

Least Material Condition (LMC) means a feature contains the least material permitted within the tolerance dimension, that is maximum hole size, minimum shaft size.

Regardless of Feature Size (RFS) applies to any size of the feature within its tolerance. This is more restrictive. For example, RFS does not allow the tolerance of the center-to-center dimension of a pair of pegs fitting into a pair of holes greater leeway if the peg diameters are smaller and/or the holes are bigger, whereas MMC does.

The diameter symbol, Ø, is used in lieu of the abbreviation DIA.

CHARACTERISTIC SYMBOLS

FORM	▱	FLATNESS
	▭	STRAIGHTNESS
	◯	ROUNDNESS
	⌭	CYLINDRICITY
	⌒	LINE PROFILE
	⌓	SURFACE PROFILE
	∠	ANGULARITY
	∥	PAREALLELISM
	⊥	PERPENDICULARITY
LOCATION	◎	CONCENTRICITY
	⊕	POSITION
	≗	SYMMETRY
RUNOUT	↗	CIRCULAR RUNOUT
	⫽↗	TOTAL RUNOUT

Figure 7–40 *Geometric characteristics symbols*

MATERIAL CONDITIONS OF DATUM SYMBOLS

Ⓜ	MAXIMUM MATERIAL CONDITIONS (MMC)
Ⓢ	REGARDLESS OF FEATURE SIZE
⌀	DIAMETER

Figure 7–41 *Material conditions of datum symbols*

Figure 7–42 *Symbol dialog box*

An application of geometric tolerancing is found in Exercise 7–1 at the end of this chapter. Geometric tolerancing is becoming widely accepted. It is highly encouraged that you study the latest drafting texts concerning the significance of the various symbols, so that you will be able to apply geometric tolerancing properly.

DRAWING CROSS MARKS FOR ARCS OR CIRCLES

As shown in Figure 7–43, the DIMCENTER command is used to draw the cross marks that indicate the center of an arc or circle.

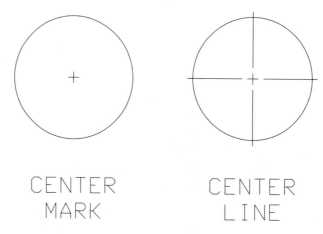

CENTER MARK   CENTER LINE

Figure 7–43 *Circles with center cross marks*

Invoke the CENTER MARK command:

Dimension toolbar	Choose the CENTER MARK command (see Figure 7–44)
Dimension menu	Choose Center Mark
Command: prompt	**dimcenter** (ENTER)

Figure 7–44 *Invoking the* CENTER MARK *command from the Dimension toolbar*

AutoCAD prompts:

Command: **dimcenter** (ENTER)
Select arc or circle:

Select an arc or circle, and AutoCAD draws the cross marks in accordance with the setting of the DIMCEN dimensioning variable.

OBLIQUE DIMENSIONING

The OBLIQUE command allows you to slant the extension lines of a linear dimension to a specified angle. The dimension line will follow the extension lines, retaining its original direction. This is useful for having the dimension stay clear of other dimensions or objects in your drawing. It is also a conventional method of dimensioning isometric drawings.

Invoke the OBLIQUE command:

Dimension menu	Choose Oblique

AutoCAD prompts:

> Command: **oblique** (ENTER)
> Select objects: *(select the dimension(s) for obliquing)*
> Enter obliquing angle (press ENTER for none): *(specify an angle or press ENTER)*

 Note: The OBLIQUE command is actually the Dimension Style Edit command in which AutoCAD automatically chooses the oblique option for you. All you have to do is select a dimension.

The extension lines of the selected dimension(s) will be slanted at the specified angle.

EDITING DIMENSION TEXT

AutoCAD allows you to edit dimensions with MODIFY commands and grip editing modes. Also, AutoCAD provides two additional MODIFY commands specifically designed to work on dimension text objects, DIMEDIT and DIMTEDIT.

DIMEDIT COMMAND

The options that are available in the DIMEDIT command allow you to replace the dimension text with new text, rotate the existing text, move the text to a new location, and if necessary, restore the text back to its home position, which is the position defined by the current style. In addition, these options allow you to change the angle of the extension lines (normally perpendicular) relative to the direction of the dimension line (by means of the Oblique option).

Invoke the DIMEDIT command:

Dimension toolbar	Choose the DIMENSION EDIT command (see Figure 7–45)
Command: prompt	**dimedit** (ENTER)

Figure 7–45 *Invoking the* DIMENSION EDIT *command from the Dimension toolbar*

AutoCAD prompts:

> Command: **dimedit** (ENTER)
> Enter type of dimension editing [Home/New/Rotate/Oblique]<Home>:
> *(press* ENTER *for the Home option or right-click and choose one of the available options from the shortcut menu)*

Home

The Home option returns the dimension text to its default position. AutoCAD prompts:

> Select objects: *(select the dimension objects and press* ENTER*)*

New

The New option allows you to change the original dimension text to the new text. AutoCAD opens the Multiline Text Editor dialog. Enter the new text and choose the OK button. AutoCAD then prompts:

> Select objects: *(select the dimension objects for which the existing text will be replaced by the new text)*

Rotate

The Rotate option allows you to change the angle of the dimension text. AutoCAD prompts:

> Specify angle for dimension text: *(specify the rotation angle for text)*
> Select objects: *(select the dimension objects for which the dimension text has to be rotated)*

Oblique

The Oblique option adjusts the obliquing angle of the extension lines for linear dimensions. This is useful to keep the dimension parts from interfering with other objects in the drawing. Also, it is an easy method by which to generate the slanted dimensions used in isometric drawings. AutoCAD prompts:

> Select objects: *(select the dimension objects)*
> Enter obliquing angle (press ENTER for none): *(specify the angle or press ENTER)*

DIMTEDIT COMMAND

The DIMTEDIT command is used to change the location of dimension text (with the Left/Right/Home options) along the dimension line and its angle (with the Rotate option).

Invoke the DIMTEDIT command:

Dimension toolbar	Choose the DIMENSION TEXT EDIT command (see Figure 7–46)
Command: prompt	**dimtedit** (ENTER)

Figure 7–46 *Invoking the* DIMENSION TEXT EDIT *command from the Dimension toolbar*

AutoCAD prompts:

Command: **dimtedit** (ENTER)
Select dimension: *(select the dimension object to modify)*

A preview image of the dimension selected is displayed on the screen, with the text located at the cursor. You will be prompted:

Specify new location for dimension text or [Left/Right/Center/Home/ Angle]: *(specify a new location for the dimension text or right-click for the shortcut menu and select one of the available options)*

By default, AutoCAD allows you to position the dimension text with the cursor, and the dimension updates dynamically as it drags.

Left
The Left option will cause the text to be drawn toward the left extension line.

Right
The Right option will cause the text to be drawn toward the right extension line.

Home
The Home option returns the dimension text to its default position.

Angle
The Angle option changes the angle of the dimension text. AutoCAD prompts:

Specify angle for dimension text: *(specify the angle)*

The angle specified becomes the new angle for the dimension text.

EDITING DIMENSIONS WITH GRIPS

If the Grips feature is set to ON, you can select an associative dimension object and its grips will be displayed at strategic points. The grips will be located at the object ends of the extension lines, the intersections of the dimension and extension lines, and at the insertion point of the dimension text. In addition to the normal grip editing of the dimension as a group (rotate, move, copy, etc.), each grip can be selected for editing the configuration of the dimension as follows: Moving the object end grip of an extension line will move that specified point, making the value change accordingly. Horizontal and vertical dimensions remain horizontal and vertical. Aligned dimensions follow the alignment of the relocated point. Moving the grip at the intersection of the dimension line and one of the extension lines causes the dimension line to be nearer to or farther from the object dimensioned. Moving the grip at the insertion point of the text does the same as the intersection grip and also permits you to move the text back and forth along the dimension line.

DIMENSION STYLES

Each time a dimension is drawn it conforms to the settings of the Dimensioning System Variables in effect at the time. The entire set of Dimensioning System Variable settings can be saved in their respective states as a Dimension Style, with a name by which it can be recalled for application to a dimension later in the drawing session or in a subsequent session. Some Dimensioning System Variables affect every dimension. For example, every time a dimension is drawn, the DIMSCALE setting determines the relative size of the dimension. But DIMDLI, the variable that determines the offsets for Baseline Dimensions, comes into effect only when a Baseline Dimension is drawn. However, when a Dimension Style is created and named, all Dimensioning System Variable settings (except DIMASO and DIMSHO) are recorded in that Dimension Style, whether or not they will have an effect. The DIMASO and DIMSHO dimensioning variable settings are saved in the drawing separate from the Dimension Styles. As stated at the beginning of this chapter: "Dimension variables change when their associated settings are changed in the Modify Dimension Style section of the Dimension Style Manager dialog box. For example, the value of the DIMEXO (extension line offset) dimensioning system variable is established by the number in the Offset From Origin text box in the Modify Dimension Style subdialog box."

DIMENSION STYLE MANAGER DIALOG BOX

AutoCAD provides a comprehensive set of subdialog boxes accessible through the Dimension Style Manager dialog box for creating new Dimension Styles and managing existing ones. In turn, these subdialog boxes compile and store Dimensioning System Variable settings. Creating Dimension Styles through use of the DIMSTYLE command's dialog boxes allows you to make the desired changes to the appearance of dimensions without having to search for or memorize the names of the Dimensioning System Variables in order to change the settings directly.

Invoke the DIMSTYLE command:

Dimension toolbar	Choose the DIMSTYLE command (see Figure 7–47)
Dimension menu	Choose Style
Command: prompt	**dimstyle** (ENTER)

Figure 7–47 *Invoking the* DIMSTYLE *command from the Dimension toolbar*

AutoCAD displays the Dimension Style Manager dialog box, shown in Figure 7–48.

Figure 7–48 *Dimension Style Manager dialog box*

Current Dimstyle:

The Current Dimstyle heading shows the name of the Dimension Style to which the next drawn dimension will conform. The name of the Current Dimstyle is recorded as the value of the DIMSTYLE dimensioning system variable.

Styles:

The **Styles:** section displays the name(s) of the current style and any other style(s) available depending on the option chosen in the **List:** selection box. If you choose one of the styles in the list, it is shown in white characters on a blue background and will be the Dimstyle acted upon when one of the buttons (Set Current, New, Modify, Override, or Compare) is chosen. It will also be the Dimstyle whose appearance is shown in the **Preview of:** viewing area.

List:

The options in the **List:** selection box are All styles and Styles in use. You can have all of the styles available in the drawing displayed or only those in use by choosing the appropriate option from this selection box. The Standard style cannot be deleted and is always available to be listed. Styles from externally referenced drawings can be listed, though they are not changeable in the current drawing.

Preview of:

The view below the **Preview of:** heading shows how dimensions will appear when drawn using the current Dimstyle.

Description

The Description section shows the difference(s) between the Dimstyle chosen in the **Styles:** section and the current Dimstyle. For example, if the current Dimstyle is Standard and the Dimstyle chosen in the **Styles:** section is named Harnessing, then the Description section might say "Standard + Angle format = 1, Fraction format = 1, Length units = 4" to indicate that the Angle format is in Degrees/Minutes/Seconds, the Fraction format is in Diagonal, and the Length units is in Architectural, these settings being different from the corresponding settings for the Standard Dimstyle.

Set Current

Choosing the **Set Current** button causes the Dimstyle chosen in the **Styles:** section to become the current Dimstyle.

New

Choosing the **New** button causes the Create New Dimension Style dialog box to appear (see Figure 7–49).

The **Create New Dimension Style** dialog box allows you to create and name a new Dimstyle.

The **New Style Name** text box is where you enter the name of the new Dimstyle you wish to create.

The **Start With:** list box lets you choose the existing Dimstyle that you would like to start with in creating your new Dimstyle. The Dimensioning System Variables will be the same as those in the **Start With:** list box until you change their settings. In many cases you may wish to change only a few Dimensioning System Variable settings.

Figure 7–49 *Create New Dimension Style dialog box*

The **Use For:** list box lets you choose the type(s) of dimensions for which to apply the new Dimstyle you will be creating.

Modify

Choosing the **Modify** button causes the Modify Dimension Style dialog box to appear (see Figure 7–50). There are tabs for six subdialog boxes to choose from: Lines and Arrows, Text, Fit, Primary Units, Alternate Units, and Tolerance.

Figure 7–50 *Modify Dimension Style dialog box with Lines and Arrows tab chosen*

Lines and Arrows

The Lines and Arrows subdialog box allows you to modify the geometry of the various elements that make up the dimensions, such as Dimension Lines, Extension Lines, Arrowheads, and Center Marks for circles. There is a preview showing how dimensions will appear when drawn using the settings as you change them.

In both the Dimension Line and Extension Line sections:

> **Color:** list boxes let you choose how the color of Dimension Lines and Extension Lines will be determined. By Layer causes the color of the line to match that of its layer. By Block causes the color of the line to match that of its block reference if it is part of a block reference. You can also choose one of the standard colors or an Other color for the line color to match.
>
> **Lineweight:** list boxes lets you choose how the lineweight of Dimension and Extension Lines will be determined. By Layer causes the lineweight of the line to match that of its layer. By Block causes the lineweight of the line to match that of its block reference if it is part of a block reference. You can also choose one of the standard lineweights or enter a value.

In the Dimension Line section only:

> **Extend beyond ticks:** text box lets you enter the distance that Dimension Lines will be drawn past Extension Lines if the arrowhead type is a tick (small diagonal line).
>
> **Baseline spacing:** text box lets you enter the distance that AutoCAD uses between Dimension Lines when they are drawn using the BASELINE DIMENSION command.
>
> **Suppress:** check boxes let you determine if one or more of the Dimension or Extension Lines is suppressed when the dimension is drawn.

In the Extension Line section only:

> **Extend beyond dim lines:** text box lets you enter the distance that Extension Lines will be drawn past Dimension Lines.
>
> **Offset from origin:** text box lets you enter the distance that AutoCAD uses between the Extension Line and the origin point specified when drawing an extension line.
>
> **Suppress:** check boxes let you determine if one or more of the Dimension or Extension Lines is suppressed when the dimension is drawn.

In the Arrowhead section:

> **1st, 2nd,** and **Leader** list boxes let you determine if and how arrowheads will be drawn at the terminations of the Extension Lines or the start

point of a Leader. Unless you specify a different type for the second line termination, it will be the same type as the first line termination. The type of arrowheads is recorded in DIMBLK if both the first and second are the same or in DIMBLK1 and DIMBLK2 if they are different..

The **Arrow size:** text box lets you enter the distance that AutoCAD uses for the length of an arrowhead when drawn in a dimension or leader.

Some of the arrowheads types included in the standard library are None, Closed, Dot, Closed Filled, Oblique, Open, Origin Indication, and Right-Angle. Figure 7–51 presents an example of each.

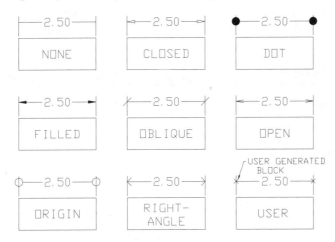

Figure 7–51 *Arrowhead types*

The **User Arrow** selection lets you use a previously saved block definition by entering its name. Figure 7–52 shows the related dialog box. The block definition should be created as though it were drawn for the right end of a horizontal Dimension Line, with the insertion point at the intersection of the Extension Line and the Dimension Line.

Figure 7–52 *Select Custom Arrow dialog box*

In the Center Marks for Circles section:

> The **Type:** list box lets you choose if and what type of center mark is drawn when a Diameter Dimension or Radius Dimension is drawn. You may choose from None, Mark, and Line. If Mark is chosen, a Center Mark will be drawn with its length determined by the value in the **Size:** text box. If Line is chosen, the Center Marks will be drawn as lines that extend outside the circle a distance determined by the value in the **Size:** text box.

Text

The Text subdialog box lets you modify the appearance, location, and alignment of Dimension Text that is included when a dimension is drawn. There is a preview showing how dimensions will appear when drawn using the settings as you change them. (see Figure 7–53)

Figure 7–53 *Modify Dimension Style dialog box with Text tab chosen*

In the Text Appearance section:

> The **Text Style:** list box lets you choose the text style to which the Dimension Text will conform.

Note: Do not confuse Text Style with Dimension Style. Dimensions are drawn in accordance with the current Dimstyle, which has as part of its configuration a Text Style to which Dimension Text will conform.

The **Color:** list box lets you choose how the color of Dimension Text will be determined. By Layer causes the color of the text to match that of its layer. By Block causes the color of the text to match that of its block reference if it is part of a block reference. You can also choose one of the standard colors for the text color to match.

The **Text height:** text box lets you enter the distance that AutoCAD uses for text height when drawing Dimension Text as part of the dimension.

The **Fraction height scale:** text box lets you enter a scale which determines the distance that AutoCAD uses for text height when drawing Dimension Text for fractions as part of the Dimension Text. This scale is the ratio of the text height for normal Dimension Text to the height of the fraction text.

The **Draw frame around text** check box, when checked, causes AutoCAD to draw the Dimension Text inside a rectangular frame.

In the Text Placement section:

The **Vertical:** list box lets you choose how the Dimension Text will be drawn in relation to the Dimension Line. The options include Centered, Above, Outside, and JIS.

The **Horizontal:** list box lets you choose how the Dimension Text will be drawn in relation to the Extension Lines. The options include Centered, At Ext Line 1, At Ext Line 2, Over Ext Line 1, and Over Ext Line 2.

The **Offset from dim line:** text box lets you enter the distance from the Dimension Line that AutoCAD uses when drawing Dimension Text as part of the dimension.

In the Text Alignment section, there are three radio buttons that let you choose how the text will be aligned in relation to the Dimension Lines. The options include Horizontal, Aligned with Dimension Line, and ISO Standard.

Fit

The Fit subdialog box lets you to determine the arrangement of the various elements of dimensions when a dimension is drawn. There is a preview showing how dimensions will appear when drawn using the settings as you change them (see Figure 7–54).

Figure 7–54 *Modify Dimension Style dialog box with Fit tab chosen*

In the Fit Options section, there are five radio buttons that let you choose which of the text or arrowheads will be drawn between the extension lines. The options include:

Choosing the **Either the Text or the Arrows, Whichever Fits Best** button causes the text to be drawn outside and the arrows inside the extension lines if there is room for the arrows only.

Choosing the **Arrows** button causes the text to be drawn outside and the text inside the extension lines if there is room for the arrows only.

Choosing the **Text** button causes the arrows to be drawn outside and the text inside the extension lines if there is room for the text only.

Choosing the **Both Text and Arrows** button causes both the text and arrows to be drawn outside if the Dimension Lines are forced outside and both to be drawn inside if space is available. If you have selected the User Defined check box and are placing diameter/radius text, you can force the leader to be drawn outside the circle/arc by specifying outside the circle/arc with the cursor.

Choosing the **Always Keep Text Between Ext Lines** button causes text to always be drawn between the Extension Lines..

The **Suppress arrows if they don't fit inside the extension lines** check box, when checked, causes AutoCAD to not draw the arrowheads between extension lines if they do not fit.

In the Text Placement section there are three radio buttons that let you choose how the text will be placed when text is not in the default position. The options include: include: Beside the Dimension Line, Over the Dimension Line, with a Leader, and Over the Dimension Line, without a Leader.

In the Scale for Dimension Features section:

The **Use overall scale of:** radio button activates the text box that lets you enter the number AutoCAD uses as a ratio of the true scale to drawn dimensions. This is useful when you wish to create different parts of the drawing at different scales but have the dimension elements uniform in size. This does not include distances, coordinates, angles, or tolerances.

The Scale section lets you increase or decrease the size of all of the dimensioning components uniformly by the factor entered. For example, say you have set the sizes of the components to plot at a scale factor of 1/4" = 1'-0" (1:48) and wish to draw a detail to true size (including the dimensions) and then scale it up to 3/4" = 1'-0" (1:16). You can set the dimensioning scale to 1/3 so that the components, when scaled up by a factor of 3, will plot at the same size as those not scaled up.

Choosing the **Scale dimensions to layout (paperspace)** radio button causes dimensions to be drawn with elements scaled to layout.

In the Fine Tuning section:

Choosing the **Place text manually when dimensioning** check box allows you to dynamically specify where Dimension Text is placed when the dimension is drawn.

Choosing the **Always draw dim line between ext lines** button causes a Dimension Line to be drawn between the extension lines regardless of the distance between Extension Lines.

Primary Units

The Primary Units subdialog box lets you to determine the appearance and format of numerical values of distances and angles when they are drawn with dimensions. There is a preview showing how dimensions will appear when drawn using the settings as you change them (see Figure 7–55).

Figure 7–55 *Modify Dimension Style dialog box with Primary Units tab chosen*

In the Linear Dimensions section:

> The **Unit format:** list box lets you choose the format of the units that AutoCAD uses when drawing Dimension Text as part of the dimension. The options include Scientific, Decimal, Engineering, Architectural, Fractional, and Windows Desktop. (see Figure 7–56). The value is recorded in DIMUNIT.

> The **Precision:** list box lets you determine to how many decimals the text will be shown if the format is in Scientific, Decimal, or Windows Desktop Units. It will show to how many decimals the inches will be shown if the format is in Engineering Units. It will show how small the fraction will be shown in Architectural or Fractional Units, and the options in the **Fractional format:** shows whether the fractions will be shown Horizontal, Diagonal, or Not Stacked. This value is recorded in DIMDEC.

Figure 7–56 *Unit Format options in the Primary Units subdialog box*

The **Decimal separator:** list box lets you choose whether the decimal will be separated by a Period, Comma, or a Space if the format is in Scientific, Decimal, or Windows Desktop Units.

The **Round off:** text box lets you enter the value to which distances will be rounded off. For example, a value of 0.5 causes dimensions to be rounded to the nearest 0.5 units.

The **Prefix:** text box allows you to include a prefix in the Dimension Text. The prefix text will override any default prefixes, such as those used in radius (R) dimensioning. The value is recorded in DIMPOST.

The **Suffix:** text box allows you to include a suffix in the Dimension Text. If you specify tolerances, AutoCAD includes the suffix in the tolerances, as well as in the main dimension. The value is recorded in DIMPOST.

In the Measurement Scale section:

The **Scale factor:** text box lets you enter the number AutoCAD uses as a ratio of the true dimension distances to drawn dimensions distances, or the linear scale factor for the linear measured distances of a dimension without affecting the components, angles, or tolerance values. For example, you are drawing with the intention of plotting at the quarter-size scale, or 3" = 1'-0". You have scaled up a detail by a factor of 4 so that it will plot to full scale. If you wish to dimension it after it has been enlarged, you can set the linear scale factor in this section of the Primary Units dialog box to .25 so that dimensioned distances will represent the dimension of the object features before it was scaled up. This method keeps the components at the same size as dimensions created without a scale change. The value is recorded in DIMLFAC. This is useful when you wish to create different parts of the drawing at different scales but have the dimension elements uniform in size. Selecting the **Apply to layout dimensions only** check box causes the ratio to be applies only to layout dimensions.

In the **Zero Suppression** section you can specify whether or not zeros are displayed, as follows:

Selecting the **Leading** check box causes zeros ahead of the decimal point to be suppressed. For example: .700 is displayed instead of 0.700.

Selecting the **Trailing** check box causes zeros behind the decimal point to be suppressed. For example: 7 or 7.25 is displayed instead of 7.000 or 7.250, respectively.

Selecting the **0 Feet** check box causes zeros representing feet to be suppressed if the Dimension Text represents inches and/or fractions only. For example: 7" or 7 1/4" is displayed instead of 0'-7" or 0'-7 1/4".

Selecting the **0 Inches** check box causes zeros representing inches to be suppressed if the Dimension Text represents feet only. For example: 7' is displayed instead of 7'-0".

In the **Angles** section:

The **Units format:** list box lets you choose the format of the units for the angular Dimension Text. Options include Decimal Degrees, Degrees/Minutes/Seconds, Grads, and Radians. The value is recorded in DIMAUNIT. In the **Zero Suppression** subsection you can specify whether or not Leading or Trailing zeros are displayed.

Alternate Units

The Alternate Units subdialog box (see Figure 7–57) is similar to the Primary Units section, except for the Display **alternate units** check box, which lets you enable (DIMALT set to 1) or disable (DIMALT set to 0) alternate units, and there is no Angle subsection. There is a Placement subsection with **After primary value** and **Before primary value** radio buttons that let you determine where to place the Alternate Units.

Figure 7–57 *Modify Dimension Style dialog box with Alternate Units tab chosen*

Tolerances

The Tolerances subdialog box (see Figure 7–58) allows you to set the tolerance method and appropriate settings. Detailed explanation is provided in the Tolerance section, earlier in this chapter.

Once the necessary changes are made for the appropriate settings, choose the *OK* button to close Modify Dimension Style dialog box and then choose the *Close* button to close the Dimension Style Manager dialog box.

Figure 7–58 *Modify Dimension Style dialog box with Tolerances tab chosen*

EDITING DIMENSIONS USING THE SHORTCUT MENU

Dimensions already drawn can be edited easily by using the right-click shortcut menu in the following manner.

First, GRIPS must be enabled by choosing Options... from the Tools menu and then selecting the Enable Grips check box from the Selection section. Once the grips have been made available, select the dimension(s) that you wish to edit. You will see the grips at the strategic points of the elements of the dimension(s) you have selected. Now simply right-click (with the cursor anywhere in the drawing area) and the shortcut menu will include a section for editing dimensions (see Figure 7–59).

The **Dim Text Position** flyout includes options for Above Dim Line, Centered, Home Text, Move Text Alone, Move With Leader, and Move With Dim Line. Choosing one of these options causes the text in the selected dimension(s) to conform to the position option chosen.

The **Precision** flyout includes a range of 0.0 to 0.000000 for Decimal dimensions and a range of 0'-0 to 0'-0 1/128 for Architectural and Fractional dimensions. Choosing one of the decimal or fractional options causes the distances/angles in the selected dimension(s) to conform to the level of precision chosen.

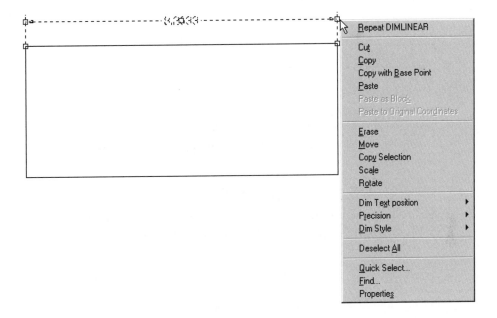

Figure 7–59 *Shortcut Menu with Dimension Editing Section*

Note: If the dimensions selected (and identified by visible grips) include styles with both decimal and fractional units, the Precision option is not available in the shortcut menu.

The **Dimstyle** flyout includes an option for Save As New Style…, and a list of available Dimstyles to choose from. If you have made some changes to settings of individual variables in a particular dimension, you can choose the Save As New Style option to create and name a new, unique Dimstyle conforming to the selected dimension's variable settings.

OVERRIDING THE DIMENSION FEATURE

The DIMOVERRIDE command allows you to change one of the features in a dimension without having to change its dimension style or create a new dimension style. For example, you may wish to have one leader with the text centered at the ending horizontal line, rather than over the ending horizontal line in the manner that the current dimension style may call for. By invoking the DIMOVERRIDE command, you can respond to the prompt with the name of the Dimensioning System Variable (DIMTAD in this case) and set the value to 0 rather than 1. Then you can select the dimension leader you wish to have overridden.

Invoke the DIMOVERRIDE command:

Command: prompt	**dimoverride** (ENTER)

AutoCAD prompts:

> Command: **dimoverride** (ENTER)
> Enter dimension variable name to override or [Clear overrides]: *(specify the Dimensioning System Variable name or right-click and choose Clear to clear overrides)*

If you specify a Dimensioning System Variable, you are prompted:

> Enter new value for dimension variable <current>: *(specify the new value)*
> Enter dimension variable name to override: *(specify another Dimension System Variable or press ENTER)*

If you press ENTER, you are prompted:

> Select objects: *(select the dimension objects)*

The dimensions selected will have the Dimensioning System Variable settings overridden in accordance with the value specified.

If you choose Clear, you are prompted:

> Select objects: *(select the dimension to clear the overrides)*

The dimensions selected will have the *Dimensioning System Variable* setting overrides cleared.

UPDATING DIMENSIONS

The DIMENSION UPDATE command permits you to make selected existing dimension(s) conform to the settings of the current dimension style.

Invoke the DIMENSION UPDATE command:

Dimension toolbar	Choose the DIMENSION UPDATE **command** (see Figure 7–60)
Dimension menu	Choose Update

Figure 7–60 *Invoking the* DIMENSION UPDATE *command from the Dimension toolbar*

AutoCAD prompts:

> Enter a dimension style option
> [Save/Restore/STatus/Variables/Apply/?] <Restore>: apply
> Select objects: *(select any dimension(s) whose settings you wish to have*
>   *updated to conform to the current dimension style).*

 Note: Note that the UPDATE command is actually the Dimension Style Edit command in which AutoCAD automatically chooses the **apply** option for you. All you have to do is select a dimension.

PROJECT EXERCISE

This project exercise provides point-by-point instructions for dimensioning the objects in the drawing shown in Figure P7–1.

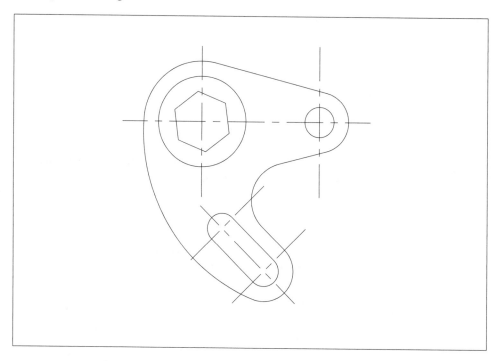

Figure P7–1 *Completed project drawing*

In this project you will use Dimension commands such as DIMLINEAR, DIMSTYLE, DIMALIGNED, DIMCONTINUE, DIMANGULAR, DIMRADIUS, and DIMDIAMETER to dimension objects.

Step 1: Start the AutoCAD program.

Step 2: Invoke the OPEN command from the Standard toolbar and select the drawing completed in the Chapter 4 Project Exercise.

Step 3: Invoke the LAYER command from the Object Properties toolbar, or select Layer from the Format menu. AutoCAD displays the Layer Properties Manager dialog box.

Create one new layer, and rename it as shown in the following table, assigning an appropriate color and linetype.

Layer Name	Color	Linetype
Dimension	Blue	Continuous

Set *dimension* as the current layer, and close the Layer Properties Manager dialog box.

Step 4: Select Style from the Dimension menu, AutoCAD displays the Dimension Style Manager dialog box.

Choose the **Modify** button to open the Modify Dimension Style dialog box. Then choose the Lines and Arrows tab. In the Dimension Line section enter **0.50** in the **Baseline spacing** text box. In the Arrowheads section enter **0.125** in the **Arrow Size:** text box, and in the Extension Line section enter **0.125** in the **Extend beyond dim lines:** text box.

Choose the Primary Units tab in the Modify Dimension Style dialog box. From the **Precision:** list box (in Linear and Angular),, select **0.00** for two decimal place precision.

Choose the Text tab in the Modify Dimension Style dialog box. In the Text Appearance section enter **0.125** in the **Text Height:** text box.

Choose **OK** to close the Modify Dimension Style dialog box.

Choose **Close** to close the Dimension Style Manager dialog box.

Step 5: From the Object Snap toolbar choose the Object Snap Settings button. The Drafting Settings dialog box will appear. In the Object Snap section set Center object snap mode to ON and the remaining object snap modes to OFF. Also make sure OSNAP toggle button is set to ON in the status bar.

Step 6: Invoke the DIMLINEAR command from the Dimension toolbar. Draw the linear dimension, as shown in Figure P7–2.

Figure P7–2 *Completed linear dimension*

538

Step 7: Invoke the ALIGNED DIMENSION command from the Dimension toolbar (see Figure P7–3) to add the aligned dimensions to the drawing.

Figure P7–3 *Invoking the DIMALIGNED command from the Dimension toolbar*

Command: **dimaligned** (ENTER)
Specify first extension line origin or <select object>: *(select the arc at PT. 1, as shown in Figure P7–4)*
Specify second extension line origin: *(select the arc at PT. 2, as shown in Figure P7–4)*
Specify dimension line location or [Mtext/Text/Angle]: *(select PT. 3 to locate the dimension, as shown in Figure P7–4)*

AutoCAD draws the dimension as shown in Figure P7–5.

Figure P7–4 *Specifying points for the aligned dimension*

Invoke the DIMCONTINUE command from the Dimension toolbar.

Command: **dimcontinue** (ENTER)
Specify a second extension line origin or [Undo<Select>]: *(select the large circle to continue the dimension, as shown in Figure P7–6 and press ENTER to complete the selection).*

Figure P7–5 *Completed aligned dimension*

Figure P7–6 *Aligned dimension*

Step 8: Draw centerlines to assist in completing the dimensioning exercise. Set Centerline as the current layer. Invoke the LINE command from the Draw toolbar, and draw lines I and 2 as shown in Figure P7–7.

Command: **line** (ENTER)
Specify first point: **7,8** (*start point for line 1*)
Specify next point or [Undo]: **@6.5<315**
Specify next point or [Undo]: (ENTER)

Command: (ENTER)
Specify first point: 7,8 (*start point for line 2*)
Specify next point or [Undo]: (*use the object snap tool "intersection" to draw the line to the intersection of the polygon, as shown in Figure P7–7*)
Specify next point or [Undo]: (enter)

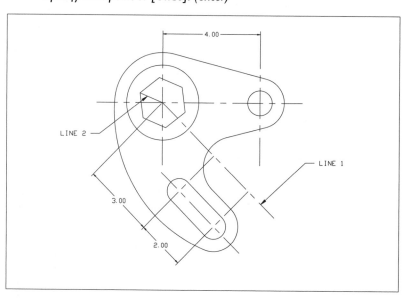

Figure P7–7 *Lines 1 and 2*

Step 9: Set Dimension as the current layer. Invoke the DIMALIGNED command from the Dimension toolbar to draw the remaining aligned dimensions for the drawing.

Command: **dimaligned** (ENTER)
Specify first extension line origin or <select object>: (*select the arc at PT. 1, as shown in Figure P7–8*)
Specify second extension line origin: (*use the object snap tool "perpendicular" to specify PT. 2, as shown in Figure P7–8*)
Specify dimension line location or [Mtext/Text/Angle]: (*specify PT. 3 to locate the dimension, as shown in Figure P7–8*)

AutoCAD draws the dimension as shown in Figure P7–9.

Command: **dimaligned** (ENTER)
Specify first extension line origin or <select object>: (*use the object snap tool "intersection" to locate PT. 1, as shown in Figure P7–10*)

Specify second extension line origin: *(use the object snap tool "perpendicular" to specify PT. 2, as shown in Figure P7–10)*

Specify dimension line location or [Mtext/Text/Angle]: *(specify PT. 3 to locate the dimension, as shown in Figure P7–10)*

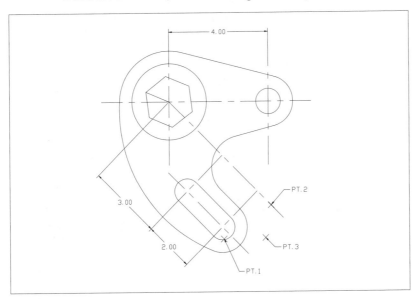

Figure P7–8 *Drawing the aligned dimension*

Figure P7–9 *Completed dimension*

Figure P7–10 *Drawing the aligned dimension*

AutoCAD draws the dimension as shown in Figure P7–11.

Figure P7–11 *Completed dimension*

Step 10: Set the OSNAP toggle button to OFF in the status bar.

Step 11: Invoke the DIMANGULAR command from the Dimension toolbar to draw the angular dimensions.

> Command: **dimangular** (ENTER)
> Select arc, circle, line, or ENTER: *(select line 1 as shown in Figure P7–12)*

Select second line: *(select line 2 as shown in Figure P7–12)*
Specify dimension arc line location or [Mtext/Text/Angle]: *(locate the dimension as at PT. 1, as shown in Figure P7–12)*

Command: **dimangular** (ENTER)
Select arc, circle, line or ENTER: *(select line 4 as shown in Figure P7–12)*
Select second line: *(select line 5 as shown in Figure P7–12)*
Specify dimension arc line location or [Mtext/Text/Angle]: *(locate the dimension as at PT. 2, as shown in Figure P7–12)*

AutoCAD draws the dimensions as shown in Figure P7–13.

Figure P7–12 *Selecting points for angular dimensions*

Figure P7–13 *Completed angular dimensions*

Step 12: Invoke the DIMRADIUS and DIMDIAMETER commands to draw the radius and diameter dimensions as shown in Figure P7–14.

Note: Before using DIMRADIUS and DIMDIAMETER, open the Dimension Style Manager dialog box from the Dimension menu. Choose the Modify button, then select the Fit subdialog box and choose **Place text manually when dimensioning** in the Fine Tuning section of the dialog box.

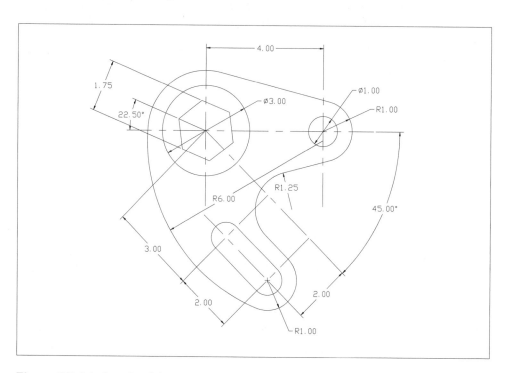

Figure P7–14 *Completed dimensions*

Step 13: Save and close the drawing.

EXERCISES 7–1 THROUGH 7–5

EXERCISE 7–1

Create the drawing shown, with all the dimensions (includes both lateral tolerancing and geometric tolerancing symbology).

Settings	Value		
1. Units	Decimal		
2. Limits	Lower left corner: –250,–100		
	Upper right corner: 100,100		
3. Grid	5		
4. Snap	5		
5. Layers	*NAME*	*COLOR*	*LINETYPE*
	Dimension	White	Continuous
	Border	Red	Continuous
	Object	Green	Continuous
	Center	Magenta	Center
	Text	Blue	Continuous
	Hidden	White	Hidden

Hints: The border can be drawn from coordinates –245.00, –95.00 to 95.00, 95.00 using the RECTANGLE command. Create the drawing to the specification.

In the **Dimension Style Manager** dialog box, choose Modify button, then select the Tolerances tab to set the Method to Limits, giving an Upper Value of 0.20 and a Lower Value of 0.10.

For each feature control frame you can select Tolerance from the Dimension menu.

In order to force the arrows inside the circle of diameter dimensions and the text outside the circle, choose Over the dimension line without a leader option in the Text Placement section of the Fit sub-dialog ox in the Dimension Style Manager dialog box.

Figure Ex7–1 *Completed drawing*

EXERCISE 7–2

Open the drawing that was completed in Exercise 4–2 and add the dimensions shown in the drawing here.

Settings	Value
1. Units	Architectural
2. Limits	Lower left corner: 10'-0",-10',-0" Upper right corner: 50'-0",35',-0"
3. Grid	12"
4. Snap	6"

Settings	Value		
5. Layers	*NAME*	*COLOR*	*LINETYPE*
	Beam	White	Continuous
	Column	White	Continuous
	Dimension	White	Continuous
	Border	Red	Continuous
	Construction	Green	Continuous
	Text	Blue	Continuous

Figure Ex7–2 *Completed drawing*

EXERCISE 7–3

Open the drawing that was completed in Exercise 6–6 and add the dimensions shown in the drawing here.

Settings	Value		
1. Units	Architectural		
2. Limits	Lower left corner: 0',0'		
	Upper right corner: 12',9'		
3. Grid	6"		
4. Snap	3"		
5. Layers	*NAME*	*COLOR*	*LINETYPE*
	Border	Red	Continuous
	Object	Green	Continuous
	Text	Blue	Continuous

Hints: Detail B can be drawn to true scale and then increased in size by means of the SCALE command.

In order to have the dimensions reflect the true sizes of features before their size was changed, set DIMLFAC to the reciprocal of the factor by which the object was scaled. For example, if the scale factor was 2, DIMLFAC should be set to 1/2.

Figure Ex7–3 *Completed drawing with dimensions*

EXERCISE 7–4

The drawing of the instrument panel in Figure Ex7–4a has all the information needed to create your drawing. Add the dimensions shown in Figure Ex7–4b with the BASELINE DIMENSION command.

Settings	Value
1. Units	Decimal
2. Limits	Lower left corner: 0,0
	Upper right corner: 12,10.5
3. Grid	0.5"
4. Snap	0.25"
5. Layers	_NAME_ _COLOR_ _LINETYPE_
	Border Red Continuous
	Object Green Continuous
	Text Blue Continuous
	Dimension White Continuous

HOLE	X-COORD.	Y-COORD.	DIA.
SCREW	0.250	0.250	0.125
SCREW	0.250	4.750	0.125
SCREW	4.250	0.250	0.125
SCREW	4.250	4.750	0.125
RPM	1.250	3.375	1.500
MPH	3.250	3.375	1.500
BATT	1.125	1.250	0.500
OIL	2.250	1.250	0.500
FUEL	3.375	1.250	0.500

Figure Ex7–4a *Layout of the drawing*

Figure Ex7–4b *Layout of the drawing with baseline dimensions*

EXERCISE 7-5

The drawing of the instrument panel in Figure Ex7–5a has all the information needed to create your drawing. Add the dimensions shown in Figure Ex7–5b with the ORDINATE DIMENSION command.

Settings	Value		
1. Units	Decimal		
2. Limits	Lower left corner: −1.5,−2.5		
	Upper right corner: 10.5,6.5		
3. Grid	0.5"		
4. Snap	0.25"		
5. Layers	*NAME*	*COLOR*	*LINETYPE*
	Border	Red	Continuous
	Object	Green	Continuous
	Text	Blue	Continuous
	Dimension	White	Continuous

HOLE	X-COORD.	Y-COORD.	DIA.
SCREW	0.250	0.250	0.125
SCREW	0.250	4.750	0.125
SCREW	4.250	0.250	0.125
SCREW	4.250	4.750	0.125
RPM	1.250	3.375	1.500
MPH	3.250	3.375	1.500
BATT	1.125	1.250	0.500
OIL	2.250	1.250	0.500
FUEL	3.375	1.250	0.500

Figure Ex7–5a *Layout of the drawing*

Figure Ex7–5b *Layout of the drawing with ordinate dimensions*

EXERCISE 7-6

Create the drawing of the Bracket as shown, with all the dimensions.

Settings	Value		
1. Units	Decimal		
2. Limits	Lower left corner: 0,0		
	Upper right corner: 8.5,11		
3. Dimscale	1		
4. Ltscale	0..5		
5. Layers	*NAME*	*COLOR*	*LINETYPE*
	Border	Red	Continuous
	Object	Green	Continuous
	Text	Blue	Continuous
	Dimension	White	Continuous
	Center	Red	Center
	Hidden	Magenta	Hidden

BRACKET
SCALE: FULL SIZE

Figure Ex7–6 *Bracket*

EXERCISE 7-7

Create the drawing of the Lock Ring as shown, with all the dimensions.

Settings	Value		
1. Units	Decimal		
2. Limits	Lower left corner: 0,0		
	Upper right corner: 11,8.5		
3. Dimscale	1		
4. Ltscale	0..5		
5. Layers	_NAME_	_COLOR_	_LINETYPE_
	Border	Red	Continuous
	Object	Green	Continuous
	Text	Blue	Continuous
	Dimension	White	Continuous
	Center	Red	Center
	Hidden	Magenta	Hidden

LOCK RING
SCALE: FULL SIZE

Figure Ex7–7 *Lock Ring*

REVIEW QUESTIONS

1. Dimension types available in AutoCAD include:

   a. Linear

   b. Angular

   c. Diameter

   d. Radius

   e. All of the above

2. The associative dimension drawn with the DIMASO variable set to ON has all of its separate parts drawn as separate objects.

   a. True

   b. False

3. The LINEAR DIMENSIONING command allows you to draw horizontal, vertical, and aligned dimensions.

   a. True

   b. False

4. To draw a linear dimension you must (1) specify the first extension line origin, (2) locate the dimension line, and then (3) specify the second extension line.

   a. True

   b. False

5. The ANGULAR DIMENSION command allows you to draw angular dimensions between two parallel lines.

   a. True

   b. False

6. By default, the dimension text for a radius dimension is preceded by:

   a. Radius

   b. Rad

   c. R

7. Using PROPERTIES (modify properties) command will allow you to override the Dimensioning System Variable settings for a single dimension, without modifying the base dimension style.

   a. True

   b. False

8. The BASELINE DIMENSION command is used to draw dimensions from a single datum baseline.

   a. True

   b. False

9. The DIMCENTER command allows you to draw center cross marks or centerlines in a circle.

   a. True

   b. False

10. You must explode a dimension before you can use the DIMTEDIT command to edit the dimension text.

    a. True

    b. False

11. The suppress option in the Lines and Arrows sub-dialog box of the Modify Dimension Style dialog box allows you to suppress only one extension line at a time.

    a. True

    b. False

12. The arrowhead section in Lines and Arrows sub-dialog box of the Modify Dimension Style dialog box allows you to change the size and style of your arrowheads.

    a. True

    b. False

13. You must use the UNITS command to determine how many decimal places will be displayed in the dimension text.

    a. True

    b. False

Directions: For each of the following questions, select which of the dimensioning sub-dialog boxes the option can be found from in the following list.

   a. Lines and Arrows Sub-dialog box

   b. Text Sub-dialog box

   c. Fit Sub-dialog box

   d. Primary Units Sub-Sub-dialog box

   e. cannot be found in the dimensioning dialog boxes

14. Arrowhead size

15. Associative/Non-associative setting

16. Text height

17. Arrowhead style

18. Text location

19. Suppressing of extension lines

20. Linear tolerance settings

21. Number of decimal places for a linear dimension

Directions: Answer the following questions based on Figure QX7–1.

Figure QX7–1 *Modify Dimension Style dialog box with Lines and Arrows tab selected*

22.   Which Dimension Variable must be changed to modify the size of Arrowheads?

   a.   DIMARO

   b.   DIMASZ

   c.   DIMTSZ

   d.   DIMDLI

23. Which Dimension Variable must be changed to modify Arrowhead types?

    a. DIMASZ   c. DIMBLK

    b. DIMTSZ   d. DIMEXE

24. Which Dimension Variable must be modified to change the color of dimension lines?

    a. DIMCLRE   c. DIMCLRL

    b. DIMCLRT   d. DIMCLRD

25. Which Dimension Variable is used to hide the first extension line?

    a. DIMEXL   c. DIMSE1

    b. DIMOFF   d. DIMEL1

26. Which Dimension Variable establishes the distance from the object to the start of the extension line?

    a. DIMEXO   c. DIMGAP

    b. DIMSEL   d. DIMSPC

27. Which Dimension Variable establishes the distance from the first dimension line to the second dimension line when using Baseline Dimensions?

    a. DIMD12   c. DIMDLI

    b. DIMSPC2   d. DIMDSPC

28. Which Dimension Variable is used to modify Architectural tick terminators?

    a. DIMARC   c. DIMTIC

    b. DIMART   d. DIMTSZ

29. Which Dimension Variable is used to hide the second dimension line?

    a. DIMSDL   c. DIM2OFF

    b. DIMHD2   d. DIMSD2

30. When one desires for the extension lines to be drawn past the dimension lines which Dimension Variable is used to perform this function?

    a. DIMEXE   c. DIMEXT

    b. DIMELD   d. DIMDLE

31. Which Dimension Variable is used to have a 2nd arrowhead that is different from the 1st?

    a. DIMARW2   c. DIMBLK2

    b. DIMABLK   d. dimaro2

CHAPTER 8

Plotting/Printing

INTRODUCTION

One task has not changed much in the transition from board drafting to CAD: obtaining a hard copy. The term "hard copy" describes a tangible reproduction of a screen image. The hard copy is usually a reproducible medium from which prints are made and can take many forms, including slides, videotape, prints, or plots. This chapter describes the most commonly used processes for getting a hard copy: plotting/printing.

In manual drafting, if you need your drawing to be done in two different scales, you physically draw the drawing for two different scales. In CAD, with minor modifications, you plot or print the same drawing in different scale factors on different sizes of paper. In AutoCAD, you can even compose your drawing in Paper Space with limits that equal the sheet size and plot it at 1:1 scale.

After completing this chapter, you will be able to do the following:

- Plan the plotted sheet
- Plot from Model Space
- Set up a Layout
- Create and modify a Layout
- Create Floating Viewports
- Scale views relative to paper space
- Control the visibility of layers within Viewports
- Plot from Layout (WYSIWYG)
- Create and Modifying Plot Style Tables
- Change the Plot Style Property for an Object or Layer
- Configure Plotters

PLANNING THE PLOTTED SHEET

Planning ahead is still required in laying out the objects to be drawn on the final sheet. The objects drawn on the plotted sheet must be arranged. At least in CAD, with its true-size capability, an object can be started without first laying out a plotted sheet. But eventually, limits, or at least a displayed area, must be determined. For schematics, diagrams, and graphs, plotted scale is of little concern. But for architectural, civil, and mechanical drawings, plotting to a conventional scale is a professionally accepted practice that should not be abandoned just because it can be circumvented.

When setting up the drawing limits, you must take the plotted sheet into consideration to get the entire view of the object(s) on the sheet. So, even with all the power of the CAD system, some thought must still be given to the concept of scale, which is the ratio of true size to the size plotted. In other words, before you start drawing, you should have an idea at what scale the final drawing will be plotted or printed on a given size of paper.

The limits should correspond to some factor of the plotted sheet. If the objects will fit on a 24" x 18" sheet at full size with room for a border, title block, bill of materials, dimensioning, and general notes, then set up your limits to (0,0) (lower left corner) and (24,18) (upper right corner). This can be plotted or printed at 1:1 scale, that is, one object unit equals one plotted unit.

Plot scales can be expressed in several formats. Each of the following five plot scales is exactly the same; only the display formats differ.

> 1/4" = 1'-0"
> 1" = 4'
> 1 = 48
> 1:48
> 1/48

A plot scale of 1:48 means that a line 48 units long in AutoCAD will plot with a length of 1 unit. The units can be any measurement system, including inches, feet, millimeters, nautical miles, chains, angstroms, and light-years, but, by default, plotting units in AutoCAD are inches.

There are four variables that control the relationship between the size of objects in the AutoCAD drawing and their sizes on a sheet of paper produced by an AutoCAD plot:

- Size of the object in AutoCAD. For simplification it will be referred to as ACAD_size.

- Size of the object on the plot. For simplification it will be referred to as ACAD_plot.

- Maximum available plot area for a given sheet of paper. For simplification it will be referred to as ACAD_max_plot.

- Plot scale. For simplification it will be referred as to ACAD_scale.

The relationship between the variables can be described by the following three algebraic formulas:

$$\text{ACAD\_scale} = \text{ACAD\_plot} / \text{ACAD\_size}$$
$$\text{ACAD\_plot} = \text{ACAD\_size} \times \text{ACAD\_scale}$$
$$\text{ACAD\_size} = \text{ACAD\_plot} / \text{ACAD\_scale}$$

EXAMPLE OF COMPUTING PLOT SCALE, PLOT SIZE, AND LIMITS

An architectural elevation of a building 48' wide and 24' high must be plotted on a 36" x 24" sheet. First, you determine the plotter's maximum available plot area for the given sheet size. This depends on the model of plotter you use.

In the case of a Summagraphics/Houston Instruments plotter, the available area for 36" x 24" is 33.5" x 21.5". Next, you determine the area needed for the title block, general notes, and other items, such as an area for revision notes and a list of reference drawings. For the given example, let's say that an area of 27" x 16" is available for the drawing.

The objective is to arrive at one of the standard architectural scales in the form of x in. = 1 ft. The usual range is from 1/16" = 1'-0" for plans of large structures to 3" = 1'-0" for small details. To determine the plot scale, substitute these values for the appropriate variables in the formula:

$$\text{ACAD\_scale} = \text{ACAD\_plot}/\text{ACAD\_size}$$
$$\text{ACAD\_scale} = 27"/48' \text{ for } X \text{ axis}$$
$$= 0.5625"/1'-0" \text{ or } 0.5625"=1'-0"$$

The closest standard architectural scale that can be used in the given situation is 1/2" = 1'-0" (0.5" = 1'-0", 1/24 or 1:24).

To determine the size of the object on the plot, substitute these values for the appropriate variables in the formula:

$$\text{ACAD\_plot} = \text{ACAD\_size} \times \text{ACAD\_scale}$$
$$\text{ACAD\_plot} = 48' \times (0.5"/1') \text{ for } X \text{ axis}$$
$$= 24" \text{ (less than the 27" maximum allowable space on the paper)}$$
$$\text{ACAD\_plot} = 24' \times (0.5"/1') \text{ for } Y \text{ axis}$$
$$= 12" \text{ (less than the 16" maximum allowable space on the paper)}$$

If instead of 1/2" = 1'-0" scale, you wish to use a scale of 3/4" = 1'-0", then the size of the object on the plot will be 48' x (0.75"/1') = 36" for the X axis. This is more than the available space on the given paper, so the drawing will not fit on the given paper size. You must select a larger paper size.

Once the plot scale is determined and you have verified that the drawing fits on the given paper size, you can then determine the drawing limits for the plotted sheet size of 33.5" x 21.5".

To determine the limits for the X and Y axes, substitute the appropriate values in the formula:

ACAD_limits (X axis)	= ACAD_max_plot/ACAD_scale
	= 33.5"/(0.5"/1'-0")
	= 67'
ACAD_limits (Y axis)	= 21.5"/(0.5"/1'-0")
	= 43'

Appropriate limits settings in AutoCAD for a 36" x 24" sheet with a maximum available plot area of 33.5" x 21.5" at a plot scale of 0.5" = 1'-0" would be:

```
lower left corner: 0,0
upper right corner: 67',43'
```

Another consideration in setting up a drawing for user convenience is to have the (0,0) coordinates at some point other than the lower left corner of the drawing sheet. Many objects have a reference point from which other parts of the object are dimensioned. Being able to set that reference point to (0,0) is very helpful. In many cases, the location of (0,0) is optional. In other cases, the coordinates should coincide with real coordinates, such as those on an industrial plant area block. In still other cases, only one set of coordinates might be a governing factor.

In this example, the 48' wide x 24' high front elevation of the building is to be plotted on a 36" x 24" sheet at a scale of 1/2" = 1'-0". It has been determined that (0,0) should be at the lower left corner of the front elevation view, as shown in Figure 8–1.

Centering the view on the sheet requires a few minutes of layout time. Several approaches allow the drafter to arrive at the location of (0,0) relative to the lower left corner of the plotted sheet or limits. Having computed the limits to be 67' wide x 43' high, the half-width and half-height (dimensions from the center) of the sheet are 33.5' and 21.5' to scale, respectively. Subtracting the half-width of the building from the half-width of the limits will set the X coordinate of the lower left corner at −9.5' (from the equation, 24' − 33.5'). The same is done for the Y coordinate −9.5' (12' − 21.5'). Therefore, the lower left corner of the limits are at (−9.5',−9.5').

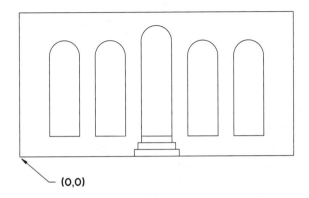

Figure 8–1 *Setting the reference point to the origin (0,0) in a location other than the lower left corner*

Appropriate limits settings in AutoCAD for a 36" x 24" sheet with a maximum available plot area of 33.5" x 21.5" by centering the view at a plot scale of 0.5" = 1'-0" (see Figure 8–2) is:

> lower left corner: −9.5',−9.5'
> upper right corner: 57.5',33.5'

 Note: The absolute X coordinates, when added (57.5' + 9.5',) equal 67', which is the width of the limits, and the absolute Y coordinates (33.5' + 9.5') equal 43', which is the height of the limits.

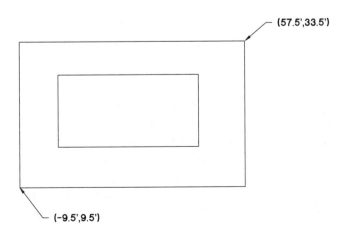

Figure 8–2 *Setting the limits to the maximum available plot area by centering the view*

SETTING FOR LTSCALE

As explained in Chapter 3, the LTSCALE system variable provides a method of adjusting the linetypes to a meaningful scale for the drawing. This sets the length of dashes and spaces in linetypes. When the value of LTSCALE is set to the reciprocal of the plot scale, the linetypes provided with AutoCAD plot out on paper at the sizes they are defined in ACAD.LIN.

$$LTSCALE = 1/ACAD\_scale$$

SETTING FOR DIMSCALE

As explained in Chapter 7, AutoCAD provides a set of dimensioning variables that control the way dimensions are drawn. The DIMSCALE dimension variable is applied globally to all dimensions that are applied to sizes or distances, as an overall scaling factor. The default DIMSCALE value is set to 1. When DIMSCALE is set to the reciprocal of the plot scale, it applies globally to all dimension variables for the plot scale factor.

$$DIMSCALE = 1/ACAD\_scale$$

If necessary, you can set individual dimensioning variables to the size that you want the dimension to appear on the paper by substituting the appropriate values in the following formula:

$$\text{size of the plotted dimvars\_value} = \text{dimvars\_value} \times ACAD\_scale \times DIMSCALE$$

As an example, here's how to determine the arrow size DIMASZ for a plot scale of 1/2" = 1'-0", DIMSCALE set to 1, and default DIMASZ of 0.18":

$$\text{size of the plotted arrow} = 0.18" \times (1/24) \times 1$$
$$= 0.0075"$$

SCALING ANNOTATIONS AND SYMBOLS

How can you determine the size at which text and symbols (blocks) will plot? As mentioned earlier, you almost always draw objects in their actual size, that is, to real-world dimensions. Even in the case of text and blocks, you place them at the real-world dimensions. In the previous example, the architectural elevation of a building 48' x 24' is drawn to actual size and plotted to a scale of 1/2" = 1'-0'. Let's say you wanted your text to plot at 1/4" high. If you were to create your text and annotations at 1/4", they would be so small relative to the elevation drawing itself that you could not read the characters.

Before you begin placing the text, you need to know the scale at which you will eventually plot the drawing. In the previous example of an architectural elevation, the plot scale is 1/2" = 1'-0" and you want the text to plot 1/4" high. You need to find a relationship between 1/4" on the paper and the size of the text for the real-world dimensions in the drawing. If 1/2" on the paper equals 12" in the model, then 1/4"-high text on the paper equals 6", so text and annotations should be drawn at 6" high in the drawing to plot at 1/4" high at this scale of 1/2" = 1'-0". Similarly, you can calculate the various text sizes for a given plot scale. Table 8–1 shows the model text size needed to achieve a specific plotted text height at some common scales.

Table 8–1 Text Size Corresponding to Specific Plotted Text Height at Various Scales

SCALE	FACTOR	PLOTTED TEXT HEIGHT								
		1/16"	3/32"	1/8"	3/16"	1/4"	5/16"	3/8"	1/2"	5/8"
1/16" = 1'-0"	192	12"	18"	24"	36"	48"	60"	66"	96"	120"
1/8" = 1'-0"	96	6"	9"	12"	18"	24"	30"	36"	48"	60"
3/16" = 1'-0"	64	4"	6"	8"	12"	16"	20"	24"	32"	40"
1/4" = 1'-0"	48	3"	4.5"	6"	9"	12"	15"	18"	24"	30"
3/8" = 1'-0"	32	2"	3"	4"	6"	8"	10"	12"	16"	20"
1/2" = 1'-0"	24	1.5"	2.25"	3"	4.5"	6"	7.5"	9"	12"	15"
3/4" = 1'-0"	16	1"	1.5"	2"	3"	4"	5"	6"	8"	10"
1" = 1'-0"	12	0.75"	1.13"	1.5"	2.25"	3"	3.75"	4.5"	6"	7.5"
1 1/2" = 1'-0"	8	0.5"	.75"	1"	1.5"	2"	2.5"	3"	4"	5"
3" = 1'-0"	4	0.25"	.375"	0.5"	0.75"	1"	1.25"	1.5"	2"	2.5"
1" = 10'	120	7.5"	11.25"	15"	22.5"	30"	37.5"	45"	60"	75"
1" = 20'	240	15"	22.5"	30"	45"	60"	75"	90"	120"	150"
1" = 30'	360	22.5"	33.75"	45"	67.5"	90"	112.5"	135"	180"	225"
1" = 40'	480	30"	45"	60"	90"	120"	150"	180"	240"	300"
1" = 50'	600	37.5"	56.25"	75"	112.5"	150"	187.5"	225"	300"	375"
1" = 60'	720	45"	67.5"	90"	135"	180"	225"	270"	360"	450"
1" = 70'	840	52.5"	78.75"	105"	157.5"	210"	262.5"	315"	420"	525"
1" = 80'	960	60"	90"	120"	180"	240"	300"	360"	480"	600"
1" = 90'	1080	67.5"	101.25"	135"	202.5"	270"	337.5"	405"	540"	675"
1" = 100'	1200	75"	112.5"	150"	225"	300"	375"	450"	600"	750"

PLOTTING FROM MODEL SPACE AND PAPER SPACE

One of the useful features of AutoCAD is the option to work on your drawing in two different environments, model space and paper space. AutoCAD 2000 allows you to plot a drawing from model space as well as paper space, also referred to as Layout. Most of the drafting and design work is created in the *3D* environment of model space, even though your objects may have been drawn only in a *2D* plane. You can plot the drawing from model space to any scale by specifying the plot scale in the plot dialog box.

Paper space is a *2D* environment used for arranging various views (floating viewports) of what was drawn in model space. It represents the paper on which you arrange the drawing prior to plotting. With AutoCAD 2000, single or multiple paper space environments (layouts) can be easily designed and manipulated. Each layout represents an individual plot output sheet, or an individual sheet in a drawing project. You can apply different scales to each floating viewport and specify different visibility for layers in the views. After arranging the views and scaling appropriately, you can plot the drawing from layout at 1:1 (full scale). Paper space allows you to plot the drawing in WYSIWYG (What You See Is What You Get) mode.

You can switch between model space and paper space by selecting the appropriate tabs provided at the bottom of the drawing window. Model space can be accessed from the Model tab and selecting one of the available Layout tabs can access paper space. You can also switch between model space and paper space by changing the value of the system variable TILEMODE. When TILEMODE is set to 1 (default), you will be working in model space; when it is set to 0, you will be working is paper space. When you are working in a layout, you can access model space by double-clicking inside one of the floating viewports. If you double-click anywhere outside the viewport, AutoCAD will switch you to paper space. You must have at least one floating viewport in paper space to see the model. By default, AutoCAD creates two layout tabs called Layout1 and Layout2. If necessary, you can create additional layouts. A detailed explanation is provided later in the chapter for creating and modifying layouts.

PLOTTING FROM MODEL SPACE

To plot/print the current drawing from the model space, invoke the PLOT command:

Standard toolbar	Choose the PLOT command (see Figure 8–3)
File menu	Choose Plot
Command: prompt	**plot** (ENTER)

Figure 8–3 *Invoking the* PLOT *command from the Standard toolbar*

AutoCAD displays the Plot dialog box, as shown in the Figure 8–4.

Figure 8–4 *Plot dialog box*

AutoCAD displays the current layout name or displays "Selected layouts" if multiple tabs are selected. If the Model tab is current when you choose Plot, the Layout Name shows "Model." Set the **Save changes to layout** toggle button to ON to save changes you make in the Plot dialog box to the layout. This option is not available if multiple layouts are selected.

The **Page setup name:** list box displays a list of any saved page setups. You can choose from one of the saved page setups to restore the plot settings. Or you can save the current settings to a named page setup and it can be applied later to plot any drawing from model space. To save the current settings, choose the **Add** button, and AutoCAD displays the User Defined Page Setups dialog box, as shown in Figure 8–5.

Figure 8–5 *User Defined Page Setups dialog box*

You can create, delete, or rename named page setups. To save the current settings, specify a name in the **New page setup name:** text box and choose the **OK** button. To delete the currently selected user-defined page setup, choose the **Delete** button; to rename the currently selected user-defined page setup, choose the **Rename** button and change the name. Choose the **Import** button to import a user-defined page setup from another drawing.

PLOT DEVICE SETTINGS

The Plot Device tab of the Plot dialog box specifies the plotter to use, a plot style table, what to plot, and information about plotting to a file.

Plotter Configuration

The **Plotter configuration** section (see Figure 8–6) displays the currently configured plotting device, the port to which it's connected, or its network location. If necessary, you can plot the drawing to a different plotting device from the list of available system plotters displayed in the **Name:** list box. The default plotter is specified in the Options dialog box. To add a new plotter to the **Name:** list box, refer to the section on "Configuring Plotters," later in this chapter.

Figure 8–6 *Plot dialog box – Plotter configuration section*

Choose the **Properties** button to change the Plotter Configuration of the currently configured plotting device. Refer to the section on "Configuring Plotters," later in this chapter, for a detailed explanation on plotter configuration.

Choose the **Hints** button to display information of the currently configured plotting device.

Plot Style Table

The **Plot style table** section (see Figure 8–7) sets the plot style table, edits the plot style table, or creates a new plot style table. Plot style tables are settings that give you control over how objects in your drawing are plotted into hard copy plots. You can control how colors are translated into line weight and how area fills are converted into shades of gray or screened colors, as well as many other output options. You can control how the plotter treats each individual object in a drawing. Refer to the section on "Creating and Modifying Plot Style Tables," later in this chapter, for a detailed explanation on creating and modifying a plot style table.

Figure 8–7 *Plot dialog box – Plot style table section*

What to Plot

The **What to plot** section (see Figure 8–8) defines what you want to plot: a Model tab or a Layout tab. The **Current Tab** selection plots the current Model or Layout tab. The **Selected Tabs** selection plots multiple preselected Model or Layout tabs. To select multiple tabs, hold down the CTRL key while selecting the tabs. If only one tab is selected, this option is not available. The **All Layout tabs** selection plots all layout tabs, regardless of which tab is selected.

Figure 8–8 *Plot dialog box – What to plot section*

The **Number of Copies:** text field specifies number of copies that are plotted.

Plot to File

The **Plot to file** section (see Figure 8–9) controls whether to create a plot file instead of using a plotter to plot the drawing. The creation of a plot file is useful if you need to use a plotter connected to another computer. You can copy the file to a floppy disk and then transfer the disk to the other computer. AutoCAD creates the plot file with .PLT as an extension to the given file name. Set the **Plot to file** toggle to ON to create a plot file. By default, the name of the plot file will be the same as the current drawing name and saved in the current folder. If you want to save it to a different file name, specify the file name in the **File name:** text field and/or save it to a different folder, specify the name of the folder in the **Location:** text field.

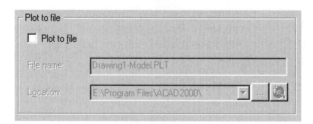

Figure 8–9 *Plot dialog box – Plot to file section*

PLOT SETTINGS

The Plot Settings tab (see Figure 8–10) of the Plot dialog box specifies paper size, drawing orientation, plot area and scale, plot offset, and plot options.

Figure 8–10 *Plot dialog box – Plot Settings tab*

Paper Size and Paper Units

The **Paper size and paper units** section allows you to select the paper size and the units to plot. Select the paper size to plot from the **Paper Size:** list box for the selected plotting device. Actual paper sizes are indicated by the width (*X* axis direction) and height (*Y* axis direction). A default paper size is set for the plotting device when you create a plotter configuration file with the Add-a-Plotter wizard. For more information about the Add-a-Plotter wizard, refer to the section on "Configuring Plotters," later in this chapter.

AutoCAD displays the name of the currently selected plot device and actual area on the paper that is used for the plot based on the current paper size. Select the appropriate radio button to plot your drawing in inches or millimeters.

Drawing Orientation

The **Drawing orientation** section (see Figure 8–11) specifies the orientation of the drawing on the paper for plotters that support landscape or portrait orientation. Select the **Portrait** radio button to orient and plot the drawing so that the short edge of the paper represents the top of the page. Select the **Landscape** radio button to orient and plot the drawing so that the long edge of the paper represents the top of the page. Set the **Plot upside-down** toggle button to ON to orient and plot the drawing upside down.

Figure 8–11 *Plot dialog box – Drawing orientation section*

Plot Area

The **Plot area** section (see Figure 8–12) specifies the portion of the drawing to be plotted.

Figure 8–12 *Plot dialog box – Plot area section*

Select the **Limits** radio button to plot the drawing to its limits. In general, this makes the lower left corner of the limits of the drawing equal to the origin of the plot. This option is available only when you plot from Model space.

Select the **Extents** radio button to plot the entire drawing. It forces the lower left corner of the rectangle that includes all objects and/or the current limits in the drawing, rather than the display, to become the origin of the plot. This option is similar to the ZOOM EXTENTS command and ensures that the entire drawing is plotted, including any objects drawn outside the limits.

Select the **Display** radio button to plot what is currently displayed on the screen. An important point to remember is that the lower left corner of the current display is the origin point of the plot. This option is useful if you want to plot only part of the drawing. Before you select this option, make sure the view you want to plot is displayed on the screen by using the ZOOM and PAN commands.

Select the **View** radio button to plot a previously saved view. If you plot a previously created view, the plot is identical to the screen image after the VIEW RESTORE command is used to bring the view on-screen. View plotting makes it easy to plot predefined areas of a drawing. You can select a named view from the list provided. If there are no saved views in the drawing, this option is not available.

Select the **Window** radio button to pick a window on the screen and plot the objects that are inside the window. The lower left corner of the window becomes the origin of the plot. This is similar to using the ZOOM WINDOW command to zoom into a specific portion of the drawing, and then using the Display option of the PLOT command to plot. To specify the Window, choose the **Window <** button. Specify the coordinates of the two diagonal points or designate the window by means of your pointing device.

Plot Scale

The **Plot scale** section (see Figure 8–13) controls the plot area. Select the plot scale from the **Scale:** list box. When plotting from Model Space the default setting is Scaled to Fit. When you select a standard scale, the scale is displayed in Custom text fields.

Figure 8–13 *Plot dialog box – Plot scale section*

Set the **Scale Lineweights** toggle to ON to plot in proportion to the plot scale and to OFF to plot the objects with assigned lineweights. Lineweights normally specify the linewidth of printed objects and are plotted with the linewidth size regardless of the plot scale. This option is available only in a Layout (Paper Space).

Plot Offset

The Plot Offset section (see Figure 8–14) specifies an offset of the plotting area from the lower left corner of the paper. In a layout, the lower left corner of a specified plot area is positioned at the lower left margin of the paper. You can offset the origin by entering a positive or negative value.

Set the **Center the plot** toggle button to ON to automatically center the plot on the paper.

Figure 8–14 *Plot dialog box – Plot offset section*

Plot Options

The **Plot options** section (see Figure 8–15) specifies options for lineweights, plot styles, and the current plot style table.

Set the **Plot object lineweights** toggle button to ON to plot the objects with assigned lineweights and to OFF to plot with the default lineweight. This option is not available if Plot Styles is set to ON.

Set the **Plot with plot styles** toggle button to ON to plot using the object plot styles that are assigned to the geometry, as defined by the plot style table. Refer to the section on "Creating and Modifying Plot Style Tables," later in this chapter, for a detailed explanation on creating and modifying a plot style table.

Set the **Plot paperspace last** toggle button to ON to plot model space geometry before paper space objects are plotted. This option is not available when plotting from Model Space.

Set the **Hide objects** toggle button to ON to plot layouts (paper space) with hidden lines removed. Hidden lines are those that normally would be obscured by objects placed in front of them. This option is not applicable to *2D* drawings.

Figure 8–15 *Plot dialog box – Plot options section (Plotting from Model Space)*

Partial Plot Preview

Choose the **Partial Plot Preview** button to quickly display a partial preview of the effective plot area relative to the paper size and printable area (see Figure 8–16). In addition, AutoCAD provides information regarding paper size, printable area, and effective area. Partial plot preview also gives advance notice of any AutoCAD warnings that may be encountered when you plot the drawing.

Figure 8–16 *Partial Plot Preview dialog box*

Full Preview

A full preview displays the drawing on the screen as it would appear when plotted. This takes much longer than the partial preview, but it takes less time than regular plot regeneration. This is quicker than plotting something incorrectly and having to start all over again.

To get a full preview of your plot, choose the **Full Preview** button. AutoCAD temporarily clears the plotting dialog boxes, draws an outline of the paper size, and displays the drawing as it would appear on the paper when it is plotted (see Figure 8–17).

The cursor changes into a magnifying glass with plus and minus signs. Holding the pick button and dragging the cursor toward the top of the screen enlarges the preview image. Dragging it toward the bottom of the screen reduces the preview image. Right-click and AutoCAD displays a shortcut menu offering additional preview options: Pan, Zoom, Zoom Window, Zoom Original, Plot, and Exit.

To end the full preview, choose the Exit option from the shortcut menu. AutoCAD returns to the Plot dialog box.

 Note: You can also access a full preview by invoking the PRINT PREVIEW command from the File menu.

Figure 8–17 *Plot Preview dialog box*

After making the necessary changes in the plot settings, choose the **OK** button. AutoCAD starts plotting and reports its progress as it converts the drawing into the plotter's graphics language by displaying the number of vectors processed.

If something goes wrong or if you want to stop immediately, choose the CANCEL button at any time. AutoCAD cancels the plotting.

PLOTTING FROM LAYOUT (WYSIWYG)

As mentioned earlier, AutoCAD allows you to plot a drawing from model space as well as paper space. The paper space environment allows you to set up multiple layouts. You can have as many layouts as you like, and each one can be set up for a different type of output. A layout is used to compose or layout your model drawing for plotting. A layout may consist of a title block, one or more viewports, and annotations. As you create a layout, you can design floating viewport configurations to visualize different details in your drawing. You can apply different scales to each view within the viewport, and specify different visibility for layers in the viewport. Layouts are accessible by choosing the Layout tab at the bottom of the drawing area.

By default, whenever you start a new drawing, AutoCAD creates two layouts, called Layout1 and Layout2. If you open a file created in a previous versions of AutoCAD, you only see one Layout tab.

SETTING UP A LAYOUT

To set up a layout, first select the Layout tab located at the bottom of the drawing area. AutoCAD displays the Page Setup dialog box by default, as shown in Figure 8–18.

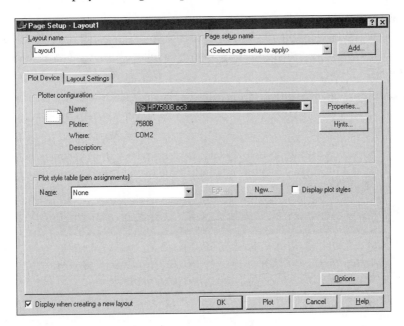

Figure 8–18 *Page Setup dialog box*

AutoCAD displays the current layout name in the **Layout name:** text field.

The **Page setup name:** list box displays a list of any saved page setups. You can choose from one of the saved page setups to restore the plot settings. Or you can save the current settings to a named page setup and it can be applied later to plot any drawing from a layout. To save the current settings, choose the **Add** button, and AutoCAD displays the User Defined Page Setups dialog box. You can create, delete, or rename named page setups.

Plot Device Settings

The Plot Device tab specifies a currently configured plotting device to plot the selected layout.

The **Plotter Configuration** section displays the currently configured plotting device, the port to which it's connected, or its network location. If necessary, you can plot the drawing to a different plotting device from the list of available system plotters displayed in the **Name:** list box. The default plotter is specified in the Options dialog box. To add a new plotter to the **Name:** list box, refer to the section on "Configuring Plotters," later in this chapter. Choose the **Properties** button to change the Plotter Configuration of the currently configured plotting device. Refer to the section on **Configuring of Plotters** for a detailed explanation on plotter configuration. Choose the **Hints** button to display information on the currently configured plotting device.

The **Plot Style Table** section sets the plot style table, edits the plot style table, or creates a new plot style table. Refer to the section on "Creating and Modifying Plot Style Tables," later in this chapter, for a detailed explanation on creating and modifying a plot style table.

Set the **Display when creating a new layout** toggle box to ON (default is set to ON) to display the Page Setup dialog box each time you select a new Layout tab to begin creating a new layout.

Choose the **Options** button to display the Plotting tab in the Options dialog box, as shown in Figure 8–19. You can view or modify the current layout and plot settings in the Options dialog box.

Figure 8–19 *Options dialog box with Plotting tab selected*

Refer to Chapter 13 for a detailed explanation of various options available in the Options dialog box.

Layout Settings

The Layout Settings tab (see Figure 8–20) specifies paper size, drawing orientation, plot area and plot scale, plot offset, and plot options.

Figure 8–20 *Plot dialog box – Layout Settings tab*

The **Paper size And paper units** section allows you to select the paper size and the units to plot. Select the paper size to plot from the **Paper Size:** list box for the selected plotting device. Actual paper sizes are indicated by the width (*X* axis direction) and height (*Y* axis direction). A default paper size is set for the plotting device when you create plotter configuration file with the Add-a-Plotter wizard. For more information about the Add-a-Plotter wizard, refer to the section on "Configuring Plotters," later in this chapter.

AutoCAD displays the name of the currently selected plot device and actual area on the paper that is used for the plot based on the current paper size. Select the appropriate radio button to plot your drawing in inches or millimeters.

The **Drawing orientation** section specifies the orientation of the drawing on the paper for plotters that support landscape or portrait orientation. Select the **Portrait** radio button to orient and plot the drawing so that the short edge of the paper represents the top of the page. Select the **Landscape** radio button to orient and plot the drawing so that the long edge of the paper represents the top of the page. Set the **Plot Upside-Down** toggle button to ON to orient and plot the drawing upside down.

The **Plot area** section specifies the portion of the drawing to be plotted.

Select the **Layout** radio button to plot everything within the margins of the specified paper size, with the origin calculated from 0,0 in the layout.

Select the **Extents** radio button to plot the extent of the current space of the drawing. It forces the lower left corner of the rectangle that includes all objects or the current limits in the drawing, rather than the display, to become the origin of the plot. This option is similar to the ZOOM EXTENTS command and ensures that the entire drawing is plotted, including any objects drawn outside the limits.

Select the **Display** radio button to plot what is currently displayed on the screen. An important point to remember is that the lower left corner of the current display is the origin point of the plot. This option is useful if you want to plot only part of the drawing. Before you select this option, make sure the view you want to plot is displayed on the screen by using the ZOOM and PAN commands.

Select the **View** radio button to plot a previously saved view. If you plot a previously created view, the plot is identical to the screen image after the VIEW RESTORE command is used to bring the view on-screen. View plotting makes it easy to plot predefined areas of a drawing. You can select a named view from the list provided. If there are no saved views in the drawing, this option is not available.

Select the **Window** radio button to specify a window on the screen and plot the objects that are inside the window. The lower left corner of the window becomes the origin of the plot. This is similar to using the ZOOM WINDOW command to zoom into a specific portion of the drawing, and then using the Display option of the PLOT command to plot. To specify the Window, choose the **Window <** button. Specify the coordinates of the two diagonal points or designate the window by means of your pointing device.

The **Plot scale** section controls the plot area. Select the plot scale from the **Scale:** list box. When plotting from Layout, the default setting is set to 1:1 (full scale). When you select a standard scale, the scale is displayed in Custom. The **Custom** selection allows you to define a user-defined scale. You can create a custom scale by entering the number of inches (or millimeters) equal to the number of drawing units in the **Custom** text fields.

Set the **Scale Lineweights** toggle to ON to plot in proportion to the plot scale and to OFF to plot the objects with assigned lineweights. Lineweights normally specify the linewidth of printed objects and are plotted with the linewidth size regardless of the plot scale.

The **Plot offset** section specifies an offset of the plotting area from the lower left corner of the paper. In a layout, the lower left corner of a specified plot area is positioned at the lower left margin of the paper. You can offset the origin by entering a positive or negative value. Set the **Center the plot** toggle button to ON to automatically center the plot on the paper.

The **Plot Options** section specifies options for lineweights, plot styles, and the current plot style table. Set the **Plot Object Lineweights** toggle button to ON to plot the objects with assigned lineweights and to OFF to plot with default lineweight. Set the **Plot With Plot Styles** toggle button to ON to plot using the object plot styles that are assigned to the geometry, as defined by the plot style table. Refer to the section on "Creating and Modifying Plot Style Tables," later in this chapter, for a detailed explanation on creating and modifying a plot style table. Set the **Plot Paperspace Last** toggle button to ON to plot model space geometry before paper space objects are plotted. Set the **Hide Objects** toggle button to ON to plot layouts (paper space) with hidden lines removed from objects. Hidden lines are those that normally would be obscured by objects placed in front of them. This option is not applicable to *2D* drawings.

Choose the **Plot** button to plot the drawing from the selected Layout or choose the **OK** button to close the Page Setup dialog box.

AutoCAD sets up a layout with a rectangular outline (shadow) indicating the paper size of the currently configured plotting device (see Figure 8–21). The margins displayed (dashed lines) within the paper indicate the printable area of the paper.

Figure 8–21 *Parts of a Layout tab*

582

 Note: The Display tab of the Options dialog box controls the display of the paper background and margins.

In addition, by default, AutoCAD creates a floating viewport and displays the drawing in the viewport. Refer to the section later in this chapter on "Creating and Modifying Floating Viewports" to create and modify floating viewports.

 Note: If necessary, you can make changes to the page setup anytime by invoking the PAGESETUP command.

CREATING AND MODIFYING A LAYOUT

The LAYOUT command allows you to create a new layout, copy an existing layout, rename a layout, delete a layout, save a layout, and make a layout current.

Invoke the NEW LAYOUT command:

Layouts toolbar	Choose NEW LAYOUT (see Figure 8–22)
Insert menu	Choose Layout > New Layout
Command: prompt	**layout** (ENTER)

Figure 8–22 *Invoke the* NEW LAYOUT *command from the Layouts toolbar*

AutoCAD prompts:

Command: **layout** (ENTER)
Enter layout option [Copy/Delete/New/Template/Rename/SAveas/Set/?]
    <set>: *(specify an option or right-click for shortcut menu and select one of the options)*

Creating a New Layout

The New option creates a new layout tab with the specified name using the default plot device. AutoCAD prompts:

Enter new Layout name <default>: *(specify a layout name or press* ENTER *to accept the default name)*

When you make the newly created layout tab current, AutoCAD displays the Page Setup dialog box by default. If necessary, you can make changes to the Plot Device settings and Layout Settings.

Note: You can also create a new layout by choosing the **New Layout** option from the shortcut menu that appears when you right-click after placing the cursor on the existing layout tab.

Copy an Existing Layout

The **Copy** option creates a new layout by copying an existing layout. AutoCAD prompts:

> Enter name of layout to copy <default>: *(specify the name of an existing layout name from which to copy)*
> Enter layout name for copy <default>: *(specify a layout name or press* ENTER *to accept the default name)*

The new layout tab is inserted after the last current layout.

 Note: You can also create a new layout by copying an existing layout. This is done by choosing the **Copy Layout** option from the shortcut menu that appears when you right-click after placing the cursor on an existing layout tab.

Creating a Layout from a Template

The **Template** option creates a new layout from an existing layout in a drawing (DWG) or template drawing (DWT) file. The layout (and all geometry) from the specified template or drawing file is inserted into the current drawing.

When you select the Template option, AutoCAD displays the Select File dialog box. Select a drawing or template file from the appropriate directory and choose the **Open** button. AutoCAD displays the Insert Layout(s) dialog box, as shown in Figure 8–23.

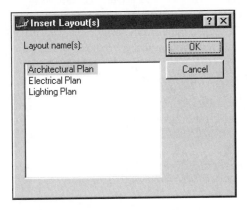

Figure 8–23 *Insert Layout(s) dialog box*

Select one or more layouts from the Insert dialog box and choose the **OK** button to close the dialog box.

Note: You can also create a new layout from an existing layout in a drawing or template drawing by choosing the **From template** option from the shortcut menu that appears when you right-click after placing the cursor on an existing layout tab.

Renaming a Layout

The **Rename** option allows you to rename an existing layout. The last current layout is used as the default for the layout to rename. Layout names must be unique. Layout names can contain up to 255 characters, and are not case-sensitive. Only the first 32 characters are displayed in the tab.

AutoCAD prompts:

> Enter layout to rename <default>: *(specify the layout name to rename)*
> Enter new layout name: *(specify a new layout name)*

Note: You can also rename an existing layout by choosing the **Rename** option from the shortcut menu that appears when you right-click after placing the cursor on an existing layout tab to be renamed. AutoCAD displays the Rename Layout dialog box. Specify a new layout name and choose the **OK** button.

Delete an Existing Layout

The **Delete** option deletes the selected layout. The most current layout is the default. AutoCAD prompts:

> Enter name of layout to delete <default>: *(press* ENTER *to accept the default layout to delete or specify the name of the layout to delete)*

Note: The Model tab cannot be deleted. To remove all the geometry from the Model tab, you must select all geometry and use the ERASE command. You can also delete layouts by choosing the **Delete** option from the shortcut menu that appears when you right-click after placing the cursor on an existing layout tab to be deleted.

Saving an Existing Layout to a Template

The **SAveas** option saves the layout to a drawing template (DWT) file. The last current layout is used as the default for the layout to save.

AutoCAD prompts:

> Enter layout to save as template <current>: *(press* ENTER *to accept the default or specify a layout name to save)*

AutoCAD displays the **Create Drawing File** dialog box. You can create a new drawing file or template drawing file with the layout. You can also save the layout to an existing drawing file or template drawing file.

Setting a Layout as Current

The **Set** option makes a layout current. AutoCAD prompts:

Enter layout to make current <last>: *(specify the name of the layout to make it current)*

You can also make a layout current by selecting the appropriate layout tab.

Listing of Layouts in the Current Drawing

The ? option lists all the layouts defined in the drawing.

CREATING FLOATING VIEWPORTS

Similar to creating tiled viewports in model space (see Chapter 3 for a detailed explanation), you can also create viewports in paper space called floating viewports. In paper space, you can create multiple, overlapping, contiguous, or separated untiled viewports as shown in Figure 8–24. By default, AutoCAD sets up one floating viewport in each layout. Consider viewports as objects with a view into model space that you can move and resize. AutoCAD treats viewports in layout like any other object, such as a line, arc, or text object. You can use any of the standard AutoCAD modify commands, such as MOVE, COPY, STRETCH, SCALE, and ERASE, to manipulate the floating viewports. For example, you can use the MOVE command to grab one viewport and move it around the screen without affecting other viewports. A viewport can be of any size and can be located anywhere in the layout. You must have at least one floating viewport to view the model.

Figure 8–24 *Floating viewports in a layout*

You can create or manipulate floating viewports only when you are in the paper space. You cannot edit the model when you manipulate the floating viewports. To edit the model you must switch to model space using one of the following methods:

- Choose the Model tab.
- Double-click over the floating viewport. On the status bar, Paper changes to Model.
- Click Paper on the status bar to return to the floating viewport that was last current.

When you are working in a layout, you can switch back and forth between model space and paper space. When you make a floating viewport in a layout current, you are then working in model space in a floating viewport. Any modification to the drawing in model space is reflected in all paper space viewports as well as tiled viewports. When you double-click outside a floating viewport, AutoCAD switches the mode to paper space. In paper space you can add annotations or other graphical objects, such as title block. Objects you add in paper space do not change the model or other layouts.

To create multiple floating viewports, open the Display Viewports Dialog box:

Layouts toolbar	Choose Display Viewports Dialog (see Figure 8–25)
View menu	Choose Viewports > New Viewports
Command: prompt	**vports** (ENTER)

Figure 8–25 *Invoking the Display Viewports Dialog box from the Layouts toolbar*

AutoCAD displays the Viewports dialog box, as shown in Figure 8–26.

Choose the name of the configuration you want to use from the **Standard viewports:** list. AutoCAD displays the corresponding configuration in the preview window. Specify the spacing between the viewports if you are creating multiple viewports in the **Viewport Spacing** text field. Select *2D* from the **Setup:** menu for *2D* viewport setup or select *3D* for *3D* viewport setup. Choose the **OK** button to create the selected viewport configuration.

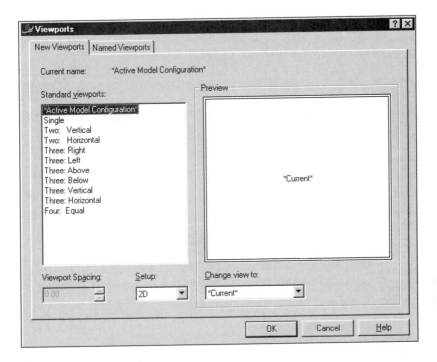

Figure 8–26 *Viewports dialog box*

AutoCAD prompts:

> Specify first corner or [Fit] <Fit>: *(specify the first corner to define selected viewport configuration or right-click for shortcut menu to select one of the available options)*
>
> Specify opposite corner: *(specify opposite corner to define selected viewport configuration)*

Select the **Fit** option to create the selected viewport configuration to fit the paper size.

You can also create floating viewports from the command version of the VPORTS command:

AutoCAD prompts:

> Command: **-vports** (ENTER)
> Specify corner of viewport or
>
> [ON/OFF/Fit/Hideplot/Lock/Object/Polygonal/Restore/2/3/4] <Fit>: *(specify the first corner to define a single viewport or right-click for shortcut menu to select one of the available options)*

If you specify the first corner to define a single viewport, AutoCAD prompts:

Specify opposite corner: *(specify the opposite corner to define the floating viewport)*

Fit Option

The **Fit** option (default) creates a single floating viewport to fit the paper size. The actual size of the viewport depends on the dimensions of the paper space view.

ON Option

The **ON** option turns on a viewport if it is set to OFF, making it active and making its objects visible. By default, all the new viewports are set to ON.

OFF Option

The OFF option turns off a viewport. When a viewport is off, its objects are not displayed, and you cannot make that viewport current. You can increase system performance by turning some viewports off or by limiting the number of active viewports. If you don't want to plot a viewport, you can turn that viewport off.

 Note: You can also turn a viewport on or off from the shortcut menu that appears when you right-click after selecting the viewport, as shown in Figure 8–27.

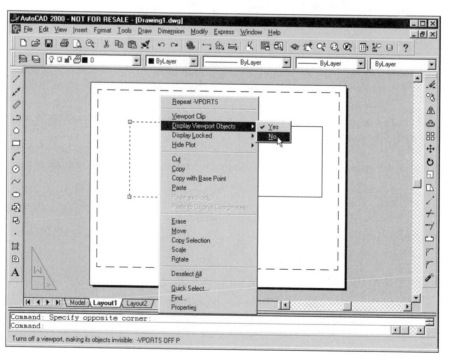

Figure 8–27 *Shortcut menu that appears after selecting a viewport*

Hideplot Option

The Hideplot option removes hidden lines from a viewport during plotting from paper space (layout).

 Note: You can also turn the hideplot option on or off for a viewport from the shortcut menu that appears when you right-click after selecting the viewport.

Lock Option

The Lock option locks selected viewports. This is similar to layer locking. AutoCAD allows you to scale views relative to paper space, which establishes a consistent scale for each displayed view. To accurately scale the plotted drawing, you must scale each view relative to paper space. If you zoom in to the geometry after setting the plot scale, you are changing the viewport scale at the same time. By locking the viewport, you can zoom in without altering the viewport scale. The zoom is performed in paper space, not in the viewport.

 Note: You can also turn the display lock on or off for a viewport from the shortcut menu that appears when you right-click after selecting the viewport.

Object Option

The **Object** option allows you to select a closed polyline, ellipse, spline, region, or circle to convert into a viewport. The polyline you specify must be closed and contain at least three vertices. It can be self-intersecting, and it can contain arcs as well as line segments.

When you select this option, AutoCAD prompts:

> Select object to clip viewport: *(select the object to convert into a floating viewport)*

AutoCAD creates a floating viewport from the selected object.

Polygonal Option

The **Polygonal** option creates an irregularly shaped viewport defined by specifying points.

When you select this option, AutoCAD prompts:

> Specify start point: *(specify start point to create an irregularly shaped viewport)*
> Specify next point or [Arc/Close/Length/Undo]: *(specify next point or choose an option)*

The available sub-options are the same as in the XCLIP command. Refer to Chapter 12 for a detailed explanation for all the available sub-options under the XCLIP command.

Restore Option

The Restore option allows you to display a saved viewport configuration. When you select this option, AutoCAD prompts:

> Enter viewport configuration name or [?]: *(Specify the name of the viewport configuration to restore)*

2 Option

The 2 option splits the current viewport in half. When you select this option, AutoCAD prompts:

> Enter viewport arrangement [Horizontal/Vertical] <Vertical>: *(select one of the available options)*
>
> Specify first corner or [Fit] <Fit>: *(specify first corner to define selected viewport configuration or right-click for shortcut menu and select one of the available options)*
>
> Specify opposite corner: *(specify opposite corner to define selected viewport configuration)*

Select the **Fit** option to create the selected viewport configuration to fit the paper size.

3 Option

The 3 option divides the current viewport into three viewports. When you select this option, AutoCAD prompts:

> Enter viewport arrangement [Horizontal/Vertical/Above/Below/Left/Right] <Right>: *(select one of the available options)*

You can select the Horizontal or Vertical option to split the current viewport into thirds by horizontal or vertical division. The other options let you split into two small ones and one large one, specifying whether the large one is to be placed above, below, or to the left or right, similar to creating tiled viewports.

4 Option

The 4 option divides the current viewport into four viewports of equal size both horizontally and vertically. When you select this option, AutoCAD prompts:

> Specify first corner or [Fit] <Fit>: *(specify first corner to define selected viewport configuration or right-click for shortcut menu)*
>
> Specify opposite corner: *(specify opposite corner to define selected viewport configuration)*

Select the **Fit** option to create the selected viewport configuration to fit the paper size.

CREATING A LAYOUT BY LAYOUT WIZARD

The Layout Wizard lets you create a new (paper space) layout for plotting. Each wizard page instructs you to specify different layout and plot settings for the new layout you are creating. Once the layout is created using the wizard, you can modify layout settings using the Page Setup dialog box.

To open the Layout Wizard, invoke the LAYOUTWIZARD command:

Insert menu	Choose Layout > Layout wizard
Command: prompt	**layoutwizard** (ENTER)

AutoCAD displays the Begin page of the Layout Wizard, as shown in Figure 8–28.

Figure 8–28 *Create Layout - Begin page*

Specify the name of the layout in the **Enter a name for the new layout you are creating** text field. Choose the **Next** button. AutoCAD displays the **Create Layout – Printer** page of the Layout Wizard, as shown in Figure 8–29.

Figure 8–29 *Create Layout - Printer page*

Select a configured plotter for the new layout from the list box. If you do not see the name of the plotter to which you want to plot, refer to the section on **Configuring Plotters** to configure a plotter. Once you have selected the plotter configuration, choose the **Next** button. AutoCAD displays the **Create Layout – Paper Size** page of the Layout Wizard, as shown in Figure 8–30.

Figure 8–30 *Create Layout - Paper Size page*

Select a paper size to be used for the layout from the list box. The paper sizes available are based on the plot device you selected. Select drawing units from one of the two radio buttons located in the **Drawing units** section. Choose the **Next** button and AutoCAD displays the **Create Layout – Orientation** page of the Layout Wizard, as shown in Figure 8–31.

Figure 8–31 *Create Layout - Orientation page*

Select the orientation of the drawing on the paper from one of the two radio buttons: Portrait or Landscape. Choose the **Next** button and AutoCAD displays the **Create Layout – Title Block** page of the Layout Wizard, as shown in Figure 8–32.

Figure 8–32 *Create Layout - Title Block page*

Select a title block from the list box to use for the layout. Select whether the title block will be inserted as a block or attached as an external reference. Choose the **Next** button and AutoCAD displays the **Create Layout – Define Viewports** page of the Layout Wizard, as shown in Figure 8–33.

Figure 8–33 *Create Layout - Define Viewports page*

Choose one of the following four available options in the Viewport section.

- None—if you do not need any floating viewports.
- Single—to create one floating viewport.
- Std. 3D Engineering Views—to create four viewports with top left set for top view, top right for isometric view, bottom left for front view, and bottom right for right side view. If necessary, you can specify the distance between the viewports in the **Spacing Between Rows:** and **Spacing Between Columns:** text fields.
- Array—to create an array of viewports. Specify the number of viewports in rows and columns in the **Rows:** and **Columns:** text fields. If necessary, you can specify the distance between the viewports in the **Spacing Between Rows:** and **Spacing Between Columns:** text fields.

Choose the **Next** button and AutoCAD displays the **Create Layout – Pick Location** page of the Layout Wizard, as shown in Figure 8–34.

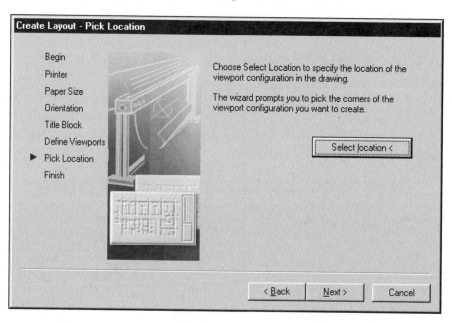

Figure 8–34 *Create Layout - Pick Location page*

Choose **Select location** button to specify the location of the viewport configuration in the drawing. The wizard prompts you to specify the corners of the viewport configuration you want to create. After specifying the location, AutoCAD displays the **Create Layout – Finish** page of the Layout Wizard, as shown in Figure 8–35.

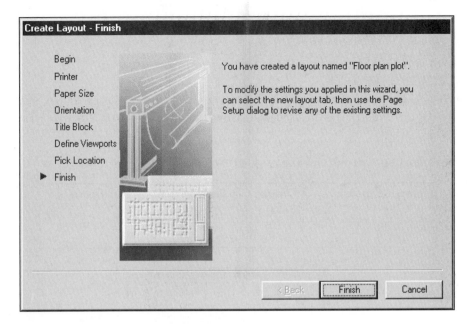

Figure 8–35 *Create Layout - Finish page*

Choose the **Finish** button to create the layout. If necessary, you can make any changes to the newly created layout by using the Page Setup dialog box.

SCALING VIEWS RELATIVE TO PAPER SPACE

AutoCAD allows you to scale views relative to paper space, which establishes a consistent scale for each displayed view. To accurately scale the plotted drawing, you must scale each view relative to paper space.

When you are working in paper space, the scale factor represents a ratio between the actual size of the model displayed in the viewports and the size of the layout. Usually the layout is plotted at a 1:1 ratio. The ratio is determined by dividing the paper space units by the model space units. The scale factor can be set with the ZOOM XP command. For example, specifying 1/24xp or 0.04167xp (1/24 = 0.01467) will display an image to a scale of 1/2" = 1'0", which is the same as 1:24 or 1/24. You can also change the plot scale of the viewport using the Viewport Scale Control on the Viewports toolbar (see Figure 8–36).

Figure 8–36 *Viewports toolbar*

Note: If necessary, after scaling the viewport you can lock the viewport with the -VPORTS LOCK command.

CLIPPING VIEWPORTS

The VPCLIP command allows you to clip a floating viewports to a user-drawn boundary. To clip a viewport, you can select an existing closed object, or specify the points of a new boundary.

To clip a floating viewport, invoke the VPCLIP command:

Viewports toolbar	Choose Clip Existing Viewport (see Figure 8–37)
Command: prompt	**vpclip** (ENTER)

Figure 8–37 *Invoking the* VPCLIP *command from the Viewports toolbar*

AutoCAD prompts:

> Select viewport to clip: *(select viewport to clip)*
> Select clipping object or [Polygonal/Delete] <Polygonal>: *(select clipping object or select one of the available options)*

If you select an object for clipping, AutoCAD converts the object into a clipping boundary. Objects that are valid as clipping boundaries include closed polylines, circles, ellipses, closed splines, and regions.

Polygonal Option

The Polygonal option allows you to create a clipping boundary. When you select the Polygonal option, AutoCAD prompts:

> Specify start point: *(specify start point for a clipping boundary)*

> Specify next point or [Arc/Close/Length/Undo]: *(specify next point or select one of the available options)*

Arc, Close, Length, and Undo sub-options are similar to the functionality of the options available for the PLINE command. Refer to Chapter 4 for a detailed explanation for the PLINE command.

Delete Option

The Delete option deletes the clipping boundary of a selected viewport. This option is available only if the selected viewport has already been clipped. If you clip a viewport that has been previously clipped, the original clipping boundary is deleted and the new clipping boundary is applied.

 Note: You can also use grips to create an irregularly shaped viewport just as you edit any object with grips.

CONTROLLING THE VISIBILITY OF LAYERS WITHIN VIEWPORTS

The VPLAYER (short for ViewPort LAYER) command controls the visibility of layers in a single viewport or in a set of viewports. This enables you to select a viewport and freeze a layer in it while still allowing the contents of that layer to appear in another viewport. See Figure 8–38, in which two viewports contain the same view of the drawing, but in one viewport the layer containing the dimensioning is on, and in another it is off, by using the VPLAYER command. To invoke the VPLAYER command, you must be working in layout (TILEMODE set to 0).

Figure 8–38 *One viewport with Dimlayer ON and another with Dimlayer OFF*

The VPLAYER command can be invoked from either model space or paper space (TILEMODE set to 0). Several options in the VPLAYER command require you to select one or more viewports in which to make your changes. AutoCAD prompts:

Enter an option [All/Select/Current] <Current>:

To accept the default option, you must be in model space; AutoCAD applies changes in the current viewport. If you choose the **Select** option and are in model space, AutoCAD temporarily switches to paper space so you can select a viewport. The All option applies your changes to all paper space viewports.

If you set TILEMODE to 1 (ON), the global layer settings take precedence over any VPLAYER settings.

Invoke the VPLAYER command:

Command: prompt	**vplayer** (ENTER)

AutoCAD prompts:

Command: **vplayer**
Enter an option [?/Freeze/Thaw/Reset/Newfrz/Vpvisdflt]: *(select one of the available options)*

? Option

The ? option displays the names of layers in a specific viewport that are frozen. When you select this option, AutoCAD prompts:

Select a viewport: *(select a viewport)*

If you are in model space, AutoCAD switches temporarily to paper space to let you select a viewport.

Freeze Option

The Freeze option allows you to specify one or more layers to freeze in the selected viewport. When you select this option, AutoCAD prompts:

Layer(s) to freeze:

You can respond to this prompt with a single layer name, a list of layer names separated by commas, or any valid wild-card specification. Then AutoCAD prompts:

Enter an option [All/Select/Current] <Current>:

Select the viewport(s) in which to freeze the selected layers.

Thaw Option

The Thaw option allows you to specify one or more layers to thaw that were frozen by the VPLAYER command in selected viewports. When you select this option, AutoCAD prompts:

Layer(s) to thaw:

You can respond to this prompt with a single layer name, a list of layer names separated by commas, or any valid wild-card specification. Then AutoCAD prompts:

Enter an option [All/Select/Current] <Current>:

Select the viewport(s) in which to thaw the selected layers.

Reset Option

The Reset option allows you to restore the default visibility setting for a layer in a given viewport. The default visibility is controlled by the Vpvisdflt option (explained later in this section). When you select this option, AutoCAD prompts:

Layer(s) to Reset:

You can respond to this prompt with a single layer name, a list of layer names separated by commas, or any valid wild-card specification. Then AutoCAD prompts:

Enter an option [All/Select/Current] <Current>:

Select the viewport(s) in which to reset the selected layers.

Newfrz (New Freeze) Option

The New Freeze option allows you to create new layers that are frozen in all viewports. If you create a new viewport, the layers that are created by the Newfrz option will be frozen by default. (The layer can be thawed in the chosen viewport by using the Thaw option.) When you select this option, AutoCAD prompts:

Enter name(s) of new layers frozen in all viewports:

You can respond to this prompt with a single layer name or a list of layer names separated by commas.

Vpvisdflt (Viewport Visibility Default) Option

The Viewport Visibility Default option allows you to set a default visibility for one or more existing layers. This default determines the frozen/thawed status of an existing layer in newly created viewports. When you select this option, AutoCAD prompts:

Enter layer name(s) to change viewport visibility:

You can respond to this prompt with a single layer name, a list of layer names separated by commas, or any wild-card specification. Then AutoCAD prompts:

Enter a viewport visibility option [Frozen/Thawed] <Thawed>:

You can respond to this prompt with a null response to set the default visibility to thaw, or enter F to set the default visibility to freeze.

You can also control the visibility of layers in viewports from the Layer Properties Manager dialog box. Invoke the LAYER command, and AutoCAD displays the Layer Properties Manager dialog box, similar to Figure 8–39.

The Active VP Freeze column (tenth column from the left) toggles between freezing and thawing of selected layer(s). Similarly, the New VP Freeze column (eleventh column from left) toggles between freezing and thawing of selected layer(s) for new viewports.

In Figure 8–39, the layers dim, OBJECT, and TEXT are frozen in the current viewport, and ELEVATION and HIDDEN are frozen in all the new viewports.

Figure 8–39 *Layer Properties Manager dialog box*

SETTING PAPER SPACE LINETYPE SCALING

Linetype dash lengths and the space lengths between dots or dashes are based on the drawing units of the model or paper space in which the objects were created. They can be scaled globally by the system variable LTSCALE factor, as explained earlier. If you want to display objects in viewports at different scales in layout (TILEMODE set to 0), the linetype objects would be scaled to model space rather than paper space by default. However, by setting paper space linetype scaling (system variable PSLTSCALE) to 1 (default), dash and space lengths are based on paper space drawing units, including the linetype objects that are drawn in model space. For example, a single linetype definition with a dash length of 0.30, displayed in several viewports with different zoom factors, would be displayed in paper space with dashes of length 0.30, regardless of the scale of the viewpoint in which it is being displayed (PSLTSCALE set to 1).

Invoke the PSLTSCALE command:

Command: prompt	**psltscale** (ENTER)

AutoCAD prompts:

> Command: **psltscale**
> Enter new value for PSLTSCALE <current>: *(specify the new value, or press* ENTER *to accept the current value)*

 Note: When you change the PSLTSCALE value to 1, the linetype objects in the viewport are not automatically regenerated. Use the REGEN or REGENALL command to update the linetypes in the viewports.

DIMENSIONING IN MODEL SPACE AND PAPER SPACE

Dimensioning can be done in both model space and paper space. There are no restrictions placed on the dimensioning commands by the current mode. It is advisable to draw associative dimensions in model space, since AutoCAD places the defining points of the dimension in the space where the dimension is drawn. If the model geometry is modified with a command such as STRETCH, EXTEND, or TRIM, the dimensions are updated automatically. In contrast, if the dimensions are drawn in paper space, the paper space dimension does not change if the model geometry is modified.

For dimensioning in model space, the DIMSCALE factor should be set to 0.0. This causes AutoCAD to compute a scale factor based on the scaling between paper space and the current model space viewport. If dimensioning that describes model geometry should be created in paper space, then the DIMLFAC dimension variable scale factor should be based on the model space viewport. It is important that the length scaling be set to a value that is appropriate for the view being dimensioned.

ATTACHING PLOT STYLE TABLES TO VIEWPORTS

Plot style tables are settings that give you control over how objects in your drawing are plotted into hard copy plots. You can control how colors are translated into line weight and how area fills are converted into shades of gray or screened colors, as well as many other output options. You can control how the plotter treats each individual object in a drawing. AutoCAD allows you to assign a plot style table to a layout. In addition, you can assign a separate plot style table to floating viewports overriding the assigned plot style table to the layout. To visualize the effect of the plot styles attached to your current layout or viewport, set **Display Plot Style** to ON in the Page Setup dialog box and then regenerate the drawing or preview the drawing by invoking the PRINT PRE-VIEW command. Refer to the section on "Creating and Modifying Plot Style Tables," later in this chapter, for a detailed explanation on creating and modifying a plot style table.

PLOTTING FROM LAYOUT

Invoke the PLOT command to plot the drawing in WYSIWYG (What You See Is What You Get). Before you plot the drawing from the layout, make sure the following tasks are completed:

Create a model drawing

Create or activate a layout

Open the page setup dialog box and set settings such as plotting device (if necessary configure plotting device), paper size, plot area, plot scale, and drawing orientation

If necessary, insert a title block or attach title block as a reference file

Create floating viewports and position them in the layout

Set the view scale of the floating viewports

Annotate or create geometry in the layout as needed.

Plot the layout

To plot the current drawing from the layout, invoke the PLOT command:

Standard toolbar	Choose the PLOT command (see Figure 8–40)
File menu	Choose Plot
Command: prompt	**plot** (ENTER)

Figure 8–40 *Invoking the* PLOT *command from the Standard toolbar*

AutoCAD displays the Plot dialog box, as shown in Figure 8–41.

Figure 8–41 *Plot dialog box – Plot Device tab*

AutoCAD displays the current layout name or displays "Selected layouts" if multiple tabs are selected. Set the **Save changes to layout** toggle button to ON to save changes you make in the Plot dialog box to the layout. This option is not available if multiple layouts are selected.

The **Page setup name:** list box displays a list of any saved page setups. Similar to plotting from model space, you can choose from one of the saved page setups to restore the plot settings. Or you can save the current settings to the named page setup by selecting the **Add** button and it can be applied later to plot any drawing from paper space.

Plot Device Settings

The Plot Device tab (see Figure 8–41) specifies the plotter to use, a plot style table, the layout or layouts to plot, and information about plotting to a file. The options that are available in the Plot Device tab are the same as those explained earlier in the section "Plotting From Model Space."

Plot Settings

The Plot Settings tab (see Figure 8–42) specifies paper size, drawing orientation, plot area and plot scale, plot offset, and plot options. The options that are available in the

604

Plot Settings tab are the same as those explained earlier in the section "Plotting From Model Space," except an additional option is provided in the **Plot area** called **Layout**. Whenever you want to plot the current layout, select the **Layout** radio button from the Plot area section. This causes AutoCAD to plot everything within the margins of the specified paper size, with the origin calculated from 0,0 in the layout. If you choose to turn off the Display paper background on the Display tab of the Options dialog box, the Layouts selection becomes Limits.

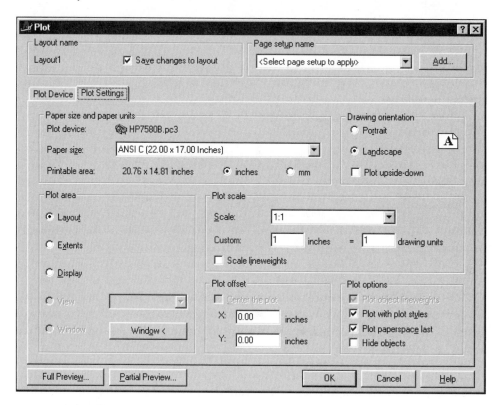

Figure 8–42 *Plot dialog box – Plot Settings tab*

Choose the **Partial Preview** button to quickly display a partial preview of the effective plot area relative to the paper size and printable area. In addition, AutoCAD provides information regarding paper size, printable area, and effective area. Partial preview also gives advance notice of any AutoCAD warnings that may be encountered when you plot the drawing.

A full preview displays the drawing on the screen as it would appear when plotted. This takes much longer than the partial preview, but it takes less time than regular plot regeneration. This is quicker than plotting incorrectly and having to start all over again.

To get a full preview of your plot, choose the **Full Preview** button. AutoCAD temporarily clears the plotting dialog boxes, draws an outline of the paper size, and displays the drawing as it will appear on the paper when it is plotted.

The cursor changes to a magnifying glass with plus and minus signs. Holding the pick button and dragging the cursor toward the top of the screen enlarges the preview image. Dragging it toward the bottom of the screen reduces the preview image. Right-click and AutoCAD displays a shortcut menu offering additional preview options: Pan, Zoom, Zoom Window, Zoom Original, Plot, and Exit.

To end the full preview, choose the Exit option from the shortcut menu. AutoCAD returns to the Plot dialog box.

 Note: You can also access a full preview by invoking the PLOT PREVIEW command from the pull-down menu File or the Print Preview icon on the Standard toolbar.

After making the necessary changes in the plot settings, choose the **OK** button. AutoCAD starts plotting and reports its progress as it converts the drawing into the plotter's graphics language by displaying the number of vectors processed.

If something goes wrong or if you want to stop immediately, choose the CANCEL button at any time. AutoCAD cancels the plotting.

CREATING AND MODIFYING PLOT STYLE TABLES

Plot style tables are settings that give you control over how objects in your drawing are plotted into hard copy plots. By modifying an object's plot style, you can override that object's color, linetype, and lineweight. You can also specify end, join, and fill styles, as well as output effects such as dithering, gray scale, pen assignment, and screening. You can use plot styles if you need to plot the same drawing in different ways.

By default, every object and layer has a plot style property. The actual characteristics of plot styles are defined in plot style tables that you can attach to a model tab and layouts within layouts. If you assign a plot style to an object, and then if you detach or delete the plot style table that defines the plot style, the plot style will not have any effect on the object.

AutoCAD provides two plot style modes: Color-Dependent and Named.

The Color-dependent plot styles are based on object color. There are 255 color-dependent plot styles. You cannot add, delete, or rename color-dependent plot styles. You can control the way all objects of the same color plot in color-dependent mode by adjusting the plot style that corresponds to that object color. Color-dependent plot style tables are stored in files with the extension .CTB.

Named plot styles work independently of object color. You can assign any plot style to any object regardless of that object's color. Named plot style tables are stored in files with the extension .STB.

By default, all the plot style table files are saved in the Acad2000\Plot styles folder.

The default plot style mode is set in the Plotting tab (Default plot style behavior for new drawings section) of the Options dialog box, as shown in Figure 8–43.

Figure 8–43 *Options dialog box (Plotting tab)*

Every time you start a new drawing in AutoCAD 2000, the plot style mode that is set in the Options dialog box is applied. Anytime you change the mode, it is applied only for the new drawings or an open drawing that has not yet been saved in AutoCAD 2000. You cannot change the mode for the current drawing.

CREATING A PLOT STYLE TABLE

AutoCAD 2000 allows you to create a named plot style table to utilize all the flexibility of named plot styles or a color-dependent plot style table to work in a color-based mode. The **Add Plot Style Table** Wizard allows you to create a plot style from scratch, modify an existing plot style table, import style properties from an acadr14.cfg file, or import style properties from an existing PCP or PC2 file.

To open the Add Plot Style Table Wizard:

Tools menu	Choose Wizards > Add Plot Style Table

AutoCAD displays the Introductory text of the Add Plot Style Table Wizard, as shown in Figure 8–44.

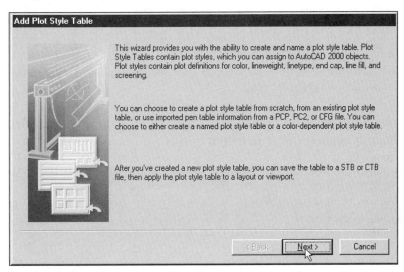

Figure 8–44 *Introductory text page of the Add Plot Style Table Wizard*

Choose the **Next** button, and AutoCAD displays the **Add Plot Style Table – Begin** page, as shown in Figure 8–45.

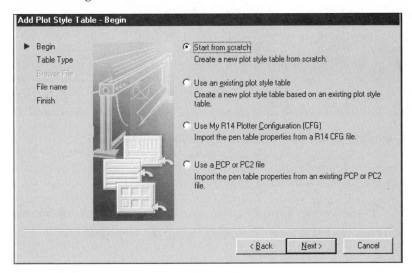

Figure 8–45 *Add Plot Style Table – Begin page*

Choose one of the following four radio buttons

- Start from Scratch—Allows you to create a new plot style from scratch.

- Use an Existing Plot Style Table—Creates a new plot style using an existing named plot style table.

- Use My R14 Plotter Configuration (CFG)—Creates a new plot style table using the pen assignments stored in the ACADR14.CFG file. Choose this option if you do not have an equivalent PCP or PC2 file.

- Use a PCP or PC2 File—Creates a new plot style table using pen assignments stored in a PCP or PC2 file

Choose the **Next** button, and AutoCAD displays the **Add Plot Style Table – Pick Plot Style Table** page, as shown in Figure 8–46.

Figure 8–46 *Add Plot Style Table – Pick Plot Style Table page*

Choose one of the following radio buttons:

- Color-Dependent Plot Style table—Creates a plot style table with 255 plot styles.

- Named Plot Style table—Creates a named plot style table.

Note: If you selected the **Use an existing style table** option to create a new plot style, AutoCAD displays the **Add Plot Style Table – Browse File Name** wizard. Specify the plot style table file name from which to create a new plot style name.

Choose the **Next** button, and AutoCAD displays the **Add Plot Style Table – File name** page, as shown in Figure 8–47.

Figure 8–47 *Add Plot Style Table – File name page*

Specify the file name in the **File name:** text field. By default the new style table is saved in the Acad2000/Plot Styles folder.

Choose the **Next** button, and AutoCAD displays **Add Plot Style Table – Finish** page, as shown in Figure 8–48.

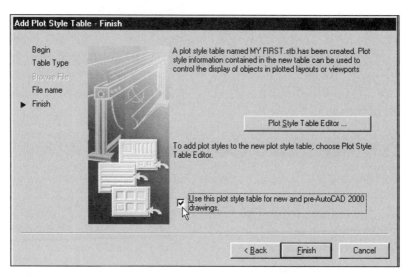

Figure 8–48 *Add Plot Style Table – Finish page*

Set the **Use this plot style table for new and pre-AutoCAD 2000 drawings** toggle button to ON to attach this plot style table to all new drawings and pre-AutoCAD 2000 drawings by default.

Choose the **Finish** button to create the plot style table and close the wizard.

MODIFYING A PLOT STYLE TABLE

AutoCAD allows you to add, delete, copy, paste, and modify plot styles in a plot style table by using the Plot Style Table Editor. You can open more than one instance of the Plot Style Table Editor at a time and copy and paste plot styles between the tables. Open the Plot Style Table Editor using any of the following methods:

- Choose Plot Style Table Editor button from the Finish screen in the Add Plot Style Table Wizard.

- Open the Plot Style Manager (File menu), right-click a CTB or STB file, and then choose Open from the shortcut menu.

- On the Plot Device tab of the Plot dialog box or Page Setup dialog box, select the plot style table you want to edit from the Plot Style Table list, and then choose the **Edit** button.

- On the Plotting tab of the Options dialog box, choose the Add or Edit Plot Style Tables button

Figure 8–49 shows an example of the Plot Style Table Editor for a named plot style table, and Figure 8–50 shows an example of the Plot Style Table Editor for a color-dependent plot style table.

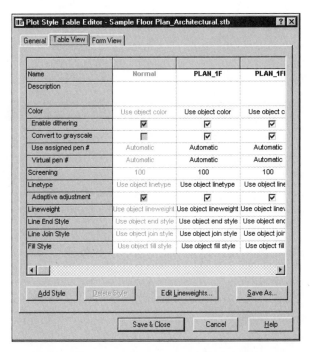

Figure 8–49 *Plot Style Table Editor for named plot style table (Table View)*

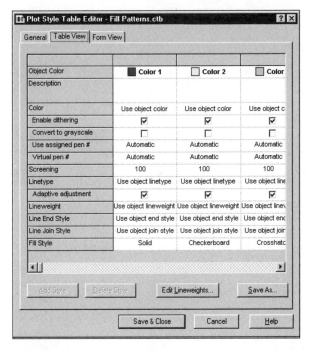

Figure 8–50 *Plot Style Table Editor for color-dependent plot style table (Table View)*

Following are the three tabs available in the Plot Style Table Editor:

- **General**—Displays the name of the plot style table, description (if any), location of the file, and version number (see Figure 8–51). You can modify the description, and apply scaling to non-ISO lines and to fill patterns.

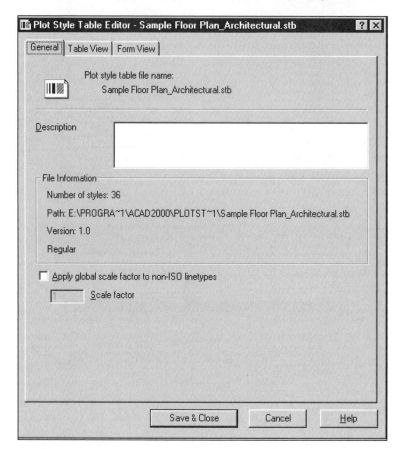

Figure 8–51 *Plot Style Table Editor – General tab*

- **Table View**—Lists entire plot styles in the plot style table and their settings in tabular form (see Figure 8–49 and Figure 8–50). The styles are displayed in columns from left to right. The setting names of each row appear at the left of the tab. By default, in the case of a named plot style table, AutoCAD sets up a style named Normal and represents an object's default properties. You cannot modify or delete the Normal style. In the case of a color-dependent plot style table, AutoCAD lists all the 255 color styles in tabular form. In general this is convenient if you have a small number of plot styles to view them in the tabular form.

- Form View—The plot style names are listed under the **Plot Styles:** list box and the settings for the selected plot style are displayed at the right side of the dialog box (see Figure 8–52).

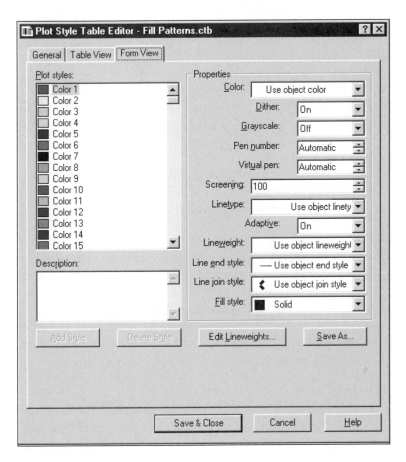

Figure 8–52 *Plot Style Table Editor – Form View*

Creating a New Plot Style

To create a new plot style, choose the **Add Style** button from thePlot Style Editor. AutoCAD adds a new style, and you can change the name to a descriptive name if necessary (cannot exceed 255 characters). You cannot duplicate names within the same plot style table.

Note: You cannot add or change the name for a plot style in the color-dependent style table.

Deleting a Pen Style

To delete a pen style, click the gray area above the plot style name in the Table View and choose the **Delete Style** button. In the Form View, select the style name from the **Plot styles:** list box and choose the **Delete Style** button.

 Note: You cannot delete a plot style from a color-dependent style table.

Adding and Modifying a Plot Style Description

AutoCAD allows you to specify a description for plot styles and modify an existing description for a plot style if necessary. The description cannot exceed 255 characters.

Assigning Plot Style Color

The **Color:** list box allows you to assign a plot style color. If you assign a color from one of the available colors, then AutoCAD overrides the object's color at plot time. By default, all of the plot styles are set to use object color.

Enabling Dithering

If you set Enable Dithering to ON, the plotter approximates colors with dot patterns, giving the impression of plotting with more colors than the ink available in the plotter. If you set the Enable Dithering to OFF, then AutoCAD maps colors to the nearest color, which limits the range of colors used for plotting. The most common reason for turning off dithering is to avoid false line typing from dithering of thin vectors and to make dim colors more visible. If the plotter does not support dithering, the dithering setting is ignored. The default setting is set to ON.

Converting to Gray Scale

If you set **Convert to grayscale** to ON, AutoCAD converts the object's colors to gray scale if the plotter supports gray scale. If you set the **Convert to grayscale** to OFF, AutoCAD uses the RGB values for the object's colors. The default setting is set to OFF.

Assigning Pen Numbers

The **Use assigned pen number** text field setting in the Plot Style Table Editor specifies which pen to use for each plot style. You can specify a pen to use in the plot style by selecting from a range of pen numbers from 1 to 32. If you specify 0, the field updates to read Automatic. AutoCAD uses the information you provided under Physical Pen Characteristics in the Plotter Configuration Editor to select the pen closest in color to the object you are plotting. The Default is set to Automatic.

Using Virtual Pens

Specify a virtual pen number in the **Virtual pen #** edit field for plotters that do not use pens but can simulate the performance of a pen plotter by using virtual pens. The default is set to Automatic or 0 to specify that AutoCAD should make the virtual pen assignment from the AutoCAD Color Index. You can specify a virtual pen number between 1 and 255. The virtual pen number setting in a plot style is used only by plotters without pens and only if they are configured for virtual pens. If this is the case, all the other style settings are ignored and only the virtual pen is used. The Default is set to Automatic.

 Note: If a plotter without pens is not configured for virtual pens, then both the virtual and the physical pen information in the plot style are ignored and all the other settings are used.

Using Screening

The **Screening** text field sets a color intensity setting that determines the amount of ink AutoCAD places on the paper while plotting. The valid range is 0 through 100. Selecting 0 reduces the color to white. Selecting 100 (default) displays the color at its full intensity. The default is set to 100%.

Assigning Plot Style Linetype

The **Linetype** list box allows you to assign a plot style linetype. If you assign a linetype from one of the available linetypes, then AutoCAD overrides the object's linetype at the plot time. By default, all the plot styles are set to use object linetype.

Adaptive Adjustment

The **Adaptive Adjustment** toggle adjusts the scale of the linetype to complete the linetype pattern. Set **Adaptive Adjustment** to ON if it is more important to have complete linetype patterns than correct linetype scaling. Set **Adaptive Adjustment** to OFF if linetype scale is important. The default is set to ON.

Assigning Plot Style Lineweight

The **Lineweight** list box allows you to assign a plot style lineweight. If you assign a lineweight from one of the available lineweights, then AutoCAD overrides the object's lineweight at the plot time. By default, all the plot styles are set to use object lineweight.

Assigning a Line End Style

The **Line End Style** list box allows you to assign a line end style. The line end style options include: Butt, Square, Round, and Diamond. If you assign a line end style from one of the available line end styles, then AutoCAD overrides the object's line end style at the plot time. By default, all the plot styles are set to use object end style.

Assigning a Line Join Style

The **Line Join Style** list box allows you to assign a line end style. The line join style options include: Miter, Bevel, Round, and Diamond. If you assign line join style from one of the available line join styles, then AutoCAD overrides the object's line join style at the plot time. By default, all the plot styles are set to use object line join style.

Assigning Fill Style

The **Fill Style** list box allows you to assign a fill style. The fill style options include: Solid, Checkerboard, Crosshatch, Diamonds, Horizontal Bars, Slant Left, Slant Right, Square Dots, and Vertical Bars. The fill style applies only to solids, plines, donuts, and *3D* faces. If you assign a fill style from one of the available fill styles, then AutoCAD overrides the object's fill style at the plot time. By default, all the plot styles are set to use object fill style.

Editing Lineweights

AutoCAD allows you edit the available lineweights by choosing the **Edit Lineweights** button. You cannot add or delete lineweights from the list.

Saving the Plot Style Table

To save the changes and close the Plot Style Table Editor, choose the **Save & Close** button. To save the changes to another plot style table, choose the **Save As** button. AutoCAD displays the Save As dialog box. Specify the file name in the **File name:** text field and choose the **Save** button to save and close the Save As dialog box.

CHANGING PLOT STYLE PROPERTY FOR AN OBJECT OR LAYER

As mentioned earlier, every object that is created in AutoCAD has a plot style property in addition to color, linetype, and lineweight. Similarly, every layer has a color, linetype, and lineweight, in addition to a plot style property. The default setting for plot styles for objects and layers are set in the Options (Plotting tab) dialog box, as shown in Figure 8–53.

The default plot style for objects can be any of the following:

- Normal—Uses the object's default properties.
- BYLAYER—Uses the properties of the layer that contains the object.
- BYBLOCK—Uses the properties of the block that contains the object.
- Named plot style—Uses the properties of the specific named plot style defined in the plot style table.

The default plot setting for an object is BYLAYER, and the initial plot style setting for a layer is Normal. When the object is plotted it retains its original properties.

If you are working in a Named plot style mode, you can change the plot style for an object or layer at any time. If you are working in a color dependent plot style mode, you cannot change the plot style for objects or layers. By default, they are set to **By Color**.

Figure 8–53 *Options (Plotting tab) dialog box*

To change the plot style for one or more objects, first select the objects (system variable PICKFIRST set to ON), and select the plot style from the **Plot Style control** from the Object Properties toolbar, as shown in Figure 8–54. If the plot style does not list the one you want to select, then choose **Other**, and AutoCAD displays the Select Plot Styles dialog box, as shown in Figure 8–55.

Figure 8–54 *Object Properties toolbar – Plot Style control*

Figure 8–55 *Select Plot Style dialog box*

Select the plot style you want to apply to the selected object(s) from the **Plot styles:** list box. If you need to select a plot style from a different plot style table, then select the plot style table from the **Active plot style table:** list box. AutoCAD lists all the available plot styles in the **Plot styles:** list box and selects the one you want to apply to the select object(s). Choose the **OK** button to close the dialog box. You can also change the plot style of the selected object(s) from the properties dialog box.

To change the plot style for a layer, open the **Layer Properties Manager** dialog box. Select the layer you want to change and select a plot style for the selected layer similar to changing color or linetype.

CONFIGURING PLOTTERS

Autodesk® Plotter Manager allows you to configure a local or network non-system plotter. In addition, you can also configure a Windows system printer with non-default settings. AutoCAD stores information about the media and plotting device in configured plot (PC3) files. The PC3 files are stored in the acad2000\plotters folder. Plot configurations are therefore portable and can be shared in an office or on a project. If you calibrate a plotter, the calibration information is stored in a plot model parameter (PMP) file that you can attach to any PC3 files you create for the calibrated plotter.

AutoCAD allows you to configure plotters for many devices and store multiple configurations for a single device. You can create several PC3 files with different output options for the same plotter. After you create a PC3 file, it's available in the list of plotter configuration names on the Plot Device tab of the Plot dialog box.

Open the Autodesk® Plotter Manager:

File menu	Choose Plotter Manager
Command: prompt	**plottermanager** (ENTER)

AutoCAD displays the Plotters window explorer, shown in Figure 8–56, listing all the plotters configured.

Figure 8–56 *Plotters window explorer (Windows 98 version)*

Double-click Add-A-Plotter Wizard, and AutoCAD displays the **Add Plotter – Introduction Page**, as shown in Figure 8–57.

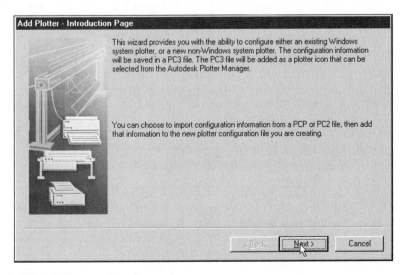

Figure 8–57 *Add Plotter – Introduction Page*

Choose the **Next** > button, and AutoCAD displays the **Add Plotter – Begin** page, as shown in Figure 8–58.

Figure 8–58 *Add Plotter – Begin page*

Choose one of the three radio buttons:

- My Computer—To configure a local non-system plotter.
- Network Plotter Server—To configure a plotter that is on the network.
- System Printer—To configure a Windows system printer. If you want to connect to a printer that is not in the list, you must first add the printer using the Windows Add Printer wizard in the Control Panel.

If you select the **My Computer** option, then the wizard prompts you to select a plotter manufacturer and model number, identify the port to which the plotter is connected, specify a unique plotter name, and choose the **Finish** button the close the wizard.

If you select the **Network Plotter Server** option, then the wizard prompts you to identify the network server, select the plotter manufacturer and model number, specify a unique plotter name, and choose the **Finish** button to close the wizard.

If you select the **System Printer** option, then the wizard prompts you to select one of the printers configured in the Windows operating system, specify a unique plotter name, and choose the **Finish** button to close the wizard.

AutoCAD saves the configuration file with PC3 file format with a unique given name in the acad2000\plotters folder (default).

If necessary, you can edit the PC3 file using the Plotter Configuration Editor. The Plotter Configuration Editor provides options for modifying a plotter's port connections and output settings including media, graphics, physical pen configuration, custom properties, initialization strings, calibration, and user-defined paper sizes. You can drag these options from one PC3 file to another.

You can open the Plotter Configuration Editor using one of the following methods:

- From the File menu, choose Page Setup. Choose Properties.

- From the File menu, choose Plot. Choose Properties.

- Double-click a PC3 file from Windows Explorer and right-click the file and choose Open. (By default, PC3 files are stored in the ACAD2000\Plotters folder.)

- Choose Edit Plotter Configuration on the Add Plotter - Finish page in the Add-a-Plotter wizard.

Figure 8–59 shows the Plotter Configuration Editor for a HP7580B plotter.

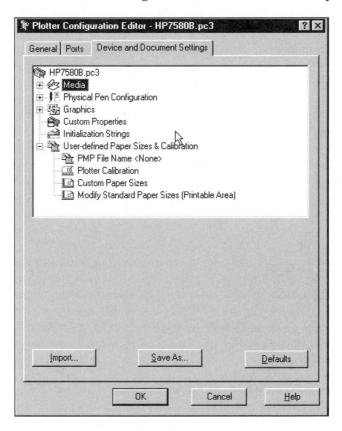

Figure 8–59 *Plotter Configuration Editor for a HP7580B plotter*

The Plotter Configuration Editor contains three tabs:

- General tab—Contains basic information about the configured plotter.
- Ports tab—Contains information about the communication between the plotting device and your computer.
- Device and Document Settings tab—Contains plotting options.

In the Device and Document Settings tab, you can change many of the settings in the configured plot (PC3) file. Following are the six areas in which you make the changes:

- Media—Specifies a paper source, size, type, and destination.
- Physical Pen Configuration—Specifies settings for pen plotters.
- Graphics—Specifies settings for printing vector graphics, raster graphics, and TrueType® fonts.
- Custom Properties—Displays settings related to the device driver.
- Initialization Strings—Sets pre-initialization, post-initialization, and termination printer strings.
- Calibration and User-Defined Paper Sizes—Attaches a plot model parameter (PMP) file to the PC3 file, calibrates the plotter, and adds, deletes, or revises custom or standard paper sizes.

The areas correspond to the categories of settings in the PC3 file you're editing. Double-click any of the six categories to view and change the specific settings. When you change a setting, your changes appear in angle brackets (<>) next to the setting name unless there is too much information to display. To save the changes to another PC3 file, choose the **Save As** button. AutoCAD displays the Save As dialog box. Specify the file name in the **File name:** text field and choose the **Save** button. To save the changes to the PC3 file and close the Plotter Configuration Editor, choose the OK button.

BATCH PLOTTING

In addition to plotting individual drawings, AutoCAD 2000 has a batch plotting utility that you can use to construct a list of AutoCAD drawings to be plotted. You can plot the drawings immediately or save them in a batch plot (BP3) file to plot later. The Batch Plot utility runs independently of AutoCAD and is available in the AutoCAD program group.

Once you have created a list of drawings to plot using the Batch Plot utility, you can attach any PC3 (plot configuration) file to each drawing. Any drawing not attached to a PC3 file plots to the device that was the default before you started the Batch Plot utility.

PROJECT EXERCISES

PROJECT 8–1

This project exercise provides point-by-point instructions for creating a plot style table, changing plot style for selected objects and plotting the drawing.

Step 1: Start the AutoCAD program.

Step 2: Open drawing PROJ2.DWG that was completed in the Chapter 2 project exercise.

Step 3: Activate Layout1 tab. By default, AutoCAD displays Page Setup dialog box similar to Figure P8–1.

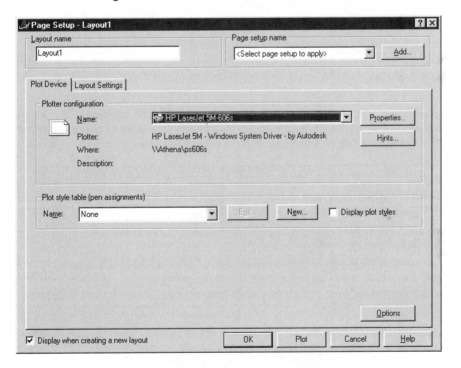

Figure P8–1 *Page Setup dialog box – Plot Device tab*

Step 4: From the Plot Device tab, select one of the available plotters in the **Plotter Configuration** section.

Step 5: Create a new plot style table by choosing the **New...** button from the Plot Style table section of the Page Setup dialog box. AutoCAD displays **Add Named Plot Style Table – Begin** wizard as shown in Figure P8–2.

Figure P8–2 *Add Named Plot Style Table – Begin wizard*

Select the **Start from scratch** radio button to create a new plot style table. Choose the **Next** button. AutoCAD displays the **Add Named Plot Style Table – File Name** wizard as shown in Figure P8–3.

Figure P8–3 *Add Named Plot Style Table – File Name wizard*

Enter **proj8-1** as the file name for the plot style table in the **File name:** edit field. Choose the **Next** button. AutoCAD displays **Add Named Plot Style Table – Finish** wizard as shown in Figure P8–4.

Figure P8–4 *Add Named Plot Style Table – Finish wizard*

To create new plot styles, choose the **Plot Style Table Editor** button. AutoCAD displays the Plot Style Table Editor as shown in Figure P8–5.

Figure P8–5 *Plot Style Table Editor*

Create four new styles, rename and assign color and lineweight for the newly created styles as listed in the following table:

Name of the style table	Color	Lineweight
COIL	RED	0.0059"
90 DEG ELL	YELLOW	0.0059"
DOOR JAMB	GREEN	0.0059"
INST PANEL	CYAN	0.0059"

Figure P8–6 shows the Plot Style Table Editor with newly created styles with the assigned color and lineweight.

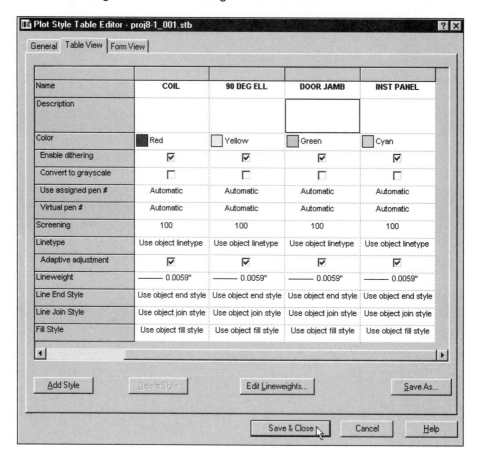

Figure P8–6 *Plot Style Table Editor with newly created plot styles.*

Choose the **Save & Close** button to save the newly created styles and close the Plot Style Table Editor. Choose the **Finish** button to close the Add Named Plot Style Table.

Step 6: Select the **Layout Settings** tab of the Page Setup dialog box and choose the paper size, drawing orientation, plot area and scale, plot offset, and plot options as shown in Figure P8–7.

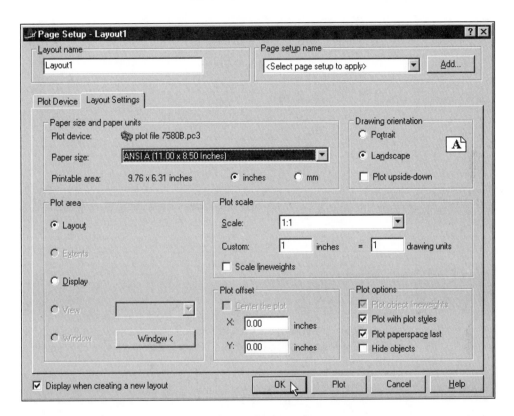

Figure P8–7 *Page Setup dialog box – Layout Settings tab.*

Choose the **OK** button to close the Page Setup dialog box. AutoCAD displays the drawing in the Layout1 as shown in Figure P8–8.

628

Figure P8–8 *Layout of the drawing in the Layout1.*

Step 7: Right-click on the Layout1 tab and select Rename from the shortcut menu. AutoCAD displays the **Rename Layout** dialog box as shown in Figure P8–9. Specify **Project 2 – Size A** in the **Name:** edit field and choose the **OK** button.

Figure P8–9 *Rename Layout dialog box.*

Step 8: Choose the **PAPER** button in the status bar to toggle to Model space.

Step 9: Select all the objects that comprise the Coil to change the plot style from the default plot style (ByLayer). Right-click and select Properties from the shortcut menu. AutoCAD displays Properties dialog box as shown in Figure P8–10.

Figure P8–10 *Properties dialog box*

Choose **Other** from the list box in the Plot Style section of the Properties dialog box. AutoCAD displays the Select Plot Style dialog box as shown in Figure P8–11.

Figure P8–11 *Select Plot Style dialog box*

Select **Coil** from the **Plot Styles:** list box and choose the **OK** button to close the dialog box. AutoCAD assigns Coil as the plot style for the selected objects. Clear the current selection of the objects.

Similarly, change the plot style for 90° ell, Door Jamb and Instrument Panel from bylayer to 90 deg ell, Door Jamb and Inst Panel plot style respectively.

Step 10: Choose the **MODEL** button in the status bar to toggle to Paper space.

Step 11: Invoke the PLOT command from the File menu. AutoCAD displays the Plot dialog box as shown in Figure P8-12. Choose the **Full Preview** button. AutoCAD displays the drawing with the applied plot styles. Right-click and select **Exit** from the shortcut menu. If necessary, make appropriate changes in the Plot dialog box. Choose the **OK** button to plot the drawing from the selected Layout.

Figure P8–12 *Plot dialog box*

AutoCAD starts plotting and reports its progress as it converts the drawing into the plotter's graphics language by displaying the number of vectors processed.

PROJECT 8-2

This project exercise provides point-by-point instructions for creating a plot style table, changing plot style for selected layers and plotting the drawing.

Step 1: Start the AutoCAD program.

Step 2: Open drawing PROJ6.DWG that was completed in the Chapter 6 project exercise.

Step 3: Activate Layout1 tab. By default, AutoCAD displays the Page Setup dialog box.

Step 4: From the Plot Device tab, select one of the available plotters that can plot on a Size D paper in the **Plotter configuration** section.

Step 5: Create a new plot style table by choosing the **New** button from the Plot Style table section of the Page Setup dialog box. AutoCAD displays the **Add Named Plot Style Table – Begin** wizard. Select the **Start from scratch** radio button to create a new plot style table. Choose the **Next** button. AutoCAD displays the **Add Named Plot Style Table – File Name** wizard. Specify **proj8–2** as the file name for the plot style table in the **File name:** edit field. Choose **Next** button. AutoCAD displays the **Add Named Plot Style Table – Finish** wizard.

To create new plot styles, choose the **Plot Style Table Editor** button. AutoCAD displays the Plot Style Table Editor.

Create four new styles, rename and assign color and lineweight for the newly created styles as listed in the following table:

Name of the style table	Color	Lineweight
BORDER	BLUE	0.012"
OBJECT	YELLOW	0.0059"
HIDDEN	GREEN	0.0059"
TEXT	CYAN	0.0059"

Figure P8–13 shows the Plot Style Table Editor with newly created styles with the assigned color and lineweight.

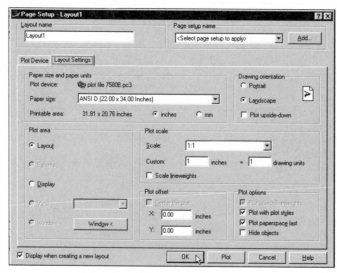

Figure P8–13 *Plot Style Table Editor with newly created plot styles.*

Choose the **Save & Close** button to save the newly created styles and close the Plot Style Table Editor. Choose the **Finish** button to close the Add Named Plot Style table.

Step 6: Select the **Layout Settings** tab of the Page Setup dialog box and choose paper size, drawing orientation, plot area and scale, plot offset, and plot options as shown in Figure P8–14.

Figure P8–14 *Page Setup dialog box – Layout Settings tab.*

Choose the **OK** button to close the Page Setup dialog box. AutoCAD by default creates a viewport and displays the drawing in the Layout1.

Step 7: Right-click on the Layout1 tab and select Rename from the shortcut menu. AutoCAD displays the **Rename Layout** dialog box. Specify **Project 6 – Size D** in the **Name:** text field and choose the **OK** button.

Step 8: Invoke the LAYER command. AutoCAD displays the Layer Properties Manager dialog box. Create a new layer and rename the layer to Viewports and assign color yellow. Set Viewports as the current layer. Change the plot style from the default to the one shown Figure P8–15 for Border, Object, Hidden, and Text layers.

Figure P8–15 *Layer Properties Manager dialog box.*

Choose the OK button to close the Layer Properties Manager dialog box.

Step 9: Invoke the ERASE command and delete the existing viewport. Invoke the -VPORTS command and create two new viewports. The first one to a size of 20 X 20 and the second to a size of 8 X 15 as shown in Figure P8–16.

Figure P8–16 *Location of the two viewports on a Size D paper.*

Step 10: Choose the **PAPER** button in the status bar to toggle to Model space.

Step 11: Activate the left viewport. Open the Viewports tool bar. To scale the drawing inside the floating viewport, select ½"=1' from Viewport Scale Control as shown in Figure P8–17.

Figure P8–17 *Viewports tool bar*

Activate the right viewport and set the Viewport Scale Control to ¾"=1'. Pan the drawing in the viewport so that it displays only the Laundry floor plan as shown in Figure P8–18.

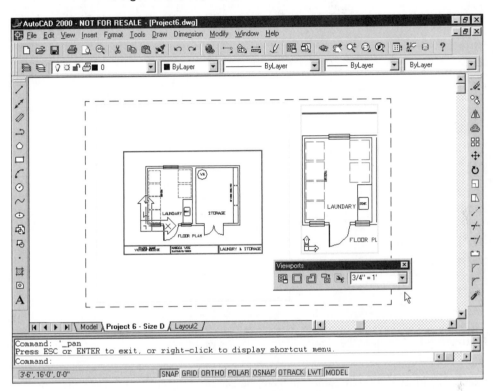

Figure P8–18 *Display of the drawing in the viewports*

Step 12: Choose the **MODEL** button in the status bar to toggle to Paper space. Set Layer 0 as the current layer and turn OFF the Viewports layer. The drawing will be displayed as shown in Figure P8–19.

636

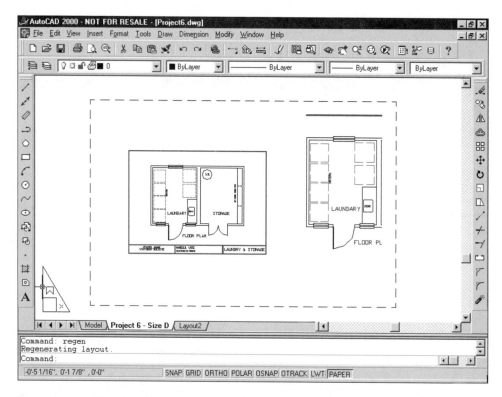

Figure P8–19 *Display of the drawing in the viewports with the Viewports layer set to OFF.*

Step 13: Invoke the PLOT command from the File menu. AutoCAD displays the Plot dialog box. Choose the **Full Preview** button. AutoCAD displays the drawing with the applied plot styles. Right-click and select **Exit** from the shortcut menu. If necessary, make appropriate changes in the Plot dialog box. Choose the **OK** button to plot the drawing from the selected Layout.

AutoCAD starts plotting and reports its progress as it converts the drawing into the plotter's graphics language by displaying the number of vectors processed.

REVIEW QUESTIONS

1. If you were to plot a drawing at a scale of 1"=60', what should you set LTSCALE to?

   a. 60

   b. 1/60

   c. 720

   d. 1/720

2. If you want to plot a drawing requiring multiple pens and you are using a single pen plotter, AutoCAD will:

   a. not plot the drawing at all

   b. pause when necessary to allow you to change pens

   c. invoke an error message

   d. plot all the drawing using the single pen

   e. none of the above

3. The drawing created at a scale of 1:1 and plotted to "Scaled to Fit" is plotted:

   a. at a scale of 1:1

   b. to fit the specified paper size

   c. at the prototype scale

   d. none of the above

4. To plot a full scale drawing at a scale of 1/4"=1', use a plot scale of:

   a. 0.25=12

   b. 0.25=1

   c. 48=1

   d. 12=0.25

   e. 24=1

5. What is the file extension assigned to all files created when plotting to a file?

   a. DWG

   b. DRW

   c. DRK

   d. PLO

   e. PLT

6. When plotting, pen numbers are assigned to:

a. colors

b. layers

c. thickness

d. linetypes

e. none of the above

You need to draw three orthographic views of an airplane whose dimensions are: wingspan of 102 feet, a total length of 118 feet, and a height of 39 feet. The drawing has to be done on a standard 12" x 9" sheet of paper. No dimensions will be added, so you will need only 1" between the views. Answer the following five questions using the information from this drawing:

7. What would be a reasonable scale for the paper plot?

a. 1=5'

b. 1=15'

c. 1=25'

d. 1=40'

8. What would be a reasonable setting of LTSCALE ?

a. 1

b. 5

c. 60

d. 25

e. 300

f. 480

9. If you were plotting from paperspace, what ZOOM scale factor would you use?

a. 1/5X

b. 1/25X

c. 1/60X

d. 1/300X

e. 1/5XP

f. 1/25XP

g. 1/60XP

h. 1/300XP

10. When inserting your border in paperspace, what scale factor should you use?

    a. 1   d. 60

    b. 5   e. 300

    c. 25

11. Which of the following options will the plot preview give you?

    a. seeing what portion of your drawing will be plotted

    b. seeing the plotted size of your drawing

    c. seeing the plotted drawing relative to the page size

    d. seeing rulers around the edge of the plotted page for size comparison

    e. none of the above

12. Which of the following determine the relationship between the size of the objects in a drawing and their sizes on a plotted copy?

    a. Size of the object in the AutoCAD drawing

    b. Size of the object on the plot

    c. Maximum available plot area

    d. Plot scale

    e. All of the above

13. AutoCAD 2000 permits plotting in which of the following environment modes?

    a. Model space

    b. Paper space

    c. layout

    d. all of the above

14. Which TILEMODE system variable setting corresponds to Model Space?

    a. 0

    b. 1

    c. either A or B

    d. none of the above

12. When starting a new drawing, how many default plotting layouts does AutoCAD 2000 create?

    a. 0   c. 2

    b. 1   d. unlimited

13. Within multiple floating viewports you can establish various scale and layer visibility settings for each individual viewport.

    a. True

    b. False

14. Paper sizes are indicated by X axis direction (drawing length) and Y axis direction (drawing width)?

    a. True

    b. False

15. Floating viewports, like lines, arcs, and text, can be manipulated using AutoCAD 2000 commands such as MOVE, COPY, STRETCH, SCALE or ERASE.

    a. True

    b. False

16. While in Paper Space both the floating viewports and the 3D model can be modified or edited.

    a. True

    b. False

17. Which of the following can be converted into a viewport?

    a. ellipse

    b. splines

    c. circles

    d. all of the above

18. Which of the following commands allows for the control of layer visibility in a specific viewport?

    a. VPORTS

    b. VPLAYER

    c. VIEWLAYER

    d. LAYERVIS

CHAPTER 9

Hatching and Boundaries

INTRODUCTION

After completing this chapter, you will be able to do the following:

- Use the BHATCH and HATCH commands.
- Fine-tune the hatching boundaries by means of the Advanced tab of the Boundary Hatch dialog box.
- Modify hatch patterns via the HATCHEDIT command.

WHAT IS HATCHING?

Drafters and designers use repeating patterns, called hatching, to fill regions in a drawing for various purposes (see Figure 9–1). In a cutaway (cross-sectional) view, hatch patterns help the viewer differentiate between components of an assembly and indicate the material of each. In surface views, hatch patterns depict material and add to the readability of the view. In general, hatch patterns greatly help the drafter/designer achieve his or her purpose, that is, communicating information. Because drawing hatch patterns is a repetitive task, it is an ideal application of computer-aided drafting.

You can use patterns that are supplied in an AutoCAD support file called ACAD.PAT, patterns in files available from third-party custom developers, or you can create your own custom hatch patterns. See Appendix E for the list of patterns supplied with ACAD.PAT.

AutoCAD allows you to fill with a solid color in addition to a hatch pattern. AutoCAD creates an associative hatch, which updates when its boundaries are modified, or a nonassociative hatch, which is independent of its boundaries. Before AutoCAD draws the hatch pattern, it allows you to preview the hatching and to adjust the definition if necessary.

CONCRETE/EARTH

GRATING

BRICK

Figure 9–1 *Examples of hatch patterns*

Hatch patterns are considered as separate objects. The hatch pattern behaves as one object; if necessary, you can separate it into individual objects with the EXPLODE command. Once it is separated into individual objects, the hatch pattern will no longer be associated with the boundary object.

Hatch patterns are stored with the drawing, so they can be updated, even if the pattern file containing the hatch is not available. You can control the display of the hatch pattern with the FILLMODE system variable. If FILLMODE is set to OFF, then the patterns are not displayed, and regeneration calculates only the hatch boundaries. By default, FILLMODE is set to ON.

The hatch pattern is drawn with respect to the current coordinate system, current elevation, current layer, color, linetype and current snap origin.

DEFINING THE HATCH BOUNDARY

A region of the drawing may be filled with a hatch pattern if it is enclosed by a boundary of connecting lines, circles, or arc objects. Overlapping boundary objects can be considered as terminating at their intersections with other boundary objects. There must not be any gaps between boundary objects, however. Figure 9–2 illustrates variations of objects and the potential boundaries that might be established from them.

Note in Figure 9–2 how the enclosed regions are defined by their respective boundaries. A boundary might include all or part of one or more objects. In addition to lines, circles, and arcs, boundary objects can include 2D and 3D polylines, 3D faces, and viewports. Boundary objects should be parallel to the current UCS. You can also hatch Blocks References that have been inserted with unequal X and Y scale factors.

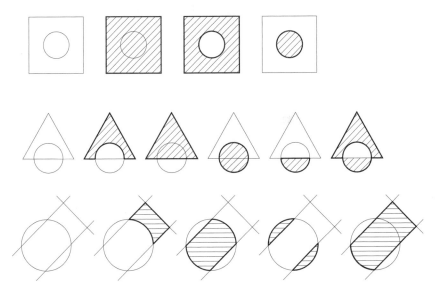

Figure 9–2 *Allowed hatching boundaries made from different objects*

BHATCH VERSUS HATCH

AutoCAD provides two commands for hatching: BHATCH and HATCH. The BHATCH command (introduced in AutoCAD Release 12) means Boundary Hatch and includes several features that greatly improve the ease of use over the HATCH command (introduced in Version 1.4). The BHATCH command automatically creates a boundary; conversely, with the HATCH command you must define a boundary or it allows you to create it. In Figure 9–3, the HATCH command selects the four lines via the Window option. These four lines comprise the hatching boundary. These four objects are valid boundary segments for use by the HATCH command. They connect at their endpoints and do not overlap. Instead of using the Window option of the object selection process, you can select the four lines individually. This may be desirable if there were unwanted objects were within a window used to select them.

In Figure 9–4, the BHATCH command permits you to select a point in the region enclosed by the four lines. Then, AutoCAD creates a polyline with vertices that coincide with the intersections of the lines. There is also an option that allows you to retain or discard the boundary when the hatching is complete.

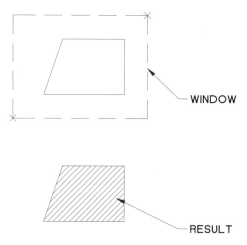

Figure 9–3 *The* HATCH *command requires the user to select objects to define the boundary*

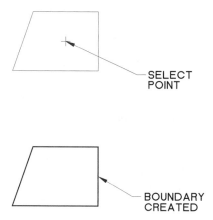

Figure 9–4 *The* BHATCH *command allows the user to pick a single point and automatically creates the boundary*

For example, if the four lines shown in Figure 9–3 had been segments of a closed polyline, then you could have selected that polyline by picking it with the cursor. Otherwise, all objects enclosing the region to be hatched must be selected when using the HATCH command, and those objects must be connected at their endpoints. For example, to use the HATCH command for the region in Figure 9–5, you would need to draw three lines (from 1 to 2, 2 to 3, and 3 to 4) and an arc from 4 to 1, select the four objects or connect them into a polyline, and select them (or it) to be the boundary.

The BHATCH command permits you to select a point in the region and have AutoCAD automatically create the needed polyline boundary. This ease of use and automation of the BHATCH command almost eliminates the need for the HATCH command except for rare, specialized applications.

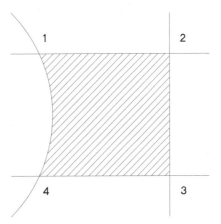

Figure 9–5 *A Region bounded by three lines and an arc*

 Note: Even though the Draw menu lists "Hatch" as the item to choose and the Draw toolbar Tooltip says "Hatch", they actually invoke the BHATCH command instead of the HATCH command. To invoke the HATCH command, you must enter **hatch** at the Command: prompt.

The dialog box used by the BHATCH command provides a variety of easy-to-select options, including a means to preview the hatching before completing the command. This saves time. Consider the variety of effects possible, such as areas to be hatched, angle, spacing between segments in a pattern, and even the pattern selected. The Preview option lets you make necessary changes without having to start over.

HATCH PATTERNS WITH THE BHATCH COMMAND

Invoke the BHATCH command:

Draw toolbar	Choose the HATCH command (see Figure 9–6)
Draw menu	Choose Hatch...
Command: prompt	**bhatch** (ENTER)

Figure 9–6 *Invoking the* BHATCH *command from the Draw toolbar*

AutoCAD displays the Boundary Hatch dialog box, similar to Figure 9–7.

Figure 9–7 *Boundary Hatch dialog box with Quick tab selected*

The **Type:** text box in the Quick tab section of the Boundary Hatch dialog box enables you to select the type of hatch pattern to be applied. There are three choices: Predefined, User-defined, or Custom.

Predefined Pattern

When the Predefined pattern type is chosen, the **Pattern:** text box lets you select a pattern from those defined in the ACAD.PAT file. Or you can select one of the available patterns by choosing the image tile located below the **Pattern:** text box. This causes the Hatch Pattern Palette dialog box to be displayed as shown in Figure 9–8. There are four tabs from which to select Predefined patterns: ANSI, ISO, Other Predefined, and Custom. Each tab displays icons representing that pattern. To select one of the patterns, choose it and then press the OK button or double-click on the icon. An example of the selected pattern is displayed in the **Swatch:** box in the Quick tab section of the dialog box. The selected pattern becomes the value of the HPNAME system variable.

In the **Scale:** and **Angle:** text boxes, you can change the scale and angle respectively. The angle 0 (zero) corresponds to the positive X axis of the current UCS and refers to the hatch pattern as shown in the Swatch box. The default scale is set to 1 and the default angle is set to 0 degrees. These settings can be changed to suit the desired appearance, as shown in Figure 9–9.

Figure 9–8 *Hatch Pattern Palette dialog box with the ANSI tab selected*

SCALE=1   SCALE=2   SCALE=1
ANGLE=0   ANGLE=0   ANGLE=90

Figure 9–9 *Hatch pattern with different scale and angle values*

The **ISO Pen Width:** list box allows you to specify ISO-related pattern scaling based on the selected pen width. The option is available only if a predefined ISO hatch pattern is selected.

The **Relative to Paper Space** check box allows you to scale the hatch pattern relative to the units in Paper Space. This can only be used in layout mode.

User-Defined Pattern

The User-defined pattern type allows you to define a simple pattern using the current linetype on the fly. You can specify a simple pattern of parallel lines or two groups of parallel lines (crossing at 90 degrees) at the spacing and angle desired. Specify the angle and spacing for the user defined pattern in the **Angle:** and **Spacing:** text boxes. To draw a second set of lines at 90 degrees to the original lines, set the **Double** check box to ON.

Custom Pattern

When the Custom pattern type is chosen, the **Custom Pattern:** text box lets you select a custom pattern from a .PAT file other than the ACAD.PAT file. Or you can select one of the available patterns by choosing the image tile located below **Custom Pattern:** text box.

Solid Fill

To create a solid fill in an enclosed area, select Solid pattern. The solid fill is drawn with the current color settings, and all pattern properties are disabled, such as scale, angle, and spacing.

The Hatching Boundary

AutoCAD provides two methods by which you can select the objects, which determines the boundary for drawing hatch patterns: Pick Points and Select Objects.

The **Pick Points** method determines a boundary from existing objects that form an enclosed area. When you choose the Pick Points button to specify a point the default hatching style is set to Normal style and the default Island Detection Method is set to Flood. The available styles, Normal, Outer, and Ignore, and available detection methods, Flood and Ray Casting, are discussed later in this chapter. It is very important to make sure the appropriate hatching style and detection methods are selected.

To invoke the Pick Points method, choose the Pick Points button in the dialog box. AutoCAD prompts:

> Select internal point: *(specify a point within the hatched area)*
> Select internal point: *(specify a point, enter* **u** *to undo the selection, or press* ENTER *to end point specification)*

See Figure 9–10 for an example of hatching by specifying a point inside a boundary.

The **Select Objects** method allows you to select specific objects for hatching. To invoke the Select Objects option, choose the Select Objects button in the dialog Box. AutoCAD prompts:

> Select objects: *(select the objects by one of the standard methods, and press* ENTER *to terminate object selection)*

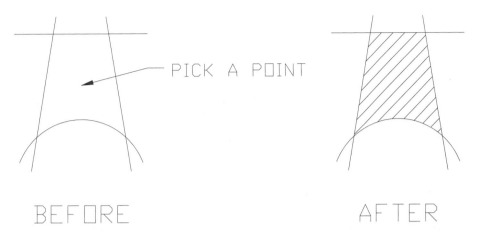

Figure 9–10 *Hatching by specifying a point*

Choosing the **View Selection** option causes AutoCAD to highlight the defined boundary set. If necessary, you can remove boundary set objects defined as a boundary set by the **Pick Points** option by choosing **Remove Islands** button. AutoCAD prompts you to select a boundary set that has to be removed from the defined boundary set. You cannot remove the outermost boundary.

The Select Objects option can be used to select an object, such as text, to cause AutoCAD not to hatch over the selected object and to leave a clear unhatched area around the text to provide better readability.

Caution must be observed when hatching over dimensioning. Dimensions are not affected by hatching as long as the DIMASO dimension variable (short for "associative dimensioning") is set to ON when the hatching is created and the dimension has not been exploded. The DIMASO system variable toggles between associative and nonassociatvie dimensioning. If the dimensions are drawn with DIMASO set to OFF (or exploded into individual objects), then the lines (dimension and extension) have an unpredictable (and undesirable) effect on the hatching pattern. Therefore, selecting in this case should be done by specifying the individual objects on the screen.

Blocks are hatched as though they are separate objects. Note, however, that when you select a block, all objects that make up the block are selected as part of the group to be considered for hatching.

If the selected items include text, shape, and/or attribute objects, AutoCAD does not hatch through these items if identified in the selection process. AutoCAD leaves an unhatched area around the text objects so they can be clearly viewed, as shown in Figure 9–11. Using the Ignore style will negate this feature so that the hatching is not interrupted when passing through the text, shape, and attribute objects.

Figure 9–11 *Hatching in an area where there is text*

 Note: When you select objects individually (after selecting the Select Objects button from the Boundary Hatch dialog box), AutoCAD no longer automatically creates a closed border. Therefore, any objects selected that will be part of the desired border must be either connected at their endpoints or a closed polyline.

When a filled solid or trace with width is selected in a group to be hatched, AutoCAD does not hatch inside that solid or trace. However, the hatching stops right at the outline of the filled object, leaving no clear space around the object as it does around text, shape, and attributes.

The **Composition** section of the Boundary Hatch dialog box controls whether the hatch is associative or nonassociative. Associative selection associates the hatch pattern element with the objects they pattern. For example, if the object is stretched, the hatch pattern expands to fill the new size. Nonassociative selections do not associate hatch patterns with the object they pattern. Figure 9–12 shows an examples of associative and nonassociative hatch patterns.

Figure 9–12 *Example of the associative/nonassociative hatch pattern*

To apply the hatch pattern settings, such as pattern type, pattern angle, pattern scale, from an existing associative pattern to another area to be hatched, choose the **Inherit Properties** button. AutoCAD prompts:

> Select associative hatch object: *(select an associative hatch pattern)*
> Select internal point: *(specify a point inside a closed area to hatch)*
> Select internal point: *(specify a point inside a closed area to hatch or press*
>   ENTER *to complete the selection)*

 Note: AutoCAD does not allow the properties of a nonassociative hatch pattern to be inherited.

To preview the hatch pattern for the selected objects, choose the **Preview** button. AutoCAD displays the currently defined boundaries with the current hatch settings. After previewing the hatch, press ENTER to return to the dialog box. If necessary, make any changes to the settings, choose **OK** button to apply the hatch pattern, or choose **Cancel** to disregard the selection.

Before you choose **OK** button, you can fine-tune additional parameters related to the boundary set if necessary; these are accessible through the Advanced tab of the Boundary Hatch dialog box (see Figure 9–13).

Figure 9–13 *Boundary Hatch dialog box with Advanced tab selected*

Advanced Settings

The **Island Detection Style** section of the Advanced Settings specifies whether objects within the outermost boundary are used as boundary objects. These internal objects are known as islands. You can select one of the three available options: Normal, Outer, and Ignore. The **Normal** style causes AutoCAD to hatch between alternate areas, starting with the outermost area. The **Outer** style hatches only the outermost area. The **Ignore** style causes AutoCAD to hatch the entire area enclosed by the outermost boundary, regardless of how you select the object, as long as its outermost objects comprise a closed polygon and are joined at their endpoints.

For example, in Figure 9–14, specifying point P1 in response to the Pick Points option results in hatching for Normal style, as shown in the upper right; Outer style, as shown in the lower left; and Ignore style, as shown in the lower right.

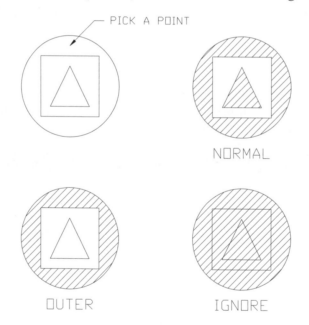

Figure 9–14 *Examples of hatching by specifying a point for Normal, Outer, and Ignore styles*

The **Object Type** section of the Advanced tab specifies whether to retain boundaries as objects and allows you to select the object type AutoCAD applies to those objects. The **Retain Boundaries** check box specifies whether the area to be hatched is to be retained. If it is to be retained then the adjacent selection box lets you select between whether the area is to be determined by a Polyline boundary or by a Region.

The **Boundary Set** section of the Advanced tab allows you to select a set of objects (called a boundary set) that AutoCAD analyzes when defining a boundary from a specified point. The selected boundary set applies only when you use the Pick Points option to create a boundary to draw hatch patterns. By default, when you use Pick Points to define a boundary, AutoCAD analyzes all objects visible in the current viewport. To create a new boundary set, choose **New** button in the Boundary Set section, and AutoCAD clears the dialog box and returns you to the drawing area to select objects from which a new boundary set will be defined. AutoCAD creates a new boundary set from those objects selected that are hatchable; existing boundary sets are abandoned. If hatchable objects are selected, they remain as a boundary set until you define a new one or exit the BHATCH command. Defining a boundary set will be helpful when you are working on a drawing that has too many objects to analyze to create a boundary for hatching.

It is important to distinguish between a boundary set and a boundary. A boundary set is the group of objects from which AutoCAD creates a boundary. As explained earlier, a boundary set is defined by selecting objects in a manner similar to selecting objects for some modify commands. The objects (or parts of them) in the group are used in the subsequent boundary. A boundary is created by AutoCAD after it has analyzed the objects (the boundary set) you have selected. It is the boundary that determines where the hatching begins and ends. The boundary consists of line/arc segments, which can be considered to be a closed polygon with segments that connect at their endpoints. If objects in the boundary set overlap, then in creating the boundary AutoCAD uses only the parts of objects in the boundary set that lie between intersections with other objects in the boundary set.

The **Island detection method** section of the Advanced tab specifies whether to include objects (called islands) within the outermost boundary as boundary objects. Choose the **Flood** radio button to include islands within the selected boundary. The hatching is performed depending on the selection of the Island detection style (Normal, Outer, and Ignore). If the **Ray Casting** radio button is chosen instead, then AutoCAD draws an imaginary line from the selected point to the nearest object and traces the boundary in the counterclockwise direction. If it cannot trace a closed boundary, AutoCAD displays an error message and will not be able to hatch.

For example, in Figure 9–15, point A is valid and point B is not when the Flood radio button is chosen. The object nearest to point A is the line that is part of a potential boundary (the square) of which point A is inside and AutoCAD considers the square as the hatch boundary. Conversely, point B is nearest a line that is part of a potential boundary (the triangle) of which point B is outside and AutoCAD displays an error message with the selected point as outside the boundary.

After making necessary adjustments in the Advanced section of the Boundary Hatch dialog box, proceed with the selection of the objects as explained earlier to create the hatch patterns.

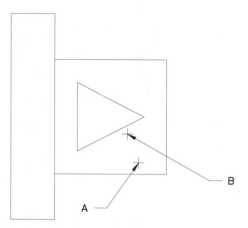

Figure 9–15 *Selecting points for Hatching*

HATCH PATTERNS WITH THE HATCH COMMAND

The HATCH command creates a nonassociative hatch or fill. As mentioned earlier, if the boundary is made of multiple objects, then their endpoints must coincide for the hatch to be created properly with the HATCH command. AutoCAD allows you to create a polyline hatch boundary with the HATCH command if you do not have a closed boundary for drawing the hatch pattern.

Invoke the HATCH command:

Command: prompt	**hatch** (ENTER)

AutoCAD prompts:

> Enter a pattern name or [?/Solid/User defined]: *(specify a predefined or custom pattern name, or right-click for the shortcut menu and select one of the avilable options)*

When you specify a predefined or custom pattern name, AutoCAD prompts:

> Specify a scale for the pattern <default>: *(specify the scale for the selected pattern)*
>
> Specify an angle for the pattern <default>: *(specify the angle for the selected pattern)*
>
> Select objects to define hatch boundary or <direct hatch>: *(select the objects to draw the hatch pattern and press the ENTER to terminate object selection)*

AutoCAD draws the hatch pattern to the selected boundary objects.

Instead of selecting the objects, if you press the ENTER (null response) in response to the "Select objects:" prompt, AutoCAD allows you to draw a polyline boundary with a combination of lines and arcs for drawing the hatch pattern. The prompt sequences are as follows:

> Command: **hatch**
> Enter a pattern name or [?/Solid/User defined]: *(specify a predefined or custom pattern name)*
> Specify a scale for the pattern <default>: *(specify the scale for the selected pattern)*
> Specify an angle for the pattern <default>: *(specify the angle for the selected pattern)*
> Select objects to define hatch boundary or <direct hatch>: (ENTER)
> Retain polyline boundary? [Yes/No] <N>: *(press ENTER to discard the hatching boundary, or enter **y** to retain the hatching boundary)*
> Specify start point: *(specify a start point for the polyline boundary)*
> Specify next point or [Arc/Close/Length/Undo]: *(specify a point for the polyline boundary, or select an option)*

The options are similar to those for the PLINE command (see Chapter 4). When you've completed the polyline boundary, the HATCH command prompts you to create additional polyline boundaries. Once you have completed creating polyline boundaries, press ENTER (null response), and AutoCAD draws the selected hatch patterns.

When you specify the pattern name, you can select one of the three styles explained earlier. To invoke the Normal style, just type the name of the style. To invoke the Outermost style, type the name of the pattern followed by a comma (,) and the letter o. To invoke the Ignore style, type the name of the pattern followed by a comma (,) and the letter i.

? Option

The ? option lists and provides a brief description of the hatch patterns defined in the ACAD.PAT file.

Solid

The Solid option specifies a solid fill. As in drawing a hatch pattern, here also you can select objects to define a boundary or draw a polyline boundary with a combination of lines and arcs for a solid fill. The boundary of a solid fill must be closed and must not intersect itself. In addition, if the hatch area contains more than one loop, the loops must not intersect.

User-Defined

The User-defined option draws a pattern of lines using the current linetype. You can specify a simple pattern of parallel lines or two groups of parallel lines (crossing at 90 degrees) at the specified spacing and angle. The prompt sequences are as follows:

```
Command: hatch
Enter a pattern name or [?/Solid/User defined] : (specify u for user defined)
Specify angle for crosshatch lines <current>: (specify the angle for
    crosshatch lines)
Specify spacing between the lines <current>: (specify the spacing for
    crosshatch lines)
Double hatch area? [Yes/No} <current>: (specify y for a second set of lines
    to be drawn at 90 degrees or n to draw simple patter of parallel lines)
Select objects to define hatch boundary or <direct hatch>: (select objects or
    press ENTER to define a polyline boundary)
```

As in drawing a hatch pattern, here also you can select objects to define a boundary or draw a polyline boundary with a combination of lines and arcs for a user-defined hatch pattern.

Note: The selection of objects for hatching must be done with awareness of how each object will affect or be affected by the HATCH command. Complex hatching of large areas can be time-consuming. Forgetting to select a vital object can change the whole effect of hatching. You can terminate hatching before it is completed by pressing the ESC key.

Hatching Base Point and Angle

Different areas hatched with the same (or a similar) pattern at the same scale and angle have corresponding lines lined up with each other in adjacent areas. This is because the families of lines were defined in the pattern(s) with the same basepoint and angle, no matter where the areas to be filled are in the drawing. This causes hatching lines to line up in adjacent hatched areas. But if you wish to offset the lines in adjacent areas, make the basepoint in one of the areas different from the basepoint in the adjacent area. Change the snap basepoint by either using the SNAP command and the Rotation option or by changing the SNAPBASE system variable. This is also useful for improving the appearance of hatching in any one area.

Changing the SNAPANG system variable or the base angle (from the SNAP/Rotate command/option) affects the angles of lines in a hatching pattern. This capability is also possible when responding to the HATCH command's "Specify angle for crosshatch lines <default>:" prompt.

Multiple Hatching

When you have finished hatching an area and have pressed the ENTER to repeat the HATCH command, AutoCAD prompts only for the objects to be selected. The optional parameters of pattern, mode, scale, and angle remain unchanged. In order to change any options, you must invoke the HATCH command by entering **hatch** at the "Command:" prompt. AutoCAD resumes prompting for the hatch options.

EDITING HATCHES

The HATCHEDIT command allows you to modify hatch patterns or choose a new pattern for an existing hatch. In addition, it allows you to change the pattern style of an existing pattern.

Invoke the HATCHEDIT command:

Modify II toolbar	Choose the EDIT HATCH command (see Figure 9–16)
Modify menu	Choose Object > Hatch...
Command: prompt	**hatchedit** (ENTER)

Figure 9–16 *Invoking the* HATCHEDIT *command from the Modify II toolbar*

AutoCAD prompts:

Select associative hatch object: *(select a hatch object)*

AutoCAD displays the Hatch Edit dialog box, similar to Figure 9–17.

Figure 9–17 *Hatch Edit dialog box*

Select one of the three options, Predefined, User-defined, or Custom, from the **Type:** list box. To change the existing pattern to a new one, select one of the predefined patterns by choosing the image tile located in the **Hatch Pattern Palette** section of the dialog box. You can also select a new pattern from the **Pattern:** list box located in the Pattern Properties section of the dialog box.

If necessary, you can change the Scale and Angle of the hatch pattern in the **Scale:** and **Angle:** text boxes. To inherit properties of an existing hatch pattern, select the **Inherit Properties** button, and then select an associative hatch pattern. The **Associative** check box controls whether or not selected hatching will be associative. If this option is selected, the modified hatch pattern is associative. If necessary, you can make changes in the settings in the Advanced section of the Hatch Edit dialog box.

Once the necessary changes are made in the Hatch Edit dialog box, choose the **OK** button to modify the selected hatch pattern.

When selecting a solid fill to modify, pick an outer edge of the hatch pattern with your pointing device or select it with the Crossing Window selection over the solid fill.

CONTROLLING THE VISIBILITY OF HATCH PATTERNS

The FILL command controls the visibility of hatch patterns in addition to the filling of multilines, traces, solids, and wide polylines.

Invoke the fill command:

Command: prompt	**fill** (ENTER)

AutoCAD prompts:

> Command: **fill**
> Enter mode [ON/OFF] <current>: *(specify **ON** to display the hatch pattern and **OFF** to turn off the display of hatch pattern)*

You have to invoke the REGEN command after changing the setting of fill to see the effect.

PROJECT EXERCISE

This project exercise provides point-by-point instructions for the objects shown in Figure P9–1. In this exercise you will apply the skills acquired in Chapters 1 through 9.

Figure P9–1 *Completed project drawing*

In this project you will do the following:

- Set up the drawing, including Limits, Units, and Layers.
- Use the LINE, PLINE, and BHATCH commands to create objects.
- Use the ARRAY command to create objects from existing objects.
- Use the FILLET and TRIM commands to modify objects.

SET UP THE DRAWING

Step 1: Start the AutoCAD program.

Step 2: To create a new drawing, invoke the NEW command from the Standard toolbar or select New from the File menu.

AutoCAD displays the Create New Drawing dialog box. Select the Start from Scratch button, and then set the Units, Limits, Snap, and Grid to the values shown in the SETTINGS/VALUES table.

Step 3: Invoke the LAYER command from the Object Properties toolbar, or select Layer from the Format menu. AutoCAD displays the Layer Properties Manager dialog box. Create 10 layers, and rename them as shown in the table, and assign appropriate colors and linetypes.

SETTINGS	VALUE		
UNITS	Architectural		
LIMITS	Lower left corner: -2'-0",-2'-6"		
	Upper right corner: 4'-0",3'-6"		
GRID	1"		
SNAP	1/2"		
LAYERS	NAME	COLOR	LINETYPE
	Border	Cyan	Continuous
	Slab	White	Continuous
	Text	Blue	Continuous
	Brick section	White	Continuous
	Concrete	White	Continuous
	Earth	White	Continuous
	Hatch boundary	Red	Continuous
	Insulation	Green	Continuous
	Sand	White	Continuous
	Wall	White	Continuous

HATCH PATTERNS	NAME	LAYER	SCALE	ANGLE
	AR-CONC	Concrete	0'-0 1/2"	0
	FLEX	Insulation	0'-3"	45
	ANSI31	Brick section	0'-6"	0
	EARTH	Earth	0'-9"	45
	AR-SAND	Sand	0'-0 1/2"	0

DRAWING THE CONCRETE SLAB, BRICK WALL, AND FRAME WALL

Step 4: Set Slab as the current layer. Invoke the PLINE command with a width of 0.125 to draw the concrete perimeter and a width of 0.0 to draw the "break" line, as shown in Figure P9–2. Figure P9–1 shows all the required dimensions to draw the layout. After you draw the layout, the drawing should look like Figure P9–2.

Figure P9–2 *Layout of the drawing*

Step 5: Set Wall as the current layer. Invoke the PLINE command to draw the bricks. The bottom brick is 2-3/8" x 3-1/2" and is 3/8" above the brick ledge.

Step 6: In order to create the arc that represents the mortar joints, invoke the CIRCLE command with the Three point option. Select the corners of the concrete and the brick that the desired arc touches, and then drag the cursor to a point that gives you a radius slightly larger than the joint gap. Then, using the TRIM command with the concrete and the brick as the objects to trim to, remove the outer part of the circle, leaving the desired arc. Invoke the MIRROR command to create an arc on the opposite edge of the brick, using the Midpoint Osnap on the bottom line of the brick as the basepoint about which to mirror. The completed drawing should look like Figure P9–3.

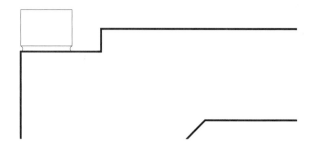

Figure P9–3 *Layout of a brick to draw the wall*

Step 7: Invoke the ARRAY command to duplicate the bricks and mortar joints and create a stack of 10 bricks. Draw a "break" line through the top brick, and use the TRIM command to trim off any brick extending past the "break" line, as shown in Figure P9–4.

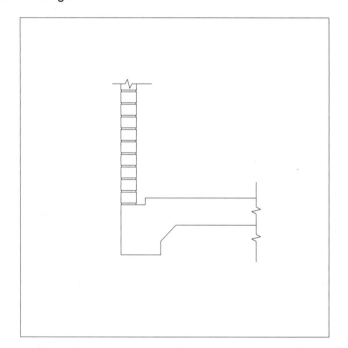

Figure P9–4 *Layout of the wall*

Step 8: Invoke the LINE command to draw the 1/2" gypsum board and the 1/2" exterior sheathing comprising the frame wall. The 2 x 4 sole plate is 1-1/2" thick and is shown by crossed lines. The heavy flashing at the bottom can be drawn with the PLINE command, with 0.125 width; then use the FILLET command to round the corner, as shown in Figure P9–5.

Figure P9–5 *Layout of the gypsum board and exterior sheathing*

Step 9: Set Hatchboundary as the current layer. Draw copies of the two faces of the bottom brick and the two bottom arcs representing the mortar joint. Set the Wall layer to OFF. The layout should look like Figure P9–6.

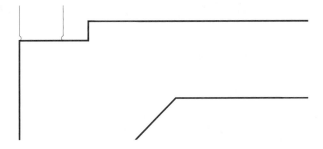

Figure P9–6 *Layout of the brick and mortar joint*

Step 10: Invoke the ARRAY command to duplicate the faces of the bricks and mortar joints into a stack of 10 double faces and double arcs. Set the Wall layer to ON and, using the "break" line through the top brick, invoke the TRIM command to trim off the brick faces extending past the "break" line. Invoke the PLINE command and trace over the parts of the "break" line between the brick faces. Set the Wall layer to OFF. The layout should look like Figure P9–7.

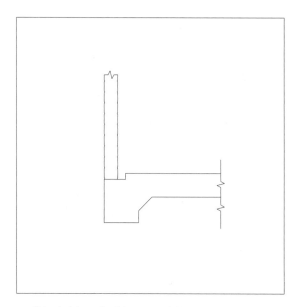

Figure P9–7 *Layout of the brick wall with mortar joint*

Step 11: Set Bricksection as the current layer. Invoke the BHATCH command and, using the ANSI31 pattern at a scale of 6 and a hatch angle of zero (0), select a point in the boundary of brick faces and mortar arcs just created. Invoke the LINE command, and draw the necessary lines as shown in Figure P9–8 to complete the boundaries for the Sand and Earth hatch patterns.

Figure P9–8 *Hatch pattern for brick faces and mortar arcs*

Step 12: Set Concrete as the current layer. Invoke the BHATCH command and, using the AR-CONC pattern at a scale of 1/2 and a hatch angle of zero (0), select a point in the boundary representing the concrete slab.

Step 13: Set Sand as the current layer. Invoke the BHATCH command and, using the AR-SAND pattern at a scale of 1/2 and a hatch angle of zero (0), select a point in the boundary representing the sand beneath the concrete slab.

Step 14: Set Earth as the current layer. Invoke the BHATCH command and, using the EARTH pattern at a scale of 9 and a hatch angle of 45, select a point in the boundary representing the earth beneath the sand and concrete slab.

Step 15: Set Insulation as the current layer. Invoke the BHATCH command and, using the FLEX pattern at a scale of 3 and a hatch angle of 45, select a point in the boundary representing the frame wall between the exterior and interior sheathings. Set the Hatchboundary layer to OFF. The layout should look as shown in Figure P9–9.

Figure P9–9 *Layout of the exterior wall section*

Step 16: Set Text as the current layer. Invoke the LEADER command, and draw the text object with the leader lines as shown in Figure P9–10.

Figure P9–10 *Layout of the exterior wall section with leader lines*

Step 17: Set Dimension as the current layer. Draw all the required dimensioning to complete the exterior wall section as shown in Figure P9–11.

Figure P9–11 *Layout of the exterior wall section with leader lines and dimensions*

EXERCISES 9–1 THROUGH 9–6

EXERCISE 9–1

Create the drawing shown in accordance with the settings given in the following table:

Settings	Value			
1. Units	Decimal			
2. Limits	Lower left corner: 0",0"			
	Upper right corner: 12",9"			
3. Grid	0.5			
4. Snap	0.25			
5. Layers	*NAME*	*COLOR*	*LINETYPE*	
	Construction	Cyan	Continuous	
	Object	White	Continuous	
	Center	Green	Center	
6. Hatch Patterns	*NAME*	*LAYER*	*SCALE*	*ANGLE*
	ANSI31	Hatch	1.0	0
	ANSI31	Hatch	1.0	90
	ANSI32	Hatch	1.0	67.5

Hints: Use the PLINE command to draw half the object without the rounded corners. Grid lines are spaced at 0.5 in the figure.

Use the FILLET command to round the corners of the object.

Use the MIRROR command to complete the object.

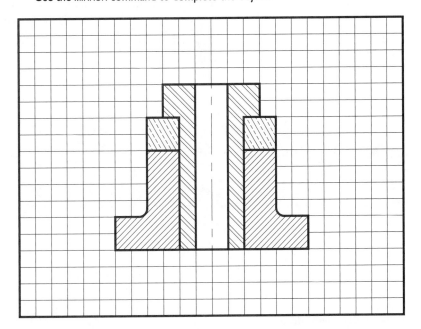

Figure Ex 9–1 *Completed project drawing (with grid added for reference)*

EXERCISE 9–2

Create the drawing shown, including dimensions, in accordance with the settings given in the following table:

Settings	Value			
1. Units	Architectural			
2. Limits	Lower left corner: 0,0			
	Upper right corner: 48',32'			
3. Grid	6"			
4. Snap	12"			
5. Layers	*NAME*	*COLOR*	*LINETYPE*	
	Border	Cyan	Continuous	
	Object	White	Continuous	
	Text	Blue	Continuous	
	Brick	White	Continuous	
	Roof	White	Continuous	
6. Hatch Patterns	*NAME*	*LAYER*	*SCALE**	*ANGLE*
	BRICK	Brick	16"	0
	AR-RSHKE	Roof	1/2"	0
	* Using a value of 16.0 adjusts the brick hatch pattern for a plotscale of 1=32 or 3/8"=1'-0" plot, yielding a 4" x 8" brick face			

Hint: Using the dimensions in Figure Ex9–2b, draw the front elevation of the house. The main concern about being able to draw hatch patterns within the proper boundaries is to make sure the boundaries are closed. Overlapping at the intersections of the segments that make up the boundary is acceptable. But AutoCAD will not complete the hatch pattern if there is a gap between any of the boundary segments.

FRONT ELEVATION

Figure Ex9–2a *Completed drawing*

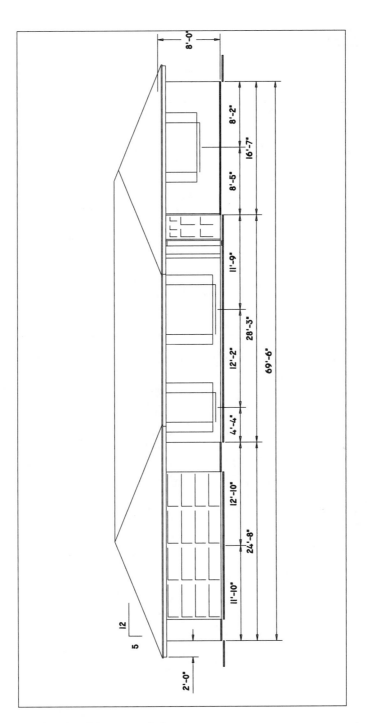

Figure Ex9–2b *Completed drawing with dimensions*

EXERCISE 9–3

Create the drawing shown, including dimensions, in accordance with the settings given in the following table:

Settings	Value			
1. Units	Decimal			
2. Limits	Lower left corner: 0,0			
	Upper right corner: 17,11			
3. Grid	0.5			
4. Snap	0.25			
5. Layers	*NAME*	*COLOR*	*LINETYPE*	
	Object	Green	Continuous	
	Center	Red	Center	
	Text	Blue	Continuous	
	Hatch	White	Continuous	
6. Hatch Pattern	*NAME*	*LAYER*	*SCALE**	*ANGLE*
	ANSI31	Hatch	1.0	0
	* Using a value of 1.0 adjusts the hatch pattern for a plotscale of 1=1 or Full scale plot			

 Hint: All three areas can be selected at once, thus creating a single hatch object.

Figure Ex9–3 *Completed project drawing*

EXERCISE 9–4

Create the drawing shown, including dimensions, in accordance with the settings given in the following table:

Settings	Value			
1. Units	Decimal			
2. Limits	Lower left corner: 0,0			
	Upper right corner: 17,11			
3. Grid	0.5			
4. Snap	0.25			
5. Layers	*NAME*	*COLOR*	*LINETYPE*	
	Object	Green	Continuous	
	Center	Red	Center	
	Hidden	Magenta	Hidden	
	Text	Blue	Continuous	
	Hatch	White	Continuous	
6. Hatch Pattern	*NAME*	*LAYER*	*SCALE**	*ANGLE*
		ANSI31	Hatch	1.00
	* Using a value of 1.0 adjusts the hatch pattern for a plot scale of 1=1 or Full scale plot			

Figure Ex9–4 *Completed drawing*

EXERCISE 9–5

Create the drawing shown, including dimensions, in accordance with the settings given in the following table:

Settings	Value			
1. Units	Decimal			
2. Limits	Lower left corner: 0,0			
	Upper right corner: 9,12			
3. Grid	0.1			
4. Snap	0.1			
5. Layers	*NAME*	*COLOR*	*LINETYPE*	
	Object	Green	Continuous	
	Center	Red	Center	
	Hidden	Magenta	Hidden	
	Text	Blue	Continuous	
	Hatch1	White	Continuous	
	Hatch2	White	Continuous	
	Knurl	White	Continuous	
6. Hatch Patterns	*NAME*	*LAYER*	*SCALE**	*ANGLE*
	ANSI31	Hatch1	1.0	0
	ANSI31	Hatch2	1.0	90
	ANSI37	Knurl	1.0	15
	* Using a value of 1.0 adjusts the hatch pattern for a plot scale of 1=1 or Full scale plot			

Hints: The angle pattern areas can use the same pattern, ANSI31, but with the hatch angles 90 degrees from each other.

The knurled pattern needs added lines to close the boundary. These can be drawn on a separate layer. That layer can be set to OFF for viewing and plotting the drawing.

0.3126

1/32" X 1/32"
CHAMFER
(TYP.)

BEAM
COMPASS
POINTER

NO. 10-FINE
THREAD

.5962

0.3125

4.4063   .4999

.2187

1/4 DIA.
NOM. SIZE
24" LONG

SECTION

Ø.375
Ø.139
Ø.1078

1.0938   .4522

.375

.8125

0.156
0.375

VERY SHARP
(.005 MAX. R.)

Figure Ex9–5 *Completed drawing*

EXERCISE 9–6

Create the drawing shown, including dimensions, in accordance with the settings given in the following table:

Settings	Value			
1. Units	Decimal			
2. Limits	Lower left corner: 0,0			
	Upper right corner: 17,11			
3. Grid	0.5			
4. Snap	0.25			
5. Layers	*NAME*	*COLOR*	*LINETYPE*	
	Object	Green	Continuous	
	Center	Red	Center	
	Hidden	Magenta	Hidden	
	Text	Blue	Continuous	
	Hatch1	White	Continuous	
6. Hatch Pattern	*NAME*	*LAYER*	*SCALE**	*ANGLE*
	ANSI31	Hatch	1.0	0
	* Using a value of 1.0 adjusts the hatch pattern for a plot scale of 1=1 or Full scale plot			

Figure Ex9–6 *Completed drawing*

REVIEW QUESTIONS

1. AutoCAD will ignore text within a crosshatching boundary.

   a. True   b. False

2. The BHATCH command allows you to create an associative hatch pattern which updates when its boundaries are modified.

   a. True   b. False

3. By default, hatch patterns are drawn at a 45 degree angle.

   a. True   b. False

4. All of the following may be used as boundaries of the HATCH command, except:

   a. ARC   d. CIRCLE

   b. LINE   e. PLINE

   c. BLOCK

5. The following are all valid AutoCAD commands except:

   a. ANGLE

   b. POLYGON

   c. BHATCH

   d. ELLIPSE

   e. MULTIPLE

6. When using the BHATCH command with a named hatch pattern, one can change

   a. the color and scale of the pattern

   b. the angle and scale of the pattern

   c. the angle and linetype of the pattern

   d. the color and linetype of the pattern

   e. the color and angle of the pattern

7. Boundary hatch patterns inserted with an asterisk "*" preceding the name of the pattern will:

   a. exclude inside objects

   b. ignore inside objects

   c. be inserted as individual objects

   d. be inserted on layer 0

   e. none of the above

8. The AutoCAD hatch feature:

   a. provides a selection of numerous hatch patterns

   b. allows you to change the color and linetype

   c. hatches over the top of text when the text is contained inside the boundary

   d. all of the above

9. The BHATCH command will allow you to create a polyline around the area being hatched and to retain that polyline upon completion of the command.

   a. True   b. False

10. Hatch patterns created with the BHATCH command can be nonassociative and associative.

    a. True   b. False

11. The HATCH command will place a hatch pattern over any text contained within the hatch boundary.

    a. True   b. False

12. If you insert a hatch pattern with an astrick (*) preceding the name, the pattern will be inserted as individual objects.

    a. True   b. False

13. The _____ command automatically defines the nearest boudary surrounding a point you have specified.

    a. HATCH   c. BOUNDARY

    b. BHATCH   d. PTHATCH

14. The BHATCH and Hatch Edit dialog boxes look the same.

    a. True   b. False

15. What default file does AutoCAD use to load hatch patterns from?

    a. ACAD.mnu   c. ACAD.dwg

    b. ACAD.pat   d. ACAD.hat

15. You can terminate the HATCH command before applying the hatch pattern by pressing the ESC.

    a. True   b. False

16. AutoCAD allows the properties of a nonassociative hatch pattern to be inherited.

    a. True   b. False

CHAPTER 10

Block References and Attributes

INTRODUCTION

The AutoCAD BLOCK command feature is a powerful design/drafting tool. The BLOCK command enables a designer to create an object from one or more objects, save it under a user-specified name, and later place it back into the drawing. When block references are inserted in the drawing they can be scaled up or down in both or either of the X and Y axes. They can also be rotated as they are inserted in the drawing. Block references can best be compared with their manual drafting counterpart, the template. Even though an inserted block reference can be created from more than one object, the block reference acts as a single unit when operated on by certain modify commands, like MOVE, COPY, ERASE, ROTATE, ARRAY, and MIRROR. You can export a block reference to become a drawing file outside the current drawing and create a symbol library from which block references are inserted into other drawings. Like the plastic template, block references greatly reduce repetitious work.

The BLOCK command can save time because you don't have to draw the same object(s) more than once. Block references save computer storage because the computer needs to store the object descriptions only once. When inserting block references, you can change the scale and/or proportions of the original object(s).

After completing this chapter, you will be able to do the following:

- Create and insert block references in a drawing
- Convert individual block references into drawing files
- Define attributes, edit attributes, and control the display of attributes
- Use the DIVIDE and MEASURE commands

CREATING BLOCKS

When you invoke the BLOCK command to create a block, AutoCAD refers to this as defining the block. The resulting definition is stored in the drawing database. The same block can be inserted as a block reference as many times as needed.

Blocks may comprise one or more objects. The first step in creating blocks is to create a block definition. In order to do this, the objects that make up the block must be visible on the screen. That is, the objects that will make up the block definition must have already been drawn so you can select them when prompted to do so during the BLOCK command.

The layer the objects comprising the block are on is very important. Objects that are on layer 0 when the block is created will assume the color, linetype, and lineweight of the layer on which the block reference is inserted. Objects on any layer other than 0 when included in the block definition will retain the characteristics of that layer, even when the block reference is inserted on a different layer. See Figure 10–1 for an example.

Figure 10–1 *Example of inserting block references drawn in different layers*

You should be careful when invoking the PROPERTIES command to change the color, linetype, or lineweight of elements of a block reference. It is best to keep the color, linetype, and lineweight of block references and the objects that comprise them in the BYLAYER state.

Examples of some common uses of blocks in various disciplines are shown in Figure 10–2.

Figure 10–2 *Examples of common uses of blocks in various disciplines*

CREATING A BLOCK DEFINITION

The BLOCK command creates a block definition for selected objects. Invoke the BLOCK command:

Draw toolbar	Choose the MAKE BLOCK command (see Figure 10–3)
Draw menu	Choose Block > Make
Command: prompt	**block** (ENTER)

Figure 10–3 *Invoking the* BLOCK *command from the Draw toolbar*

AutoCAD displays the Block Definition dialog box, similar to the Figure 10–4.

Figure 10–4 *Block Definition dialog box*

Block Name

Specify the block name in the **Name:** text box. The block name can be up to 255 characters long and may contain letters, numbers, and the special characters $ (dollar), - (hyphen), and _ (underscore). To list the block names in the current drawing, click the down arrow to the right of the **Name:** text box. AutoCAD lists the blocks in the current drawing.

Base Point

In the Base Point section, you can specify the insertion point for the block.

The insertion point specified during the creation of the block becomes the basepoint for future insertions of this block as a block reference. It is also the point about which the block reference can be rotated or scaled during insertion. When determining where to locate the base insertion point it is important to consider what will be on the drawing before you insert the block reference. Therefore, you must anticipate this preinsertion state of the drawing. It is sometimes more convenient for the insertion point to be somewhere off of the object than on it.

You can specify the insertion point on the screen, or you can specify *X*, *Y*, and *Z* coordinates of the insertion point in the **X:**, **Y:**, and **Z:** text boxes, respectively, located in the Base Point section of the Block Definition dialog box. To specify the basepoint on the screen, choose the **Pick point** button located in the Base point section of the dialog box. AutoCAD prompts:

Specify insertion point: *(specify the insertion point)*

Once you have specified the insertion point, the Block Definition dialog box reappears.

Selecting Objects

To select objects to include in the block definition, choose the **Select Objects** button located in the Objects section of the Block Definition dialog box. AutoCAD prompts:

Select objects: *(select objects using one of the AutoCAD object selection methods, and press* ENTER *to complete object selection)*

Once the objects are selected, the Block Definition dialog box reappears. You can also use the **Quick Select** option to define the selection of objects by clicking the icon to the right of **Select Objects** in the Objects section of the Block Definition dialog box. Choose one of the three radio buttons located in the Objects section to specify whether to retain or delete the selected objects or convert them to a block reference after you create the block.

If the **Retain** radio button is chosen, then the objects selected to be included in the block definition will not be deleted from the drawing after the block definition is created. They will remain in place as separate objects.

If the **Convert to Block** radio button is chosen, then the block definition created from the selected objects will become a block reference inserted into the drawing at the location that the block definition was created.

If the **Delete** radio button is chosen, then the objects selected to be included in the block definition will be deleted from the drawing after the block definition is created.

Preview Icon

In the Preview Icon section, if the **Create Icon From Block Geometry** radio button is chosen, then an icon will be displayed in the Preview Icon area on the right. The icon will simulate the appearance of the block definition. If the **Do Not Include An Icon** radio button is chosen, then no icon will be displayed.

Insert Units

Select one of the available units from the **Insert Units:** option menu. If the block is dragged from the AutoCAD DesignCenter (for detailed explanation on DesignCenter, see Chapter 12), it will be scaled to the units in the Insert Units list box.

Description

The **Description** text box can be used to enter a description of the block.

Choose the **OK** button to create the block definition with the given name. If the given name is the same as an existing block in the current drawing, AutoCAD displays a warning like that shown in Figure 10–5.

Figure 10–5 *Warning dialog box regarding block definition*

To redefine the block, choose the **Yes** button in the Warning dialog box. The block with that same name is then redefined. Once the drawing is regenerated, any insertion of this block reference on the drawing redefines to the new block definition with this name.

Choose the **No** button in the Warning dialog box to cancel the block definition. Then, to create a new block definition, specify a different block name in the **Name:** text box of the Block Definition dialog box and choose **OK**.

You can also create a block definition from selected objects by invoking the BLOCK command preceded with a hyphen from the Command: prompt, as shown in the following command sequence:

> Command: **-block** (ENTER)
> Enter block name or [?]: *(specify a block name, or enter ? to list the block names in the current drawing)*
> Specify insertion base point: *(specify the insertion base point)*
> Select objects: *(select the objects to include in the block definition, and press ENTER to complete object selection)*

AutoCAD creates the block with the given name and then erases the objects that make up the definition from the screen. You can restore the objects by entering the OOPS command immediately after the BLOCK command.

INSERTING BLOCK REFERENCES

You can insert previously defined blocks into the current drawing by invoking the INSERT command. If there is no block definition with the specified name in the current drawing, AutoCAD searches the drives and directories on the path for a drawing of that name and inserts it instead.

Note: If blocks were created and stored in a template drawing, and you make your new drawing equal to the template, those blocks will be in the new drawing ready to insert. Any drawing inserted into the current drawing will bring with it all of its block definitions, whether they have been inserted or are only stored as definitions.

Invoke the INSERT command:

Draw toolbar	Choose the INSERT BLOCK command (see Figure 10–6)
Insert menu	Choose Insert
Command: prompt	**insert** (ENTER)

Figure 10–6 *Invoking the* INSERT *command from the Draw toolbar*

AutoCAD displays the Insert dialog box, similar to Figure 10–7.

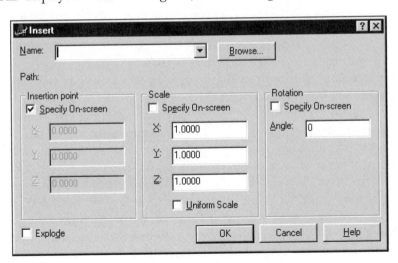

Figure 10–7 *Insert dialog box*

Specify a block name in the **Name:** text box. Or choose the down arrow to display a list of blocks defined in the current drawing (see Figure 10–8) and select the block you wish to insert.

Figure 10–8 *Listing of blocks defined in the current drawing*

Insertion Point

The Insertion point section of the Insert dialog box allows you to specify the insertion point for inserting a copy of the block definition. You can specify the insertion point in terms of *X*, *Y*, and *Z* coordinates in the **X:, Y:** and **Z:** text boxes when the **Specify On-screen** check box is set to OFF. If you prefer to specify the insertion point on screen, then the **Specify On-screen** check box is set to ON.

Scale

The Scale section of the Insert dialog box allows you to specify the scale for the inserted block. The default scale factor is set to 1 (Full scale). You can specify a scale factor between 0 and 1 to insert the block reference smaller than the original size of the block and specify more than 1 to increase the size from the original size. If necessary, you can specify different *X* and *Y* scale factors for insertion of the block reference. If you specify a negative scale factor, then AutoCAD inserts a mirror image of the block about the insertion point. As a matter of fact, if −1 were used for both *X* and *Y* scale factors, it would "double mirror" the object, the equivalent of rotating it 180 degrees. If you prefer to specify the scale factor on screen, then set the **Specify On-screen** check box to ON. If you check the **Uniform Scale** check box, then you can enter a value only in the **X:** text box. The *Y* and *Z* scales will be the same as that entered for the *X* scale.

Rotation Angle

The Rotation angle section of the Insert dialog box allows you to specify the rotation angle for the inserted block. To rotate the block reference, specify a positive or negative angle, referencing the block in its original position. If you prefer to specify the rotation angle on screen, then set the **Specify On-screen** check box to ON.

Explode

The Explode check box allows you to insert the block reference as a set of individual objects rather than as a single unit.

To specify a drawing file to insert as a block definition, enter the drawing file name in the **Name:** text box. Or choose the **Browse** button to display a standard file dialog box and select the appropriate drawing file.

 Note: The name of the last block reference inserted during the current drawing session is remembered by AutoCAD. The name becomes the default for subsequent use of the INSERT Command.

Choose the **OK** button to insert the selected block.

You can also insert a block by invoking the INSERT command preceded with a hyphen, as shown in the following command sequence:

> Command: **-insert** (ENTER)
> Enter block name or [?]: *(specify a block name, or enter ? to list the block names in the current drawing)*
> Specify insertion point or [Scale/X/Y/Z/Rotate/PScale/PX/PY/PZ/PRotate]: *(specify the insertion point to insert the block reference or right-click for the shortcut menu and select one of the avilable options)*
> Enter X scale factor, specify opposite corner, or [Corner/XYZ] <1>: *(specify a scale factor for the X axis, specify a point, or press ENTER to accept the default scale factor)*
> Enter Y scale factor <use X scale factor>: *(specify a scale factor for the Y axis, specify a point, or press ENTER to accept the same scale factor as for the X axis)*
> Specify rotation angle <0>: *(specify a rotation angle, specify a point, or press ENTER to accept the default rotation angle)*

Instead of specifying the insertion point to insert the block reference, you can invoke one of the following options at the "Insertion point:" prompt to preset the scale and rotation of a block reference before you specify its position:

Scale The Scale option sets the scale factor for the X, Y, and Z axes.

Xscale The Xscale option sets the X scale factor.

Yscale The Yscale option sets the Y scale factor.

Zscale The Zscale option sets the Z scale factor.

Rotate The Rotate option sets the rotation angle.

PScale The PScale option sets the scale factor for the X, Y, and Z axes to control the display of the block reference as it is dragged into position.

PXscale The PXscale option sets the scale factor for the *X* axis to control the display of the block reference as it is dragged into position.

PYscale The PYscale option sets the scale factor for the *Y* axis to control the display of the block reference as it is dragged into position.

PZscale The PZscale option sets the scale factor for the *Z* axis to control the display of the block reference as it is dragged into position.

PRotate The PRotate option sets the rotation angle of the block reference as it is dragged into position.

After you provide the appropriate information to the selected option, AutoCAD returns to the "Insertion point:" prompt. AutoCAD continues with the scale factor and rotation angle prompts. A copy of the specified block is inserted, with its defined insertion point located at the specified point in the current drawing to the specified scale and rotation angle.

NESTED BLOCKS

Blocks can contain other blocks. That is, when using the BLOCK command to combine objects into a single object, one or more of the selected objects can themselves be blocks. And the blocks selected can have blocks nested within them. There is no limitation to the depth of nesting. You may not, however, use the name of any of the nested blocks as the name of the block being defined. This would mean that you were trying to redefine a block, using its old definition in the new.

Any objects within blocks (as nested blocks) that were on layer 0 when made into a block will assume the color, linetype and lineweight of the layer on which the block reference is inserted. If an object (originally on layer 0 when included in a block definition) is in a block reference that has been inserted on a layer other than layer 0, it will retain the color, linetype and lineweight of the layer it was on when its block was included in a higher-level block. For example, you draw a circle on layer 0 and include it in a block named Z1. Then, you insert Z1 on layer R, whose color is red. The circle would then assume the color of layer R (in this case it will be red). Create another block called Y3 by including the block Z1. If you insert block reference Y3 on a layer whose color is blue, the block reference Y3 will retain the current color of layer R (in this case it will be red) instead of taking up the color of blue.

EXPLODE COMMAND

The EXPLODE command causes block references, hatch patterns, and associative dimensioning to be turned into the separate objects from which they were created. It also causes polylines/polyarcs and multilines to separate into individual simple line and arc objects. The EXPLODE command causes 3D polygon meshes to become 3Dfaces, and 3D polyface meshes to become 3Dfaces and simple line and point objects. When an object is exploded, the new, separate objects are created in the space (model or paper) of the exploded objects.

Invoke the EXPLODE command:

Modify toolbar	Choose the EXPLODE command (see Figure 10–9)
Modify menu	Choose Explode
Command: prompt	**explode** (ENTER)

Figure 10–9 *Invoking the* EXPLODE *command from the Modify toolbar*

AutoCAD prompts:

> Command: **explode**
> Select objects: *(select objects to explode, and press* ENTER *to complete object selection)*

You can use one or more object selection methods. The object selected must be eligible for exploding, or an error message will appear. An eligible object may or may not change its appearance when exploded.

POSSIBLE CHANGES CAUSED BY THE EXPLODE COMMAND

A polyline segment having width will revert to a zero-width line and/or arc. Tangent information associated with individual segments is lost. If the polyline segments have width or tangent information, the EXPLODE command will be followed by the message:

> Exploding this polyline has lost (width/tangent) information.
> The UNDO command will restore it.

Individual elements within blocks that were on layer 0 when created (and whose color was BYLAYER) but were inserted on a layer with a color different than that of layer 0 will revert to the color of layer 0.

Attributes are special text objects that, when included in a block definition, take on the values (names and numbers) specified at the time the block reference is inserted. The power and usage of attributes are discussed later in this chapter. To understand the effect of the EXPLODE command on block references that include attributes, it is sufficient to know that the fundamental object from which an attribute is created is called an attribute definition. It is displayed in the form of an attribute tag before it is included in the block.

An attribute within a block will revert to the attribute definition when the block reference is exploded and will be represented on the screen by its tag. The value of the attribute specified at the time of insertion is lost. The group will revert to those elements created by the ATTDEF command prior to combining them into a block via the BLOCK command.

In brief, an attribute definition is turned into an attribute when the block in which it is a part is inserted; conversely, an attribute is turned back into an attribute definition when the block reference is exploded.

EXPLODING BLOCK REFERENCES WITH NESTED ELEMENTS

Block references containing other blocks and/or polylines are separated for one level only. That is, the highest-level block reference will be exploded, but any nested blocks or polylines will remain block references or polylines. They in turn can be exploded when they come to the highest level.

Viewport objects in a block definition cannot be turned on after being exploded unless they were inserted in paper space.

Block references with equal X, Y, and Z scales explode into their component objects. Block references with unequal X, Y, and Z scales (nonuniformly scaled block references) might explode into unexpected objects.

 Note: Block references inserted via the MINSERT command or external references and their dependent blocks cannot be exploded.

MULTIPLE INSERTS OF BLOCK REFERENCES

The MINSERT (multiple insert) command is used to insert block references in a rectangular array. The total pattern takes the characteristics of a block, except the group cannot be exploded. This command works similar to the rectangular ARRAY command.

Invoke the MINSERT command:

Command: prompt	**minsert** (ENTER)

AutoCAD prompts:

Command: **minsert**
Enter block name or [?] <y>: *(enter name of the block)*
Specify insertion point or [Scale/X/Y/Z/Rotate/PScale/PX/PY/PZ/PRotate]:
    *(specify the insertion point or right-click for shortcut menu and select one of
    the avilable options)*

Enter X scale factor, specify opposite corner, or [Corner/XYZ] <1>:
  (enter a number or specify a point)
Enter Y scale factor <use X scale factor>: *(enter a number or press* ENTER*)*
Specify rotation angle <0>: *(enter a number or specify a point)*
Enter number of rows (—) <default>: *(enter the number of rows)*
Enter number of columns (||||) <default>: *(enter the number of columns)*
Enter distance between rows or specify unit cell (—): *(specify the distance
  between rows)*
Specify distance between columns (||||):*(specify the distance between columns)*

The row/column spacing can be specified by the delta-X/delta-Y distances between two points picked on the screen. For example, if, in response to the "Distance" prompt, you selected points 2,1 and 6,4 for the first and second points, respectively, the row spacing would be 3 (4 – 1) and the column spacing would be 4 (6 – 2).

INSERTING UNIT BLOCK REFERENCES

Groups of objects often need to be duplicated within a drawing. We showed earlier how the AutoCAD BLOCK command makes this task easier. The task of transferring blocks or groups of objects to another drawing is demonstrated in the section on the WBLOCK command. This section covers additional aspects of creating blocks in anticipation of inserting them later with differing scale factors (sometimes with *X* and *Y* unequal). This is referred to as a unit block.

Doors and windows are a few of the objects that can be stored as blocks in a symbol library. But blocks can be used in different ways to suit differing situations. The following examples are offered as procedures that are used without customizing menus or AutoLISP routines to enhance the process. It should be noted that these procedures can be improved either with customization or possibly with some variations involving standard commands and features. Also, the symbology and names of items are subject to variation.

You can employ a variety of symbols to represent windows in an architectural plan view. Horizontal sliding windows may need to be distinguished from single-hung windows, as shown in Figure 10–10. You may also wish to have more than one option as to how you will insert a window. You may wish to make its insertion point the center of the window sometimes or one of its edges at other times.

Figure 10–10 *Two different windows shown as block references*

Because there are windows of varying widths, you might have to create a separate block for each width. A group of windows might be 1'-0", 1'-6", 2'-0", 2'-8", 3'-0", 3'-4", 4'-0", 5'-4", 6'-0", and 8'-0", and some widths in between. You would have to make a block for each width in addition to the different types. The symbol for the single-hung 2'-8"–wide window could be drawn with the snap set a 0.5 (1/2") to the dimensions shown in Figure 10–11. The objects in Figure 10–11 could be saved as a block named WDW32 (for a 32"-wide window). This window could be inserted from its corner, as shown, if you have established that intersection in the wall in which it is to be drawn.

The preceding method requires a separate block for each window width. Another method is to make a drawing of a window in which the X and/or Y dimension is 1 unit, in anticipation of making the final desired dimension the X and/or Y scale factor during insertion. The following Unit Block symbol for the window can be used for a window of any width, as shown in Figure 10–12. This group of objects might be made into a block named WDW1. Then, to use it for a 2'-8" window, the INSERT command preceded with a hyphen would be:

Command: **-insert**
Enter block name or [?] <y>: **wdw1**
Specify insertion point or [Scale/X/Y/Z/Rotate/PScale/PX/PY/PZ/PRotate]:
   (select insertion point)
Enter X scale factor, specify opposite corner, or [Corner/XYZ] <1>: **32**
Enter Y scale factor <use X scale factor>: **1**
Specify rotation angle <0>: (ENTER)

Figure 10–11 *Single-hung 2'–8"–wide window*

Note that when you specify a value for the X scale factor, AutoCAD assumes you wish to apply the same value to the Y scale factor, making the resulting shape of the object(s) proportional to the unit block from which it was generated. That is why the Y scale factor defaults to the X scale factor. Therefore, if you wish to insert the block reference with an X scale factor different from the Y scale factor, you must input a Y scale factor even if it is just a factor of 1. The block WDW1 can be used for a window of any width by making the desired width (in inches) the X scale factor and then using 1 as the *Y* scale factor. Figure 10–13 shows the window inserted in its location with the proper *X* scale factor. Also shown is the cleanup that can be done with the BREAK command.

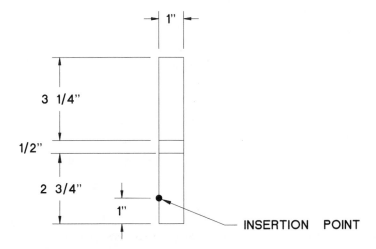

Figure 10–12 *Creating a block for a window symbol*

Another variation on the unit block WDW1 would be to use the same shape for center insertion. It would be drawn the same, but the insertion point would be chosen as shown in Figure 10–14. Figure 10–15 shows the window inserted.

Figure 10–13 *Inserting a unit block reference with a large X scale factor*

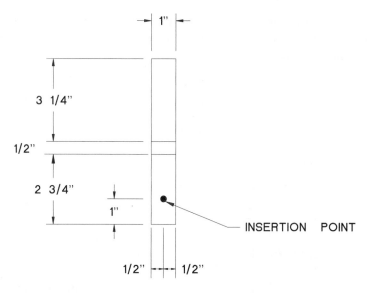

Figure 10–14 *Using a unit block reference with a specified insertion point*

Figure 10–15 *The unit block reference inserted as a window*

Window symbols in plan view lend themselves to the application of the one-way scaling unit block reference. As demonstrated, they vary in only the X scale and not the Y scale from one size to another. Doors, however, present a special problem when we try to apply the unit block method. A symbol for a 2'-8"–wide door might look like that in Figure 10–16. The 2'-8"–wide door is drawn half open, that is, swung at 45 degrees to the wall. Therefore, if you used the same approach on the door and on the window unit block, some problems would be encountered.

First, a base block 1-unit–wide simplified door symbol would not be practical. For instance, if the unit block reference in Figure 10–17 were inserted with an X scale factor of 32 and a Y scale factor of 1, it would look like Figure 10–18. Even though the 4" lines that represent the jambs are acceptable, the "door" part of the symbol will not retain its 45-degree swing if not inserted with equal X and Y scale factors. Equal X and Y scale factors present another problem. Using a Y scale factor other than 1 would make the 4"-wide jamb incorrect. Therefore, combining the jambs with the door in the same symbol presents problems that might be impossible to overcome when trying to use a unit block approach to permit one block to serve for all sizes of doors.

Figure 10–16 *Creating a unit block symbol for a door*

Figure 10–17 *Creating a unit block of a door symbol*

Figure 10–18 *A unit block reference of the door inserted with X and Y scale factors*

SEPARATE UNIT BLOCKS FOR ONE SYMBOL

The solution to making the jambs adaptable to using these two (jambs and door) into two block references, is shown in Figure 10–19.

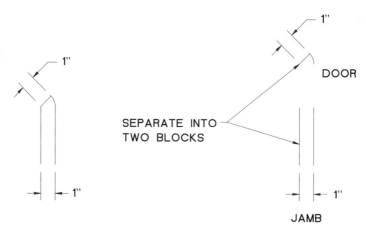

Figure 10–19 *Creating one symbol from two unit block references*

Now you can insert the block references separately, as follows (see Figure 10–20):

    Command: -insert
    Enter block name or [?] <y>: jmb
    Specify insertion point or [Scale/X/Y/Z/Rotate/PScale/PX/PY/PZ/PRotate]:
        (specify a point)
    Enter X scale factor, specify opposite corner, or [Corner/XYZ] <1>: 32
    Enter Y scale factor <use X scale factor>: 1
    Specify rotation angle <0>: (ENTER)

    Command: -insert
    Enter block name or [?] <y>: dr
    Specify insertion point or [Scale/X/Y/Z/Rotate/PScale/PX/PY/PZ/PRotate]:
        (specify a point)
    Enter X scale factor, specify opposite corner, or [Corner/XYZ] <1>: 32
    Enter Y scale factor <use X scale factor>: (ENTER)
    Specify rotation angle <0>: (ENTER)

Figure 10–20 *Two unit block references inserted to create one symbol*

Note: When you intend to apply the concept of the unit block, whether or not it is to be scaled uniformly (X scale equal to Y scale), be sure that shapes and sizes of all items in the symbol will be correct at their new scale.

COLUMNS AS UNIT BLOCKS

Another unit block application is when both *X* and *Y* scale factors are changed, but not at the same factor. The column shown in Figure 10–21 is made into a block named COL1 and is inserted for a 10 x 8–wide flange symbol as follows:

> Command: **-insert**
> Block name: **col1**
> Insertion point: *(specify point)*
> X scale factor <1>: **8**
> Y scale factor (default=X>): **10**
> Rotation angle: (ENTER)

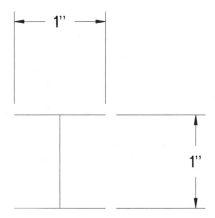

← 1" →

1"

Figure 10–21 *Creating a unit block of a column symbol*

This can also be done for a 16 x 12 column rotated at 90 degrees, as shown in Figure 10–22.

> Command: **-insert**
> Enter block name or [?] <y>: **col1**
> Specify insertion point or [Scale/X/Y/Z/Rotate/PScale/PX/PY/PZ/PRotate]:
>   *(specify a point)*
> Enter X scale factor, specify opposite corner, or [Corner/XYZ] <1>: **12**
> Enter Y scale factor <use X scale factor>: **16**
> Specify rotation angle <0>: **90**

Note the 90-degree rotation. But don't forget that the *X* and *Y* scale factors are applied to the block reference in the respective *X* and *Y* directions that were in effect when it was created, not to the *X* and *Y* directions after a rotated insertion.

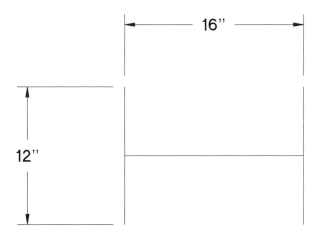

Figure 10–22 *The 16 x 12 column symbol inserted and rotated 90 degrees*

CREATING ONE DRAWING FROM ANOTHER

The WBLOCK command permits you to group objects in a manner similar to the BLOCK command. But, in addition, WBLOCK exports the group to a file, which, in fact, becomes a new and separate drawing. The new drawing (created by the WBLOCK command) might consist of a selected block reference in the current drawing. Or it might be made up of selected objects in the current drawing. You can even export the complete current drawing to the new drawing file. Whichever of these you choose to WBLOCK, the new drawing assumes the layers, linetypes, styles, and other environmental items, such as system variable settings of the current drawing.

Invoke the WBLOCK command:

Command: prompt	**wblock** (ENTER)

AutoCAD displays the Write Block dialog box, similar to Figure 10–23, where you will provide a drawing name in the **File name:** text of the Destination section of the dialog box as you do when you begin a new drawing. Make sure appropriate folder is selected in the **Location:** text box. You should enter only the file name and not the extension. AutoCAD appends the .DWG extension automatically. The name should comply with the operating system requirements of valid characters and should be unique for drawings on the specified folder/path. Otherwise you will get the message:

Drawingname.dwg already exists. Do you want to replace it?

Choose the Yes button only if you wish to overwrite the existing drawing with the same name. Otherwise choose the No button and enter a different, unique name for the drawing.

If you wish to use an existing block, the entire drawing, or existing objects in the current drawing to create a new drawing, choose one of the radio buttons in the Source section of the Write Block dialog box.

Figure 10–23 *Write Block dialog box*

Block

If you wish to create a drawing to be a duplicate of an existing block in the current drawing, then in the Source section of the Write Block dialog box, the **Block** radio button should be selected. Select the name of the desired block from the selection box to the right of the **Block** radio button. By default, the name of the new drawing will be the same as the block chosen. You can enter a different name for the new drawing in the **File Name:** text box. You can also enter a path to the new drawing in the **Location:** text box.

Entire Drawing

If you wish to create a drawing to be a duplicate of the entire current drawing, then in the Source section of the Write Block dialog box, the **Entire Drawing** radio button should be selected. Enter a name for the new drawing in the **File Name:** text box and a path to the new drawing in the **Location:** text box.

Objects

If you wish to create a drawing from objects to be selected in the current drawing, then in the Source section of the Write Block dialog box, the **Objects** radio button should be selected. You must select one or more objects to be exported as a new drawing. The object selection process is the same as when you are creating a block with the BLOCK command, described earlier in this chapter. Enter a name for the new drawing in the **File Name:** text box and a path to the new drawing in the **Location:** text box.

Insert Units

When the new file is inserted as a block, it will be scaled to the units in the **Insert Units** list box.

Choose the OK button to create the drawing.

One advantage of using the WBLOCK command when the entire drawing is written to a file is that all of the unused blocks, layers, linetypes, and other unused named objects are not written. That is, the drawing is automatically purged. This means that unused items will not be written to the new drawing file. For example, unused items include block definitions that have not been inserted, noncurrent layers that have no objects drawn on them, and styles that are not being used. This can be useful if you just wish to clean up a cluttered drawing, especially one that has had other drawing files inserted into it, each bringing with it various unused named objects.

BASE COMMAND

The BASE command allows you to establish a base insertion point for the whole drawing in the same manner that you specify a base insertion point when using the BLOCK command to combine elements into a block. The purpose of establishing this basepoint is primarily so that the drawing can be inserted into another drawing by way of the INSERT command and having the specified basepoint coincide with the specified insertion point. The default basepoint is the origin (0,0,0). You can specify a 2D point, and AutoCAD will use the current elevation as the base Z coordinate. Or you can specify the full 3D point.

Invoke the BASE command:

Draw menu	Block > Base
Command: prompt	**base** (ENTER)

AutoCAD prompts:

> Command: **base**
> Enter base point <current>: *(specify a point, or press ENTER to accept the default)*

ATTRIBUTES

Attributes can be used for automatic annotation during insertion of a block reference. Attributes are special text objects that can be included in a block definition and must be defined beforehand and then selected when you are creating a block definition.

Attributes have two primary purposes:

> The first use of attributes is to permit annotation during insertion of the block reference to which the attributes are attached. Depending on how you define an attribute, it either appears automatically with a preset (constant) text string or it prompts you (or other users) for a string to be written as the block reference is inserted. This feature permits you to insert each block reference with a string of preset text or with its own unique string.

> The second (perhaps the more important) purpose of attaching attributes to a block reference is to have extractable data about each block reference stored in the drawing database file. Then, when the drawing is complete (or even before), you can invoke the ATTEXT (short for "attribute extract") command to have attribute data extracted from the drawing and written to a file in a form that database-handling programs can use. You can have as many attributes attached to a block reference as you wish. As just mentioned, the text string that makes up an attribute can be either constant or user-specified at the time of insertion.

A DEFINITION WITHIN A DEFINITION

When creating a block, you select objects to be included. Objects such as lines, circles, and arcs are drawn by means of their respective commands. Normal text is drawn with the TEXT, or MTEXT command.

As with drawing objects, attributes must be drawn before they can be included in the block. This is complicated, and it requires additional steps to place them in the drawing; AutoCAD calls this procedure defining the attribute. Therefore, an attribute definition is simply the result of defining an attribute by means of the ATTDEF command. The attribute definition is the object that is selected during the BLOCK command. Later, when the block reference is inserted, the attributes that are attached to it and the manner in which they become a part of the drawing are a result of how you created (defined) the attribute definition.

VISIBILITY AND PLOTTING

If an attribute is to be used only to store information, then you can, as part of the definition of the attribute, specify whether or not it will be visible. If you plan to use an attribute with a block as a note, label, or callout, you should consider the effect of scaling (whether equal or unequal X/Y factors) on the text that will be displayed. The scaling factor(s) on the attribute will be the same as on the block reference. Therefore, be sure that it will result in the size and proportions desired. You should also be aware of the effect of rotation on visible attribute text. Attribute text that is defined as horizontal in a block will be displayed vertically when that block reference is inserted with a 90-degree angle of rotation.

Note that, like any other object in the drawing, the attribute must be visible on the screen (or would be if the plotted view were the current display) for that object to be eligible for plotting.

TAG, VALUE, PROMPT, AND DEFAULT

Four components associated with attributes should be understood before attempting a definition. The purpose of each is described as follows:

Tag

An attribute definition has a tag, just as a layer or a linetype has a name. The tag is the identifier of the attribute definition and is displayed where this attribute definition is located, depicting text size, style, and angle of rotation. The tag cannot contain spaces. Two attribute definitions with the same tag should not be included in the same block. Tags appear in the block definition only, not after the block reference is inserted. However, if you explode a block reference, the attribute value (described herein) changes back into the tag.

If multiple attributes are used in one block, each must have a unique tag in that block. This restriction is similar to each layer, linetype, block, and other named object having a unique name within one drawing. An attribute's tag is its identifier at the time that attribute is being defined, before it is combined with other objects and attributes by the BLOCK command.

Value

The value of an attribute is the actual string of text that appears (if the visibility mode is set to ON) when the block reference (of which it is a part) is inserted. Whether visible or not, the value is tied directly to the attribute, which, in turn, associates it with the block reference. It is this value that is written to the database file. It might be a door or window size or, in a piping drawing, the flange rating, weight, or cost of a valve or fitting. In an architectural drawing the value might represent the manufacturer, size, color, cost or other pertinent information attached to a block representing a desk.

 Note: When an extraction of attribute data is performed, it is the value of an attribute that is written to a file, but it is the tag that directs the extraction operation to that value. This will be described in detail in the later section on "Extracting Attributes."

Prompt

The prompt is what you see when inserting a block reference with an attribute whose value is not constant or preset. During the definition of an attribute, you can specify a string of characters that will appear in the prompt area during the insertion of the block reference to prompt you to enter the appropriate value. What the prompt says to you during insertion is what you told it to say when you defined the attribute.

Default

You can specify a default value for the attribute during the definition procedure. Then, during insertion of the block reference, it will appear behind the prompt in brackets, i.e., <default>. It will automatically become the value if ENTER is pressed in response to the prompt for a value.

ATTRIBUTE COMMANDS

The four primary commands to manage Attributes are:

> ATTDEF—attribute definition
>
> ATTDISP—attribute display
>
> ATTEDIT—attribute edit
>
> ATTEXT—attribute extract

As explained earlier, the ATTDEF command is used to create an attribute definition. The attribute definition is the object that is selected during the BLOCK command.

The ATTDISP command controls the visibility of the attributes.

The ATTEDIT command provides a variety of ways to edit without exploding the block reference.

The ATTEXT command allows you to extract the data from the drawing and have it written to a file in a form that database-handling programs can use, as shown in the following table:

DOORS						
SIZE	**THKNS**	**CORE**	**FINISH**	**LOCKSET**	**HINGES**	**INSET**
3070	1.750	SOLID	PAINT	PASSAGE	4 x 4	3/4
3070	1.750	SOLID	VARNISH	KEYED	4 x 4	20 x 20
2868	1.375	HOLLOW	PAINT	PRIVACY	3 x 3	3/4

ROOM FINISHES					
NAME	**WALL**	**CEILING**	**FLOOR**	**BASE**	**REMARKS**
LIVING	GYPSUM	GYPSUM	CARPET	NONE	PAINT
FAMILY	PANEL	ACOUSTICAL	TILE	STAIN	STAIN
BATH	PAPER	GYPSUM	TILE	COVE	4'_CERAMIC_TILE
GARAGE	GYPSUM	GYPSUM	CONCRETE	NONE	TAPE_FLOAT_ONLY

You can include an attribute in the WDW block to record the size of the window. A suggested procedure would be to zoom in near the insertion point and create an attribute definition with a tag that reads WDW-SIZE, as shown in Figure 10–24.

If, during the insertion of the WDW block reference, you respond to the prompt for the SIZE attribute with 2054 for a 2'-0"–wide x 5'-4"–high window, the resulting block reference object would be as shown in Figure 10–25, with the normally invisible attribute value shown here for illustration purposes. Figure 10–26 shows the result of the attribute being inserted with unequal scale factors.

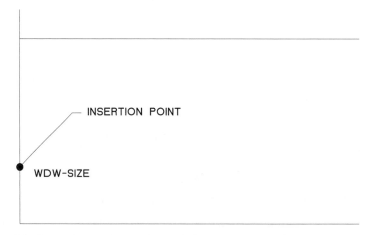

Figure 10–24 *Create the attribute definition before defining the block*

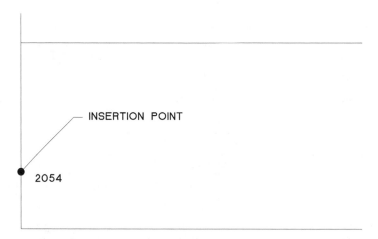

Figure 10–25 *The attribute value (2054) visible with the inserted block*

Even though the value displayed is distorted, the string is not affected when extracted to a database file for a bill of materials.

Figure 10–26 *The attribute in a block reference with exaggerated, unequal scale factors*

To solve the distortion and rotation problem, if you wish to have an attribute displayed for rotational purposes, you can create a block that contains attributes only or only one attribute. Then it can be inserted at the desired location and rotated for readability to produce the results shown in Figure 10–27.

Attribute definitions would be created as shown in Figure 10–27 and inserted as shown in Figure 10–28.

Figure 10–27 *Attributes created as separate blocks*

Figure 10–28 *Attribute inserted separately for the window*

The insertion points selected would correspond to the midpoint of the outside line that would result from the insertion of the WDW block reference. The SIZE attributes could be defined into blocks called WDWSIZE and inserted separately with each WDW block reference, thereby providing both annotation and data extraction.

 Note: The main caution in having an attribute block reference separate from the symbol (WDW) block reference is in editing. Erasing, copying, and moving the symbol block reference without the attribute block reference could mean that the data extraction results in the wrong quantity.

There is another solution to the problem of a visible attribute not being located or rotated properly in the inserted block reference. If the attribute is not constant, you can edit it independently after the block reference has been inserted by way of the ATTEDIT command. It permits changing an attribute's height, position, angle, value, and other properties. The ATTEDIT command is covered in detail in the later section on "Editing Attributes," but it should be noted here that the height editing option applies to the X and Y scales of the text. Therefore, for text in the definition of a block that was inserted with unequal X and Y scale factors, you will not be able to edit its proportions back to equal X and Y scale factors.

CREATING AN ATTRIBUTE DEFINITION

The ATTDEF command allow you to create an attribute definition. The ATTDEF command allows you to create an attribute definition through a dialog box; the -ATTDEF command allows you to create an attribute definition at the Command: prompt.

Invoke the ATTDEF command:

Draw menu	Block > Define Attributes
Command: prompt	**attdef** (ENTER)

AutoCAD displays the Attribute Definition dialog box, as shown in Figure 10–29.

Set one or more of the available modes to ON in the **Mode** section of the Attribute Definition dialog box.

Invisible Setting the Invisible mode to ON causes the attribute value *not* to be displayed when the block reference is completed. Even if visible, the value will not appear until the insertion is completed. Attributes needed only for data extraction should be invisible, to quicken regeneration and to avoid cluttering your drawing. You can use the ATTDISP command to override the Invisible mode setting. Setting the Invisible mode to Y (yes, or ON) does not affect the visibility of the tag in the attribute definition.

Figure 10–29 *Attribute Definition dialog box*

Constant If the Constant mode is set to ON, you must enter the value of the attribute while defining it. That value will be used for that attribute every time the block reference to which it is attached is inserted. There will be no prompt for the value during insertion, and you cannot change the value.

Verify If the Verify mode is set to ON, you will be able to verify its value when the block reference is inserted. For example, if a block reference with three (nonconstant value) attributes is inserted, once you have completed all prompt/value sequences that have displayed the original defaults, you will be prompted again, with the latest values as new defaults, giving you a second chance to be sure the values are correct before the INSERT command is completed. Even if you press ENTER to accept an original default value, it also appears as the second-chance default. If, however, you make a change during the verify sequence, you will *not* get a third chance, that is, a second verify sequence.

Preset If the Preset mode is set to ON, the attribute automatically takes the value of the default that was specified at the time of defining the attribute. During a normal insertion of the block reference, you will not be prompted for the value. You must be careful to specify a default during the ATTDEF command or the attribute value will be blank. A block consisting of only attributes whose defaults were blank when Preset modes were set to ON could be inserted, but it would not display anything and cannot be purged from the drawing. The only adverse effect would be that of adding to the space taken in memory. One way to get rid of a nondisplayable block reference like this is to use a visible entity to create a block with the same name, thereby redefining it to something that can be edited, that is, erased and subsequently purged.

You can duplicate an attribute definition with the COPY command, and use it for more than one block. Or you can explode a block reference and retain one or more of its attribute definitions for use in subsequent blocks.

The **Attribute** section of the Attribute Definition dialog box allows you to set attribute data. Enter the attribute's tag, prompt, and default value in the text boxes.

The attribute's tag identifies each occurrence of an attribute in the drawing. The tag can contain any characters except spaces. AutoCAD changes lowercase letters to uppercase.

The attribute's prompt appears when you insert a block reference containing the attribute definition. If you do not specify the prompt, AutoCAD uses the attribute tag as the prompt. If you turn on the Constant mode, the Prompt field is disabled.

The default Value specifies the default attribute value. This is optional, except if you turn on the Constant mode, for which the default value needs to be specified.

The **Insertion Point** section of the dialog box allows you to specify a coordinate location for the attribute in the drawing, either by choosing the Pick Point button to specify the location on the screen or by entering coordinates in the text boxes provided.

The **Text Options** section of the dialog box allows you to set the justification, text style, height, and rotation of the attribute text.

The **Align Below Previous Attribute** check box allows you to place the attribute tag directly below the previously defined attribute. If you haven't previously defined an attribute definition, this option is unavailable.

Choose the OK button to define the attribute definition.

After you close the Attribute Definition dialog box, the attribute tag appears in the drawing. Repeat the procedure to create additional attribute definitions.

You can also define attributes by invoking ATTDEF command preceded with a hyphen from the Command: prompt, as shown in the following command sequence:

Command: **-attdef**
Current attribute modes: Invisible=N Constant=N Verify=N Preset=N
Enter an option to change [Invisible/Constant/Verify/Preset] :
  (ENTER **i**, **c**, **v**, *or* **p** *to change the current settings for attribute modes)*
Enter attribute tag name: *(specify the attribute tag name, with no spaces or exclamation marks)*
Enter attribute prompt: *(specify the text for the prompt line, or press* ENTER *to accept the attribute tag as the attribute prompt)*
Enter default attribute value: *(specify a value for the default attribute value, or press* ENTER*)*

This last prompt appears unless the Constant mode is set to Y (yes, or ON), in which case the prompt will be as follows:

Attribute value: *(specify a value for the attribute value)*

After the foregoing ATTDEF prompts have been answered, you will be prompted to place the tag, in the same manner as you would for placing text, except AutoCAD will create the tag in place of a text string. Subsequent attributes can be placed in a manner similar to that for placing lines of text, using the insertion point and line spacing as left justified, centered, aligned, or right justified lines of text. Simply press ENTER to invoke a repeating attribute definition.

Figure 10–30 shows a pre-drawn 2'-0"–wide door made into a block reference with attributes.

Figure 10–30 *Pre-drawn door with attributes*

 Note: In this example the block is drawn to true size with jambs so it can be inserted with X and Y scale factors equal to 1 (one). But, the attributes will all be invisible, because the door might be inserted at a rotation incompatible with acceptable text orientations. The attributes can also be very small, located at a point that will be easy to find if they must be made visible in order to read (Figure 10–31).

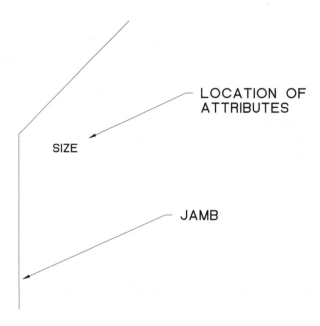

LOCATION OF
ATTRIBUTES

SIZE

JAMB

Figure 10–31 *Pre-drawn jamb block with attributes*

The Attributes are entered as follows (Figures 10–32 and 10–33):

Command: **-attdef**
Attribute modes—Invisible:N Constant:N Verify:N Preset:N
Enter an option to change [Invisible/Constant/Verify/Preset] <done>: **i**
Attribute modes—Invisible:Y Constant:N Verify:N Preset:N
Enter an option to change [Invisible/Constant/Verify/Preset] <done>: **c**
Attribute modes—Invisible:Y Constant:Y Verify:N Preset:N
Enter an option to change [Invisible/Constant/Verify/Preset] <done>:
    (ENTER)
Enter attribute tag name: **size**
Enter attribute value: **26681.375** *(for 2'-6" x 6'-8" x 1-3/8")*
Current text style: "Standard" Text height: 0.2000
Specify start point of text or [Justify/Style]: *(specify point)*
Specify height <0.2000>: **1/16**
Specify rotation angle of text <0>: (ENTER)

Command: (ENTER)
Attribute modes—Invisible:Y Constant:Y Verify:N Preset:N
Enter an option to change [Invisible/Constant/Verify/Preset] <done>: **c**
Attribute modes—Invisible:Y Constant:N Verify:N Preset:N
Enter an option to change [Invisible/Constant/Verify/Preset] <done>:
    (ENTER)
Enter attribute tag name: **matl**

Enter attribute prompt: **material**
Enter default attribute value: **mahogany**
Specify start point of text or [Justify/Style]: *(specify point)*
Command: (ENTER)

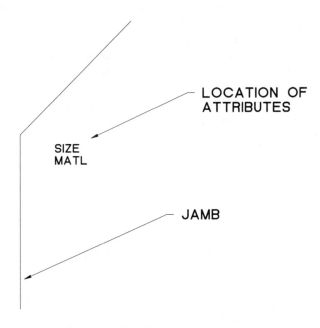

Figure 10–32 *Pre-drawn jamb block with character attributes*

Figure 10–33 *Pre-drawn jamb block with character/numeric attributes defined*

Note: The values given for the various attributes can be written just as any string of text is written. You should note that these strings, when written to a database-handling file, might eventually need to be interpreted as numbers rather than characters. The difference is primarily of concern to the person who will use the data in a database-handling program. So, if you are not familiar with data types, such as numeric, character, and date, you might wish to consult with someone (or study a book on databases) if you are entering the values. For example, an architectural distance such as 12'-6 1/2" may need to be written in decimal feet (12.54) without the apostrophe or in decimal inches (150.5) without the inch mark if it is going to be used mathematically once extracted.

You can continue to define additional attributes in the same manner for the L.H./R.H. SWING, PAINT/VARNISH FINISH, Type of HINGE, Type of LOCKSET, and so on (see Figure 10–34).

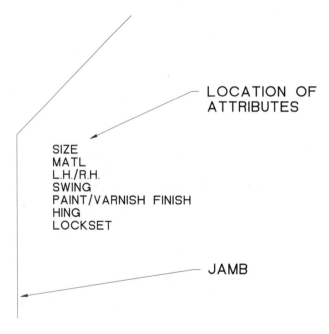

Figure 10–34 *Pre-drawn jamb with additional attributes defined*

INSERTING A BLOCK REFERENCE WITH ATTRIBUTES

Blocks with attributes may be inserted in a manner similar to that for inserting regular block references. If there are any nonconstant attributes, you will be prompted to enter the value for each. You may set the system variable called ATTREQ to 0 (zero), thereby suppressing the prompts for attribute values. In this case the values will either be blank or be set to the default values if they exist. You can later use either the DDATTE or ATTEDIT command to establish or change values.

You can set the ATTDIA system variable to a nonzero value, which will display a dialog box for attribute value input.

CONTROLLING THE DISPLAY OF ATTRIBUTES

The ATTDISP command controls the visibility of attributes. Attributes will normally be visible if the Invisible mode is set to N (normal) when they are defined.

Invoke the ATTDISP command:

View menu	Display > Attribute Display
Command: prompt	**attdisp** (ENTER)

AutoCAD prompts:

> Command: **attdisp**
> Enter attribute visibility setting [Normal/ON/OFF] <Normal>: *(choose one of the available options, or press ENTER to accept the default option)*

Responding ON makes all attributes visible; OFF makes all attributes invisible. The Normal option displays the attributes as you created them. The ATTMODE system variable is affected by the ATTDISP setting. If REGENAUTO is set to ON, changing the ATTDISP setting causes drawing regeneration.

EDITING ATTRIBUTES

Unlike other objects in an inserted block reference, attributes can be edited independently of the block reference. You can, however, edit groups of attributes collectively. This permits you to insert a block reference with generic attributes; that is, the default values can be used in anticipation of changing them to the desired values later. Or you can copy an existing block reference that may need only one or two attribute changes to make it correct for its new location. And, of course, there is always the chance that either an error was made in entering the value or design changes necessitate subsequent changes.

Editing an attribute is accomplished by invoking the ATTEDIT command. The ATTEDIT command edits individual, nonconstant attribute values associated with a specific block reference, whereas the -ATTEDIT command edits both attribute values and attribute properties individually or globally, independent of the block reference.

Invoke the ATTEDIT command:

Modify II toolbar	Choose the EDIT ATTRIBUTE command (see Figure 10–35)
Modify menu	Choose Attribute > Single
Command: prompt	**attedit** (ENTER)

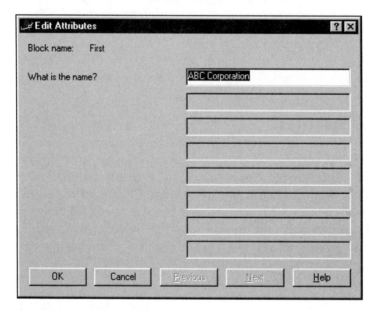

Figure 10–35 *Invoking the* EDIT ATTRIBUTE *command from the Modify II toolbar*

AutoCAD prompts:

> Command: **attedit**
> Select block reference: *(select the block reference)*

AutoCAD displays the Edit Attributes dialog box, similar to the one shown in Figure 10–36. Selecting objects that are not block references or block references that contain no attributes will cause an error message to appear.

Figure 10–36 *Edit Attributes dialog box*

The dialog box lists all the attributes defined with values for the selected block reference. Using the pointing device, you can select values to be changed in the dialog box. Enter the new values and accept the changes by clicking the OK button or pressing ENTER. Choosing the CANCEL button terminates the command, returning all values to their original state.

You can use the ATTEDIT command to look at the values of a selected attribute without making changes. Or you can employ repeated editing or repeated inquiry looks at attribute values by modifying the ATTEDIT command with the MULTIPLE command, as follows:

> Command: **multiple**
>
> Enter command name to repeat: **attedit**

To change attribute properties such as position, height, and style, invoke the -ATTEDIT command. The -ATTEDIT command provides a variety of ways to specify attributes to be edited. It also allows various properties of the selected attributes to be edited. It should be noted that attributes with constant values cannot be edited in this manner.

Invoke the ATTEDIT command (preceded with a hyphen):

Modify menu	Choose Attribute > Global
Command: prompt	**-attedit** (ENTER)

AutoCAD prompts:

> Command: **-attedit**
> Edit attributes one at a time? [Yes/No] <Y>:

Responding **Y** permits you to edit visible attributes individually. You can limit those attributes eligible for selection by specifying the block name, tag, or value. In addition to an attribute's value, other properties that can be changed during this one-at-a-time mode include its position, height, and angle of rotation.

Responding **N** permits global (or mass) editing of attributes. Again, you can limit those eligible for editing to a specified block name, tag, or value. However, this mode permits editing of attribute values only and no other properties.

After responding **Y** or **N**, you will be prompted to specify eligible attributes as follows:

> Enter block name specification <*>:
> Enter attribute tag specification <*>:
> Enter attribute value specification <*>:

For any attribute to be eligible for editing, the block name, attribute tag, and attribute value must all match the name, tag, and value specified. Use of "*" or "?" wild-card characters allows more than one string to match. Pressing ENTER in response to the "Block name specification:" prompt defaults to the asterisk, which makes all attributes eligible with regard to the block name restriction. Attributes still have to match tag and value responses in order to be eligible for editing.

Global Editing

If you responded to the "Edit attributes one at a time? <Y>:" prompt with **N** (no), thereby choosing global editing, and set the block name, tag, and value limitations, the following prompts appear:

> Performing global editing of attribute values.
> Edit only attributes visible on screen? [Yes/No] <Y>:

An N reply switches AutoCAD to the text window with the following message:

> Drawing must be regenerated afterwards.

When the editing for this command is completed (and if AUTOREGEN is set to ON) a drawing regeneration occurs.

If at the "Edit only Attributes visible on screen? <Y>:" prompt you respond Y (or default to Y by pressing ENTER), you will be prompted as follows:

> Select Attributes:

Attributes may now be selected either by specifying them on the screen, or by the window, crossing, or last method. Eligible attributes are then highlighted and the following prompts appear:

> Enter string to change: *(select a string to change)*
> Enter new string: *(enter a new text string)*

 Note: The changes you specify will affect a group of attributes all at one time. Take care that unintended changes do not occur.

The responses to the "String to change:" prompt cause AutoCAD to search eligible selected attribute strings for matching strings. Each matching string will be changed to your response to the "New string:" prompt. For example, blocks named WDW20, having tags named SIZE and values of 2054 (for 2'-0" x 5'-4"), can be changed to 2060 by the following sequence:

> Enter string to change: **54**
> Enter new string: **60**

If you did not limit the attributes by block name, tag, or value, you might unintentionally change a window whose SIZE is 5440 (for 5'-4" x 4'-0") to 6040.

If you respond to "Enter string to change:" by pressing ENTER, it will cause any response to "New string:" to be placed ahead of all eligible attribute value strings. For example, if you specified only the block name WDW20 with an attribute value of 2054 to be eligible for editing, you can make an addition to the value by the following sequence:

> Enter string to change: (ENTER)
> Enter new string: **dbl hung**

The value 2054 will be changed to read DBL HUNG 2054. Be sure to add a space behind the G if you do not want the result to be DBL HUNG2054.

Editing Attributes One at a Time

If you responded to "Edit attributes one at a time? [Yes/No] <Y>:" by pressing ENTER (defaulting to Y), and have specified the eligible attributes through the block name, tag, and value sequences, you will be prompted:

> Select Attributes:

You may select attributes by specifying them on the screen or by the window, crossing, or with the last method. The eligible selected attributes with nonconstant values will be marked sequentially with an X. The next prompt is as follows:

> Enter an option [Value/Position/Height/Angle/Style/Layer/Color/Next]
> <N>: *(choose one of the available options)*

Angle is not an option for an attribute defined as fitted text, nor are the angle and height options for aligned attributes. Entering any option's initial letter permits a change relative to that option, followed by the repeated prompt for the list of options until you default to N for the next attribute.

The Value option, if chosen, will prompt you as follows:

> Enter type of value modification [Change/Replace] <R>:

Defaulting to R causes the following prompt:

> Enter new attribute value:

Any string entered will become the new value. Even a null response will become a null (blank) value. Entering C at the "Enter type of value modification [Change/Replace] <R>:" prompt will cause the following prompts to appear:

> Enter string to change:
> Enter new string:

Responses to these prompts follow the same rules described in the previous section on global editing. In addition to the Position/Height/Angle options normally established during attribute definition, you can change those preset properties such as Style/Layer/Color with the -ATTEDIT command.

EXTRACTING ATTRIBUTES

Extracting data from a drawing is one of the most innovative features in CAD. Paper copies of drawings have long been used to communicate more than just how objects look. In addition to dimensions, drawings tell builders or fabricators what materials to use, quantities of objects to make, manufacturers' names and models of parts in an assembly, coordinate locations of objects in a general area, and what types of finishes to apply to surfaces. But, until computers came into the picture (or pictures came into the computer), extracting data from manual drawings involved making lists (usually by hand) while studying the drawing, often checking off the data with a marker. The AutoCAD attribute feature and the ATTEXT command combine to allow complete, fast, and accurate extraction of (1) data consciously put in for the purpose of extraction, (2) data used during the drawing process, and (3) data that AutoCAD maintains about all objects (block references in this case).

The CAD drawing in Figure 10–37 shows a piping control set. The 17 valves and fittings are a fraction of those that might be on a large drawing. Each symbol is a block with attributes attached to it. Values that have been assigned to each attribute tag record the type (TEE, ELL, REDUCER, FLANGE, GATE VALVE, or CONTROL VALVE), size (3", 4", or 6"), rating (STD or 150#), weld (length of weld), and many other vital bits of specific data. Keeping track of hundreds of valves, fittings, and even cut lengths of pipe is a time-consuming task subject to omissions and errors if done manually, even if the drawing is plotted from CAD. Just as important as extracting data from the original drawing is the need to update a list of data when the drawing is changed. Few drawings, if any, remain unchanged. The AutoCAD attribute feature makes the job fast, thorough, and accurate. Examples of some of the block references with attribute definition are shown in Figures 10–38 through 10–40.

These are three examples of block references as defined and inserted, each attribute having been given a value during the INSERT command. Remember, the value that will be extracted is in accordance with how the template specifies the tags. Figure 10–41 shows the three block references inserted, with corresponding attribute values in tabular form. By using the ATTEXT command, a complete listing of all valves and fittings in the drawing can be written to a file, as shown in Table 10–1.

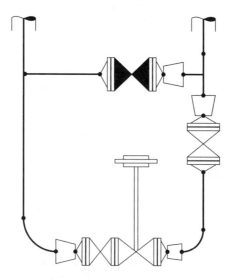

Figure 10–37 *Piping control set*

ENTITIES THAT WILL BE
INCLUDED IN EACH BLOCK

RED

TEXT ITEMS REPRESENT
ATTRIBUTE TAGS

TYPE
SIZE
RATING
BLTQUAN
BLTSIZE
GSKETMATL
GSKETSIZE
WELD

Figure 10–38 *Reducer with attribute definition*

ENTITIES THAT WILL BE
INCLUDED IN EACH BLOCK

FLANGE

TEXT ITEMS REPRESENT
ATTRIBUTE TAGS

TYPE
SIZE
RATING
BLTQUAN
BLTSIZE
GSKETMATL
GSKETSIZE
WELD

Figure 10–39 *Flange with attribute definition*

ENTITIES THAT WILL BE
INCLUDED IN EACH BLOCK

VALVE

TEXT ITEMS REPRESENT
ATTRIBUTE TAGS

TYPE
SIZE
RATING
BLTQUAN
BLTSIZE
GSKETMATL
GSKETSIZE
WELD

Figure 10–40 *Valve with attribute definition*

BLOCKS INSERTED WITH
VALUES ASSIGNED TO
ATTRIBUTES

TEXT ITEMS REPRESENT
ATTRIBUTE TAGS

	REDUCER	RFWN FLG	GT VLV
TYPE	REDUCER	RFWN FLG	GT VLV
SIZE	6X4	4"	4"
RATING	STD	150#	150#
BLTQUAN	-	8	8
BLTSIZE	-	5/8"x3"	5/8"x3"
GSKETMATL	-	COMP	COMP
GSKETSIZE	-	1/8"	1/8"
WELD	34.95	14.14	-

Figure 10–41 *Inserted block references with corresponding attribute values*

Table 10–1 A List of Inventory as Specified by the ATTEXT Command and Tags

TYPE	SIZE	RATING	BLTQUAN	BLTSIZE	GSKETMATL	GSKETSIZE	WELD
TEE	6"	STD	4	5/8"X3"	COMP	1/8"	62.44
ELL	6"	STD	4	5/8"X3"	COMP	1/8"	41.63
ELL	4"	STD	4	5/8"X3"	COMP	1/8"	14.14
RED	6X4	STD	8	5/8"X3"	COMP	1/8"	34.95
RED	6X4	STD	8	5/8"X3"	COMP	1/8"	34.95
RED	6X3	STD	4	5/8"X3"	COMP	1/8"	31.81
RED	4X3	STD	8	5/8"X3"	COMP	1/8"	25.13
FLG	4"	150#	8	5/8"X3"	COMP	1/8"	14.14
FLG	4"	150#	8	5/8"X3"	COMP	1/8"	14.14
FLG	4"	150#	8	5/8"X3"	COMP	1/8"	14.14
FLG	4"	150#	8	5/8"X3"	COMP	1/8"	14.14
FLG	3"	150#	4	5/8"X3"	COMP	1/8"	11.00
FLG	3"	150#	4	5/8"X3"	COMP	1/8"	11.00
GVL	4"	150#	16	5/8"X3"	COMP	1/8"	11.00
GVL	4"	150#	16	5/8"X3"	COMP	1/8"	11.00
GVL	3"	150#	8	5/8"X3"	COMP	1/8"	11.00
CVL	3"	150#	8	5/8"X3"	COMP	1/8"	11.00

The headings above each column in Table 10–1 are for your information only. These will not be written to the extract file by the ATTEXT command. They signify the tags whose corresponding values will be extracted.

When operated on by a database program, this file can be used to sort valves and fittings by type, size, or other value. The scope of this book is too limited to cover database applications. But generating a file like Table 10–1 that a database program can use is the important linkage between computer drafting and computer management of data for inventory control, material takeoff, flow analysis, cost, maintenance, and many other applications. The CAD drafter can apply the ATTEXT feature to perform this task.

Files

Using the ATTEXT command involves concepts of computer applications other than CAD (although not necessarily more advanced). Some fundamental understanding of the computer's operating system is needed. The operating system is used to manipulate and store files.

Many different types of files or groups of files are involved in using the ATTEXT command.

EXE and *DLL:* AutoCAD program files are being used to perform the attribute extraction.

DWG: The drawing file is a data file that contains the block references and their associated attributes whose values are eligible for extraction.

TXT: A template file must be created to tell AutoCAD what type of data to extract. A line editor or word processor is used to create the TXT template file.

TXT or *DXF:* The extraction process creates a *FILENAME.TXT* file containing the data in accordance with the instructions received from the template file.

A set of database program files usually can operate on the extracted file.

Database

Extracted data can be manipulated by a database application program. The telephone directory, a database, is an alphabetical listing of names, each followed by a first name (or initial), an address, and a phone number. A listing of pipe, valves, and fittings in a piping system can be a database if each item has essential data associated with it, such as its size, flange rating, weight, material of manufacture, product that it handles, cost, and location within the system, among many others.

The two elementary terms used in a database are the record and the field. A record is like a single listing in the phone book, made up of a name and its associated first name, address, and phone number. The name Jones with its data is one record. Another Jones with a different first name (or initials) is another record. Another Jones with the same first name or initials at a different address is still another record. Each listing is a record. The types of data that may be in a record come under the heading of a field. Name is a field. All the names in the list come under the name field. Address is a field. Phone number is a field. And even though some names may have first names and some may not, first name is a field. It is possible to take the telephone directory that is listed alphabetically by name, feed it into a computer database program, and generate the same list in numerical order by phone number. You can generate a partial list of all the Jones's sorted alphabetically by the first name. Or you can generate a list of everyone who lives on Elm Street. The primary purpose of the ATTEXT command is to generate the main list that includes all of the desired objects to which the database program manipulations can be applied.

Creating a Template

The template is a file saved in ASCII format; it lists the fields that specify the tags and determine which block references will have their attribute data extracted. The template must be a file on an accessible path with the extension of .TXT. When you use a text editor or word processor in the ASCII mode, you must add this extension when you name the file.

Field Name

The field name must correspond exactly to the attribute tag if you wish for that attribute's value to be extracted. If the attribute tag is called type, then there must be a field name in the template called type. A field name in the template called ratings will not cause a tag name called rating to have its values written to the extract file.

Character-Numeric

The template tells the ATTEXT command to classify the data written to a particular field either as numeric or as character type. Characters (a, b, c, A, #, ", etc.) are always character type, but numbers do not always have to be numeric type. Characters occupy less memory space in the computer. Sometimes numbers, like an address and phone number, are better stored as characters unless they are to be operated on by mathematical functions (addition, subtraction, etc.). The only relative significance of numbers as characters is their order (1 2 3...) for sorting purposes by the database program. Characters (a b c...) have that same significance. The template contains two elements for each field. The first is the field name; the second is the character-numeric element.

Numbers in strategic spaces in the character-numeric element of the template file specify the number of spaces to allow for the values to be written in the extract file. Others also specify how many decimal places to carry numeric values.

The format is as follows:

fieldname	Nwwwddd	for numeric values with decimal allowance
fieldname	Nwww000	for numeric values without decimal allowance
fieldname	Cwww000	for character values

Each line in the template is one field. The w's and d's are to be filled in with the necessary digits when the template file is created. The order of fields listed in the template does not have to coincide with the order that attribute tags appear in a block reference. Any group of fields, in any order, is acceptable. The only requirement for a block reference to be eligible for extraction by the ATTEXT command is that there be at least one tag–field match.

A template for the example in Figure 10–40 would be written as follows:

TYPE	C008000
SIZE	C008000
RATING	C006000
BLTQUAN	N004000
BLTSIZE	C010000
GSKETMATL	C006000
GSKETSIZE	C006000
WELD	N006002

 Note: Word processors add coded characters (often hidden) to files unless specifically set up to write in the ASCII format. These coded characters are not acceptable in the template. Be sure that the text editor or word processor being used is in the proper mode. Also, do not use the TABy key to line up the second column elements, but key in the necessary spaces, because the TAB key involves coded characters.

From the extracted file in Table 10–1, a database program can generate a sorted list whose items correspond to selected values. The procedure (again, to explain would take another book entirely) might list only 3" flanges, or all 3" fittings, or all reducers, or any combination of available records required corresponding to tag–field association.

BL:xxxxxx Nonattribute Fields Available

If necessary, additional data is also automatically stored with each inserted block reference. The suggested format in a template file and a description of each (in parentheses) follows:

Field Name	Field Format	Description
BL:LEVEL	Nwww000	(block nesting level)
BL:NAME	Cwww000	(block name)
BL:X	Nwwwddd	(X coordinate of block insertion point)
BL:Y	Nwwwddd	(Y coordinate)
BL:Z	Nwwwddd	(Z coordinate)
BL:NUMBER	Nwww000	(block counter; same for all members of a MINSERT)
BL:HANDLE	Cwww000	(block's handle; same for all members of a MINSERT)
BL:LAYER	Cwww000	(block insertion layer name)
BL:ORIENT	Nwwwddd	(block rotation angle)

Field Name	Field Format	Description
BL:XSCALE	Nwwwddd	(*X* scale factor of block)
BL:YSCALE	Nwwwddd	(*Y* scale factor)
BL:ZSCALE	Nwwwddd	(*Z* scale factor)
BL:XEXTRUDE	Nwwwddd	(*X* component of block's extrusion direction)
BL:YEXTRUDE	Nwwwddd	(*Y* component)
BL:ZEXTRUDE	Nwwwddd	(*Z* component)

 Note: The comments in parentheses are for your information and must not be included in the template file.

The first column lists the name of each field, for example: BL:ORIENT. The field format begins with a C or an N, denoting character or numeric data, respectively. The next three digits denote the width of the field; that is, how many spaces are to be allowed in the extract file for values to be written under this particular field. If the value to be written under this field for any record is too long for the width allowed, AutoCAD truncates the data written, proceeds with the extraction, and displays the following error message:

**Field overflow in record <record number>

The last three digits in the field format denote the number of decimal places to which numeric values will be written. Character fields should have zeros in these three places. When fields are specified as numeric, the attribute values must be numbers, or AutoCAD will display an error message.

You can use one of the two commands available to extract attributes. To extract attribute objects through the attributes extraction dialog box, use the ATTEXT command. To extract objects at the "Command:" line, use the -ATTEXT command.

Invoke the ATTEXT command:

Command: prompt	**attext** (ENTER)

AutoCAD displays the Attribute Extraction dialog box, similar to Figure 10–42.

Choose one of the three radio buttons available in the File Format section of the Attribute Extraction dialog box.

The **Comma Delimited Format** (CDF) format generates a file containing no more than one record for each block reference in the drawing. The values written under the fields in the extract files are separated by commas, with the character fields enclosed in single quotes.

The **Space Delimited Format** (SDF) format writes the values lined up in the widths allowed. It is possible for adjacent mixed fields (characters and numeric) not to have spaces between them. It may be necessary to add dummy fields to provide spaces in these cases.

Figure 10–42 *Attribute Extraction dialog box*

The **AutoCAD Drawing Interchange File** format is called DXF. Unlike the DXFOUT command, extraction files generated by the ATTEXT command contain only block reference, attribute, and end-of-sequence objects.

Specify the Template Filename in the **Template File** edit box if you have selected either CDF or SDF file formats. AutoCAD appends the .TXT file type. If you choose the DXF file format, the **Template File...** button and edit box are disabled.

Specify the output extract file name in the **Output File...** text box. AutoCAD appends the .TXT file type for CDF and SDF file formats and DXX for DXF files. DOS permits a file name of con, for writing to the screen, or a file name of prn, for writing to a printer. In this case, a printer must be connected and ready to print.

Choose the **Select Objects <** button to select the objects to be included in the extraction of attributes. The dialog box will disappear, and AutoCAD allows you to select the attribute entities. After you have done so, the dialog box reappears, and the number of objects you selected is displayed after the **Number found:** label.

Choose the OK button to close the dialog box and extract the attributes from the selected block references. The extracted information is stored in the output file specified in the Attribute Extraction dialog box. You can import the output file into a database program or open the output file with the help of a text editor and analyze the extracted information.

As mentioned earlier, you can invoke the ATTEXT command preceded with a hyphen as shown in the following command sequence:

```
Command: -attext
Enter extraction type or enable object selection [Cdf/Sdf/Dxf/Objects]
   <C>:
```

Choose one of the three options: CDF, SDF, or DXF. You can also choose the Objects option, which allows you to select the objects for extraction of their attributes. Once selection is complete, the prompt reverts to the CDF, SDF, or DXF prompt without the Objects option. Then AutoCAD prompts for the template file name and extract file name, respectively.

Duplications

The example extract file listed some records whose values in every field were the same. This would serve no purpose in a telephone directory but in a list of objects in a drawing it is possible that the only difference between two objects is their location in the drawing. In a bill of materials, the purchasing agent is not concerned with where, but how many duplicate objects there are. If it were essential to distinguish every object, the BL:X, BL:Y, and BL:Z fields could be included in the template to identify each object. Multiple insertions of the same block in the same location could still be distinguished by their BL:HANDLE field if the handles system variable were set to ON, or by their BL:NUMBER if they were inserted with the MINSERT command.

Also, to the experienced pipe estimator (or astute novice) there are other duplications that are possible if not taken into account. Counting bolts and weld lengths for every fitting and valve could result in twice the quantities required. Mating flanges, each having 4 bolt holes, only require 4 bolts to assemble. An ell welded to a tee likewise requires the specified weld length only once. Therefore, Attribute value quantities associated with corresponding tags should take this and similar problems in mating assemblies into account.

If the same fittings (in the form of block references with attributes) are shown in more than one view in a drawing, some mechanism should be provided to prevent duplication of quantities. As you can see, advanced features (attributes) that provide solutions to complex problems (data extraction) often require carefully planned implementation.

Changing Objects in a Block Reference without Losing Attribute Values

Sometimes it might be desirable to make changes to the objects within block references that have been inserted with attributes and have had values assigned to the attributes. Remember that the values of the attributes can be edited with the ATTEDIT command. But, in order to change the geometry of a block reference, it is normally required that you explode the block reference, make the necessary changes, and then redefine the block. As long as the attribute definitions keep the same tags in the new definition, the redefined blocks that have already been inserted in the drawing will retain those attributes with the original definitions.

Another method of having a new block definition applied to existing block references is to create a new drawing with the new block definition. This can be done by using the WBLOCK command to create a drawing that conforms to the old definition and then to edit that drawing. Or you can start a new drawing with the name of the block that you wish to change.

In order to apply the new definition to the existing block reference, you can call up the drawing from the OPEN command, and then use the INSERT command with the **Block Name:** option. For example, if you wish to change the geometry of a block named Part_1, and the changes have been made and stored in a separate drawing with the same name (Part_1), the sequence of prompts is as follows:

```
Command: -insert
Enter block name or [?] <x>: part_1=
Block "part_1" already exists. Redefine it? [Yes/No] <N>: y
Block "Part_1" redefined
Regenerating model.
Specify insertion point or [Scale/X/Y/Z/Rotate/PScale/PX/PY/PZ/PRotate]:
    (ENTER)
```

It is not necessary to specify an insertion point or to respond to scale factor and rotation angles. Simply inserting the block reference with the = (equals) behind the name will cause the definition of the drawing to become the new definition of the block with the same name residing in the current drawing.

REDEFINING A BLOCK AND ATTRIBUTES

The ATTREDEF command allows you to redefine a block reference and updates associated attributes. Invoke the ATTREDEF command:

Command: prompt	**attredef** (ENTER)

AutoCAD prompts:

```
Command: attredef
Enter name of the block you wish to redefine: (specify the block name to
    redefine)
Select objects for new Block...
Select objects: (select objects for the block to redefine and press ENTER)
Specify insertion base point of new Block: (specify the insertion basepoint of
    the new block)
```

New attributes assigned to existing block references are given their default values. Old attributes in the new block definition retain their old values. Old attributes not included in the new block definition are deleted from the old block references.

DIVIDING OBJECTS

The DIVIDE command causes AutoCAD to divide an object into equal-length segments, placing markers at the dividing points. Objects eligible for application of the DIVIDE command are the line, arc, circle, ellipse, spline, and polyline. Selecting an object other than one of these will cause an error message to appear, and you will be returned to the "Command:" prompt.

Invoke the DIVIDE command:

Draw menu	Point > Divide
Command: prompt	**divide** (ENTER)

AutoCAD prompts:

Command: **divide**
Select the object to divide: *(select a line, arc, circle, ellipse, spline, or polyline)*
Enter the number of segments or [Block]: *(specify the number of segments)*

You may respond with an integer from 2 to 32767, causing points to be placed along the selected object at equal distances but not actually separating the object. The Object Snap NODe can snap at the divided points. Logically, there will be one less point placed than the number entered, except in the case of a circle. The circle will have the first point placed at the angle from the center of the current snap rotation angle. A closed polyline will have the first point placed at the first point drawn in the polyline. The total length of the polyline will be divided into the number of segments entered without regard to the length of the individual segments that make up the polyline. An example of a closed polyline is shown in Figure 10–43.

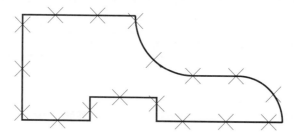

Figure 10–43 *The DIVIDE command as used with a closed polyline*

 Note: It is advisable to set the PDSIZE and PDMODE system variables to values that will cause the points to be visible.

The **Block** option allows a named block reference to be placed at the dividing points instead of a point. The sequence of prompts is as follows:

Command: **divide**
Select object to divide: *(select a line, arc, circle, ellipse, spline, or polyline)*
Enter the number of segments or [Block]: **block** *(or just* **b***)*
Enter name of block to insert: *(enter the name of the block)*
Align block with object? [Yes/No] <Y>: *(press* ENTER *to align the block
    reference with the object, or enter* **n***, for not to align with the block reference)*
Enter the number of segments: *(specify the number of segments)*

If you respond with **No** or **N** to the "Align block with object?" prompt, all of the block references inserted will have a zero angle of rotation. If you default to **Yes**, the angle of rotation of each inserted block reference will correspond to the direction of the linear part of the object at its point of insertion or to the direction of a line tangent to a circular part of an object at the point of insertion.

MEASURING OBJECTS

The MEASURE command causes AutoCAD to divide an object into specified-length segments, placing markers at the measured points. Objects eligible for application of the MEASURE command are the line, arc, circle, ellipse, spline, and polyline. Selecting an object other than one of these will cause an error message to appear and you will be returned to the "Command:" prompt.

Invoke the MEASURE command:

Draw menu	Point > Measure
Command: prompt	**measure** (ENTER)

AutoCAD prompts:

Command: **measure**
Select object to measure: *(select a line, arc, circle, ellipse, spline, or polyline)*
Specify length of segment or [Block]: *(specify the length of the segment, or
    enter* **b** *for the Block option)*

If you reply with a distance, or show AutoCAD a distance by specifying two points, the object is measured into segments of the specified length, beginning with the closest endpoint from the selected point on the object. The **Block** option allows a named block reference to be placed at the measured point instead of a point. The sequence of prompts is as follows:

Command: **measure**
Select object to measure: *(select a line, arc, circle, ellipse, spline, or polyline)*
Specify length of segment or [Block]: **block** *(or just* **b***)*
Enter name of block to insert: *(enter the name of the block)*
Align block with object? [Yes/No] <Y>: *(press* ENTER *to align the block
    reference with the object, or enter* **n***, for not to align with the block reference)*
Specify length of segment: *(specify the length of segments)*

PROJECT EXERCISE

This project exercise provides point-by-point instructions for creating the objects shown in Figure P10–1. In this exercise you will apply the skills acquired in Chapters 1 through 10.

Figure P10–1 *Completed project drawing*

In this project you will learn to do the following:

- Set up the drawing, including Limits, and Layers.
- Use the LINE, PLINE, CIRCLE, ARC, and SOLID commands to create objects.
- Use the BLOCK command to save objects for later insertion.
- Use the ATTDEF command to create an attribute and include it in the block definition. This enables you to attach a text string to the block reference when it is inserted.
- Use the INSERT command to insert the block references created earlier.

SET UP THE DRAWING AND DRAW A BORDER

Step 1:   Start the AutoCAD program.

Step 2:   To create a new drawing, invoke the NEW command from the Standard toolbar or choose New from the File menu. AutoCAD displays the **Create New Drawing** dialog box. Choose the **Start From Scratch** option, and choose **OK** to create a new drawing.

Step 3:   Invoke the LIMITS command, and set the limits to the lower left and upper right coordinates as shown in the SETTINGS/VALUES table. Set the grid and snap values as shown also.

SETTINGS	VALUE		
UNITS	Decimal		
LIMITS	Lower left corner: 0,0		
	Upper right corner: 18,12		
GRID	0.5		
SNAP	0.25		
LAYERS	*NAME*	*COLOR*	*LINETYPE*
	Blocks	Red	Continuous
	Border	Cyan	Continuous
	Equipment	Green	Continuous
	Instrument	Magenta	Continuous
	Pipeline	Blue	Continuous
	Text	White	Continuous

Step 4:   Invoke the LAYER command from the Object Properties toolbar, or choose Layer from the Format menu. AutoCAD displays the Layer Properties Manager dialog box. Create six layers, rename them as shown in the table, and assign the appropriate color and linetype.

Step 5:   Set Border as the current layer and set the Ortho, Grid, and Snap options all to ON. Invoke the ZOOM ALL command to display the entire limits on screen.

Step 6:   Invoke the RECTANGLE command from the Draw toolbar to draw the border (17" x 11"), as shown in Figure P10–2.

> Command: **rectangle**
> Specify first corner point or [Chamfer/Elevation/Fillet/Thickness/
>   Width]: **.5,.5**
> Specify other corner point: **@17,11**

Figure P10–2 *Border for piping flowsheet*

Step 7:   Set Blocks as the current layer.

Step 8:   Invoke the ZOOM command and use the Window option to zoom in on a small area of the display, as shown in Figure P10–3.

> Command: **zoom**
> Specify corner of window, enter a scale factor (nX or nXP), or
> [All/Center/Dynamic/Extents/Previous/Scale/Window] <real time>:
>   **window**
> Specify first corner: **3,3**
> Specify opposite corner: **6,6**

Figure P10–3 *Zoom in on a portion of the display*

Step 9: Open the Drafting Settings dialog box from the Tools menu, and set the grid to 0.125 and the snap to 0.0625, and set the grid and snap to ON.

Step 10: Draw the gate valve as shown in Figure P10–4.

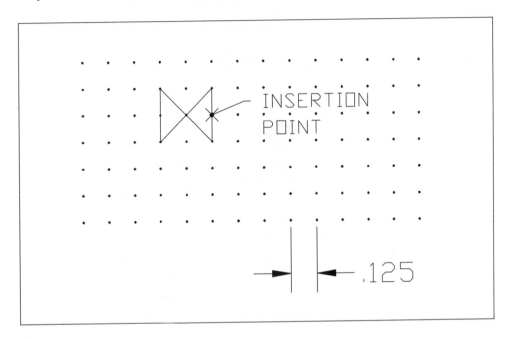

INSERTION POINT

.125

Figure P10–4 *The gate valve*

Step 11: Invoke the BLOCK command from the Draw toolbar, as shown in Figure P10–5.

Figure P10–5 *Invoking the* BLOCK *command from the Draw toolbar*

AutoCAD displays the Block Definition dialog box shown in Figure P10–6.

Figure P10–6 *Block Definition dialog box*

Enter **GATE** in the **Block name:** text box. Select the **Pick Point** button. AutoCAD prompts:

> Specify insertion base point: *(specify the insertion point as shown in Figure P10–4)*

Once you specify the insertion point, AutoCAD redisplays the Block Definition dialog box. Choose the **Select Objects** button. AutoCAD prompts:

> Select objects: *(select all the objects that comprise the gate valve)*

Once you select the objects, AutoCAD redisplays the Block Definition dialog box. Choose the **Delete** radio button in the Objects section of the dialog box. Choose the **OK** button to create the block called GATE and close the Block Definition dialog box.

Step 12: Draw the remaining symbols required for the piping flow sheet on the appropriate layers, as shown in the Figure P10–7.

Figure PI0–7 *Required blocks for the piping flowsheet drawing*

Create individual blocks for all the symbols by providing appropriate insertion points and block names (as shown in uppercase letters in Figure P10–7). If necessary, refer to Step 11 for the Step-by-Step procedure to create a block.

Before you create a block for the instrument bubble, define an attribute.

Invoke the ATTDEF command. AutoCAD displays the Attribute Definition dialog box. Set appropriate attributes and the required prompts, as shown in Figure P10–8. Choose the **Pick Point** button to place the attribute in the center of the instrument bubble, as shown in Figure P10–7.

Once you have defined the attribute, create the block for the instrument bubble. Make sure to include the attribute tag as part of the block when selecting objects.

Figure P10–8 *Attribute Definition dialog box*

Step 13: Invoke the ZOOM ALL command to display the entire drawing.

Step 14: Open the Drafting Settings dialog box from the Tools menu and set the Grid to 0.25, the Snap to 0.125, and set the Grid and Snap to ON.

Step 15: Set Equipment as the current layer.

Step 16: Draw the vertical and horizontal vessels and the boxes for the exchangers as shown in Figure P10–9. Following are the prompt sequences to draw the vessels and boxes for the exchangers.

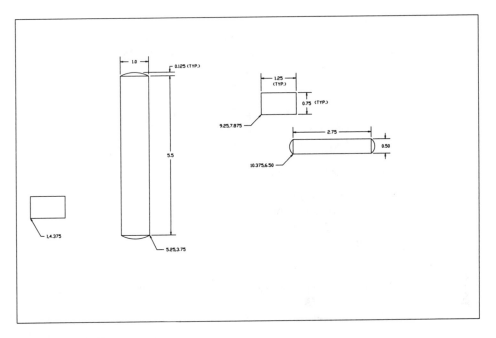

Figure P10–9 *Equipment layout*

To draw the vertical vessel:

> Command: **rectangle**
> Specify first corner point or [Chamfer/Elevation/Fillet/Thickness/
>   Width]: **5.25,3.75**
> Specify other corner point: **@–1,5.5**

> Command: **arc**
> Specify start point of arc or [CEnter]: **4.25,3.75**
> Specify second point of arc or [CEnter/ENd]: **4.75,3.625**
> Specify end point of arc: **5.25,3.75**

> Command: **arc**
> Specify start point of arc or [CEnter]: **5.25,9.25**
> Specify second point of arc or [CEnter/ENd]: **4.75,9.375**
> Specify end point of arc: **4.25,9.25**

To draw the horizontal vessel:

> Command: **rectangle**
> Specify first corner point or [Chamfer/Elevation/Fillet/Thickness/
>   Width]: **10.375,6.5**
> Specify other corner point: **@2.75,0.50**

Command: **arc**
Specify start point of arc or [CEnter]: **10.375,7**
Specify second point of arc or [CEnter/ENd]: **10.25,6.75**
Specify end point of arc: **10.375,6.5**

Command: **arc**
Specify start point of arc or [CEnter]: **13.125,6.5**
Specify second point of arc or [CEnter/ENd]: **13.25,6.75**
Specify end point of arc: **13.125,7**

To draw the boxes for the exchangers:

Command: **rectangle**
Specify first corner point or [Chamfer/Elevation/Fillet/Thickness/
    Width]: **9.25,7.875**
Specify other corner point: **@1.25,0.75**

Command: **rectangle**
Specify first corner point or [Chamfer/Elevation/Fillet/Thickness/
    Width]: **1,4.375**
Specify other corner point: **@1.25,0.75**

Step 17: Insert the exchangers and pump as shown in Figure P10–10 by invoking the
INSERT BLOCK command from the Draw toolbar (see Figure P10–11).

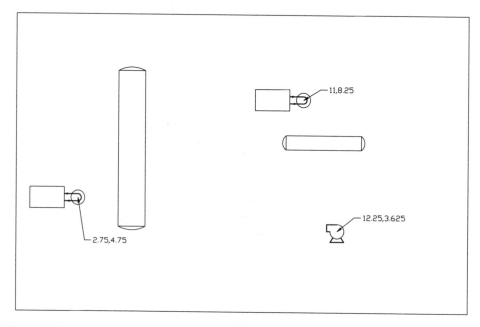

Figure P10–10 *Layout with exchangers and pump*

Figure P10–11 *Invoking the* INSERT BLOCK *command from the Draw toolbar*

AutoCAD displays the Insert dialog box shown in Figure P10–12.

Figure P10–12 *Insert dialog box*

Select EXCH block name from the **Name:** list box. Set the **Specify On-screen** check box to OFF in the **Insertion point** section of the dialog box. In the Insertion Point section of the dialog box, enter **2.75** in the **X:** text box and **4.75** in the **Y:** text box. In the Scale section of the dialog box, enter **1.00** in the **X:** text box and **1.00** in the **Y:** text box. In the Rotation section of the dialog box, enter **0** in the **Angle:** text box. Choose the **OK** button to insert the block reference and close the dialog box.

Similarly, insert the EXCH block reference again at the insertion point **11,8.25**, with scale set to **1.00** and rotation to **0** degrees. Insert the PUMP block reference at the insertion point **12.25,3.625**, with scale set to **1.00** and rotation set to **0** degrees.

Step 18: Insert the NOZ block reference at the appropriate locations, as shown in Figure P10–13.

Draw a horizontal line (representing a tray) in the vessel just below the highest side, and then copy it three times, as shown in Figure P10–13.

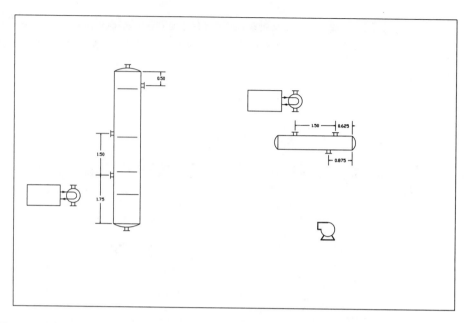

Figure P10–13 *Layout with nozzles*

Step 19: Set Pipeline as the current layer.

Step 20: Lay out the pipelines as shown in Figure P10–14 by invoking the LINE command.

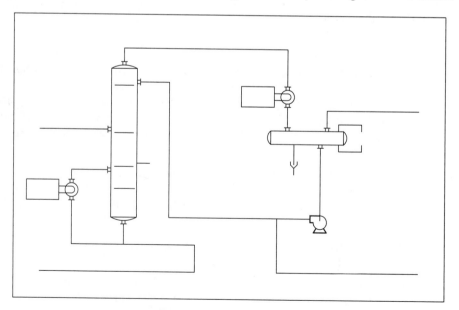

Figure P10–14 *Layout with pipelines*

Step 21: Insert the flowsheet symbols at the appropriate places by invoking the INSERT command, as shown in Figure P10–15. Make sure to provide appropriate attribute values while inserting the instrument bubble block reference. After inserting the block references, use the BREAK command to break out the pipeline passing through the valve symbols.

Figure P10–15 *Layout with flowsheet symbols*

Step 22: Set Instruments as the current layer.

Step 23: Invoke the PLINE command to draw LINE 1, LINE 2, and LINE 3 instrument air lines as shown in Figure P10–16.

Figure P10–16 *Layout with instrument air lines*

Step 24: Invoke the DIVIDE command to insert the block reference tic on the instrument air lines as shown in Figure P10–17. AutoCAD prompts:

> Command: **divide**
> Select object to divide: *(select LINE 1)*
> Enter the number of segments or [Block]: **b**
> Enter name of block to insert: **tic**
> Align block with object? [Yes/No] <Y>: **n**
> Enter the number of segments: **5**
>
> Command: **divide**
> Select object to divide: *(select LINE 2)*
> Enter the number of segments or [Block]: **b**
> Enter name of block to insert: **tic**
> Align block with object? [Yes/No] <Y>: **n**
> Enter the number of segments: **7**
>
> Command: **divide**
> Select object to divide: *(select LINE 3)*
> Enter the number of segments or [Block]: **b**
> Enter name of block to insert: **tic**
> Align block with object? [Yes/No] <Y>: **n**
> Enter the number of segments: **8**

Figure P10–17 *Layout with tic marks on the instrument air lines*

Step 25: Set Text as the current layer.

Step 26: Invoke the TEXT command to add the text as shown in Figure P10–18. The larger text height is set to 0.18 and the smaller text height is set to 0.125.

Figure P10–18 *Layout with text*

Step 27: Save the drawing as CH10-PROJ and exit the AutoCAD program.

Congratulations! You have just successfully applied several AutoCAD concepts in creating a simple schematic drawing.

EXERCISES 10–1 THROUGH 10–4

EXERCISE 10–1

Create the drawing according to the settings given in the following table. Create all the required blocks by referring to the piping flowsheet symbols in Figure Ex10–1b. (Set the grid to 0.125 and the snap to 0.0625 to draw the symbols.)

Settings	Value		
1. Units	Decimal		
2. Limits	Lower left corner: 0,0		
	Upper right corner: 12,9		
3. Grid	0.5		
4. Snap	0.125		
5. Layers	*NAME*	*COLOR*	*LINETYPE*
	Blocks	Red	Continuous
	Border	Cyan	Continuous
	Equipment	Green	Continuous
	Lines	Magenta	Continuous
	Text	White	Continuous

Hints: Note that the insertion points are located on the symbol where they join the objects to which they are connected.

In order to attach a nozzle to the curved top of the reactor, one of its legs will have to overlap. Then the block reference will have to be exploded before part of it can be trimmed.

Figure Ex10–1a

Figure Ex10–1b *Piping Flowsheet Symbols*

EXERCISE 10-2

Create the drawing according to the settings given in the following table. Create all the required blocks by referring to the electrical symbols in Figure Ex10–2b. (Set the grid to 0.125 and the snap to 0.0625 to draw the symbols.)

Settings	Value		
1. Units	Decimal		
2. Limits	Lower left corner: 0,0		
	Upper right corner: 12,9		
3. Grid	0.5		
4. Snap	0.125		
5. Layers	*NAME*	*COLOR*	*LINETYPE*
	Blocks	Red	Continuous
	Border	Cyan	Continuous
	Lines	Magenta	Continuous
	Text	White	Continuous
	Construction	Red	Continuous

Hints: Note that the insertion points are located on the symbol where they join the objects to which they are connected.

Draw the circuits with continuous lines on the Construction layer. Insert the block references by using the Osnap mode Nearest. Then set the Construction layer to OFF and draw the lines between the symbols, connecting to their endpoints.

Figure Ex10–2a *Electrical Schmatic*

Figure Ex10–2b *Electrical symbols*

EXERCISE 10-3

Create the drawing according to the settings given in the following table. Create all the blocks used in this drawing (define attributes where necessary). Create all required blocks by referring to the electrical symbols in Figure Ex10–2b. (Set the grid to 0.125 and the snap to 0.0625 to draw the symbols.)

Settings	Value		
1. Units	Decimal		
2. Limits	Lower left corner: 0,0		
	Upper right corner: 12,9		
3. Grid	0.5		
4. Snap	0.125		
5. Layers	*NAME*	*COLOR*	*LINETYPE*
	Blocks	Red	Continuous
	Border	Cyan	Continuous
	Lines	Magenta	Continuous
	Text	White	Continuous
	Construction	Red	Continuous

Hints: Note that the insertion points are located on the symbol where they join the objects to which they are connected.

Draw the circuits with continuous lines on the Construction layer. Insert the block references by using the Osnap mode Nearest. Then set the Construction layer to OFF and draw the lines between the symbols, connecting to their endpoints.

Figure Ex10–3 *Electrical schematic*

EXERCISE 10–4

Create the drawing according to the settings given in the following table. Create all the required blocks by referring to the architectural symbols in Figure Ex10–4b.

Settings	Value
1. Units	Architectural
2. Limits	Lower left corner: 0'-0",0'-0" Upper right corner: 64'-0",48'-0"
3. Grid	1'-0"
4. Snap	6"

Settings	Value		
5. Layers	*NAME*	*COLOR*	*LINETYPE*
	Blocks	Red	Continuous
	Border	Cyan	Continuous
	Floor plan	Magenta	Continuous
	Text	White	Continuous
	Construction	Red	Continuous

Hint: Defining attributes is useful for labeling the desks as they are inserted.

Figure Ex10-4 *Office Layout*

Figure Ex10–4b *Architectural Symbols*

REVIEW QUESTIONS

1. If you used "BL??7* as the pattern for listing blocks, which of the following block names would be listed?

    a. BL7

    b. BLOCK739

    c. BLUE727

    d. BLACK

    e. none of the above

2. It is possible to force invisible attributes to display on a drawing.

    a. True

    b. False

3. Which of the following forms of attribute extraction require template files?

    a. CDF   d. a and b

    b. SDF   e. none of the above

    c. DXF

4. The maximum number of characters for a block name is:

    a. 8   d. 31

    b. 16   e. 43

    c. 23

5. A block is:

    a. a rectangular-shaped figure available for insertion into a drawing

    b. a single element found in a block formation of a building drawn with AutoCAD

    c. one or more objects stored as a single object for later retrieval and insertion

    d. none of the above

6. A block reference cannot be exploded if it:

    a. consists of other blocks (nested)

    b. has a negative scale factor

    c. has been moved

    d. has different X and Y scale factors

    e. none of the above

7. A new drawing may be created from an existing block by using the command:

   a. WBLOCK   d. SAVEAS

   b. NEW   e. none of the above

   c. BLOCK=

8. All of the following can be exploded, except:

   a. block references

   b. associative dimensions

   c. polylines

   d. Block references inserted with MINSERT command

   e. none of the above

9. To insert a block reference called "TABLE" and have the block reference converted to an individual object as it is inserted, what should you use for the block name?

   a. /TABLE   d. TABLE

   b. *TABLE   e. none of the above

   c. ?TABLE

10. The command used to write all or just part of a drawing out to a new drawing file is:

   a. SAVEAS   d. FILE

   b. BLOCK   e. WBLOCK

   c. DXFOUT

11. To return a block reference back to its original objects, use:

   a. EXPLODE   d. UNDO

   b. BREAK   e. STRETCH

   c. CHANGE

12. To identify a new insertion point for a drawing file which will be inserted into another drawing, invoke the:

   a. BASE command   d. WBLOCK command

   b. INSERT command   e. DEFINE command

   c. BLOCK command

13. MINSERT places multiple copies of an existing block similar to the command:

   a. ARRAY   d. COPY

   b. MOVE   e. MIRROR

   c. INSERT

14. If one drawing is to be inserted into another drawing and editing operations are to be performed on the inserted drawing, you must first:

    a. use the PEDIT command

    b. EXPLODE the inserted drawing

    c. UNDO the inserted drawing

    d. nothing, it can be edited directly

    e. none of the above

15. The BASE command:

    a. can be used to move a block reference

    b. is a subcommand of PEDIT

    c. will accept 3D coordinates

    d. allows one to move a dimension baseline

    e. none of the above

16. Attributes are associated with:

    a. objects

    b. block references

    c. text

    d. layers

    e. shapes

17. To merge two drawings, use:

    a. INSERT

    b. MERGE

    c. BIND

    d. BLOCK

    e. IGESIN

18. The WBLOCK command can be used to create a new block for use:

    a. in the current drawing

    b. in an existing drawing

    c. in any drawing

    d. only in a saved drawing

    e. none of the above

19. The DIVIDE command causes AutoCAD to:

    a. divide an object into equal length segments

    b. divide an object into two equal parts

    c. break an object into two objects

    d. all of the above

20. One cannot explode:

    a. polylines containing arcs

    b. blocks containing polylines

    c. dimensions incorporating leaders

    d. block references inserted with different X, Y, and Z scale factors

    e. none of the above

21. The DIVIDE command will:

    a. place points along a line, arc, polyline, or circle

    b. accept 1.5 as segment input

    c. place markers on the selected object and separate it into different segments

    d. divide any object into the equal number of segments

22. When using the -WBLOCK command to create a new block, one should respond to the block name with:

    a. a block name   d. a file name

    b. an equals sign   e. none of the above

    c. an asterisk

23. The WBLOCK command creates:

    a. a drawing file

    b. a collection of blocks

    c. a symbol library

    d. an object file

    e. none of the above

24. The MEASURE command causes AutoCAD to divide an object

    a. into specified length segments

    b. into equal length segments

    c. into two equal parts

    d. all of the above

25. A command used to edit attributes is:

    a. DDATTE   d. ATTFILE

    b. EDIT   e. none of the above

    c. EDITATT

26. The WBLOCK command:

  a. means "Window Block" and allows for the use of a window to define a block

  b. allows you to send a previously defined block to a file, thus creating a drawing file of the block

  c. is used in lieu of the INSERT command when the block is stored in a separate disk file

  d. none of the above

27. Attributes are defined as the:

  a. database information displayed as a result of entering the LIST command

  b. X and Y values which can be entered when inserting a block reference

  c. coordinate information of each vertex found along a SPLINE object

  d. none of the above

Directions: Using the drawing created in Exercise 10-2, answer the following questions.

28. If the NPN Transistor object were drawn and inserted into the drawing in the orientation shown for item on the left, what scale factor would be required to insert the NPN Transistor in the orientation for the item shown on the right?

  a. x = 1, y = 1

  b. x = -1, y = -1

  c. x = 1, y = -1

  d. x = -1, y = 1

29. If the Resistor object were drawn and inserted into the drawing in the orientation shown for item R2 what rotation angle would be required to insert the Resistor in the orientation for item R1?

  a. 90

  b. -180

  c. 270

  d. 320

30. How many Resistors have been inserted into the drawing in the position in which it was originally created?

  a. 5

  b. 7

  c. 10

  d. 12

31. How many uniquely created symbols have been inserted into the drawing?

    a. 5   c. 25

    b. 7   d. 27

32. The _____ command returns a block back into its original objects.

    a. CHANGE

    b. BREAK

    c. RETURN

    d. EXPLODE

33. A new drawing may be created from an existing block by using what command?

    a. BMAKE

    b. BLOCK

    c. WBLOCK

    d. NEW

34. What is the maximum number of times a block can be inserted into a drawing?

    a. 25

    b. 50

    c. 100

    d. unlimited

35. Which command allows you to delete uninserted blocks from the block definition table?

    a. ID

    b. PURGE

    c. LIST

    d. DBLIST

36. How many attributes can be attached to a block?

    a. 1   c. 10

    b. 5   d. unlimited

37. What is the string of text in an attribute called?

    a. name   c. text

    b. tag   d. value

CHAPTER 11

External References

INTRODUCTION

One of the most powerful time-saving features of AutoCAD is the ability to have one drawing become part of a second drawing while maintaining the integrity and independence of the first one. AutoCAD lets you display or view the contents of as many as 32,000 other drawing files while working in your current drawing file. This feature is provided by the XREF command, short for external reference.

After completing this chapter, you will be able to do the following:

- Attach and detach reference files
- Change the path for reference files
- Load and unload reference files
- Decide whether to attach or overlay an external reference file
- Clip external reference files
- Control dependent symbols
- Edit external references
- Manage external references
- Use the BIND command to add dependent symbols to the current drawing

EXTERNAL REFERENCES

Prior to Release 11, existing AutoCAD drawings could be combined in only one way, by means of the INSERT command, to insert one drawing into another. When one drawing is inserted into another, it is actually a duplicate of the inserted drawing that becomes a part of the drawing into which it is inserted. The data from the inserted drawing is added to the data of the current drawing. Once the duplicate of a drawing is inserted, no link or association remains between the original drawing from which the inserted duplicate came and the drawing it has been inserted into.

The INSERT command and the XREF command now give users a choice of two methods for combining existing drawing files. The external reference (XREF) feature does not obsolete the INSERT feature; users can decide which method is more appropriate for the current application.

When a drawing is externally referenced (instead of inserted as a block), the user can view and object snap to the external reference from the current drawing, but each drawing's data is still stored and maintained in a separate drawing file. The only information in the reference drawing that becomes a permanent part of the current drawing is the name of the reference drawing and its folder path. If necessary, externally referenced files can be scaled, moved, copied, mirrored, or rotated by using the AutoCAD modify commands. You can control the visibility, color, and linetype of the layers belonging to an external drawing file. This lets you control which portions of the external drawing file are displayed, and how. No matter how complex an external reference drawing may be, it is treated as a single object by AutoCAD. If you invoke the MOVE command and point to one line, for example, the entire object moves, not just the line you pointed to. You cannot explode the externally referenced drawing. A manipulation performed on an external reference will not affect the original drawing file, because an external reference is only an image, however scaled or rotated.

Borders are an excellent example of drawing files that are useful as external reference files. The objects that make up a border will use considerable space in a file, and commonly amount to around 20,000 bytes. If a border is drawn in each drawing file, this would waste a large amount of disk space when you multiply 20,000 bytes by 100 drawing files. If external reference files are used correctly, they can save 2 MB of disk space in this case.

Accuracy and efficient use of drawing time are other important design features that are enhanced through external reference files. When an addition or change is made to a drawing file that is being used as an external reference file, all drawings that use the file as an external reference will reflect the modifications. For example, if you alter the title block of a border, all the drawing files that use that border as an external reference file will automatically display the title block revisions. (Can you imagine accessing 100 drawing files to correct one small detail?) External reference files will save time and ensure the drawing accuracy required to produce a professional product. Figure 11–1 shows a drawing that externally references four other drawings to illustrate the doorbell detail.

There is a limit of 32,000 external references you can add to a drawing. In practice, this represents an unlimited number. If necessary, you can nest them so that loading one external reference automatically causes another external reference to be loaded. When you attach a drawing file as an external reference file, it is permanently attached until it is detached or bound to the current drawing. When you load the drawing with external references, AutoCAD automatically reloads each external reference drawing file; thus, each external drawing file reflects the latest state of the referenced drawing file.

```
Command: xref
?/Bind/Detach/Path/Reload/<Attach>: a
Xref to Attach:
```

Figure 11–1 *This drawing appears to be a single drawing but references four other drawings*

The XREF command, when combined with the networking capability of AutoCAD, gives the project manager powerful features to cope with the problems of file management. The project manager instantaneously sees the work of the departments and designers working on aspects of the contract. If necessary, you can overlay a drawing where appropriate, track the progress, and maintain document integrity. At the same time, departments need not lose control over individual designs and details.

In AutoCAD you can control the display of the external reference file by means of clipping, so you can display only a specific section of the reference file.

EXTERNAL REFERENCES AND DEPENDENT SYMBOLS

The symbols that are carried into a drawing by an external reference are called dependent symbols, because they depend on the external file, not on the current drawing, for their characteristics. The symbols have arbitrary names and include blocks, layers, linetypes, text styles, and dimension styles.

When you attach an external reference drawing, AutoCAD automatically renames the xref's dependent symbols. AutoCAD forms a temporary name for each symbol by combining its original name with the name of the xref itself. The two names are separated by the vertical bar (|) character. Renaming the symbols prevents the xref's objects from taking on the characteristics of existing symbols in the drawing.

For example, you created a drawing called PLAN1 with layers 0, First-fl, Dim, and Text, in addition to blocks Arrow and Monument. If you attach the PLAN1 drawing as an external reference file, the layer First-fl will be renamed as PLAN1|first-fl, Dim as PLAN1|dim, and Text as PLAN1|text, as shown in Figure 11–2. Blocks Arrow and Monument will be renamed as PLAN1|Arrow and PLAN1|Monument. The only exceptions to renaming are unambiguous defaults like layer 0 and linetype continuous. The information on layer 0 from the reference file will be placed on the active layer of the current drawing when the drawing is attached as an external reference of the current drawing. It takes on the characteristics of the current drawing.

Figure 11–2 *Layer Properties Manager*

This prefixing is carried to nested xrefs. For example, if the external file PLAN1 included an xref named "Title" that has a layer Legend, it would get the symbol name PLAN1|Title|Legend if PLAN1 were attached to another drawing.

This automatic renaming of an xref's dependent symbols has two benefits:

- It allows you to see at a glance which named objects belong to which external reference file.

- It allows dependent symbols to have the same name in both the current drawing and an external reference, and coexist without any conflict.

The AutoCAD commands and dialog boxes for manipulating named objects do not let you select an xref's dependent symbols. Usually, dialog boxes display these entries in lighter text.

For example, you cannot insert a block that belongs to an external reference drawing in your current drawing, nor can you make a dependent layer the current layer and begin creating new objects.

You can control the visibility of the layers (ON/OFF, Freeze/Thaw) of an external reference drawing and, if necessary, you can change the color and linetype. When the VISRETAIN system variable is set to 0 (default), any changes you make to these settings apply only to the current drawing session. They are discarded when you end the drawing. If VISRETAIN is set to 1, then the current drawing visibility, color, and linetype for xref dependent layers take precedence. They are saved with the drawing and are preserved during xref reload operations.

There may be times when you want to make your xref data a permanent part of your current drawing. To make an xref drawing a permanent part of the current drawing, use the Bind option of the XREF command. With the Bind option, all layers and other symbols, including the data, become part of the current drawing. This is similar to inserting a drawing via the INSERT command.

If necessary, you can make dependent symbols such as layers, linetypes, text styles, and dim styles part of the current drawing by using the XBIND command instead of binding the whole drawing. This allows you to work with the symbol just as if you had defined it in the current drawing.

ATTACHING AND MANIPULATING WITH THE XREF COMMAND

The XREF command provides various options for attaching and detaching external references files.

Invoke the XREF command:

Reference toolbar (see	Choose the EXTERNAL REFERENCE command Figure 11-3)
Insert menu	Choose External Reference
Command: prompt	**xref** (ENTER)

Figure 11-3 *Invoking the* EXTERNAL REFERENCE *command from the Reference toolbar*

AutoCAD displays the Xref Manager dialog box, similar to Figure 11–4.

 Note: Enter **-xref** and press ENTER at the "Command:" prompt for all the options available at the Command: prompt level for attaching and manipulating external reference drawings. The options are the same ones available in the Xref Manager dialog box.

AutoCAD provides two options for listing attached external reference drawings: the **List View** and the **Tree View**. By default, the List View option (see Figure 11–4) displays a list of the attached reference files and their associated data. To sort a column alphabetically, select the column heading. A second click sorts it in reverse order. To resize a column's width, select the separator between columns and drag the pointing device to the right or left.

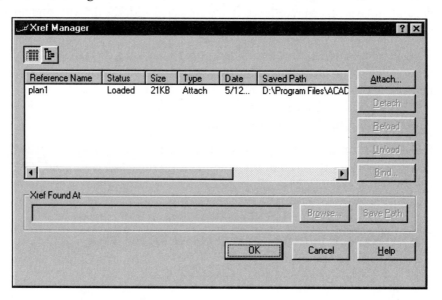

Figure 11–4 *Xref Manager dialog box*

The external reference file's **Name** does not have to be the same as its original file name. To rename the external file, double-click the name or press F2. AutoCAD allows you to rename the file. The new name can be up to 31 characters and must be without any spaces.

The **Status** column displays the state of the external reference file, which can be Loaded, unLoaded, Unreferenced, Unresolved, Orphaned, Reload, or Not found. A detailed discussion of these states is provided later in the chapter.

The **Size** column shows the file size of the corresponding external reference drawing. The size is not displayed if the external reference is unloaded, not found, or unresolved.

The **Type** column indicates whether the external reference is an attachment or an overlay.

The **Date** column displays the last date the associated drawing was modified. The date is not displayed if the external reference is unloaded, not found, or unresolved.

The **Saved Path** column shows the saved path of the associated external reference file.

Selecting any field highlights the external file's reference name.

You can also display the information as a Tree View. To do so, click the Tree View button located at the top left of the dialog box, or press F4. To switch back to the List View, press F3.

In a Tree View listing, AutoCAD displays a hierarchical representation of the external reference in alphabetical order, as shown in Figure 11–5. Tree View shows the level of nesting relationship of the attached external references, whether they are attached or overlaid, whether they are loaded, unloaded, marked for reload or unload, or not found, unresolved, or unreferenced.

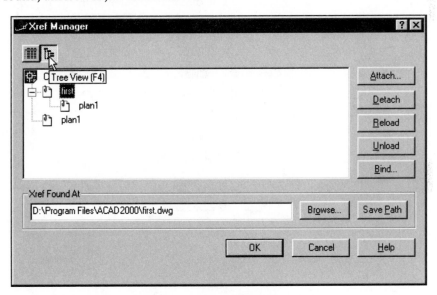

Figure 11–5 *Xref Manager dialog box with Tree View listing*

ATTACHING EXTERNAL REFERENCE DRAWINGS

The **Attach** option allows you attach a drawing as an external reference. Use it when you want to attach a new external reference file or a copy of the external reference file already attached to the current drawing. If you attach a drawing that itself contains an attached external reference, the attached external reference appears in the current drawing. If another person is currently editing the external reference drawing, the drawing attached is based on the most recently saved version.

To attach a drawing as an external reference, invoke the Attach option:

Xref Manager dialog box	Choose the Attach button
Reference toolbar	Choose the EXTERNAL REFERENCE ATTACH command (see Figure 11–6)
Command: prompt	**xattach** (ENTER)

Figure 11–6 *Invoking the EXTERNAL REFERENCE ATTACH command from the Reference toolbar*

AutoCAD displays the Select Reference File dialog box. Select the drawing file from the appropriate directory to attach to the current drawing. AutoCAD displays External Reference dialog box, as shown in Figure 11–7.

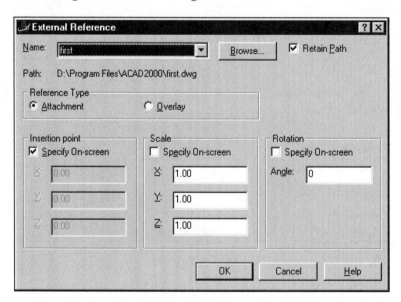

Figure 11–7 *External Reference dialog box*

Once an external reference drawing is attached to the current drawing, the external reference drawing name is added to the list box located next to the **Browse** button. When an attached external reference file name is selected from the list box, its path is displayed below.

In the **Reference Type** section of the dialog box, select one of the two available options: **Attachment** or **Overlay**. When you attach an external reference in the Attachment mode, the external reference will be included in the drawing when the drawing itself is attached as an external reference to another drawing. If you attach an external reference in the Overlay mode, in contrast, an overlaid external reference is not included in a drawing when the drawing itself is attached as an external reference or overlaid external reference to another drawing.

For example, PLAN-A drawing is attached as an overlaid external reference to PLAN-B. Then PLAN-B is attached as an external reference to PLAN-C. PLAN-A is not seen in PLAN-C because it is overlaid in PLAN-B, not attached as an external reference. But, if PLAN-A is attached as an external reference in PLAN-B, and in turn PLAN-B is attached in PLAN-C, both PLAN-A and PLAN-B will be seen in PLAN-C.

So, the only behavioral difference between overlays and attachments is how nested references are handled. Overlaid external references are designed for data sharing. If necessary, you can change the status from Attach to Overlay, or vice versa, by double-clicking the field in the Type column in the Xref Manager dialog box or by selecting the appropriate radio button in the Reference Type section of the External Reference dialog box.

The **Retain Path** check box in the External Reference dialog box determines whether or not the full path to the external reference is saved. If Retain Path is set to ON, then the external reference path is saved in the drawing database; if it is set to OFF, the name of the external reference drawing is saved without a path in the database. AutoCAD searches for the external reference drawing in the AutoCAD Support File Search Path and in the paths associated with the Files tab of the Options dialog box.

In the **Insertion Point** section of the External Reference dialog box, you can specify the insertion point of the external reference.

In the **Scale** section of the External Reference dialog box, you can specify the insertion point, X Scale Factor, Y Scale Factor, and Z Scale Factor of the external reference.

In the **Rotation** section of the External Reference dialog box, you can specify the rotation angle of the external reference.

The Insertion Point, Scale, and Rotation features are similar to those in the insertion of a block, explained in Chapter 10.

DETACHING EXTERNAL REFERENCE DRAWINGS

The **Detach** option allows you detach one or more external reference drawings from the current drawing. Only the external reference drawings attached or overlaid directly to the current drawing can be detached. You cannot detach an external reference drawing referenced by another external reference drawing. If the external reference is currently being displayed as part of the current drawing, it disappears when you detach it.

To detach an external reference drawing from the current drawing, first select the drawing name from the displayed list in the Xref Manager dialog box, and then select the **Detach** button. AutoCAD detaches the selected external reference drawing(s) from the current drawing. You can select several files at once by using the standard Windows methods (holding down SHIFT or CTRL while selecting).

RELOADING EXTERNAL REFERENCE DRAWINGS

The **Reload** option allows you to update one or more external reference drawings any time while the current drawing is in AutoCAD. When you load a drawing into AutoCAD, it automatically reloads any external references attached. The Reload option has been provided to reread the external drawing from the external drawing file whenever it is desirable to do so from within AutoCAD. The Reload option is helpful in a network environment to get the latest version of the reference drawing while you are in an AutoCAD session.

To reload external reference drawing(s), first select the drawing name(s) from the displayed list in the Xref Manager dialog box, and then select the **Reload** button. AutoCAD reloads the selected external reference drawing(s). You can select several files at once by using the standard Windows methods (holding down SHIFT or CTRL while selecting).

UNLOADING EXTERNAL REFERENCE DRAWINGS

The **Unload** option allows you to unload one or more external reference drawings from the current drawing. Unlike the Detach option, the Unload option merely suppresses the display and regeneration of the external reference definition, to help current session editing and improve performance. This option can also be useful when a series of external reference drawings needs to be viewed during a project on an as-needed basis. Rather than having the referenced files displayed at all times, you can reload the drawing when you require the information.

To unload external reference drawing(s), first select the drawing name(s) from the displayed list in the Xref Manager dialog box, then select the **Unload** button. AutoCAD turns off the display of the selected external reference drawing(s). You can select several files at once by using the standard Windows methods (holding down SHIFT or CTRL while selecting).

The effects of Unload and Reload take effect when you close the dialog box.

BINDING EXTERNAL REFERENCE DRAWINGS

The **Bind** option allows you to make your external reference drawing data a permanent part of the current drawing. To bind external reference drawing(s), first select the drawing name(s) from the displayed list in the Xref Manager dialog box, then select the **Bind** button. AutoCAD displays the Bind Xrefs dialog box shown in Figure 11–8. Select one of the two available bind types: Bind or Insert. You can select several files at once by using the standard Windows methods (holding down SHIFT or CTRL while selecting).

Figure 11–8 *Bind Xrefs dialog box*

With Bind, the external reference drawing becomes an ordinary block in your current drawing. It also adds the dependent symbols to your drawing, letting you use them as you would any other named objects. In the process of binding, AutoCAD renames the dependent symbols. The vertical bar symbol (|) is replaced with three new characters: a $, a number, and another $. The number is assigned by AutoCAD to ensure that the named object will have a unique name.

For example, if you bind an external reference drawing named PLAN1 that has a dependent layer PLAN1|FIRST-FL, AutoCAD will try to rename the layer to PLAN1$0$FIRST-FL. If there is already a layer by that name in the current drawing, then AutoCAD tries to rename the layer to PLAN1$1$FIRST-FL, incrementing the number until there is no duplicate.

If you do not want to bind the entire external reference drawing, but only specific dependent symbols, such as a layer, linetype, block, dimension style, or text style, then you can use the XBIND command, explained later in this chapter, in the section on "Adding Dependent Symbols to the Current Drawing."

With Insert, the external reference drawing is inserted in the current drawing just like inserting a drawing with the INSERT command. AutoCAD adds the dependent symbols to the current drawing by stripping off the external reference drawing name.

For example, if you insert an external reference drawing named PLAN1 that has a dependent layer PLAN1|FIRST-FL, AutoCAD will rename the layer to FIRST-FL. If there is already a layer by that name in the current drawing, then the layer FIRST-FL would assume the properties of the layer in the current drawing.

CHANGING THE PATH

To change to a different path or file name for the currently selected external reference file, choose the **Browse** button in the Xref Manager dialog box. AutoCAD displays the Select New Path dialog box, in which you can specify a different path or file name.

SAVING THE PATH

To save the path as it appears in the Saved Path field of the currently selected external reference file (and displayed in the **Xref Found At** text box), select the **Save Path** button. AutoCAD saves the path of the currently selected external reference file.

After making the necessary changes in the Xref Manager dialog box, choose the **OK** button to keep the changes and close the dialog box.

ADDING DEPENDENT SYMBOLS TO THE CURRENT DRAWING

The XBIND command lets you permanently add a selected subset of external reference-dependent symbols to your current drawing. The dependent symbols include the block, layer, linetype, dimension style, and text style. Once the dependent symbol is added to the current drawing, it behaves as if it were created in the current drawing and saved with the drawing when you ended the drawing session. While adding the dependent symbol to the current drawing, AutoCAD removes the vertical bar symbol (|) from each dependent symbol's name, replacing it with three new characters: a $, a number, and another $ symbol.

For example, you might want to use a block that is defined in an external reference. Instead of binding the entire external reference with the Bind option of the XREF command, it is advisable to use the XBIND command. With the XBIND command, the block and the layers associated with the block will be added to the current drawing. If the block's definition contains a reference to an external reference, AutoCAD binds that xref and all its dependent symbols as well. After binding the necessary dependent symbols, you can detach the external reference file.

Invoke the XBIND command:

Reference toolbar	Choose the EXTERNAL REFERENCE BIND command (see Figure 11–9)
Modify menu	Choose Object > External Reference > Bind
Command: prompt	**xbind** (ENTER)

Figure 11–9 *Invoking the* EXTERNAL REFERENCE BIND *command from the Reference toolbar*

AutoCAD displays the Xbind dialog box, similar to the Figure 11–10.

Figure 11–10 *Xbind dialog box*

 Note: To invoke the XBIND command at the "Command:" prompt level, enter **-xbind** and press ENTER at the "Command:" prompt. The available options are the same ones as in the Xbind dialog box.

On the left side of the Xbind dialog box AutoCAD lists the external reference files currently attached to the current drawing. Double-click on the name of the external reference file, and AutoCAD expands the list by listing the dependent symbols. Select the dependent symbol from the list and choose the **Add->** button. AutoCAD moves the selected dependent symbol into the Definitions to Bind list. If necessary, return it back to the external reference dependent list from the Definitions to Bind list by choosing the **<-Remove** button after selecting the appropriate dependent symbol.

Click the **OK** button to bind the definitions to the current drawing.

CONTROLLING THE DISPLAY OF EXTERNAL REFERENCES

The XCLIP command allows you to control the display of unwanted information by clipping the external reference drawings and blocks. Clipping does not edit or change the external reference or block, it just prevents part of the object from being displayed. The defined clipping boundary can be visible or hidden. You can also define the front and back clipping planes.

The clipping boundary consists of planar straight-line segments. You can also generate the clip boundary from polylines. If the polyline is fit-curved or has arcs, the clip boundary is displayed as if the polyline had been decurved prior to being used as a clip boundary.

The XCLIP command can be applied to one or more external references or blocks. If you set the clip boundary to OFF, the entire external reference or block is displayed. If you subsequently set the clip boundary to ON, the clipped drawing is displayed again. If necessary, you can delete the clipping boundary; AutoCAD redisplays the entire external reference or block. In addition, AutoCAD also allows you to generate a polyline from the clipping boundary.

Invoke the XCLIP command:

Reference toolbar command	Choose the EXTERNAL REFERENCE CLIP (see Figure 11–11)
Modify menu	Choose Object > Clip
Command: prompt	**xclip** (ENTER)

Figure 11–11 *Invoking the* EXTERNAL REFERENCE CLIP *command from the Reference toolbar*

AutoCAD prompts:

> Command: **xclip**
> Select objects: *(select one or more external references and/or blocks to be included in the clipping and press* ENTER *to complete the selection)*
> Enter clipping option [ON/OFF/Clipdepth/Delete/generate Polyline/New boundary] <New> *(right-click and select one of the available options from the shortcut menu)*

CREATING A NEW CLIPPING BOUNDARY

The **New** option (default) allows you to define a rectangular or polygonal clip boundary or generates a polygonal clipping boundary from a polyline. AutoCAD prompts:

> Specify clipping boundary:
> [Select polyline/Polygonal/Rectangular] <Rectangular>: *(right-click and select one of the available options from the shortcut menu)*

The **Rectangular** option (default) allows you to define a rectangular boundary by specifying the opposite corners of a window. The clipping boundary is applied in the current UCS and is independent of the current view.

The **Select polyline** option defines the boundary by using a selected polyline. The polyline can be open or closed, can be made of straight-line segments, but cannot intersect itself.

The **Polygonal** option defines a polygonal boundary by specifying points for the vertices of a polygon.

Once the clipping boundary is defined, AutoCAD displays only the portion of the drawing that is within the clipping boundary and then exits the command.

If you already have a clipping boundary of the selected external reference drawing, and you invoke the New Boundary option, then AutoCAD prompts:

> Delete old boundary(s)? [Yes/No] <Yes>: *(select one of the two available options)*

If you choose Yes, the entire reference file is redrawn and the command continues; if you choose No, the command sequence is terminated.

 Note: The display of the boundary border is controlled by the XCLIPFRAME system variable. If it is set to I (ON), then AutoCAD displays the boundary border; if it is set to 0 (OFF), then AutoCAD does not display the boundary border.

CONTROLLING THE DISPLAY OF THE CLIPPED BOUNDARY

The ON/OFF option controls the display of the clipped boundary. The OFF option displays all of the geometry of the external reference or block, ignoring the clipping boundary. The ON option displays the clipped portion of the external reference or block only.

SETTING THE FRONT AND BACK CLIPPING PLANES

The **Clip depth** option sets the front and back clipping planes on an external reference or block. Objects outside the volume defined by the boundary and the specified depth are not displayed.

DELETING THE CLIPPING BOUNDARY

The **Delete** option removes the clipping boundary for the selected external reference or block. To turn off the clipping boundary temporarily, use the OFF option explained earlier. The Delete option erases the clipping boundary and the clipdepth and displays the entire reference file.

 Note: The ERASE command cannot be used to delete clipping boundaries.

GENERATING A POLYLINE

AutoCAD draws a polyline coincident with the clipping boundary. The polyline assumes the current layer, linetype, and color settings. When you delete the clipping boundary, AutoCAD deletes the polyline. If you need to keep a copy of the polyline, then invoke the **Generate Polyline** option. AutoCAD makes a copy of the clipping boundary. You can use the PEDIT command to modify the generated polyline, and then redefine the clipping boundary with the new polyline. To see the entire external reference while redefining the boundary, use the OFF option to turn off the clipping boundary.

EDITING REFERENCE FILES/XREF EDIT CONTROL

You can edit block references and external references while working in a drawing session by means of the REFEDIT command. This is referred to as in-place reference editing. If you select a reference for editing and it has attached xrefs or block definitions, the nested references and the reference are displayed and available for selection in the Reference Edit dialog box.

 Note: You can edit only one reference at a time. Block references inserted with the MINSERT command cannot be edited. You cannot edit a reference file if it is use by someone else.

You can also display the attribute definitions for editing if the block reference contains attributes. The attributes become invisible and their definitions can be edited along with the reference geometry. Attributes of the original reference remain unchanged when the changes are saved back to the block reference. Only subsequent insertions of the block will be affected by the changes.

Invoke the REFEDIT:

Modify menu	Choose In-Place Xref And Block Edit > Edit Reference
Command: prompt	**refedit** (ENTER)

AutoCAD prompts:

Command: **refedit**
Select reference: *(select external references and/or blocks to be included in the editing)*

AutoCAD displays the Edit Reference dialog box as shown in Figure 11–12.

Figure 11–12 *Edit Reference dialog box*

Objects selected that belong to any nested references cause all the references to become candidates for editing. Select the specific reference you want to edit by choosing **Next** in the Reference Edit dialog box to cycle through the references displayed. Or you can also select a reference name in the list. This will lock the reference file to prevent other users from opening the file. To display layer and symbol names with a prefix of $#$, select **Enable Unique Layer and Symbol names**. To make attribute definitions in block references available for selection when editing blocks with attributes, select **Display Attribute Definitions for Editing**. Choose OK button to close the Reference Edit dialog box.

AutoCAD prompts:

> Select nested objects: *(selected the objects that you would like to edit in the reference)*

The objects you choose are temporarily extracted for modification in the current drawing and become the *working set*. The working set objects stand out so they can be distinguished from other objects. All other objects not selected appear faded. You can now perform modifications on the working set objects.

 Note: Make sure **Reference Edit fading intensity** is set to appropriate setting in the Options dialog box (Display tab).

ADDING/REMOVING OBJECTS FROM THE WORKING SET

If a new object is created while editing a reference, it is usually added to the working set automatically. However, if making changes to objects outside the working set causes a new object to be created, it will not be added to the working set.

Objects removed from the working set are added to the host drawing and removed from the reference when the changes are saved back. Objects created or removed are automatically added to or deleted from the working set. You can tell whether an object is in the working set or not by the way it is displayed on the screen; a faded object is not in the working set. When a reference is being edited, the **Refedit** toolbar is displayed as shown in Figure 11–13.

Figure 11–13 *The Refedit toolbar*

To add objects to the working set, invoke **Add Objects To The Working Set** from the Refedit toolbar and select objects to be added. You can only select items when the type of space (Model or Paper) is in effect that was in effect when the REFEDIT command was initiated. To remove objects from the working set, invoke **Remove Objects From The Working Set** from the Refedit toolbar and select objects to be removed. As mentioned earlier, if you remove objects from the working set and save changes, the objects are removed from the reference and added to the current drawing. Any changes you make to objects in the current drawing (not in the xref or block) are not discarded. Once the modifications are complete, invoke **Save Reference Edits** to save the changes to the reference file and to discard the changes, invoke **Discard Reference Edits** command.

MANAGING EXTERNAL REFERENCES

Several tools are available to help in the management and tracking of external references.

One of the tracking mechanisms is an external ASCII log file that is maintained on each drawing that contains external references. This file, which AutoCAD generates and maintains automatically, has the same name as the current drawing and a file extension .XLG. You can examine the file with any text editor and/or print it. The log file registers each Attach, Bind, Detach, and Reload of each external reference for the current drawing. AutoCAD writes a title block to the log file that contains the name of the current drawing, the date and time, and the operation being performed. Once a log file has been created for a drawing, AutoCAD continues to append to it. The log file is always placed in the same folder as the current drawing. The log file is maintained only if the XRECTL system variable is set to 1. The default setting for XRECTL is 0.

External references are also reported in response to the ? option of the XREF command and the BLOCK command. Because of the external reference feature, the contents of a drawing may now be stored in multiple drawing files. This means that new backup procedures are required to handle drawings linked in external reference partnerships. Three possible solutions are:

1. Make the external reference drawing a permanent part of the current drawing prior to archiving with the Bind option of the XREF command.

2. Modify the current drawing's path to the external reference drawing so that they are both stored in the same folder, and then archive them together.

3. Archive the folder location of the external reference drawing with the drawing which references it. Tape backup machines do this automatically.

In AutoCAD, a combination of demand loading and the saving of drawings with indexes helps you increase the performance of drawing with external references. In conjunction with the XLOADCTL and INDEXCTL system variables, demand loading provides a method of displaying only those parts of the referenced drawing that are necessary.

The XLOADCTL system variable controls whether demand loading is set to ON or OFF and whether it opens the original drawing or a copy. If XLOADCTL is set to 0, then

AutoCAD turns off demand loading, and the entire reference file is loaded. If XLOADCTL is set to 1, then AutoCAD turns on the demand loading, and the reference file is kept open. AutoCAD loads only the objects that are necessary to display on the current drawing. AutoCAD places a lock on all reference drawings that are set for demand loading. Other users can open those reference drawings, but they cannot save changes to them. If XLOADCTL is set to 2, then AutoCAD turns on demand loading and a copy of the reference file is opened. AutoCAD makes a temporary copy of the externally referenced file and demand-loads the temporary file. Other users are allowed to edit the original drawing. When you disable demand loading, AutoCAD reads in the entire reference drawing regardless of layer visibility or clip instances.

The INDEXCTL system variable determines whether Layer, Spatial, or Layer & Spatial indexes are created when a drawing file is saved. Using layer and spatial indexes increases performance when AutoCAD is demand-loading external references. If INDEXCTL is set to 0, then indexes are not created; if INDEXCTL is set to 1, a layer index is created. The layer index maintains a list of objects that are on specific layers, and with demand loading it determines which objects need to be read in and displayed. If INDEXCTL is set to 2, then a spatial index is created. The spatial index organizes lists of objects based on their location in 3-dimensional space, and it determines which objects lie within the clip boundary and reads only those objects into the current session. If INDEXTCTL is set to 3, both layer and spatial indexes are created and saved with the drawing. If you intend to take full advantage of demand loading, then INDEXCTL should be set to 3.

 Note: If the drawing you are working on is not going to be referenced by another drawing, it is recommended that you set the INDEXCTL to 0 (OFF).

AutoCAD provides another system variable, VISRETAIN, to control the visibility of layers in the external reference drawing. If the VISRETAIN system variable is set to 0 (OFF), any changes you make to settings of the external reference drawing's layers, such as ON/OFF, Freeze/Thaw, Color, and Linetype, apply to the current drawing session only. If VISRETAIN is set to 1 (ON), then any changes you make to settings of the external reference drawing's layers take precedence over the external reference layer definition.

IMAGES

The IMAGE command allows you insert a raster or bit-mapped bi-tonal, 8-bit gray, 8-bit color, or 24-bit color image file into the drawing. The image formats that can be inserted into AutoCAD include BMP, TIFF, RLE, JPG, PCX, FLIC, GIF, IG4, IG5, RIC, PCT, CALSI, and TGA. More than one image can be displayed in any viewport, and the number and size of images is not limited.

Invoke the IMAGE command:

Reference toolbar	Choose the IMAGE command (see Figure 11–14)
Insert menu	Choose Raster Image
Command: prompt	**image** (ENTER)

Figure 11–14 *Invoking the* IMAGE *command from the Reference toolbar*

AutoCAD displays the Image Manager dialog box, similar to Figure 11–15.

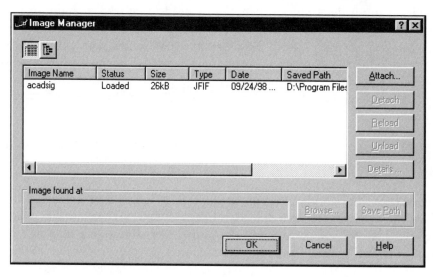

Figure 11–15 *Image Manager dialog box*

The Image Manager dialog box looks a lot like the Xref Manager dialog box. In the Image Manager dialog box, AutoCAD lists the images attached to the current drawing. The information provided in the list box of the Image Manager dialog box, such as Image name, Status, Size, Type, Data, and Saved path, is similar to the information provided in the Xref Manager dialog box. You can switch between List View and Tree View, as in the Xref Manager dialog box, by clicking the two buttons at the top left of the Image Manager dialog box.

ATTACHING AN IMAGE TO THE CURRENT DRAWING

The Attach option allows you to attach an image object to the current drawing. Invoke the Attach option:

Image Manager dialog box	Choose the **Attach** button
Reference toolbar	Choose the IMAGE ATTACH command (see Figure 11–16)
Command: prompt	**imageattach** (ENTER)

Figure 11–16 *Invoking the* IMAGE ATTACH *command from the Reference toolbar*

AutoCAD displays the Image dialog box. Select the image file from the appropriate directory and choose the **Open** button. AutoCAD displays the Select Image File dialog box, similar to Figure 11–17.

Figure 11–17 *Select Image File dialog box*

To select a different image choose the **Browse** button to display the Select Image File dialog box. Select the appropriate image file to attach to the current drawing.

Note: Once you have attached your first image file to the current drawing, then whenever you invoke the Attach option, AutoCAD displays the Select Image File dialog box instead of the Image dialog box.

Once an image file is attached to the current drawing, the image file name is added to the list box located next to the **Browse** button. When an attached image file name is selected from the list box, its path is displayed below.

The **Retain Path** check box determines whether or not the full path to the image file is saved. If Retain Path is set to ON, then the image file path is saved in the drawing database; if it is set to OFF, the name of the image file is saved without a path in the database. AutoCAD searches for the image file in the AutoCAD Support File Search Path and in the paths associated with the Files tab of the Options dialog box.

In the **Insertion Point** section of the Image dialog box, you can specify the insertion point of the external reference.

In the **Scale** section of the Image dialog box, you can specify X Scale Factor, Y Scale Factor, and Z Scale Factor of the external reference.

In the **Rotation** section of the Image dialog box, you can specify the rotation angle of the external reference.

The Insertion Point, Scale, and Rotation features are similar to those in the insertion of a block, explained in Chapter 10.

Choose the **Details** button to display the image information for the selected image file. This information includes image resolution in horizontal and vertical units, image size by width and height in pixels, and image size by width and height in the current selected units.

DETACHING AN IMAGE FROM THE CURRENT DRAWING

The **Detach** option removes the selected image definitions from the drawing database and erases all the associated image objects from the drawing and from the display. To detach an image file from the current drawing, first select the image file name from the displayed list in the Image Manager dialog box, and then select the **Detach** button. AutoCAD detaches the selected image file from the current drawing.

RELOADING AN IMAGE INTO THE CURRENT DRAWING

The **Reload** option loads the most recent version of an image. To reload the image file to the current drawing, first select the image file name from the displayed list in the Image Manager dialog box, and then select the **Reload** button. AutoCAD reloads the selected image file into the current drawing.

UNLOADING AN IMAGE FROM THE CURRENT DRAWING

The **Unload** option unloads the image data from working memory without erasing the image objects from the drawing. It is highly recommended that you unload images that are no longer needed for editing. By unloading the images, you can improve performance by reducing the memory requirement for AutoCAD. To unload the image file from the current drawing, first select the image file name from the displayed list in the Image Manager dialog box, and then select the **Unload** button. AutoCAD unloads the selected image file from the current drawing.

DETAILS OF THE IMAGE FILE

The Details option provides detailed information about the selected image, including the image name, saved path, active path, file creation date and time, file size and type, color, color depth, width and height in pixels, resolution, default size in units, and a preview image. To display the detailed information of the selected image file, first select the image file name from the displayed list in the Image Manager dialog box, then select the **Details** button. AutoCAD displays the detailed information of the selected image file in the Image File Details dialog box, similar to the Figure 11–18.

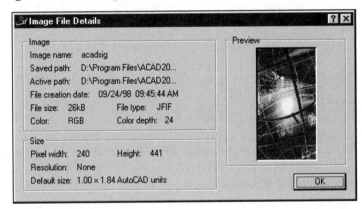

Figure 11–18 *Image File Details dialog box*

CHANGING THE PATH

To change to a different path or file name for the currently selected image file, choose the **Browse** button in the Image Manager dialog box. AutoCAD displays the Image dialog box, in which you can select a different path or file name.

SAVING THE PATH

To save the path as it appears in the Saved Path field of the currently selected image file (and displayed in the **Image Found At** text box), select the **Save Path** button. AutoCAD saves the path of the currently selected image file.

After making the necessary changes in the Image Manager dialog box, choose the **OK** button to keep the changes and close the dialog box.

CONTROLLING THE DISPLAY OF THE IMAGE OBJECTS

The Imageclip Command

The IMAGECLIP command allows you to control the display of unwanted information by clipping the image object; this is similar to the use of the XCLIP command for external references and blocks.

Invoke the IMAGECLIP command:

Reference toolbar 11–19)	Choose the IMAGE CLIP command (see Figure
Modify menu	Choose Object > Image Clip
Command: prompt	**imageclip** (ENTER)

Figure 11–19 *Invoke the* IMAGE CLIP *command from the Reference toolbar*

AutoCAD prompts:

Command: **imageclip**
Select image to clip: *(select the image to clip)*
Enter image clipping option [ON/OFF/Delete/New boundary] <New>:
    (right-click and select one of the available options from the shortcut menu)

Creating a New Clipping Boundary

The **New** boundary option (default) allows you to define a rectangular or polygonal clip boundary. AutoCAD prompts:

Enter clipping type [Polygonal/Rectangular] <Rectangular>: *(right-click and select one of the available options from the shortcut menu)*

The **Rectangular** option (default) allows you to define a rectangular boundary by specifying the opposite corners of a window. The rectangle is always drawn parallel to the edges of the image.

The **Polygonal** option defines a polygonal boundary by specifying points for the vertices of a polygon.

If you already have a clipping boundary of the selected image, and you invoke the New Boundary option, then AutoCAD prompts:

Delete old boundary(s)? [Yes/No] <Yes>: *(select one of the two available options)*

If you choose Yes, the entire reference file is redrawn and the command continues; if you choose No, the command sequence is terminated.

Once the clipping boundary is defined, AutoCAD displays only the portion of the image that is within the clipping boundary and then exits the command.

Note: The display of the boundary border is controlled by the IMAGEFRAME system variable. If it is set to 1 (ON), then AutoCAD displays the boundary border; if it is set to 0 (OFF), then AutoCAD does not display the boundary border.

Controlling the Display of the Clipped Boundary

The **ON/OFF** option controls the display of the clipped boundary. The OFF option displays all of the image, ignoring the clipping boundary. The ON option displays the clipped portion of the image only.

ADJUSTING THE IMAGE SETTINGS

The IMAGEADJUST command controls the brightness, contrast, and fade values of the selected image. Invoke the IMAGEADJUST command:

Reference toolbar	Choose the IMAGE ADJUST command (see Figure 11–20)
Modify menu	Choose Object > Image > Adjust
Command: prompt	**imageadjust** (ENTER)

Figure 11–20 *Invoking the* IMAGE ADJUST *command from the Reference toolbar*

AutoCAD prompts:

Command: **imageadjust**
Select image to adjust: *(select an image object)*

AutoCAD displays the Image Adjust dialog box, similar to Figure 11–21.

Figure 11–21 *Image Adjust dialog box*

You can adjust the Brightness, Contrast, and Fade within the range of 0 to 100.

Select the **Reset** button to reset values for the brightness, contrast, and fade parameters to the default settings of 50, 50, and 0, respectively.

ADJUSTING THE DISPLAY QUALITY OF IMAGES

The IMAGEQUALITY command controls the display quality of images. The quality setting affects display performance. A high-quality image takes longer to display. Changing the setting updates the display immediately without causing a regeneration. Images are always plotted using a high-quality display.

Invoke the IMAGEQUALITY command:

Reference toolbar	Choose the IMAGE QUALITY command (see Figure 11–22)
Modify menu	Choose Object > Image > Quality
Command: prompt	**imagequality** (ENTER)

Figure 11–22 *Invoking the* IMAGE QUALITY *command from the Reference toolbar*

AutoCAD prompts:

> Command: **imagequality**
> Enter image quality setting [High/Draft] <current>: *(right-click and select one of the available options from the shortcut menu)*

The High option produces a high-quality image on screen, and the Draft option produces a lower-quality image on screen.

CONTROLLING THE TRANSPARENCY OF AN IMAGE

The TRANSPARENCY command controls whether the background pixels in an image are transparent or opaque. Invoke the TRANSPARENCY command:

Reference toolbar	Choose the IMAGE TRANSPARENCY command (see Figure 11–23)
Modify menu	Choose Object > Image > Transparency
Command: prompt	**transparency** (ENTER)

Figure 11–23 *Invoking the* IMAGE TRANSPARENCY *command from the Reference toolbar*

AutoCAD prompts:

> Command: **transparency**
> Select image: *(select the images, and press* ENTER *to complete selection)*
> Enter transparency mode [ON/OFF] <current>: *(select one of the two available options)*

The ON option turns transparency on so that objects beneath the image are visible. The OFF option turns transparency off so that objects beneath the image are not visible.

CONTROLLING THE FRAME OF AN IMAGE

The IMAGEFRAME command controls whether the background pixels in an image are transparent or opaque.

Invoke the IMAGEFRAME command:

Reference toolbar	Choose the IMAGE FRAME command (see Figure 11–24)
Modify menu	Choose Object > Image > Frame
Command: prompt	**imageframe** (ENTER)

Figure 11–24 *Invoking the* IMAGE FRAME *command from the Reference toolbar*

AutoCAD prompts:

> Command: **imageframe**
> Select image: *(select the images, and press* ENTER *to complete selection)*
> Enter image frame setting [ON/OFF] <current>: *(right-click and select one of the available options from the shortcut menu)*

The ON option turns Imageframe on so that the frames around images are visible. The OFF option turns Imageframe off so that frames around images are not visible.

PROJECT EXERCISE

This project exercise provides point-by-point instructions for creating two drawings and then using them as external references in the third drawing, shown in Figure P11–1.

Figure P11–1 *Completed project drawing*

Step 1:   Start the AutoCAD program.

Step 2:   Create a new drawing with the parameters given in the following table.

SETTINGS	VALUE		
UNITS	Architectural		
LIMITS	Lower left corner: 0,0		
	Upper right corner: 28'-0",35'-0"		
GRID	1'		
SNAP	6"		
LAYERS	*NAME*	*COLOR*	*LINETYPE*
	Centerline	Green	Center
	Object	White	Continuous
	Text	Blue	Continuous
	Border	Red	Phantom

Set Centerline as the current layer and set up LTSCALE factor to 32.

Step 3:   Invoke the LINE command from the Draw toolbar and draw the centerlines as shown in Figure P11–2.

Command: **line**
Specify first point: **2',0**
Specify next point or [Undo]: **@35'<90**
Specify next point or [Undo]: (ENTER)
Command: **line**
Specify first point: **0,5'**
Specify next point or [Undo]: **@23'<0**
Specify next point or [Undo]: (ENTER)

Invoke the OFFSET command from the Modify toolbar, and offset the vertical line 14' to the right and the horizontal line 25' upward.

Figure P11-2 *Centerlines*

Step 4: Set Object as the current layer, and use the PLINE command with the width set to 1" to draw the steel columns at the points shown in Figure P11–3. The steel columns are 12" × 12".

Note: Draw one column and then copy it to the other locations. Make sure the grid and snap are set to ON.

Figure P11–3 *Steel columns*

Step 5:   Set Text as the current layer, and invoke the TEXT command to draw in the coordinate callouts as shown in Figure P11–4. Set the text height to 8"; this will cause the text to be plotted out 0.25" high when plotted at 3/8" = 1'-0".

Figure P11–4 *Add text to the drawing*

Step 6: Save the drawing with the name "Piperack" and close the session.

Step 7: Create another new drawing with the parameters given earlier in the settings table. Set Centerline as the current layer.

Step 8: Invoke the LINE command from the Draw toolbar and draw the centerlines as shown in Figure P11–5.

Command: **line**
Specify first point: **12'6,8'6**
Specify next point or [Undo]: **@10'<0**
Specify next point or [Undo]: (ENTER)
Command: **line**
Specify first point: **19',7'**
Specify next point or [Undo]: **@12'<90**
Specify next point or [Undo]: (ENTER)

Invoke the OFFSET command from the Modify toolbar and offset the horizontal centerline 9' upward.

Figure P11–5 *Pump centerlines*

Step 9:   Set Object as the current layer and use the LINE command to draw the pump foundation as shown in Figure P11–6.

Command: **line**
Specify first point: **14'6,6'6**
Specify next point or [Undo]: **@5'6<0**
Specify next point or [Undo]: **@4'<90**
Specify next point or [Close/Undo]: **@5'6<180**
Specify next point or [Close/Undo]: **c**

Command: **line**
Specify first point: **14'6,15'6**
Specify next point or [Undo]: **@5'6<0**
Specify next point or [Undo]: **@4'<90**
Specify next point or [Close/Undo]: **@5'6<180**
Specify next point or [Close/Undo]: **c**

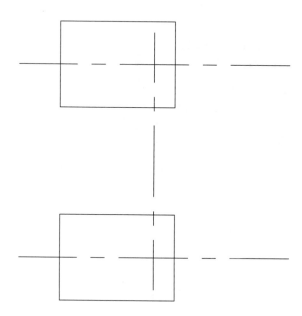

Figure P11–6 *Pump foundations*

Step 10: Set Text as the current layer and invoke the TEXT command from the Draw toolbar to draw in the coordinate callouts as shown in Figure P11–7. Set the text height to 8". This will cause the text to be plotted out 0.25" high when plotted at 3/8=1'-0".

Figure P11–7 *Add text to the drawing*

Step 11: Save the drawing with the name "Pump" and close the session.

Step 12: Create another new drawing with the parameters given earlier in the settings table. Set Border as the current layer.

Step 13: Invoke the LINE command from the Draw toolbar and draw the border line as shown in Figure P11–8.

> Command: **line**
> Specify first point: **0,0**
> Specify next point or [Undo]: **@28'<0**
> Specify next point or [Undo]: **@35'<90**
> Specify next porin or [Close/Undo]: **@28'<180**
>
> Specify next point or [Close/Undo]: **c**

Figure P11–8 *Border drawing*

Step 14: Attach the pipe rack drawing as a reference file to the current drawing with an insertion point of 0,0, a scale factor of 1.0, and a rotation angle of 0 degrees.

After attaching the pipe rack drawing, your drawing will appear as shown in Figure 11–9.

Figure P11–9 *Attach the pipe rack drawing*

Step 15: Attach the pump drawing as a reference file to the current drawing with an insertion point of 0,0, a scale factor of 1.0, and a rotation angle of 0 degrees.

After you attach the pump drawing, your drawing will appear as shown in Figure PI1–10.

Figure PI1–10 *Attach the pump drawing*

Step 16:   Complete the title block as shown in Figure PI1–11, with your name and the date filled in. Save the drawing as PROJCH11, and close the session.

Figure P11–11 *Completed drawing*

EXERCISES 11–1 THROUGH 11–5

EXERCISE 11–1

Create a new drawing according to the parameters given in the table, and draw the column layout as shown in Figure Ex11–1a. Add the necessary dimensions and save the drawing as EX11–1a.

Settings	Value		
1. Units	Architectural		
2. Limits	Lower left corner: –10'-0",–10'-0"		
	Upper right corner: 50'-0",35'-0"		
3. Grid	12"		
4. Snap	6"		
5. Layers	*NAME*	*COLOR*	*LINETYPE*
	Border	Red	Continuous
	Beams	Cyan	Continuous
	Columns	Green	Continuous
	Text	Blue	Continuous
	Dimension	White	Continuous

Hint: The columns on the corners are 8" x 8", and the remaining columns are 10" x 10". They can be drawn as polylines with a width of 0.5.

Figure Ex11–1a *Layout of the columns*

Create another new drawing according to the parameters given in the table. Attach the file EX11–1a as a reference file. Draw the walls and doors as shown in Figure Ex11–1b. Add text for the view name and for labeling the doors. Add the necessary dimensions, and save the drawing as EX11–1b.

Figure Ex11–1b *Layout of the walls, doors, and columns*

Open the drawing EX11–1a and make the necessary changes for column spacing as shown in Figure Ex11–1c. Save the drawing.

Open the drawing EX11–1b; the changes in the column layout show up automatically. The walls and doors will have to be altered to conform to the revised external reference column layout. Make the necessary changes, including eliminating one of the doors, as shown in Figure Ex11–1d. The wall along the column line on the left has to be relocated, and the dimension of the remaining door in the left bay changed from 8'-0" to 10'-0". Save the drawing as Ex11–1d.

Figure Ex11–1c *Layout of the column spacing (updated)*

Figure Ex11–1d *Layout of the columns (final revision)*

EXERCISE 11–2

Open the drawing of the laundry and storage floor plan, P6-1.dwg, that was completed in Chapter 6. Set the Construction, Hidden, and Text layers to OFF. The drawing should appear as shown in Figure Ex11–2a. Save the drawing with the name EX11–2a.

Settings	Value		
1. Units	Architectural		
2. Limits	Lower left corner: –5'-0",–6'-0"		
	Upper right corner: 24'-4",15'-4"		
3. Grid	6"		
4. Snap	2"		
5. Layers	*NAME*	*COLOR*	*LINETYPE*
	Border	Red	Continuous
	Columns	Green	Continuous
	Text	Blue	Continuous
	Dimension	White	Continuous
	Electrical	Magenta	Continuous
	Lighting	Red	Continuous

Hint: The electrical outlet symbols (110V and 220V) can each be drawn one time and then copied to the other locations. The curved line from the switch to the light fixture can be drawn on the same layer as the switch and the fixture. You can then use the Properties button on the standard toolbar and change the Linetype to hidden.

Figure Ex11–2a *Laundry and storage floor plan with the Construction, Dim, Hidden, and Text layers set to OFF*

Create a new drawing according to the parameters given in the table. Attach the file EX11–2a as a reference file. Turn on the layer Ex11-2a hidden to show squares on the left side of the wall. Draw the lighting and electrical symbols as shown in Figure Ex11–2b. Save the drawing as EX11–2b.

Figure Ex11–2b *Layout of the walls, doors, and columns*

EXERCISE 11–3

Open the drawing *EX5-4.dwg* that was completed in Chapter 5. Erase Detail A, Detail B, and the details of the stair riser. Also erase the interior footings (the 24 double rectangles inside the perimeter grade beams), the dimensions, and two interior lines of the five perimeter lines. Change the linetype of the perimeter grade beam lines to hidden linetype. The drawing should appear as shown in Figure Ex11–3a. Save the drawing with the name EX11–3a.

Settings	Value
1. Units	Architectural
2. Limits	Lower left corner: 0'-0",0'-0" Upper right corner: 60'-0", 45'-0"
3. Grid	12"
4. Snap	6"

Settings	Value		
5. Layers	*NAME*	*COLOR*	*LINETYPE*
	Construction	Cyan	Continuous
	Border	Red	Continuous
	Object	Green	Continuous
	Text	Blue	Continuous
	Dim	White	Continuous

Hint: The foundation layout can be used from the external reference as the basis for the walls and stair for the roofing plan. Once the roofing plan is completed, the layers in the external reference can be turned off. If any later changes are made in the foundation layout, they will automatically show up in the roofing plan, which can then be revised to suit the changes.

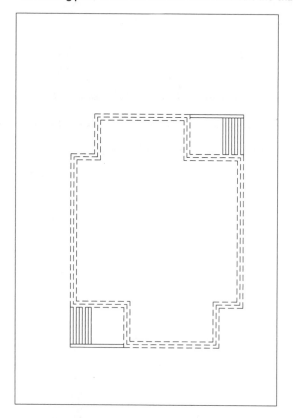

Figure Ex11–3a *Foundation plan*

Create a new drawing according to the parameters given in the table. Attach the file EX11–3a as a reference file, with an insertion point of 0,0, a scale factor of 1.0, and a rotation angle of 0 degrees. Draw the roof as shown in Figure Ex11–3b. Save the drawing as EX11–3b.

Figure Ex11–3b *Roof layout*

EXERCISE 11–4

Create a new drawing according to the parameters given in the table, and draw the floor plan as shown in Figure Ex11–4a. Do *not* add the dimensions, which are provided here only for reference. Save the drawing as EX11–4a.

Settings	Value		
1. Units	Architectural		
2. Limits	Lower left corner: –2'-0",–2'-0"		
	Upper right corner: 70'-0",50'-0"		
3. Grid	6"		
4. Snap	2"		
5. Layers	*NAME*	*COLOR*	*LINETYPE*
	Walls	Red	Continuous
	Ductwork	White	Continuous
	Text	Blue	Continuous

Hint: The ductwork can be drawn with the MLINE command.

Figure Ex11–4a *Floor plan*

Create a new drawing according to the parameters given in the table. Attach the file EX11–4a as a reference file. Draw the A/C duct layout as shown in Figure Ex11–4b. Save the drawing as EX11–4b.

A/C DUCT LAYOUT

Figure Ex11–4b *Floor plan with the A/C duct layout*

EXERCISE 11–5

Create a new drawing according to the parameters given in the table, and draw the vessel shown in Figure Ex11–5a. Save the drawing with the name EX11–5a.

Settings	Value		
1. Units	Architectural		
2. Limits	Lower left corner: 0,0		
	Upper right corner: 28'-0",35'-0"		
3. Grid	1'		
4. Snap	6"		
5. Layers	*NAME*	*COLOR*	*LINETYPE*
	Centerline	Green	Center
	Object	White	Continuous
	Text	Blue	Continuous
	Border	Red	Phantom

Hint: This exercise uses a drawing that already has external references attached to it, and those external references must also be accessible for this exercise.

Figure Ex11–5 *Vessel layout*

Create another new drawing according to the parameters given in the table, and draw an octagonal foundation as shown in Figure Ex11–5b. Do *not* add the dimensions, which are provided here only for reference.

2" DIA. BOLTS
ON 9'-1 1/2"
BOLT CIRCLE

4'-0"

FOUNDATION PLAN

Figure Ex11–5b *Octagonal foundation*

Attach the file Ex11–5a as a reference file, with an insertion point of 0,0, a scale factor of 1.0, and a rotation of 0 degrees. The drawing should appear as shown in Figure Ex11–5c. Save the drawing as EX11–5c.

VESSEL C-101
84" DIA.

Figure Ex11–5c *Vessel in an octagonal foundation*

Open the drawing of the vessel and pump plan, PROJCH11.DWG, the project exercise completed earlier in this chapter. Attach the file Ex11–5c as a reference file, with an insertion point of 7'-0",8'-0", a scale factor of 1.0, and a rotation of 0 degrees. The drawing should appear as shown in Figure Ex11–5d. Save the drawing as EX11–5d.

Figure Ex11–5d *Layout of the vessel and pump with a vessel in an octagonal foundation*

REVIEW QUESTIONS

1. If an externally referenced drawing called "FLOOR.DWG" contains a block called "TABLE" and is permanently bound to the current drawing, the new name of the block is:

   a. FLOOR0TABLE

   b. FLOOR|TABLE

   c. FLOOR$0TABLE

   d. FLOOR_TABLE

   e. FLOOR$|$TABLE

2. If an externally referenced drawing called "FLOOR.DWG" contains a block called "TABLE," the name of the block is listed as:

   a. FLOOR0TABLE

   b. FLOOR|TABLE

   c. FLOOR$0TABLE

   d. FLOOR_TABLE

   e. FLOOR$|$TABLE

3. The maximum number of files which can be externally referenced into a drawing is:

   a. 32   d. 32,000

   b. 1,024   e. only limited by memory

   c. 8,000

4. If you want to retain, from one drawing session to another, any changes you make to the color or visibility of layers in an externally referenced file, the system variable that controls this is:

   a. XREFRET   d. VISRETAIN

   b. RETXREF   e. these changes cannot be saved from one session to another

   c. XREFLAYER

5. XREFs are converted to blocks if you detach them.

   a. True   b. False

6. When detaching XREFs from your drawing, it is acceptable to use wild cards to specify which XREF should be detached.

   a. True   b. False

7. Overlaying XREFs rather than attaching them causes AutoCAD to display the file as a bit map image, rather than a vector based image.

   a. True   b. False

8. To make a reference file a permanent part of the current drawing database, use the XREF command with the:

   a. Attach option   d. Reload option

   b. Bind option   e. Path option

   c. ? option

9. The XREF command is invoked from the toolbar:

   a. Draw

   b. Modify

   c. External Reference

   d. Any of the above

   e. None of the above

10. The Attach option of the XREF command is used to:

    a. bind the external drawing to the current drawing

    b. attach a new external reference file to the current drawing

    c. reload an external reference drawing

    d. all of the above

11. The following are the dependent symbols that can be made a permanent part of your current drawing, except:

    a. Blocks

    b. Dimstyles

    c. Text Styles

    d. Linetypes

    e. Grid and Snap

12. Including an image file in a drawing will incorporate the image similar to the way a drawing file is merged by using:

    a. INSERT

    b. XREF

    c. WBLOCK

13. Which of the following is not a valid file type to use with the IMAGE command?

    a. BMP   d. JPG

    b. TIF   e. GIF

    c. WMF

14. Which of the following parameters can be adjusted on a bitmapped image?

    a. Brightness   d. all of the above

    b. Contrast   e. none of the above

    c. Fade

Directions: Open PROJCH11.dwg and attach file EX11–5C.dwg as a reference file in order to answer the following questions:

15. What is the measured horizontal distance from the centerline of P.S. #1 to the centerline of P.S. # 2?

    a. 10'-0"   c. 20'-0"

    b. 15'-0"   d. 25'-0"

16. What is the measured horizontal distance from the centerline of Pump 1B to the centerline of Pump 1C?

    a. 3'-0"   c. 9'-0"

    b. 6'-0"   d. 12'-0"

17. What is the measured angular distance between Nozzle B and Nozzle D on Vessel C-101?

    a. 45 degrees   c. 135 degrees

    b. 90 degrees   d. 145 degrees

18. The measured horizontal distance from the centerline of P.S. #1 to the centerline of Nozzle D on Vessel C-101 is _____.

19. What is the measured horizontal distance from the columns located at 2'-0" to the centerline of the pump discharge ?

    a. 12'-0"   c. 17'-0"

    b. 15'-0"   d. 19'-0"

20. The combined measured area of the pump foundations is _____.

21. The combined measured area of Vessel C-101 is _____.

22. The combined measured area of the Project boundaries is _____.

23. The combined measured area of the bolt circle on Vessel C-101 is _____.

CHAPTER 12

AutoCAD DesignCenter

INTRODUCTION

AutoCAD DesignCenter, a new feature in AutoCAD 2000, makes it much easier to manage content within your drawing. Content includes blocks, external references, layers, raster images, linetypes, layouts, text styles, dimension styles, and custom content created by third-party applications. You can now manage content between your drawing and other sources such as other drawings, whether currently open, stored on any drive, or even elsewhere on a network or somewhere on the Internet. The AutoCAD DesignCenter provides a program window with a specialized drawing file-handling section. It allows you to drag and drop content and images into your current drawing or attach a drawing as an external reference.

After completing this chapter, you will be able to do the following:

- Open, undock, move, resize, dock, and close AutoCAD DesignCenter.
- Locate drawings, files, and their content in a manner similar to Windows Explorer.
- Use the DesignCenter Palette and Tree View.
- Preview images, drawings, content, and their written descriptions.
- Customize and use Autodesk *Favorites* folder.
- Manage blocks, layers, xrefs, layouts, dimstyles, textstyles, and raster images.
- Manage web-based content and custom content from third-party applications.
- Create shortcuts to drawings, folders, drives, the network, and Internet locations.
- Browse sources for content and drag and drop (copy) into drawings.

CONTENT

As mentioned above, content includes block definitions, external references, layer names and compositions, raster images, linetypes, layouts, text styles, dimension styles, and custom content created by third-party applications. For example, a layer of one name may have the same properties as a layer of another name. But, because its name is part of its composition, it can be identified as a unique layer. Also, two items of content can have the same name if they are not the same type of content. For example, you can name both a text style and a dimension style "architectural".

 Note: You can only view the name of an item if content in the Palette along with a raster image if it is a block or image and a description of one has been written. The item itself still resides in its container. Through the DesignCenter, you can drag and drop copies of the item's definition into your current drawing, but you cannot edit the item itself from the DesignCenter.

CONTENT TYPE

Content types are only types or categories and not the items themselves. Lines, circles, and other objects are not included in what are considered content types, although a block may be made up of such objects. To view a particular item of content you must first select the name of its content type (under a drawing name) in the Tree View or double-click the name of its content type in the Palette. For example, if you were trying to locate an Xref named "1st floor plan", you could search through drawing names and select the content type named "xref" under each drawing name. Under each drawing name there will be a content type named "xref", but under each content type named "xref" there may or may not be listed any xrefs.

CONTAINER

The primary container is the drawing. It contains the blocks, images, linetypes, and other definitions that are most commonly sought to add to your current drawing. A folder can be considered a container because it contains files, and a drawing is a file. An image can also be a file in a folder. An image is normally not a container like a drawing. Drawings and images can be dragged and dropped into your current drawing from the Palette. So, a drawing can be both content and a container.

Figure 12–1 shows an example for a content (block named BR 5), content type (Block), and Container (drawing named 1st floor plan).

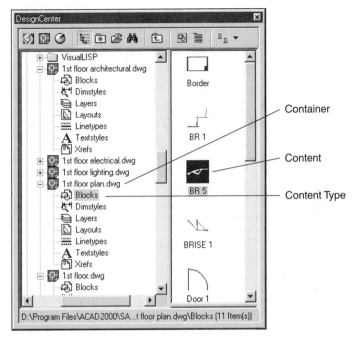

Figure 12–1 *An example showing a content, content type, and container from the DesignCenter window*

OPENING THE DesignCenter WINDOW

The AutoCAD DesignCenter is a window rather than a dialog box. It is like calling up a special program that runs along with AutoCAD and expedites file managing and drawing content handling tasks.

Invoke the AutoCAD DesignCenter command:

Standard toolbar	Choose the **AUTOCAD DESIGNCENTER** command (see Figure 12–2)
Tools menu	Choose AutoCAD DesignCenter
Command: prompt	**adcenter** (ENTER)

Figure 12–2 *Invoking the AutoCAD DesignCenter from the Standard toolbar*

AutoCAD displays the AutoCAD DesignCenter Window, similar to Figure 12–3.

Figure 12–3 *AutoCAD DesignCenter Window*

POSITIONING THE DesignCenter WINDOW

DOCKED POSITION

The default position of the DesignCenter is docked at the left side of the drawing area. This is where it is located when AutoCAD is started for the first time, as shown in Figure 12–4. However, once the DesignCenter has been repositioned and the drawing session is closed, the next time you open AutoCAD it will be at its relocated position.

FLOATING (UNDOCKED) POSITION

You can undock the DesignCenter by double-clicking on its title bar or border. Or you can drag the window into the drawing area until its drag preview image clears its present position, as shown in Figure 12–5.

Figure 12–4 *DesignCenter docked on the left side of the drawing area*

Figure 12–5 *DesignCenter floating (undocked) in the drawing area*

REDOCKING

To redock the DesignCenter, double-click the title bar or border or drag the DesignCenter to the left or right side of the drawing area. It cannot be docked at the top or bottom of the screen. When undocked, the DesignCenter can be resized by holding down the pick button over an edge or corner of the window and, when the cursor becomes a double arrow, dragging the edge or corner with the pointing device.

PARTS OF THE DesignCenter WINDOW

The DesignCenter consists of five major areas: Toolbar, Palette pane, Tree View pane, Preview pane, and Description pane. The Palette and Toolbar are always displayed as shown in Figure 12–6. The Tree View, Preview, and Description panes can be turned off and on when desired.

Figure 12–6 *Parts of the DesignCenter window*

TOOLBAR

The toolbar contains buttons designed especially to help you select and manage the type of content desired within the DesignCenter window, as shown in Figure 12–6.

PALETTE

The Palette or content pane is AutoCAD DesignCenter's primary area for displaying the names and icons representing content, as shown in Figure 12–6.

TREE VIEW

The **Tree View** or navigation pane is an optional area on the left side of the DesignCenter window for displaying files such as drawings and images and their paths. Choose the Tree View button in the toolbar, as shown in Figure 12–7, to display the Tree View.

Figure 12–7 *Invoking the Tree View from the DesignCenter toolbar*

The Tree View also displays content types. The Tree View displays multiple levels and operates in a manner similar to the Windows Explorer. Figure 12–8 shows the DesignCenter window with the tree view window display on and off.

 Note: The Tree View can be used to navigate up and down through the hierarchy of networks, drives, folders, files, and content type. But, you cannot drag and drop to, from, or within the Tree View.

Figure 12–8 *AutoCAD DesignCenter window with view window display on (left) and off (right)*

PREVIEW

The Preview pane displays a raster image on the Preview pane of the selected item of content if the item is a drawing, block, or image, as shown in Figure 12–6.

DESCRIPTION

The Description pane displays a written description of the item of content selected on the Palette if a description has been written, as shown in Figure 12–6.

TOOLBAR BUTTONS

The buttons on the toolbar are used to manage the DesignCenter panes and what is being displayed in them, as shown in Figure 12–9.

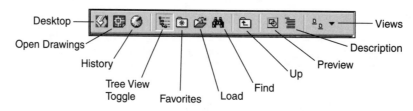

Figure 12–9 *AutoCAD DesignCenter showing toolbar buttons*

TREE VIEW, PREVIEW, AND DESCRIPTION

The buttons for the Tree View, Preview, and Description simply cause their respective panes to be displayed or not displayed, depending on their current status.

FAVORITES

The **Favorites** button causes the Autodesk *Favorites* feature to display icons in the Palette representing shortcuts to frequently used files. These are files for which you have previously set up shortcut icons to make them quickly accessible. The Tree View displays and highlights the subfolder *Favorites* of the folder Autodesk, as shown in Figure 12–10.

Figure 12–10 *The Tree View with the Autodesk Favorites folder selected*

Note: Only the shortcuts are in the *Favorites* folder. The files themselves remain in their original locations

A drawing or image file or a folder can be added to *Favorites* by right-clicking on the file name or folder name in the Palette or Tree View and choosing **Add To Favorites** from the shortcut menu. To view and organize items in *Favorites*, choose Organize Favorites from the shortcut menu. An Autodesk window is displayed for managing items in *Favorites*.

When the *Favorites* folder is highlighted in the Tree View, the Palette displays the icons, which are shortcuts to other files or folders. Double-click on the icon for the desired folder or file.

After you have double-clicked the desired icon in the Palette, the selected file or folder will be displayed in the Tree View and it will be highlighted.

And, in turn, the Palette will display the contents of the highlighted folder or file. Figure 12–11 shows the selection of the *Favorites* folder and corresponding selection.

Figure 12–11 *Selection of the Favorites selection*

LOAD

The **Load** button causes the **Load DesignCenter Palette** dialog box to be displayed, as shown in Figure 12–12. It is similar to the Windows File Manager dialog box. From it you can select a drawing file whose contents will be loaded onto the Palette pane.

Figure 12–12 *Load DesignCenter Palette dialog box*

FIND

The **Find** button provides a means to find containers, content types, and content similar to the Windows Find feature.

UP

Each time the **Up** button is selected it causes the selection in the Palette or Tree View to move up one level in the hierarchy of content, content types, containers, folders, drives, and other locations in the same manner as the Up button in the Windows Explorer and other file manager dialog boxes.

VIEWS

The Views and down arrow combination button allows you to choose between the four optional modes of displaying content: **Large Icons, Small Icons, List,** and **Details**. Selecting the left part of the Views button (with the icon) causes the mode to cycle to the next mode of display and the icon will change to reflect the mode in effect. Selecting the down arrow on the right will display the shortcut menu from which you can select one of the four modes, as the four type of views in Figures 12–13, 12–14, 12–15, and 12–16 show. If the Palette has drawings, xrefs, blocks, or images, the large icon will include a small raster image of the item of content.

Figure 12–13 *The Palette with the Large Icons display mode in effect*

Figure 12–14 *The Palette with the Small Icons display mode in effect*

Figure 12–15 *The Palette with the List display mode in effect*

Figure 12–16 *The Palette with the Details display mode in effect*

Whenever the Tree View pane is displayed, there are three additional buttons available on the AutoCAD DesignCenter toolbar: **Open Drawings, Desktop,** and **History.** These are shown in Figure 12–17.

Desktop

Open Drawings

History

Figure 12–17 *Desktop, Open Drawings, and History buttons in DesignCenter toolbar*

The **Open Drawings** button displays a list in the Tree View of drawings that are currently open, as shown in Figure 12–18. When you select one of the drawings, its content types will be displayed in the Palette pane. Or you can double-click on the file name and it will display the content types at one level below the name you double-clicked on. Selecting one of the content types causes content of that type (if any exists in the drawing) to be listed in the Palette pane.

Figure 12–18 *DesignCenter displaying the open drawings*

The **Desktop** button causes the Tree View to display the drives, folders, subfolders, and files that are located on your computer's desktop, as shown in Figure 12–19. This is the type of view you normally see when you open Windows Explorer. From there you can step through the path(s) necessary to the desired location. Selection of the Desktop button is needed whenever you want to return from Open drawings tree view and History view (explained later this section) to display the drives, folders, subfolders, and files that are located on your computer's desktop.

Figure 12–19 *DesignCenter displaying the Desktop*

The **History** button displays the last 20 items accessed through AutoCAD DesignCenter, including their paths, as shown in Figure 12–20.

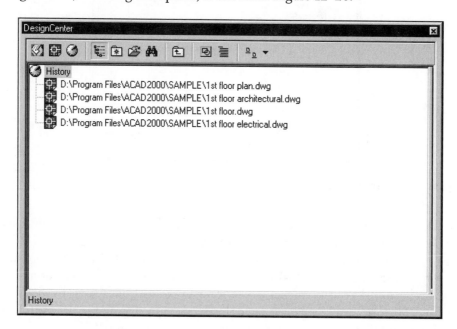

Figure 12–20 *DesignCenter with History of drawings accessed*

VIEWING CONTENT

The names of content type, such as blocks, linetypes, linestyles, and so on, and containers such as drawings, image files, folders, drives, networks, and Internet locations, can be viewed in the Tree view as well as Palette pane. Names of content can be viewed in the Palette pane but not in the Tree View pane. Either pane can be used to move up and down through the path from the drive to the item of content. It is usually quicker to navigate the path in the Tree View because of its ability to display multiple levels of hierarchy. When the container (a folder for drawings/images and a drawing/content type for blocks, images, and other items) appears in the Tree View, select it and then view the name of the items of content (if any) in the Palette pane.

USING THE TREE VIEW

As mentioned earlier, in the Tree View you can view the content type and container, but not the actual content. You can use the Tree View to display the icon for an item of content in the Palette. Just doing this will not, however, cause a raster image to be displayed in the Preview panel, as shown in using the Palette in this section.

The folder named Sample that comes with the AutoCAD 2000 program contains numerous drawings which will be used in this section as examples. First, select the Up button as many times as necessary to display the level (in the Tree View pane) that includes the branch leading through the folder named ACAD2000 to the sub-folder named Sample. This, again, might require going all the way back to a particular drive. Then click the plus sign or double-click the folder name or drive name to display the folders within a folder. Figure 12–21 shows the folder named SAMPLE highlighted in the Tree View. Note that the list of content folders is displayed in the Palette pane.

Figure 12–21 *The Tree View showing the drawing named 1st floor plan in the folder named Sample*

Select the 1st floor plan in the Tree View. Figure 12-22 shows the drawing named 1st floor plan highlighted in the Tree View. Note that the list of content types is displayed in the Palette pane.

Figure 12–22 *The Tree View with the content type Blocks highlighted*

Next, double-click on the drawing named "1st floor plan" in the Tree View or select the box with the plus in it to the left of the drawing name. The list of content types will be listed below the drawing name, as well as in the Palette pane. To display the names of individual items of content, blocks for example, select the content type Blocks in the Tree View, as shown in Figure 12–23.

By choosing one of the block icons displayed in the Palette pane, AutoCAD displays the corresponding preview image and description, if any, in the Preview pane and Description pane, respectively. Figure 12–24 shows the selection of a block named BR 5, and its corresponding preview image and description are shown in the preview pane and in the Description pane.

Figure 12–23 *An example of the Tree View displaying container and content type and the Palette pane displaying content*

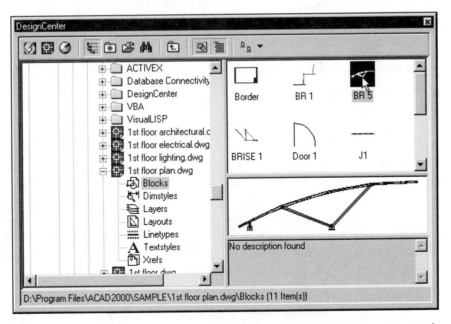

Figure 12–24 *Selection of a block named BR 5 and the corresponding preview image and description are shown in the preview pane and in the Description pane*

The AutoCAD 2000 program has a folder named Express, which contains numerous bitmap images. You can use the Tree View to display the names of these images in a similar manner as you used it to display content types. However, when you select Express, the list of bitmap (.BMP) files are listed in the Palette pane, as shown in Figure 12–25. You cannot double-click Express and have the list displayed in the Tree View.

Figure 12–25 *The Tree View with the folder Express highlighted and listing the files in the Palette pane*

USING THE PALETTE PANE

As mentioned earlier, the Palette Pane is used for displaying Content. It can also display Containers and Content type. A container is a source of content such as a drawing. In the hierarchy of what can be displayed, containers are one step higher than content types. Content types are one step higher than content. From drawing files and other containers, the next step up is a folder. Then, progressing up through the folders (if there is a hierarchy of folders), next is the drive, which may lead to a network, a Web site on the Internet, a floppy diskette, or a CD-ROM.

Note: The palette only displays items of the same level that are members of a single container one level above them. The Palette does not display more than one level at a time. For example, content types such as blocks, xrefs, and layers are members of one drawing. If there are any blocks in the drawing, then they are listed when you have selected that drawing's content type called Blocks. Several drawings may be members of a particular folder. Folders are members of one drive or folders may actually be subfolders of a folder one level up. Remember, in the palette, only one level of the hierarchy is displayed and all items shown are members of the same component one level above. For displaying more than one level at a time, use the Tree View.

You can navigate up and down through the hierarchy of drives, folders, drawings, content types, and content in the Palette pane. It is not as easy as in the Tree View, however, because only one level is displayed at a time. If the content is a drawing, block, or image, you can view a raster image of it in the Preview pane. Simply select the item in the Palette. Remember that you can enlarge the Preview pane for viewing, even if only temporarily. As mentioned earlier, you can double-click the title bar to dock or undock the DesignCenter.

For example, there is a block named "Border" in the drawing named "1st floor" in the folder named Sample. To get to the block named "Border" in the Palette pane, as described for the example in the Tree View above, select the Up button as many times as necessary to display the level that includes the branch leading to the folder named "1st floor". This might require going all the way back to a particular drive. Then double-click the sequence of drive and folders on the path to the drawing named "1st floor" in the Palette, and then double-click on "1st floor" icon. When you double-click the icon for "1st floor", the icons for content types will be displayed, including blocks, dimstyles, layers, layouts, linetypes, textstyles, and xrefs, as shown in Figure 12–26.

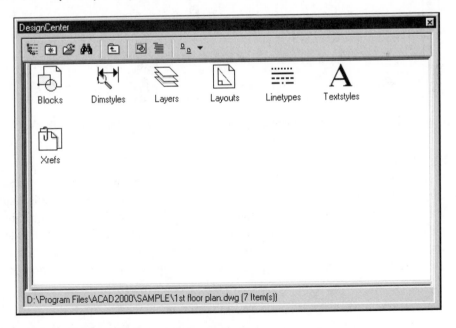

Figure 12–26 *The Palette pane with the display of the content types*

AutoCAD also displays the path to the selected drawing at the bottom of the DesignCenter window, as shown in Figure 12–26. Double-click on the Blocks icon, and AutoCAD displays the icons (when the display is set to Large icons) representing all the blocks in the selected drawing. Choose one of the block icons (do not double-click), and AutoCAD displays corresponding raster images and descriptions in the Preview pane and description pane respectively. Figure 12–27 shows the selection of the block named "border" and the corresponding display of its raster image and description. You can enlarge the Preview pane to get a bigger picture. Note, however, that it is a raster image, and sharp details are not available in close-up views.

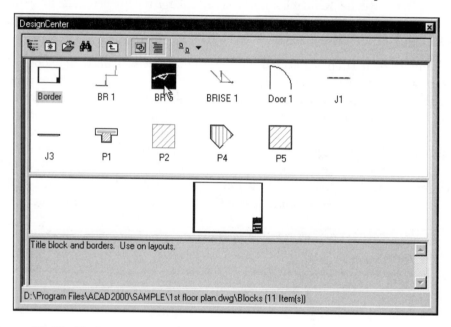

Figure 12–27 *The Preview pane displaying the raster image of the block named "Border"*

VIEWING IMAGES

If the item of content is a bitmap image, its name can be displayed in the Palette pane by selecting the name of its container in the Tree View. For example, there is a sub-folder named Express in the folder named ACAD2000. When Express is highlighted in the Tree View, you can choose the bitmap file named Laylck16.bmp in the Palette pane and the icon resembling a lock is displayed in the Preview pane, as shown in Figure 12–28. Also, there is a description that can be viewed in the description pane.

All the images in the sub-folder have a file extension of .BMP. In another sub-folder named Textures in the folder named ACAD2000, there are files with the extension .TGA. You can view these files in the same manner as the .BMP file in the previous example, as shown in Figure 12–29.

Figure 12–28 *Display of the image named "Laylck16bmp" with its raster image in the Preview pane and its corresponding description in the Description pane*

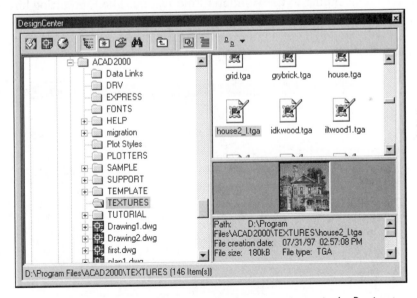

Figure 12–29 *The image named "House_2ltga" with its raster image in the Preview pane and corresponding description in the Description pane*

Note: When you have selected the name of a folder in the Tree View, if there are any files in that folder that are viewable, their names will appear in the Palette pane. Some folders may have other files in them, but if they are not viewable, their names will not appear.

LOAD

As described in the section on "Toolbar Buttons," the Load button causes the **Load DesignCenter Palette** dialog box to be displayed, from which you can select a drawing file whose contents will be loaded onto the Palette pane. You can also load a drawing or folder from the Find dialog box by dragging and dropping it into the Palette or Tree View pane. Or you can right-click and select Load Into Palette from the shortcut menu.

 Note: The Load feature does not load the item selected into your drawing. It only loads its icon into the Palette, as shown in the next section on adding content to drawings.

REFRESHING THE PALETTE AND TREE VIEW

When file manipulations are made to the Palette or Tree View, they do not always show up immediately on the screen. If this happens, you can right-click in one of the panes and choose Refresh from the shortcut menu. The views will be updated.

SEARCHING FOR CONTENT

As mentioned earlier, the Find feature provides a means to locate containers, content type, and content similar to the Windows Find feature. The Find dialog box is displayed by selecting the **Find** button on the toolbar, or you can right-click in the Palette or Tree View and choose Find from the shortcut menu. AutoCAD displays the Find dialog box, similar to Figure 12–30.

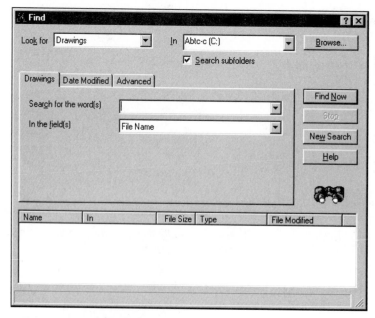

Figure 12–30 DesignCenter Find dialog box

From the **Look for** list box in the Find dialog box, select the type of content you wish to find. Available options include Blocks, Dimstyles, Drawings, Drawings and Blocks, Layers, Layouts, Linetypes, Textstyles, and Xrefs.

From the **In** list box, select the location for searching. Available options include My Computer, local hard drives (C:) (and any others), 3½ Floppy (A:), and any other drives.

The **Browse** button causes the **Browse For Folder** dialog box to be displayed, which has a file manager window as shown in Figure 12–31. Here you can specify a path by stepping through the levels to the location which you wish to search. When the final location is highlighted, select **OK** and the path to this location is displayed in the **In** list box.

Figure 12—31 *Browse For Folder dialog box*

The number of tab(s) that will be displayed in the Find dialog box depends on the type of content selected in the **Look for** list box. Each of the content types will have a tab that corresponds to the type of content selected. If Drawings is selected, there are two additional tabs: **Date Modified** and **Advanced**.

In the **Drawings** tab, the **Search for the word(s)** text box lets you enter the word(s), such as a drawing name or author, to determine what to search for. The **In the field(s)** text box lets you select a field type to search. Fields include File Name, Title, Subject, Author, and Keywords.

In the **Date Modified** tab (available only when Drawings is selected in the **Look for** text box of the Find dialog box) you can specify a search of drawing files by the date they were modified as shown in Figure 12–32. The search will include all files that comply with other filters if the **All files** radio button is selected. If the **Find all files created or modified:** radio button is selected, you can limit the search to the range of dates specified by selecting one of the following secondary radio buttons. The **between** and **and** radio button lets you limit the search to drawings modified between the dates entered. The **during the previous months** radio button lets you limit the search to drawings modified during the number of previous months entered. The **during the previous days** radio button lets you limit the search to drawings modified during the number of previous days entered.

Figure 12—32 *Display of the Date Modified tab in the Find dialog box*

In the **Advanced** tab (available only when Drawings is selected in the **Look For** text box of the Find dialog box) you can specify a search of files by additional parameters as shown in Figure 12–33. In the **Containing:** text box you can specify a file by one of four options: Block Name, Block and Drawing Description, Attribute Tag, and Attribute Value. The search will then be limited to items containing the text entered in the **Containing text:** text box. The **Size is:** text boxes let you limit the search to drawings that are At Least or At Most as many KBs (size of the drawing file) as the number entered into the second text box.

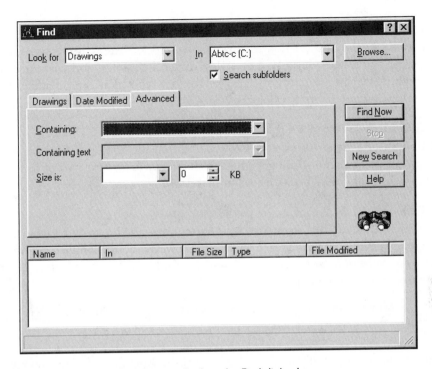

Figure 12—33 *Display of the Advanced tab in the Find dialog box*

All of the other tabs (Blocks, Dimstyles, Drawings and Blocks, Layers, Layouts, Linetypes, Textstyles, and Xrefs) have a **Search for the name** text box in which you can enter the name of the item you wish to find. This can be a block name, dimstyle, or one of the other content types.

To the right of the tab section are four buttons: **Find Now, Stop, New Search,** and **Help**.

Once the parameters have been specified, such as the path, content type, and date modified, choosing the **Find Now** button initiates the search. Select the **Stop** button to terminate the search. Choosing the **New Search** lets you specify new parameters for another search. Choosing the **Help** button causes the **AutoCAD Command Reference** (Help) dialog box to be displayed.

Each time a search is performed, the text entered in the **Search for the name** text box or the **Search for the word(s)** text box is saved. If you wish to repeat the same search again, select the down arrow next to the text box and select the text associated with the search you wish to repeat.

ADDING CONTENT TO DRAWINGS

As discussed in the introduction to this chapter, blocks, external references, layers, raster images, linetypes, layouts, text styles, dimension styles, and custom content created by third-party applications are the content types that can be added to the current drawing in session by using the AutoCAD DesignCenter.

LAYERS, LINETYPES, TEXT STYLES, AND DIMENSION STYLES

A definition of a layer or linetype or a style created for text or dimensions can be dragged into the current drawing from the Palette. It will become part of the current drawing as though it had been created in that drawing.

BLOCKS

A block definition can be inserted into the current drawing by dragging its icon from the Palette into the drawing area. You cannot, however, do this while a command is active. It must be done at the command prompt. Only one block definition at a time can be inserted from the Palette.

 Note: If you double-click a block and it has nested blocks, the hierarchy is flattened.

There are two methods of inserting blocks from the AutoCAD DesignCenter. One method uses *Autoscaling*, which scales the block reference as needed based on the comparison of the units in the source drawing to the units in the target drawing. Another method is to use the Insert dialog box to specify the insertion point, scale, and rotation.

 Note: When dragging and dropping using the automatic scaling method, the dimension values inside the blocks will not be true.

When you drag a block definition from the Palette or the Find dialog box into your drawing, you can release the button on the pointing device (drop) when the block is at the desired location. This is useful when the desired location can be specified with a running object snap mode in effect. The block will be inserted with the default scale and rotation.

To invoke the Insert dialog box, double-click the block icon or right-click the block definition in the Palette or Find dialog box and then select Insert Block from the shortcut menu. From the Insert dialog box, specify the Insertion Point, Scale, and Rotation, or select Specify On-Screen. You can select Explode to have the block definition exploded on insertion.

RASTER IMAGES

A raster image such as a digital photo, print screen capture saved in a paint program as a bitmap, or a company logo can be copied into the current drawing by dragging its icon from the Palette into the drawing area. Then specify the Insertion Point, Scale, and Rotation. Or you can right-click the image icon and choose Attach Image from the shortcut menu.

EXTERNAL REFERENCES

To attach an xref to the current drawing, drag its icon from the Palette or Find dialog with the right button on the pointing device into the drawing area. Then release the button and select Attach from the shortcut menu. The Attach Xref dialog box is displayed, from which you can choose between Attachment or Overlay in the Reference Type option. From the Attach Xref dialog box, specify the Insertion Point, Scale, and Rotation, or select Specify On-Screen.

NAME CONFLICTS

When copying, inserting, or attaching content into a drawing that already has an item of the content type with the same name, AutoCAD will display a warning. The item is not added to the drawing. If the item is a block or xref, AutoCAD checks to determine if the name is already listed in the database. The warning "Duplicate definition of [object][name] ignored" is displayed. If the xref exists in the drawing, the warning "Xref [name] has already been defined. Using existing definition" is displayed. If the item with the duplicate name is a layer, linetype, or other item that is not a block or xref, the warning "Add [object] operation performed. Duplicate definitions will be ignored" is displayed.

OPENING DRAWINGS

To open a drawing being displayed in the Palette, right-click on its icon and choose Open In Window from the shortcut menu. You can also drag and drop the icon in the drawing area. Be sure to drop the icon in an area that is clear of another drawing. If necessary, resize or minimize any other drawing(s) first.

EXITING AutoCAD DesignCenter

You can exit AutoCAD DesignCenter by either selecting the X in the upper right corner of the DesignCenter Window or entering **adcclose** at the AutoCAD command: prompt.

DRAWING EXERCISES 12–1 THROUGH 12–4

EXERCISE 12–1

The drawing shown is typical of an architectural furniture symbols library. Create the drawing shown according to the settings given in the following table.

Settings	Value		
1. Units	Architectural		
2. Limits	Lower left corner: 0",0"		
	Upper right corner: 34',44'		
3. Grid	12		
4. Snap	12		
5. Layers	*NAME*	*COLOR*	*LINETYPE*
	Furniture	Green	Continuous
	Object	Blue	Continuous
	Text	Red	Continuous
	Dim	Yellow	Continuous
	Const	Magenta	Continuous
	Border	White	Continuous
	Walls	Blue	Continuous
6. Blocks	As indicated in Figure 12–1		

Hints: To draw objects at their proper size, the grid is spaced at 12". Use this to approximate the furniture dimensions.

Figure Ex12–1 *Furniture Blocks drawing*

EXERCISE 12-2

Create the drawing shown according to the settings given in the following table.

Settings	Value		
1. Units	Architectural		
2. Limits	Lower left corner: 0",0"		
	Upper right corner: 44',34'		
3. Grid	12		
4. Snap	12		
5. Layers	NAME	COLOR	LINETYPE
	Walls	Yellow	Continuous
	Furniture	Green	Continuous
	Object	Blue	Continuous
	Text	Red	Continuous
	Dim	Yellow	Continuous
	Const	Magenta	Continuous
	Border	White	Continuous
	Walls	Blue	Continuous

Use the AutoCAD Design Center to drag and drop blocks, layers, etc. from the drawing created in Exercise 12–1 to create the drawing as shown in Ex12–2.

Figure Ex12–2 *Completed drawing*

EXERCISE 12–3

The drawing shown is typical of an architectural kitchen appliances symbols library. Create the drawing shown according to the settings given in the following table.

Settings	Value		
1. Units	Architectural		
2. Limits	Lower left corner: 0",0"		
	Upper right corner: 34',44'		
3. Grid	12		
4. Snap	12		
5. Layers	*NAME*	*COLOR*	*LINETYPE*
	Kitchen	Green	Continuous
	Walls	Yellow	Continuous
	Furniture	Green	Continuous
	Object	Blue	Continuous
	Text	Red	Continuous
	Dim	Yellow	Continuous
	Const	Magenta	Continuous
	Border	White	Continuous
6. Blocks	As indicated in Figure 12–3		

Hints: To draw objects at their proper size, the grid is spaced at 12". Use this to approximate the appliance dimensions.

Figure Ex12–3 *Architectural Kitchen blocks drawing*

EXERCISE 12–4

Create the drawing shown according to the settings given in the following table.

Settings	Value		
1. Units	Architectural		
2. Limits	Lower left corner: 0",0"		
	Upper right corner: 22',17'		
3. Grid	6		
4. Snap	6		
5. Layers	*NAME*	*COLOR*	*LINETYPE*
	Walls	Yellow	Continuous
	Kitchen	Green	Continuous
	Walls	Yellow	Continuous
	Furniture	Green	Continuous
	Object	Blue	Continuous
	Text	Red	Continuous
	Dim	Yellow	Continuous
	Const	Magenta	Continuous
	Border	White	Continuous

Use the AutoCAD Design Center to drag and drop blocks, layers, etc. from the drawing created in Exercise 12–3 to create the drawing as shown in Ex12–4.

Figure Ex12–4 *Completed drawing*

REVIEW QUESTIONS

1. What command is used to invoke the AutoCAD 2000 Design Center?

    a. DSNCEN

    b. DGNCEN

    c. ADCENTER

    d. DGNCTR

2. Which of the following is considered to be drawing content types?

    a. blocks

    b. layers

    c. linetypes

    d. all the above

3. AutoCAD 2000 allows items to be directly edited from the Design Center.

    a. True

    b. False

4. Which of the following are not drawing content types?

    a. lines

    b. circles

    c. arcs

    d. all of the above

5. Within the Design Center a drawing is considered to be a _____?

    a. content

    b. container

    c. folder

    d. a & b

    e. a & c

6. Invoking the AutoCAD 2000 Design Center command opens the Design Center dialog box.

    a. True

    b. False

7. The default position for the Design Center is in the lower right corner.

a. True

b. False

8. Which of the following areas of the Design Center displays the names and icons representing content?

a. Toolbar

b. Palette

c. Tree

d. Preview Pane

e. Description Pane

9. Large Icons, Small Icons, List and Details are four optional modes of displaying content using which of the following buttons?

a. LOAD

b. FIND

c. UP

d. VIEWS

10. Within the Tree View Pane which button can display the previous twenty items and their paths accessed through the AutoCAD 2000 Design Center?

a. Open Drawings

b. Desktop

c. History

d. None of the above

CHAPTER 13

Utility Commands

INTRODUCTION

After completing this chapter, you will be able to do the following:

- Partial Open and Load drawings
- Manage Drawing Properties
- Use the geometric calculator to perform geometric calculations
- Manage named objects
- Delete unused named objects
- Use the utility display commands
- Use object properties
- Use X,Y,Z filters
- Use the SHELL command
- Set up a drawing by means of the MVSETUP utility
- Use the TIME and AUDIT commands
- Customize AutoCAD settings
- Export and Import data
- Use object linking and embedding
- Use script and slide commands

PARTIAL OPEN/PARTIAL LOAD

The PARTIAL OPEN command allows you to work with just part of a drawing by loading geometry only from specific views or layers. When a drawing is partially open, specified and named objects are loaded. Named objects include blocks, layers, dimension styles, linetypes, layouts, text styles, viewport configurations, UCSs, and views. You can load additional information into the current drawing as long as the drawing is partially open.

Invoke the PARTIAL OPEN command:

Standard toolbar	Choose the OPEN command (see Figure 13–1) and then select Partial Open from the Select File dialog box.
Files menu	Choose Open and then select Partial Open from the Select File dialog box
Command: prompt	**partialopen** (ENTER)

Figure 13–1 *Invoking the* OPEN *command from the Standard toolbar*

AutoCAD displays the Select File dialog box. When a drawing name is entered in the **File Name** text box, select the **Partial Open** button as shown in Figure 13–2. If the system variable FILEDIA is set to 0 (zero), there will be a prompt on the command line in response to the OPEN command.

 Note: AutoCAD will display the AutoCAD Macro Virus Protection dialog box if the drawing you open contains macros. If the current drawing is partially open, PARTIALOAD loads additional data into a drawing file

Figure 13–2 *Choose the Partial Open button from the Select File dialog box*

AutoCAD displays the Partial Open dialog box as shown in Figure 13–3.

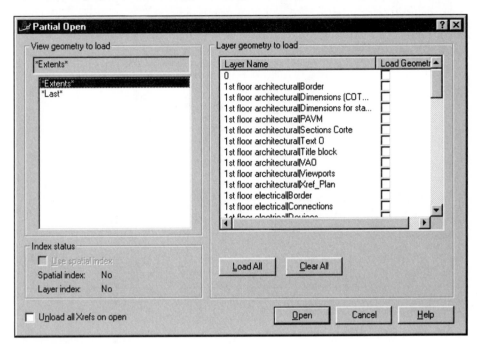

Figure 13–3 *The Partial Open dialog box*

In the **Partial Open** dialog box, select a view. The default view is Extents. Only geometry from model space views that are saved in the current drawing can be loaded. Select layer(s). No layer geometry is loaded into the drawing if you do not select any layers to load. However, all drawing layers will exist in the drawing. Even if the geometry from a view is specified to load and no layer is specified to load, no geometry is loaded. If you don't want to load xrefs, but the drawing contains them, select **Unload All Xrefs** on Open. When the view(s) and layer(s) have been specified, choose **Open** to partially open the drawing.

The PARTIAL LOAD command allows you to load additional geometry into a current partially-loaded drawing by loading geometry only from specific views or layers.

Invoke the PARTIAL LOAD command:

Files menu	Choose Partial Load
Command: prompt	**partiaload** (ENTER)

AutoCAD displays the Partial Load dialog box as shown in Figure 13–4.

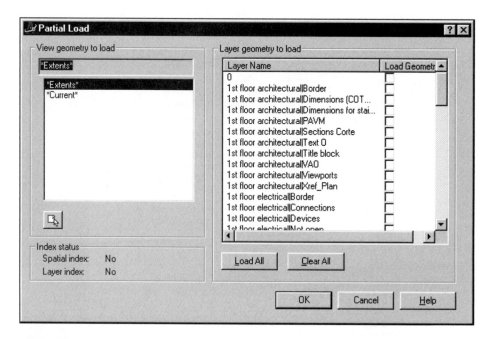

Figure 13–4 *The Partial Load dialog box*

In the **Partial Load** dialog box, select a view. The default view is Extents. Only geometry from model space views that are saved in the current drawing can be loaded. Select layer(s). No layer geometry is loaded into the drawing if you do not select any layers to load. However, all drawing layers will exist in the drawing. Even if the geometry from a view is specified to load, and no layer is specified to load, no geometry is loaded. When the view(s) and layer(s) have been specified, choose **OK** to partially load the geometry from the specified layers.

DRAWING PROPERTIES

Drawing Properties allow you to keep track of your drawings by having properties assigned to them. The properties you assign will identify the drawing by its title, author, subject, and keywords for the model or other data. Hyperlink addresses or paths can be stored along with ten custom properties.

Invoke the DRAWING PROPERTIES command:

Files menu	Choose Drawing Properties
Command: prompt	**dwgprops** (ENTER)

AutoCAD displays the Drawing Properties dialog box as shown in Figures 13–5.

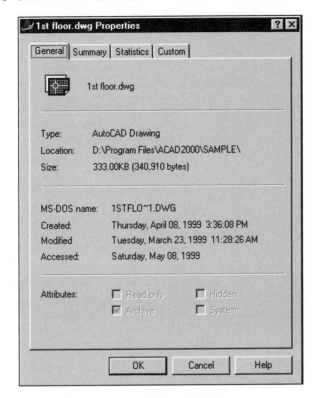

Figure 13–5 *The Drawing Properties dialog box with the General tab displayed*

The **Drawing Properties** dialog box has four tabs: the General tab, Summary tab, Statistics tab, and Custom tab.

The **General** tab (see Figure 13–5) displays the drawing type, location, size, and other information. These come from the operating system. These fields are read-only. However, the attributes options are made available by the operating system if you access file properties through Windows Explorer.

The **Summary** tab (see Figure 13–6) lets you enter a title, subject, author, keywords, comments, and a hyperlink base for the drawing. Keywords for drawings sharing a common property will help in your search. For a hyperlink base you can specify a path to a folder on a network drive or an Internet address.

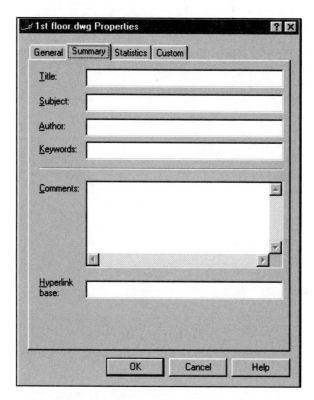

Figure 13–6 *The Drawing Properties dialog box with the Summary tab displayed*

The **Statistics** tab (see Figure 13–7) displays information such as the dates files were created and last modified. You can search for all files created at a certain time.

The **Custom** tab (see Figure 13–8) lets you specify up to ten custom properties. Enter the names of the custom fields in the left column, and the value for each custom field in the right column.

 Note: Properties entered in the Drawing Properties dialog box are not associated with the drawing until you save the drawing.

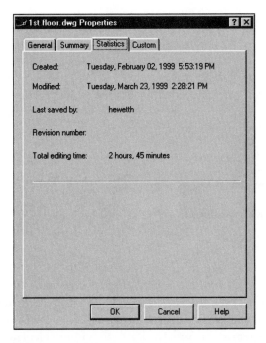

Figure 13–7 *The Drawing Properties dialog box with the Statistics tab displayed*

Figure 13–8 *The Drawing Properties dialog box with the Custom tab displayed*

THE GEOMETRIC CALCULATOR (CAL)

The CAL command is an on-line calculator that evaluates real, integer, and vector expressions. The CAL command, when entered at the "Command:" prompt (or 'cal transparently when prompted for input) switches you to the calculator mode. When the "Command:" prompt is preceded by double angle brackets (>>), AutoCAD is awaiting input, which must be in the acceptable calculator format. For example, you can solve a simple algebraic equation by entering the following:

> Command: **cal**
> Initializing. . .
> >>Expression: **A=pi*4^2** (ENTER)

producing the following response:

> 50.2655

This sequence might be used to derive the area of a circle with a radius of 4. The variable A and the equal sign are not required to arrive at the solution. Entering pi*4^2 will suffice. However, the A= serves to give a user-defined name (the letter A in this case) to a variable so that the result of the equation can be set as the value of the named variable for later use in AutoLISP or in another calculator expression. In the built-in calculator, "pi" has been assigned a preset value of approximately 3.141592654, which is the ratio of a circle's circumference to its diameter. The "*" and the "^" are the symbols for multiply and exponent, respectively, as is noted later in the listing of arithmetic operators you may use in calculator expressions.

You may also enter the built-in calculator transparently (while in the middle of another command), as shown here during the LINE command and in Figure 13–9:

> Command: **line**
> Specify first point: **'cal**
> >>Expression: **(mid+end)/2**
> >>Select entity for MID snap: *(select one line)*
> >>Select entity for END snap: *(select other line)*
> Specify next point or [Undo]: (ENTER)

This expression causes AutoCAD to pause while the user selects points by object snapping to a midpoint and an endpoint. Then AutoCAD uses the midpoint between those two points as the starting point of the line to be drawn.

Responses to the ">>Expression:" and other calculator (with leading >>'s) prompts vary according to the desired function.

In addition to using CAL (and 'CAL transparently) while in AutoCAD, you can use CAL as an operator within an AutoLISP expression, followed by a calculator expression enclosed in double quotes.

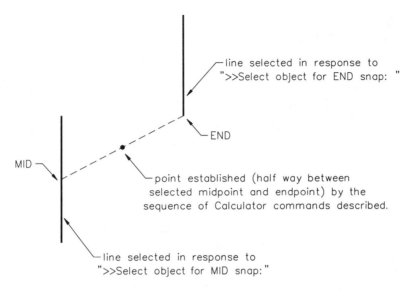

Figure 13–9 *Using the* CAL *command transparently during the* LINE *command*

CAL can be considered an AutoLISP subroutine with an argument of a calculator expression-string. In the following example, CAL finds the midpoint between two other midpoints of parallel and equal-length lines 1 and 2, as shown in Figure 13–10.

Command: **(setq pt1 (cal "(mid+mid)/2"))**
>>Select entity for MID snap: *(select one line)*
>>Select entity for MID snap: *(select other line)*
(4.5 5.0 0.0)

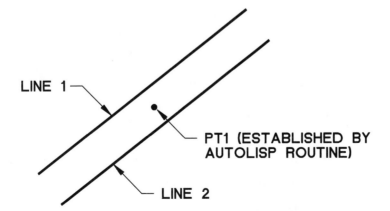

Figure 13–10 CAL *finds the midpoint between two other midpoints parallel and equal-length lines 1 and 2*

Things you can do directly or interactively (with AutoLISP) through this programmable calculator include the following:

- Use expressions similar to common algebra to solve equations.
- Name and store values directly (or indirectly through expressions) to variables for later use in expressions, by AutoLISP or in AutoCAD commands.
- Derive information about the existing geometry, or use available data to generate or analyze geometric constructs such as points, vectors, components (X, X, and Z coordinates) of a point, coordinate systems, and axes.
- Apply Object Snap modes.

NOTATIONS AND FORMAT

Calculator expressions are entered in a format similar to algebra, known as infix notation. Operational precedences are as follows:

1. Left to right for operators of equal precedence
2. Parentheses, from innermost out
3. Exponents first; multiplication and division second; addition and subtraction last

Numeric operators that can be used include the following:

+	add
–	subtract
*	multiply
/	divide
^	raise to a power
[ ]	signifies expression

Vector operators that can be used include the following:

+	add vectors
–	subtract vectors
*	multiply a vector by a real number or scaler product of vectors
/	divide a vector by a real number
&	vector product of vectors
[ ]	signifies expression

Point/vector operators and symbols that can be used include the following:

[ ]	encloses point or vector
,	separates coordinate values
<	precedes angle
@	precedes distance
*	denotes WCS (overrides UCS)

Architectural distances can be entered as either feet'-inches" or feet'inches".

Angles can be entered in degrees/minutes/seconds as degrees d, minutes', seconds", radians r, or gradians g. Degrees must be entered as 0d if the angle in d/m/s is less than one degree.

Points and vectors are entered by enclosing the coordinate values in square brackets, separated by commas. Zero values can be omitted as long as the commas are there. For example, the following values are valid point/vectors:

[,,]	equals	[0,0,0]
[8,9,]	equals	[8,9,0]
[,,7]	equals	[0,0,7]

Coordinate systems can be entered in the following formats:

System	Format	Example
polar	=[dist<angle]	[7<45]
cylindrical	=[dist<angle,z]	[3<0.57r,1.0]
spherical	=[dist<angle1,angle2]	[7<30<60]
relative	=[@x,y,z]	[@1.0,2.5,3.75]
WCS	=[*x,y,z]	[*5,6,20]

Conversion of point/vectors between the User Coordinate System and the World Coordinate System can be achieved by the following functions:

w2u converts from the WCS to the UCS

u2w converts from the UCS to the WCS

A point/vector expression can include arithmetic operators, as the following example shows:

[4+3<45]
[3<3.14/6,0.5*2]

Other point/vector operators include the following:

sin, asin	sine and arcsine of angle
cos, acos	cosine and arccosine of angle
tang, atan	tangent and arctangent of angle
ln	natural log of the number
log	base-10 log of the number
exp	natural exponent of the number
exp10	base-10 exponent of the number
sqr, sqrt	square and square root
abs	absolute value
round	number rounded to nearest integer
trunc	integer with decimal portion removed
r2d	converts angle in radians to degrees
d2r	converts angle in degrees to radians
pi	quotient of circumference divided by diameter
vec	obtains the vector between two points
vec1	obtains a unit vector between two points

Coordinate values of point/vectors can be obtained singly and in pairs by using one of the following operators:

xyof(point)	returns new point with X and Y coordinates of point
xzof(point)	returns new point with X and Z coordinates of point
yzof(point)	returns new point with Y and Z coordinates of point

xof(point)	returns new point with X coordinate of point
yof(point)	returns new point with Y coordinate of point
zof(point)	returns new point with Z coordinate of point
rxof(point)	returns real that is X coordinate of point
ryof(point)	returns real that is Y coordinate of point
rzof(point)	returns real that is Z coordinate of point

Cur

The CUR operator pauses and prompts the user to pick a point with the cursor, using that point in the expression.

@

The @ symbol causes the last point to be used in the expression. This is similar to the relative coordinate method of specifying points.

Object Snap

Using the first three letters of any of the Object Snap modes in an expression causes AutoCAD to pause and prompt the user for a point subject to the Object Snap mode specified.

Finding Points

You can obtain a point on a line by entering the two endpoints and either a distance or a scaler value, as follows:

pld(pt1,pt2,distance) returns a point that is the given distance from pt1 in the direction of pt2 from pt1.

pld([1,2,3], [5,5,3],3.5) obtains pt3, as shown in Figure 13–11.

plt(pt1,pt2,proportion) returns a point that is in the direction of pt2 from pt1, and its distance from pt1 will be the product of the proportion multiplied by the distance between pt1 and pt2.

plt(pt1,pt2, 0.5) obtains pt 3, as shown in Figure 13–12.

plt(pt1, pt2, 7/8) obtains pt3, as shown in Figure 13–13.

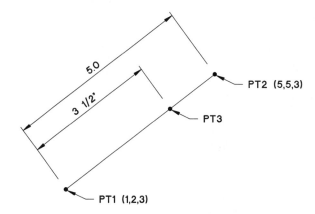

Figure 13–11 *Obtaining a point on a line by entering the two endpoints*

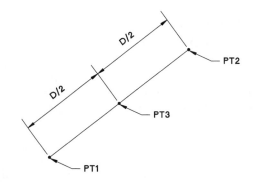

Figure 13–12 *Obtaining a point on a line by entering the two endpoints and a distance*

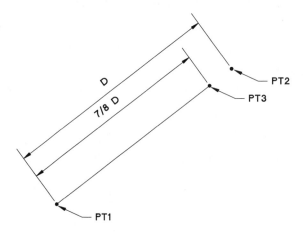

Figure 13–13 *Obtaining a point on a line by entering the two endpoints and a scaler value*

ADDITIONAL GEOMETRIC FUNCTIONS

Special points, vectors, distances, radii, and angles can be derived by using the calculator functions described in this section.

Rotation

Rotation of a point about an axis can be achieved with the ROT command.

> **rot(p,origin,angle)** returns the point *p* rotated *ang* angle about an axis through the origin in the *y* direction, as shown in Figure 13–14.

> **rot(p,Axpt1,Axpt2,ang)** returns the point *p* rotated *ang* angle about an axis through Axpt1 and Axpt2, as shown in Figure 13–15.

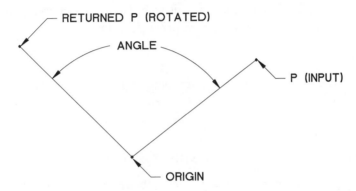

Figure 13–14 *Rotation of a point about an axis through the origin in the Y direction*

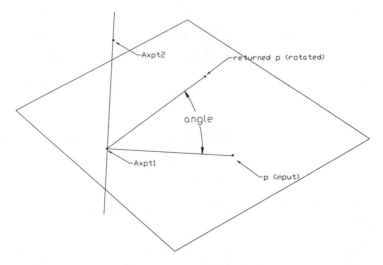

Figure 13–15 *Rotation of a point about an axis through Axpt1 and Axpt2*

Intersection Point

Ill(pt1,pt2, pt3,pt4) returns the intersection point between two lines (p1,p2) and (p3,p4). AutoCAD considers all points three-dimensional.

ilp(p1,p2,p3,p4,p5) returns the intersection point between a line (p1,p2) and a plane passing through three points (p3,p4,p5).

Distance

dpl(pt1,pt2,pt3) returns the distance (shortest) between point pt1 and a line through pt2 and pt3.

dpp(pt1,pt2,pt3,pt4) returns the distance (shortest) between point pt1 and a plane defined by the three points pt2, pt3, and pt4.

Radius

RAD pauses for the user to select an arc, circle, or *2D* polyline arc segment, and returns its radius.

Angle

ang(v) returns the angle between the *X* axis and a line defined by the vector v.

ang(pt1,pt2) returns the angle between the *X* axis and a line defined by points pt1 and pt2.

ang(apex,pt1,pt2) returns the angle between lines defined by apex to pt1 and apex to pt2.

ang(apex,pt1,pt2,pt3) returns the angle between lines defined by apex to pt1 and apex to pt2 and measured counterclockwise about the axis defined by apex to pt3.

Normal Vector

nor returns the unit normal vector (*3D*) of a selected arc, circle, or polyline arc segment.

nor(v) returns the unit normal vector (*2D*) to the vector v.

nor(pt1,pt2) returns the unit normal vector (*2D*) to the line through pt1 and pt2.

nor(pt1,pt2,pt3) returns the unit normal vector (*3D*) to the plane defined by the three points pt1, pt2, and pt3.

SHORTCUT OPERATIONS

Following are the shortened versions of calculator operators:

dee	=	dist(end,end)
ille	=	ill(end,end,end,end)
mee	=	(end+end)/2
nee	=	nor(end,end)
vee	=	vec(end,end)
vee1	=	vec1(end,end)

MANAGING NAMED OBJECTS

The RENAME command allows you to change the names of blocks, dimension styles, layers, linetypes, text styles, views, User Coordinate Systems, or viewport configurations.

Invoke the RENAME command:

Format menu	Choose Rename
Command: prompt	**rename** (ENTER)

AutoCAD displays the Rename dialog box, shown in Figure 13–16

Figure 13–16 *Rename dialog box*

 Note: Except for the layer named 0 and the linetype named Continuous, you can change the name of any of the named objects.

In the **Named Objects** list box, select the object name you want to change. The **Items** list box displays the names of all objects that can be renamed. To change the object's name, pick the name in the Items list box or enter it into the **Old Name:** text box. Enter the new name in the **Rename To:** text box, and select the **Rename To:** button to update the object's name in the Items list box. To close the dialog box, choose the OK button.

DELETING UNUSED NAMED OBJECTS

The PURGE command is used to selectively delete any unused named objects.

Invoke the PURGE command:

File menu	Choose Drawing Utilities > Purge
Command: prompt	**purge** (ENTER)

AutoCAD prompts:

Command: **purge**

If you select Purge from the Drawing Utilities in the File menu, a submenu will be displayed from which you can select All or one of the types of named objects to purge from the drawing.

If you choose the All option, AutoCAD prompts:

Enter type of unused objects to purge [Blocks/Dimstyles/LAyers/LTypes/ Plotstyles/SHapes/textSTyles/Mlinestyles/All]: *(right-click and select one of the available options to purge from the shortcut menu)*

If you have chosen an option from the submenu other than All, or if you specify one of the available options at the command prompt, AutoCAD prompts:

Enter name(s) to purge <*>: *(specify the name(s) of the object(s) to purge, or with wild-card characters specify the objects to purge)*
Verify each name to be purged? [Yes/No] <Y>: *(press ENTER for AutoCAD to prompt for verification of each symbol name displayed before it is purged, or enter n, for AutoCAD to purge without verification)*

Depending on the response, AutoCAD prompts for verification of each symbol name before it is purged.

The PURGE command removes only one level of reference. For instance, if a block has nested blocks, the PURGE command removes the outer block definition only. To remove the second-, third-, or deeper-level blocks within blocks, you must repeat the

PURGE command until there are no referenced objects. You can use the PURGE command at any time during a drawing session.

Individual shapes are part of a .SH*X* file. They cannot be renamed, but references to those that are not being used can be purged. Views, User Coordinate Systems, and viewport configurations cannot be purged, but the commands that manage them provide options to delete those that are not being used.

Note: The PURGE command cannot be used in a partially opened drawing.

COMMAND MODIFIER—MULTIPLE

MULTIPLE is not a command, but when used with another AutoCAD command, it causes automatic recalling of that command when it is completed. You must press ESC to terminate this repeating process. Here is an example of using this modifier to cause automatic repeating of the ARC command:

> Command: **multiple arc** (ENTER)
> Enter command name to repeat: **arc** (ENTER)

You can use the MULTIPLE command modifier with any of the draw, modify, and inquiry commands. PLOT, however, will ignore the MULTIPLE command modifier.

UTILITY DISPLAY COMMANDS

The utility display commands include VIEW, REGENAUTO, DRAGMODE, and BLIPMODE.

SAVING VIEWS

The VIEW command allows you to give a name to the display in the current viewport and have it saved as a view. You can recall a view later by using the VIEW command and responding with the name of the view desired. This is useful for moving back quickly to needed areas in the drawing without having to resort to zoom and pan.

Invoke the VIEW command:

View menu	Choose Named Views...
Command: prompt	**view** (ENTER)

AutoCAD displays the View dialog box, shown in Figure 13–17.

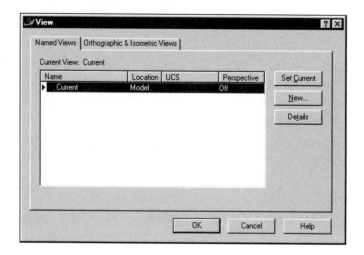

Figure 13–17 *View dialog box*

There are two tabs in the View dialog box: **Named Views** and **Orthographic & Isometric Views**. In the Named Views tab, AutoCAD lists any saved view(s) in the Views list box. You can right-click on a view name and select one of the options in the shortcut menu. Options include **Set Current, New, Delete,** and **Details.** The Set Current option makes the highlighted view the current view. The Rename option highlights the view name for editing. It can be edited in the same manner as you would a file name after clicking it once and then once again a second or so after the first click. The Delete option deletes the highlighted option. The Details option causes a dialog box to be displayed, described later in this section.

To create a new view, choose the **New...** button. AutoCAD displays the **New View** dialog box as shown in Figure 13–18.

Figure 13–18 *New View dialog box*

In the **New View** dialog box, you can enter a name for the new view in the **View Name:** text box. There are two radio buttons for specifying the area for the new view: **Current Display** and **Define Window**. If the Current Display radio button is selected the view will include the current display as the new view. If the Define Window radio button is selected the **Define New Window** button will be highlighted. Select the **Define New Window** button and you will be prompt to specify diagonally opposite corners of a window that will define the area for the new view.

From the **UCS Settings** section of the New View dialog box, the available Coordinate Systems (User or World) can be selected in the **UCS Name:** list box. You can select the **Save UCS With View** check box and have the coordinate system that is displayed in the **USC Name:** list box saved with the new view. Select **OK** to close the New View dialog box.

Selecting the **Set Current** button causes the view that is highlighted in the Current View list box to become the current view.

Selecting the **Details** button causes the **View Details** dialog box to be displayed, as shown in Figure 13–19.

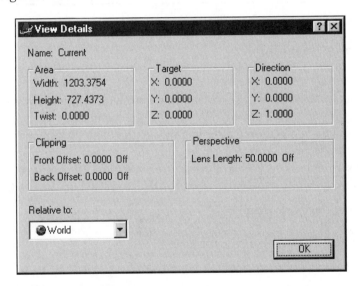

Figure 13–19 *View Details dialog box*

The View Details dialog box shows data for the view's Area, Target, Direction, Clipping and Perspective and which Coordinate system the view is Relative to.

The Orthographic & Perspective Views Tab presents other options. In the **Current View:** list box of the Orthographic & Perspective Views tab is listed six orthographic views (Top, Bottom, Front, Back, Left, and Right) and four isometric views (Southwest, Southeast, Northeast, and Northwest). Selecting one of the standard orthographic or isometric views and then selecting the **Set Current** button and the **OK** button will cause the view to be taken from the direction specified by the view name. For example, if you are viewing the front view of a building and select Top, your view will be from the traditional plan view direction. It will not, of course, automatically cut away the roof and ceiling (if the roof and ceiling are part of the *3D* drawing) and give you the horizontal section that you expect to see in a plan view.

CONTROLLING THE REGENERATION

The REGENAUTO command controls automatic regeneration. When the REGENAUTO is set to ON, AutoCAD drawings regenerate automatically. When it is set to OFF, then you may have to regenerate the drawing manually to see the current status of the drawing.

Invoke the REGENAUTO command:

Command: prompt	**regenauto** (ENTER)

AutoCAD prompts:

Command: **regenauto**
Enter mode [ON/OFF] <ON>: *(select one of the two available options)*

When you set REGENAUTO to OFF, what you see on the screen may not always represent the current state of the drawing. When changes are made by certain commands, the display will be updated only after you invoke the REGEN command. But waiting time can be avoided as long as you are aware of the status of the display. Turning the REGENAUTO setting back to ON will cause a regeneration. If a command should require regeneration while REGENAUTO is set to OFF, you will be prompted:

About to regen, proceed? <Y>

Responding with No will abort the command.

Regeneration during a transparent command will be delayed until a regeneration is performed after that transparent command. The following message will appear:

REGEN QUEUED

CONTROLLING THE DRAWING OF OBJECTS

The DRAGMODE command controls the way dragged objects are displayed. Certain draw and modify commands display highlighted dynamic (cursor-following) representations of the objects being drawn or edited. This can slow down the drawing process if the objects are very complex. Setting DRAGMODE to OFF turns off dragging.

Invoke the DRAGMODE command:

Command: prompt	**dragmode** (ENTER)

AutoCAD prompts:

> Command: **dragmode**
> Enter new value [ON/OFF/Auto] <Auto>: *(right-click and select one of the three available options from the shortcut menu)*

When DRAGMODE is set to OFF, all calls for dragging are ignored. Setting DRAGMODE to ON allows dragging by use of the DRAG command modifier. Setting DRAGMODE to Auto (default) causes dragging wherever possible.

When DRAGMODE is set to ON the DRAG command modifier can be used wherever dragging is permitted. For example, during the MOVE prompt you can use the following:

> Command: **move**
> Select objects: *(select the objects)*
> Specify base point or displacement: *(specify the base point)*
> Specify second point of displacement or
> <use first point as displacement>: **drag**

At this point the selected objects follow the cursor movement.

CONTROLLING THE DISPLAY OF MARKER BLIPS

The BLIPMODE command controls the display of marker blips. When BLIPMODE is set to ON, a small cross mark is displayed when points on the screen are specified with the cursor or by entering their coordinates. After you edit for a while, the drawing can become cluttered with these blips. They have no effect other than visual reference and can be removed at any time by using the REDRAW, REGEN, ZOOM, or PAN commands. Any other command requiring regeneration causes the blips to be removed. When BLIPMODE is set to OFF, the blips marks are not displaced.

Invoke the BLIPMODE command:

Command: prompt	**blipmode** (ENTER)

AutoCAD prompts:

> Command: **blipmode**
> Enter mode [ON/OFF] <OFF>: *(right-click and select one of the two available options from the shortcut menu)*

CHANGING THE DISPLAY ORDER OF OBJECTS

The DRAWORDER command allows you to change the display order of objects as well as images. This will ensure proper display and plotting output when two or more objects overlay one another. For instance, when a raster image is attached over an existing object, AutoCAD obscures them from view. With the help of the DRAWORDER command you can make the existing object display over the raster image.

Invoke the DRAWORDER command:

Modify II toolbar	Choose the DRAWORDER command (see Figure 13–20)
Tools menu	Choose Display Order
Command: prompt	**draworder** (ENTER)

Figure 13–20 *Invoking the* DRAWORDER *command from the Modify II toolbar*

AutoCAD prompts:

> Command: **draworder**
> Select objects: *(select the objects for which you want to change the display order, and press ENTER to complete object selection)*
> Enter object ordering option [Above object/Under object/Front/Back] <Back>: *(right-click and select one of the available options from the shortcut menu)*
> Select reference object: *(select the reference object for changing the order of display)*

When multiple objects are selected for reordering, the relative display order of the objects selected is maintained.

Above Object

The Above Object option moves the selected object(s) above a specified reference object.

Under Object

The Under Object option moves selected object(s) below a specified reference object.

Front

The Front option moves selected object(s) to the front of the drawing order.

Back

The Back option moves selected object(s) to the back of the drawing order.

 Note: The DRAWORDER command terminates when selected object(s) are reordered. The command does not continue to prompt for additional objects to reorder.

OBJECT PROPERTIES

There are three important properties that control the appearance of objects: color, linetype, and lineweight. You can specify the color, linetype, and lineweight for the objects to be drawn with the help of the LAYER command, as explained in Chapter 3. You can do the same thing by means of the COLOR, LINETYPE, and LINEWEIGHT commands.

COLOR COMMAND

The COLOR command allows you to specify a color for the objects to be drawn, separate from the layer color.

Invoke the COLOR control command:

Object Properties toolbar	Select one of the colors available from the option menu (see Figure 13–21)
Format menu	Choose Color...
Command: prompt	**color** (ENTER)

Figure 13–21 *Selecting one of the available colors from the option menu in the Object Properties toolbar*

AutoCAD displays the Select Color dialog box. The color may be entered as a standard name (red, green, cyan, yellow, magenta, blue, white, or green) or by the number code (1 through 255). Or you can respond with BYLAYER or BYBLOCK. BYLAYER is the default. If you reply with a standard name or number code, this becomes the current color. All new objects you create are drawn with this color, regardless of which layer is current, until you again set the color to BYLAYER or BYBLOCK. BYLAYER causes the objects drawn to assume the color of the layer on which it is drawn. BYBLOCK causes objects to be drawn in white until selected for inclusion in a block definition. Subsequent insertion of a block that contains objects drawn under the BYBLOCK option causes those objects to assume the color of the current setting of the COLOR command. You can use the PROPERTIES command to change the color of the existing objects.

 Note: As noted in Chapter 3, the options to specify colors by both layer and the COLOR command can cause confusion in a large drawing, especially one containing blocks and nested blocks. You are advised not to mix the two methods of specifying colors in the same drawing.

LINETYPE COMMAND

The LINETYPE command allows you to draw lines with different dash/dot/space combinations. It is used to load linetype definitions from a library or to let you create custom linetypes.

A linetype must exist in a library file and be loaded before you can apply it to an object or layer. Standard linetypes are in the library file called ACAD.LIN and are not loaded with the LAYER command. You must load the linetype before you assign it to a specific layer.

Linetypes are combinations of dashes, dots, and spaces. Customized linetypes permit "out of line" objects in a linetype such as circles, wavy lines, blocks, and skew segments.

Dash, dot, and space combinations eventually repeat themselves. For example, a six-unit-long dash, followed by a dot between two one-unit-long spaces, repeats itself according to the overall length of the line drawn and the LTSCALE setting.

Lines with dashes (not all dots) usually have dashes at both ends. AutoCAD automatically adjusts the lengths of end dashes to reach the endpoints of the adjoining line. Intermediate dashes will be the lengths specified in the definition. If the overall length of the line is not long enough to permit the breaks, the line is drawn continuous.

There is no guarantee that any segments of the line fall at a particular location. For example, when placing a centerline through circle centers, you cannot be sure that the short dashes will be centered on the circle centers as most conventions call for. To achieve this effect, the short and long dashes have to be created by either drawing them individually or by breaking a continuous line to create the spaces between the dashes. This also creates multiple in-line lines instead of one line of a particular linetype. Or you can use the DIMENSION command Center option to place the desired mark.

Individual linetype names and definitions are stored in one or more files whose extension is .LIN. The same name may be defined differently in two different files. Selecting the desired one requires proper responses to the prompts in the Load option of the LINETYPE command. If you redefine a linetype, loading it with the LINETYPE command will cause objects drawn on layers assigned to that linetype to assume the new definition.

Mastering the use of linetypes involves using the LAYER command, the LINETYPE command, the LTSCALE command, and knowing which files contain the linetype definition(s) desired. Also, with the LINETYPE command you can define custom linetypes.

Invoke the LINETYPE command:

Object Properties toolbar	Choose Other... from Linetype Control (see Figure 13–22)
Format menu	Choose Linetype...
Command: prompt	**linetype** (ENTER)

Figure 13–22 *Invoking the* LINETYPE *command from the Object Properties toolbar*

AutoCAD displays the Linetype Manager dialog box, similar to Figure 13–23.

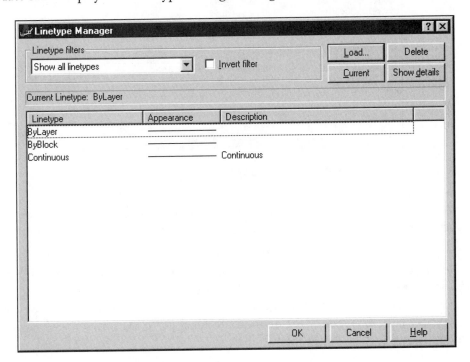

Figure 13–23 *Linetype Manager dialog box*

AutoCAD lists the available linetypes for the current drawing and displays the current linetype setting next to the **Current:** button. By default, it is set to ByLayer. To change the current linetype setting, select the appropriate linetype from the Linetype list box and select the **Current:** button. All new objects you create will be drawn with the selected linetype, irrespective of the layer you are working with, until you again set the linetype to ByLayer or ByBlock. ByLayer causes the object drawn to assume the linetypes of the layer on which it is drawn. ByBlock causes objects to be drawn in Continuous linetype until selected for inclusion in a block definition. Subsequent insertion of a block that contains objects drawn under the ByBlock option will cause those objects to assume the linetype of the block.

To load a linetype explicitly into your drawing, choose the **Load** button. AutoCAD displays the Load or Reload Linetypes dialog box, as shown in Figure 13–24.

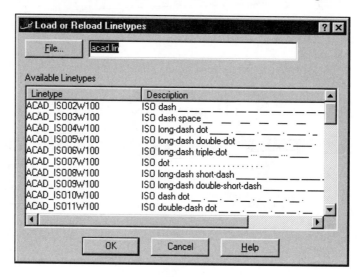

Figure 13–24 *Load or Reload Linetypes dialog box*

By default, AutoCAD lists the available linetypes from the ACAD.LIN file. Select the linetype to load from the Available Linetypes list box, and choose the **OK** button.

If you need to load linetypes from a different file, choose the **File** button in the Load or Reload Linetypes dialog box. AutoCAD displays the Select Linetype File dialog box. Select the appropriate linetype file and choose the **OK** button. In turn, AutoCAD lists the available linetypes from the selected linetype file in the Load or Reload Linetypes dialog box. Select the appropriate linetype to load from the Available Linetypes list box, and choose the **OK** button.

To delete a linetype that is currently loaded in the drawing, first select the linetype from the list box in the Linetype Manager dialog box, and then select the **Delete** button. You can delete only linetypes that are not referenced in the current drawing. You cannot delete linetype Continuous, ByLayer, or ByBlock. Deleting is the same as using the PURGE command to purge unused linetypes from the current drawing.

To display additional information of a specific linetype, first select the linetype from the Linetype list box in the Linetype Manager dialog box, and then choose the **Show Details** button. AutoCAD displays an extension of the dialog box, listing additional settings, as shown in Figure 13–25.

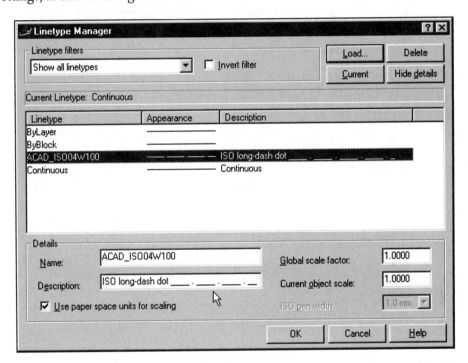

Figure 13–25 *Extended Linetype Manager dialog box*

The **Name:** and **Description:** edit fields display the selected linetype name and description, respectively.

The **Global Scale Factor:** text field displays the current setting of the LTSCALE factor. The **Current Object Scale:** text field displays the current setting of the CELTSCALE factor. If necessary, you can change the values of the LTSCALE and CELTSCALE system variables.

The **ISO Pen Width:** box sets the linetype scale to one of a list of standard ISO values. The resulting scale is the global scale factor multiplied by the object's scale factor.

Setting the **Use Paper Space Units For Scaling** check box to ON scales linetypes in paper space and model space identically.

After making the necessary changes, click the **OK** button to keep the changes and close the Linetype Manager dialog box.

 Note: As noted in Chapter 3, the options to specify linetypes by both the LAYER and the LINETYPE commands can cause confusion in a large drawing, especially one containing blocks and nested blocks. You are advised not to mix the two methods of specifying linetypes in the same drawing.

LINEWEIGHT COMMAND

The LINEWEIGHT command allows you to specify a lineweight for the objects to be drawn, separate from the assigned lineweight for the layer.

Invoke the LINEWEIGHT command:

Object Properties toolbar	Select one of the lineweights available from the option menu (see Figure 13–26)
Format menu	Choose Lineweight...
Command: prompt	**lineweight** (ENTER)

Figure 13–26 *Selecting one of the available lineweights from the option menu in the Object Properties toolbar*

If you select the Lineweight from the Format menu or type LINETWEIGHT at the Command: prompt, AutoCAD displays the Lineweight Settings dialog box, as shown in Figure 13–27.

Figure 13–27 *Lineweight Settings dialog box*

AutoCAD displays current lineweight at the bottom of the dialog box. To change the current lineweight, select one of the available lineweights from the Lineweights list box. Bylayer is the default. All new objects you create are drawn with the current lineweight, regardless of which layer is current, until you again set the lineweight to Bylayer, Byblock, or Default. Bylayer causes the objects drawn to assume the lineweight of the layer on which it is drawn. Byblock causes objects to be drawn in default until selected for inclusion in a block definition. Subsequent insertion of a block that contains objects drawn under the Byblock option causes those objects to assume the Lineweight of the current setting of the Lineweight command. The Default selection causes the objects to be drawn to the default value as set by the LWDEFAULT system variable and defaults to a value of 0.01 inches or 0.25 mm. You can also set the default value from the Default option menu located in the right side of the dialog box. The lineweight value of 0 plots at the thinnest lineweight available on the specified plotting device and is displayed at one pixel wide in model space. You can use the PROPERTIES command to change the lineweight of the existing objects.

The **Units for Listing** section specifies whether lineweights are displayed in millimeters or inches. You can also set Units for Listing by using the LWUNITS system variable.

The **Display Lineweight** toggle controls whether lineweights are displayed in the current drawing. If it is set to ON, lineweights are displayed in model space and paper space. AutoCAD regeneration time increases with lineweights that are represented by more than one pixel. If it is set to OFF, AutoCAD performance improves. Performance slows down when working with lineweights set to ON in a drawing.

The **Adjust Display Scale** controls the display scale of lineweights on the Model tab. On the Model tab, lineweights are displayed in pixels. Lineweights are displayed using a pixel width in proportion to the real-world unit value at which they plot. If you are using a high-resolution monitor, you can adjust the lineweight display scale to better display different lineweight widths. The Lineweight list reflects the current display scale. Objects with lineweight that are displayed with a width of more than one pixel may increase AutoCAD regeneration time. If you want to optimize AutoCAD performance when working in the Model tab, set the lineweight display scale to the minimum value or turn off lineweight display altogether.

Choose the **OK** button to close the dialog box and keep the changes in the settings.

Note: As noted in Chapter 3, the options to specify lineweights by both the LAYER and the LINEWEIGHT commands can cause confusion in a large drawing, especially one containing blocks and nested blocks. You are advised not to mix the two methods of specifying lineweights in the same drawing.

X, Y, AND Z FILTERS—AN ENHANCEMENT TO OBJECT SNAP

The AutoCAD filters feature allows you to establish a *2D* point by specifying the individual (*X* and *Y*) coordinates one at a time in separate steps. In the case of a *3D* point you can specify the individual (*X*, *Y*, and *Z*) coordinates in three steps. Or you can specify one of the three coordinate values in one step and a point in another step, from which AutoCAD extracts the other two coordinate values for use in the point being established.

The filters feature is used when being prompted to establish a point, as in the starting point of a line, the center of a circle, drawing a node with the POINT command, or specifying a base point or second point in displacement for the MOVE or COPY command, to mention just a few.

Note: During the application of the filters feature there are steps where you can input either single coordinate values or points, and there are steps where you can input only points. It is necessary to understand these restrictions and options and when one type of input is more desirable than the other.

When selecting points during the use of filters, you need to know which coordinates of the specified point are going to be used in the point being established. It is also essential to know how to combine Object Snap modes with those steps that use point input.

The filters feature is actually an enhancement to either the object snap or the @ (last point) feature. The AutoCAD ability to establish a point by snapping to a point on an existing object is one of the most powerful features in AutoCAD, and being able to have AutoCAD snap to such an existing point and then filter out selected coordinates for use in establishing a new point adds to that power. Therefore, in most cases you will not use the filters feature if it is practical to enter in all of the coordinates from the keyboard, because entering in all the coordinates can be done in a single step. The filters feature is a multistep process, and each step might include substeps, one to specify the coordinate(s) to be filtered out and another to designate the Object Snap mode involved.

FILTERS WITH @

When AutoCAD is prompting for a point, the filters feature is initiated by entering a period followed by the letter designation for the coordinate(s) to be filtered out. For example, if you draw a point starting at (0,0) and use the relative polar coordinate response @3<45 to determine the endpoint, you can use filters to establish another point whose X coordinate is the same as the X coordinate of the end of the line just drawn. It works for Y and Z coordinates and combinations of XY, XZ, and YZ coordinates also. The following command sequence shows how to apply filter to a line that needs to be started at a point whose X coordinate is the same as that of the end of the previous line and whose Y coordinate is 1.25. The line will be drawn horizontally 3 units long. The sequence is as follows:

```
Command: line
Specify first point: 0,0
Specify next point or [Undo]: @3<45
Specify next point or [Undo]: (ENTER)

Command: line (or ENTER)
Specify first point: .x
of @
(need YZ): 0,1.25
Specify next point or [Undo]: @3<0
```

Entering .x initiates the filters feature. AutoCAD then prompts you to specify a point from which it can extract the X coordinate. The @ (last point) does this. The new line has a starting point whose X coordinate is the same as that of the last point drawn. By using the filters feature to extract the X coordinate, that starting point will be on an imaginary vertical line through the point specified by @ in response to the "of" prompt.

When you initiate filters with a single coordinate (.x in our example) and respond with a point (@), the prompt that follows asks for a point also. From it (the second point specified) AutoCAD extracts the other two coordinates for the new point.

Even though the prompt is for "YZ," the point may be specified in *2D* format as 0,1.25 (the X and Y coordinates), from which AutoCAD takes the second value as the needed *Y* coordinate. The *Z* coordinate is assumed to be the elevation of the current coordinate system.

You can use the two-coordinate response to initiate filters. Then specify a point, and all that AutoCAD requires is a single value for the final coordinate. An example of this follows.

> Command: **line**
> Specify first point: **0,0**
> Specify next point or [Undo]: **@3<45**
> Specify next point or [Undo]: (ENTER)
>
> Command: **line** (or ENTER)
> Specify first point: **.xz**
> of **@**
> (need Y): **1.25**
> Specify next point or [Undo]: **@3<0**

You can also specify a point in response to the "(need Y):" prompt:

> (need Y): **0,1.25** *(or pick a point on the screen)*

In this case, AutoCAD uses the *Y* coordinate of the point specified as the *Y* coordinate of the new point.

Remember, it is an individual coordinate in *2D* (one or two coordinates in *3D*) of an existing point that you wish AutoCAD to extract and use for the new point. In most cases you will be object snapping to a point for the response. Otherwise, if you knew the value of the coordinate needed, you would probably enter it at the keyboard.

FILTERS WITH OBJECT SNAP

Without filters, an Object Snap (Osnap) mode establishes a new point to coincide with one on an existing object. With filters, an Osnap mode establishes selected coordinates of a new point to coincide with corresponding coordinates of one on an existing object.

Extracting one or more coordinate values to be applied to corresponding coordinate values of a point that you are being prompted to establish is shown in the following example.

In Figure 13–28, a 2.75" by 7.1875" rectangle has a 0.875"-diameter hole in its center. A board drafter would determine the center of a square or rectangle by drawing diagonals and centering the circle at their intersection. AutoCAD drafters (without filters) could do the same, or they might draw orthogonal lines from the midpoint of a horizontal line and from the midpoint of one of the vertical lines to establish a centering intersection. The following command sequence shows steps in drawing a rectangle with a circle in the center using filters.

Figure 13–28 *Extracting coordinate values to be applied to corresponding coordinate values*

Command: **line**
Specify first point: *(select a point)*
Specify next point or [Undo]: **@2.75<90**
Specify next point or [Undo]: **@7.1875<0**
Specify next point or [Close/Undo]: **@2.75<270**
Specify next point or [Close/Undo]: **c**

Command: **circle**
Specify center point for circle or [3P/2P/Ttr (tan tan radius)]: **.x**
of **mid**
of (select line 1)
(need YZ): **mid**
of (select line 2)
Specify radius of circle or [Diameter]: **d**
Specify diameter of circle <previous diameter >: **.875**

SHELL COMMAND

The SHELL command allows you to execute operating system programs without leaving AutoCAD. You can execute any operating system program as long as there is sufficient memory to execute.

Invoke the SHELL command:

Command: prompt	**shell** (ENTER)

AutoCAD prompts:

Command: **shell**
OS Command: (invoke one of the available operating system's utility
   programs)

When the utility program is finished, AutoCAD takes you back to the "Command:" prompt. If you need to execute more than one operating system program, give a null response to the "OS command:" prompt. AutoCAD responds with the appropriate operating system prompt. You can now enter as many operating system commands as you wish. When you are finished, you may return to AutoCAD by entering EXIT. It will take you back to the "Command:" prompt.

If there is not enough free memory for the SHELL command, the following message will appear:

Shell error: insufficient memory for command.

Where there is insufficient memory, you can execute the SH command instead of the SHELL command. SH requires less memory than the SHELL command and can be used to access internal DOS commands, such as DIR, COPY, and TYPE. If the need arises, you can adjust the amount of memory required for the SH and SHELL commands by modifying the ACAD.PGP file. For additional information, see Chapter 18 on "Customizing AutoCAD."

In the Windows 95, Windows 98, and Windows NT 4.xx operating system, instead of invoking the SHELL command, you can switch between programs by selecting the appropriate program from the task bar. Depending on the memory available in the computer, you can open multiple programs and switch between the programs by selecting appropriate open programs from the task bar.

 Note: Do not delete the AutoCAD lock files or temporary files created for the current drawing when you are at the operating system prompt.

SETTING UP A DRAWING

The MVSETUP command is used to control and set up the view(s) of a drawing, including the choice of standard plotted sheet sizes with a border, the scale for plotting on the selected sheet size, and multiple viewports. MVSETUP is an AutoLISP routine that can be customized to insert any type of border and title block.

Options and associated prompts depend on whether the TILEMODE system variable is set to ON (1) or OFF (0). When TILEMODE is set to ON, Tiled Viewports is enabled. When TILEMODE is set to OFF, the Floating Viewports menu item is enabled. Other paper space–related drawing setup options are available.

Invoke the MVSETUP command:

| Command: prompt | **mvsetup** (ENTER) |

AutoCAD prompts depend whether you are working on Model tab or Layout tab. If you are working on Model tab, then AutoCAD prompts:

> Enable paper space? [No/Yes] <Y>: *(press* ENTER *to enable paper space, and AutoCAD changes the* TILEMODE *setting to 0, or enter n and press* ENTER *to stay in the* TILEMODE *setting of 1)*

If you are working in Layout tab, then AutoCAD prompts:

> Enter an option [Align/Create/Scale viewports/Options/Title block/Undo]:
> *(right-click and select one of the available options from the shortcut menu)*

The following is the procedure for setting up the drawing when you are working on Model tab. AutoCAD prompts:

> Enable paper space? [No/Yes] <Y>: **n** *(ENTER)*
> Enter units type [Scientific/Decimal/Engineering/Architectural/Metric]:
> *(select a unit type. Depending on the units selected, AutoCAD lists the available scales. Select one of the available scales, or you can even specify a custom scale factor)*
> Enter the scale factor: *(specify a scale factor)*
> Enter the paper width: *(specify the paper width at which the drawing will be plotted)*
> Enter the paper height: *(specify the paper height at which the drawing will be plotted)*

AutoCAD sets up the appropriate limits to allow you to draw to full scale and also draws a bounding box enclosing the limits. Draw the drawing to full scale; when you are ready to plot, specify the scale mentioned earlier to plot the drawing.

Following is the procedure for setting up the drawing when you are working in the Layout tab. AutoCAD prompts:

> Enter an option [Align/Create/Scale viewports/Options/Title block/Undo]:

The options are explained here in the order that is logical to complete the drawing setup.

Title Block

The Title Block option allows you to select an appropriate title block. AutoCAD promtps:

> Enter title block option [Delete objects/Origin/Undo/Insert] <Insert>:
> *(right-click and select one of the available options from the shortcut menu)*

The Insert option (default) allows you to insert one of the available standard title blocks. AutoCAD lists the available title blocks and prompts you to select one of the available title blocks as follows:

0. None
1. ISO A4 Size(mm)
2. ISO A3 Size(mm)
3. ISO A2 Size(mm)
4. ISO A1 Size(mm)
5. ISO A0 Size(mm)
6. ANSI-V Size(in)
7. ANSI-A Size(in)
8. ANSI-B Size(in)
9. ANSI-C Size(in)
10. ANSI-D Size(in)
11. ANSI-E Size(in)
12. Arch/Engineering (24 x 36in)
13. Generic D size Sheet (24 x 36in)

Enter number of title block to load or [Add/Delete/Redisplay]: *(select one of the available title blocks and press* ENTER, *and AutoCAD inserts a border and title block, as shown in Figure 13–29, or enter an option)*

Figure 13–29 *Title block and border for ANSI-B Size*

The **Add** option allows you to add a title block drawing to the available list.

The **Delete** option allows you to delete an entry from the available list.

The **Redisplay** option redisplays the list of title block options.

Create

The Create option allows you to establish viewports. AutoCAD prompts:

> Enter option [Delete objects/Create viewports/Undo] <Create>: *(press*
> ENTER *to create viewports, and AutoCAD lists the available viewport layout*
> *options, or enter an option)*

The **Create viewports** option (default) allows you to create multiple viewports to one of the standard layouts. AutoCAD prompts:

> Available Mview viewport layout options:
>
> 0: None
> 1: Single
> 2: Std. Engineering
> 3: Array of Viewports
>
> Enter layout number to load or [Redisplay]: *(select one of the available*
> *layouts)*

The **None** selection creates no viewports. The **1** selection creates a single viewport whose size is determined during subsequent prompts responses. The **2** selection creates four viewports with preset viewing angles by dividing a specified area into quadrants. The size is determined by responses to subsequent prompts. The **3** selection creates a matrix of viewports along the X and Y axes.

The **Delete Objects** option deletes the existing viewports.

The **Undo** option reverses operations performed in the current MVSETUP session.

Scale Viewports

The Scale Viewports option adjusts the scale factor of the objects displayed in the viewports. The scale factor is specified as a ratio of paper space to model space. For example, 1:48 is 1 paper space unit for 48 model space units (scale for 1/4" = 1'0").

Options

The Options option lets you establish several different environment settings that are associated with your layout. AutoCAD prompts:

> Enter an option [Layer/LImits/Units/Xref] <exit>: *(enter an option)*

The **Layer** option permits you to specify a layer for placing the title block.

The **LImits** option permits you to specify whether or not to reset the limits to drawing extents after the title block has been inserted.

The **Units** option permits you to specify whether sizes and point locations will be translated to inch or millimeter paper units.

The **Xref** option permits you to specify whether the title block is to be inserted or externally referenced.

Align

The Align option causes AutoCAD to pan the view in one viewport so that it aligns with a basepoint in another viewport. Whichever viewport the other point moves to becomes the active viewport. AutoCAD prompts:

Enter an option [Angled/Horizontal/Vertical alignment/Rotate view/Undo]:

The **Angled** option causes AutoCAD to pan the view in a viewport in a specified direction.

The **Horizontal** option causes AutoCAD to pan the view in one viewport, aligning it horizontally with a basepoint in another viewport.

The **Vertical Alignment** option causes AutoCAD to pan the view in one viewport, aligning it vertically with a basepoint in another viewport.

The **Rotate View** option causes AutoCAD to rotate the view in a viewport around a basepoint.

Undo

The Undo option causes AutoCAD to undo the results of the current MVSETUP command.

TIME COMMAND

The TIME command displays the current time and date related to your current drawing session. In addition, you can find out how long you have been working in AutoCAD. This command uses the clock in your computer to keep track of the time functions and displays to the nearest millisecond using 24-hour military format.

Invoke the TIME command:

Tools menu	Choose Inquiry > Time
Command: prompt	**time** (ENTER)

The following listing is displayed in the text screen followed by a prompt:

Current time: Sunday, April 25, 1999 at 9:46:10:110 PM
Times for this drawing:

```
Created: Sunday, April 25, 1999 at 9:26:29:370 PM
Last updated: Sunday, April 25, 1999 at 9:26:29:370 PM
Total editing time: 0 days 00:19:40.790
Elapsed timer (on): 0 days 00:19:40.790
Next automatic save in: 0 days 01:40:24.455
Enter option [Display/ON/OFF/Reset]:
```

The first line gives today's date and time.

The third line gives the date and time the current drawing was initially created. The drawing time starts when you initially begin a new drawing. If the drawing was created by means of the WBLOCK command, it is the date and time the command was executed that is displayed here.

The fourth line gives the date and time the drawing was last updated. Initially set to the drawing creation time, this is updated each time you use the END or SAVE command.

The fifth line gives the length of time you are in AutoCAD. This timer is continuously updated by AutoCAD while you are in the program, excluding plotting and printer plot time. This timer cannot be stopped or reset.

The sixth line provides information about the stopwatch timer. You can turn this timer ON or OFF and reset it to zero. This timer is independent of other functions.

The seventh line provides information about when the next automatic save will take place.

Display Option
The Display option redisplays the time functions, with updated times.

ON Option
The ON option turns the stopwatch timer to ON, if it is OFF. By default it is ON.

OFF Option
The OFF option turns the stopwatch timer to OFF and displays the accumulated time.

Reset Option
The Reset option resets the stopwatch timer to zero.

To exit the TIME command at the prompt, give a null response or press ESC.

AUDIT COMMAND

The AUDIT command serves as a diagnostic tool to correct any errors or defects in the database of the current drawing. AutoCAD generates an extensive report of the problems, and for every error detected AutoCAD recommends action to correct it.

Invoke the AUDIT command:

File menu	Choose Drawing Utilities >Audit
Command: prompt	**audit** (ENTER)

AutoCAD prompts:

> Command: **audit**
> Fix any errors detected? [Yes/No] <N>: *(specify y for yes or n for no)*

If you respond with Y or Yes, AutoCAD will fix all the errors detected and display an audit report with detailed information about the errors detected and fixing them. If you answer with N or No, AutoCAD will just display a report and will not fix any errors.

In addition, AutoCAD creates an ASCII report file (AUDITCLT system variable should be set to ON) describing the problems and the actions taken. It will save the file in the current directory, using the current drawing's name with the file extension .ADT. You can use any ASCII editor to display the report file on the screen or print it on the printer, respectively.

 Note: If a drawing contains errors that the AUDIT command cannot fix, open the drawing with the RECOVER command to retrieve the drawing and correct its errors.

OBJECT LINKING AND EMBEDDING (OLE)

Object linking and embedding (OLE) is a Microsoft Windows feature that combines various application data into one compound document. AutoCAD has client as well as server capabilities. As a client, AutoCAD permits you to have objects from other Windows applications either embedded in or linked to your drawing.

When an object is inserted into an AutoCAD drawing from an application that supports OLE, the object can maintain a connection with its source file. If you insert an embedded object into AutoCAD (client), it is no longer associated with the source (server). If necessary, you can edit the embedded data from inside the AutoCAD drawing by using the original application. But at the same time, this editing does not change the original file.

If, instead, you insert an object as a linked object into AutoCAD (client), the object remains associated with its source (server). When you edit a linked object in AutoCAD by using the original application, the original file as well as the object inserted into AutoCAD change.

Linked or embedded objects appear on the screen in AutoCAD and can be printed or plotted using Windows system drivers.

Let's look at an example of object linking between an AutoCAD (server) drawing and Microsoft Word (client). Figure 13–30 shows a drawing of a desk, a computer, and a chair, that contains various attribute values. We are going to link this drawing to a Microsoft Word document.

From the Edit menu select the **Copy** option and AutoCAD prompts you to select objects. Select the computer, the table, and the chair, and press ENTER. This will copy the selection to the Windows clipboard. Minimize the AutoCAD program.

Instead of selecting specific objects, you can copy the current view into the Windows clipboard by invoking the COPYLINK command from the Edit menu.

Figure 13–30 *Drawing of a desk, a computer, and a chair with attribute values*

Open the Microsoft Word program from the Desktop by double-clicking the Word program icon. The Microsoft Word program is displayed, as shown in Figure 13–31.

From the Edit menu in Microsoft Word, select **Paste Special...**, and Word displays the Paste Special dialog box shown in Figure 13–32. Select the **Paste Link** button. This will insert the AutoCAD drawing object into Microsoft Word, as shown in Figure 13–33.

Minimize the Word program and maximize AutoCAD (or if the AutoCAD program is not open, double-click the drawing image in the Word document, which will launch the AutoCAD program with the image drawing open). Edit the values of the attributes in the computer block to 450, 13.0 GB, 21", which represent a Pentium 450 computer with a 13.0 GB hard drive and a 21" monitor.

Switch back to Word, and from the Edit menu select the **Links** option. Word will display the Links dialog box shown in Figure 13–34.

Figure 13–31 *Microsoft Word program*

Figure 13–32 *Paste Special dialog box*

Figure 13–33 *Microsoft Word document with the AutoCAD drawing*

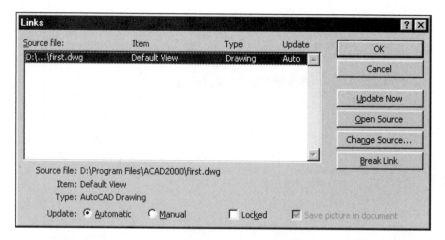

Figure 13–34 *Links dialog box*

Select the **Update Now** button and then choose the OK button. The image in the Word document is updated, as shown in Figure 13–35.

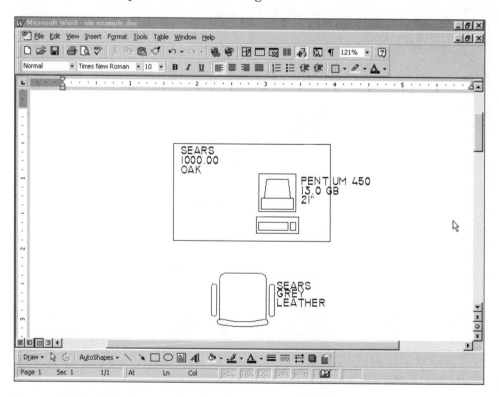

Figure 13–35 *AutoCAD drawing updated in Word as the client*

In our example, AutoCAD is the server and Microsoft Word is the client.

Conversely, you can place a linked object into AutoCAD, where AutoCAD is the client and another application is the server. Let's look an example in which AutoCAD is the client and Excel is the server.

Start the Excel program and create a spreadsheet. Copy the contents into the Windows clipboard.

From the Edit menu in AutoCAD, select the **Paste Special** option. AutoCAD displays the Paste Special dialog box. Select the **Paste Link** radio button and Microsoft Excel Worksheet from the list box, and click the **OK** button.

The Excel spreadsheet will be linked to the drawing, as shown in Figure 13–36. AutoCAD is now the client and Microsoft Excel is the server.

To edit the spreadsheet, double-click anywhere on the spreadsheet, which in turn will launch Excel with the spreadsheet document open. Any changes made to the spreadsheet will be reflected in the drawing. Figure 13–37 shows the changes that were made in the spreadsheet.

Figure 13–36 *An Excel spreadsheet in the AutoCAD drawing*

Figure 13–37 *AutoCAD drawing showing the changes made in the Excel spreadsheet*

Here is another example in which an AutoCAD drawing is the client, for both embedding from Word and linking from Excel.

Figure 13–38 shows both an AutoCAD screen and an Excel spreadsheet, in which the spreadsheet is being used for area calculations. The AREA cells are formulas that calculate the product of the corresponding WIDTH and LENGTH cells. In turn, the TOTAL cell is the sum of the AREA cells.

Figure 13–39 shows the same AutoCAD screen and a Word document that was used for typing the GENERAL NOTES, which were in turn embedded into the AutoCAD drawing.

Figure 13–40 shows how the cell value in the WIDTH for ROOM1 has been changed, resulting in changes in the AREA and TOTAL cells. Because this object (the spreadsheet consisting of four columns and seven rows, including the title) was paste linked into the AutoCAD drawing, the linked object automatically reflects the changes.

Figure 13–38 *AutoCAD drawing screen and an Excel spreadsheet*

Figure 13–39 *AutoCAD drawing and a Word document*

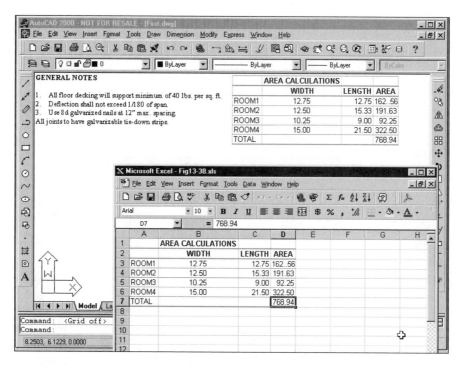

Figure 13–40 *Changes shown in the spreadsheet and AutoCAD screen*

AutoCAD displays the OLE Properties dialog box as shown in Figure 13–41 whenever you insert an OLE object into a drawing.

Figure 13–41 *OLE Properties dialog box*

If necessary, you can change the size of the object by changing the height and width in drawing units, or enter a percentage of the current height or width in the Scale section. You can also change the text size and OLE plot quality. If you need to change the OLE properties after pasting the objects into the current drawing, first select the OLE object, and from the shortcut menu select properties or at the Command: prompt invoke the OLESCALE command. AutoCAD displays the OLE Properties dialog box.

CUSTOM SETTINGS WITH THE OPTIONS DIALOG BOX

The **Options** dialog box (in AutoCAD Release 14 it was called the Preference dialog box) allows you to customize the AutoCAD settings. AutoCAD allows you to save and restore a set of custom preferences called a profile. A profile can include preference settings that are not saved in the drawing, with the exception of pointer and printer driver settings. By default, AutoCAD stores your current settings in a profile named <Unnamed Profile>.

To open the Options dialog box, invoke the OPTIONS command:

Tools menu	Choose Options
Command: prompt	**options** (ENTER)

AutoCAD displays the Options dialog box, similar to Figure 13–42. From the Options dialog box, the user can control various aspects of the AutoCAD environment. The Options dialog box has nine tabs; to make changes to any of the sections, select the corresponding tab from the top of the Options dialog box.

Figure 13–42 *Options dialog box (with the Files tab selected)*

FILES

The Files tab of the Options dialog box, shown in Figure 13–42, specifies the directory in which AutoCAD searches for support files, driver files, project files, template drawing file location, temporary drawing file location, temporary external reference file location, and texture maps. It also specifies the location of menu, help, log, text editor, and dictionary files.

When you choose the **Browse...** button, AutoCAD displays the Browse for Folder or Select a File dialog box, depending on what you selected from the list.

When you choose the **Add** button, AutoCAD adds a search path for the selected folder.

When you choose the **Remove** button, AutoCAD removes the selected search path or file.

When you select the **Move Up** button, AutoCAD moves the selected search path above the preceding search path.

When you select the **Move Down** button, AutoCAD moves the selected search path below the following search path.

When you select the **Set Current** button, AutoCAD makes the selected project or spelling dictionary current.

DISPLAY

The Display tab of the Options dialog box, shown in Figure 13–43, controls preferences that relate to AutoCAD performance.

Figure 13–43 *Options dialog box (with the Display tab selected)*

Window Elements

The Window Elements section controls the parameters of the AutoCAD drawing window. The **Display Scroll Bars In Drawing Window** check box (set to OFF by default) specifies whether to display scroll bars at the bottom and right sides of the drawing window. The **Display Screen Menu** check box (set to OFF by default) specifies whether to display the screen menu on the right side of the drawing window. The **Text Lines In Command Line Window:** text field (set to 3 by default) specifies the number of lines of text you want to display in the docked command window. You can set the value to any integer from 1 to 100. Choose the **Colors** button and AutoCAD displays the AutoCAD Window Colors dialog box, which can be used to set the colors for drawing area, screen menu, text window, and command line. Choose the **Fonts** button and AutoCAD displays the Graphics Window Font dialog box, which can be used to specify the font AutoCAD uses for the screen menu and command line and in the text window.

Display Resolution

The Display Resolution section lets you set the resolution in the following text boxes:

- Arc And Circle Smoothness
- Segments In A Polyline Curve
- Rendered Object Smoothness
- Contour Lines Per Surface.

Layout Elements

The Layout Elements section has check boxes to toggle on and off the following:

- Display Layout And Model Tabs
- Display Margins
- Display Paper Background
- Display Paper Shadow
- Show Page Setup Dialog For New Layouts
- Create Viewport In New layouts

Display Performance

The Display Performance section has check boxes to toggle on and off the following:

- Pan and Zoom With Raster Image
- Highlight Raster Image Frame Only
- True Color Raster Images And Rendering
- Apply Solid Fill
- Show Text Boundary Frame Only
- Show Silhouettes In Wireframe

OPEN AND SAVE

The Open And Save tab of the Options dialog box, shown in Figure 13–44, lets you determine formats and parameters for drawings, external references, and ObjectARX Applications as they are opened or saved.

Figure 13–44 *Options dialog box (with the Open And Save tab selected)*

File Save

In the File Save section, the **Save As:** text box lets you select a default save format when you invoke the SAVEAS command. Choose from the following formats:

- AutoCAD 2000 Drawing (*.dwg)
- AutoCAD R14/LT98/LT97 Drawing (*.dwg)
- AutoCAD R13/LT95 Drawing (*.dwg)
- AutoCAD Drawing Template File (*.dwt)
- AutoCAD 2000 DXF (*.dxf)
- AutoCAD R14/LT98/LT97 DXF (*.dxf)
- AutoCAD R13/LT95 DXF (*.dxf)
- AutoCAD R12/LT2 DXF (*.dxf)

External References (Xref)

The External References (Xref) section has a text box from which you can select Disabled, Enabled, or Enabled With Copy. **Disabled** means that demand loading is not on in the current drawing. Someone else can open and edit an xref file except as it is being read into the current drawing. **Enabled** means that demand loading is ON in the current drawing. No one else can edit an xref file while the current drawing is open. However, someone else can reference the xref file. **Enabled with Copy** means that demand loading is ON in the current drawing. An xref file can still be opened and edited by someone else. AutoCAD only uses a copy of the xref file, treating it as a completely separate file from the original xref.

The **Retain Changes To Xref Layers** check box controls the properties of layers in xrefs to be saved as they are changed in the current drawing for reloading later. Another check box entitled **Allow Other Users To Refedit Current Drawing** allows the current drawing to be edited while being referenced by other drawing(s).

File Safety Precautions

The File Safety Precautions section helps to detect errors and avoid losing data. The **Automatic Save** check box with the **Minutes Between Saves** text box allow you to determine if and for what intervals periodic automatic saves will be performed. The **Create Backup Copy with Each Save** check box allows you to determine whether a backup copy is created when you save the drawing. The **Full-Time CRC Validation** check box allows you to determine whether a cyclic redundancy check (CRC) is performed when an object is read into the drawing. Cyclic redundancy check is a mechanism for error-checking. If you suspect a hardware problem or AutoCAD error is causing your drawings to be corrupted, set this check box on. The **Maintain a Log File** check box allows you to determine whether the contents of the text window are written to a log file. Use the Files tab in the Options dialog box to specify the name and location of the log file. The **File Extension for Temporary Files** text box allows you to specify an extension for temporary files on a network. The default extension is .AC$.

ObjectARX Applications

The ObjectARX Applications section controls parameters for AutoCAD Runtime Extension applications and proxy graphics. The **Demand Load ObjectARX Apps** text box let you select when and if a third-party application is demand-loaded when a drawing has custom objects that were created in that application. The **Disable Load On Demand** option turns off demand-loading. The **Custom Object Detect** option selection demand-loads the source application when you open a drawing that contains custom objects. It does not demand-load the application when you invoke one of the application's commands. The **Command Invoke** option selection demand-loads the source application when you invoke the application's command. This setting does not demand-load the application when you open a drawing that contains custom objects. The **Object Detect and Command Invoke** option selection demand-loads the source application when you open a drawing that contains custom objects or when you invoke one of the application's commands.

The **Proxy Images for Custom Objects** text box controls how custom objects in the drawings are displayed. The **Do Not Show Proxy Graphics** option, when selected, causes custom objects in drawings not to be displayed. The **Show Proxy Graphics** option, when selected, causes custom objects in drawings to be displayed. The **Show Proxy Bounding Box** option, when selected, causes a box to be displayed in place of custom objects in drawings. The **Show Proxy Information Dialog Box** check box, when checked, causes a warning to be displayed when you open a drawing that contains custom objects.

PLOTTING

The **Plotting** tab of the Options dialog box, shown in Figure 13–45, lets you select the parameter for plotting your drawing.

Figure 13–45 *Options dialog box (with the Plotting tab selected)*

Default Plot Settings For New Drawings

The Default Plot Settings For New Drawings section determines plotting parameters for new drawings. It will also determine settings for drawings created in earlier releases of AutoCAD prior to AutoCAD 2000 that have never been saved in AutoCAD 2000 format. The **Use As Default Output Device** text box determines the default output device for new drawings and for drawings created in an earlier release of AutoCAD that have never been saved in AutoCAD 2000 format. The list displays all plotter configuration files (PC3) that are found in the plotter configuration search path. It also displays all system printers configured in the system. The **Use Last Successful Plot Settings** check box, when checked, uses the settings of the last successful plot for the current settings. The **Add or Configure Plotters** buttons lets you add or configure a plotter from the Plotters program window. See the chapter on plotting for adding and configuring plotters.

General Plot Options

The General Plot Options section lets you set general parameters such as paper size, system printer alert parameters, and OLE objects. Under **When Changing The Plot Device:**, the **Keep the Layout Paper Size If Possible** radio button, when selected, applies the paper size in the Layout Settings tab in the Page Setup dialog box provided the selected output device is able to plot to this paper size. If it cannot, AutoCAD displays a warning message and uses the paper size specified either in the plotter configuration file (PC3) or in the default system settings if the output device is a system printer. The **Use the Plot Device Paper Size** radio button applies the paper size in either the plotter configuration file (PC3) or in the default system settings if the output device is a system printer.

The **System Printer Spool Alert** list box selection determines if a warning will be displayed if the plotted drawing is spooled through a system printer because of an input or output port conflict. The **Always Alert (And Log Errors)** option, when selected, displays a warning and always logs an error when the plotted drawing spools through a system printer. The **Alert First Time Only (And Log Errors)** option, when selected, displays a warning once and always logs an error when the plotted drawing spools through a system printer. The **Never Alert (And Log First Error)** option, when selected, will never display a warning and logs only the first error when the plotted drawing spools through a system printer. The **Never Alert (Do Not Log Errors)** option, when selected, never displays a warning or logs an error when the plotted drawing spools through a system printer.

The **OLE Plot Quality** list box selection determines plotted OLE objects' quality. Options include Line Art, Text, Graphics, Photograph, and High Quality Photograph.

The **Use OLE Application When Plotting OLE Objects** check box starts the application that creates the OLE object when you plot a drawing with OLE objects. This will help optimize quality of OLE objects.

Default Plot Style Behavior

The Default Plot Style Behavior for new drawings section is used to set up parameters for plot styles in all drawings. A plot style is a set of parameters defined in a plot style table and used when the drawing is plotted. See the Chapter 8 on plotting . The **Use Color Dependent Plot Styles** radio button, when selected, disables the plot style list on the Object Properties toolbar. This is the default setting. The **Use Named Plot Styles** radio button, when selected, enables the list.

When the **Use Color Dependent Plot Styles** is selected, color-dependent plot styles are used in both new drawings and drawings created in earlier versions of AutoCAD. Color-dependent plot styles utilize numbers from the AutoCAD color index to create a plot style table. This is a file with a .CTB file extension. Each color is identified by a number ranging from 1 to 255 or a name. You can assign color numbers to individual pens on a pen plotter. This can allocate specific property settings in the plotted drawing. If this option is selected, a plot style is created for each color setting.

To change the default plot style behavior for a drawing, select the **Use Named Plot Styles** before opening or creating a drawing. Changing the default settings using the Options dialog box affects only new drawings or drawings created in a release of AutoCAD prior to ACAD2000 and have never been saved in AutoCAD 2000 format. This setting is saved with the drawing. Once a drawing is saved with Use Color Dependent Plot Styles as the default, you can change the default to Use Named Plot Styles with a migration utility. But, once a drawing is saved with Use Named Plot Styles as the default, it cannot be changed to Use Color Dependent Plot Styles.

The **Default Plot Style Table** list box selection allows you to determine which plot style table will be attached to new drawings. A plot style table is a file. It has a file extension of .CTB or .STB and it defines plot styles. If you are using color-dependent plot styles, this option lists the value of None along with all color dependent plot style tables found in the search path. This option lists all named plot styles tables when named plot styles are being used

The **Default Plot Style for Layer 0** list box selection determines the plot style for Layer 0 for new drawings or drawings created in a release of AutoCAD prior to ACAD2000 and have never been saved in AutoCAD 2000 format. The Normal style and any other style defined in the currently loaded plot style table are listed.

The **Default Plot Style for Objects** list box selection determines the default plot style assigned for new objects. It includes Normal, BYBLOCK, and BYLAYER, and any other style defined in the currently loaded plot style table.

The **Add or Edit Plot Style Tables** button causes the **Plot Styles** program window to be displayed for managing (creating and editing) plot style tables. See the Chapter 8 on plotting.

SYSTEM

The System tab of the Options dialog box, shown in Figure 13–46, has sections for managing the *3D* Graphics Display, Pointing Devices, dbConnect Options, and General Options.

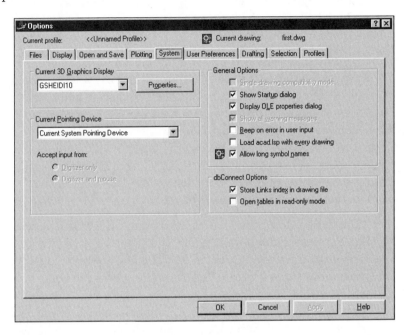

Figure 13–46 *Options dialog box (with the System tab selected)*

Current 3D Graphics Display

The Current 3D Graphics Display section text box allows you to specify the current display from the list. The **Properties** button causes the **3d Graphics System Configuration** dialog box to be displayed.

In the **3d Graphics System Configuration** dialog box the **Adaptive Degradation** section allows you to determine to what degree the view can degrade to maintain the speed of the view manipulation. For example, selecting Wireframe will allow the display to speed up. The **Flat Shaded** check box, when checked, allows the display to change to flat shaded. The **Wireframe** check box, when checked, allows the display to change to wireframe. The **Bounding Box** check box, when checked, allows the display to change to a bounding box which will display in place of objects. The **Maintain Speed FPS** text box lets you specify the display speed in frames per second.

The **Backfaces** section determines whether to display back faces in *3D*. The **Discard Back Faces** check box, when checked, causes back faces to not be drawn.

The **Material** section determines whether to display materials attached to objects in *3D*. The **Enable Materials** check box, when checked, causes materials in *3D* to be displayed.

The **Isolines** section determines how isolines in *3D* are to be displayed. The **Isolines Always on Top** check box, when checked, causes isolines to be displayed for front and back faces in all shade modes except hidden.

The **Dynamic Tessellation** section allows you to determine the smoothness of the objects in a drawing. The **Dynamic Tessellation** check box, when checked, determines that tessellations will be drawn. The **Surface Tessellation** and **Curve Tessellation** slide bars let you set the tessellation from **Low Detail (Less Memory)** to **High Detail (More Memory)** for surface and curves respectively. The **Number of Tessellations to Cache** text box allows you to specify whether the number of tessellations in the cache is 1 or more. When the number is 1, all viewports have the same tessellation. Increasing the number of tessellations can reduce the occurrence of regenerations when more than one viewport is being used.

The **Acceleration** section determines whether software or hardware acceleration is used in *3D*. The **Software** radio button, when selected, uses the software graphics system to perform all of the drawing tasks in *3D*. The **Hardware** radio button, when selected, uses the hardware graphics card to perform most of the drawing tasks in *3D*. This speeds up the drawing time. The **Use Geometry Acceleration (Single Precision)** check box, when checked, causes geometry acceleration to be used. The **Enable Anti-aliasing lines** check box, when checked, causes anti-aliasing lines to be used. Lines will appear smoother.

Current Pointing Device

The Current Pointing Device section of the **System** tab determines the parameter for the pointing device(s) being used. The text box lists available pointing device drivers from which to choose. The **Current System Pointing Device** option causes the system pointing device to be the current pointing device. The **Wintab Compatible Digitizer** ADI 4.2 by Autodesk option causes the Wintab Compatible Digitizer to be the current pointing device. The **Accept Input From:** options determines if the digitizer only is active or both the mouse and digitizer are active. These are determined by the **Digitizer Only** and the **Digitizer And Mouse** radio buttons.

General Options

The General Options section of the **System** tab has check boxes relating to a variety of options for system parameter settings. The **Single-Drawing Compatibility Mode** check box, when checked, causes AutoCAD to open only one drawing at a time (Single-drawing Interface or SDI). Otherwise AutoCAD can open multiple drawing sessions (Multi-drawing Interface or MDI). The **Show Startup Dialog** check box, when checked, causes the **Startup** dialog box to be displayed when you start AutoCAD. The **Display OLE Properties Dialog** check box, when checked, causes the **OLE Properties** dialog box to be displayed when inserting OLE objects into AutoCAD drawings. The **Show All Warning Messages** check box, when checked, causes all dialog boxes that include a Don't Display This Warning Again option to be displayed. Dialog boxes with warning options will be displayed regardless of previous settings specific to each dialog box. The **Beep on Error in User Input** check box, when checked, causes an alarm beep when AutoCAD detects an invalid entry. The **Load acad.lsp with Every Drawing** check box, when checked, causes AutoCAD to load the acad.lsp file into every drawing. The **Allow Long Symbol Names** check box, when checked, allows you to use up to 255 characters for named objects.

dbConnect Options

The dbConnect Options section of the **System** tab lets you determine parameters affecting database connectivity. The **Store Links Index in Drawing File** check box, when checked, causes the database index to be stored within the AutoCAD drawing file. The **Open Tables in Read-Only Mode** check box, when checked, causes database tables to be opened in Read-only mode within the AutoCAD drawing file.

USER PREFERENCES

The User Preferences tab of the Options dialog box, shown in Figure 13–47, controls options that optimize the way you work in AutoCAD.

Figure 13–47 *Options dialog box (with the User Preferences tab selected)*

Windows Standard Behavior

The Windows Standard Behavior section of the **User Preferences** tab lets you apply Windows techniques and methods in AutoCAD. The **Windows Standard Accelerator Keys** check box, when checked, arranges for Windows standards to be applied in interpreting keyboard accelerators (for example, CTRL+C equals COPYCLIP). Otherwise, AutoCAD standards will be applied for keyboard accelerators (for example, CTRL+C equals Cancel; CTRL+V toggles among the viewports). The **Shortcut Menus in Drawing Area** check box, when checked, causes shortcut menus to be displayed when the pointing device is right-clicked. Otherwise, right-clicking is the same as pressing ENTER. The **Right-Click Customization** button causes the Right-Click Customization dialog box to be displayed.

The Right-Click Customization dialog box determines whether right-clicking in the drawing area displays a shortcut menu or is the same as pressing ENTER. This will allow you to have a right-click invoke ENTER while a command is active. You can also disable the following Command shortcut menus Options:

The **Default Mode** section of the Right-Click Customization dialog box determines the effect of right-clicking when no objects are selected. The **Repeat Last Command** radio button causes right-clicking to be the same as pressing ENTER. The **Shortcut Menu** radio button causes right-clicking to display a shortcut menu when applicable.

The **Edit Mode** section of the Right-Click Customization dialog box determines the effect of right-clicking when one or more objects are selected. The **Repeat Last Command** radio button causes right-clicking to be the same as pressing ENTER. The **Shortcut Menu** radio button causes right-clicking to display the **Edit** shortcut menu.

The **Command Mode** section of the Right-Click Customization dialog box determines the effect of right-clicking when a command is in progress. The **Enter** radio button causes right-clicking to be the same as pressing ENTER when a command is in progress. The **Shortcut Menu: Always Enabled** radio button causes right-clicking to display the Command shortcut menu. The **Shortcut Menu: Enabled When Command Options Are Present** radio button causes right-clicking to display the Command shortcut menu to be displayed only when options are currently available from the command line. Otherwise, right-clicking is the same as pressing ENTER.

AutoCAD DesignCenter

The AutoCAD DesignCenter section of the **User Preferences** tab determines parameters relating to the AutoCAD DesignCenter. Under **When Insert Units Are Not Defined, Assume:**, the **Source Content Units** text box lets you select from approximately 21 different type of units to apply to the source contents. These include Unspecified-Unitless, Inches, Feet, Miles, Millimeters, Centimeters, Meters, Kilometers, Microinches, Mils, Yards, Angstroms, Nanometers, Microns, Decimeters, Decameters, Hectometers, Gigameters, Astronomical Units, Light Years, and Parsecs. The **Target Drawing Units** text box lets you select from the same sets of units to apply to the contents when inserted into the target.

Hyperlink

The Hyperlink section of the **User Preferences** tab determines display properties settings of hyperlinks. The **Display Hyperlink Cursor and Shortcut Menu** check box, when checked, causes the hyperlink cursor and shortcut menu to be displayed when the cursor is over an object that contains a hyperlink. The **Display Hyperlink Tooltip** check box, when checked, causes the hyperlink tooltip to be displayed when the cursor is over an object that contains a hyperlink.

Priority for Coordinate Data Entry

The Priority for Coordinate Data Entry section of the **User Preferences** tab determines how input of coordinate data affects AutoCAD's actions. The **Running Object Snap** radio button, when selected, causes running object snaps to be used at all times instead of specific coordinates. The **Keyboard Entry** radio button, when selected, causes the coordinates that you enter to be used at all times and overrides running object snaps. The **Keyboard Entry Except Scripts** radio button, when selected, causes the specific coordinates that you enter to be used rather than running object snaps, except in scripts.

Object Sorting Methods

The Object Sorting Methods section of the **User Preferences** tab controls the order in which objects are sorted. The **Object Selection** check box, when checked, causes AutoCAD to sort selectable objects from those created first to those created last. The latest object will be selected when two objects are chosen at the same time. Otherwise the selection is random. The **Object Snap** check box, when it is checked and a selection is made using Object Snap, causes AutoCAD to sort selectable objects from those created first to those created last. The latest object will be selected when two objects are chosen at the same time The **Redraws** check box, when checked, causes AutoCAD to sort objects from those created first to those created last when redrawing objects during a REDRAW or REDRAWALL. The **Regens** check box, when checked, causes AutoCAD to sort objects from those created first to those created last when redrawing objects during a REGEN or REGENALL. The **Plotting** check box, when checked, causes AutoCAD to sort objects from those created first to those created last when redrawing objects when plotting. The **PostScript Output** check box, when checked, causes AutoCAD to sort objects from those created first to those created last when redrawing objects during PostScript output.

Lineweight Settings

The **Lineweight Settings** button causes the Lineweight Settings dialog box to be displayed. This allows you to set lineweight options.

DRAFTING

The Drafting tab of the Options dialog box, shown in Figure 13–48, lets you customize drafting options in AutoCAD. In this tab are sections for **AutoSnap Settings, AutoSnap Marker Size, AutoTrack Settings, Alignment Point Acquisition,** and **Aperture size.** The options available in these sections are explained in Chapter 3 in the section on Drafting Settings.

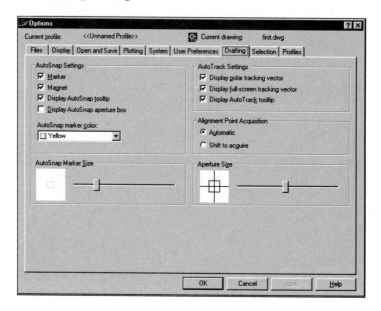

Figure 13–48 *Options dialog box (with the Drafting tab selected)*

SELECTION

The Selection tab of the Options dialog box, shown in Figure 13–49, lets you customize selection options in AutoCAD. In this tab are sections for **Selection Modes, Pickbox Size, Grips,** and **Grip Size.** The options available for the Selection Modes are explained in Chapter 5 in the section on Object Selection Modes. The options available for Grips are explained in Chapter 6 in the section on Editing With Grips.

PROFILES

The Profiles tab of the Options dialog box, shown in Figure 13–50, lets you manage profiles. A profile is a named and saved group of environment settings. This profile can be restored as a group when desired. AutoCAD stores your current options in a profile named Unnamed Profile. AutoCAD displays the current profile name, as well as the current drawing name, in the Options dialog box. The profile data is saved in the system registry and can be written to a text file (an ARG file). AutoCAD organizes essential data and maintains changes in the registry as necessary.

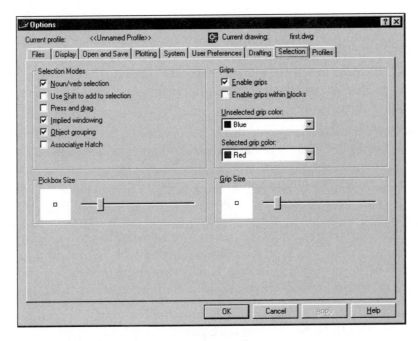

Figure 13–49 *Options dialog box (with the Selection tab selected)*

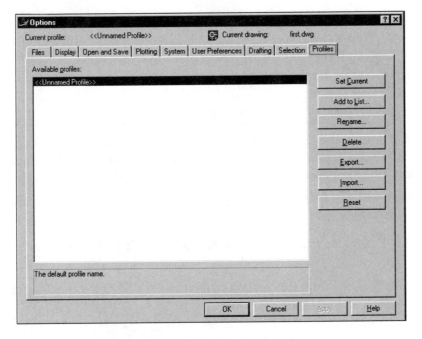

Figure 13–50 *Options dialog box (with the Profiles tab selected)*

A profile can be exported or imported to and from different computers. If changes have been made to your current profile during an AutoCAD session and you want to save them in the ARG file, the profile must be exported. After the profile with the current profile name has been exported, AutoCAD updates the ARG file with the new settings. Then the profile can be imported again into AutoCAD, thus updating your profile settings.

The **Set Current** makes the profile that is highlighted in the **Available Profiles** list box the current profile. Choosing the **Add To List** button lets you name and save the current environment settings as a profile. The **Rename** button lets you rename the highlighted profile. The **Delete** button lets you delete the highlighted profile. The **Export** button causes the Export Profiles dialog box to be displayed. This is a file manager dialog box in which the highlighted profile can be saved to the path you specify. The **Import** button causes the Import Profiles dialog box to be displayed. This is a file manager dialog box in which you can select a profile from a saved path to be imported. The **Reset** button causes the highlighted profile to be reset. The default profile name is listed in the description pane at the bottom of the dialog box.

After making all the necessary changes, click the **Apply** button to apply the changes. Then choose the **OK** button to close the Options dialog box.

SAVING OBJECTS IN OTHER FILE FORMATS (EXPORTING)

The EXPORT command allows you to save a selected object in other file formats, such as .BMP, .DXF, .DWF, .SAT, .3DS, and .WMF. Invoke the EXPORT command:

File menu	Choose Export
Command: prompt	**export** (ENTER)

AutoCAD displays the **Export Data** dialog box, similar to Figure 13–51. In the **Save As Type:** list box, select the format type in which you wish to export objects. Enter the file name in the **File Name:** edit box. Select the **Save** button, and AutoCAD prompts:

> Select objects: *(select the objects to export, and press ENTER to complete object selection)*

AutoCAD exports the selected objects in the specified file format using the specified file name.

Table 13–1 lists the format types available in AutoCAD for exporting the current drawing.

Figure 13–51 *Export Data dialog box*

Table 13–1 Exportable Format Types

Format Type	Description
3DS	3D Studio file
BMP	Device-independent bitmap file
DWG	AutoCAD 2000 drawing file (same as invoking the WBLOCK command)
DXX	AutoCAD attribute extract DXF file (same as invoking the ATTEXT command)
EPS	Encapsulated PostScript file
SAT	ACIS solid-object file
STL	Solid object stereo-lithography file
WMF	Windows metafile

IMPORTING VARIOUS FILE FORMATS

The IMPORT command allows you to import various file formats, such as .3DS, .DXF, .EPS, .SAT, and WMF, into AutoCAD. Invoke the IMPORT command:

Insert toolbar	Select IMPORT command (see Figure 13–52)
Command: prompt	**import** (ENTER)

Figure 13–52 *Invoking the* IMPORT *command from the Insert toolbar*

AutoCAD displays the Import File dialog box, similar to Figure 13–53. In the **Files of Type:** list box, select the format type you wish to import into AutoCAD. Select the file from the appropriate directory from the list box and pick the **Open** button. AutoCAD imports the file into the AutoCAD drawing.

Figure 13–53 *Import File dialog box*

Table 13–2 lists the format types available to import into AutoCAD.

Table 13–2 Format types available to import into AutoCAD

Format Type	Description
3DS format	Imports a 3D Studio file
DXF format	Imports drawing interchange file
EPS format	Imports Encapsulated PostScript file
SAT format	Imports ACIS solid object file
WMF format	Imports Windows Metafile

SLIDES AND SCRIPTS

Slides are quickly-viewable, non-editable views of a drawing or parts of a drawing. There are two primary uses for slides. One is to have a quick and ready picture to display symbols, objects, or written data for informational purposes only. The other very useful application of slides is to be able to display a series of pictures, organized in a prearranged sequence for a timed slide show. This is a very useful tool for demonstrations to clients or in a showroom. This feature supplements the time-consuming calling up of views required when using the ZOOM, PAN, or other display commands. The "slide show" is implemented through the SCRIPT command (described later in this chapter).

It should be noted that a slide merely masks the current display. Any cursor movement or editor functions employed while a slide is being displayed affects the current drawing under the slide and not the slide itself.

MAKING A SLIDE

The current display can be made into a slide with the MSLIDE command. The current viewport becomes the slide while you are working in model space. The entire display, including all viewports, becomes the slide when using MSLIDE while you are working in paper space. The MSLIDE command takes a picture of the current display and stores it in a file, so be sure it is the correct view.

Invoke the MSLIDE command:

Command: prompt	**mslide** (ENTER)

The **Create Slide File** dialog box appears, as shown in Figure 13–54. The default is the drawing name, which can be used as the slide file name by pressing ENTER. Or you can type any other name, as long as you are within the limitations of the operating system file-naming conventions. AutoCAD automatically appends the extension .SLD. Only objects that are visible in the screen drawing area (or in the current viewport when in model space) are made into the slide.

If you plan to show the slide on different systems, you should use a full-screen view with a high-resolution display for creating the slide.

Figure 13–54 *Create Slide File dialog box*

SLIDE LIBRARIES

The SLIDELIB program provided with AutoCAD is used to create a slide library file in conjunction with a list of slide file names. Each slide file name must correspond to an actual slide whose extension is .SLD. The list of file names is in a separate file, written in ASCII format, with each slide file name on a separate line by itself. The file names in the list may or may not include the .SLD extension, but they must include the path (drive and/or folder) for the SLIDELIB command to access the proper file for inclusion in the library.

A file named SLDS001 could read as follows:

> pic_1
>
> A:pic_2.sld
>
> c:\dwgs\pic_3.sld
>
> b:\other\pic_6

Each line represents a slide name, some with extensions, some without. Each one has a different path. When the SLIDELIB command is used to address this file, all the slides listed will be included in the library you specify. The paths are not included when used with the SLIDELIB program. If there were a PIC_2.SLD on another path, it would not be included or it would conflict with the one specified for its particular library file. This allows for control of which slides will be used if the same slide file name exists in several drive/directory locations. The SLIDELIB program appends .SLB to the library file it creates from the slide list.

The preceding listing file can be used to create a library file named ALLPICS.SLB using the form:

> slidelib *library* [< *slidelist*]

From the DOS window enter the following:

> **slidelib allpics** [< slds001]

 Note: In the DOS window, the proper path for each of the three elements of the function must either be entered or you must be on the same folder/directory as the element(s).

VIEWING A SLIDE

The VSLIDE command displays a slide in the current viewport.

Invoke the VSLIDE command:

| Command: prompt | **vslide** (ENTER) |

The **Select Slide File** dialog box appears. This is similar to most file-managing dialog boxes. Select the slide to display in the current viewport. If you have stored the slide in a library file and are calling it up for view, you can use the library file keyname followed by the slide name in parentheses. For example, if a slide named PIC_1 (actually stored under the filename pic_1.sld) is in a library file called ALLPICS.SLB, you can use the following format:

Command: **vslide**
Slide file <default>: **allpics(pic_1)**

 Note: The library file lists only the address of each slide. It can be on a path or directory other than that of the slide(s) listed in it. It is necessary to specify the path to the library (.SLB file), but it is not necessary to specify the path in front of the slide name. For example, the ALLPICS.SLB file might be on the working directory, while the slide PIC_1 is on drive F in the directory SL1, and the slide PIC_2 is on drive F in the directory SL2. The sequences to call them into view would be as follows:

Command: **vslide**
Slide file <default>: **allpics(pic_1)**

Command: **vslide**
Slide file <default>: **allpics(pic_2)**

If the ALLPICS.SLB file is on drive F in the subdirectory SLB, the sequences are as follows:

Command: **vslide**
Slide file <default>: **f:\slb\allpics(pic_1)**

Command: **vslide**
Slide file <default>: **f:\slb\allpics(pic_2)**

SCRIPTS

Of the many means available to enhance AutoCAD through customization, scripts are among the easiest to create. Scripts are similar to the macros that can be created to enhance word processing programs. They permit you to combine a sequence of commands and data into one or two entries. Creating a script, like most enhancements to AutoCAD, requires that you use a text editor to write the script file (with the extension .SCR), which contains the instructions and data for the SCRIPT command to follow.

Because script files are written for use at a later time, you must anticipate the conditions under which they will be used. Therefore, familiarity with sequences of prompts that will occur and the types of responses required is necessary to have the script function properly. Writing a script is a simple form of programming.

The Script Text

A script text file must be written in ASCII format. That is, it must have no embedded print codes or control characters that are automatically written in files when created with a word processor in the document mode. If you are not using one of the available line editors, be sure you are in the nondocument, programmer, or ASCII mode of your word processor when creating or saving the file. Save the script with the file extension .SCR.

Each command can occupy a separate line, or you can combine several command/data responses on one line. Each space between commands and data is read as an ENTER, just as pressing the SPACEBAR is read while in AutoCAD. The end of a line of text is also the same as an ENTER.

Spaces and Ends of Lines in Script Files

The following script file contains several commands and data. The commands are GRID, LINE, CIRCLE, LINE, and CIRCLE again. The data are the response ON, coordinates (such as 0,0 and 5,5), and distances, such as radius 3.

```
GRID ON LINE 0,0 5,5  CIRCLE 3,3 3
LINE 0,5 5,0
CIRCLE 5,2.5 I
```

The first line includes the GRID, LINE, and CIRCLE commands, and their responses. Note the two spaces after 5,5; these are required to simulate the double ENTER. Not obvious is the extra space following the 5,0 response in the second LINE command. This extra space and the invisible CR-LF (carriage-return, linefeed) code that ends every line in a text file combine to simulate pressing the SPACEBAR twice. This is necessary, again, to exit the LINE command.

Some text editors automatically remove blank spaces at the end of text lines. To guard against that, an alternative is to have a blank line indicate the second ENTER, as follows:

```
GRID ON LINE 0,0 5,5 CIRCLE 3,3 3
LINE 0,5 5,0

CIRCLE 5,2.5 I
```

Invoke the SCRIPT command:

Tools menu	Choose Run Script
Command: prompt	**script** (ENTER)

AutoCAD displays the Select Script File dialog box. Select the appropriate script file from the list box and choose the **Open** button. AutoCAD executes the command sequence from the script file.

Changing Block Definitions with a Script File

Using a script to perform a repetitive task is illustrated in the following example. This application also offers some insight on changing the objects in an inserted block with attributes without affecting the attribute values.

Figure 13–55 shows a group of drawings all of which utilize a common block with attribute values in one insertion that are different from the attribute values of those in other drawings. In this case, the border/title block is a block named BRDR. It was originally drawn with the short lines around and outside of the main border line. It was discovered that these lines interfered with the rollers on the plotter and needed to be removed. The BRDR block definition is shown in Figure 13–56. Remember, the inserted block has different attribute values in each drawing, such as drawing number, date, and title.

Figure 13–55 *Drawings utilizing common block and attribute values that are different from those of other drawings*

You want to change objects in the block but maintain the attribute values as they are. There are two approaches to redefinition. One is to find a clear place in the drawing and insert the block with an asterisk (*). This is the same as inserting and exploding

Figure 13–56 *A block definition*

the block. Then you make the necessary changes in the objects and make the revised group into a block with the same block name. This redefines all insertions of blocks with that same name in the drawing. In this case there is only one insertion. You must be attentive to how any changes to attributes might affect the already inserted block of that name.

The second method is to use the WBLOCK command to place a copy of the block in a file with the same name. This makes a new and separate drawing of the block. Then you exit the current drawing, call up the newly created drawing, make the required changes in the objects, and end the drawing that was created by the WBLOCK command. Then you re-enter the drawing in which the block objects need to be changed. You now use the INSERT command and respond with "blockname=" and have the block redefined without losing the attribute values. For example, if the block name is BRDR, the sequence would be as follows:

```
Command: insert
Enter block name or [?]: brdr=
Block "b" already exists. Redefine it? [Yes/No] <N>: y
```

Block BRDR redefined
Regenerating drawing
Specify insertion point or [Scale/X/Y/Z/Rotate/PScale/PX/PY/PZ/PRotate]:
   (press ENTER *to cancel the command sequence*)

The key to this sequence is the equals sign (=) following the block name. This causes AutoCAD to change the definition of the block named BRDR to be that of the drawing named BRDR, although it maintains the attribute values as long as attribute definitions remain unchanged.

If the preceding procedure must be repeated many times, this is where a script file can be employed to automate the process. In the following example, we show how to apply the script to a drawing named PLAN_1. The sequence included an ENTER as it was described to be used while in AutoCAD. This expedited the operation by not actually having the block inserted, but only having its definition brought into the drawing. Because an ENTER during the running of a script file causes the script to terminate, those keystrokes cannot be in the middle of a script; besides, invoking in a SCRIPT command requires using the AutoLISP function "(command)." The script can be written in a file (called BRDRCHNG.SCR for this example) in ASCII format as follows:

```
INSERT
BRDR=
0,0        (the 0,0 is followed by six spaces)
ERASE L (the L is followed by one space)
REDRAW
```

Now, from the "Command:" prompt, you can apply the script to drawing PLAN_1 by responding as follows:

Command: script

There are several important aspects of this sequence:

Lines 1–2 This is where you might enter if you were not in the SCRIPT command and have the definition of the block named BRDR take on that of the drawing BRDR without actually having to continue with the insertion in the drawing.

Line 3 In this case, 0,0 as the insertion point is arbitrary because the inserted block is going to be erased anyway. Special attention is given to the six spaces following the insertion point. These are the same as pressing SPACEBAR or ENTER six times in response to the "X-scale," "Y-scale," and "Rotation angle" prompts, and the number of spaces that follow must correspond to the number of attributes that require responses for values. In this example there were three attributes. Again, the fact that the responses are null is immaterial, because this insertion will not be kept.

Line 4 The ERASE L is self-explanatory, but do not forget that after the "L" you must have another space (press SPACEBAR again) to terminate the object selection process and complete the ERASE command.

Line 5 The REDRAW command is not really required except to show the user for a second time that the changes have been made before the script ends.

UTILITY COMMANDS FOR A SCRIPT
Following are the utility commands that may be used within a script file.

Specifying the Delay
The DELAY command causes the script to pause for the number of milliseconds that have been specified by the command. To set the delay in the script for five seconds, you would write:

    DELAY 5000

Resuming the Script
The RESUME command causes the script to resume running after the user has pressed either or to interrupt the script. It may be entered as follows:

    Command: resume

Graphscr and Textscr
The GRAPHSCR and TEXTSCR commands are used to flip or toggle the screen to the graphics or text mode, respectively, during the running of the script. Each is simply entered as a command in the script, as follows:

    Command: **graphscr**
or
    Command: **textscr**

These two screen toggle commands can be used transparently by preceding them with an apostrophe.

Repeating the Script
The RSCRIPT command, when placed at the end of a script, causes the script to repeat itself. With this feature you can have a slide show run continuously until terminated by ESC key.

A repeating demonstration can be set up to show some sequences of commands and responses as follows:

```
GRID ON
LIMITS 0,0 24,24
ZOOM A
CIRCLE 12,12 4
DELAY 2000
COPY L M 12,12 18,12 12,18 6,12 12,6 (an extra space at the end)
DELAY 5000
ERASE W 0,0 24,24 (an extra space at the end)
DELAY 2000
LIMITS 0,0 12,9
ZOOM A

GRID OFF
TEXT 1,1 .5 0 THAT'S ALL FOLKS!
ERASE L (an extra space at the end)
RSCRIPT
```

This script file utilizes the DELAY and RSCRIPT subcommands. Note the extra spaces where continuation of some actions must be terminated.

The SCRIPT command can be used to show a series of slides, as in the following sequence:

```
VSLIDE SLD_A
VSLIDE *SLD_B
DELAY 5000
VSLIDE
VSLIDE *SLD_C
DELAY 5000
VSLIDE
DELAY 10000
RSCRIPT
```

This script uses the asterisk (*) before the slide name prior to the delay. This causes AutoCAD to load the slide, ready for viewing. Otherwise, there would be a blank screen between slides while the next one is being loaded. The RSCRIPT command repeats the slide show.

REVIEW QUESTIONS

1. All of the following can be renamed using the RENAME command, except:

    a. a current drawing name

    b. named views within the current drawing

    c. block names within the current drawing

    d. text style names within the current drawing

2. The PURGE command can be used:

    a. after an editing session

    b. at the beginning of the editing session

    c. at any time during the editing session

    d. a and b only

3. To calculate the product of 4 times 21.5 times pi and store it for use with later AutoCAD commands, you should respond to the ">>Expression" prompt of the CAL command with:

    a. A=* 4 21.5 pi   d. A= (* 4 21.5 pi)

    b. A= 4 * 21.5 * 3.14   e. none of the above

    c. A= 4 * 21.5 * pi

4. To use the product from the previous question to respond to a scale factor in the SCALE command you would type:

    a. %A   d. A

    b. ^A   e. (A)

    c. !A

5. The most common setting for the current drawing color is:

    a. Red   d. White

    b. Bylayer   e. none of the above

    c. Byblock

6. Which of the following are not valid color names in AutoCAD?

    a. Brown   d. Magenta

    b. Red   e. none of the above (i.e. all are valid)

    c. Yellow

7. To change the background color for the graphics drawing area, you should use:

   a. COLOR

   b. BGCOLOR

   c. PREFERENCES

   d. SETTINGS

   e. CONFIG

8. AutoCAD release 14 can be used to edit a drawing saved as an ACAD2000 drawing.

   a. True

   b. False

9. ACAD2000 can be used to edit a drawing that is saved in an AutoCAD release 14 drawing.

   a. True

   b. False

10. All the following items can be purged from a drawing file, except:

    a. Text styles

    b. Blocks

    c. System variables

    d. Views

    e. Linetypes

11. The following can be deleted with the PURGE command, except:

    a. blocks not referenced in the current drawing

    b. linetypes that are not being used in the current drawing

    c. layer 0, if it is not being use

    d. none of the above (i.e. all can be purged)

12. With OLE (object linking and embedding), you can copy or move information from one application to another while retaining the ability to edit the information in the original application.

    a. True

    b. False

13. If the TIME command is not turned off during lunch break, the TIME command will:

    a. include the lunch break time

    b. exclude the lunch break time

    c. turn itself off after 10 minutes of inactivity

    d. automatically subtract one hour for lunch

    e. none of the above

14. The VIEW command:

   a. serves a purpose similar to the PAN command

   b. will restore previously saved views of your drawing

   c. is normally used on very small drawings

   d. none of the above

15. When using a .X point filter, AutoCAD will request that you complete the point selection process by entering:

   a. An X coordinate

   b. A Y and Z coordinate

   c. An OSNAP mode

   d. Nothing, .X is not a valid filter

16. The geometric calculator function MEE will:

   a. calculate the midpoint of a selected line

   b. calculate the midpoint between the endpoints of any two objects

   c. Return the node name for the computer you are working on

   d. average a string of numbers

   e. none of the above

17. To load a script file called "SAMPLE.SCR", use:

   a. SCRIPT, then SAMPLE   c. LOAD, SCRIPT, then SAMPLE

   b. LOAD, then SAMPLE   d. none of the above

18. The AutoCAD command used for viewing a slide is:

   a. VSLIDE   d. SLIDE

   b. VIEWSLIDE   e. MSLIDE

   c. SSLD

19. If a REDRAW is performed while viewing a slide:

   a. the command will be ignored

   b. the current slide will be deleted

   c. the current drawing will be displayed

   d. AutoCAD will load the drawing the slide was created from

   e. none of the above

20. A script file is identified by the following extension:

   a. SCR   d. SPT

   b. BAK   e. none of the above

   c. DWK

21. A slide file is identified by the following extension:

   a. SLD   d. SLU

   b. SCR   e. none of the above

   c. SLE

22. Slides can be removed from the display with the command:

   a. ZOOM ALL   d. both a and b

   b. REGEN   e. none of the above

   c. OOPS

23. The SCRIPT command cannot be used to:

   a. insert blocks

   b. create layers

   c. place text

   d. create another script file

   e. none of the above (i.e. all are possible)

24. To create a slide library, you must:

   a. use the AutoCAD command SLIDELIB

   b. add slides to the library one at a time

   c. create an ASCII text file listing all slides to be included in the library

   d. have all slide files in a common directory

25. To cause a script file to execute in an infinite loop, you should place what command at the end of the file?

   a. REPEAT   d. BEGIN

   b. RSCRIPT   e. none of the above

   c. GOTO:START

26. If AutoCAD is executing in an infinite script file loop, how can you terminate the loop?

    a. press BACKSPACE   d. press F1

    b. press CTRL + C   e. none of the above

    c. press ALT + C

27. What command permits a user to work with just a portion of a drawing by loading geometry from specific views or layer?

    a. pload   c. partialopen

    b. partopn   d. pltopn

28. What command permits additional geometry to be loaded into the current partially loaded drawing?

    a. PLOAD   c. PARTADD

    b. PARTIAL LOAD   d. ADDGEO

29. Can the PURGE command be used in partially open drawings?.

    a. Yes

    b. No

30. What term can be added to draw, modify, or inquiry commands which will cause them to automatically repeat?

    a. redo   d. multiple

    b. repeat   e. no such option

    c. return

31. When one of more objects overlay each other which command is used to control their order of display?

    a. DWGORDER   c. VIEWORDER

    b. DRAWORDER   d. ARRANGE

32. Which command permits you to execute operating system programs without exiting AutoCAD 2000?

    a. RUN   c. OPSYS

    b. EXOPRG   d. SHELL

CHAPTER 14

Internet Utilities

INTRODUCTION

The Internet is the most important way to convey digital information around the world. You are probably already familiar with the best-known uses of the Internet: e-mail (electronic mail) and the Web (short for "World Wide Web"). E-mail lets users exchange messages and data at very low cost. The Web brings together text, graphics, audio, and video in an easy-to-use format. Other uses of the Internet include FTP (file transfer protocol, for effortless binary-file transfer), Gopher (presents data in a structured, subdirectory-like format), and Usenet, a collection of more than 10,000 news groups.

AutoCAD allows you to interact with the Internet in several ways. You can launch a Web browser from within AutoCAD. AutoCAD can create DWF (short for "Drawing Web Format") files for viewing drawings in two-dimensional format on Web pages. AutoCAD can open and insert drawings from, and save drawings to the Internet.

After completing this chapter, you will be able to do the following:

- Launch the default Web browser
- Open drawings from the Internet
- Save drawings to the Internet
- Create and Use hyperlinks
- Creating and viewing DWF files

LAUNCHING THE DEFAULT WEB BROWSER

The BROWSER command lets you start a Web browser from within AutoCAD. By default, the BROWSER command uses whatever brand of Web browser program is registered in your computer's Windows operating system. AutoCAD prompts for the URL (short for "uniform resource locator"). The URL is the Web site address, such as http://www.autodesk.com. (More about URLs in the next section.) The BROWSER command can be used in scripts, toolbar or menu macros, and AutoLISP routines to access the Internet automatically.

Invoke the BROWSER command:

Web toolbar	Choose the BROWSE THE WEB command (see Figure 14–1)
Command: prompt	**browser** (ENTER)

Figure 14–1 *Invoking the* BROWSE THE WEB *command from the Web toolbar*

AutoCAD prompts:

> Command: **browser**
> Enter Web location (URL) <default location>: *(specify a new location, or press* ENTER *to accept the default location)*

You can specify a default URL in the **Options** dialog box. Select the Files tab, choose Menu, Help, and Miscellaneous File Name, select Default Browser Location (see Figure 14–2) and specify the default URL Web location. When you invoke the BROWSER command and press ENTER to accept the default location, AutoCAD launches the Web browser, which contacts the Web site. Figure 14–3 shows Netscape Communicator and the Autodesk Web site.

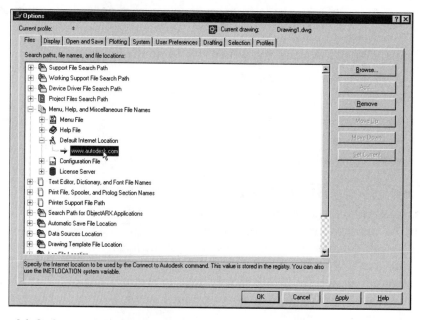

Figure 14–2 *Options dialog box with Files tab selected*

Figure 14–3 *Netscape Communicator displaying the Autodesk Web site*

If your computer does not have Internet access, the default URL location <drive:path\ACAD2000\Home.htm> can be accessed through the BROWSER command. This file contains information about AutoCAD's Internet Features and hyperlinks to Autodesk Web sites.

UNIFORM RESOURCE LOCATOR

As mentioned earlier, the file-naming system of the Internet is known as URL, short for "uniform resource locator." The URL system allows you to find any resource (a file) on the Internet. Example resources include a text file, a Web page, a program file, an audio or movie clip—in short, anything you might also find on your own computer. The primary difference is that these resources are located on somebody else's computer. A typical URL looks like the examples shown in Table 14–1.

The "http://" prefix is not required. Most of today's Web browsers automatically add in this routing prefix, which saves you a few keystrokes. URLs can access several different kinds of resources—such as Web sites, e-mail, or news groups—but always take on the following format:

    scheme://netloc

The "scheme" accesses the specific resource on the Internet, including those listed in Table 14–2.

The "://" characters indicate a network address. Table 14–3 lists the formats recommended for specifying URL-style file names with the BROWSE command. *Servername* is the name of the server, such as "www.autodesk.com." The *pathname* is the same name as a subdirectory or folder name. The *drive:* is the driver letter, such as C: or D:. A local file is a file located on your computer. The *localhost* is the name of the network host computer.

Table 14–1 Example URLs

Example	Meaning
http://www.autodesk.com	Autodesk primary Web site
http://data.autodesk.com	Autodesk data publishing Web site
news://adesknews.autodesk.com	Autodesk news server
ftp://ftp.autodesk.com	Autodesk FTP server

Table 14–2 URL Prefix Meanings

Scheme Prefix	Meaning
file://	Files on your computer's hard drive or local network
ftp://	File transfer protocol (downloading files)
http://	Hypertext transfer protocol (Web sites)
mailto://	Electronic mail (e-mail)
news://	Usenet news (news groups)
telnet://	Telnet protocol
gopher://	Gopher protocol

Table 14–3 Recommended Formats for Specifying URL-Style File Names

Drawing Location	Template URL
Web site	http://servername/pathname/filename
FTP site	ftp://servername/pathname/filename
Local file	file:///drive:/pathname/filename file:///drive:/pathname/filename file://\\localPC\pathname\filename file:////localPC/pathname/filename
Network file	file://localhost/drive:/pathname/filename file://localhost/drive:/pathname/filename

OPENING AND SAVING DRAWINGS FROM THE INTERNET

AutoCAD 2000 allows you to open and save drawing files from the Internet or an intranet. You can also attach externally referenced drawings stored on the Internet/ intranet to drawings stored locally on your system. Whenever you open a drawing file from an Internet or intranet location, it is first downloaded into your computer and opened in the AutoCAD drawing area. Then you can edit the drawing and save it, either locally or back to the Internet or intranet location for which you have appropriate access privileges.

To open an AutoCAD drawing from an Internet/intranet location, invoke the OPEN command:

Standard toolbar	Choose the OPEN command (see Figure 14–4)
File menu	Choose Open
Command: prompt	**open** (ENTER)

Figure 14–4 *Invoking the* OPEN *command from the Standard toolbar*

AutoCAD displays the Select File dialog box, as shown in Figure 14–5.

Select File								? X

Look in: Sample

Activex
Database Connectivity
DesignCenter
Vba
VisualLISP
1st floor architectural.dwg

1st floor electrical.dwg
1st floor lighting.dwg
1st floor plan.dwg
1st floor.dwg
Campus.dwg
City base map.dwg

Preview

File name: _____ Open Partial Open... Find File...

Files of type: Drawing (*.dwg) Cancel Locate

☐ Open as read-only   ☐ Select Initial View

Figure 14–5 *Select File dialog box*

In the **File name:** text field, specify the URL to the file you wish to open, and choose the **Open** button to open the drawing from the specified Internet/intranet location. Be sure to specify the transfer protocol, such as ftp:// or http://, and the file extension, such as DWG or DWT. You can also choose the **Search the Web** button to open the Browse the Web dialog box. From there you can navigate to the Internet location where the file is stored.

 Note: AutoCAD Internet functionality is only available when you have installed AutoCAD 2000 with full installation or have Microsoft Internet Explorer 4.0, Netscape Navigator 4.0 or a later version installed on your system.

To save an AutoCAD drawing to an Internet/intranet location, invoke the SAVEAS command:

File menu	Choose Save As
Command: prompt	**saveas** (ENTER)

AutoCAD displays the Save Drawing As dialog box, as shown in Figure 14–6.

Figure 14–6 *Save Drawing As dialog box*

In the **File name:** text field, specify the URL to the file you wish to save and choose the **Save** button to save the drawing to the specified Internet/intranet location. Be sure to specify the transfer protocol and file extension (such as DWG or DWT). You can also choose the **Search the Web** button to open the Browse the Web dialog box. From there you can navigate to the Internet location where the file is to be saved.

Note: You can save the drawing to a specific Internet location by using the FTP protocol only.

To attach an xref to a drawing stored on the Internet/Intranet location, invoke the XREF command:

Reference toolbar	Choose the EXTERNAL REFERENCE ATTACH command (see Figure 14–7)
Insert menu	Choose External Reference...
Command: prompt	**xref** (ENTER)

Figure 14–7 *Invoking the* EXTERNAL REFERENCE ATTACH *command from the Reference toolbar*

AutoCAD displays the Select Reference File dialog box, as shown in Figure 14–8.

Figure 14–8 *Select Reference File dialog box*

In the **File name:** text field, specify the URL to the file you wish to open and choose the **Open** button to attach the drawing as a reference file from the specified Internet/intranet location. Be sure to specify the transfer protocol, such as ftp:// or http://, in the URL. You can also choose the **Search the Web** button to open the Browse the Web dialog box. From there you can navigate to the Internet location where the file is stored.

WORKING WITH HYPERLINKS

AutoCAD 2000 allows you to create hyperlinks that provide jumps to associated files. Hyperlinks provide a simple and powerful way to quickly associate a variety of documents with an AutoCAD drawing. For example, you can create a hyperlink that opens another drawing file from the local drive, network drive, or from an Internet web site. You can also specify a named location to jump to within a file, such as a view name in an AutoCAD drawing, or a bookmark in a word processing program. You can attach hyperlinks to any graphical object in an AutoCAD drawing.

AutoCAD allows you to create both *Absolute* and *Relative* hyperlinks in your AutoCAD drawings. Absolute hyperlinks store the full path to a file location, whereas Relative hyperlinks store a partial path to a file location, relative to a default URL or name of the directory you specify using the HYPERLINKSBASE system variable.

You can also specify the relative path for a drawing in the Drawing properties (Summary tab) dialog box (as shown in Figure 14–9), opened from File menu.

Figure 14–9 *Drawing properties dialog box (Summary tab)*

Whenever you attach a hyperlink to an object, AutoCAD provides a cursor feedback when you position the cursor over the object. To activate the hyperlink, first select the object (make sure the PICKFIRST system variable is set to 1). Right-click to display the shortcut menu and activate the link from the Hyperlink sub-menu, as shown in Figure 14–10.

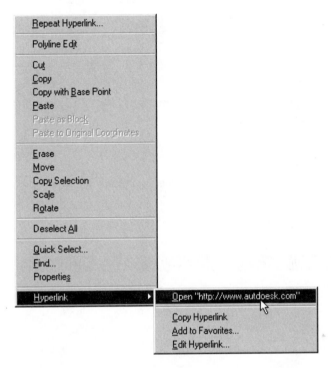

Figure 14–10 *Hyperlink sub-menu from the shortcut menu*

When you create a hyperlink that points to an AutoCAD drawing template (DWT) file, AutoCAD creates a new drawing file based on the selected template rather than opening the actual template when you activate the hyperlink. With this method there is no risk of accidentally overwriting the original template.

When you create a hyperlink that points to an AutoCAD named view and activate the hyperlink, the named view that was created in the model space is restored in the Model tab, and the named view that was created in paper space is restored in the Layout tab.

CREATING A HYPERLINK

To create an hyperlink, invoke the HYPERLINK command:

Standard toolbar	Choose the INSERT HYPERLINK command (see Figure 14–11)
Insert menu	Choose Hyperlink
Command: prompt	**hyperlink** (ENTER)

Figure 14–11 *Invoking the* INSERT HYPERLINK *command from the Standard toolbar*

AutoCAD prompts:

> Command: **hyperlink**
> Select objects: *(select one or more objects to attach the hyperlink and press* ENTER*)*

AutoCAD displays the Insert Hyperlink dialog box, as shown in Figure 14–12.

Figure 14–12 *Insert Hyperlink dialog box*

Specify the URL or path with the name of the file that you wish to have associated with the selected objects in the **Link to file or URL:** text field, or choose the **Browse** button. AutoCAD displays the Browse the Web – Select Hyperlink dialog box, as shown in Figure 14–13.

Figure 14–13 *Browse the Web – Select Hyperlink dialog box*

Navigate to the location of the file that you wish to have associated with the hyperlink and choose the **Open** button.

The **Path:** text field displays the path to the file associated with the hyperlink. If **Use Relative Path for Hyperlink** is selected, only the file name is listed. If **Use Relative Path for Hyperlink** is cleared, the full path and the file name are listed.

The **Base:** text field displays the default base path for all relative path hyperlinks in the current drawing. As mentioned earlier, you can specify the default path using the HYPERLINKBASE system variable. If you haven't set HYPERLINKBASE for the current drawing, the directory of the current drawing is listed and becomes the default base path for relative hyperlinks.

If desired, specify a named location to jump to in the **Named location in file:** text field.

 Note: To specify a named location for another application, refer to that application's documentation for named location conventions.

If necessary, specify a description for the newly created hyperlink in the **Hyperlink description:** text field.

The **Use Relative Path for Hyperlink** check box toggles the use of a relative path for the current drawing. If this option is selected, the full path to the linked file is not stored with the hyperlink. AutoCAD sets the relative path to the value specified by the HYPERLINKBASE system variable or, if this variable is not set, the current drawing path. If this option is not selected, the full path to the associated file is stored with the hyperlink.

 Note: AutoCAD Release 14 does not support Relative path for a hyperlink. If you save the drawing in Release 14 format, you will not be able to execute any relative hyperlinks. But, subsequently, if you save the drawing in AutoCAD 2000 format, the Relative hyperlinks are restored.

Choose the OK button to create the hyperlink to the selected objects.

EDITING A HYPERLINK

To edit an hyperlink, invoke the HYPERLINK command:

Standard toolbar	Choose the HYPERLINK command (see Figure 14–14)
Insert menu	Choose Hyperlink...
Command: prompt	**hyperlink** (ENTER)

Figure 14–14 *Invoking the* HYPERLINK *command from the Standard toolbar*

AutoCAD prompts:

Command: **hyperlink**
Select objects: *(select one or more objects to edit the hyperlink and press* ENTER*)*

AutoCAD displays the Edit Hyperlink dialog box, similar to Figure 14–15. Make necessary changes in the **Link to file or URL:** and/or **Named location in file:** text fields. Choose the OK button to accept the changes.

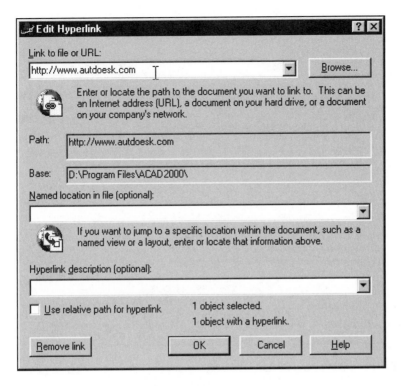

Figure 14–15 *Edit Hyperlink dialog box*

REMOVING THE HYPERLINK

To remove the hyperlink, invoke the HYPERLINK command:

Standard toolbar	Choose the HYPERLINK command
Insert menu	Choose Hyperlink…
Command: prompt	**hyperlink** (ENTER)

AutoCAD prompts:

> Command: **hyperlink**
> Select objects: *(select one or more objects from which to remove the hyperlink and press* ENTER*)*

AutoCAD displays the Edit Hyperlink dialog box, as shown in Figure 14–16.

Figure 14–16 *Choosing the Remove Link button from the Edit Hyperlink dialog box*

Click the **Remove link** button. AutoCAD removes the hyperlink from the selected objects and closes the Edit Hyperlink dialog box.

HYPERLINKS IN BLOCKS

AutoCAD 2000 allows you to attach hyperlinks to blocks, including nested objects contained within blocks. If the blocks contain any relative hyperlinks, the relative hyperlinks adopt the relative base path of the current drawing when you insert them.

Whenever you attach a hyperlink to a block reference, AutoCAD provides a cursor feedback when you position the cursor over the inserted block. To activate the hyperlink, first select the block reference (make sure the PICKFIRST system variable is set to 1). Right-click to display the shortcut menu and activate the hyperlink associated with the currently selected block element from the Hyperlink sub-menu.

Whenever you include objects that have hyperlinks in a block definition, you can activate the hyperlink from any of the block references. If you attach a hyperlink to a block reference, then you will have the choice of activating the block hyperlink or selected object hyperlink. To remove or edit hyperlinks of the objects within the block, you must explode the block reference, then proceed with the removing or editing of hyperlinks. You can remove or edit the hyperlink that was attached to a block reference without exploding the block.

Note: When a hyperlink is attached to a block reference that already contains an object with a hyperlink, the cursor feedback for that block will only show the block hyperlink. The object hyperlink can still be accessed through the Hyperlink sub-menu as previously described.

CONVERTING RELEASE 14 ATTACHED URLS TO HYPERLINKS

Whenever you open a Release 14 drawing that has attached URLs in AutoCAD 2000, they are automatically converted to hyperlinks with no description. If you save an AutoCAD 2000 drawing to Release 14 format, hyperlinks in the drawing are converted to Release 14 attached URLs. Hyperlink descriptions are not available in Release 14, but if you save them again in AutoCAD 2000 format, the hyperlink descriptions are restored.

DRAWING WEB FORMAT

To display AutoCAD drawings on the Internet, Autodesk established a new file format called Drawing Web Format, or .DWF for short. The .DWF file has several benefits and some drawbacks. The .DWF file is compressed and is as small as 1/8 of the original .DWG drawing file so that it takes less time to transmit over the Internet, particularly with relatively slow telephone modem connections. The .DWF format is more secure, since the original drawing is not being displayed. Another user cannot tamper with the original .DWG file.

To view a .DWF file on the Internet, your Web browser needs a plug-in. A plug-in is a software extension that lets a Web browser handle a variety of file formats. Autodesk makes the .DWF plug-in freely available from its Web site. It's a good idea to check the following URL regularly for updates to the .DWF plug-in, which is updated about twice a year at www.autodesk.com. (The exact location of the .DWF plug-in is not given, since Autodesk changes its Web site around every year. Simply use the Web site's Index to search for the plug-in.)

USING EPLOT TO CREATE DWF FILES

AutoCAD 2000's ePlot feature allows you to create DWF files that can be opened, viewed, and plotted with an Internet browser and Autodesk WHIP! 4.0 plug-in. (See the explanation of the WHIP! Plug-in later in this chapter.)

WHIP! 4.0 plug-in supports real-time panning and zooming as well as control over the display of layers, named views, and embedded hyperlinks. With ePlot, you can specify a variety of settings, such as pen assignments, rotation, and paper size, all of which control the appearance of plotted DWF files. With ePlot you can also create DWF files that have rendered images and multiple viewports displayed in Layout tab.

Note: By default, AutoCAD 2000 plots all objects with a lineweight of 0.06 inches, even if you haven't specified lineweight values in the Layer Properties Manager. If you want to plot without any lineweight, clear the **Plot Object Lineweights** option from the Plot Settings tab in the Plot dialog box.

AutoCAD 2000 provides two preconfigured ePlot PC3 (plotter configuration) files (ePlot.pc3 and classic.pc3) that you can use to create DWF files. The ePlot.pc3 configuration file creates DWF files with a white drawing background and a paper boundary. The classic.pc3 configuration file creates DWF files with a black drawing background and resembles AutoCAD Release14 DWFs. If necessary, you can create additional plotter configurations to create DWF files.

To create a .DWF file, invoke the PLOT command:

Standard toolbar	Choose the PLOT command
File menu	Choose Plot
Command: prompt	**plot** (ENTER)

AutoCAD displays the Plot dialog box similar to Figure 14–17.

Figure 14–17 *Plot dialog box*

In the **Plot Device** tab (Figure 14–17), select ePlot.pc3 or classic.pc3 from the **Name:** list box (Plotter configuration section) to create a DWF file. Specify a file name in the

File name: text field (Plot to file section) and location or network folder to plot the file in the **Location:** text field. You can only plot DWF files to the Internet using the FTP protocol.

Make necessary changes to paper size and paper units, Drawing orientation, Plot area, Plot scale, Plot offset, Plot options, and choose the **OK** button to create the DWF file.

Setting Custom Properties to Create DWF files

AutoCAD 2000 allows you to fine-tune custom plotting properties such as resolution, file compression, background color, inclusion of paper boundary, and so on, to create DWF files.

To modify custom plotting properties, open the Plot dialog box and choose the Plot Device tab. Make sure to select an ePlot plotting device and then choose the **Properties** button. AutoCAD displays the Plotter Configuration Editor dialog box, as shown in Figure 14–18.

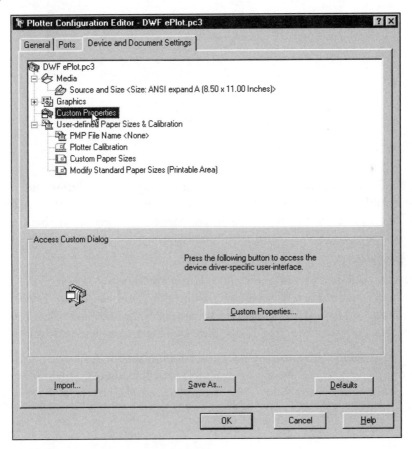

Figure 14–18 *Plotter Configuration Editor dialog box*

Choose the Device and Document Settings tab, and then select Custom Properties from the tree window (see Figure 14–18). Choose the **Custom Properties...** button. AutoCAD displays the DWF Properties dialog box, as shown in Figure 14–19.

Figure 14–19 *DWF Properties dialog box*

In the **Resolution** section, specify the resolution using the Resolution slider. By default, the resolution of the DWF files is set to medium. The higher the resolution of the DWF file, the greater the precision and larger the file size. If there are large numbers of objects in the drawing, then it is recommended to create a DWF file with high resolution.

Specify a file compression option from the **Format** section. By default, DWF files are created in a compressed binary format. If necessary, you can create DWF files as uncompressed binary format or ASCII format (uncompressed text) files. You do not lose any data by compressing the DWF files. It is the recommended format.

Specify background color from the **Background color shown in viewer:** option menu. In addition, specify toggle settings for **Include Layer information, Include Scale And Measurement Information, Show Paper Boundaries,** and **Convert .DWG Hyperlink Extensions To .DWF**.

After making necessary changes, choose the **OK** button to close the DWF Properties dialog box. Choose the **OK** button to save the settings and close the Plotter Configuration Editor dialog box. AutoCAD displays the changes to the **Changes to a Printer Configuration File** dialog box, as shown in Figure 14–20.

Figure 14–20 *Changes to a Printer Configuration File dialog box*

There are two available options. If you want to apply the changes in the settings to the current plot, then select the **Apply changes to the current plot only** radio button. If you want to save the settings and apply them to all plots, select the **Save changes to the following file** radio button and specify the name of the file in the edit field. Choose the OK button to close the Printer Configuration File dialog box.

Choose the **OK** button to create the DWF file and close the Plot dialog box.

VIEWING DWF FILES
AutoCAD itself cannot display DWF files, nor can DWF files be converted back to DWG format without using file translation software from a third-party vendor. In order to view a DWF file, you need a Web browser with a special plug-in that allows the browser to interpret the file correctly. Autodesk has named their DWF plug-in WHIP!, short for "Windows High Performance." The plug-in should not be confused with the AutoCAD display driver, even though both are named Whip!

Figure 14–21 shows an example of DWF displayed in Netscape 4.1 browser.

Figure 14–21 *Netscape Browser displaying DWF file*

Autodesk updates the DWF plug-in approximately twice a year. Each update includes some new features. In summary, all versions of the Whip! plug-in perform the following functions:

- View DWF files created by AutoCAD within a browser.

- Right-clicking the DWF image displays a shortcut menu listing available commands.

- Real-time pan and zoom lets you change the view of the DWF file as quickly as a drawing file in AutoCAD.

- Embedded hyperlinks let you display other documents and files.

- File compression means that a DWF file appears in your Web browser faster than the equivalent DWG drawing file would.

- Print the DWF file alone or with the entire Web page.

- Work with Netscape Navigator v3.x or Microsoft Internet Explorer v3.x. A separate plug-in is required, depending upon which of the two browsers you use.

As this book was being written, DWF plug-in Release 4 was current. Here are some of its features:

- Views ePlots created by AutoCAD 2000.
- Allows you to "drag and drop" a DWG file from a Web site into AutoCAD as a new drawing or as a block.
- Views percentage- or pixel-specified DWF files; views a DWF file in a specified browser frame; views a named view stored in the DWF file; and can specify a view using *X,Y* coordinates.
- Real time Pan and Zoom.
- Layer Controls—turn layers in DWF on and off.
- ActiveX Automation—use WHIP! With Visual Basic, VBA, and VBScript.
- Sends DWF file information to CGI scripts.
- Supports Netscape Communicator v4.0 (previously named Navigator) and Microsoft Internet Explorer v4.0 Web browsers. A separate plug-in is required for each browser.
- Supports raster images in DWF files; renders images as a purple outline during pans and zooms to help speed up the display.
- Printing—full DWF or HTML page output capability.
- Hyperlink Display—enables easy navigation to associated drawings.
- ASCII DWF Support—view DWF drawings written in ASCII.
- Java—customize and control WHIP! with LiveConnect, Java, or JavaScript.
- Enhanced Navigation—Microsoft IntelliMouse support.
- Supports TrueType fonts.

To use the WHIP! plug-in with a Web browser, your computer should have a fast CPU, such as a Pentium or Pentium II. Your computer must be running Windows 95, Windows 98, or Windows NT v4.0. The computer should display at least 256 colors. The Web browser must be at least Netscape v3.0x or above or Explorer v3.0x or above.

If you don't know whether the DWF plug-in is installed in your Web browser, select Help > About Plug-ins from the browser's menu bar. You may need to scroll through the list of plug-ins to find something like this:

```
WHIP!
File name: C:\INTERNET\NETSCAPE\PROGRAM\plugins\npdwf.dll
Autodesk Drawing Web Format File
Mime Type        Description            Suffixes    Enabled
Drawing/x-dwf    Drawing Web Format file    dwf       Yes
```

If the plug-in is not installed or is an older version, then you need to download it from Autodesk's Web site at www.autodesk.com/products/. For Netscape users, the file is quite large at 4.5 MB and takes about a half-hour to download using a typical 28.8-Kbps modem. The file you download from the Autodesk Web site is a self-extracting installation file. After the download is complete, start the program and follow the instructions on the screen.

 Note: If your computer has an older version of the DWF plug-in for Netscape, you must uninstall it before installing the newer version. If the Netscape Web browser is running, close it before installing the DWF plug-in.

For Internet Explorer users, the DWF plug-in is an ActiveX control. Explorer's auto-download feature automatically installs the control the first time your browser accesses a Web page that includes a DWF file. The download time is about 10 minutes with a 28.8-Kbps modem.

DWF PLUG-IN COMMANDS

To display the DWF plug-in's commands, move the cursor over the DWF image, and select one of the available commands from the shortcut menu, such as PAN, ZOOM, NAMED VIEWS.

Pan

PAN is the default command. Press the left mouse button and move the mouse. The cursor changes to an open hand to signal that you can pan the view around the drawing. This is the same as real-time panning in AutoCAD.

Zoom

ZOOM is like the ZOOM REALTIME command in AutoCAD. The cursor changes to a magnifying glass. Hold down the left mouse button and move the cursor up (to zoom in) or down (to zoom out).

Zoom rectangle

ZOOM RECTANGLE is the same as AutoCAD ZOOM WINDOW command . The cursor changes to a plus sign. Click the left mouse button at one corner, and then drag the rectangle to specify the size of the new view.

Fit to window

FIT TO WINDOW is the same as the AutoCAD ZOOM EXTENTS command. You see the entire drawing.

Layers

Selecting LAYERS opens the **Layers** dialog box, a non-modal dialog box that allows you to select layers. From here you can turn layers on and off. (A non-modal dialog box remains on the screen; unlike AutoCAD modal dialog boxes, you do not need to close a non-modal dialog box to continue working.)

Named views

NAMED VIEWS works only when the original DWG drawing file contains named views created with the VIEW command. Selecting Named Views displays a non-modal dialog box that allows you to select a named view.

Location

Selecting LOCATION opens the **Current Location** non-modal dialog box. When you move the cursor it displays the current coordinate location in X and Y.

Highlight urls

HIGHLIGHT URLS displays a highlight box around all objects and areas with URLs in the image. This helps you see where the hyperlinks are. To read the associated URL, pass the cursor over a highlighted area and look at the URL on the browser's status line. (A shortcut method is to hold down the SHIFT key, which causes the URLs to highlight until you release the SHIFT key).

Print

The PRINT option prints the DWF image. To print the entire Web page (including the DWF image), use the browser's Print button.

Saveas

SAVEAS saves the DWF file in one of three formats to your computer's hard drive: DWF, BMP (Windows bitmap), or DWG. Saving in DWG (AutoCAD drawing file) format works only when a copy of the DWG file is available at the same subdirectory as the DWF file. You cannot use the SAVEAS command until the entire DWF file has finished being transmitted to your computer.

About whip!

ABOUT WHIP! displays information about the DWF file, including the DWF file version, description, author, creator, copywright, keywords, source file name, creation time, modification time, source creation time, source modification time, current view left, current view right, current view bottom, and current view top.

Forward

FORWARD is similar to choosing the browser's Forward button. When the DWF image is in a frame, only that frame goes forward.

Back

BACK is similar to choosing the browser's Back button. It works differently when the DWF image is displayed in a frame.

DRAG-AND-DROP

The DWF plug-in allows you to perform several drag-and-drop functions. Drag-and-drop is when you use the mouse to drag an object from one application to another.

Hold down CTRL to drag a DWF file from the browser into AutoCAD. Recall that AutoCAD cannot translate a DWF file into the DWG drawing format. For this reason, this form of drag-and-drop works only when the originating DWG file exists in the same subdirectory as the DWF file.

Another drag-and-drop function is to drag a DWF file from the Windows Explorer (or File Manager) into the Web browser. This causes the Web browser to load the DWF plug-in and display the DWF file. Once the file is displayed, you can execute all of the commands listed in the previous section.

Finally, you can also drag-and-drop a DWF file from the Windows Explorer (or File Manager) into AutoCAD. This causes AutoCAD to launch another program that is able to view the DWF file. This does not work if you have no other software on your computer system capable of viewing DWF files.

EMBEDDING A DWF FILE

To let others view your DWF file over the Internet, you need to embed the DWF file in a Web page. There are several approaches to embedding a DWF file in a Web page.

The quickest way is to use the <embed> tag in a very simple manner. <Embed> is an HTML (hypertext markup language) code for embedding an object in a Web page.

```
<embed src="filename.dwf">
```

Replace "filename.dwf" with the name of the DWF file. Remember to keep the quotation marks in place. To specify the size of the DWF image on the Web page, include the Width and Height options.

```
<embed width=800 height=600 name=description src="filename.dwf">
```

Replace 800 and 600 with any appropriate numbers, such as 100 and 75 for a "thumbnail" image, 300 and 200 for a small image, or 640 and 480 for a medium-size image. These width and height values are measured in pixels.

The Name option displays a textual description of the image when the browser does not load images. For example, you might replace the description with the DWF file name.

If you use Netscape Navigator, you can include a description of where to get the WHIP! plug-in if it is not available with the Web browser.

```
<embed width=800 height=600 name=description src="filename.dwf"
pluginspage=http://www.autodesk.com/products/autocad/whip/whip.htm>
```

If you use Microsoft Internet Explorer, you must add the <object> and <param> tags.

```
<object width=600 height=400 classid="clsid:b2be75f3-9197-11cf-abf4-
   08000996e931"
codebase="ftp://ftp.autodesk.com/pub/autocad/plugin/
   whip.cab#Version=2,0,0,0">
<param name="description" value="filename.dwf">
<embed width=600 height=400 name=description src="filename.dwf"
pluginspage=http://www.autodesk.com/products/autocad/whip/whip.htm>
  </object>
```

PROJECT EXERCISE

In this project, you use AutoCAD HYPERLINK command to attach a drawing, a text document, and a Web document. In addition, if your computer can access the Internet, this project has you open a drawing from the Autodesk Press Web site.

Step 1: Start AutoCAD and open **1st floor plan.dwg** from AutoCAD2000\Sample folder.

Step 2: Choose Model tab to display the drawing in the Model space. Invoke the HYPERLINK command from the Standard toolbar as shown in Figure P14–1.

Figure P14–1 *Invoking the* HYPERLINK *command from the Standard toolbar*

AutoCAD prompts:

Command: **hyperlink**
Select objects: (select the left most column as shown in Figure P14–2
    and press ENTER)

Figure P14–2 *Selection of the object for attaching an hyperlink*

AutoCAD displays Insert Hyperlink dialog box as shown in Figure P14–3.

Figure P14–3 *Insert Hyperlink dialog box*

Choose the **Browse** button. AutoCAD displays **Browse the Web – Select Hyerlink** dialog box. Select **City base map.dwg** from the AutoCAD 2000\Sample folder. Choose the **OK** button to close the dialog box. AutoCAD attaches the selected drawing file as an hyperlink to the selected object. If you move the cursor over the column, AutoCAD displays a tool tip "City base map.dwg".

Similarly, attach License.rtf document and home.htm found located in the AutoCAD 2000 folder to the columns as shown in Figure P14–2.

Now you have attached a drawing, a text document, and a Web document to objects in the drawing.

Step 3: Select the column to which the drawing is attached, right-click and from the shortcut menu select Open "City base map.dwg". AutoCAD opens the City base map drawing. To display both the drawings, select Tile Vertically from Window menu. Similarly, you can also open the License.rtf and home.htm documents.

Step 4: To open a drawing from the Autodesk Press Web site, invoke the OPEN command from the Standard tool bar. Before you invoke the OPEN command, ensure that you have a live connection to the internet.

AutoCAD displays Select File dialog box. Choose the Search the Web button as shown in Figure P14–4.

Figure P14–4 *Select File dialog box*

AutoCAD displays Browse the Web dialog box similar to Figure P14–5.

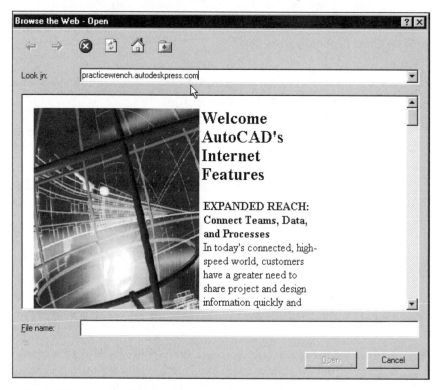

Figure P14–5 *Browse the Web dialog box*

Type **practicewrench.autodeskpress.com** in the **Look in:** text field and press ENTER. After a few seconds, AutoCAD displays the web site. In the **File name:** text field, type **wrench.dwg** and press ENTER. AutoCAD begins transferring the file. Depending on the speed of your internet connection, this will take between a couple of seconds and a half-minute. AutoCAD displays the Wrench drawing as shown in Figure P14–6, it looks just like any drawing you open from your computer's hard drive.

Figure P14–6 *Wrench.dwg opened from the Internet*

Save the drawing in your local drive and exit AutoCAD.

REVIEW QUESTIONS

1. The command to invoke the Internet browser in AutoCAD is:

   a. INTERNET

   b. BROWSER

   c. HTTP

   d. WWW

   e. none of the above

2. DWF is short for:

   a. DraWing Format.

   b. Drawing Web Format.

   c. DXF Web Format.

3. The purpose of DWF files is to view:

   a. 2D drawings on the Internet.

   b. 3D drawings on the Internet.

   c. 3D drawings in another CAD system.

   d. all of the above.

4. URL is short for:

   a. Union Region Lengthen.

   b. Earl.

   c. Useful Resource Line.

   d. Uniform Resource Locator.

5. Which of the following URLs are valid:

   a. www.autodesk.com

   b. http://www.autodesk.com

   c. all of the above.

   d. none of the above.

6. What is the purpose of a URL?

   a. To access files on computers, networks, and the Internet.

   b. It is a universal file-naming system for the Internet.

   c. To create a link to another file.

   d. All of the above.

7. FTP is short for:

   a. Forwarding Transfer Protocol

   b. File Transfer Protocol

   c. File Transference Protocol

   d. File Transfer Partition

8. URLs are used in an AutoCAD drawing to browse the Internet.

   a. True   b. False

9. The purpose of URLs is to let you create _____ between files.

   a. backups

   b. links

   c. copies

   d. partitions

10. You can attach a URL to rays and xlines.

    a. True   b. False

11. Compression in the DWF file causes it to take _____ time to transmit over the Internet.

    a. more

    b. less

    c. all of the above

    d. none of the above

12. A "plug-in" lets a Web browser:

    a. plug into the Internet

    b. display a file format

    c. log in to the Internet

    d. display a URL

13. A Web browser can view DWG drawing files over the Internet.

    a. True   b. False

14. The WHIP! plug-in and the WHIP! display driver are the same.

    a. True   b. False

15. Which AutoCAD command is used to start a Web browser?

    a. WEBSTART

    b. WEBDWG

    c. BROWSERS

    d. LAUNCH

16. Before a URL can be accessed you must always type the prefix "http://".

    a. True   b. False

17. Which AutoCAD command is used to open an AutoCAD drawing file from an Internet site?

    a. LAUNCH

    b. OPEN

    c. START

    d. LAUNCHFILE

18. Once an Internet AutoCAD drawing file has been modified it can only be saved back to the Internet site.

    a. True   b. False

20. AutoCAD creates hyperlinks as being _____.

    a. Absolute

    b. Relative

    c. Polar

    d. Only a or b

21. Hyperlinks can only be attached to text objects within a drawing.

    a. True   b. False

22. The system variable PICKFIRST must be set to a value of 1 before a hyperlink can be attached.

    a. True   b. False

23. DWF files are compressed to as much as _____ of the original .DWG file size.

    a. 1/8

    b. 1/4

    c. 1/2

    d. equal to

CHAPTER 15

AutoCAD 3D

INTRODUCTION

After completing this chapter, you will be able to do the following:

- Define a User Coordinate System
- View in *3D* using the VPOINT, DVIEW, and 3DORBIT commands
- Create *3D* objects
- Use the REGION command
- Use the 3DPOLY and 3DFACE commands
- Create meshes
- Edit in *3D* using the ALIGN, ROTATE3D, MIRROR3D, 3DARRAY, EXTEND, and TRIM commands
- Create solid shapes: solid box, solid cone, solid cylinder, solid sphere, solid torus, and solid wedge
- Create solids from existing *2D* objects and regions
- Create solids by means of revolution
- Create composite solids
- Edit *3D* solids using the CHAMFER, FILLET, SECTION, SLICE, and INTERFERE commands, in addition to Face editing, body editing, and edge editing.
- Obtain the mass properties of a solid
- Place a multiview in paper space
- Generate Views in viewports
- Generate profiles

WHAT IS *3D*?

In *two-dimensional* drawings you have been working in a single plane with two axes, X and Y. In *three-dimensional* drawings you work with the Z axis, in addition to the X and Y axes, as shown in Figure 15–1. Plan views, sections, and elevations represent only two dimensions. Isometric, perspective, and axonometric drawings, on the other hand, represent all three dimensions. For example, to create three views of a cube, the cube is simply drawn as a square with thickness and then viewed along each of the three axes. Drawing in this manner is referred to as extruded *2D*. Only objects that are extrudable can be drawn by this method. Other views are achieved by rotating the viewpoint or the object, just as if you were physically holding the cube. You can get an isometric or perspective view by simply changing the viewpoint.

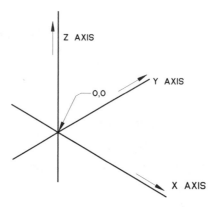

Figure 15–1 *The X, Y, and Z axes for a 3D drawing*

Whether you realize it or not, all the drawings you have done in previous chapters were created by AutoCAD in true *3D*. What this means is that every line, circle, or arc that you have drawn, even if you think you have drawn it in *2D*, is really stored with three coordinates. By default, AutoCAD stores the *Z* value as your current elevation with a thickness of zero. What you think of now as *2D* is really only one of an infinite number of views of your drawing in *3D* space.

Drawing objects in *3D* provides three major advantages:

- An object is drawn once and then can be viewed and plotted from any angle (viewpoint).

- A *3D* object holds mathematical information that can be used in engineering analysis such as finite element analysis and computer numerical control (CNC) machinery.

- Shading and rendering enhances the visualization of an object.

There are two major limitations to working in *3D*:

1. Whenever you want to input *3D* coordinates whose *Z* coordinate is different from the current construction plane's elevation, you have to use the keyboard instead of your pointing device. One exception is to Object Snap to an object not in the current construction plane. The input device (mouse or digitizer) can supply AutoCAD with only two of the three coordinates at a

time. Three-dimensional-input devices exist, but there is no practical support by AutoCAD for them at this time. So you are limited to using the keyboard.

2. Determining where you are in relationship to an object in *3D* space is difficult.

COORDINATE SYSTEMS

AutoCAD provides two types of coordinate systems. One is a single fixed coordinate system called the World Coordinate System, and the other is an infinite set of user-defined coordinate systems available through the User Coordinate System.

The World Coordinate System (WCS) is fixed and cannot be changed. In this system (when viewing the origin from 0,0,1), the *X* axis starts at the point 0,0,0, and values increase as the point moves to the operator's right; the *Y* axis starts at 0,0,0, and values increase as the point moves to the top of the screen; and finally, the *Z* axis starts at the 0,0,0 point, and values get larger as it comes toward the user. All drawings from previous chapters are created with reference to the WCS. The WCS is still the basic system in virtually all *2D* AutoCAD drawings. However, because of the difficulty in calculating *3D* points, the WCS is not suited for many *3D* applications.

The User Coordinate System (UCS) allows you to change the location and orientation of the *X, Y,* and *Z* axes to reduce the number of calculations needed to create *3D* objects. The UCS command lets you redefine the origin of your drawing and establish the positive *X* and the positive *Y* axes. New users think of a coordinate system simply as the direction of positive *X* and positive *Y.* But once the directions of *X* and *Y* are defined, the direction of *Z* will be defined as well. Thus, the user has only to be concerned with *X* and *Y.* As a result, when you are drawing in *2D,* you are also somewhere in *3D* space. For example, if a sloped roof of a house is drawn in detail using the WCS, each endpoint of every object on the inclined roof plane must be calculated. On the other hand, if the UCS is set to the same plane as the roof, each object can be drawn as if it were in the plan view. You can define any number of UCSs within the fixed WCS and save them, assigning each a user-determined name. But at any given time, only one coordinate system is current and all coordinate input and display is relative to it. This is unlike a rotated snap grid in which direction and coordinates are based on the WCS. If multiple viewports are active, AutoCAD 2000 allows you to assign a different UCS to each viewport. Each UCS can have a different origin and orientation for various construction requirements.

RIGHT-HAND RULE

The directions of the *X, Y,* and *Z* axes change when the UCS is altered; hence, the positive rotation direction of the axes may become difficult to determine. The right-hand rule helps in determining the rotation direction when changing the UCS or using commands that require object rotation.

To remember the orientation of the axes, do the following:

1. Hold your right hand with the thumb, forefinger, and middle finger pointing at right angles to each other, as shown in Figure 15–2.

2. Consider the thumb to be pointing in the positive direction of the X axis.

3. The forefinger points in the positive direction of the Y axis.

4. The middle finger points in the positive direction of the Z axis.

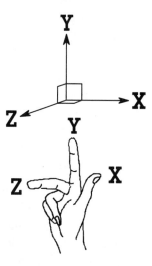

Figure 15–2 *The correct hand position when using the right-hand rule*

SETTING THE DISPLAY OF THE UCS ICON

The UCS icon provides a visual reminder of how the UCS axes are oriented, where the current UCS origin is, and the viewing direction relative to the UCS *XY* plane. AutoCAD displays a different coordinate system icon in paper space and in model space, as shown in Figure 15–3.

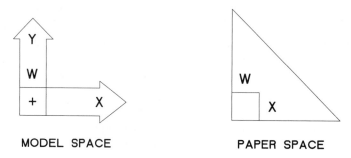

MODEL SPACE   PAPER SPACE

Figure 15–3 *The UCS icon for model space and for paper space*

The *X* and *Y* axis directions are indicated with arrows labeled appropriately, and the *Z* axis is indicated by the placement of the icon. The icon displays a W in the *Y* axis arrow if the current coordinate system is the World Coordinate System, and a "+" appears at the base of the arrows if the icon is placed at the origin of the current coordinate system. When looking straight up or down on the *Z* plane, the icon seems flat. When viewed at any other angle, the icon looks skewed. The orientation of the *Z* axis is indicated further by the presence or absence of a box at the base of the arrows that create the icon. If the box is visible, you are looking down on (from the positive *Z* side of) the *XY* plane; if the box is not present, the bottom of the *XY* plane is being viewed (from the negative *Z* side of it). See Figure 15–4 for different orientations of the UCS icon.

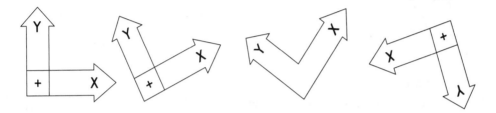

Figure 15–4 *The UCS icon in different orientations*

 Note: If the viewing angle comes within 1 degree of the XY plane, the UCS icon will change to a "broken pencil," as shown in Figure 15–5. When this icon is showing in a view, it is recommended that you avoid trying to use the cursor to specify points in that view, because the results may be unpredictable.

Figure 15–5 *The UCS icon becomes a broken pencil when the viewing angle is within 1 degree of the XY plane*

The display and placement on the origin of the UCS icon is handled by the UCSICON command. Invoke the UCSICON command:

View menu	Choose Display > UCS icon
Command: prompt	**ucsicon** (ENTER)

AutoCAD prompts:

> Command: **ucsicon** (ENTER)
> Enter an option [ON/OFF/All/Noorgin/Origin]<default>: *(select one of the available options or right-click for shortcut menu and select one of the available options)*

ON Option

The ON option allows you to set the icon to ON if it is OFF in the current viewport.

OFF Option

The OFF option allows you to set the icon to OFF if it is ON in the current viewport.

Noorigin Option

The Noorigin option (default setting) tells AutoCAD to display the icon at the lower left corner of the viewport, regardless of the location of the UCS origin. This is like parking the icon in the lower left corner.

ORigin Option

The Origin option forces the icon to be displayed at the origin of the current coordinate system.

 Note: If the origin is off screen, the icon is displayed at the lower left corner of the viewport.

All Option

The All option determines whether the options that follow affect all of the viewports or just the current active viewport. This option is issued before each and every option if you want to affect all viewports. For example, to turn ON the icon in all the viewports and display the icon on the origin, the following sequence of prompts is displayed:

> Command: **ucsicon**
> Enter an option [ON/OFF/All/Noorigin/Origin] <default>: **a**
> Enter an option [ON/OFF/Noorigin/Origin]: **on**
> Command: (ENTER)
> Enter an option [ON/OFF/All/Noorigin/Origin]: **a**
> Enter an option [ON/OFF/Noorigin/Origin]: **or**

DEFINING A NEW UCS

The UCS is the key to almost all 3D operations in AutoCAD, as mentioned earlier. Many commands in AutoCAD are traditionally thought of as 2D commands. They are effective in *3D* because they are always relative to the current UCS. For example, the ROTATE command rotates in only the *X* and *Y* directions; if you want to rotate an object in the *Z* direction, change your UCS so that *X* or *Y* is now in the direction of what was previously *Z*. Then you could use the *2D* ROTATE command.

The UCS command lets you redefine the origin in your drawing. Broadly, you can define the origin by five methods:

- Specify a data point for an origin, specify a new XY plane by providing three data points, or provide a direction for the Z axis.

- Define an origin relative to the orientation of an existing object.

- Define an origin by selecting a face.

- Define an origin by aligning with the current viewing direction.

- Define an origin by rotating the current UCS around one of its axes.

Invoke the UCS command:

UCS toolbar	Choose the UCS command (see Figure 15–6)
Tools menu	Choose New UCS
Command: prompt	**ucs** (ENTER)

Figure 15–6 *Invoking the* UCS *command from the UCS toolbar*

AutoCAD prompts:

Command: **ucs**
Enter an option [New/Move/orthoGraphic/Prev/Restore/Save/Del/Apply/?/
     World]<World>: *(select the new option or right-click for the shortcut menu
     and select New option)*

Specify origin of new UCS or [Zaxis/3point/OBject/Face/View/X/Y/Z]
     <0,0,0>: *(select one of the available options to define a new origin or right-
     click for the shortcut menu and select one of the available options)*

Origin Option

The Origin option defines a new UCS by shifting the origin of the current UCS, leaving the directions of the X, Y, and Z axes unchanged. You can also invoke the UCS Origin command:

UCS toolbar	Choose the UCS Origin command (see Figure 15–7)
Tools menu	Choose New UCS > Origin

Figure 15-7 *Invoking the UCS Origin command from the UCS toolbar*

AutoCAD prompts:

> **Specify new Origin point (0,0,0):** *(Specify a new origin point relative to the origin of the current UCS, as shown in Figure 15-8)*

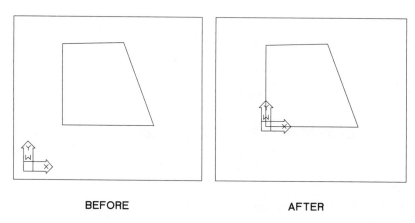

BEFORE   AFTER

Figure 15-8 *Specifying a new origin point relative to the origin of the current UCS*

ZAxis Option

The ZAxis option allows you to define an origin by specifying a point and the direction for the *Z* axis. AutoCAD arbitrarily, but consistently, sets the direction of the *X* and *Y* axes in relation to the given *Z* axis. You can also invoke the UCS ZAxis command:

UCS toolbar	Choose the UCS Z Axis Vector command (see Figure 15-9)
Tools menu	Choose New UCS > Z Axis Vector

Figure 15-9 *Invoking the UCS Z Axis Vector command from the UCS toolbar*

AutoCAD prompts:

> Specify new Origin point (0,0,0): *(specify a point to define a new origin point)*
> Specify point on positive portion of the Z axis <default>: *(specify the direction for the Z axis)*

If you give a null response to the second prompt, the Z axis of the new coordinate system will be parallel to the previous one. This is similar to using the Origin option.

3point Option

The 3point option is the easiest and most often used option for controlling the orientation of the UCS. This option allows the user to select three points to define the origin and the directions of the positive *X* and *Y* axes. The origin point acts as a base for the UCS rotation, and when a point is selected to define the direction of the positive *X* axis, the direction of the *Y* axis is limited because it is always perpendicular to the *X* axis. When the *X* and *Y* axes are defined, the *Z* axis is automatically placed perpendicular to the *XY* plane. You can also invoke the UCS 3point command:

UCS toolbar	Choose the UCS 3 point command (see Figure 15–10)
Tools menu	Choose New UCS > 3 point

Figure 15–10 *Invoking the UCS 3 point command from the UCS toolbar*

AutoCAD prompts:

> Specify new Origin point (0,0,0): *(specify the origin point)*
> Specify Point on positive portion of the X-axis <default>: *(specify a point for the positive X axis)*
> Specify Point on positive-Y portion of the UCS XY plane <default>: *(specify a point for the positive Y axis)*

Specify a data point and the direction for the positive *X* and *Y* axes, as shown in Figure 15–11. The points must not form a straight line. If you give a null response to the first prompt, the new UCS will have the same origin as the previous UCS. If you give a null response to the second or third prompt, then that axis direction will be parallel to the corresponding axis in the previous UCS.

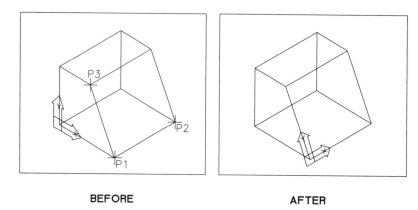

BEFORE   AFTER

Figure 15–11 *Specifying a data point and the direction for the positive X and Y axes using the 3Point Option*

OBject Option

The OBject option lets you define a new coordinate system by selecting an object. The actual orientation of the UCS depends on how the object was created. When the object is selected, the UCS origin is placed at the first point used to create the object (in the case of a line, it will be the closest endpoint; for a circle, it will be the center point of the circle); the X axis is determined by the direction from the origin to the second point used to define the object. And the Z axis direction is placed perpendicular to the XY plane in which the object sits. Table 15–1 lists the location of the origin and its X axis for different types of objects.

Table 15–1 Location of the Origin and Its X Axis for Different Types of Objects

Object	Method of UCS Determination
Line	The endpoint nearest the specified point becomes the new UCS origin. The new X axis is chosen so that the line lies in the XZ plane of the new UCS.
Circle	The circle's center becomes the new UCS origin, and the X axis passes through the point specified.
Arc	The arc's center becomes the new UCS origin, and the X axis passes through the endpoint of the arc closest to the pick point.
2D polyline	The polyline's start point becomes the new UCS origin, with the X axis extending from the start point to the next vertex.
Solid	The first point of the solid determines the new UCS origin, and the X axis lies along the line between the first two points.
Dimension	The new UCS origin is the middle point of the dimension text, and the direction of the X axis is parallel to the X axis of the UCS in effect when the dimension was drawn.

You can also invoke the UCS Object command:

UCS toolbar	Choose the UCS Object command (see Figure 15–12)
Tools menu	Choose New UCS > Object

Figure 15–12 *Invoking the UCS Object command from the UCS toolbar*

AutoCAD prompts:

> Select object to align UCS: *(identify an object to define a new coordinate system)*

Identify an object to define a new coordinate system, as shown in Figure 15–13.

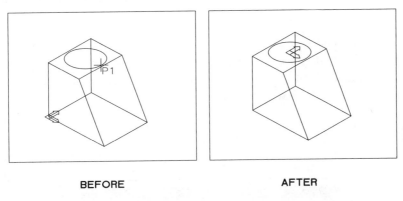

BEFORE   AFTER

Figure 15–13 *Defining a new coordinate system by the UCS Object command*

Face Option

The Face option aligns the UCS to the selected face of a solid object. To select a face, click within the boundary of the face or on the edge of the face. The face is highlighted and the UCS *X* axis is aligned with the closest edge of the first face found. You can also invoke the UCS Face command:

UCS toolbar	Choose the UCS Face command (see Figure 15–14)
Tools menu	Choose New UCS > Face

Figure 15–14 *Invoking the UCS Face command from the UCS toolbar*

AutoCAD prompts:

> Select face of solid object: *(select the face of the solid object)*
> Enter an option [Next/Xflip/Yflip] <accept>: *(select one of the available options or right-click for the shortcut menu and select one of the available options)*

The **Next** option locates the UCS on either the adjacent face or the back face of the selected edge.

The **Xflip** option rotates the UCS 180 degrees around the *X* axis.

The **Yflip** option rotates the UCS 180 degrees around the *Y* axis.

The **Accept** option accepts the location. The prompt repeats until you accept a location.

View Option

The View option places the *XY* plane parallel to the screen, and makes the *Z* axis perpendicular. The UCS origin remains unchanged. This method is used mainly for labeling text, which should be aligned with the screen rather than with objects. You can also invoke the UCS View command:

UCS toolbar	Choose the UCS View command (see Figure 15–15)
Tools menu	Choose New UCS > View

Figure 15–15 *Invoking the UCS View command from the UCS toolbar*

X/Y/Z Rotation Option

The X/Y/Z rotation option lets you define a new coordinate system by rotating the *X*, *Y*, and *Z* axes independently of each other. You can show AutoCAD the desired angle by specifying two points, or you can enter the rotation angle from the keyboard. In either case, the new angle is specified relative to the *X* axis of the current UCS. See Figures 15–16, 15–17, and 15–18 for examples of rotating the UCS around the *X*, *Y*, and *Z* axes, respectively.

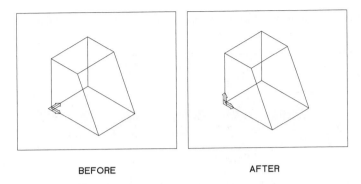

Figure 15–16 *Example of rotating the UCS around the X axis*

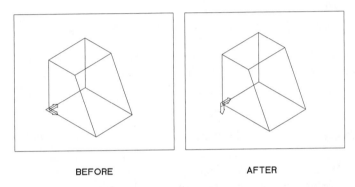

Figure 15–17 *Example of rotating the UCS around the Y axis*

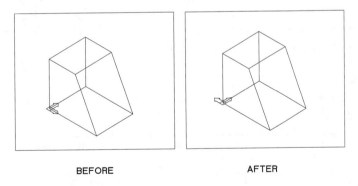

Figure 15–18 *Example of rotating the UCS around the Z axis*

Previous Option

The Previous option is similar to the Previous option of the ZOOM command. AutoCAD saves the last 10 coordinate systems in both model space and paper space. You can step back through them by using repeated Previous options. You can also invoke the UCS Previous command:

UCS toolbar	Choose the UCS Previous command (see Figure 15–19)

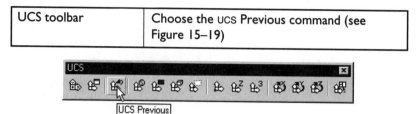

Figure 15–19 *Invoking the* UCS *Previous command from the UCS toolbar*

Restore Option

The Restore option allows you to restore any previously saved UCS.

 Note: You can also restore a previously saved UCS by invoking the DDUCS command, which in turn displays a dialog box listing the previously saved UCS.

Save Option

The Save option allows you to save the current UCS under a user-defined name.

Delete Option

The Delete option allows you to delete any saved UCS.

? Option

The ? option lists the name of the UCS you specify, the origin, and the X, Y, and Z axes for each saved coordinate system, relative to the current UCS. To list all the UCS names, accept the default, or you can specify wild cards.

World Option

The World option returns the drawing to the WCS. You can also invoke the UCS World command:

UCS toolbar	Choose the UCS World command (see Figure 15–20)
Tools menu	Choose New UCS > World

Figure 15–20 *Invoking the* UCS *World command from the UCS toolbar*

SELECTING A PREDEFINED ORTHOGRAPHIC UCS

In addition to various methods explained earlier for defining a new UCS, AutoCAD's UCS dialog box allows you to change the UCS to one of the available standard orthographic settings. In addition, the UCS dialog box lists saved user coordinate systems and allows you to modify UCS icon settings and UCS settings saved with a viewport. Invoke the UCSMAN command to open UCS dialog box:

UCS toolbar	Choose the Display UCS dialog command (see Figure 15–21)
Tools menu	Choose Orthographic UCS > Preset...

Display UCS Dialog

Figure 15–21 *Invoking the Display UCS dialog command from the UCS toolbar*

AutoCAD displays the UCS dialog box with the Orthographic UCSs tab selected, as shown in Figure 15–22.

Figure 15–22 *UCS dialog box with the Orthographic UCSs tab selected*

AutoCAD displays the name of the current UCS view at the top of the dialog box. If the UCS setting has not been saved and named, the current UCS reads UNNAMED. AutoCAD lists the standard orthographic coordinate systems in the current drawing. The orthographic coordinate systems are defined relative to the UCS specified in the **Relative to:** list box. By default it is set to World coordinate system. To set the UCS to one of the orthographic coordinate systems, select one of the six listed names and choose the **Set Current** button, double-click on the listed name, or right-click and select Set Current from the shortcut menu. The Depth field lists the distance between the orthographic coordinate system and the parallel plane passing through the origin of the UCS base setting (stored in the UCSBASE system variable). To change the depth, double-click on the Depth field, and AutoCAD displays **Orthographic UCS Depth** dialog box. Make the necessary changes to the depth and choose the **OK** button to close the dialog box.

To set the UCS to one of the saved UCSs, select the Named UCSs tab, and AutoCAD lists the saved UCS in the current drawing, similar to Figure 15–23. Select one of the listed UCS saved names and choose the **Set Current** button, double-click on the listed name, or right-click and select Set Current from the shortcut menu.

Figure 15–23 *UCS dialog box with the Named UCSs tab selected*

The Settings tab of the UCS dialog box displays and allows you to modify UCS icon settings and UCS settings saved with a viewport, as shown in Figure 15–24.

Figure 15–24 *UCS dialog box with the Settings tab selected*

The **UCS icon** settings section of the dialog box specifies the UCS settings for the current viewport.

The **On** toggle button controls the display of the UCS icon in the current viewport.

The **Display at UCS Origin point** toggle button controls the display of the UCS icon at the origin of the current coordinate system for the current viewport. If this option is set to off, or if the origin of the coordinate system is not visible in the viewport, the UCS icon is displayed at the lower left corner of the viewport.

The **Apply all to active viewports** toggle button applies the UCS icon settings to all active viewports in the current drawing.

The **UCS settings** section specifies the UCS settings for the current viewport.

The **Save UCS with viewport** toggle button saves the coordinate system setting with the viewport. If this option is set to off, the viewport reflects the UCS of the viewport that is current.

The **Update View to Plan When UCS Is Changed** toggle button restores plan view when the coordinate system in the viewport changes. Plan view is restored when the dialog box is closed and the selected UCS setting is restored.

Choose the **OK** button to close the UCS dialog box and accept the changes.

You can also set the current UCS to one of the saved UCS or to one of the standard orthographic UCS from the UCS II toolbar drop down menu, as shown in Figure 15–25.

Figure 15–25 *UCS II toolbar*

VIEWING A DRAWING FROM PLAN VIEW

The PLAN command provides a convenient means of viewing a drawing from plan view. The definition of a plan means that you are at positive Z and looking perpendicularly down on the XY plane, with X to the right and Y pointing up. You can select the plan view of the current UCS, a previously saved UCS, or the WCS.

Invoke the PLAN command:

View menu	Choose 3D views > Plan view
Command: prompt	**plan** (ENTER)

AutoCAD prompts:

> Command: **plan**
> Enter an option [Current ucs>/Ucs/World] <current>: *(select one of the available options or right-click for the shortcut menu and select one of the available options)*

Current UCS Option

The Current UCS option displays the plan view of the current UCS. This is the default option.

Ucs Option

The Ucs option displays a plan view of a previously saved UCS. When you select this option, AutoCAD prompts for a name of the UCS.

World Option

The World option displays the plan view of the World Coordinate System.

VIEWING IN 3D

Until now, you have been working on the plan view, or the *XY* plane. You have been looking down at the plan view from a positive distance along the *Z* axis. The direction from which you view your drawing or model is called the viewpoint. You can view a drawing from any point in model space. From your selected viewpoint, you can add objects, modify existing objects, or suppress the hidden lines from the drawing.

The VPOINT, DVIEW, and 3DORBIT commands are used to control viewing of a model from any point in model space.

VIEWING A MODEL BY MEANS OF THE VPOINT COMMAND

To view a model in *3D*, you may have to change the viewpoint. The location of the viewpoint can be controlled by means of the VPOINT command. The default viewpoint is 0,0,1; that is, you are looking at the model from 0,0,1 (on the positive *Z* axis above the model) to 0,0,0 (the origin).

Invoke the VPOINT command:

View menu	Choose 3D Views > VPOINT
Command: prompt	**vpoint** (ENTER)

AutoCAD prompts:

> Command: **vpoint**
> Specify a view point or [Rotate] <display compass and tripod>: *(select one of the available options or right-click for the shortcut menu and select one of the available options)*

The default method requires you to enter *X*, *Y*, and *Z* coordinates from the keyboard. These coordinates establish the viewpoint. From this viewpoint, you will be looking at the model in space toward the model's origin. For example, a 1,–1,1 setting gives you a –45-degree angle projected in the *XY* plane and 35.264-degree angle above the *XY* plane (top, right, and front views); looking at the model origin (0,0,0). You can set the viewpoint to any *X, Y, Z* location. Table 15–2 lets you experiment with the rotation of *3D* objects.

Table 15–2 Various Viewpoint Settings for Rotating *3D* objects

Viewpoint Setting	Displayed View(s)	Viewpoint Setting	Displayed View(s)
0,0,1	Top	–1,–1,1	Top, Front, Left side
0,0,–1	Bottom	1,1,1	Top, Rear, Right side
0,–1,0	Front	–1,1,1	Top, Rear, Left side
0,1,0	Rear	1,–1,–1	Bottom, Front, Right side
1,0,01	Right side	–1,–1,–1	Bottom, Front, Left side
–1,0,0	Left side	1,1,–1	Bottom, Rear, Right side
1,–1,1	Top, Front, Right side	–1,1,–1	Bottom, Rear, Left side

If instead of entering coordinates you give a null response (press ENTER or the SPACEBAR), a compass and axes tripod appear on the screen, as shown in Figure 15–26. The compass, in the upper right of the screen, is a *2D* representation of a globe. The center point of the circle represents the north pole (0,0,1), the inner circle represents the equator, and the outer circle represents the south pole (0,0,–1), as shown in Figure 15–27. A small cross is displayed on the compass. You can move the cross with your pointing device. If the cross is in the inner circle, you are above the equator looking down on your model. If the cross is in the outer circle, you are looking from beneath your drawing, or from the Southern Hemisphere. Move the cross, and the axes tripod rotates to conform to the viewpoint indicated on the compass. When you achieve the desired viewpoint, press the pick button on your pointing device or press ENTER. The drawing regenerates to reflect the new viewpoint position.

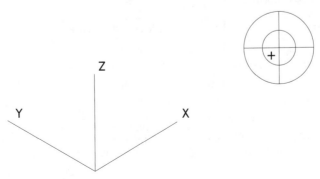

Figure 15–26 *The* VPOINT *command's compass and axes tripod*

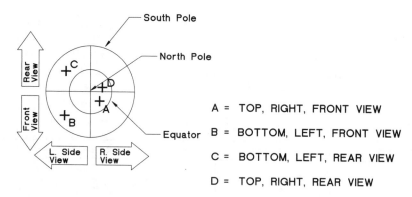

Figure 15–27 *Components of the* VPOINT *command's compass and its poles*

Rotate Option

The Rotate option allows you to specify the location of the viewpoint in terms of two angles. The first angle determines the rotation in the *XY* plane from the *X* axis (0 degrees) clockwise or counterclockwise. The second angle determines the angle from the *XY* plane up or down. When you select the Rotate option, AutoCAD prompts:

> Enter angle in X-Y plane from X axis <current>: (specify the angle in the XY plane from the X axis)
> Enter angle from X-Y plane <current>: (specify the angle from the XY plane)

AutoCAD regenerates to reflect the new viewpoint position.

Viewpoint Presets

AutoCAD provides Viewpoint Presets dialog box when you invoke the DDVPOINT command. The dialog box lets you set a *3D* viewing direction by specifying an angle from the *X* axis and an angle from the *XY* plane. This is similar to using the Rotate option of the VPOINT command.

Invoke the DDVPOINT command:

View menu	Choose 3D Views > Viewpoint Presets...
Command: prompt	**ddvpoint** (ENTER)

AutoCAD displays the Viewpoint Presets dialog box, similar to the one shown in Figure 15–28. Specify viewing angles from the image tile, or enter their values in the text boxes. You specify the view direction relative to the current UCS or the WCS, the viewing angles are updated accordingly. The new angle is indicated by the white arm; the current viewing angle is indicated by the red arm. By selecting the **Set to Plan View** button, you can set the viewing angles to display the plan relative to the selected coordinate system.

Figure 15–28 *Viewpoint Presets dialog box*

 Note: By default, AutoCAD always places the model to your current viewpoint position in reference to the WCS, not the current UCS. If necessary, you can change the WORLDVIEW system variable from 1 (default) to 0; AutoCAD then places the model in reference to the UCS for your current viewpoint position. It is recommended that you keep the WORLDVIEW set to 1 (default). Regardless of the WORLDVIEW setting, you are always looking through your viewpoint to the WCS origin.

VIEWING A MODEL BY MEANS OF THE DVIEW COMMAND

The DVIEW command is an enhanced VPOINT command. Here you visually move an object around on the screen, dynamically viewing selected objects as the view changes. The DVIEW command provides either parallel or perspective views, whereas the VPOINT command provides only parallel views. In the case of a parallel view, parallel lines always remain parallel, whereas in perspective view, parallel lines converge from your view to a vanishing point. Figures 15–29 and 15–30 show parallel and perspective views, respectively, of a model. The viewing direction is the same in each case.

Figure 15–29 *The DVIEW command displaying a model with parallel projection*

Figure 15–30 *The DVIEW command displaying the same model with perspective projection*

Invoke the DVIEW command:

Command: prompt	**dview** (ENTER)

AutoCAD prompts:

> Command: **dview**
> Select objects or (use DVIEWBLOCK): *(select objects or press ENTER)*

All or any part of the objects in the drawing can be selected for viewing during the DVIEW command process. But once you exit the DVIEW command, all objects in the drawing are represented in the new view created. If your drawing is too large to display quickly in the DVIEW display, small portions can be selected and used to orient the entire drawing. The purpose of this is to save time on slower machines and still give dynamic rotation so that you can quickly and effortlessly adjust the view of your object before you begin working with it.

If you give a null response (press ENTER or SPACEBAR) to the "Select objects:" prompt, AutoCAD provides you with a picture of a *3D* house. Whatever you do to the *3D* house under DVIEW will be done to your current drawing when you exit the DVIEW command.

Each time you exit the DVIEW command, AutoCAD performs an unconditional regeneration. No matter what the current setting for REGENAUTO is, AutoCAD automatically performs a regeneration.

After selecting the objects, AutoCAD prompts with the following options:

> [CAmera/TArget/Distance/POints/PAn/Zoom/TWist/CLip/Hide/Off/Undo]:
> *(select one of the available options or right-click for the shortcut menu and select one of the available options)*

CAmera Option

The Camera option is one of the six options that adjusts what is seen in the view. With the Camera option, the drawing is stationary while the camera moves. It can move up and down (above or below) or it can move around the target to the left or right (clockwise or counterclockwise). When you are moving the camera, the target is fixed.

When you select the Camera option, AutoCAD prompts:

> Specify camera location, or enter angle from *X-Y* plane [or Toggle (angle in)] <current>:

You can specify the amount of rotation you want by positioning the graphics cursor in the graphics area. When you move the cursor, you will see the object begin to rotate dynamically, and the AutoCAD status line displays a continuous readout of the new angle. Move the cursor to the desired angle and then press the pick button. Or you could type the desired angle from the keyboard. Either way, you have selected an angle of view above or below the target.

Next, AutoCAD prompts for the desired rotation angle of the camera around the target:

> Specify camera location, or enter angle in *X-Y* plane from *X* axis or [Toggle (angle from)] <current>:

You can move the camera 180 degrees clockwise and 180 degrees counterclockwise around the target. You specify the angle by using the cursor to specify a point on the screen. Or you could type in the desired angle from the keyboard. Either way, you have selected an angle of view around the target clockwise or counterclockwise.

The **Toggle angle** option allows you to move between two angle input modes.

AutoCAD takes you back to the 11-option prompt of the DVIEW command. When the new angle of view is correct, you exit the command sequence by giving a null response (press ENTER or the SPACEBAR). This takes you back to the "Command:" prompt.

When you exit DVIEW, your entire drawing will rotate to the same angle of view as the few objects that you selected.

The following command sequence shows an example of using the Camera option of the DVIEW command.

> Command: **dview**
> Select objects or <use DIVEWBLOCK>: *(select the objects)*
> Enter option [CAmera/TArget/Distance/POints/PAn/Zoom/TWist/CLip/ Hide/Off/Undo]: **ca**
> Specify camera location, or enter angle from *X-Y* plane or [Toggle (angle in)] <current>: **45**
> Specify camera location, or enter angle in the *X-Y* plane from *X* axis or [Toggle (angle from)] <current>: **45**
> Enter option [CAmera/TArget/Distance/POints/PAn/Zoom/TWist/CLip/ Hide/Off/Undo]: (ENTER)

982

TArget Option

The TArget option is similar to the CAmera option, but in this case the target is rotated around the camera. The camera remains stationary except for maintaining its lens on the target point. The prompts are similar to those for the CAmera option. There may seem to be no difference between the CAmera and TArget options, but there is a difference in the actual angle of view. For instance, if you elevate the camera 75 degrees above the target, you are then looking at the target from the top down. On the other hand, if you raise the target 75 degrees above the camera, you are then looking at the target from the bottom up. The angles of view are reversed. The real difference comes when you are typing in the angles rather than visually picking them.

The following command sequence shows an example of using the TArget option of the DVIEW command.

```
Command: dview
Select objects or <use DVIEWBLOCK>: (select the objects)
Enter option [CAmera/TArget/Distance/POints/PAn/Zoom/TWist/CLip/
    Hide/Off/Undo]: ta
Specify camera location, or enter angle from X-Y plane or [Toggle (angle
    in)] <current>: 75
Specify camera location, or enter angle in the X-Y plane from X axis or
    [Toggle (angle from)] <current>: 75
Enter option [CAmera/TArget/Distance/POints/PAn/Zoom/TWist/CLip/
    Hide/Off/Undo]: (ENTER)
```

Distance Option

The Distance option creates a perspective projection from the current view. The only information required for this option is the distance from the camera to the target point. Once AutoCAD knows the distance, it will apply the correct perspective. When perspective viewing is on, a box icon appears on the screen, as shown in Figure 15–31, in place of the UCS icon. Some commands (like ZOOM and PAN) will not work while perspective is on. You turn on the perspective just for visual purposes or for plotting.

Figure 15–31 *The perspective box icon appears when perspective mode is on*

When you select the Distance option, AutoCAD prompts:

Specify a new camera—target distance <current>:

In addition to the prompt, you also see a horizontal bar at the top of the screen. The bar goes from 0x to 16x. These are factor distances times your current distance from the object. Moving the slider cursor toward the right increases the distance between the target and the camera. Moving the slider cursor toward the left reduces the distance between the target and the camera. The current distance is represented by 1x. For instance, moving the slider cursor to 3x makes the new distance three times the previous distance. Or you could also type the desired distance in the current linear units from the keyboard.

The following command sequence shows an example of using the Distance option of the DVIEW command.

Command: **dview**
Select objects or <use DVIEWBLOCK>: *(select the objects)*
Enter option [CAmera/TArget/Distance/POints/PAn/Zoom/TWist/CLip/
    Hide/Off/Undo]: **d**
Specify a new camera—target distance <current>: **75**
Enter option [CAmera/TArget/Distance/POints/PAn/Zoom/TWist/CLip/
    Hide/Off/Undo]: (ENTER)

Off Option

The Off option turns off the perspective view. The following command sequence shows an example of using the Off option of the DVIEW command.

Command: **dview**
Select objects or <use DVIEWBLOCK>: *(select the objects)*
Enter option [CAmera/TArget/Distance/POints/PAn/Zoom/TWist/CLip/
    Hide/Off/Undo]: **o**
Enter option [CAmera/TArget/Distance/POints/PAn/Zoom/TWist/CLip/
    Hide/Off/Undo]: (ENTER)

 Note: To turn the perspective on again, select the Distance option and press ENTER for all the defaults. There is no option called On for turning on the perspective view.

POints Option

The Points option establishes the location of the camera as well as the target points. This gives AutoCAD the basic information needed to create the view. The location of the camera and target points must be specified in a parallel projection. If perspective is set to ON, AutoCAD temporarily turns it off while you specify the new location for camera and target points, and then redisplays the image in perspective.

When you select the POints option, AutoCAD prompts:

Specify target point: *(specify the target location)*
Specify camera point: *(specify the camera location)*

After the locations are defined, the screen shows the new view immediately.

The following command sequence shows an example of using the Points option of the DVIEW command.

> Command: **dview**
> Select objects or <use DVIEWBLOCK>: *(select the objects)*
> Enter option [CAmera/TArget/Distance/POints/PAn/Zoom/TWist/CLip/
>   Hide/Off/Undo]: **po**
> Specify target point: *(specify a point)*
> Specify camera point: *(specify a point)*
> Enter option [CAmera/TArget/Distance/POints/PAn/Zoom/TWist/CLip/
>   Hide/Off/Undo]: (ENTER)

PAn Option

The Pan option allows you to view a different location of the model by specifying the pan distance and direction. This option is similar to the regular PAN command. The following command sequence shows an example of using the Pan option of the DVIEW command.

> Command: **dview**
> Select objects or <use DVIEWBLOCK>: *(select the objects)*
> Enter option [CAmera/TArget/Distance/POints/PAn/Zoom/TWist/CLip/
>   Hide/Off/Undo]: **pa**
> Specify displacement base point: *(specify a point)*
> Specify second point: *(specify a point)*
> Enter option [CAmera/TArget/Distance/POints/PAn/Zoom/TWist/CLip/
>   Hide/Off/Undo]: (ENTER)

Zoom Option

The Zoom option lets you zoom in on a portion of the model. This option is similar to the regular AutoCAD ZOOM Center command, with the center point lying at the center of the current viewport. This option is controlled by a scale factor value.

When you select the Zoom option, AutoCAD prompts:

> Specify zoom scale factor <current>: *(specify the zoom scale factor)*

In addition to the prompt, you see a horizontal bar at the top of the screen. The slider bar lets you specify a zoom scale factor, with 1x being the current zoom level. Any value greater than 1 increases the size of the objects in the view; any decimal value less than 1 decreases the size.

 Note: When the perspective is set to ON by Distance option, the Zoom option prompts for a lens size rather than a zoom factor, but the effect is similar. The larger the lens size, the closer the object.

TWist Option

The TWist option rotates or twists the view. It allows you to rotate the image around the line of sight at a given angle from zero, with zero being to the right. The angle is measured counterclockwise.

The following command sequence shows an example of using the TWist option of the DVIEW command.

```
Command: dview
Select objects or <use DVIEWBLOCK>: (select the objects)
Enter option [CAmera/TArget/Distance/POints/PAn/Zoom/TWist/CLip/
    Hide/Off/Undo]: tw
Specify view twist angle <current>: (select a point)
Enter option [CAmera/TArget/Distance/POints/PAn/Zoom/TWist/CLip/
    Hide/Off/Undo]: (ENTER)
```

CLip Option

The CLip option hides portions of the object in view so that the interior of the object can be seen or parts of the complex object can be more clearly identified.

The CLip option has three suboptions: Back, Front, and Off.

The **Back** suboption eliminates all parts of the object in view that are located beyond the designated point along the line of sight.

The **Front** suboption eliminates all parts of the object in view that are located between the camera and the front clipping plane.

The **Off** suboption turns off front and back clipping.

The following command sequence shows an example of using the CLip option of the DVIEW command.

```
Command: dview
Select objects or <use DVIEWBLOCK>: (select the objects)
Enter option [CAmera/TArget/Distance/POints/PAn/Zoom/TWist/CLip/
    Hide/Off/Undo]: cl
Enter clipping option [Back/Front/Off] <current>: b
Specify distance from target or [On/Off] <current>: (specify the distance, or
    turn on and off the previously defined clipping plane)
Enter option [CAmera/TArget/Distance/POints/PAn/Zoom/TWist/CLip/
    Hide/Off/Undo]: (ENTER)
```

Hide Option

The Hide option is similar to the regular AutoCAD HIDE command.

Undo Option

The Undo option will undo the last DVIEW operation. You use it to step back through multiple DVIEW operations.

To exit the Undo option in the DVIEW command, press ENTER without selecting any of the available options and AutoCAD returns you to the "Command:" prompt. It is the default option of the DVIEW command.

USING 3DORBIT

The 3DORBIT command allows you to view the model interactively in the current viewport. By using the pointing device you can manipulate the view of the model. You can view the entire model or any object in your model from different points around it.

Invoke the 3DORBIT command:

3D orbit toolbar	Choose the 3D Orbit command (see Figure 15–32)
View menu	Choose 3D Orbit
Command: prompt	**3dorbit** (ENTER)

Figure 15–32 *Invoking the 3DORBIT command from the 3D orbit toolbar*

AutoCAD prompts:

Command: **3dorbit**
Press ESC or ENTER to exit, or right-click to display shortcut-menu.

AutoCAD displays an arcball—as shown in Figure 15–33—a circle divided into four quadrants by smaller circles. While 3DORBIT is active, the point (target) that you are viewing stays stationary and the camera moves around the target. By default, the center of the arcball is the target point.

Figure 15–33 *Display of the arcball with a 3D model*

Click and drag the cursor to rotate the view. When you move your cursor over different parts of the arcball, the cursor icon changes.

When you move the cursor inside the arcball a small sphere encircled by two lines is the cursor icon displayed, as shown in Figure 15–34. By clicking and dragging, you can manipulate the view freely. You can drag horizontally, vertically, and diagonally.

When you move the cursor outside the arcball a circular arrow around a small sphere is the cursor icon displayed, as shown in Figure 15–34. Clicking outside the arcball and dragging the cursor around the arcball moves the view around an axis that extends through the center of the arcball, perpendicular to the screen.

When you move the cursor over one of the smaller circles on the left or right of the arcball a horizontal ellipse around a small sphere is the cursor icon displayed, as shown in Figure 15–34. Clicking and dragging from either of these points rotates the view around the vertical or *Y* axis that extends through the center of the arcball. The *Y* axis is represented on the cursor by a vertical line.

When you move the cursor over one of the smaller circles on the top or bottom of the arcball a vertical ellipse around a small sphere is the cursor icon displayed, as shown in Figure 15–34. Clicking and dragging from either of these points rotates the view around the horizontal or *X* axis that extends through the center of the arcball. The *X* axis is represented on the cursor by a horizontal line.

Figure 15–34 *Various cursor icons displayed when the* 3DORBIT *command is invoked*

Panning and Zooming in the *3D* Orbit View

You can pan and zoom while you are viewing the model with 3DORBIT command. It works similar to the REALTIME PAN and REALTIME ZOOM commands (see Chapter 3 for a detailed explanation). Select pan and zoom options from the shortcut menu while 3DORBIT is active, from the *3D* Orbit toolbar, or enter **3dpan** and **3dzoom** at the Command: prompt to invoke the pan and zoom options of the 3DORBIT command, respectively.

Using Projection Options in the *3D* Orbit View

AutoCAD allows you to display a perspective or a parallel projection of the view while 3DORBIT is active. The **perspective** view changes the view so that lines parallel to each other converge at a point. Objects appear to recede into the distance while parts of the objects appear larger and closer to you. The shapes are somewhat distorted when the object is very close. This view correlates most closely to what your eye sees.

The **parallel** view changes the view so that two parallel lines never converge at a single point. The shapes in the drawing always remain the same and do not appear distorted when they are closer. You can select perspective and parallel options from the shortcut menu while 3DORBIT is active.

Shading Objects in the *3D* Orbit View

AutoCAD allows you to view shaded view while 3DORBIT is active. You can change the way objects are shaded using the different shading modes. The available modes are: Wireframe, hidden, Flat Shaded, Gourand Shaded, Flat Shaded – Edges On, Gourand Shaded – Edges On. You can select one of the available shading options from the shortcut menu while 3DORBIT is active. See Chapter 16 for a detailed explanation of various shading options.

 Note: Shading is applied even after exiting the 3DORBIT command. Use SHADEMODE to change the shading mode when 3DORBIT is not active.

Adjusting Clipping Planes in the *3D* Orbit View

AutoCAD allows you to set clipping planes for the objects in *3D* orbit view. Objects or parts of the objects that are beyond a clipping plane cannot be seen in the view. Open the Adjust Clipping Planes window by selecting Adjust Clipping Planes from the shortcut menu while 3DORBIT is active, from the *3D* Orbit toolbar, or type **3dclip** at the Command: prompt. In the Adjust Clipping Planes window, there are two clipping planes, front and back. The front and back clipping planes are represented as lines at the top and bottom of the Adjust Clipping Planes window. You can adjust the lines with your pointing device and you can use toolbar buttons or the options from the shortcut menu to choose the clipping plane that you want to adjust.

If clipping planes are on when you exit the *3D* orbit view, they remain on in the *2D* and *3D* view. The only way to turn off the clipping planes is to remove the check mark from the Front Clipping On or Back Clipping On in the shortcut menu while 3DORBIT is active.

Using Continuous Orbit in the *3D* Orbit View

AutoCAD allows you to display the model in a continuous motion. To start the continuous motion, first select the Continuous orbit option for the shortcut menu when 3DORBIT is active. Then click and drag in the direction that you want the continuous orbit to move. Then release the pick button. The orbit continues to move in the direction that you indicated with your pointing device. The speed of the orbit rotation is determined by the speed with which you move the pointing device. If necessary, you can change the direction by clicking and dragging in a new direction. To stop continuous orbit, choose Pan, Zoom, Orbit, or Adjust Clipping Planes from the shortcut menu.

Resetting to Preset Views in the *3D* Orbit View

AutoCAD allows you to reset the view to the view that was current when your first entered the *3D* orbit view or set to one of the preset views, such as top, front, side, isometric, and so on. You can choose the reset view or one of the preset views from the shortcut view when 3DORBIT is active.

To exit 3DORBIT command, press ESC or ENTER.

WORKING WITH MULTIPLE VIEWPORTS IN 3D

As mentioned in Chapter 3, AutoCAD allows you to set multiple viewports to provide different views of your model. For example, you might set up viewports that display top, front, right side, and isometric views. You can do so with the help of the VPOINT, DIVEW, or 3DORBIT commands. To facilitate editing objects in different views, you can define a different UCS for each view. Each time you make a viewport current, you can begin drawing using the same UCS you used the last time that viewport was current.

> The UCSVP system variable controls the setting for saving the UCS in the current viewport.

> When UCSVP is set to 1 (default setting) in a viewport, the UCS last used in that viewport is saved with the viewport and is restored when the viewport is made current again.

> When UCSVP is set to 0 in a viewport, its UCS is always the same as the UCS in the current viewport.

For example, you might set up four viewports: top view, front view, right side view, and isometric view. If you set the UCSVP system variable to 0 in the isometric viewport and to 1 in the top view, front view, and right side view, when you make the front viewport current, the isometric viewport's UCS reflects the UCS front viewport. Likewise, making the top viewport current switches the isometric viewport's UCS to match that of the top viewport.

CREATING *3D* OBJECTS

As mentioned earlier, there are several advantages to drawing objects in *3D*, including viewing the model at any angle, automatic generation of standard and auxiliary *2D* views, rendering and hidden-line removal, interference checking, and engineering analysis.

AutoCAD supports three types of *3D* modeling: wireframe, surface, and solid.

The wireframe model consists of only points, lines, and curves that describe the edges of the object. In AutoCAD you can create a wireframe model by positioning *2D* (planar) objects anywhere in *3D* space. In addition, AutoCAD provides additional commands, such as 3DPOLY, for creating a wireframe model.

The surface model is more sophisticated than the wireframe model. It defines not only the edges of a *3D* object but also its surfaces. The AutoCAD surface modeler defines faceted surfaces by using a polygonal mesh. It is possible to create a mesh to a flat or curved surface by locating the boundaries or edges of the surface.

Solid modeling is the easiest type of *3D* modeling. Solids are the unambiguous and informationally complete representation of the shape of a physical object. Fundamentally, solid modeling differs from wireframe or surface modeling in two ways:

- The information is more complete in the solid model.

- The method of construction of the model itself is inherently straightforward.

In wireframe or surface modeling, objects are created by positioning lines or surfaces in *3D* space. In solid modeling, you build the model as you would with building blocks; from beginning to end, you think, draw, and communicate in *3D*. One of the main benefits of solid modeling is its ability to be analyzed. You can calculate the mass properties of a solid object, such as its mass, center of gravity, surface area, and moments of inertia.

Each modeling type uses a different method for constructing *3D* models, and the use of each editing method varies among model types. It is recommended that you not mix modeling methods. It is possible in AutoCAD to convert between model types from solids to surfaces and from surfaces to wireframe; however, you cannot convert from wireframe to surfaces or surfaces to solids.

2D DRAW COMMANDS IN 3D SPACE

You can use most of the draw commands discussed in previous chapters with a *Z* coordinate value. But *2D* objects such as polylines, circles, arcs, and solids are constrained to the *XY* plane of the current UCS. For these objects, the *Z* value is accepted only for the first coordinate to set the elevation of the *2D* object above or below the current plane. When you specify a point by using an Object Snap mode, it assumes the *Z* value of the point to which you snapped.

SETTING ELEVATION AND THICKNESS

You can create new objects by first setting up a default elevation (*Z* value). Subsequently, all the objects drawn assume the current elevation as the *Z* value whenever a *3D* point is expected but you supply only the *X* and *Y* values. The current elevation is maintained separately in model space and paper space.

Similarly, you create new objects with extrusion thickness by presetting a value for the thickness. Subsequently, all the objects drawn, such as lines, polylines, arcs, circles, and solids, assume the current thickness and extrude in their *Z* direction. For example, you can draw a cylinder by drawing a circle with preset thickness, or you can draw a cube simply by drawing a square with preset thickness.

Note: Thickness can be positive or negative. Thickness is in the Z axis direction for 2D objects. For 3D objects that can accept thickness, it is always relative to the current UCS. They will appear oblique if they do not lie in or parallel the current UCS. If thickness is added to a line drawn directly in the Z direction, the line appears to extend beyond its endpoint in the positive or negative thickness direction. Text and dimensions ignore the thickness setting.

To set the default elevation and thickness, invoke the ELEV command:

Command: prompt	**elev** (ENTER)

AutoCAD prompts:

Command: **elev**
Specify new default elevation <current>: *(specify the elevation, or press*
     ENTER *to accept the current elevation setting)*
Specify new default thickness <current>: *(specify the thickness, or press*
     ENTER *to accept the current thickness setting)*

For example, the following are the command sequences to draw a six-sided polygon at zero elevation with a radius of 2.5 units and a height of 4.5 units, and to place a cylinder at the center of the polygon with a radius of 1.0 unit at an elevation of 2.0 units with a height of 7.5 units, as shown in Figure 15–35.

Command: **elev**
New current elevation <0.0000>: (ENTER)
New current thickness <0.0000>: **4.5**

Command: **polygon** *(draw a polygon with a radius of 2.5 units)*
Command: **elev**
New current elevation <0.0000>: **2.0**
New current thickness <0.0000>: **7.5**
Command: **circle** *(draw a circle with a radius of 1.0 unit)*

MODEL (PLAN VIEW)   MODEL (VPOINT 1,-1,1)

Figure 15–35 *Specifying a new current elevation and thickness*

You can change the thickness and elevation (Z coordinate) of the existing objects by invoking the PROPERTIES command.

CREATING A REGION OBJECT

Regions are *2D* areas you create from closed shapes or loops. The REGION command creates a region object from a selection set of objects. Closed polylines, lines, curves, circular arcs, circles, elliptical arcs, ellipses, and splines are valid selections. Once you create a region, you can extrude it with the EXTRUDE command to make a *3D* solid. You can also create a composite region with the UNION, SUBTRACTION, and INTERSECTION commands. If necessary, you can hatch a region with the BHATCH command.

AutoCAD converts closed *2D* and planar *3D* polylines in the selection set to separate regions and then converts polylines, lines, and curves that form closed planar loops. If more than two curves share an endpoint, the resultant region might be arbitrary. Each object retains its layer, linetype, and color. AutoCAD deletes the original objects after converting them to regions and, by default, does not hatch the regions.

Invoke the REGION command:

Draw toolbar	Choose the REGION command (see Figure 15–36)
Draw menu	Choose Region
Command: prompt	**region** (ENTER)

Figure 15–36 *Invoking the REGION command from the Draw toolbar*

AutoCAD prompts:

Command: **region**
Select objects: *(select the objects, and press ENTER to complete object selection)*

DRAWING 3D POLYLINES

The 3DPOLY command draws polylines with independent *X, Y,* and *Z* axis coordinates using the continuous linetype. The 3DPOLY command works similar to the PLINE command, with a few exceptions. Unlike the PLINE command, 3DPOLY draws only straight-line segments without variable width. Editing a *3D* polyline with the PEDIT command is similar to editing a *2D* polyline, except for some options. *3D* Polylines cannot be joined, curve-fit with arc segments, or given a width or tangent.

Invoke the 3DPOLY command:

Draw menu	Choose the 3DPOLY command
Command: prompt	**3dpoly** (ENTER)

AutoCAD prompts:

> Command: **3dpoly**
> Specify start point of polyline: *(specify a point)*
> Specify endpoint of line or [Undo]: *(specify a point or select an option)*
> Specify endpoint of line or [Undo]: *(specify a point or select an option)*
> Specify endpoint of line or [Close/Undo]: *(specify a point or select one of the available options)*

The available options are similar to those for the PLINE command, described earlier in Chapter 4.

CREATING 3D FACES

When you create a *3D* model, it is often necessary to have solid surfaces for hiding and shading. These surfaces are created with the 3DFACE command. The 3DFACE command creates a solid surface, and the command sequence is similar to that for the SOLID command. Unlike the SOLID command, you can give differing *Z* coordinates for the corner points of a face, forming a section of a plane in space. Unlike the SOLID command, a 3DFACE is drawn from corner to corner clockwise or counterclockwise around the object (and it does not draw a "bow tie"). A *3D* face is a plane defined by either three or four points used to represent a surface. It provides a means of controlling which edges of a *3D* face will be visible. You can describe complex, *3D* polygons using multiple *3D* faces, and you can tell AutoCAD which edges you want to be drawn. If you have an object with curved surfaces, then the 3DFACE command is not suitable. One of the mesh commands is more appropriate, as explained later in the chapter.

Invoke the 3DFACE command:

Surfaces toolbar	Choose the 3DFACE command (see Figure 15–37)
Draw menu	Choose Surfaces > 3D Face
Command: prompt	**3dface** (ENTER)

Figure 15–37 *Invoking the* 3DFACE *command from the Surfaces toolbar*

AutoCAD prompts:

> Command: **3dface**
> Specify first point or [Invisible]: *(specify the first point or enter i)*

Specify the first point, and AutoCAD prompts you for the second, third, and fourth points in sequence. Then AutoCAD closes the face from the fourth point to the first point and prompts for the third point. If you give a null response to the prompt for the third point, AutoCAD closes the *3D* face with three sides, terminates the command, and takes you to the "Command:" prompt.

If you want to draw additional faces in one command sequence, the last two points of the first face become the first two points for the second face. And the last two points of the second face become the first two points of the third face, and so on. You have to be very careful in drawing several faces in one command sequence, since AutoCAD does not have an Undo option that works inside the 3DFACE command. A single mistake can cause the entire face to be redrawn. For this reason, it is a good idea to draw *3D* faces one at a time.

For example, the following command sequence demonstrates the placement of *3D* faces, as shown in Figure 15–38.

> Command: **3dface**
> Specify first point or [Invisible]: *(select point A1)*
> Specify second point or [Invisible]: *(select point A2)*
> Specify third point or [Invisible] <exit>: *(select point A3)*
> Specify fourth point or [Invisible]: *(select point A4)*
> Specify third point or [Invisible] <exit>: *(select point A5)*
> Specify fourth point or [Invisible]: *(select point A6)*
> Specify third point or [Invisible] <exit>: *(select point A7)*
> Specify fourth point or [Invisible]: i
> Specify third point or [Invisible] <exit>: *(select point A1)*
> Specify fourth point or [Invisible]: *(select point A2)*
> Specify third point or [Invisible] <exit>: (ENTER)

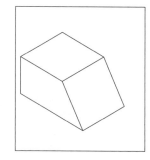

MODEL   MODEL WITH 3D FACES   MODEL-HIDDEN LINES REMOVED

Figure 15–38 *Drawing a 3D object with the* 3DFACE *command*

The surface created, as shown in Figure 15–38, required four faces to cover it. Some of the edges are overlapping, which is not acceptable when viewing the object. The 3DFACE command allows face edges to be "invisible." To create an invisible edge, the letter i must be entered at the prompt for the first point of the edge to be invisible, and then the point can be entered.

The following command sequence shows the placement of *3D* faces for invisible edges, as shown in Figure 15–38.

```
Command: 3dface
Specify first point or [Invisible]: (select point A1)
Specify second point or [Invisible]: (select point A8)
Specify third point or [Invisible] <exit>: (select point A5)
Specify fourth point or [Invisible]: (select point A4)
Specify third point or [Invisible] <exit>: (select point A3)
Specify fourth point or [Invisible]: (select point A6)
Specify third point or [Invisible] <exit>: i (ENTER)
Specify third point or [Invisible] <exit>: (select point A7)
Specify fourth point or [Invisible]: i (ENTER)
Specify fourth point or [Invisible]: (select point A8)
Specify third point or [Invisible] <exit>: (select point A1)
Specify fourth point or [Invisible]: i (ENTER)
Specify fourth point or [Invisible]: (select point A2)
Specify third point or [Invisible] <exit>: (select point A3)
Specify fourth point or [Invisible]: (select point A4)
Specify third point or [Invisible] <exit>: (ENTER)
```

3DFACE commands ignore thickness. The SPLFRAME system variable controls the display of invisible edges in *3D* faces. If SPLFRAME is set to a nonzero value, all invisible edges of *3D* faces are displayed.

Controlling the Visibility of a *3D* Face

The EDGE command allows you to change the visibility of *3D* face edges. You can selectively set the edges to ON/OFF.

Invoke the EDGE command:

Surfaces toolbar	Choose the EDGE command (see Figure 15–39)
Draw menu	Choose Surfaces > Edge
Command: prompt	edge (ENTER)

Figure 15–39 *Invoking the* EDGE *command from the Surfaces toolbar*

AutoCAD prompts:

> Command: **edge**
> Specify edge of 3dface to toggle visibility or [Display]: *(select the edges and press* ENTER *to complete selection, or select the Display option)*

The **Display** option highlights invisible edges of *3D* faces so you can change the visibility of the edges. AutoCAD prompts:

> Enter selection method for display of hidden edges [Select/All] <All>:

The default option displays all the invisible edges. Once the edges are displayed, AutoCAD allows you to change the status of the visibility.

The **Select** option allows you to selectively identify hidden edges to be displayed. Then, if necessary, you can change the status of the visibility.

Note: All edges are visible regardless of the visibility setting if the SPLFRAME system variable is set to 1 (ON). If necessary, you can also use the PROPERTIES command to modify 3D faces.

CREATING MESHES

A *3D* mesh is a single object. It defines a flat surface or approximates a curved one by placing multiple *3D* faces on the surface of an object. It is a series of lines consisting of columns and rows. AutoCAD lets you determine the spacing between rows (M) and columns (N).

It is possible to create a mesh for a flat or curved surface by locating the boundaries or edges of the surface. Surfaces created in this fashion are called geometry-generated surfaces. Their size and shape depend on the boundaries used to define them and on the specific formula (or command) used to determine the location of the vertices between the boundaries. AutoCAD provides four different commands to create geometry-generated surfaces: RULESURF, REVSURF, TABSURF, and EDGESURF. The differences between these types of meshes depend on the types of objects connecting the surfaces. In addition, AutoCAD provides two additional commands for creating polygon mesh: 3DMESH and PFACE. The key to using meshes effectively is to understand the purpose and requirement of each type of mesh and to select the appropriate one for the given condition.

CREATING A FREE-FORM POLYGON MESH

You can define a free-form *3D* polygon mesh by means of the 3DMESH command. Initially, it prompts you for the number of rows and columns, in terms of mesh M and mesh N, respectively. Then it prompts for the location of each vertex in the mesh. The product of M x N gives the number of vertices for the mesh.

Invoke the 3DMESH command:

Surfaces toolbar	Choose the 3DMESH command (see Figure 15–40)
Draw menu	Choose Surfaces > 3D Mesh
Command: prompt	**3dmesh** (ENTER)

Figure 15–40 *Invoking the* 3DMESH *command from the Surfaces toolbar*

AutoCAD prompts:

> Command: **3dmesh**
> Enter size of mesh in M direction: *(specify an integer value between 2 and 256)*
> Enter size of mesh in N direction: *(specify an integer value between 2 and 256)*

The points for each vertex must be entered separately, and the M value can be considered the number of lines that will be connected by faces, while the N value is the number of points each line consists of. Vertices may be specified as *2D* or *3D* points, and may be any distance from each other.

The following command sequence creates a simple 5 x 4 polygon mesh. The mesh is created between the first point of the first line, the first point of the second line, and so on, as shown in Figure 15–41.

> Command: **3dmesh**
> Enter size of mesh in M direction: **5**
> Enter size of mesh in N direction: **4**
> Specify location for vertex (0,0): *(select point A1)*
> Specify location for vertex (0,1): *(select point A2)*
> Specify location for vertex (0,2): *(select point A3)*
> Specify location for vertex (0,3): *(select point A4)*
> Specify location for vertex (1,0): *(select point B1)*
> Specify location for vertex (1,1): *(select point B2)*
> Specify location for vertex (1,2): *(select point B3)*
> Specify location for vertex (1,3): *(select point B4)*
> Specify location for vertex (2,0): *(select point C1)*
> Specify location for vertex (2,1): *(select point C2)*
> Specify location for vertex (2,2): *(select point C3)*
> Specify location for vertex (2,3): *(select point C4)*
> Specify location for vertex (3,0): *(select point D1)*

Specify location for vertex (3,1): *(select point D2)*
Specify location for vertex (3,2): *(select point D3)*
Specify location for vertex (3,3): *(select point D4)*
Specify location for vertex (4,0): *(select point E1)*
Specify location for vertex (4,1): *(select point E2)*
Specify location for vertex (4,2): *(select point E3)*
Specify location for vertex (4,3): *(select point E4)*

LINE DIAGRAM

LINE DIAGRAM WITH 3D FACES   MODEL AT VPOINT 1,-1,1

Figure 15–41 *Creating a 3D mesh*

Note: Specifying 3D mesh of any size can be time-consuming and tedious. It is preferable to use one of the commands for geometry-generated surfaces, such as RULESURF, REVSURF, TABSURF, or EDGESURF. The 3DMESH command is designed primarily for AutoLISP and ADS applications.

CREATING A 3D POLYFACE MESH

The PFACE command allows you to construct a mesh of any topology you desire. This command is similar to the 3DFACE command, but it creates surfaces with invisible interior divisions. You can specify any number of vertices and *3D* faces, unlike the other meshes. Producing this kind of mesh lets you conveniently avoid creating many unrelated *3D* faces with the same vertices.

AutoCAD first prompts you to pick all the vertex points, and then you can create the faces by entering the vertex numbers that define their edges.

Invoke the PFACE command:

Command: prompt	**pface** (ENTER)

AutoCAD prompts:

> Command: **pface**
> Specify location for vertex 1: *(specify a point)*

One by one, specify all the vertices used in the mesh, keeping track of the vertex numbers shown in the prompts. You can specify the vertices as *2D* or *3D* points and place them at any distance from one another. Enter a null response (press ENTER) after specifying all the vertices, and AutoCAD prompts for a vertex number that has to be assigned to each face. You define any number of vertices for each face, and enter a null response (press ENTER). AutoCAD prompts for the next face. After all the vertex numbers for all the faces are defined, enter a null response (press ENTER), and AutoCAD draws the mesh.

The following command sequence creates a simple polyface for a given six-sided polygon, with a circle of 1" radius drawn at the center of the polygon at a depth of −2, as shown in Figure 15–42.

> Command: **pface**
> Specify location for vertex 1: *(select point A1)*
> Specify location for vertex 2 or <define faces>: *(select point A2)*
> Specify location for vertex 3 or <define faces>: *(select point A3)*
> Specify location for vertex 4 or <define faces>: *(select point A4)*
> Specify location for vertex 5 or <define faces>: *(select point A5)*
> Specify location for vertex 6 or <define faces>: *(select point A6)*
> Specify location for vertex 7 or <define faces>: (ENTER)
> Face 1, Vertex 1:
> Enter a vertex number or [Color/Layer]: 1
> Face 1, Vertex 2:
> Enter a vertex number or [Color/Layer] <new face>: 2
> Face 1, Vertex 3:
> Enter a vertex number or [Color/Layer] <new face>: 1
> Face 1, Vertex 4:
> Enter a vertex number or [Color/Layer] <new face>: 2
> Face 1, Vertex 5:
> Enter a vertex number or [Color/Layer] <new face>: 1
> Face 1, Vertex 6:
> Enter a vertex number or [Color/Layer] <new face>: 2
> Face 1, Vertex 7:
> Enter a vertex number or [Color/Layer] <new face>: (ENTER)
> Face 2, Vertex 1:
> Enter a vertex number or [Color/Layer]: (ENTER)

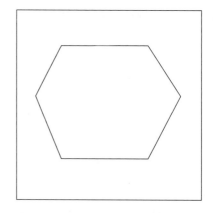

POLYGON WITH PFACE   POLYGON WITH CIRCLE HIDDEN

Figure 15–42 *Creating a polyface for a given six-sided polygon with a circle at the center*

If necessary, you can make an edge of the polyface mesh invisible by entering a negative number for the beginning vertex of the edge. By default, the faces are drawn on the current layer and with the current color. However, you can create the faces in layers and colors different from the original object. You can assign a layer or color by responding to the "Enter a vertex number or [Color/Layer]:" prompt with **L** for layer or **C** for color. AutoCAD then prompts for the name of the layer or color, as appropriate. It will continue with the prompts for vertex numbers. The layer or color you enter is used for the face you are currently defining and for any subsequent faces created.

 Note: Specifying the layer or color within the PFACE command does not change object properties for subsequent commands. Specifying PFACE of any size can be time-consuming and tedious. It is preferable to use one of the commands for geometry-generated surfaces, such as RULESURF, REVSURF, TABSURF, or EDGESURF. The PFACE command is designed primarily for AutoLISP and ADS applications.

CREATING A RULED SURFACE BETWEEN TWO OBJECTS

The RULESURF command creates a polygon mesh between two objects. The two objects can be lines, points, arcs, circles, 2D polylines, or 3D polylines. If one object is open, such as a line or an arc, the other must be open too. If one is closed, such as a circle, so must the other be. A point can be used as one object, regardless of whether the other object is open or closed. But only one of the objects can be a point.

RULESURF creates an M x N mesh, with the value of mesh M a constant 2. The value of mesh N can be changed depending on the required number of faces. This can be done with the help of the SURFTAB1 system variable. By default, SURFTAB1 is set to 6.

The following command sequence shows how to change the value of SURFTAB1 from 6 to 20:

Command: **surftab1**
Enter new value for SURFTAB1 <6>: **20**

Invoke the RULESURF command:

Surfaces toolbar	Choose the RULESURF command (see Figure 15–43)
Draw menu	Choose Surfaces > Ruled Surface
Command: prompt	**rulesurf** (ENTER)

Figure 15–43 *Invoking the* RULESURF *command from the Surfaces toolbar*

AutoCAD prompts:

Command: **rulesurf**
Current wire frame density: SURFTAB1=<current>
Select first defining curve: *(select the first defining curve)*
Select second defining curve: *(select the second defining curve)*

Identify the two objects to which a mesh has to be created. See Figure 15–44, in which an arc (A1–A2) and a line (A3–A4) were selected and a mesh was created with SURFTAB1 set to 15. Two lines (B1–B2 and B3–B4) were selected and a mesh was created with SURFTAB1 set to 20. A cone was created by drawing a circle at an elevation of 0 and a point (C1) at an elevation of 5, followed by the application of the RULESURF command with SURFTAB1 set to 20.

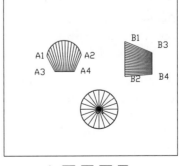

Figure 15–44 *Creating ruled surfaces with the* RULESURF *command*

Note: When you identify the two objects, make sure to select on the same side of the objects, left or right. If you pick the left side of one of the sides and the right side of the other, you will get a bow-tie effect.

CREATING A TABULATED SURFACE

The TABSURF command creates a surface extrusion from an object with a length and direction determined by the direction vector. The object is called the defining curve and can be a line, arc, circle, 2D polyline, or 3D polyline. The direction vector can be a line or open polyline. The endpoint of the direction vector nearest the specified point will be swept along the path curve, describing the surface. Once the mesh is created, the direction vector can be deleted. The number of intervals along the path curve is controlled by the SURFTAB1 system variable, similar to the RULESURF command. By default, SURFTAB1 is set to 6.

Invoke the TABSURF command:

Surfaces toolbar	Choose the TABSURF command (see Figure 15–45)
Draw menu	Choose Surfaces > Tabulated Surface
Command: prompt	**tabsurf** (ENTER)

Figure 15–45 *Invoking the* TABSURF *command from the Surfaces toolbar*

AutoCAD prompts:

> Command: **tabsurf**
> Select object for path curve: *(select the path curve)*
> Select object for direction vector: *(select the direction vector)*

The location at which the direction vector is selected determines the direction of the constructed mesh. The mesh is created in the direction from the selection point to the nearest endpoint of the direction vector. In Figure 15–46, a mesh was created with SURFTAB1 set to 16 by identifying a polyline as the path curve and the line as the direction vector.

Note: The length of the 3D mesh is the same as that of the direction vector.

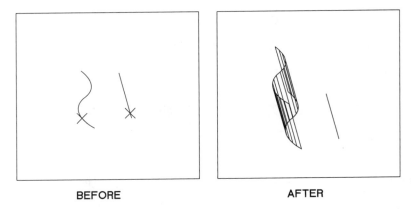

BEFORE   AFTER

Figure 15–46 *Creating a tabulated surface with the* TABSURF *command*

CREATING A REVOLVED SURFACE

The REVSURF command creates a *3D* mesh that follows the path defined by a path curve and is rotated around a centerline. The object used to define the path curve may be an arc, circle, line, *2D* polyline, or *3D* polyline. Complex shapes consisting of lines, arcs, or polylines can be joined into one object using the PEDIT command, and then you can create a single rotated mesh instead of several individual meshes.

The centerline can be a line or polyline that defines the axis around which the faces are constructed. The centerline can be of any length and at any orientation. If necessary, you can erase the centerline after the construction of the mesh. Thus it is recommended that you make the axis longer than the path curve so that it is easy to erase after the rotation.

In the case of REVSURF, both the mesh M size as well as mesh N are controlled by the SURFTAB1 and SURFTAB2 system variables, respectively. The SURFTAB1 value determines how many faces are placed around the rotation axis and can be an integer value between 3 and 1024. The SURFTAB2 determines how many faces are used to simulate the curves created by arcs or circles in the path curve. By default, SURFTAB1 and SURFTAB2 are set to 6.

The following command sequence shows how to change the value of SURFTAB1 from 6 to 20 and that of SURFTAB2 from 6 to 15:

> Command: **surftab1**
> Enter new value for SURFTAB1 <6>: **20**
>
> Command: **surftab2**
> Enter new value for SURFTAB1 <6>: **15**

Invoke the REVSURF command:

Surfaces toolbar	Choose the REVSURF command (see Figure 15–47)
Draw menu	Choose Surfaces > Revolved Surface
Command: prompt	**revsurf** (ENTER)

Figure 15–47 *Invoking the* REVSURF *command from the Surfaces toolbar*

AutoCAD prompts:

> Command: **revsurf**
> Current wire frame density: SURFTAB1=<current>
>   SURFTAB2=<current>
> Select object to revolve: *(select the path curve)*
> Select object that defines the axis of revolution: *(select the axis of revolution)*
> Specify Start angle<0>: *(specify the start angle, or press* ENTER *to accept the default angle)*
> Specify Included angle (+=ccw,-=cw)<360>: *(specify the included angle, or press* ENTER *to accept the default)*

For the "Specify Start Angle:" prompt, it does not matter if you are going to rotate the curve 360 degrees (full circle). If you want to rotate the curve only a certain angle, you must provide the start angle in reference to three o'clock (the default) and then indicate the angle of rotation in the counterclockwise (positive) or clockwise (negative) direction. See Figure 15–48, in which a mesh was created with SURFTAB1 set to 16 and SURFTAB2 set to 12, by identifying a closed polyline as the path curve and the vertical line as the axis of revolution, and then rotated 360 degrees.

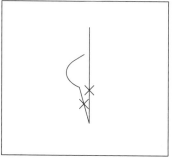

BEFORE   AFTER

Figure 15–48 *Creating a meshed surface with the* REVSURF *command*

CREATING AN EDGE SURFACE WITH FOUR ADJOINING SIDES

The EDGESURF command allows a mesh to be created with four adjoining sides defining its boundaries. The only requirement for EDGESURF is that the mesh has exactly four sides. The sides can be lines, arcs, or any combination of polylines and polyarcs. Each side must join the adjacent one to create a closed boundary.

In EDGESURF, both the mesh M size and mesh N can be controlled by the SURFTAB1 and SURFTAB2 system variables, respectively, just as in REVSURF.

Invoke the EDGESURF command:

Surfaces toolbar	Choose the EDGESURF command (see Figure 15–49)
Draw menu	Choose Surfaces > Edge Surface
Command: prompt	**edgesurf** (ENTER)

Figure 15–49 *Invoking the* EDGESURF *command from the Surfaces toolbar*

AutoCAD prompts:

```
Command: edgesurf
Current wire frame density: SURFTAB1=<current>
    SURFTAB2=<current>
Select edge 1 for surface edge: (select the first edge)
Select edge 2 for surface edge: (select the second edge)
Select edge 3 for surface edge: (select the third edge)
Select edge 4 for surface edge: (select the fourth edge)
```

When picking four sides, you must be consistent in picking the beginning of each polyline group. If you pick the beginning of one side and the end of another, the final mesh will cross and look strange. See Figure 15–50, in which a mesh was created with SURFTAB1 set to 25 and SURFTAB2 set to 20 by identifying four sides.

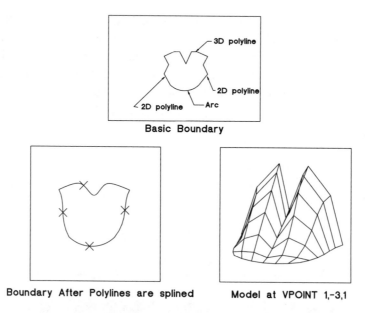

Basic Boundary

Boundary After Polylines are splined   Model at VPOINT 1,-3,1

Figure 15–50 *Creating a meshed surface with the* EDGESURF *command*

EDITING POLYMESH SURFACES

As with blocks, polylines, hatch, and dimensioning, you can explode a mesh. When you explode a mesh it separates into individual *3D* faces. Meshes can also be altered by invoking the PEDIT command, similar to editing polylines using the PEDIT command. Most of the options under the PEDIT command can be applied to meshes, except giving width to the edges of the polymesh. For a detailed explanation of the PEDIT command, refer to Chapter 5.

EDITING IN *3D*

This section describes how to perform various *3D* editing operations, such as aligning, rotating, mirroring, arraying, extending, and trimming.

ALIGNING OBJECTS

The ALIGN command allows you to translate and rotate objects in *3D* space regardless of the position of the current UCS. Three source points and three destination points define the move. The ALIGN command lets you select the objects to move, and then subsequently prompts for three source points and three destination points.

Invoke the ALIGN command:

Modify menu	Choose 3D Operation > Align
Command: prompt	**align** (ENTER)

AutoCAD prompts:

> Command: **align**
> Select objects: *(select the objects)*

Select the objects to move and press ENTER; AutoCAD then prompts for three source points and three destination points. Temporary lines are displayed between the matching pairs of source and destination points. If you enter all six points, the move consists of a translation and two rotations based on the six points. The translation moves the 1st source point to the 1st destination point. The first rotation aligns the line defined by the 1st and 2nd source points with the line defined by the 1st and 2nd destination points. The second rotation aligns the plane defined by the three source points with the plane defined by the three destination points.

If instead of entering three pairs of points you enter two pairs of points, the transformation reduces to a translation from the 1st source point to the 1st destination point and a rotation such that the line passing through the two source points aligns with the line passing through the two destination points. The transformation occurs in either *2D* or *3D*, depending on your response to the following prompt:

> **<2d> or 3d transformation:**

If you enter 2d or press ENTER, the rotation is performed in the *XY* plane of the current UCS. If you enter 3d, the rotation is in the plane defined by the two destination points and the 2nd source point.

If you enter only one pair of points, the transformation reduces to a simple translation from the source to the destination point. This is similar to using the AutoCAD regular MOVE command without the dynamic dragging.

ROTATING OBJECTS ABOUT A *3D* OBJECT

The ROTATE3D command lets you rotate an object about an arbitrary *3D* axis.

Invoke the ROTATE3D command:

Modify menu	Choose 3D Operation > Rotate 3D
Command: prompt	**rotate3d** (ENTER)

AutoCAD prompts:

> Command: **rotate3d**
> Select objects: *(select objects)*

AutoCAD lists the options for selecting the axis of rotation:

> Specify first point on axis or define axis by [Object/Last/View/Xaxis/Yaxis/ Zaxis/ 2points]: *(specify a point or select one of the available options)*

2points Option

The 2points option prompts you for two points. The axis of rotation is the line that passes through the two points, and the positive direction is from the first to the second point.

Axis by Entity Option

The Axis by Entity option lets you select an object and then derives the axis of rotation based on the type of object selected. Valid objects include: line, circle, arc, and pline.

Last Option

The Last option specifies the last-used axis. If there is no last axis, a message to that effect is displayed and the axis selection prompt is redisplayed.

View Option

The View option prompts you to select a point. The axis of rotation is perpendicular to the view direction and passes through the selected point. The positive axis direction is toward the viewer.

X/Y/Zaxis Option

The X/Y/Z axis option prompts you to select a point. The axis of rotation is parallel to the standard axis of the current UCS and passes through the selected point.

Once you have selected the axis of rotation, AutoCAD prompts:

> Specify by Rotation angle or [Reference]: *(specify the rotation angle, or enter r, for reference)*

AutoCAD rotates the selected object(s) to the specified rotation angle. The **Reference** option allows you to specify the current orientation as the reference angle or to show AutoCAD the angle by pointing to the two endpoints of a line to be rotated and then specifying the desired new rotation. AutoCAD automatically calculates the rotation angle and rotates the selected object appropriately.

See Figure 15–51 for an example of rotating a cylinder around the Z axis.

BEFORE   AFTER

Figure 15–51 *Rotating a cylinder about the Z axis with the* ROTATE3D *command*

MIRRORING ABOUT A PLANE

The MIRROR3D command lets you mirror a selected object about a plane.

Invoke the MIRROR3D command:

Modify menu	Choose *3D Operation > Mirror 3D*
Command: prompt	**mirror3D** (ENTER)

AutoCAD prompts:

Command: **mirror3d**
Select objects: *(select the objects)*

Select the objects to mirror and press ENTER. AutoCAD then lists the options for selecting the mirroring plane:

Specify first point of mirror plane [3 points] or [Object/Last/Zaxis/View/
XY/YZ/XZ/3points] <3points>: *(select one of the options to specify the
mirroring plane)*
Delete source objects? <No>: *(enter **y** to delete the objects or **n** not to delete
the objects)*

3points Option

The 3points option prompts you for three points. The mirroring plane is the plane that passes through the three selected points.

Plane by Object Option

The Plane by Object option lets you select an object, and the mirroring plane is aligned with the plane of the object selected. Valid objects include: circle, arc, and pline.

Last Option

The Last option specifies the last used plane. If there is no last plane, a message is displayed to that effect and the plane selection prompt is redisplayed.

Zaxis Option

The Zaxis option prompts you to select two points. The mirroring plane is the plane specified by a point on the plane and a point on the plane's normal (perpendicular to the plane).

View Option

The View option prompts you to select a point. The mirroring plane is created perpendicular to the view direction and passes through the selected point.

XY/YZ/XZ Option

The XY/YZ/XZ option prompts you to select a point. The mirroring plane is created parallel to the standard plane of the current UCS and passes through the selected point.

See Figure 15–52 for an example of mirroring a cylinder aligned with the plane of the object selected (pline).

BEFORE   AFTER

Figure 15–52 *Mirroring a cylinder about a polyline object*

CREATING A 3D ARRAY

The 3DARRAY command is used to make multiple copies of selected objects in either rectangular or polar array in *3D*. In the rectangular array, specify the number of columns (*X* direction), the number of rows (*Y* direction), the number of levels (*Z* direction), and the spacing between columns, rows, and levels. In the polar array, specify the number of items to array, the angle that the arrayed objects are to fill, the start point and endpoint of the axis about which the objects are to be rotated, and whether or not the objects are rotated about the center of the group.

Invoke the 3DARRAY command:

Modify menu	Choose *3D Operation > Array 3D*
Command: prompt	**3darray** (ENTER)

AutoCAD prompts:

> Command: **3Darray**
> Select objects: *(select the objects to array)*
> Enter type of array [Rectangular or Polar array] (R): *(select one of the two available options)*

Rectangular Array

To generate a rectangular array, enter **R** (for rectangular array) and AutoCAD prompts:

> Enter the number of rows(—)<1>: *(specify the number of rows or press ENTER)*
> Enter the number of columns(||||)<1>: *(specify the number of columns or press ENTER)*

Number of levels(...)<1>: *(specify the number of levels or press ENTER)*
Specify the distance between rows(—)<1>: *(specify a distance)*
Specify the distance between columns(—)<1>: *(specify a distance)*
Specify the distance between levels(—)<1>: *(specify a distance)*

Any combination of whole numbers of columns, rows, and levels may be entered. AutoCAD includes the original object in the number you enter. An array must have at least two columns, two rows, or two levels. Specifying one row requires that more than one column be specified, or vice versa. Specifying one level creates a *2D* array. Column, row, and level spacing can be different from one another. They can be entered separately when prompted, or you can select two points and let AutoCAD measure the spacing. Positive values for spacing generate the array along the positive *X, Y*, and *Z* axes. Negative values generate the array along the negative *X, Y*, and *Z* axes.

Polar Array

To generate a polar array, enter **p** (for polar array) and AutoCAD prompts:

Enter the number of items in the array: *(specify the number of items in the array; include the original object)*
Specify the angle to fill (+=ccw, -=cw) <360>: *(specify an angle, or press ENTER for 360 degrees)*
Rotate arrayed objects? [Yes/No] <Y>: *(enter **y** to rotate the objects as they are copied, or enter **n** not to rotate the objects as they are copied)*
Specify center point of array: *(specify a point)*
Specify second point on axis of rotation: *(specify a point for the axis of rotation)*

EXTENDING AND TRIMMING IN 3D

AutoCAD allows you to extend an object by means of the EXTEND command (explained in Chapter 4) to any object in *3D* space or to trim an object to any other *3D* space by means of the TRIM command (explained in Chapter 4), regardless of whether the objects are on the same plane or parallel to the cutting or boundary edges. Before you select an object to extend or trim in *3D* space, specify one of the three available projection modes: None, UCS, or View. The **None** option specifies no projection. AutoCAD extends/trims only objects that intersect with the boundary/cutting edge in *3D* space. The **UCS** option specifies projection onto the *XY* plane of the current UCS. AutoCAD extends/trims objects that do not intersect with the boundary/cutting objects in *3D* space. The **View** option specifies projection along the current view direction. the PROJMODE System variable allows you to set one of the available projection modes. You can also set the projection mode by selecting the Project option available in the EXTEND command.

In addition to specifying the projection mode, you have to specify one of the two available options for the edge. The edge determines whether the object is extended/trimmed to another object's implied edge or only to an object that actually intersects it in *3D* space. The available options are Extend and No Extend. The **Extend** option extends the boundary/cutting object/edge along its natural path to intersect another object or its implied edge in *3D* space. The **No Extend** option specifies that the object is extended/trimmed only to a boundary/cutting object/edge that actually intersects it in *3D* space. The EXTEDGE system variable allows you to set one of the available modes. You can also set the Edge by selecting the **Edge** option available in the EXTEND command.

CREATING SOLID SHAPES

As mentioned earlier, solids are the most informationally complete and least ambiguous of the modeling types. It is easier to edit a complex solid shape than to edit wireframes and meshes.

You create solids from one of the basic solid shapes: box, cone, cylinder, sphere, torus, or wedge. The user-defined solids can be created by extruding or revolving *2D* objects and regions to define a *3D* solid. In addition, you can create more complex solid shapes by combining solids together by performing a Boolean operation—union, subtraction, or intersection.

Solids can be further modified by filleting and chamfering their edges. AutoCAD provides commands for slicing a solid into two pieces or obtaining a *2D* cross-section of a solid.

Like meshes, solids are displayed as a wireframe until you hide, shade, or render them. AutoCAD provides commands to analyze solids for their mass properties (volume, moments of inertia, center of gravity, etc.). AutoCAD allows you to export data about a solid object to applications such as NC (numerical control) milling and EXTEDGE FEM (finite element method) analysis. If necessary, you can use the AutoCAD EXPLODE command to explode solids into mesh and wireframe objects.

 Note: The ISOLINES system variable controls the number of tessellation lines used to visualize curved portions of the wireframe. The default value for ISOLINES is set to 4.

CREATING A SOLID BOX

The BOX command creates a solid box or cube. The base of the box is defined parallel to the current UCS by default. The solid box can be drawn by one of two options: by providing a center point or a starting corner of the box.

Invoke the BOX command:

Solids toolbar	Choose the BOX command (see Figure 15–53)
Draw menu	Choose Solids > Box
Command: prompt	**box** (ENTER)

Figure 15–53 *Invoking the* BOX *command from the Solids toolbar*

AutoCAD prompts:

> Command: **box**
> Specify corner of box or [CEnter]<0,0,0>: *(specify a point, or type* **c** *for the Center option)*

First, by default, you are prompted for the starting corner of the box. Once you provide the starting corner, the box's dimensions can be entered in one of three ways.

The default option lets you create a box by locating the opposite corner of its base rectangle first, and then its height. The following command sequence defines a box, as shown in Figure 15–54, using the default option:

> Command: **box**
> Specify corner of box or [CEnter]<0,0,0>: **3,3**
> Specify corner or [Cube/Length]: **7,7**
> Specify height: **4**

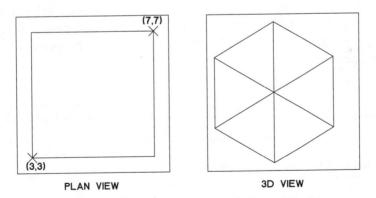

PLAN VIEW   3D VIEW

Figure 15–54 *Creating a solid box using the default option of the* BOX *command*

The **Cube** option allows you to create a box in which all edges are of equal length. The following command sequence defines a box using the Cube option:

> Command: **box**
> Specify corner of box or [CEnter]<0,0,0>: **3,3**
> Specify corner or [Cube/Length]: **c**
> Specify length: **3**

The **Length** option lets you create a box by defining its length, width, and height. The following command sequence defines a box using the Length option:

```
Command: box
Specify corner of box or [CEnter]<0,0,0>: 3,3
Specify corner or [Cube/Length]: l
Specify length: 3
Specify width: 4
Specify height: 3
```

Center Option

The Center option allows you to create a box by locating its center point. Once you locate the center point, a line rubberbands from this point to help you visualize the size of the rectangle. Then AutoCAD prompts you to define the size of the box by entering one of the following options:

```
Specify corner or [Cube/Length]: (select one of the available options)
```

 Note: Once you create a box you cannot stretch it or change its size. However, you can extrude the faces of a box with the SOLIDEDIT command.

CREATING A SOLID CONE

The CONE command creates a cone, either round or elliptical. By default, the base of the cone is parallel to the current UCS. Solid cones are symmetrical and come to a point along the Z axis. The solid cone can be drawn two ways: by providing a center point for a circular base or by selecting the elliptical option to draw the base of the cone as an elliptical shape.

Invoke the CONE command:

Solids toolbar	Choose the CONE command (see Figure 15–55)
Draw menu	Choose Solids > Cone
Command: prompt	cone (ENTER)

Figure 15–55 *Invoking the CONE command from the Solids toolbar*

AutoCAD prompts:

```
Command: cone
Current wire frame density ISOLINES=4
```

*Specify center point for base of cone or [Elliptical] <0,0,0>: (specify a point, or type **e** for the Elliptical option)*

By default, AutoCAD prompts you for the center point of the base of the cone and assumes the base to be a circle. Subsequently, you are prompted for the radius (or enter **D**, for diameter). Enter the appropriate value and then it prompts for the apex/ height of the cone. The height of the cone is the default option, and it allows you to set the height of the cone, not the orientation. The base of the cone is parallel to the current base plane. The apex option, in contrast, prompts you for a point. In turn, it sets the height and orientation of the cone. For example, the following command sequence lists the steps in drawing a cone, as shown in Figure 15–56, using the default option:

Command: **cone**
Current wire frame density ISOLINES=4

Specify center point for base of cone or [Elliptical] <0,0,0>: **5,5**
Specify radius base of cone or [Diameter]: **3**
Specify height of cone or [Apex]: **4**

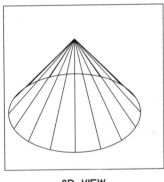

PLAN VIEW   3D VIEW

Figure 15–56 *Creating a solid cone using the default option of the* CONE *command*

Elliptical Option

Selecting the Elliptical option indicates that the base of the cone is an ellipse. The prompts are identical to the regular AutoCAD ELLIPSE command. For example, the following command shows steps in drawing a cone using the Elliptical option:

Command: **cone**
Current wire frame density ISOLINES=4

Specify center point for base of cone or [Elliptical] <0,0,0>: **e**
Specify axis endpoint of ellipse for base of cone or [Center]: **3,3**
Specify second axis endpoint of ellipse for base of cone: **6,6**
Specify length of other axis for base of cone: **5,7**
Specify height of cone or [Apex]: **4**

CREATING A SOLID CYLINDER

The CYLINDER command creates a cylinder of equal diameter on each end and similar to an extruded circle or an ellipse. The solid cylinder can be by means of one of two options: by providing a center point for a circular base, or by selecting the elliptical option to draw the base of the cylinder as an elliptical shape.

Invoke the CYLINDER command:

Solids toolbar	Choose the CYLINDER command (see Figure 15–57)
Draw menu	Choose Solids > Cylinder
Command: prompt	**cylinder** (ENTER)

Figure 15–57 *Invoking the CYLINDER command from the Solids toolbar*

AutoCAD prompts:

Command: **cylinder**
Current wire frame density ISOLINES=4

Specify center point for base of cylinder or [Elliptical] <0,0,0>: *(specify a point, or type* **e** *for the Elliptical option)*

The prompts are identical to those for a cone. For example, the following command sequence lists the steps in drawing a cylinder, as shown in Figure 15–58, using the default option:

Command: **cylinder**
Current wire frame density ISOLINES=4

Specify center point for base of cylinder or [Elliptical] <0,0,0>: **5,5**
Specify radius for base of cylinder or [Diameter]: **3**
Specify height of cylinder or [Center of other end]: **4**

PLAN VIEW   3D VIEW

Figure 15–58 *Creating a solid cylinder using the default option of the CYLINDER command*

CREATING A SOLID SPHERE

The SPHERE command creates a *3D* body in which all surface points are equidistant from the center. The sphere is drawn in such a way that its central axis is coincident with the *Z* axis of the current UCS.

Invoke the SPHERE command:

Solids toolbar	Choose the SPHERE command (see Figure 15–59)
Draw menu	Choose Solids > Sphere
Command: prompt	**sphere** (ENTER)

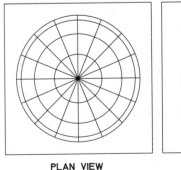

Figure 15–59 *Invoking the* SPHERE *command from the Solids toolbar*

AutoCAD prompts:

Command: **sphere**
Current wire frame density ISOLINES=4
Specify center of sphere <0,0,0>: *(specify a point)*

First, AutoCAD prompts for the center point of the sphere; then you can provide the radius or diameter to define a sphere.

For example, the following command sequence shows steps in drawing a sphere, as in Figure 15–60:

Command: **sphere**
Current wire frame density ISOLINES=4
Specify center of sphere <0,0,0>: **5,5**
Specify radius of sphere or [Diameter]: **3**

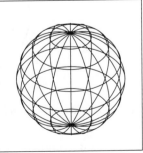

PLAN VIEW   3D VIEW

Figure 15–60 *Creating a solid sphere using the default option of the* SPHERE *command*

CREATING A SOLID TORUS

The TORUS command creates a solid with a donut-like shape. If a torus were a wheel, the center point would be the hub. The torus is created lying parallel to and bisected by the *XY* plane of the current UCS.

Invoke the TORUS command:

Solids toolbar	Choose the TORUS command (see Figure 15–61)
Draw menu	Choose Solids > Torus
Command: prompt	**torus** (ENTER)

Figure 15–61 *Invoking the* TORUS *command from the Solids toolbar*

AutoCAD prompts:

Command: **torus**
Current wire frame density ISOLINES=4

Specify center of torus<0,0,0>: *(specify a point)*

AutoCAD prompts for the center point of the torus and then subsequently for the diameter or radius of the torus and the diameter or radius of the tube, as shown in Figure 15–62. You can also draw a torus without a center hole if the radius of the tube is defined as greater than the radius of the torus. A negative torus radius would create a football-shaped solid.

Figure 15–62 *Creating a solid torus with a center hole using the* TORUS *command*

For example, the following command sequence lists the steps in drawing a torus, as shown in Figure 15–63:

Command: **torus**
Current wire frame density ISOLINES=4

Specify center of torus<0,0,0>: **5,5**
Specify radius of torus or [Diameter]: **3**
Specify radius of tube or [Diameter]: **0.5**

PLAN VIEW   3D VIEW

Figure 15–63 *Creating a torus by specifying the baseplane and central axis direction using the* TORUS *command*

CREATING A SOLID WEDGE

The WEDGE command creates a solid like a box that has been a cut in half diagonally along one face. The face of the wedge is always drawn parallel to the current UCS, with the sloped face tapering along the Z axis. The solid wedge can be drawn by one of two options: by providing a center point of the base or by providing starting corner of the box.

Invoke the WEDGE command:

Solids toolbar	Choose the WEDGE command (see Figure 15–64)
Draw menu	Choose Solids > Wedge
Command: prompt	**wedge** (ENTER)

Figure 15–64 *Invoking the* WEDGE *command from the Solids toolbar*

AutoCAD prompts:

> Command: **wedge**
> Specify first corner of wedge or [CEnter] <0,0,0>: *(specify a point, or type c for the corner of the wedge)*

First, by default you are prompted for the starting corner of the box. Once you provide the starting corner, AutoCAD prompts:

> Specify corner or [Cube/Length]: *(specify corner of the wedge or select one of the available options)*

The wedge dimensions can be specified by using one of the three options. The **specify corner** option lets you create a wedge by locating first the opposite corner of its base rectangle and then its height. The **Cube** option allows you to create a wedge in which all edges are of equal length. The **Length** option lets you create a box by defining its length, width, and height.

Center Option

The Center option allows you to create a wedge by first locating its center point. Once you locate the center point, a line rubber-bands from this point to help you visualize the size of the rectangle. Then AutoCAD prompts you to define the size of the box by entering one of the following options:

> Specify corner or [Cube/Length]: *(specify corner of the wedge or select one of the available options)*

CREATING SOLIDS FROM EXISTING 2D OBJECTS

The EXTRUDE command creates a unique solid by extruding circles, closed polylines, polygons, ellipses, closed splines, donuts, and regions. Because a polyline can have virtually any shape, the EXTRUDE command allows you to create irregular shapes. In addition, AutoCAD allows you to taper the sides of the extrusion.

Note: A polyline must contain at least 3 but not more than 500 vertices and none of the segments can cross each other. See Figure 15–65 for examples of shapes that cannot be extruded. If the polyline has width, AutoCAD ignores the width and extrudes from the center of the polyline path. If a selected object has thickness, AutoCAD ignores the thickness.

EXAMPLES OF SHAPES THAT CANNOT BE EXTRUDED (SHOWN IN PLAN VIEW)

Figure 15–65 *Shapes (shown in plan view) that cannot be extruded using the* EXTRUDE *command*

Invoke the EXTRUDE command:

Solids toolbar	Choose the EXTRUDE command (see Figure 15–66)
Draw menu	Choose Solids > Extrude
Command: prompt	**extrude** (ENTER)

Figure 15–66 *Invoking the EXTRUDE command from the Solids toolbar*

AutoCAD prompts:

> Command: **extrude**
> Current wire frame density ISOLINES=4
>
> Select objects: *(select the objects to extrude and press ENTER to complete the selection)*
>
> Specify Height of Extrusion or [Path]: *(specify height of extrusion or select path option)*

Height of Extrusion Option

The Height of Extrusion option (default) allows you to specify the distance for extrusion. Specifying a positive value extrudes the objects along the positive Z axis of the current UCS, and a negative value extrudes along the negative Z axis.

Path Option

The Path option allows you to select the extrusion path based on a specified curve object. All the profiles of the selected object are extruded along the chosen path to create solids. Lines, circles, arcs, ellipses, elliptical arcs, polylines, or splines can be paths. The path should not lie on the same plane as the profile, nor should it have areas of high curvature. The extruded solid starts from the plane of the profile and ends on a plane perpendicular to the path's endpoint. One of the endpoints of the path should be on the plane of the profile. Otherwise, AutoCAD moves the path to the center of the profile.

Once you specify the Height of Extrusion and path appropriately, AutoCAD prompts:

> Specify angle of taper for extrucsion <0>: *(specify the angle)*

Specify an angle between –90 and +90 degrees, or press ENTER or SPACEBAR to accept the default value of 0 degrees. If you specify 0 degrees as the taper angle, AutoCAD extrudes a *2D* object perpendicular to its *2D* plane, as shown in Figure 15–67. Positive angles taper in from the base object; negative angles taper out.

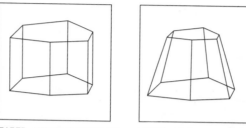

TAPER ANGLE 0 DEGREES   TAPER ANGLE 15 DEGREES

Figure 15–67 *Creating a solid with the* EXTRUDE *command with 0 degrees and with 15 degrees of taper angle*

 Note: It is possible for a large taper angle or a long extrusion height to cause the object, or portions of the object, to taper to a point before reaching the extrusion height.

CREATING SOLIDS BY MEANS OF REVOLUTION

The REVOLVE command creates a unique solid by revolving or sweeping a closed polyline, polygon, circle, ellipse, closed spline, donut, and region. Polylines that have crossing or self-intersecting segments cannot be revolved. The REVOLVE command is similar to the REVSURF command. The REVSURF command creates a surface of revolution, whereas REVOLVE creates a solid of revolution. The REVOLVE command provides several options for defining the axis of revolution.

Invoke the REVOLVE command:

Solids toolbar	Choose the REVOLVE command (see Figure 15–68)
Draw menu	Choose Solids > Revolve
Command: prompt	**revolve** (ENTER)

Revolve

Figure 15–68 *Invoking the* REVOLVE *command from the Solids toolbar*

AutoCAD prompts:

Current wire frame density ISOLINES=4
Command: **revolve**
Select objects: *(select the objects to extrude and press* ENTER *to complete the selection)*
Specify start point for axis of revolution –or define axis by [Object/X (axis)/Y(axis)]: *(specify start point for axis of revolution or select one of the available options)*

Start Point of Axis Option

The Start point of axis option (default) allows you to specify two points for the start point and the endpoint of the axis, and the positive direction of rotation is based on the right-hand rule.

Object Option

The Object option allows you select an existing line or single polyline segment that defines the axis about which to revolve the object. The positive axis direction is from the closest to the farthest endpoint of this line.

X axis Option

The *X* axis option uses the positive *X* axis of the current UCS as the axis of the revolution.

Y axis Option

The *Y* axis option uses the positive *Y* axis of the current UCS as the axis of the revolution.

Once you specify the axis of revolution, AutoCAD prompts:

Specify angle of revolution <full circle>: *(Specify the angle for revolution)*

The default is for a full circle. You can specify any angle between 0 and 360 degrees.

CREATING COMPOSITE SOLIDS

As mentioned earlier in this chapter, you can create a new composite solid or region by combining two or more solids or regions via boolean operations. Although the term Boolean implies that only two objects can be operated upon at once, AutoCAD lets you select many solid objects in a single Boolean command. There are three basic Boolean operations that can be performed in AutoCAD:

Union

Subtraction

Intersection

The UNION, SUBTRACTION, and INTERSECTION commands let you select both the solids and regions in a single use of the commands, but solids are combined with solids, and regions combined only with regions. Also, in the case of regions you can make composite regions only with those that lie in the same plane. This means that a single command creates a maximum of one composite solid, but might create many composite regions.

UNION OPERATION

Union is the process of creating a new composite object from one or more original objects. The union operation joins the original solids or regions in such a way that there is no duplication of volume. Therefore, the total resulting volume can be equal to or less than the sum of the volumes in the original solids or regions. The UNION command performs the union operation.

Invoke the UNION command:

Solids Editing toolbar	Choose the UNION command (see Figure 15–69)
Modify menu	Choose Solids Editing > Union
Command: prompt	**union** (ENTER)

Figure 15–69 *Invoking the* UNION *command from the Solids Editing toolbar*

AutoCAD prompts:

Command: **union**
Select objects: *(select the objects to extrude and press* ENTER *to complete the selection)*

You can select more than two objects at once. The objects (solids or regions) can be overlapping, adjacent, or nonadjacent.

For example, the following command sequence shows steps in creating a composite solid by joining two cylinders, as shown in Figure 15–70.

Command: **union**
Select objects: *(select cylinders A and B and press* ENTER*)*

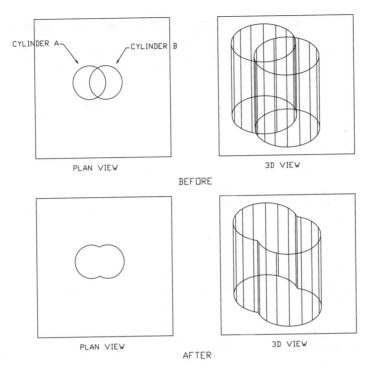

Figure 15–70 *Creating a composite solid by joining two cylinders using the* UNION *command*

SUBTRACTION OPERATION

Subtraction is the process of forming a new composite object by starting with one object and removing from it any volume that it has in common with a second object. In the case of solids, they are created by subtracting the volume of one set of solids from another set. If the entire volume of the second solid is contained in the first solid, then what is left is the first solid minus the volume of the second solid. However, if only part of the volume of the second solid is contained within the first solid, then only the part that is duplicated in the two solids is subtracted. Similarly, in the case of regions, they are created by subtracting the common area of one set of existing regions from another set. The SUBTRACT command performs the subtraction operation.

Invoke the SUBTRACT command:

Solids Editing toolbar	Choose the SUBTRACT command (see Figure 15–71)
Modify menu	Choose Solids Editing > Subtract
Command: prompt	**subtract** (ENTER)

Figure 15–71 *Invoking the* SUBTRACT *command from the Solids Editing toolbar*

AutoCAD prompts:

> Command: **subtract**
> Select solids and regions to subtract from...
> Select objects: *(select the objects from which you will subtract other objects and press* ENTER*)*

You can select one or more objects as source objects. If you select more than one, they are automatically unioned. After selecting the source objects, press ENTER or the SPACEBAR, and AutoCAD prompts you to select the objects to subtract from the source object.

> Select solids and regions to subtract...
> Select objects: *(select the objects to subtract and press* ENTER*)*

If necessary, you can select one or more objects to subtract from the source object. If you select several, they are automatically unioned before they are subtracted from the source object.

 Note: Objects that are neither solids nor regions are ignored.

For example, the following command sequence shows steps in creating a composite solid by subtracting cylinder B from A, as shown in Figure 15–72.

> Command: **subtract**
> Select objects: *(select cylinder A and press* ENTER *or the* SPACEBAR*)*
> Objects to subtract from them...
> Select objects: *(select cylinder B and press* ENTER *or the* SPACEBAR*)*

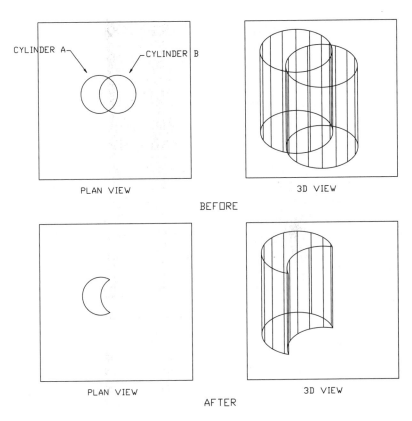

Figure 15–72 *Creating a composite solid by subtracting cylinder B from cylinder A using the* SUBTRACT *command*

INTERSECTION OPERATION

Intersection is the process of forming a composite object from only the volume that is common to two or more original objects. In the case of solids, you can create a new composite solid by calculating the common volume of two or more existing solids. Whereas in the case of regions, it is done by calculating the overlapping area of two or more existing regions. The INTERSECT command performs the intersection operation.

Invoke the INTERSECT command:

Solids Editing toolbar	Choose the INTERSECT command (see Figure 15–73)
Modify menu	Choose Solids Editing > Intersect
Command: prompt	**intersect** (ENTER)

Figure 15–73 *Invoking the* INTERSECT *command from the Solids Editing toolbar*

AutoCAD prompts:

Command: **intersect**
Select objects: *(select the objects for intersection and press* ENTER *to complete the selection)*

For example, the following command sequence shows the steps in creating a composite solid by intersecting cylinder A with cylinder B, as shown in Figure 15–74:

Command: **intersect**
Select objects: *(select cylinders A and B and press* ENTER *or the* SPACEBAR*)*

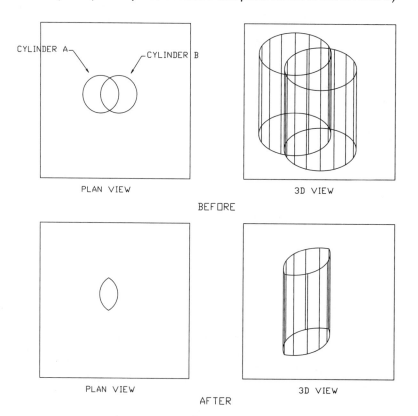

Figure 15–74 *Creating a composite solid by intersecting cylinder A from cylinder B using the* INTERSECT *command*

EDITING *3D* SOLIDS

AutoCAD makes the work of creating solids a little easier by providing editing tools, including chamfering or filleting the edges, creating a cross-section through a solid, creating a new solid by cutting the existing solid and removing a specified side, and creating a composite solid from the interference of two or more solids. In addition, AutoCAD 2000 provides additional editing tools such as extrude faces, move faces, offset faces, delete faces, rotate faces, taper faces, color faces, copy faces, color and copy edges, imprint, clean, separate solids, shell and check. If necessary, you can always use the AutoCAD modify and construct commands, such as MOVE, COPY, ROTATE, SCALE, and ARRAY to edit solids.

CHAMFERING SOLIDS

The CHAMFER command (explained in Chapter 4) can also be used to bevel the edges of an existing solid object.

Invoke the CHAMFER command:

Modify toolbar	Choose the CHAMFER command (see Figure 15–75)
Modify menu	Choose Chamfer
Command: prompt	**chamfer** (ENTER)

Figure 15–75 *Invoke the* CHAMFER *command from the Modify toolbar*

AutoCAD prompts:

> Command: **chamfer**
> Select first line or [Polyline/Distances/Angle/Trim/Method]: *(select an edge on a* 3D *solid)*

If you pick an edge that is common to two surfaces, AutoCAD highlights one of the surfaces and prompts:

> Base surface selection

> Enter surface selection option [Next/OK (current)] <OK>:

If this is the surface you want, press ENTER or SPACEBAR to accept it. If it is not, enter **N** (for next) to highlight the adjoining surface and then press ENTER or the SPACEBAR. AutoCAD prompts:

Specify base surface chamfer distance <default>: *(specify a distance, or press the* ENTER *key or* SPACEBAR *to accept the default)*
Specify other surface chamfer distance <default>: *(specify a distance, or press the* ENTER *key or* SPACEBAR *to accept the default)*

Once you provide the chamfer distances, AutoCAD prompts:

Select an edge or [Loop]: Select the edges of the highlighted surface you want chamfered, and then press ENTER or the SPACEBAR.

The **Loop** option allows you to select one of the edges on the base surface, and AutoCAD automatically selects all edges on the base surface for chamfering.

The following command sequence draws a chamfer for a solid object, as shown in Figure 15–76.

Command: **chamfer**
Select first line or [Polyline/Distances/Angle/Trim/Method]: *(select the edge)*
Enter surface selection option [Next/OK (current)] <OK>: (ENTER)
Specify base surface chamfer distance <default>: **0.25**
Specify other surface chamfer distance <default>: **0.5**
Select an edge or [Loop]: *(select the first edge)*
Loop/<Select edge>: *(select the second edge)*

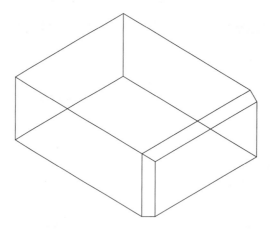

Figure 15–76 *Example of chamfering a solid surface*

FILLETING SOLIDS

The FILLET command (explained in Chapter 4) can also be used to round the edge of an existing solid object.

Invoke the FILLET command:

Modify toolbar	Choose the FILLET command (see Figure 15–77)
Modify menu	Choose Fillet
Command: prompt	**fillet** (ENTER)

Figure 15–77 *Invoking the* FILLET *command from the Modify toolbar*

AutoCAD prompts:

> Command: **fillet**
> Select first object or [Polyline/Radius/Trim]: *(select an edge in a 3D solid)*

If necessary, you can select multiple edges; but you must select the edges individually after specifying the radius for the fillet. AutoCAD prompts:

> Enter fillet radius <default>: *(specify radius for fillet, or press* ENTER *or*
>   SPACEBAR *to accept the default)*
> Select an edge or [Chain/Radius]: *(select addition edges and press* ENTER *to*
>   *complete the selection)*

The following command sequence draws a fillet for a solid object, as shown in Figure 15–78.

> Command: **fillet**
> Select first object or [Polyline/Radius/Trim]: *(select the solid edge)*
> Enter fillet radius <default>: **0.5**
> Select an edge or [Chain/Radius]: (ENTER)

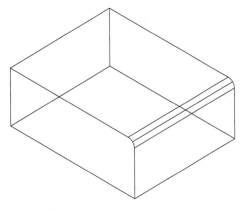

Figure 15–78 *Example of filleting a solid surface*

SECTIONING SOLIDS

The SECTION command creates a cross-section of one or more solids. The cross-section is created as one or more regions. The region is created on the current layer and is inserted at the location of the cross-section. If necessary, you can use the MOVE command to move the cross-section.

Invoke the SECTION command:

Solids toolbar	Choose the SECTION command (see Figure 15–79)
Draw menu	Choose Solids > Section
Command: prompt	**section** (ENTER)

Figure 15–79 *Invoking the SECTION command from the Solids toolbar*

AutoCAD prompts:

> Command: **section**
> Select objects: *(select the objects from which you want the cross-section to be generated)*
> Specify first point on Section plane by [Object/Zaxis/View/XY/YZ/ZX/3points] <3points>: *(specify one of the three points to define a plane or select one of the available options)*

3points Option

The 3points option (default) allows you to define a section plane by locating three points. The first point is the origin, the second point determines the positive direction of the X axis for the section plane, and the third point determines the positive Y axis of the section plane. This option is similar to the 3point option of the AutoCAD UCS command.

Object Option

The Object option aligns the sectioning plane with a circle, ellipse, circular or elliptical arc, *2D* spline, or *2D* polyline segment.

Zaxis Option

The Zaxis option defines the section plane by locating its origin point and a point on the Z axis (normal) to the plane.

View Option

The View option aligns the section plan with the viewing plane of the current viewport. Specifying a point defines the location of the sectioning plane.

XY Option

The XY option aligns the sectioning plane with the *XY* plane of the current UCS. Specifying a point defines the location of the sectioning plane.

YZ Option

The YZ option aligns the sectioning plane with the *XY* plane of the current UCS. Specifying a point defines the location of the sectioning plane.

ZX Option

The ZX option aligns the sectioning plane with the *XY* plane of the current UCS. Specifying a point defines the location of the sectioning plane.

Figure 15–80 shows a hatched cross-section produced with the SECTION command.

 Note: The section must be hatched using hatching techniques of Chapter 9.

Figure 15–80 *Creating a 2D hatched cross-section using the* SECTION *command*

SLICING SOLIDS

The SLICE command allows you to create a new solid by cutting the existing solid and removing a specified portion. If necessary, you can retain both portions of the sliced solid(s) or just the portion you specify. The sliced solids retain the layer and color of the original solids.

Invoke the SLICE command:

Solids toolbar	Choose the SLICE command (see Figure 15–81)
Draw menu	Choose Solids > Slice
Command: prompt	**slice** (ENTER)

Figure 15–81 *Invoking the SLICE command from the Solids toolbar*

AutoCAD prompts:

Command: **slice**
Select objects: *(select the objects to create a new solid by slicing)*

After selecting the objects, press ENTER or SPACEBAR. AutoCAD prompts you to define the slice plane:

Specify first point on slicing plane by [Object/Zaxis/View/XY/YZ/ZX/
3points] <3points>: *(specify one of the three points to define a plane or
select one of the available options)*

The options are the same as those for the SECTION command explained earlier in this chapter.

After defining the slicing plane, AutoCAD prompts you to indicate which part of the cut solid is to be retained:

Specify a Point on desired side of the plane or [Keep Both sides]:

The default option allows you to select with your pointing device the side of the slice that has to be retained in your drawing.

The **Both sides** option allows you to retain both portions of the sliced solids.

Figure 15–82 shows two parts of a solid model that have been cut using the SLICE command and moved apart using the MOVE command.

Figure 15–82 *Cutting a solid model into two parts using the SLICE command*

SOLID INTERFERENCE

The INTERFERE command checks the interference between two or more solids and creates a composite solid from their common volume.

There are two ways to determine the interference between solids:

- Select two sets of solids. AutoCAD determines the interference between the first and second sets of solids.

- Select one set of solids instead of two. AutoCAD determines the interference between all of the solids in the set. They are checked against each other.

Invoke the INTERFERE command:

Solids toolbar	Choose the INTERFERE command (see Figure 15–83)
Draw menu	Choose Solids > Interfere
Command: prompt	**interfere** (ENTER)

Figure 15–83 *Invoking the* INTERFERE *command from the Solids toolbar*

AutoCAD prompts:

Command: **interfere**
Select the first set of solids: *(select the first set of solids and press* ENTER *or* SPACEBAR*)*
Select the second set of solids: *(select the second set of solids or press* ENTER*)*

The second selection set is optional. Press ENTER or SPACEBAR if you do not want to define the second selection set. If the same solid is included in both the selection sets, it is considered part of the first selection set and ignored in the second selection set. AutoCAD highlights all interfering solids and prompts:

Create interference solids? [Yes/No] <N>:

Entering y creates and highlights a new solid on the current layer that is the intersection of the interfering solids. If there are more than two interfering solids, AutoCAD prompts:

Highlight pairs of interfering solids? [Yes/No] <N>:

If you specify y for yes, and if there is more than one interfering pair, AutoCAD prompts:

Enter an option [Next pair/exit[<Next>: *(Specify **x** or **n**)*

Pressing ENTER cycles through the interfering pairs of solids, and AutoCAD highlights each interfering pair of solids. Enter **x** to complete the command sequence.

EDITING FACES OF 3D SOLIDS

AutoCAD 2000 allows you to edit solid objects by extruding faces, copying faces, offsetting faces, moving faces, rotating faces, tapering faces, coloring faces, and deleting faces.

Extruding Faces

AutoCAD allows you to extrude selected faces of a *3D* solid object to a specified height or along a path. Specifying a positive value extrudes the selected face in its positive direction (usually outward); and a negative value extrudes in the negative direction (usually inward). Tapering the selected face with a positive angle tapers the face inward, and a negative angle tapers the face outward. Tapering the selected face to 0 degrees extrudes the face perpendicular to its plane.

Face extrusion along a path is based on a path curve, such as lines, circles, arcs, ellipses, elliptical arcs, polylines, or splines.

Invoke the EXTRUDE FACES command to extrude selected faces:

Solids Editing toolbar	Choose the EXTRUDE FACES command (see Figure 15–84)
Modify menu	Choose Solids Editing > Extrude faces
Command: prompt	**solidedit** (ENTER)

Figure 15–84 *Invoking the* EXTRUDE FACES *command from the Solids Editing toolbar*

AutoCAD prompts:

```
Command: solidedit
Enter a solids editing option [Face/Edge/Body/Undo/eXit] <eXit>: f
Enter a face editing option [Extrude/Move/Rotate/Offset/Taper/Delete/
    Copy/coLor/Undo/eXit] <eXit>: e
Select faces or [Undo/Remove]: (select faces and press ENTER to complete the
    selection)
Specify height of extrusion or [Path]: (specify height of extrusion or select
    path option to extrude along a path)
Specify angle of taper for extrusion <0>: (press ENTER to accept the default
    angle or specify the angle for taper for extrusion)
```

> Enter a face editing option [Extrude/Move/Rotate/Offset/Taper/Delete/
> Copy/coLor/Undo/eXit] <eXit>: *(select the eXit option to exit the face
> editing)*
> Enter a solids editing option [Face/Edge/Body/Undo/eXit] <eXit>: *(select
> the eXit option to exit solids editing)*

Figure 15–85 shows an example of a solid model in which one of the faces is extruded by a positive value with a 15 degree tapered angle with the EXTRUDE FACES command.

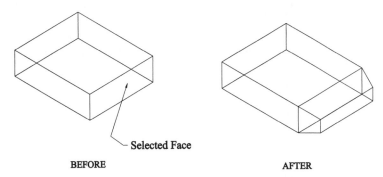

Figure 15–85 *An example in extruding one of the faces in a solid model.*

Copying Faces

The SOLIDEDIT command allows you to copy selected faces of a *3D* solid object. AutoCAD copies selected faces as regions or bodies. Prompts are similar to the regular COPY command.

Invoke the COPY FACES command to copy selected faces:

Solids Editing toolbar	Choose the COPY FACES command (see Figure 15–86)
Modify menu	Choose Solids Editing > Copy faces
Command: prompt	**solidedit** (ENTER)

Figure 15–86 *Invoking the* COPY FACES *command from the Solids Editing toolbar*

AutoCAD prompts:

> Command: **solidedit**
> Enter a solids editing option [Face/Edge/Body/Undo/eXit] <eXit>: **f**

Enter a face editing option [Extrude/Move/Rotate/Offset/Taper/Delete/ Copy/coLor/Undo/eXit] <eXit>: **c**

Select faces or [Undo/Remove]: *(select faces and press* ENTER *to complete the selection)*

Specify a base point or displacement: *(specify a base point)*

Specify a second point of displacement: *(specify a second point of displacement or press* ENTER *to consider the original selection point as a base point)*

Enter a face editing option [Extrude/Move/Rotate/Offset/Taper/Delete/ Copy/coLor/Undo/eXit] <eXit>: *(select the eXit option to exit the face editing)*

Enter a solids editing option [Face/Edge/Body/Undo/eXit] <eXit>: *(select the eXit option to exit solids editing)*

Figure 15–87 shows an example of a solid model in which one of the faces is copied by the COPY FACES command.

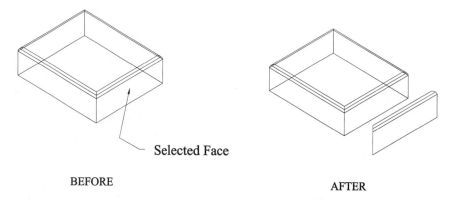

Selected Face

BEFORE

AFTER

Figure 15–87 *An example of copying one of the faces in a solid model.*

Offsetting Faces

AutoCAD allows you to uniformly offset selected faces of a *3D* solid object by a specified distance. New faces are created by offsetting existing ones inside or outside at a specified distance from their original positions. Specifying a positive value increases the size or volume of the solid; a negative value decreases the size or volume of the solid.

Invoke the OFFSET FACES command to offset selected faces:

Solids Editing toolbar	Choose the OFFSET FACES command (see Figure 15–88)
Modify menu	Choose Solids Editing > Offset faces
Command: prompt	**solidedit** (ENTER)

Figure 15–88 *Invoking the* OFFSET FACES *command from the Solids Editing toolbar*

AutoCAD prompts:

> Command: **solidedit**
> Enter a solids editing option [Face/Edge/Body/Undo/eXit] <eXit>: **f**
> Enter a face editing option [Extrude/Move/Rotate/Offset/Taper/Delete/
>   Copy/coLor/Undo/eXit] <eXit>: **o**
> Select faces or [Undo/Remove]: *(select faces and press* ENTER *to complete the*
>   *selection)*
> Specify the offset distance: *(specify the offset distance)*
> Enter a face editing option [Extrude/Move/Rotate/Offset/Taper/Delete/
>   Copy/coLor/Undo/eXit] <eXit>: *(select the eXit option to exit the face*
>   *editing)*
> Enter a solids editing option [Face/Edge/Body/Undo/eXit] <eXit>: *(select*
>   *the eXit option to exit solids editing)*

Figure 15–89 shows an example of a solid model in which one of the faces is offset (positive value) by the OFFSET FACES command.

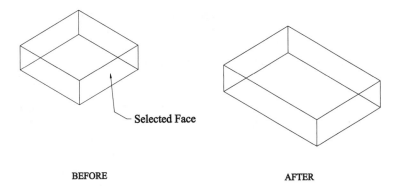

Selected Face

BEFORE   AFTER

Figure 15–89 *An example of offsetting one of the faces (positive value) in a solid model.*

Moving Faces

The SOLIDEDIT command allows you to move selected faces of a *3D* solid object. You can move holes from one location to another location in a *3D* solid. Prompts are similar to the regular MOVE command.

Invoke the MOVE FACES command to move selected faces:

Solids Editing toolbar	Choose the MOVE FACES command (see Figure 15–90)
Modify menu	Choose Solids Editing > Move faces
Command: prompt	**solidedit** (ENTER)

Figure 15–90 *Invoking the* MOVE FACES *command from the Solids Editing toolbar*

AutoCAD prompts:

Command: **solidedit**
Enter a solids editing option [Face/Edge/Body/Undo/eXit] <eXit>: **f**
Enter a face editing option [Extrude/Move/Rotate/Offset/Taper/Delete/
    Copy/coLor/Undo/eXit] <eXit>: **m**
Select faces or [Undo/Remove]: *(select faces and press* ENTER *to complete the
    selection)*
Specify a base point or displacement: *(specify a base point)*
Specify a second point of displacement: *(specify a second point of
    displacement or press* ENTER *to consider the original selection point as a base
    point)*
Enter a face editing option [Extrude/Move/Rotate/Offset/Taper/Delete/
    Copy/coLor/Undo/eXit] <eXit>: *(select the eXit option to exit the face
    editing)*
Enter a solids editing option [Face/Edge/Body/Undo/eXit] <eXit>: *(select
    the eXit option to exit solids editing)*

Figure 15–91 shows an example of a solid model in which an elliptical cylinder is moved by the MOVE FACES command.

BEFORE

AFTER

Figure 15–91 *An example of moving the elliptical cylinder in a solid model.*

Rotating Faces

The SOLIDEDIT command allows you to rotate selected faces of a *3D* solid object by choosing a base point to relative or absolute angle. All *3D* faces rotate about a specified axis.

Invoke the ROTATE FACES command to rotate selected faces:

Solids Editing toolbar	Choose the ROTATE FACES command (see Figure 15–92)
Modify menu	Choose Solids Editing > Rotate faces
Command: prompt	**solidedit** (ENTER)

Figure 15–92 *Invoking the* ROTATE FACES *command from the Solids Editing toolbar*

AutoCAD prompts:

    Command: **solidedit**
    Enter a solids editing option [Face/Edge/Body/Undo/eXit] <eXit>: **f**
    Enter a face editing option [Extrude/Move/Rotate/Offset/Taper/Delete/
        Copy/coLor/Undo/eXit] <eXit>: **r**
    Select faces or [Undo/Remove]: (select faces and press ENTERto complete the
        selection)
    Specify an axis point or [Axis by object/View/Xaxis/Yaxis/Zaxis]
        <2points>: (specify an axis point or select one of the available options)
    Enter a face editing option [Extrude/Move/Rotate/Offset/Taper/Delete/
        Copy/coLor/Undo/eXit] <eXit>: (select the eXit option to exit the face
        editing)
    Enter a solids editing option [Face/Edge/Body/Undo/eXit] <eXit>: (select
        the eXit option to exit solids editing)

Figure 15–93 shows an example of a solid model in which elliptical cylinder is rotated by the ROTATE FACES command.

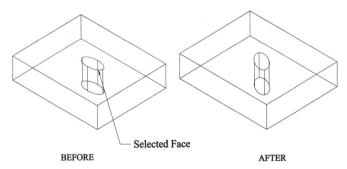

BEFORE   AFTER

Figure 15–93 *An example of rotating elliptical cylinder of a solid model.*

Tapering Faces

The SOLIDEDIT command allows you to taper selected faces of a *3D* solid object with a draft angle along a vector direction. Tapering the selected face with a positive angle tapers the face inward, and a negative angle tapers the face outward.

Invoke the TAPER FACES command to taper selected faces:

Solids Editing toolbar	Choose the TAPER FACES command (see Figure 15–94)
Modify menu	Choose Solids Editing > Taper faces
Command: prompt	**solidedit** (ENTER)

Figure 15–94 *Invoking the* TAPER FACES *command from the Solids Editing toolbar*

AutoCAD prompts:

Command: **solidedit**
Enter a solids editing option [Face/Edge/Body/Undo/eXit] <eXit>: **f**
Enter a face editing option [Extrude/Move/Rotate/Offset/Taper/Delete/
    Copy/coLor/Undo/eXit] <eXit>: **t**
Select faces or [Undo/Remove]: *(select faces and press* ENTER *to complete the
    selection)*
Specify the base point: *(specify the base point)*
Specify another point along the axis of tapering: *(specify a point to define the
    axis of tapering)*
Specify the taper angle: *(specify the taper angle and press* ENTER *to continue)*
Enter a face editing option [Extrude/Move/Rotate/Offset/Taper/Delete/
    Copy/coLor/Undo/eXit] <eXit>: *(select the eXit option to exit the face
    editing)*
Enter a solids editing option [Face/Edge/Body/Undo/eXit] <eXit>: *(select
    the eXit option to exit solids editing)*

Figure 15–95 shows an example of a solid model in which the cylinder tapered angle is changed by the TAPER FACES command.

Coloring Faces

The SOLIDEDIT command allows you to change color of selected faces of a *3D* solid object. You can choose a color from the Select Color dialog box. Setting a color on a face overrides the color setting for the layer on which the solid object resides.

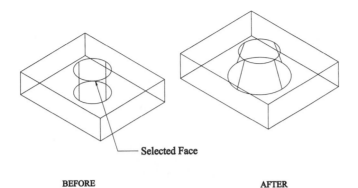

Selected Face

BEFORE   AFTER

Figure 15–95 *An example of change in a tapered angle of the cylinder in a solid model.*

Invoke the COLOR FACES command to change the color of selected faces:

Solids Editing toolbar	Choose the COLOR FACES command (see Figure 15–96)
Modify menu	Choose Solids Editing > Color faces
Command: prompt	**solidedit** (ENTER)

Color Faces

Figure 15–96 *Invoking the COLOR FACES command from the Solids Editing toolbar*

AutoCAD prompts:

> Command: **solidedit**
> Enter a solids editing option [Face/Edge/Body/Undo/eXit] <eXit>: **f**
> Enter a face editing option [Extrude/Move/Rotate/Offset/Taper/Delete/
>   Copy/coLor/Undo/eXit] <eXit>: **l**
> Select faces or [Undo/Remove]: *(select faces and press ENTER to complete the
>   selection)*

AutoCAD displays the **Select Color** dialog box. Select the color to change for selected faces and choose the **OK** button. AutoCAD prompts:

> Enter a face editing option [Extrude/Move/Rotate/Offset/Taper/Delete/
>   Copy/coLor/Undo/eXit] <eXit>: *(select the eXit option to exit the face
>   editing)*
> Enter a solids editing option [Face/Edge/Body/Undo/eXit] <eXit>: *(select
>   the eXit option to exit solids editing)*

AutoCAD changes the color of the selected faces of the *3D* solid model.

Deleting Faces

The SOLIDEDIT command allows you to delete selected faces, holes, and fillets of a *3D* solid object.

Invoke the DELETE FACES command to delete selected faces:

Solids Editing toolbar	Choose the DELETE FACES command (see Figure 15–97)
Modify menu	Choose Solids Editing > Delete faces
Command: prompt	**solidedit** (ENTER)

Figure 15–97 *Invoking the* DELETE FACES *command from the Solids Editing toolbar*

AutoCAD prompts:

Command: **solidedit**
Enter a solids editing option [Face/Edge/Body/Undo/eXit] <eXit>: **f**
Enter a face editing option [Extrude/Move/Rotate/Offset/Taper/Delete/
    Copy/coLor/Undo/eXit] <eXit>: **d**
Select faces or [Undo/Remove]: *(select faces and press* ENTER *to complete the selection)*

 Note: If the object is a simple box, AutoCAD responds with: Modeling Operation Error – Gap cannot be filled. (This informs us that there is no way to join the opposite sides to fill the gap.

Enter a face editing option [Extrude/Move/Rotate/Offset/Taper/Delete/
    Copy/coLor/Undo/eXit] <eXit>: *(select the eXit option to exit the face editing)*
Enter a solids editing option [Face/Edge/Body/Undo/eXit] <eXit>: *(select the eXit option to exit solids editing)*

AutoCAD deletes the selected faces of the *3D* solid model.

EDITING EDGES OF *3D* SOLIDS

AutoCAD 2000 allows you to copy individual edges and change color of edges on a *3D* solid object. The edges are copied as lines, arcs, circles, ellipses or spline objects.

Copying Edges

The SOLIDEDIT command allows you to copy selected edges of a *3D* solid object. AutoCAD copies selected edges as lines, arcs, circles, ellipses, or splines. Prompts are similar to the regular COPY command.

Invoke the COPY EDGES command to copy selected edges:

Solids Editing toolbar	Choose the COPY EDGES command (see Figure 15–98)
Modify menu	Choose Solids Editing > Copy edges
Command: prompt	**solidedit** (ENTER)

Figure 15–98 *Invoking the COPY EDGES command from the Solids Editing toolbar*

AutoCAD prompts:

> Command: **solidedit**
> Enter a solids editing option [Face/Edge/Body/Undo/eXit] <eXit>: **e**
> Enter an edge editing option [Copy/coLor/Undo/eXit] <eXit>: **c**
> Select edges or [Undo/Remove]: *(select edges and press ENTER to complete the selection)*
> Specify a base point or displacement: *(specify a base point)*
> Specify a second point of displacement: *(specify a second point of displacement or press ENTER to consider the original selection point as a base point)*
> Enter an edge editing option [Copy/coLor/Undo/eXit] <eXit>: *(select the eXit option to exit the edge editing)*
> Enter a solids editing option [Face/Edge/Body/Undo/eXit] <eXit>: *(select the eXit option to exit solids editing)*

Figure 15–99 shows an example of a solid model in which the edges (right side of the model) are copied by the COPY EDGES command.

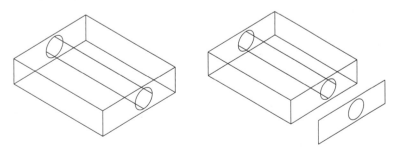

BEFORE   AFTER

Figure 15–99 *An example of copying edges of a solid model.*

Coloring Edges

The SOLIDEDIT command allows you to change color of selected edges of a *3D* solid object. You can choose a color from the Select Color dialog box. Setting a color on an edge overrides the color setting for the layer on which the solid object resides.

Invoke the COLOR EDGES command to change the color of selected edges:

Solids Editing toolbar	Choose the COLOR EDGES command (see Figure 15–100)
Modify menu	Choose Solids Editing > Color edges
Command: prompt	**solidedit** (ENTER)

Figure 15–100 *Invoking the COLOR EDGES command from the Solids Editing toolbar*

AutoCAD prompts:

> Command: **solidedit**
> Enter a solids editing option [Face/Edge/Body/Undo/eXit] <eXit>: **e**
> Enter an edge editing option [Copy/coLor/Undo/eXit] <eXit>: **l**
> Select edges or [Undo/Remove]: *(select faces and press ENTER to complete the selection)*

AutoCAD displays the **Select Color** dialog box. Select the color to change for selected faces and choose the **OK** button. AutoCAD prompts:

> Enter an edge editing option [Copy/coLor/Undo/eXit] <eXit>: *(select the eXit option to exit the edge editing)*
> Enter a solids editing option [Face/Edge/Body/Undo/eXit] <eXit>: *(select the eXit option to exit solids editing)*

AutoCAD changes the color of the selected edges of the *3D* solid model.

IMPRINTING SOLIDS

AutoCAD allows you to have an imprint of an object on the selected solid. The object to be imprinted must intersect one or more faces on the selected solid in order for imprinting to be successful. Imprinting is limited to the following objects: arcs, circles, lines, *2D* and *3D* polylines, ellipses, splines, regions, bodies, and *3D* solids.

Invoke the IMPRINT command to have an imprint of an object on the selected object:

Solids Editing toolbar	Choose the IMPRINT command (see Figure 15–101)
Modify menu	Choose Solids Editing > Imprint
Command: prompt	**solidedit** (ENTER)

Figure 15–101 *Invoking the IMPRINT command from the Solids Editing toolbar*

AutoCAD prompts:

Command: **solidedit**
Enter a solids editing option [Face/Edge/Body/Undo/eXit] <eXit>: **b**
Enter a body editing option [Imprint/seParate solids/Shell/cLean/Check/ Undo/eXit] <eXit>: **i**
Select a 3D solid: *(select a 3D solid object)*
Select an object to imprint: *(select an object to imprint)*
Delete the source object <N>: *(press ENTER to delete the source objects or enter* **y** *to keep the source object)*
Select an object to imprint: *(select another object to imprint or press ENTER to complete the selection)*
Enter a body editing option [Imprint/seParate solids/Shell/cLean/Check/ Undo/eXit] <eXit>: *(select the eXit option to exit body editing)*
Enter a solids editing option [Face/Edge/Body/Undo/eXit] <eXit>: *(select the eXit option to exit solids editing)*

AutoCAD creates imprint of the selected object.

SEPARATING SOLIDS

AutoCAD separates solids from a composite solid. But it cannot separate solids if the composite 3D solid object shares a common area or volume. After separation of the 3D solid, the individual solids retain the layers and colors of the original.

Invoke the SEPARATE command to separate solids:

Solids Editing toolbar	Choose the SEPARATE command (see Figure 15–102)
Modify menu	Choose Solids Editing > Separate
Command: prompt	**solidedit** (ENTER)

Figure 15–102 *Invoking the* SEPARATE *command from the Solids Editing toolbar*

AutoCAD prompts:

> Command: **solidedit**
> Enter a solids editing option [Face/Edge/Body/Undo/eXit] <eXit>: **b**
> Enter a body editing option [Imprint/seParate solids/Shell/cLean/Check/
>   Undo/eXit] <eXit>: **p**
> Select a 3D solid: *(select a 3D solid object and press ENTER to complete the
>   selection)*
> Enter a body editing option [Imprint/seParate solids/Shell/cLean/Check/
>   Undo/eXit] <eXit>: *(select the eXit option to exit body editing)*
> Enter a solids editing option [Face/Edge/Body/Undo/eXit] <eXit>: *(select
>   the eXit option to exit solids editing)*

AutoCAD separates the selected composite solid.

SHELLING SOLIDS

AutoCAD creates a shell or a hollow thin wall with a specified thickness from the selected 3D solid object. AutoCAD creates new faces by offsetting existing ones inside or outside their original positions. AutoCAD treats continuously tangent faces as single faces when offsetting. A positive offset value creates a shell in the positive face direction; a negative value creates a shell in the negative face direction.

Invoke the SHELL command to create a shell:

Solids Editing toolbar	Choose the SHELL command (see Figure 15–103)
Modify menu	Choose Solids Editing > Shell
Command: prompt	**solidedit** (ENTER)

Figure 15–103 *Invoking the* SHELL *command from the Solids Editing toolbar*

AutoCAD prompts:

> Command: **solidedit**

Enter a solids editing option [Face/Edge/Body/Undo/eXit] <eXit>: **b**
Enter a body editing option [Imprint/seParate solids/Shell/cLean/Check/
   Undo/eXit] <eXit>: **s**
Select a 3D solid: *(select a 3D solid object)*
Remove faces or [Undo/Add/ALL]: *(select faces to be excluded from shelling
   and press* ENTER *to complete the selection)*
Specify the shell offset value: *(specify the shell offset value)*
Enter a body editing option [Imprint/seParate solids/Shell/cLean/Check/
   Undo/eXit] <eXit>: *(select the eXit option to exit body editing)*
Enter a solids editing option [Face/Edge/Body/Undo/eXit] <eXit>: *(select
   the eXit option to exit solids editing)*

AutoCAD creates a shell with the specified thickness from the *3D* solid object.

CLEANING SOLIDS

AutoCAD allows you to remove edges or vertices if they share the same surface or
vertex definition on either side of the edge or vertex. All redundant edges, imprinted
as well as used, on the selected *3D* solid object are deleted.

Invoke the CLEAN command to clean solids:

Solids Editing toolbar	Choose the CLEAN command (see Figure 15–104)
Modify menu	Choose Solids Editing > Clean
Command: prompt	**solidedit** (ENTER)

Clean

Figure 15–104 *Invoking the* CLEAN *command from the Solids Editing toolbar*

AutoCAD prompts:

Command: **solidedit**
Enter a solids editing option [Face/Edge/Body/Undo/eXit] <eXit>: **b**
Enter a body editing option [Imprint/seParate solids/Shell/cLean/Check/
   Undo/eXit] <eXit>: **l**
Select a 3D solid: *(select a 3D solid object and press* ENTER *to complete the
   selection)*
Enter a body editing option [Imprint/seParate solids/Shell/cLean/Check/
   Undo/eXit] <eXit>: *(select the eXit option to exit body editing)*
Enter a solids editing option [Face/Edge/Body/Undo/eXit] <eXit>: *(select
   the eXit option to exit solids editing)*

AutoCAD removes the selected edges or vertices of the selected *3D* model.

CHECKING SOLIDS

AutoCAD checks to see if the selected solid object is a valid *3D* solid object. With a *3D* solid model, you can modify the object without incurring ACIS failure error messages. If the selected solid *3D* model is not valid, you cannot edit the object.

Invoke the CHECK command to clean solids:

Solids Editing toolbar	Choose the CHECK command (see Figure 15–105)
Modify menu	Choose Solids Editing > Check
Command: prompt	**solidedit** (ENTER)

Figure 15–105 *Invoking the* CHECK *command from the Solids Editing toolbar*

AutoCAD prompts:

> Command: **solidedit**
> Enter a solids editing option [Face/Edge/Body/Undo/eXit] <eXit>: **b**
> Enter a body editing option [Imprint/seParate solids/Shell/cLean/Check/
>   Undo/eXit] <eXit>: **c**
> Select a *3D* solid: *(select a* 3D *solid object and press* ENTER *to complete the*
>   *selection)*
> Enter a body editing option [Imprint/seParate solids/Shell/cLean/Check/
>   Undo/eXit] <eXit>: *(select the eXit option to exit body editing)*
> Enter a solids editing option [Face/Edge/Body/Undo/eXit] <eXit>: *(select*
>   *the eXit option to exit solids editing)*

AutoCAD checks the solid *3D* model and displays with an appropriate information of the selected solid.

MASS PROPERTIES OF A SOLID

The MASSPROP command calculates and displays the mass properties of selected solids and regions. The mass properties displayed for solids are mass, volume, bounding box, centroid, moments of inertia, products of inertia, radii of gyration, and principal moments with corresponding principal directions. The mass properties are calculated based on the current UCS.

Invoke the MASSPROP command:

Inquiry toolbar	Choose the MASSPROP command (see Figure 15–106)
Tools menu	Choose Inquiry > Mass Properties
Command: prompt	**massprop** (ENTER)

Figure 15–106 *Invoking the* MASSPROP *command from the Inquiry toolbar*

AutoCAD prompts:

> Command: **massprop**
> Select objects: *(select the objects and press* ENTER *to complete the selection)*

The MASSPROP command displays the object mass properties of the selected objects in the text window, as shown in Figure 15–107. Then, AutoCAD prompts:

> Write analysis to a file? [Yes/No] <N>: *(type* **y** *to create a file or* **n** *for not to create a file)*

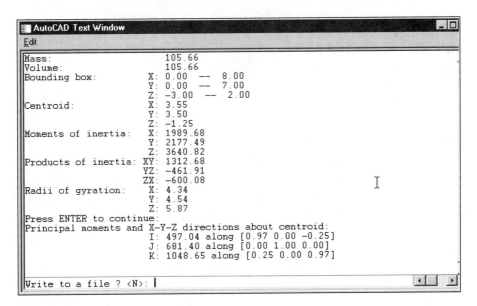

Figure 15–107 *Mass properties listing*

If you enter y, AutoCAD prompts for a file name and saves the file in ASCII format.

 Note: You can also use the AutoCAD LIST and AREA commands to obtain information about individual solid(s) (coordinates) and the area(s) of the solid(s), respectively.

HIDING OBJECTS

The HIDE command hides objects (or displays them in different colors) that are behind other objects in the current viewport. Complex models are difficult to read in wireframe form, and benefit in clarity when the model is displayed with hidden lines removed. HIDE considers circles, solids, traces, wide polyline segments, 3D faces, polygon meshes, and the extruded edges of objects with a thickness to be opaque surfaces hiding objects that lie behind them. The HIDE command remains active only until the next time the display is regenerated. Depending on the complexity of the model, hiding may take from a few seconds to even several minutes.

Invoke the HIDE command:

Render toolbar	Choose the HIDE command (see Figure 15–108)
View menu	Choose Hide
Command: prompt	**hide** (ENTER)

Figure 15–108 *Invoking the HIDE command from the Render toolbar*

AutoCAD prompts:

Command: **hide**

There are no prompts to be answered. The current viewport goes blank for a period of time, depending on the complexity of the model, and is then redrawn with hidden lines removed temporarily. Hidden-line removal is lost during plotting unless you specify that AutoCAD remove hidden lines in the plotting configuration.

PLACING MULTIVIEWS IN PAPER SPACE

The SOLVIEW command creates untiled viewports using orthographic projection to lay out orthographic views and sectional views. View-specific information is saved with each viewport as you create it. The information that is saved is used by the SOLDRAW command, which does the final generation of the drawing view. The SOLVIEW command automatically creates a set of layers that the SOLDRAW command uses to place the visible lines, hidden lines, and section hatching for each view. In addition, the SOLVIEW command creates a specific layer for dimensions that are visible in individual viewports. The SOLVIEW command applies the following conventions in naming the layers:

Layer	Name
For visible lines	VIEW NAME-VIS
For hidden lines	VIEW NAME-HID
For dimensions	VIEW NAME-DIM
For sections	VIEW NAME-HAT6

The viewport objects are drawn on the VPORTS layer. The SOLVIEW command creates the VPORTS layer if it does not already exist. All the layers created by SOLVIEW are assigned the color white and the linetype continuous. The stored information is deleted and updated when you run the SOLDRAW command, so do not draw any objects on these layers.

Invoke the SETUP VIEW command:

Solids toolbar	Choose the SETUP VIEW command (see Figure 15–109)
Draw menu	Choose Solids > Setup > View
Command: prompt	**solview** (ENTER)

Figure 15–109 *Invoking the* SETUP VIEW *command from the Solids toolbar*

AutoCAD prompts:

> Command: **solview**
> Enter an option [Ucs/Ortho/Auxiliary/Section]: *(select one of the available options)*

Ucs Option

The Ucs option creates a profile view relative to a User Coordinate System. AutoCAD prompts:

> Enter an option [Named/World/?/Current] <Current>: *(select one of the available options)*

The **Named** option uses the *XY* plane of a named UCS to create a profile view. After prompting for the name of the view, AutoCAD prompts as follows:

> Enter view scale <1.0>: *(specify the scale factor for the view to be displayed)*
> Specify View center: *(specify the center location for the viewport to be drawn in the paper space and press ENTER)*
> Specify first corner of viewport: *(specify a point for one corner of the viewport)*
> Specify opposite corner of viewport: *(specify a point for the opposite corner of the viewport)*
> Enter view name: *(specify a name for the newly created view)*

The **World** option uses the *XY* plane of the WCS to create a profile view. The prompts are the same as just described for the Named option.

The **?** option lists the names of existing User Coordinate Systems. After the list is displayed, press any key to return to the first prompt.

The **Current** option, which is the default option, uses the *XY* plane of the current UCS to create a profile view. The prompts are the same as explained earlier for the Named option.

If no untiled viewports exist in your drawing, the UCS option allows you to create an initial viewport from which other views can be created.

Ortho Option

The Ortho option creates a folded orthographic view from an existing view. AutoCAD prompts:

> Specify side of viewport to project: *(select the one of the edges of a viewport)*
> Specify view center: *(specify the center location for the viewport to be drawn in the paper space and press ENTER)*
> Specify first corner of viewport: *(specify a point for one corner of the viewport)*
> Specify opposite corner of viewport: *(specify a point for the opposite corner of the viewport)*
> Enter view name: *(specify a name for the newly created view)*

Auxiliary Option

The Auxiliary option creates an auxiliary view from an existing view. An auxiliary view is one that is projected onto a plane perpendicular to one of the orthographic views and inclined in the adjacent view. AutoCAD prompts:

Specify first point of inclined plane: *(specify a point)*
Specify second point of incline plane: *(specify a point)*
Specify side to view from: *(specify a point that determines the side from which you will view the plane)*
Specify view center: *(specify the center location for the viewport to be drawn in the paper space and press* ENTER*)*
Specify first corner of viewport: *(specify a point for one corner of the viewport)*
Specify opposite corner of viewport: *(specify a point for the opposite corner of the viewport)*
Enter view name: *(specify a name for the newly created view)*

Section Option

The Section option creates a sectional view of solids with crosshatching. AutoCAD prompts:

Specify first point of cutting plane: *(specify a point to define the first point of the cutting plane)*
Specify second point of cutting plane: *(specify a point to define the second point of the cutting plane)*
Specify side to view from: *(specify a point to define the side of the cutting plane from which to view)*
Enter view scale: *(specify the scale factor for the view to be displayed)*
Specify view center: *(specify the center location for the viewport to be drawn in the paper space and press* ENTER*)*
Specify first corner of viewport: *(specify a point for one corner of the viewport)*
Specify opposite corner of viewport: *(specify a point for the opposite corner of the viewport)*
Enter view name: *(specify a name for the newly created view)*

To exit the command sequence, press ENTER.

GENERATING VIEWS IN VIEWPORTS

The SOLDRAW command generates sections and profiles in viewports that have been created with the SOLVIEW command. Visible and hidden lines representing the silhouette and edges of solids in the viewports are created and then projected to a plane perpendicular to the viewing direction. AutoCAD deletes any existing profiles and sections in the selected viewports, and new ones are generated. In addition, AutoCAD freezes all the layers in each viewport, except those required to display the profile or section.

Invoke the SETUP DRAWING command:

Solids toolbar	Choose the SETUP DRAWING command (see Figure 15–110)
Draw menu	Choose Solids > Setup > Drawing
Command: prompt	**soldraw** (ENTER)

Figure 15–110 *Invoking the* SETUP DRAWING *command from the Solids toolbar*

AutoCAD prompts:

> Command: **soldraw**
> Select viewports to draw...
> Select objects: *(select the viewports to be drawn)*

AutoCAD generate views in the selected viewports.

GENERATING PROFILES

The SOLPROF command creates a profile image of a solid, including all of its edges, according to the view in the current viewport. The profile image is created from lines, circles, arcs, and/or polylines. SOLPROF will not give correct results in perspective view; it is designed for parallel projections only.

The SOLPROF command will work only when you are working in layout tab and you are in model space.

Invoke the SETUP PROFILE command:

Solids toolbar	Choose the SETUP PROFILE command (see Figure 15–111)
Draw menu	Choose Solids > Setup > Profile
Command: prompt	**solprof** (ENTER)

Figure 15–111 *Invoking the* SETUP PROFILE *command from the Solids toolbar*

AutoCAD prompts:

> Command: **solprof**
> Select objects: *(select one or more objects)*

Select one or more solids and press ENTER. The next prompt lets you decide the placement of hidden lines of the profile on a separate layer:

> Display hidden profile lines on separate layer? [Yes/No] <Y>:

Enter **Y** or **N**. If you answer **Y** (default option), two block inserts are created—one for the visible lines in the same linetype as the original and the other for hidden lines in the hidden linetype. The visible lines are placed on a layer whose name is PV-(viewport handle of the current viewport). The hidden lines are placed on a layer whose name is PH-(viewport handle or the current viewport). If these layers do not exist, AutoCAD will create them. For example, if you create a profile in a viewport whose handle is 6, then visible lines will be placed on a layer PV-6 and hidden lines on layer PH-6. To control the visibility of the layers, you can turn the appropriate layers on and off.

The next prompt determines whether *2D* or *3D* entities are used to represent the visible and hidden lines of the profile:

> Project profile lines onto a profile? [Yes/No] <Y>:

Enter **Y** or **N**. If you answer **Y** (default option), AutoCAD creates the visible and hidden lines of the profile with *2D* AutoCAD entities; **N** creates the visible and hidden lines of the profile with *3D* AutoCAD entities.

Finally, AutoCAD asks if you want tangential edges deleted. A tangential edge is an imaginary edge at which two facets meet and are tangent. In most of the drafting applications, the tangential edges are not shown. The prompt sequence is as follows for deleting the tangential edges:

> Delete tangential edges? [Yes/No] <Y>:

Enter **Y** to delete the tangential edges and **N** to retain them.

PROJECT EXERCISE

This project creates the bracket shown in Figure P15–1. The bracket is drawn entirely by means of AutoCAD solid-modeling features. Follow the steps, and you will be able to build the model by using various commands available in AutoCAD solid modeling.

Figure P15–1 *Creating a bracket using solid modeling*

Step 1: Begin a new drawing using the Quick Setup Wizard. Set Units to Decimal and Area to 22 by 17.

Step 2: Create the following layers with appropriate colors and linetypes:

Layer Name	Color	Linetype
Object	Red	Continuous
Border	Green	Continuous
Dim	Blue	Continuous
Viewports	Cyan	Continuous

Step 3: Invoke the VPORTS command to create four equal viewports. AutoCAD displays the Viewports dialog box, as shown in Figure P15–2.

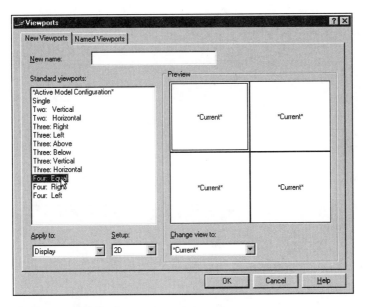

Figure P15–2 *Viewports dialog box*

Select Four: Equal from the **Standard viewports:** list box and *3D* from the
Setup option menu, as shown in Figure P15–2. Choose the **OK** button to
create the viewports and close the Viewports dialog box. AutoCAD creates
four viewports with orthographic views and sets the appropriate UCS orientation, as shown in Figure P15–3.

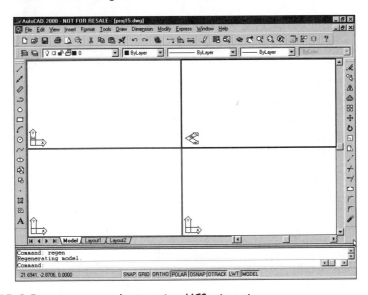

Figure P15–3 *Four viewports and appropriate UCS orientation*

Step 4: Make the upper right viewport current. Invoke the **Display UCS Dialog** from the UCS toolbar, as shown in Figure P15–4.

Figure P15–4 *Invoke the Display UCS Dialog from the UCS toolbar*

AutoCAD displays the UCS dialog box, as shown in Figure P15–5.

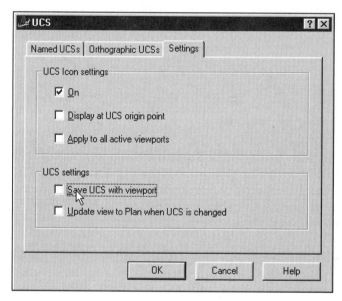

Figure P15–5 *UCS dialog box*

Select the Settings tab and set the **Save UCS with viewport** toggle button to off, as shown in Figure P15–5. Choose the **OK** button to close the UCS dialog box.

Step 5: Set Object as the current layer.

Begin the layout of the drawing by drawing four boxes using the BOX command:

Command: **box**
Specify corner of box or [CEnter] <0,0,0>: **0,0,-2**
Specify corner or [Cube/Length]: **l**
Length: **8**
Width: **7**
Height: **1**

Command: **box**
Specify corner of box or [CEnter] <0,0,0>: **0,0,-1**
Specify corner or [Cube/Length]: **l**
Length: **3**
Width: **7**
Height: **l**

Command: **box**
Specify corner of box or [CEnter] <0,0,0>: **5,0,-3**
Specify corner or [Cube/Length]: **l**
Length: **3**
Width: **7**
Height: **l**

Command: **box**
Specify corner of box or [CEnter] <0,0,0>: **2.5,3.25,-1**
Specify corner or [Cube/Length]: **l**
Length: **.75**
Width: **.5**
Height: **2**

The preceding box constructions form the basic shape of the bracket, as shown in Figure P15–6.

Figure P15–6 *Creating the basic shape of the bracket*

Step 6: Invoke the CYLINDER command to create a cylinder:

> Command: **cylinder**
> Specify center point for base of cylinder or [Elliptical] <0,0,0>: **1.5,3.5**
> Specify radius for base of cylinder or [Diameter]: **1.25**
> Specify height of cylinder or [Center of other end]: **2**

Step 7: Invoke the WEDGE command to create a wedge as shown in Figure P15–7:

> Command: **wedge**
> Specify first corner of wedge or [CEnter] <0,0,0>: **3.25,3.25,-1**
> Specify corner or [Cube/Length]: **l**
> Length: **3.75**
> Width: **.5**
> Height: **2**

Figure P15–7 *Creating the basic shape with cylinder and wedge of the bracket*

Step 8: Invoke the UCS command to define UCS as follows:

> Command: **ucs**
> Enter an option New/Move/orthoGraphic/Prev/Restore/Save/Del/Apply/ ?/World]
> <World>: **n**
> Specify origin of new UCS or [ZAxis/3point/OBject/Face/View/X/Y/Z] <0,0,0>: **3**
> Specify new origin point <0,0,0>: *(select point 1 by using the object snap ENDpoint, as shown in Figure P15–8)*

Specify point on positive portion of X-axis: *(select point 2 by using the object snap ENDpoint, as shown in Figure P15–8)*

Specify point on positive-Y portion of the UCS XY plane: *(select point 3 by using the object snap ENDpoint, as shown in Figure P15–8)*

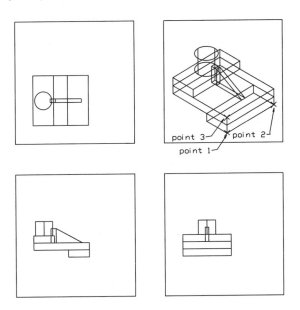

Figure P15–8 *Defining a UCS by 3 points*

Step 9: Invoke the PLINE command to draw a polyline to the given coordinates, as shown in Figure P15–9:

Command: **pline**
Specify start point: **3.5,2**
Current line-width is 0.0000
Specify next point or [Arc/Close/Halfwidth/Length/Undo/Width]: **@1<180**
Specify next point or [Arc/Close/Halfwidth/Length/Undo/Width]: **@0.5<270**
Specify next point or [Arc/Close/Halfwidth/Length/Undo/Width]: **@-0.5,-1**
Specify next point or [Arc/Close/Halfwidth/Length/Undo/Width]: **@0.5<-90**
Specify next point or [Arc/Close/Halfwidth/Length/Undo/Width]: **@1.5<0**
Specify next point or [Arc/Close/Halfwidth/Length/Undo/Width]: **c** (ENTER)

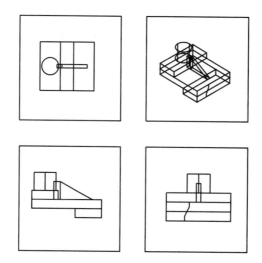

Figure P15–9 *Drawing a polyline to specified coordinates*

Step 10: Invoke the REVOLVE command to revolve the polyline just created into a solid, as shown in Figure P15–10:

> Command: **revolve**
> Current wire frame density: ISOLINES=4
> Select objects: **1**
> Select objects: (ENTER)
> Specify start point for axis of revolution or define axis by [Object/X (axis)/Y (axis)]: **3.5,2**
> Specify endpoint of axis: **@2<270**
> Specify angle of revolution <360>: **180**

Figure P15–10 *Revolving a polyline into a solid*

Step 11: Set the UCS to world coordinates systems and invoke the SPHERE command to create two spheres.

> Command: **ucs**
> Enter an option New/Move/orthoGraphic/Prev/Restore/Save/Del/Apply/
>   ?/World]
> <World>: (ENTER)
> Command: **sphere**
> Specify center of sphere <0,0,0>: **1.5,1.125,-0.5**
> Specify radius of sphere or [Diameter]: **1**

Copy the sphere to a displacement of 0,4.75, as shown in Figure P15–11.

> Command: **copy**
> Select objects: **1**
> Select objects: (ENTER)
> Specify base point or displacement, or [Multiple]: **0,0**
> Specify second point of displacement or <use first point as
>   displacement>: **0,4.75**

Figure P15–11 *Copying a sphere to a specified displacement*

Step 12: Invoke the CONE command to draw two cones, as shown in Figure P15–12.

> Command: **cone**
> Specify center point for base of cone or [Elliptical] <0,0,0>: **1.5,**
>   **1.125,-2**
> Specify radius for base of cone or [Diameter]: **0.75**
> Specify height of cone or [Apex]: **-3**

Copy the cone to a displacement of 0,4.75.

project **EXERCISE**

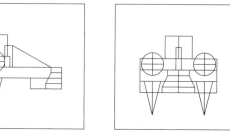

Figure P15–12 *Drawing two cones using the* CONE *command*

Step 13: Starting at 0,0,–5, create a box that is 3 x 7 x 2, as shown in Figure P15–13.

Figure P15–13 *Creating a box with starting point 0,0,–5 and dimensions of 3 x 7 x 2*

Step 14: Create a 0.5-radius cylinder, centered at 1.5,3.5,–2, to a height of 4, as shown in Figure P15–14.

Figure P15–14 *Creating a 0.5 cylinder centered at 1.5,3.5,–2 and with a height of 4*

Step 15: Create a cylinder with radius 1, centered at 1.5,3.5,1.75 to a height of 0.25, as shown in Figure P15–15.

Figure P15–15 *Creating a 0.5 cylinder centered at 1.5,3.5,1.75, with a height of 0.25 and a radius of 1*

Step 16: Create a cylinder with radius 0.25, centered at 6.5,1.0,–3 to a height of 2, as shown in Figure P15–16.

Copy the cylinder to a displacement of 0,4.

Figure P15–16 *Creating a cylinder centered at 6.5,1.0,–3, with a height of 2 and a radius of 0.25*

Step 17: Invoke the TORUS command to create a torus, as shown in Figure P15–17:

> Command: **torus**
> Specify center of torus <0,0,0>: **1.5,3.5,1.5**
> Specify radius of torus or [Diameter]: **1.25**
> Specify radius of tube or [Diameter]: **0.25**

Step 18: Select the connected boxes (except the box that was drawn in Step 13), the wedge, the large cylinder, the spheres, and the cones for use with the UNION command.

> Command: **union**
> Select objects: *(select the boxes, wedge, large cylinder, spheres, and cones, and press ENTER)*

Figure P15–17 *Creating a torus using the* TORUS *command*

Step 19: Select the resulting solid in response to the first SUBTRACT prompt, and then selecting the remaining primitives to be subtracted from it.

> Command: **subtract**
> Select solids and regions to subtract from ...
> Select objects: *(select the solid resulting from STEP 18)*
> Select objects: (ENTER)
> Select solids and regions to subtract ...
> Select objects: *(select the remaining primitives)*
> Select objects: (ENTER)

The drawing should look like Figure P15–18.

Figure P15–18 *Subtracting the primitives from the newly created solid using the* SUBTRACT *command*

Step 20: Select the faces, as shown in Figure P15–19, for the chamfer and fillet. Use the CHAMFER and FILLET commands with 0.25 as the chamfer values and the radii on the respective selected objects. The end result should look like Figure P15–20.

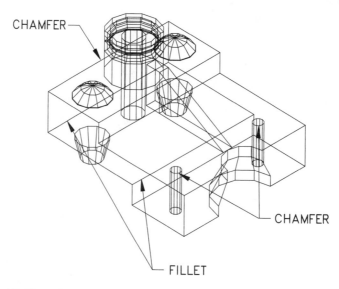

CHAMFER

CHAMFER

FILLET

Figure P15–19 *Using the* CHAMFER *and* FILLET *commands to chamfer and fillet the faces*

Figure P15–20 *The solid after chamfering and filleting the faces*

Step 21: Make the upper right viewport active. Invoke the HIDE command, and the result is as shown in Figure P15–21.

Figure P15–21 *Completed solid after using the HIDE command*

Step 22: Set Viewports as the current layer. Invoke the LAYOUTWIZARD command to create a layout to plot model. AutoCAD displays the Create Layout - Begin dialog box, as shown in Figure P15–22.

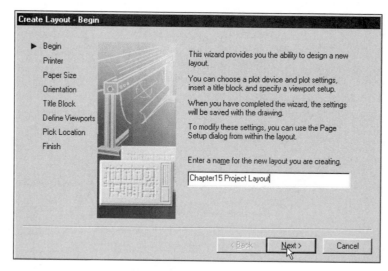

Figure P15–22 *Create Layout - Begin dialog box*

Type **Chapter15 Project Layout** in the **Enter a name for the new layout you are creating** edit field, as shown in Figure P15–22. Choose the **Next >** button, and AutoCAD displays the Create Layout - Printer dialog box, as shown in Figure P15–23.

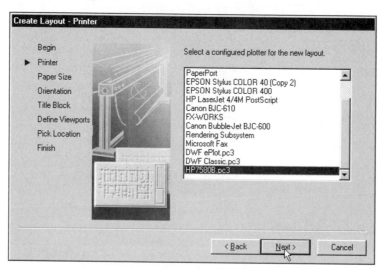

Figure P15–23 *Create Layout - Printer dialog box*

Select a printer from the **Select a configured for a new layout** list box. In Figure 15–23, hp7580b.pc3 is selected. Choose the **Next >** button, and AutoCAD displays the Create Layout – Paper Size dialog box, as shown in Figure P15–24.

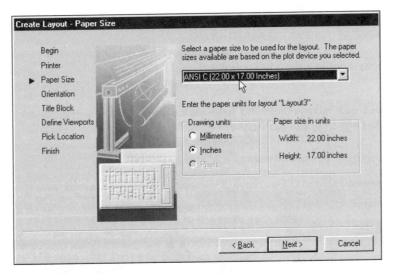

Figure P15–24 *Create Layout – Paper Size dialog box*

Select a paper size to be used in plotting. In Figure 15–24, ANSI C (22.00 × 17.00 Inches) is selected. Choose the **Next >** button, and AutoCAD displays the Create Layout – Orientation dialog box, as shown in Figure P15–25.

Figure P15–25 *Create Layout – Orientation dialog box*

Select the **Landscape** radio button, as shown in Figure P15–25. Choose the **Next >** button, and AutoCAD displays the Create Layout – Title Block dialog box, as shown in Figure P15–26.

Figure P15–26 *Create Layout – Title Block dialog box*

Select the appropriate title block for the selected paper size. In Figure P15–26, **ANSI C title block.dwg** is selected. Choose the **Next >** button, and AutoCAD displays the Create Layout – Define Viewports dialog box, as shown in Figure P15–27.

Figure P15–27 *Create Layout – Define Viewports dialog box*

Select the **Std. 3D Engineering Views** radio button in the Viewport setup section of the dialog box, as shown in Figure P15–27. Choose the **Next >** button, and AutoCAD displays the Create Layout – Pick Location dialog box, as shown in Figure P15–28.

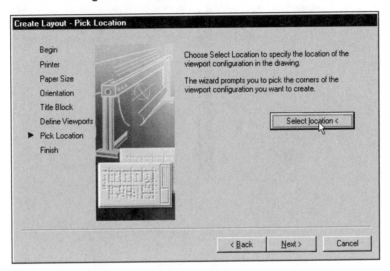

Figure P15–28 *Create Layout – Pick Location dialog box*

Choose the **Select Location <** button.

AutoCAD prompts:

> Specify first corner: **1.5,2.5**
> Specify opposite corner: **20,14**

AutoCAD displays the Create Layout – Finish dialog box. Choose the **Finish** button. AutoCAD displays 3 orthographic views and an isometric view in four viewports.

Set the Layer object as the current layout and turn off the viewport layer.

Invoke the PLOT command and plot the drawing.

Step 22: Save and close the drawing.

EXERCISES 15–1 THROUGH 15–5

EXERCISE 15–1

Layout the objects shown in 3D form. Create the drawings to the given dimension. Display the drawing with VPOINT in four different views. Select the HIDE command for one of the views.

Figure Ex15–1

EXERCISE 15–2

Figure Ex15–2

EXERCISE 15–3

Figure Ex15–3

EXERCISE 15–4

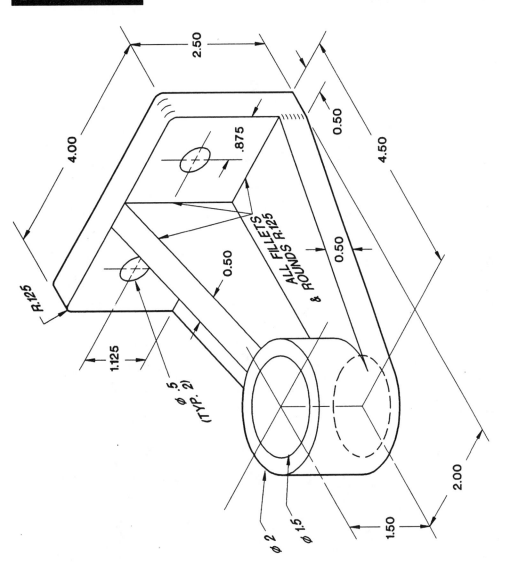

2.50

4.00

0.50

4.50

.875

0.50

ALL FILLETS
& ROUNDS R.125

R.125

0.50

1.125

Ø .5
(TYP. 2)

Ø 2

Ø 1.5

2.00

1.50

mechanical EXERCISE

Figure Ex15–4

EXERCISE 15-5

3.75

1.75

1.25

1.00

2.00

1

.5

R .75

1.75

2.0

0.75

0.50

3.00

4.75

1

Ø .75(TYP. 2)

Figure Ex15-5

REVIEW QUESTIONS

1. What type of coordinate system does AutoCAD use?

   a. Lagrangian System

   b. Right hand system

   c. Left hand system

   d. Maxwellian system

2. If the origin of the current coordinate system is off screen, then the UCSICON, if it is set to Origin, will:

   a. appear in the lower left corner of the screen

   b. attempt to be at the origin, and will therefore not be on screen

   c. appear at the center of the screen

   d. force AutoCAD to perform a ZOOM so that it will be on screen

3. Which of the following UCS options will perform a translation only on the coordinate system?

   a. Rotate

   b. Z-axis

   c. Origin

   d. View

   e. none of the above

4. Which of the following UCS options will perform only a rotation of the coordinate system?

   a. Rotate

   b. Z-axis

   c. Origin

   d. View

   e. none of the above

5. Which of the following UCS options will perform both a rotation and a translation of the coordinate system?

   a. Rotate   d. View

   b. Z-axis   e. none of the above

   c. Origin

6. The command used to generate a perspective view of a *3D* model is:

   a. VPOINT

   b. PERSPECT

   c. VPORT

   d. DVIEW

   e. none of the above

7. The VPOINT command will allow you to specify the viewing direction in all of the following methods *except*:

   a. 2 points

   b. 2 angles

   c. I point

   d. dynamically dragging the X, Y, and Z axes

   e. none of the above (that is, all are valid)

8. Objects can be assigned a negative thickness.

   a. True   b. False

9. Regions:

   a. are *3D* objects

   b. can be created from any group of objects

   c. can be used in conjunction with Boolean operations

   d. cannot be crosshatched

10. Which of the following mesh commands is primarily used in a user-defined function, rather than from the keyboard?

    a. 3DMESH

    b. RULESURF

    c. TABSURF

    d. EDGEMESH

    e. 3DFACE

11. The PFACE command:

    a. generates a mesh

    b. requests the vertexes of the object

    c. can create an opaque (will hide) flat polygon

    d. all of the above

12. To align two objects in *3D* space, invoke the command called:

    a. ALIGN

    b. ROTATE

    c. 3DROTATE

    d. TRANSFORM

    e. 3DALIGN

13. *3D* arrays are limited to rectangular patterns (that is, rows, columns, and layers).

    a. True   b. False

14. The UNION command will allow you to select more than two solid objects concurrently.

    a. True   b. False

15. Which of the following is not a standard solid primitive shape used by AutoCAD?

    a. SPHERE

    b. BOX

    c. DOME

    d. WEDGE

    e. CONE

16. Fillets on solid objects are limited to planar edges.

    a. True   b. False

17. A command which will split a solid object into two solids is:

    a. TRIM

    b. SLICE

    c. CUT

    d. SPLIT

    e. DIVIDE

18. To display a more realistic view of a *3D* object, use:

    a. HIDE

    b. VIEW

    c. DISPLAY3D

    d. MAKE3D

    e. VIEWEDIT

19. To force invisible edges of *3D* faces to display, what system variable should be set to 1?

    a. 3DEGE

    b. 3DFRAME

    c. SPLFRAME

    d. EDGEFACE

    e. EDGEVIEW

20. Advantages of solid modeling include:

    a. creation of objects which are manufacturable

    b. interfaces with Computer Aid Manufacturing

    c. analysis of physical properties of objects

    d. all of the above

21. The SECTION command:

    a. creates a region

    b. creates a polyline

    c. crosshatches the area where a plane intersects a solid

    d. is an alias for the HATCHcommand

    e. will give you a choice of either a or b

22. Which AutoCAD command permits a 3D model to be viewed interactively in the current viewport?

    a. 3DVIEW

    b. 3DPAN

    c. 3DROTATE

    d. 3DORBIT

23. When the cursor is positioned inside the arcball of the 3DORBIT command the 3D object can be viewed _____?

    a. horizontally

    b. vertically

    c. diagonally

    d. all of the above

24. When the cursor is positioned outside the arcball of the 3DORBIT command the 3D object can be viewed _____?

    a. horizontally

    b. vertically

    c. diagonally

    d. all of the above

    e. both a or b

25. Which of the following is not an option in the 3DORBIT command?

    a. pan

    b. zoom

    c. tilt

    d. none of the above

26. AutoCAD allows you to display a perspective or a parallel projection of the view while 3DORBIT is active.

    a. True

    b. False

27. AutoCAD allows you to view shaded view while 3DORBIT is active.

    a. True

    b. False

28. Which of the following AutoCAD commands allows you to set clipping planes for the objects in 3D orbit view?

    a. CLIPLANE

    b. 3DCLIP

    c. 3DFACED

    d. CLIPVIEW

29. The Continuous Orbit option of 3DORBIT permits the 3D model to be set into motion, even in shaded mode?

    a. True

    b. False

30. AutoCAD 2000 allows you to edit solid objects by _____.

   a. extruding faces

   b. copying faces

   c. offsetting faces

   d. moving faces

   e. all the above

31. When offsetting a face of a 3D solid a positive value will increase the volume of the solid.

   a. True   b. False

32. A hole in a 3D solid model can be relocated using the MoveFaces option within the SOLIDEDIT command.

   a. True   b. False

33. Colors of individual edges of a 3D solid can be changed using SOLIDEDIT?

   a. True   b. False

34. When an edge is copied using the SOLIDEDIT command it is copied as an _____.

   a. Arc

   b. Spline

   c. Circle

   d. Line

   e. all of the above

35. When a composite 3D solid is separated into individual solids each solid element will retain the layer and color settings of the original solid.

   a. True   b. False

36. AutoCAD refers to the removal of specified edges or verticies as _____.

   a. Extraction

   b. Cleaning

   c. Removing

   d. Erasing

CHAPTER 16

Rendering

INTRODUCTION

Shading or rendering turns your three-dimensional model into a realistic (eye-catching) image. The AutoCAD SHADEMODE command allows you to produce quick, shaded models. However, the AutoCAD RENDER command gives you more control over the appearance of the final image. You can add lights and control lighting in your drawing and define the reflective qualities of surfaces in the drawing, making objects dull or shiny. You can create the rendered image of your 3D model entirely within AutoCAD.

After completing this chapter, you will be able to do the following:

- Render a 3D model
- Create and modify lighting—ambient light, distant light, point light, and spotlight
- Create and modify a scene
- Create and modify a material
- Save and replay an image

SHADING A MODEL

The AutoCAD SHADEMODE command lets you produce a shaded picture of the 3D model in the current Viewport. It also provides various shading and wireframe options, and you can even edit shaded objects without regenerating the drawing. You have little control over lighting, and AutoCAD uses a single light that is logically placed just over your right shoulder in the current viewport.

Invoke the SHADEMODE command:

View menu	Choose Shade
Command: prompt	**shademode** (ENTER)

AutoCAD prompts:

Command: **shademode**
Enter option [2D wireframe/3D wireframe/Hidden/Flat/Gouraud/
fLat+edges/gOuraud+edges] <Flat+Edges>: *(right-click and select one of
the available options from the shortcut menu)*

2D Wireframe

The 2D Wireframe option displays the objects using lines and curves to represent the boundaries. Raster and OLE objects, linetypes, and lineweights are visible.

3D Wireframe

The 3D Wireframe option displays the objects using lines and curves to represent the boundaries. Raster and OLE objects, linetypes, and lineweights are visible. AutoCAD also displays a shaded *3D* user coordinate system (UCS) icon.

Hidden

The Hidden option displays the objects using the *3D* wireframe representation and hides the lines representing the back faces. Text objects are ignored unless it is given a thickness. The thickness can be as large a value as you want or as small as 0.000001 units (or any value other than 0).

Flat

The Flat option displays the objects shaded between the polygon faces. The objects appear flatter and less smooth than Gouraud shaded objects. Materials that have been applied to objects are shown.

Gouraud

The Gouraud option displays the shaded objects and smooths the edges between polygon faces, giving the objects a smooth, realistic appearance. Materials that have been applied to objects are shown.

Flat with edges on

The flat+edges option (default) displays the objects as a combination of Flat and Wireframe options. The objects are flat shaded with the wireframe showing through.

Gouraud with edges on

The gouraud+edges option displays the objects as a combination of Gouraud and Wireframe options. The objects are Gouraud shaded with the wireframe showing through.

Figure 16–1 *A 3D model rendered by AutoCAD (courtesy Autodesk)*

RENDERING A MODEL

The AutoCAD render facility allows you to create realistic models from your AutoCAD drawings. With the AutoCAD RENDER command you can adjust lighting factors, material finishes, and camera placement. All these options give you a great deal of flexibility. If you wish to take rendering "to the max," you may wish to purchase 3D Studio Max to gain the greatest flexibility and photorealism.

The tools available for rendering allow you to adjust the type and quality of the rendering, set up lights and scenes, and save and replay images. But you can always use the RENDER command without any other AutoCAD render setup. By default, RENDER uses the current view if no scene or selection is specified. If there are no lights specified, the RENDER command assumes a default over-the-shoulder distant light source with an intensity of 1.

Invoke the RENDER command:

Render toolbar	Choose the RENDER command (see Figure 16–2)
View menu	Choose Render > Render
Command: prompt	**render** (ENTER)

Figure 16–2 *Invoking the RENDER command from the Render toolbar*

AutoCAD displays the Render dialog box, as shown in Figure 16–3.

Figure 16–3 *Render dialog box*

The **Rendering Type:** list box lists the available rendering types, including AutoCAD Render, Photo Real, Photo Raytrace, and AutoVision (if installed). The default type is set to AutoCAD Render.

The **Scene to Render** list box lists the scenes available in the current drawing, including the current view, from which you can select the scene/view for rendering.

The **Rendering Procedure** section of the dialog box sets the default value for rendering. Three settings are available:

> The **Query for Selections** toggle controls whether or not to display a prompt to select objects to render.

> The **Crop Window** toggle controls the prompts for whether or not to pick an area on the screen for rendering.

> The **Skip Render Dialog** toggle controls whether or not to skip the Render dialog box and render the current view.

The **Light Icon Scale:** text field controls the size of the light blocks inserted in the drawing. The default is set to a scale factor of 1.0. If necessary, you can change the value to some other real number to rescale the blocks. Overhead, Direct, and Sh_spot are the blocks affected by the scale factor.

The **Smoothing Angle:** text field sets the angle at which AutoCAD interprets an edge. The default value is 45 degrees. Angles greater than 45 degrees are considered edges, and those less than 45 degrees as smoothed.

The **Rendering Options** section of the dialog box controls the rendering display. Four main settings plus additional options are available:

> The **Smooth Shade** option yields a smooth appearance. Depending on the object's surfaces and the direction of the screen's lights, smooth shading adds a cleaner, more realistic appearance to a rendering.

> The **Apply Materials** option applies surface materials you have defined and attaches them by color or to specific object(s). If the Apply Materials option is not selected, then all objects in the drawing assume the color, ambient, reflection, transparency, refraction, bump map, and roughness attribute values defined for the *GLOBAL* material.

> The **Shadows** option generates shadows when selected. This option is applicable only when Photo Real or Photo Raytrace rendering is selected.

> The **Render Cache** option specifies that rendering information be written to a cache file on the hard disk. As long as the drawing or view is unchanged, the cached file is used for subsequent renderings, eliminating the need for AutoCAD to retessellate.

> For fine-tuning the Rendering quality, choose the **More Options...** button. AutoCAD displays the Windows Render Options dialog box. The options available vary depending on whether you have selected Render, Photo Real, Photo Raytrace, or a third-party application as your rendering type.

The Destination section of the dialog box controls the image output setting. Three options are available:

> The **Viewport** option allows AutoCAD to render to the current viewport.

> The **File** option renders to a file. When the File option is selected, you must choose the **More Options...** button to select the file type, set the colors in the output file, and set the postscript options, if necessary.

> The **Render Window** option renders the model in a window, as shown in Figure 16–4.

The OPEN command (available from the File menu) in the Render window allows you to open three types of files: bitmap *(.BMP)* files and clipboard .dib, .rle *(.CLP)* files. The SAVE command (available from the File menu) in the render window allows you to save an image to a bitmap file. If necessary, you can also use the PRINT command (available from the File menu) to print the image. The COPY command (available from the Edit menu) can be used to copy an image from the active render window to the clipboard. The OPTIONS command (available from the File menu) displays the Windows Render Options dialog box, as shown in Figure 16–5, from which you can select aspect ratios (Size in Pixels) and resolutions (Color Depth) for bitmap images.

Figure 16–4 *Render window*

The **Sub Sampling** section of the Render dialog box controls the rendering time and image quality without abandoning effects such as shadows by rendering a fraction of all pixels. The available ratios include 1:1 (for best quality) to 8:1 (fastest).

The **Background...** button allows you to set the background for your scene. Choose the Background button to display the Background dialog box, as shown in Figure 16–6. Following are the four options available to select the type of background for rendering:

> The **Solid** option selects a one-color background. Use the controls in the **Colors** section to specify the color.

> The **Gradient** option lets you specify a two- or three-color gradient background. Use the Colors section controls and the Horizon, Height, and Rotation controls on the lower right of the dialog box to define the gradient.

The **Image** option enables you to use a bitmap *(.BMP)* file for the background. Manipulate the controls in the **Image** and **Environment** sections to define the bitmap.

The **Merge** option enables you to use the current AutoCAD image as the background.

Figure 16–5 *Windows Render Options dialog box*

Figure 16–6 *Background dialog box*

The **Fog/Depth Cue...** button in the Render dialog box provides settings for visual cues for the apparent distance of objects. Choose the **Fog/Depth Cue...** button to display the Fog/Depth Cue dialog box, as shown in Figure 16–7.

Figure 16–7 *Fog/Depth Cue dialog box*

The **Enable Fog** toggle sets the fog to ON and OFF without affecting the settings in the dialog box. The **Fog Background** toggle controls whether or not to apply the fog to the background as well as to the geometry. The **color controls** section of the dialog box controls whether AutoCAD uses the red-green-blue color system or the hue-lightness-saturation color system. The **Near/Far Distance** section of the dialog box defines where the fog starts and ends. Each value is a percentage of the distance from the camera to the back clipping plane. The Near/Far Fog Percentage section of the dialog box defines the percentage of fog at the near and far distances, ranging from 0% fog to 100% fog.

After making all the necessary changes in the Render dialog box, choose the **Render** button. AutoCAD renders the selected objects in the selected destination window.

SETTING UP LIGHTS

AutoCAD's render facility gives you great control over four types of lights in your renderings:

Ambient light can be thought of as background light that is constant and distributed equally among all objects.

Distant light gives off a fairly straight beam of light that radiates in one direction. Another property of distant light is that its brilliance remains constant, so an object close to the light receives as much light as a distant object.

Point light can be thought of as a ball of light. A point light radiates beams of light in all directions. Point lights also have more natural characteristics. Their brilliance may be diminished as the light moves away from its source. An object that is near a point light appears brighter; an object that is farther away will appear darker.

Spotlight is very much like the kind of spotlight you might be accustomed to seeing at a theater or auditorium. Spotlights produce a cone of light toward a target that you specify.

Invoke the LIGHTS command:

Render toolbar	Choose the LIGHTS command (see Figure 16–8)
View menu	Choose Render > Lights
Command: prompt	**light** (ENTER)

Figure 16–8 *Invoking the* LIGHTS *command from the Render toolbar*

AutoCAD displays the Lights dialog box, as shown in Figure 16–9.

Figure 16–9 *Lights dialog box*

CREATING A NEW LIGHT

In the Lights dialog box, select one of the three available light types from the list box located to the right of the **New...** button. Then choose the **New...** button. Depending on the type of light selected, AutoCAD displays the New Point Light dialog box (shown in Figure 16–10), New Distant Light dialog box (shown in Figure 16–11), or the New Spotlight dialog box (shown in Figure 16–12).

Figure 16–10 *New Point Light dialog box*

Figure 16–11 *New Distant Light dialog box*

Figure 16–12 *New spotlight dialog box*

Light Name

Specify the name of the light in the **Light Name:** text box. The name must be eight or fewer characters.

Intensity

The slider bar located below the **Intensity:** text box changes the brightness of the light, with 0 turning the light off. Distant light intensity values may range from 0 to 1. Point lights have a more complex intensity setting. This setting can be any real number. The factors that control the maximum intensity are the extents of the drawing and the current rate of falloff. Spotlight intensity factors are the same as for point lights except that the falloff is always inverse linear.

Position

The **Modify <** and **Show...** buttons located in the **Position** section let you modify or look at the X,Y,Z coordinate location of the light and its target. (Only available with Point and Spot Lights.)

Color

The **Color** section controls the current color of the light. To set the color you can either adjust the Red, Green, and Blue slider bars or choose the **Select Custom Color...** button (Only available with Point and Spot Lights.).

Attenuation

The setting of the attenuation controls how light diminishes over distance. The attenuation applies to both point light as well as spotlight. Select one of the three radio buttons. Selection of **None** sets no attenuation. Objects far from the point light are as bright as objects close to the light. Selection of **Inverse Linear** falloff decreases the light intensity linearly as the distance increases. For example, an object 10 units away from the light source will be 1/10 as illuminated as an object adjacent to the light source. Selection of **Inverse Square** falloff decreases the light intensity by the inverse of the squared distance. An object 10 units away from the light source receives 1/100 the amount of light of an item adjacent to the light source. This function provides rapid falloff so that a point light can be more localized.

Azimuth and Altitude

The **Azimuth and Altitude** edit boxes in the New Distant Light dialog box specify the position of the distant light by using the site-based coordinates. Azimuth can be set at any value between −180 and 180. Altitude can be set at any value between 0 and 90.

Light Source Vector

The **Light Source Vector** section in the New Distant Light dialog box displays the coordinates of the light vector that result from the light position you set using Azimuth and Altitude. You can also enter values directly in the edit boxes. AutoCAD updates the corresponding Azimuth and Altitude values.

Hotspot and Falloff

The **Hotspot and Falloff** edit fields in the New Spotlight dialog box specify the angle that defines the brightness cone of light and the full cone of light, respectively.

Once you set the appropriate values in the dialog box, choose the OK button to create a new light and close the dialog box. AutoCAD lists the name of the newly created light in the **Lights:** list box of the Lights dialog box.

AMBIENT LIGHT

The Ambient Light slider bar allows you to adjust the intensity of the ambient (background) light from a value of 0 (off) to 1 (bright). The **Color** sub-section controls the current color of the ambient light. To set the color you can adjust either the Red, Green, or Blue slider bars or choose the **Select Custom Color...** button or Select from ACI. Choose the OK button to accept the changes and close the Lights dialog box.

MODIFYING A LIGHT

In the Lights dialog box, first select the light name to modify from the **Lights:** list box. Then choose the **Modify...** button. AutoCAD displays the appropriate Modify dialog box, depending on the light type selected. Make the necessary changes, and choose the OK button to accept the changes. You cannot change a light type. For example, you cannot make a point light a distant light. But you can delete the point light and insert a new distant light in the same location.

DELETING A LIGHT

First, select the light name to delete from the **Lights:** list box. Then choose the **Delete** button. AutoCAD deletes the selected light.

SELECTING A LIGHT

The **Select <** button allows you to select a light from the screen. AutoCAD temporarily dismisses the dialog box while you specify a light on screen using the pointing device. The Lights dialog box returns, with the selected light highlighted in the **Lights:** list.

SETTING UP A SCENE

Inserting and adjusting lights allows you to render images from an unlimited number of viewpoints. If necessary, you can save a certain combination of lights and a particular view as a scene, which you can recall at anytime. A scene represents a particular view of the drawing together with one or more lights. Making a scene avoids recreating a particular set of conditions every time you need to render that image. The VPOINT and DVIEW commands are used to control viewing of a model from any point in model space, and the LIGHT command allows you to add one or more lights to the model or modify lights. The SCENE command allows you to save a scene. You can have an unlimited number of scenes in a drawing.

 Note: The VIEW command allows you to save a view but doesn't save the lights, whereas the scene can include both the view and the light positions.

Invoke the SCENES command:

Render toolbar	Choose the SCENES command (see Figure 16–13)
View menu	Choose Render > Scene
Command: prompt	**scene** (ENTER)

Figure 16–13 *Invoking the* SCENES *command from the Render toolbar*

AutoCAD displays the Scenes dialog box, as shown in Figure 16–14.

Figure 16–14 *Scenes dialog box*

AutoCAD lists the scenes available in the current drawing in the **Scenes:** list box.

CREATING A NEW SCENE

The New... button in the Scenes dialog box allows you to add a new scene to the current drawing. When you choose the **New...** button, AutoCAD displays the New Scene dialog box, as shown in Figure 16–15. (A similar dialog box is displayed when you select the **Modify...** button, except that dialog box is called Modify Scene).

Figure 16–15 *New Scene dialog box*

Scene Name

Specify the name of the scene in the **Scene Name:** text box, which may be up to eight characters long.

Views

The **Views** list box displays the list of the views in the current drawing. *CURRENT* is the current view in the active viewport. The active view in the current scene is highlighted. Selecting another view makes it the new view of the scene. You can have only one view in a scene.

Lights

The **Lights** list box displays the list of lights in the current drawing. *ALL* represents all the lights in the drawing. When you select the *ALL* option, all the lights in the drawing are added to the scene. The lights in the current scene are highlighted. Holding down the Ctrl key while selecting a nonhighlighted light adds that light to the scene. Selecting a highlighted light deselects that light and removes it from the scene.

 Note: You can create a scene with no lights, so that in this case the only lighting in the scene is ambient light.

Choose the **OK** button to create a new scene.

MODIFYING AN EXISTING SCENE

The **Modify...** button in the Scenes dialog box allows you to modify the selected scene. AutoCAD displays the Modify Scene dialog box and allows you to add or delete views and lights or change the name of the Scene.

DELETING A SCENE

The **Delete** button in the Scenes dialog box deletes the selected scene from the current drawing.

MATERIALS

The RMAT command gives you the power to modify the light reflection characteristics of the objects you will render. By modifying these characteristics, you make objects appear rough or shiny. These finish characteristics are stored in the drawing via surface property blocks. The drawing contains one surface property block for each finish you create, an attribute from the name, and AutoCAD color index (ACI) if assigned. You can modify materials by manipulating ambient, diffuse, specular, and roughness factors.

Invoke the RMAT command:

Render toolbar	Choose the MATERIALS command (see Figure 16–16)
View menu	Choose Render > Materials
Command: prompt	**rmat** (ENTER)

Figure 16–16 *Invoking the* MATERIALS *command from the Render toolbar*

AutoCAD displays the Materials dialog box, as shown in Figure 16–17.

The **Materials:** list box lists the available materials. The default for objects with no other material attached is *GLOBAL*.

Materials Library

Choosing the **Materials Library...** button displays the Materials Library dialog box, shown in Figure 16–18, from which you can select a material.

The Materials Library dialog box allows you to import a predefined material from an .MLI materials library into the current drawing. The **Current Drawing:** list box in the dialog box lists the materials currently in the drawing. The **Current Library** list box lists the materials available in the library file. If necessary, you can preview a sample of the material selected in the list. The sample can be applied to a sphere or square. You can preview only one material at a time. To import materials from the **Current Library** list box into the current drawing, first select the materials you want to import from the **Current Library** list box and then click the **<-Import** button. AutoCAD adds the selected materials to the **Current Drawing:** list box. Choose OK button to close the Materials Library dialog box, and AutoCAD returns control to the Materials dialog box.

Figure 16–17 *Materials dialog box*

Figure 16–18 *Materials Library dialog box*

Selecting Objects

The **Select** < button allows you to select an object to display the attached material. AutoCAD temporarily removes the Materials dialog box and displays the graphics area so you can select an object and display the attached material. After you select the object, the Materials dialog box reappears highlighting the material applied to the object, with the method of attachment displayed at the bottom of the dialog box.

Modify Material

The **Modify...** button allows you to modify a material by displaying the Modify Standard Material dialog box.

Creating Duplicate Material

The **Duplicate...** button duplicates a selected material and displays the New Standard Material dialog box. Make the necessary changes, and save the material with a new material name.

Creating a New Material

The **New...** button allows you to create a new material by displaying the New Standard Material dialog box.

Attaching a Material Property to an Object

The **Attach** < button displays the graphics area so you can select an object and attach the current material to it.

Detaching Material Property from an Object

The **Detach** < button displays the graphics area so you can select an object and detach the material from it.

Attaching AutoCAD Color Index

The **By ACI...** button displays the Attach by AutoCAD Color Index (ACI) dialog box, from which you can select the available ACI to attach to a material.

Attaching Material Property to a Layer

The **By Layer...** button displays the Attach by Layer dialog box, from which you can select a layer by which to attach a material.

Choose the OK button to accept the changes and close the Materials dialog box.

SETTING PREFERENCES FOR RENDERING

The Rendering Preferences dialog box allows you to establish the default settings for rendering. Invoke the RPREF command:

Render toolbar	Choose the RENDER PREFERENCES command (see Figure 16–19)
View menu	Choose Render > Preferences
Command: prompt	**rpref** (ENTER)

Figure 16–19 *Invoking the* RENDER PREFERENCES *command from the Render toolbar*

AutoCAD displays the Rendering Preferences dialog box, as shown in Figure 16–20. The available settings in the Rendering Preferences dialog box are same as in the Render dialog box. For a detailed explanation of the available settings, refer to the earlier section on "Rendering a Model."

Figure 16–20 *Rendering Preferences dialog box*

SAVING AN IMAGE

You can save the contents of the frame buffer to a BMP, TIFF, or TGA file format by invoking the SAVEIMG command.

Invoke the SAVEIMG command:

Tools menu	Choose Display Image > Save
Command: prompt	**saveimg** (ENTER)

AutoCAD displays the Save Image dialog box, as shown in Figure 16–21.

Figure 16–21 *Save Image dialog box*

Format

Choose the file format for the output image. You can save an image in any one of the three industry standard file formats: BMP, TGA, or TIFF.

Portion

The Portion section specifies the portion of the image to be rendered.

Choose the OK button to save the image. AutoCAD displays the Image File dialog box. Select the appropriate directory in which to save the file, and type the file name in the **File name:** text field. Choose the **Save** button to save the image to the given file name.

VIEWING AN IMAGE

The REPLAY command allows you to load a BMP, TIFF, or TGA file into the frame buffer for viewing, converting, or using as a background for a composite rendering.

Invoke the REPLAY command:

Tools menu	Choose Display Image > View
Command: prompt	**replay** (ENTER)

AutoCAD displays the Replay File dialog box. Select a file or enter a file name. After you select an image file, choose the OPEN button, and an Image Specifications dialog box is displayed, as shown in Figure 16–22.

Figure 16–22 *Image Specifications dialog box*

Image
The image tile in the Image Specifications dialog box lets you select a smaller part of the image you want to display. The default size of the image in the image tile reflects the entire display size in pixel measurement, with offset set to 0,0, the lower left corner of the image. To resize the image, specify two points, one for the lower left corner and the other for the upper right corner of the image in the image tile. AutoCAD automatically draws a box to mark the bounds of the reduced image area and updates the values in the X and Y Image Size coordinates.

Screen
The screen tile lets you adjust the offset location of the selected, sized image in relation to your screen. The tile displays the size of your screen or your current viewport. To change the offset, select a point in this tile to offset the center of the image to that

point. AutoCAD automatically redraws the image size boundaries to mark the new offsets and updates the values in the X and Y Screen Offset coordinates.

Reset

Resets the size and offset values to the original values.

Choose the **OK** button to close the Image Specifications dialog box.

STATISTICS

The STATISTICS command gives detailed information on your last rendering. This can be useful for diagnosing problems with your drawing.

Invoke the STATISTICS command:

Render toolbar	Choose the STATISTICS command (see Figure 16–23)
View menu	Choose Render > Statistics
Command: prompt	**stats** (ENTER)

Figure 16–23 *Invoking the* STATISTICS *command from the Render toolbar*

AutoCAD displays the Statistics dialog box. If necessary, you can save the information in the Statistics dialog box to a file.

REVIEW QUESTIONS

1. Text is ignored by the SHADE command.

   a. True

   b. False

3. Approximately where is the light source for the SHADE command?

   a. pure ambient light (i.e. all around)

   b. just behind your right shoulder

   c. directly behind the object

   d. in the upper right corner of the screen

4. It is possible to make a slide of a shaded image.

   a. True

   b. False

5. If you save your drawing while a shaded view is displayed, the preview of the drawing will appear shaded.

   a. True

   b. False

6. Which of the following is not a valid type of light you can add to a drawing for the RENDER command?

   a. point

   b. spot

   c. distant

   d. ambient

   e. none of the above (i.e. all are valid)

7. It is possible to make a slide of a rendered image.

   a. True

   b. False

8. If there is a single point light source in your drawing, objects will be darker if you set the attenuation for that light to:

   a. none

   b. inverse Linear

   c. inverse Square

   d. both b and c

9. Distant light sources do not attenuate (i.e. get dimmer with distance).

   a. True

   b. False

10. The SCENE command operates much like the VIEW command, but it also saves all the lights you have added to the drawing.

    a. True

    b. False

11. The maximum number of scenes which can be saved in a drawing is:

    a. 8

    b. 64

    c. 256

    d. 32,000

    e. unlimited (except by memory)

12. Material finishes can be assigned based on the color of the object in AutoCAD.

    a. True

    b. False

13. Which of the following file types is not available to the SAVEIMG command:

    a. BMP

    b. GIF

    c. TGA

    d. TIF

14. By default all lights emit white light.

    a. True

    b. False

15. The smoothing angle is the minimum angle:

    a. between faces which AutoCAD will ignore

    b. between the viewing angle and the face which AutoCAD will display the face

    c. between the incident light and the face which AutoCAD will still reflect the light

    d. none of the above

16. Both 2D and 3D models can be rendered using the SHADEMODE command?

    a. True

    b. False

17. The image option allows which type of files to be used as a background image?

    a. .GIF

    b. .BMP

    c. .JPG

    d. .TIF

18. Which AutoCAD command is used to control the position from which a 3D model is viewed in model space?

    a. VPORTS

    b. VIEWRES

    c. RENDER

    d. VPOINT

19. Which AutoCAD command is used to adjust the light reflection characteristics of an object?

    a. REFL

    b. LGTRFL

    c. LIGHTREF

    d. RMAT

20. Which of the following depicts the point at which a light's value fades to zero?

    a. fade

    b. highlight

    c. falloff

    d. diminish

CHAPTER 17

The Tablet and Digitizing

INTRODUCTION

The AutoCAD program consists of several major components in the form of program files, along with many other supporting files. One major part of the program, the menu files (in the forms of filename.MNU, filename.MNR, filename.MNS, filename.MNL, and filename.MNC), determine how various devices work with the program. The devices controlled by the menu include the buttons on the pointing device (mouse or tablet puck), the toolbars, pull-down menus, function box keys (not commonly used), and tablet. This chapter covers configuring and using the tablet menu.

First, using the tablet part of any menu requires that the AutoCAD program be installed and configured for the particular make and model of digitizing tablet that is properly connected to the computer.

 Note: The installation configuration (in ACAD2000) permits using a mouse as well as a digitizing tablet. But because a mouse and a tablet are both pointing devices, you must choose one or the other at any one time. This prestartup configuration is not to be confused with configuring the installed tablet with the TABLET command while in AutoCAD.

Second, a preprinted template (paperlike overlay to place on the digitizer surface) with up to four rectangular menu areas that has properly arranged columns and rows of pick areas within each menu area should be used. Each pick area corresponds to a command or line of programming in the filename.MNU menu file in effect. It is also necessary to set aside a screen area on the tablet if you wish to control the screen cursor with the tablet's puck, as shown in Figure 17–1.

After completing this chapter, you will be able to do the following:

- Configure the tablet menu
- Calibrate the tablet for digitizing

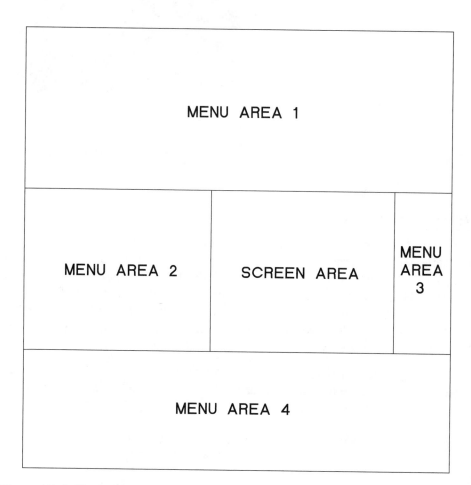

Figure 17–1 *The four screen areas of the overlay*

TABLET OPERATION

When a filename.MNU file, a digitizing tablet, and an overlay have been installed and properly set up to work together, you can use the digitizer's attached puck on the tablet surface to achieve the following results.

Normal operation The most common use of a tablet is to allow the user to move the puck and press the pick button while pointing to one of the various commands or symbols on a preprinted overlay, and be able to invoke that particular command or initiate a program (perhaps written by that user) that will draw the object(s) that that symbol represents. In addition, when the puck is moved within the overlay's designated screen area, the screen cursor will mimic the puck movement, thereby permitting the user to specify points or select objects on the screen.

Mouse movement Some tablets have an option that causes the puck to emulate mouse-type movement rather than absolute movement. Mouse emulation means that if you pick the puck up off of the tablet surface and move it to another place in the screen area, the screen cursor does not move. The cursor moves only with puck movement while it is on the tablet and in the screen area. Absolute (normal) tablet-puck operation means that, once configured, each point in the tablet's screen area corresponds to only one point on the screen. So, while in the normal mode, if you pick up the puck and put it down at another point in the tablet screen area, the cursor will immediately move to the screen's corresponding point.

Paper copying By switching Tablet mode to ON, you can cause points on the tablet to correspond to drawing coordinates rather than to screen pixel locations as it does in the normal or mouse operations just described. This allows you to fix a drawing (like a map) on the tablet surface, select two points on the map, and specify their coordinate locations on the map, after which the puck movement around the map will cause screen cursor movement to correspond to the same coordinates in the computer-generated drawing. Options and precautions for using this feature (referred to as *digitizing*) are discussed later in this chapter.

TABLET CONFIGURATION

The intended procedure is to have a preprinted template (the overlay) arranged on a sheet that can be fixed to the tablet. Tablet menu areas can then be configured to coincide with the template. Although you could try to configure a bare tablet, it would be difficult to select the required points for rectangular menu areas and also impractical to try and place a template on the tablet after it was configured in such a manner. However, if a tablet has been configured for one template, you can use another template in the same location without reconfiguring as long as the areas are the same. One benefit of this is being able to change from one set of icons/symbols or commands to another set without having to configure again. However, a change in the menu must be made in order to accommodate changes in the template, even if the configuration is the same.

The ACAD.MNC (compiled from ACAD.MNU) menu file supports a multibutton pointing device and the tablet overlay that is provided with the AutoCAD program package. That overlay is approximately 11" x 11" and has four areas for selecting icon/commands and a screen area, as shown in Figure 17–1. The pointing device (the puck furnished with every tablet) usually has three or more buttons (the menu supports up to a 10-button puck or mouse), and the cursor movement on the screen mimics the puck's movement in the tablet's configured screen area.

CUSTOM MENUS

Chapter 18 describes how to customize a menu file. Most of the explanations and examples refer to the screen menu, primarily because of its complexity. The same principles of customizing the tablet portion of a menu can be applied.

TABLET COMMAND

The TABLET command is used to switch between digitizing paper drawings and normal command/icon/screen area selections on a configured overlay. The TABLET command is also used to calibrate a paper drawing for digitizing or to configure the overlay to suit the current menu.

Invoke the TABLET command:

Tools menu	Choose Tablet
Command: prompt	**tablet** (ENTER)

AutoCAD prompts:

Command: **tablet**
Enter an option [ON/OFF/CAL/CFG]: *(right-click and select one of the available options from the shortcut menu)*

TABLET CONFIGURATION

The **CFG** option is used to set up the individual tablet menu areas and the screen pointing area. At this time a preprinted overlay should have been fixed to the tablet. Its menu areas should suit the menu you wish to use. The sequence of prompts is as follows:

Command: **tablet**
Option (ON/OFF/CAL/CFG): **cfg**
Enter number of tablet menus desired (0-4) <0>:

Select the number of individual menu areas desired (with a limit of 4). The next prompt asks:

Digitize upper left corner of menu area n:
Digitize lower left corner of menu area n:
Digitize lower right corner of menu area n:

The "n" refers to tablet menu areas of the corresponding tablet number in the menu. If the three corners you digitize do not form a right angle (90 degrees), you will be prompted to try again. Individual areas may be skewed on the tablet and with respect to each other, but such an arrangement usually does not provide the most efficient use of total tablet space. Tablet areas should not overlap.

The next prompts are:

> Enter the number of columns for menu area n:
> Enter the number of rows for menu area n:

Enter the numbers from the keyboard. The area will be subdivided into equal rectangles determined by the row and column values you have entered. If the values you enter do not correspond to the overlay row/column values, the results will be unpredictable when trying to use the tablet. Remember also that the overlay must suit the menu being used. After the menus areas have been specified, AutoCAD prompts:

Do you want to respecify the Fixed Screen Pointing Area? [Yes/No] <N>:

If you respond with **y**, AutoCAD prompts:

> Digitize lower left corner of Fixed Screen pointing area: *(specify a point with the pick button)*

> Digitize upper right corner of Fixed Screen pointing area: *(specify a point with the pick button)*

AutoCAD then prompts:

> Do you want to specify the Floating Screen Pointing area? [Yes/No] <N>:

If you respond with **y**, AutoCAD prompts:

> Do you want the Floating Screen Area to be the same size as the Fixed Screen Pointing Area? [Yes/No] <Y>:

If you respond with **n**, AutoCAD prompts:

> Digitize lower-left corner of the Floating Screen pointing area: *(specify a point with the pick button)*

> Digitize upper-right corner of the Floating Screen pointing area: *(specify a point with the pick button)*

In order to toggle the Floating Screen Area ON and OFF, press F12.

AutoCAD prompts:

> The F12 Key will toggle the Floating Screen Area ON and OFF. Would you also like to specify a button to toggle the Floating Screen Area? [Yes/No] <N>:

If you respond with y, AutoCAD prompts:

Press any non-pick button on the digitizer puck that you wish to designate as the toggle for the Floating Screen Area.

The standard AutoCAD overlay is installed as follows:

Command: **tablet**
Option (ON/OFF/CAL/CFG): **cfg**
Enter number of tablet menus desired (0-4) <default>: **4**
Do you want to realign tablet menu areas? <N>: **y** *(if required)*

At this time, digitize areas 1 through 4, as shown in Figure 17–2. The values for columns and rows must be entered as follows:

MENU AREA	COLUMN	ROW
1	25	9
2	11	9
3	9	7
4	25	7

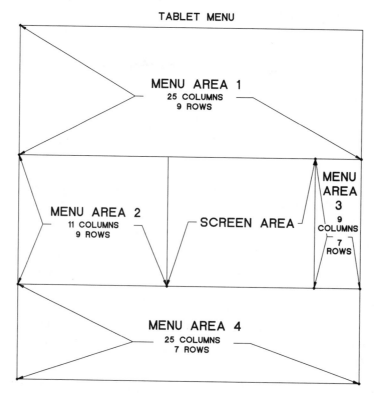

Figure 17–2 *Digitizing the screen areas of the tablet menu*

These values are for the ACAD.MNC menu. Other menus may vary.

ON/OFF OPTION

The default setting of the Tablet mode is OFF. The OFF setting does not incapacitate the tablet as you might think; rather, it means that you are not going to use the tablet for digitizing (making copies of paper drawings). With the Tablet mode set to OFF you can use the tablet to select command/icons in the areas programmed accordingly. You may also use the puck in the screen area of the tablet to control the screen cursor.

In order to digitize paper drawings, you must respond to the prompt as follows:

> Command: **tablet**
> Option (ON/OFF/CAL/CFG): **on**

Most systems have a toggle key to switch the Tablet mode ON and OFF. With many PCs, the toggle is either F10 or CTRL + T.

CALIBRATION

If the tablet has been calibrated already, the last calibration coordinates will still be in effect. If not, or if you wish to change the calibration (necessary when you move the paper drawing on the tablet), you can respond as follows:

> Option (ON/OFF/CAL/CFG): **cal**
> Digitize point #1: *(digitize the first known point)*

The point you select on the paper drawing must be one whose coordinates you know. The next prompt asks you to enter the actual paper drawing coordinates of the point you just digitized:

> Enter coordinates point for #1: *(enter those known coordinates)*

You are then prompted to digitize and specify coordinates for the second known point:

> Digitize point for #2: *(digitize the second known point)*
> Enter coordinates point #2: *(enter those known coordinates)*
> Digitize point #3 (or press ENTER to end): *(digitize the third known point
>   or press ENTER)*

An example of a drawing that might be digitized is a map, as shown in Figure 17–3.

If, for example, the map in Figure 17–3 has been printed on an 11" x 17" sheet and you wish to digitize it on a 12" x 12" digitizer, you can overlay and digitize on one-half of the map at a time. You may use the coordinates 10560,2640 and 7920,5280 for two calibrating points. But because X coordinates increase toward the left, you must consider them as negative values in order to make them increase to the right. Therefore, in calibrating the map, you may use coordinates –10560,2640 and –7920,5280 to calibrate the first half and coordinates –7920,2640 and –5280,5280 for the second half.

The points on the paper should be selected so that the X values increase toward the right and the Y values increase upward.

Once calibration has been initiated in a particular space (model or paper), turning on the Tablet mode must be done while in that particular space.

Figure 17–3 *An example of a drawing digitized as a map*

TRANSFORMATION OPTIONS

Tablet Calibration can be done by one of several transformation methods. These include Orthogonal, Affine, Projective, and Multiple-Point. The method you choose may depend on the condition of the map/drawing that is to be digitized and the desired accuracy.

Orthogonal

This option involves two points. It results in uniform scaling and rotation. The translation is arbitrary. Orthogonal (two-point) translation is most suitable for tracing paper drawings that are dimensionally and rotationally accurate (right angles are not skew). It is advisable to use this option for long, narrow applications (a pipeline, for example).

Affine

This option involves three points and can be applied to a paper drawing with scale factors, rotation, and right angle representations that are not to an acceptable accuracy in two dimensions. This can be applied to drawings with parallel lines that are represented parallel, but with an *X* direction scale/*Y* direction scale differential that is out of tolerance. Right angles may not be represented by right angles.

Whether or not you should use the Affine option depends on whether or not those lines that should be parallel are represented by parallel lines on the paper drawing. You can check the report displayed (in the form of a table) when you have digitized at least three points. If the RMS error (described in this section) is small, then the Affine option should be acceptable.

Projective

This option involves four points, to simulate a translation comparable to a perspective in which points from one plane converge while passing through another plane (skew to the first) to one point of view. This option is applicable to copying paper sheets with irregularities that differ from one area to another (known as rubber-sheeting) and parallel lines that are not always represented as parallel. However, lines do project as lines.

Multiple-Point

The Multiple-Point option will accept more than four points to be digitized and specified. AutoCAD will use the "least squares" method to optimize the calibration to compensate for imperfections in the paper drawing. The accuracy of this method is proportional to the number of points used.

THE CALIBRATION TABLE

If you use three or more points during calibration, AutoCAD computes the range or error (if any) and reports the results and displays information designed to help you determine if the paper copying process is acceptable. An example of a calibration table follows:

Four Calibration Points			
Transformation type:	Orthogonal	Affine	Projective
Outcome of fit:	Success	Success	Exact
RMS Error:	143'-6.2"	73'-4.1"	
Standard deviation:	62'-2.7"	1'-7.8"	
Largest residual:	193'-1"	74'-1"	
At point:	2	3	
Second largest residual:	177'-9"	7'-7"	
At point:	3	2	

OUTCOME OF FIT

Cancelled Cancelled occurs only with projective transformations. It indicates the fit has been cancelled.

Exact The number of points used was exactly correct and the transformation defined from them was valid.

Failure Points selected, though the correct number, were probably collinear or coincidental.

Impossible Insufficient points were selected for the transformation type under which "Impossible" is reported.

Success The transformation defined was valid and more points were used than was required.

RMS Error

If the transformation is reported as a "Success," then the RMS (root mean square) error is reported. This is the square root of the average of the squares of the distances (called residuals) of each selected point from their respective targets.

Standard Deviation

This reports the standard deviation of all residuals (the distance each point misses its target).

Point(s)/Residual(s)

The two points whose residuals (see "RMS Error" and "Standard Deviation") are largest and second largest are reported along with their respective residual values.

TABLET MODE AND SKETCHING

Sketching while in Tablet mode operates similar to sketching with a mouse or on a tablet with the Tablet mode off. The difference is that the entire tablet surface is used for digitizing while the Tablet mode is on, making the maximum area available for tracing but making the pull-down menus inaccessible.

EDITING SKETCHES

Once sketched lines have been recorded and the SKETCH command has been terminated, you can use regular editing commands (like COPY, MOVE, ERASE) to edit the individual line segments or sketched polylines (discussed next) just as though they had been drawn by the LINE or PLINE command. In the case of sketched polylines, the PEDIT command can be used for editing.

SKETCHING IN POLYLINES

You can cause AutoCAD to make the created sketch segments into polylines instead of lines by setting the SKPOLY system variable to a nonzero value.

LINETYPES IN SKETCHING

You should use the Continuous linetype while sketching, whether with regular lines or polylines.

REVIEW QUESTIONS

1. What filename extension is used to store the menu command entries for a tablet menu?

   a. .TAB

   b. .MNU

   c. .CMD

   d. .PGP

   e. .INI

2. How many different tablet menu areas can be specified on a digitizing pad?

   a. 1

   b. 2

   c. 3

   d. 4

   e. 5

3. When aligning a tablet menu with the digitizing pad, how many points are required?

   a. 1

   b. 2

   c. 3

   d. 4

   e. 5

4. To toggle the digitizing pad between menu functions and paper copying, you press:

   a. {F10}

   b. {F11}

   c. {F12}

   d. {CTRL} + {O}

   e. none of the above

5. The minimum number of points required to calibrate a digitizing pad using an Affine calibration is:

   a. 1

   b. 2

   c. 3

   d. 4

   e. 5

6. If you have an isometric drawing on paper and wished to digitize an orthographic view of the top, what type of calibration would you use?

   a. 2 point (Orthogonal)

   b. 3 point (Affine)

   c. 4 point (Projective)

   d. cannot be done

7. If you have photograph of a building and wished to digitize the front elevation, what type of calibration would you use?

    a. 2 point (Orthogonal)

    b. 3 point (Affine)

    c. 4 point (Projective)

    d. cannot be done

8. What does RMS stand for in RMS error?

    a. Real Measure Statistic

    b. Root Mean Square

    c. Radical Motion Setting

    d. ReMainder Sum

    e. none of the above

9. If you select three points in a straight line to calibrate the digitizing pad, what "Outcome of Fit" will AutoCAD report for an Affine fit:

    a. Canceled   d. Impossible

    b. Exact   e. Success

    c. Failure

10. When using the SKETCH command, what system variable determines if lines segments or polylines?

    a. SKLINE   d. LINESEG

    b. SKPOLY   e. POLYGEN

    c. SKTYPE

11. What command will allow you to generate freehand lines when digitizing a paper drawing?

    a. DIGITIZE   d. FREEHAND

    b. TABLET   e. DRAW

    c. SKETCH

12. The option of the TABLET command which allows for the configuration of a tablet menu is:

    a. ON   d. CFG

    b. OFF   e. MENU

    c. CAL

13. The RMS error will always be lower (or equal) for a projective fit versus an orthogonal fit when 6 points are selected.

    a. True   b. False

14. AutoCAD's default menu template has how many menu areas?

    a. 1   d. 4

    b. 2   e. 5

    c. 3

15. To toggle the floating screen area on and off press which of the following?

    a. F10   d. CTRL + F

    b. F11   e. none of the above

    c. F12

16. The three corners that are selected to define the tablet menu area must form a _____ degree angle?

    a. 30   c. 60

    b. 45   d. 90

17. Which of the following is used to toggle the tablet mode on and off?

    a. CTRL + B

    b. CTRL + M

    c. CTRL + T

    d. none of the above

18. Which of the following is not a tablet transformation method?

    a. Orthogonal

    b. Affine

    c. Polygonal

    d. Projective

    e. Multi-Point

19. The Projective transformation option allows copying paper with irregularities from one area to another. This type of copying is known as _____.

    a. Rubber banding

    b. Rubber sheeting

    c. Rubber stamping

    d. Applique

Customizing AutoCAD

INTRODUCTION

Off the shelf, AutoCAD is extremely powerful. But, like many popular engineering and business software programs, it does not automatically do all things for all users. It does—probably better than any software available—permit users to make changes and additions to the core program to suit individual needs and applications. Word processors offer a feature by which you can save a combination of many keystrokes and invoke them at any time with just one or a combination of two keystrokes. This is known as a *macro*. Spreadsheet and database management programs (as well as other types of programs) have their own library of user functions that can also be combined and saved as user-named, custom-designed commands. These programs also allow you to create and save standard blank forms for later use, to be filled out as needed. Using these features to make your copy of a generic program unique and more powerful for your particular application is known as *customizing*.

Customizing AutoCAD can include several facets requiring various skill levels. The topics in this chapter include creating command aliases, linetypes, shapes, and hatch patterns; menu customization; and command macros. A text editor that lets you save a file in ASCII or plain text format such as Microsoft Windows NOTEPAD.EXE is needed in order to customize AutoCAD menus. A macro is a sequence of commands executed by a single user selection from a custom menu. Macros can be written that help maintain certain standards in design and drafting conventions, allowing the user to concentrate on the project at hand.

After completing this chapter, you will be able to do the following:

- Create command aliases
- Create, modify, and understand the AutoCAD Menu Structure
- Create custom linetypes, hatch patterns, and fonts

EXTERNAL COMMANDS AND ALIASES

AutoCAD allows you to run certain programs, including internal and external DOS commands. Windows brings true multitasking to AutoCAD by making it possible to use AutoCAD simultaneously with word processors, database and spreadsheet programs, and many other applications. You can also include aliases in the ACAD.PGP file for regular AutoCAD commands. An alias is nothing but a nickname.

Command aliasing provides an alternate keystroke for invoking a command, not options of the command. The ZOOM command has a Windows option; however, you cannot define an alias "ZW" for ZOOM WINDOW. There is an alias for ZOOM, the letter "Z". Prior to modifying the ACAD.PGP file, it is recommended that you make a backup copy of the file, such as XACAD.PGP, so that if you make a mistake, you can restore the original version. Following is an extract from the ACAD.PGP file.

```
;  — Command Aliases —
;  You can abbreviate frequently-used AutoCAD commands by defining
;  aliases for them in the command alias section of acad.pgp.
;  You can create a command alias for any AutoCAD command,
;  device driver command, or external command.
;  Recommendation: back up this file before editing it.
;  External command format:
;  <Command name>,[<DOS request>],<Bit flag>,[*]<Prompt>,
;  The bits of the bit flag have the following meanings:
;  Bit 1: if set, don't wait for the application to finish
;  Bit 2: if set, run the application minimized
;  Bit 4: if set, run the application "hidden"
;  Bit 8: if set, put the argument string in quotes
;
;  Fill the "bit flag" field with the sum of the desired bits.
;  Bits 2 and 4 are mutually exclusive; if both are specified,
;  only
;  the 2 bit is used. The most useful values are likely to be 0
;  (start the application and wait for it to finish), 1 (start the
;  application and don't wait), 3 (minimize and don't wait), and 5
;  (hide and don't wait). Values of 2 and 4 should normally be
;  avoided,
;  as they make AutoCAD unavailable until the application has
;  completed.
;
;  Bit 8 allows commands like DEL to work properly with filenames
;  that
;  have spaces such as "long filename.dwg".  Note that this will
;  interfere
;  with passing space-delimited lists of file names to these same
;  commands.
;  If you prefer multiple file support to using long file names,
```

```
;  turn off
;  the "8" bit in those commands.
;  Examples of external commands for command windows:
CATALOG, DIR /W,    0,File specification: ,
DEL,DEL,      0,File to delete: ,
DIR, DIR,     0,File specification: ,
EDIT,START EDIT,    1,File to edit: ,
SH,, 1,*OS Command: ,
SHELL,,      1,*OS Command: ,
START,START,         1,*Application to start: ,
TYPE,TYPE,  0,File to list: ,
;  Examples of external commands for Windows:
;  See also the STARTAPP AutoLISP function for an alternative
;  method.
EXPLORER, START EXPLORER, 1,,
NOTEPAD,   START NOTEPAD,  1,*File to edit: ,
PBRUSH,    START PBRUSH,   1,,
;  Command alias format:
;  <Alias>,*<Full command name>
;  Sample aliases for AutoCAD commands:
;  These examples include most frequently used commands.
3A,   *3DARRAY
3F,   *3DFACE
3P,   *3DPOLY
A,    *ARC
AA,   *AREA
AL,   *ALIGN
AP,   *APPLOAD
AR,   *ARRAY
AAD, *ASEADMIN
AEX, *ASEEXPORT
ALI, *ASELINKS
ASQ, *ASESQLED
ARO, *ASEROWS
ASE, *ASESELECT
AT,  *DDATTDEF
-AT, *ATTDEF
ATE, *DDATTE
-ATE,         *ATTEDIT
B,    *BMAKE
-B,  *BLOCK
BH,  *BHATCH
BO,  *BOUNDARY
```

```
-BO, *-BOUNDARY
BR,  *BREAK
C,   *CIRCLE
CH,  *PROPERTIES
```

The External Commands are defined at the top of the ACAD.PGP file.

The format for a command line is as follows:

```
<Command name>,<executable>,<Memory reserve>,[*]<Prompt>,
   <Return code>
```

An example of lines in an ACAD.PGP file for specifying external commands is as follows:

```
SH,,0, *OS Command: ,0
SHELL,,0, *OS Command: ,0
TYPE,TYPE, 0, File to type: ,0
CATALOG,DIR/W,0, File specification: ,0
DEL,DEL, 0, File to erase: ,0
DIR,DIR, 0, File specification: ,0
EDIT,, 0, File to edit: ,0
```

The command name (to be entered at the "Command:" prompt) should not be the same as an AutoCAD command and should be in uppercase characters.

The executable string is sent to the operating system as the name of a command. It can contain parameters and switches.

The memory amount must contain a number (usually 0). It maintains compatibility with previous versions of AutoCAD.

The prompt (optional) is used to inform the operator if additional input is necessary. If the prompt is preceded by an asterisk (*), then the user response may contain spaces. The response must be terminated by pressing ENTER. Otherwise, pressing SPACEBAR or ENTER will terminate the response.

The return code is a bit-coded specification. The number you specify will represent one or more of the bitcodes. For example, if you specify 3, then bitcodes 1 and 2 will be in effect. If you specify 5, then bitcodes 1 and 4 will be in effect. The values are as follows:

0: Return to text screen.

1: Load DXB file. This causes a file named $cmd.dxb to be loaded into the drawing at the end of the command.

2: Construct block from DXB file. This causes the response to the prompt to become the name of a block to be added to the drawing, consisting of objects in the $cmd.dxb file written by the file command. This code must be used in conjunction with bitcode 1. This may not be used to redefine a previously defined block.

4: Restore text/graphics mode. If this bitcode is included, the mode you were in (text or graphics) will be returned to when the command is completed; otherwise, you will be in text mode.

The format to define an alias is as follows:

```
<Alias>,*<Full command name>
```

The abbreviation preceding the comma is the character or characters to be entered at the "Command:" prompt. The asterisk (*) must precede the command you wish to be invoked. It may be a standard AutoCAD command name, a custom command name that has been defined in and loaded with AutoLISP or ADS, or a display or machine driver command name. Aliases cannot be used in scripts. You can prefix a command in an alias with the hyphen that causes a command line version to be used instead of a dialog box, as shown here:

```
BH, *-BHATCH
```

See Appendix H for list of available aliases from the default ACAD.PGP file.

CUSTOMIZING MENUS

When you launch the standard version of AutoCAD 2000, you are presented with the standard set of menus and toolbars. Selecting an item from a menu or toolbar might execute a command, an AutoLISP routine, or a macro, or cause another menu to be displayed. Menus are user definable and are created/edited using text editors. Menu files also define the functionality and appearance of the menu area. If you perform an application-specific task on a regular basis that requires multiple steps to accomplish this task, you can place this in a menu macro and have AutoCAD complete all the required procedures in a single step while pausing for input if necessary. Menu macros are similar to script files (files ending with *.SCR). Script files are also capable of executing many commands in sequence but have no decision-making capability and cannot pause for interactive user input.

AutoCAD2000 supports the following kinds of menus:

- Pull-down and shortcut menus
- Screen menus
- Image tile menus
- Pointing device menus
- Tablet menus

- Toolbars
- Keyboard accelerators
- Help strings and tool tips
- Menu groups

MENU FILE TYPES

Prior to AutoCAD Release 13 for Windows, there were two types of menu files: ASCII text files ending with the file extension .MNU, and a compiled version of the same file with the extension .MNX. The ACAD.MNU file is provided with the AutoCAD program, and if, when you launch AutoCAD, the ACAD.MNX file is not present or any changes were made to the ACAD.MNU file, AutoCAD automatically compiles and creates a new ACAD.MNX. If you created your own custom menu—for example, CUSTOM.MNU—then upon loading the menu, AutoCAD automatically creates CUSTOM.MNX. The ACAD.MNU and ACAD.MNX files remained applicable in the DOS and UNIX versions of AutoCAD Release 14. Because of the additional functionality introduced in the Windows version of AutoCAD Release 14, a new menu scheme was introduced.

The following table lists the menu files used by AutoCAD for current releases:

Menu File Type	Description
.MNU	Template menu file, ASCII text
.MNC	Compiled menu file. This binary file contains the command strings and menu syntax that defines the functionality and appearance of the menu.
.MNR	Menu resource file. This binary file contains the bitmaps used by the menu.
.MNS	Source menu file (AutoCAD generated)

AutoCAD 2000 comes with the ACAD.MNU file. This is located in the "ACAD2000 \SUPPORT" folder, assuming AutoCAD is installed on the C: drive and AutoCAD 2000 is the subdirectory. If your location is different, substitute the drive and path location for your situation. The .MNS file is similar in structure to the .MNU file and is created automatically and dynamically updated by AutoCAD when you add new toolbars. The .MNC file compiles when you reload AutoCAD or dynamically update the menu. The .MNR file is created automatically when .MNS is completed and when buttons on the toolbars are created, changed, or added. The .MNR menu resource file is used for storing bitmaps (*.BMP) for the icons and is a binary file.

When you add toolbar information to the menu, the *.MNS file is the file that is updated. The toolbars are denoted in the ***TOOLBARS major section. If you delete the *.MNS file, you will lose the newly created toolbars, because AutoCAD creates a new *.MNS file based on the *.MNU file. If you want to add the toolbars to your *.MNU file, copy the ***TOOLBARS section from your *.MNS file and paste it into your *.MNU file. If you delete your *.MNS file, AutoCAD rebuilds it based on the information found in the *.MNU file. AutoCAD has another file that is constantly updated, called the ACAD.INI file. Windows applications contain many *.INI files. This is where the initialization and configuration is kept for Windows applications. When you move toolbars and dock menus, the information is read to the ACAD.INI file so that when you launch AutoCAD, the menus are located in their new positions.

MENU FILE STRUCTURE

Menus are divided into sections relating to specific menu areas. Menu sections can contain submenus that you can reference and display as needed. The command strings and macro syntax that define the result of a menu selection are called menu macros. The Windows NOTEPAD editor program can only open a file size less than 64K in memory. To view the contents of the ACAD.MNU file, use the Windows WordPad program, which is located in the Accessories group. Prior to opening the ACAD.MNU file, copy it and save it as XACAD.MNU. Then if something should happen, you can restore the original. It is important that files remain as text files without any print or formatting codes.

Following are the major sections in menu files for DOS and Windows versions (major sections are denoted by ***):

***BUTTONSn_Pointing device button menu, where n is either 1 or 2.

***BUTTONS1_The normal menu used by a mouse or tablet cursor (puck).

***BUTTONS2_Holding down SHIFT and pressing the right mouse button (or on a three-button mouse, pressing the middle button) activates the menu.

***AUXn_Auxiliary device button, where n is either 1 or 2.

***POPn_Pull-down and cursor menus, where n is a number from 0 to 16 (*Note:* 0 is used only for cursor menus).

***POP0_The cursor menu, which follows the crosshairs. Pressing SHIFT and the left mouse button (or pressing button 3 on a digitizing puck) activates this menu.

***POP1 through ***POP16_The pull-down menus. The AutoCAD regular menu file does not use all the available pull-down menus.

***SCREEN_Screen menu area, which slowly and surely is disappearing from AutoCAD for Windows. It may not be supported in future versions.

***TABLETn_Tablet menu area, where n is a number from 1 to 4.

***TABLET1 through ***TABLET4_The four menu areas of a digitizing tablet.

***TOOLBARS_Toolbar menus are the most flexible and easily customized method of interface now available for the AutoCAD user. Groups of buttons can be placed around or on the screen in changeable proportions in toolbars. These buttons can be used to invoke commands, macros, or user-defined programs that can make drawing easier and more accurate.

You can use the Find feature of Windows WordPad to locate the major sections.

Menu system does not have to use all of the preceding major menu sections. AutoCAD for Windows has additional major sections explained later in the chapter.

Following is the menu code taken from the ACAD.MNU file.

```
***POP1
**FILE
ID_MnFile     [&File]
ID_New        [&New...\tCtrl+N]^C^C_new
ID_Open       [&Open...\tCtrl+O]^C^C_open
ID_FILE_CLOSE [&Close]
ID_PartialOp  [$(if,$(eq,$(getvar,fullopen),0),,~)Pa&rtial
    Load]^C^C_partialload
              [--]
ID_Save       [&Save\tCtrl+S]^C^C_qsave
ID_Saveas     [Save &As...]^C^C_saveas
ID_Export     [&Export...]^C^C_export
              [--]
ID_PlotSetup  [Pa&ge Setup...]^C^C_pagesetup
ID_PlotMgr    [Plotter &Manager...]^C^C_plottermanager
ID_PlotStyMgr [Plot St&yle Manager...]^C^C_stylesmanager
ID_Preview    [Plot Pre&view]^C^C_preview
ID_Print      [&Plot...\tCtrl+P]^C^C_plot
              [--]
ID_MnDrawing  [->Drawing &Utilities]
ID_Audit      [&Audit]^C^C_audit
ID_Recover    [&Recover...]^C^C_recover
              [--]
ID_MnPurge    [->&Purge]
ID_PurgeAll     [&All]^C^C_purge _a
                [--]
ID_PurgeLay     [&Layers]^C^C_purge _la
ID_PurgeLin     [Li&netypes]^C^C_purge _lt
ID_PurgeTxt     [&Text Styles]^C^C_purge _st
```

```
ID_PurgeDim      [&Dimension Styles]^C^C_purge _d
ID_PurgeMln      [&Multiline Styles]^C^C_purge _m
ID_PurgeBlk      [&Blocks]^C^C_purge _b
ID_PurgePlt      [&Plot Styles]^C^C_purge _p
ID_PurgeShp      [<-<-&Shapes]^C^C_purge _sh
ID_SendMail   [Sen&d...]
ID_Props      [Drawing Propert&ies...]^C^C_dwgprops
              [--]
ID_MRU        [Drawing History]
              [--]
ID_APP_EXIT   [E&xit]
```

POP1 is the major section, which is the File (first) menu on the menu bar of AutoCAD. All the words that begin with ID_ are menu tag names. Labels are enclosed in square brackets []. The ampersand (&) is how you underscore the following letter, which allows menu selections from the keyboard, with the ALT + "letter" combination common to all Windows programs. The label (or letters between the brackets) is what appears in the pull-down menu. The first label in the menu is what appears on the menu bar as "File"; all other labels appear on the menu itself. This is only for POP menus 1 to 16 and IMAGE menus. The text following the closing bracket is the command macro that AutoCAD will execute.

Here is the syntax:

```
[&Open...\tCtrl+O]^C^C_open
```

The ^C^C is the cancel command, which is followed by the open command. The symbol ^ before the C is a caret, which is SHIFT + 6 on the keyboard. This combination of characters executes the ESC sequence to cancel any previous command. Note that there are two of them. This is because some AutoCAD commands require the user to press ESC twice before canceling a command. After the cancellation, the OPEN command will execute. This is an example of a simple macro.

MENU MACRO SYNTAX

Following is a partial list of the codes you will encounter in menu macros:

Syntax	Description
***	Denotes major sections of the menu
**	Denotes subsections located between major sections
[ ]	Menu label
;	Semicolon; equivalent to pressing ENTER on the keyboard
space	A space character; equivalent to pressing ENTER on the keyboard

Syntax	Description
\	Pause for user input
'	Issue a command transparently while in another command
*	Repeat a command until user cancels
+	Allows the long macros to be continued onto the next line

The following example macro will create a layer called "EL_OFFEQ" (Electrical Office Equipment), assign the color "RED", and make it the current layer:

```
[EL_OFFEQ]^C^C-LAYER;M;EL-OFFEQ;C;RED;;;
```

This is the equivalent of typing the -LAYER command, selecting the Make option, typing "EL-OFFEQ" as the desired layer name, selecting the Color option to assign the color Red, and finally, pressing ENTER three times to exit the command. Selecting this macro will execute all of this in one operation. Use the Make option of the -LAYER command in case the layer does not yet exist. If the layer does exist it will become the current layer. Note that there is no space after the ^C^C.

EXAMPLE OF CREATING A MENU FILE

Let's go through the sequence of steps in creating a menu file incorporating all the facets of menu sections. In order to create the menu file, first create the blocks with attributes shown in Figure 18–1, and save the drawing in your current working directory.

Figure 18–1 shows three blocks: a computer, a table, and a chair. Create the objects on layer 0 and the attribute definitions on layer ATTRIB. Make layer 0 "white" and layer ATTRIB "yellow." Define the attributes as visible, and verify. Following is the list of attribute tags with the corresponding block names:

Figure 18–1 *Drawing showing the three blocks required to complete the menu exercise*

Block Name	Attribute Flag
COMPUTER	CPU
	HD
	MONITOR
TABLE	MANUF
	COST
	FINISH
CHAIR	MANUF
	COLOR
	MATL

Save the blocks as separate drawing files, COMP.DWG, CHAIR.DWG, and TABLE1.DWG, by using the WBLOCK command.

Create a menu system file called TEST.MNU that will insert various items of furniture into a drawing using either a standard set of values or a custom set of values. The purpose of the menu system is to insert the blocks with a single pick from a menu that will automatically create the standard layer for insertion of the block, allow for custom values (if the Custom option is selected), and set layer 0 as the current layer.

Invoke the AutoCAD program from the Windows file manager. Begin a new drawing, and launch the Windows NOTEPAD editor from the Accessories group. Type the following code, and save the file as TEST.MNU in your current working directory.

```
***POP1
**Furniture
[&Furniture]
[->&Standard Furniture]
  [&Computer]^C^C^CATTREQ;0;-LAYER;M;EL_OFFEQ;COLOR;1;;;+
-INSERT;COMP;\;;\ATTREQ;1;-LAYER;SET;0;;
  [&Table]^C^C^CATTREQ;0;-LAYER;M;FR_OFF;COLOR;3;;;+
-INSERT;TABLE1;\;;\ATTREQ;1;-LAYER;SET;0;;
  [<-C&hair]^C^C^CATTREQ;0;-LAYER;M;FR_OFF;COLOR;3;;;+
-INSERT;CHAIR;\;;\ATTREQ;1;-LAYER;SET;0;;
[->Custo&m Furniture]
  [Com&puter]^C^C^CATTDIA;0;-LAYER;M;EL_OFFEQ;COLOR;1;;;+
-INSERT;COMP;\;;\\\\\\\-LAYER;S;0;;ATTREQ;1
  [Tab&le]^C^C^CATTDIA;0;-LAYER;M;FR_OFF;COLOR;3;;;+
-INSERT;TABLE1;\;;\\\\\\\-LAYER;S;0;;ATTREQ;1
  [<-Chai&r]^C^C^CATTDIA;0;-LAYER;M;FR_OFF;COLOR;3;;;+
-INSERT;CHAIR;\;;\\\\\\\-LAYER;S;0;;ATTREQ;1
[--]
[Layer EL_OFFEQ]^C^C^C-LAYER;M;EL_OFFEQ;COLOR;1;;;
[Layer FR_OFF]^C^C^C-LAYER;M;FR_OFF;COLOR;3;;;
[--]
[Layer 0]^C^C^C-LAYER;SET;0;;
```

Go back to the AutoCAD program, and load the TEST.MNU menu file by typing MENU at the "Command:" prompt and then pressing ENTER or SPACEBAR. The Select Menu File dialog box appears, as shown in Figure 18–2.

Select the TEST.MNU file from the appropriate drive and directory, and choose the Open button. TEST.MNU will replace ACAD.MNU. The right mouse button on your tablet puck/mouse (or ENTER) will not respond to any action, because in the TEST.MNU file the ***BUTTONS1 major section is not defined. Only the pull-down menu Furniture is displayed, as shown in Figure 18–3.

Figure 18–2 *Select Menu File dialog box*

To test the menu, select the Computer option from the Standard Furniture cascading menu. AutoCAD prompts for an insertion point and rotation angle. From the Custom Furniture cascading menu, select the Computer option. AutoCAD prompts for the insertion point and rotation angle, and in addition prompts for three attribute values. If you come across an error message, go back to the TEST.MNU file, check the syntax, and make the necessary changes. Reload the TEST.MNU file. Be sure you enter the full name and extension when you specify TEST.MNU so that the source file TEST.MNS and compiled file TEST.MNC are updated with the latest changes or corrections. Invoke the commands again, and make sure it is working properly. Let's examine some of the unique features of the TEST.MNU file line by line.

```
***POP1
```

Indicates that this is a pull-down menu located at position 1.

```
[&Furniture]
```

Figure 18–3 *Pull-down menu Furniture*

This is the title that will appear on the menu bar. The ampersand before the F will underline the letter F, allowing the menu to be selected using the ALT + F combination.

```
[->&Standard Furniture]
```

This is the beginning of the cascading menu, as noted by the "->." The letter S will be underscored and "Standard Furniture" will appear as the first item in the "Furniture" pull-down menu.

```
[&Computer]^C^C^CATTREQ;0;-LAYER;M;EL_OFFEQ;COLOR;1;;;+
-INSERT;COMP;\;;\ATTREQ;1;-LAYER;SET;0;;
```

This is the first item available in the cascaded menu. The ^C's will cancel any previous command. It is advisable to use three ^C combinations, because some commands require more than two exits to terminate. Following the ^C is ATTREQ, a setvar variable whose function is to control attribute requests. If it has a value of 0, the block will be inserted without prompting for attributes using standard values, which, in this case, is what we want. Note the use of the semicolon, which is equivalent to pressing

ENTER. Next, the -LAYER command is invoked, creating an "EL_OFFEQ" layer and setting the color to "RED" (1 is the code number for the color RED), followed by three returns to terminate the layer command. The hyphen prefix causes the prompts to be invoked at the Command: prompt instead of using the dialog box. The use of the "+" symbol at the end of the line allows the code to continue onto the next line. There are no spaces between the final return ";" and the "+". The macro continues by invoking the -INSERT command and inserting a block called COMP. The first "\" backslash character after the block name causes a pause, which allows the user to pick an insertion point. The two semicolons (;;) that follow the backslash character accept the default scale factors. The next backslash character is for the user to pick a rotation angle. The ATTREQ setvar variable is reset to the default value of 1. The final command restores layer 0 as the current layer.

```
[->Custo&m Furniture]
[Com&puter]^C^C^CATTDIA;0;-LAYER;M;EL_OFFEQ;COLOR;1;;;+
INSERT;COMP;\;;\\\\-LAYER;S;0;;ATTDIA;1
```

This code is for a custom Computer option from the Custom Furniture cascading menu. In the Custom menu, the intent is to allow the operator to insert a block and at the same time allow the user to input nonstandard values for the attributes of the block. ATTDIA (Attribute dialog box), which is a setvar variable, is set to 0. This will disable the Attribute Dialog Box from popping up on the screen. Instead, the user can input attribute values from the "Command:" prompt. The LAYER command creates the "EL_OFFEQ" layer, whose color is set to "RED." The "+" is used to allow the continuation of the command onto the next line. The INSERT command is invoked, and the COMP block is inserted, pausing for the user to pick an insertion point. The macro is followed by two semicolons, ";;", which accept the default scale factor. The following seven backslash characters, "\\\\\\\", causes a pause for the operator to pick the rotation angle of the block and input the three attribute values for the block followed by the three verifications of the attribute values. The final command is to restore layer 0 as the current layer.

 Note: It is considered good programming practice to reset the system variables to their default values.

```
[Layer EL_OFFEQ]^C^C^C-LAYER;M;EL_OFFEQ;COLOR;1;;;
[Layer FR_OFF]^C^C^C-LAYER;M;FR_OFF;COLOR;3;;;
[-]
[Layer 0]^C^C^C-LAYER;SET;0;;
```

This macro creates new layers if they do not already exist and assigns the appropriate colors. The [—] code draws a line (separator) on the pull-down menu to group similar elements visually.

MENUGROUPS AND PARTIAL MENU LOADING

One of the problems with the TEST.MNU file menu is that in order to use it, you have to replace the AutoCAD default menu file (ACAD.MNU) even though TEST.MNU is incomplete, containing only one major section. One method around this problem is to rename the major section in TEST.MNU from ***POP1 to ***POP11 and to append this menu to the AutoCAD default ACAD.MNU file. Then you can use all the available standard menu commands in addition to the custom menu. The ACAD.MNU file that comes standard with AutoCAD uses ***POP1 to ***POP10. This has been the traditional approach, which causes rather large menu files just to use the additional functionality. If AutoCAD changes the default menu file, the user will sometimes be forced to make necessary changes to additional functionality to accommodate the ACAD.MNU file changes. Windows NOTEPAD can only edit a file that is less than 64K in size. You would have to use some other editor to append the ***POP11 to the copy of the standard AutoCAD menu. But AutoCAD provides the ability for partial loading of menus so you can mix and match the functionality of various menus. Changes in one menu will not affect the other menus. AutoCAD achieves this new functionality with the addition of a new major section group called ***MENUGROUP. Each menu file can only have one major section called MENUGROUP.

The ***MENUGROUP= major section is how AutoCAD tracks which menus are loaded and referenced. A MENUGROUP string definition can be up to 32 alphanumeric characters (spaces and punctuation marks are not allowed). The ***MENUGROUP= label must precede all menu definitions that use the name-tag mechanism, which will be covered later. Open the TEST.MNU file in the NOTEPAD editor, make the following changes, and save the file as TEST.MNU to demonstrate partial menu loading.

```
//This is a test menu file and it demonstrates the basic new
//functionality of the new menu name tag syntax.
***MENUGROUP=test
***POP1
**Furniture
ID_Furn [&Furniture]
ID_Furns [->&Standard Furniture]
ID_Comps [&Computer]^C^C^CATTREQ;0;-LAYER;M;EL_OFFEQ;COLOR;1;;;+
INSERT;COMP;\;;\ATTREQ;1;-LAYER;SET;0;;
ID_Tables [&Table]^C^C^CATTREQ;0;-LAYER;M;FR_OFF;COLOR;3;;;+
INSERT;TABLE1;\;;\ATTREQ;1;-LAYER;SET;0;;
ID_Chairs [<-C&hair]^C^C^CATTREQ;0;-LAYER;M;FR_OFF;COLOR;3;;;+
INSERT;CHAIR;\;;\ATTREQ;1;-LAYER;SET;0;;
ID_Furnc [->Custo&m Furniture]
ID_Compc [Com&uter]^C^C^CATTDIA;0;-LAYER;M;EL_OFFEQ;COLOR;1;;;+
```

```
INSERT;COMP;\;;\\\\-LAYER;S;0;;ATTDIA;1
ID_Tablec   [Tab&le]^C^C^CATTDIA;0;-LAYER;M;FR_OFF;COLOR;3;;;+
INSERT;TABLE1;\;;\\\\-LAYER;S;0;;ATTDIA;1
ID_Chairc   [<-Chai&r]^C^C^CATTDIA;0;-LAYER;M;FR_OFF;COLOR;3;;;+
INSERT;CHAIR;\;;\\\\-LAYER;S;0;;ATTDIA;1
[—]
[Layer EL_OFFEQ]^C^C^C-LAYER;M;EL_OFFEQ;COLOR;1;;;
[Layer FR_OFF]^C^C^C-LAYER;M;FR_OFF;COLOR;3;;;
[—]
ID_Lyr0 [Layer 0]^C^C^C-LAYER;SET;0;;
```

The text that follows the // is considered comment lines by the AutoCAD program and is ignored in menu file compilation.

The name tags before the labels in the menu begin with "ID_". This can be any string; however, AutoCAD recommends using "ID_" as part of the string. The other requirement is that name tags be unique. Details on how name tags are used are provided later in the chapter. To partially load the menus, invoke Customize Menus from the Tools menu. AutoCAD displays the Menu Customization dialog box, as shown in Figure 18–4.

Choose the **Browse...** button, and AutoCAD displays the Select Menu file dialog box. Select the TEST.MNU file from the appropriate drive and directory, and choose the **Open** button to close the dialog box. AutoCAD displays the name of the menu file selected in the **File Name:** edit box. Choose the **Load** button. TEST.MNU will be added to the **Menu Groups:** list box. The functionality of the TEST.MNU menu is added to the AutoCAD menu. If the TEST.MNU file had toolbars, then they would have appeared on the screen. To display the TEST.MNU group as part of the AutoCAD menu bar, first select TEST.MNU from the **Menu Groups:** list box. Then choose the **Menu Bar** tab, as shown in Figure 18–5. AutoCAD displays the Menu Bar page, as shown in Figure 18–6.

The Menu Bar page of the dialog box shows the Furniture item menu on the left side and the AutoCAD menu bar on the right side. Select Furniture on the left side and Help on the right side, and then choose the **Insert>>** button. AutoCAD inserts the Furniture item before the Help item on the right side of the dialog box. Choose the **Close** button to accept the configuration. AutoCAD adds the Furniture menu item to the standard AutoCAD menu bar between the Tools and Help menus, complete with its own functionality, as shown in Figure 18–7.

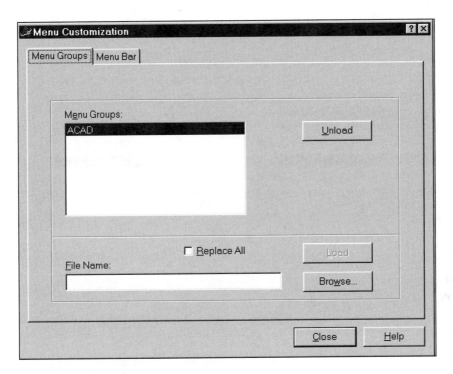

Figure 18–4 *Menu Customization dialog box*

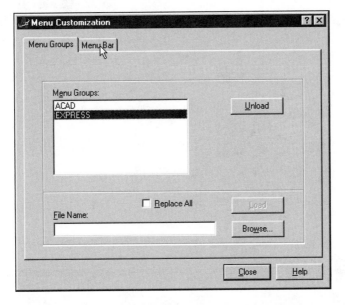

Figure 18–5 *Choosing the Menu Bar tab in the Menu Customization dialog box*

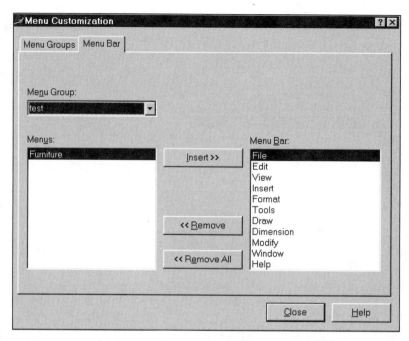

Figure 18–6 *Menu Bar page of the Menu Customization dialog box*

Figure 18–7 *Furniture menu item as part of the AutoCAD standard menu*

Partial menu loading gives you the ability to blend menus together and achieve the functionality as if the menu system were one menu. Test the menu items from the AutoCAD menu and from the Furniture menu. The system works as if dealing with one menu. To unload the TEST.MNU menugroup, open the Menu Customization dialog box (Menu Groups page), select the "test" menu group from the **Menu Groups:** list box, and then choose the **Unload** button.

HELPSTRINGS

In the menu file, you can also include help strings that will be displayed in the status bar whenever a command is selected from the pull-down menu. Open the TEST.MNU file in the NOTEPAD editor and add the ***HELPSTRINGS section, as shown below. Save the file as TEST.MNU to demonstrate the display of help strings.

```
//This is a test menu file and it demonstrates the basic new
//functionality of the new menu name tag syntax.
***MENUGROUP=test
***POP1
**Furniture
ID_Furn [&Furniture]
ID_Furns [->&Standard Furniture]
ID_Comps [&Computer]^C^C^CATTREQ;0;-LAYER;M;EL_OFFEQ;COLOR;1;;;+
INSERT;COMP;\;;\ATTREQ;1;-LAYER;SET;0;;
ID_Tables [&Table]^C^C^CATTREQ;0;-LAYER;M;FR_OFF;COLOR;3;;;+
INSERT;TABLE1;\;;\ATTREQ;1;-LAYER;SET;0;;
ID_Chairs [<-C&hair]^C^C^CATTREQ;0;-LAYER;M;FR_OFF;COLOR;3;;;+
INSERT;CHAIR;\;;\ATTREQ;1;-LAYER;SET;0;;
ID_Furnc [->Custo&m Furniture]
ID_Compc [Com&uter]^C^C^CATTDIA;0;-LAYER;M;EL_OFFEQ;COLOR;1;;;+
INSERT;COMP;\;;\\\\-LAYER;S;0;;ATTDIA;1
ID_Tablec [Tab&le]^C^C^CATTDIA;0;-LAYER;M;FR_OFF;COLOR;3;;;+
INSERT;TABLE1;\;;\\\\-LAYER;S;0;;ATTDIA;1
ID_Chairc [<-Chai&r]^C^C^CATTDIA;0;-LAYER;M;FR_OFF;COLOR;3;;;+
INSERT;CHAIR;\;;\\\\-LAYER;S;0;;ATTDIA;1
[-]
[Layer EL_OFFEQ]^C^C^C-LAYER;M;EL_OFFEQ;COLOR;1;;;
[Layer FR_OFF]^C^C^C-LAYER;M;FR_OFF;COLOR;3;;;
[-]
ID_Lyr0 [Layer 0]^C^C^C-LAYER;SET;0;;
***HELPSTRINGS
ID_Comps [Standard Computer]
ID_Tables [Standard Table]
ID_Chairs [Standard Chair]
ID_Compc [Custom Computer]
ID_Tablec [Custom Table]
ID_Chairc [Custom Chair]
```

The text inside the square brackets next to the tag name will appear on the status bar when the command is invoked from the menu bar. For example, selecting the Standard Computer option from the Custom Furniture menu causes the text inside the square brackets to appear on the status bar, as shown in Figure 18–8.

Figure 18–8 *Status bar displaying the helpstring*

To load the menu with changes, select Customize Menu from the Tools menu. From the Menu Group unload the TEST.MNU file if it is already loaded, and then reload the updated file with changes. Select the **Browse...** button, and AutoCAD displays the Select Menu file dialog box. Select the TEST.MNU file from the appropriate directory, and click the **Open** button to return to the Customize Menu dialog box. AutoCAD displays the name of the menu file selected in the **Menu Groups:** edit box. Select TEST in the edit box and then select the **Load** button. A warning is displayed that states "Loading of a template menu file (MNU file) overwrites and redefines the menu source file (MNS file), which results in the loss of any toolbar customization changes that have been made". You are then asked, "Continue loading MNU file?" Select **Yes**. TEST.MNU will be added to the **Menu Groups:** list box. The functionality of the TEST.MNU menu is added to the AutoCAD menu.

To display the TEST.MNU group as part of the AutoCAD menu bar, first select TEST.MNU from the **Menu Groups:** list box. Then choose the **Menu Bar** tab. In the **Menu Group:** pull-down text box, select TEST. The **Menus:** text box on the left side shows the Furniture menu and the default AutoCAD list of menu bar items in the **Menu Bar:** text box on the right side. Select the Furniture item on the left side and Help on the right side, and then select the **Insert>>** button. AutoCAD inserts the Furniture item before the Help item on the right side of the dialog box. Choose the **Close** button to accept the configuration. AutoCAD adds the Furniture menu item to the standard AutoCAD menu bar between the Tools and Help menus, complete with its own functionality. Test the Furniture menu to see whether the help string appears on the status bar.

CREATING AND MODIFYING TOOLBARS

The easiest way to create toolbars while you are working in AutoCAD is to let AutoCAD add the code automatically to the *.MNS file. If you want to add this information to the *.MNU file, copy and paste this information into the *.MNU file and then delete the *.MNS file. AutoCAD will rebuild the *.MNS file based on the information found in the *.MNU file.

In this section, you will create three new toolbars, as shown in Figure 18–9, as part of the TEST.MNU file: one for custom and standard computers, one for custom and standard tables, and one for custom and standard chairs. Make sure the TEST menu file is still partially loaded into the AutoCAD menu.

To create a new toolbar, select Toolbars from the View menu. AutoCAD displays the Toolbars dialog box, as shown in Figure 18–10.

From the Toolbars dialog box, select the **New...** button. AutoCAD displays the New Toolbar dialog box, as shown in Figure 18–11.

From the **Menu Group:** drop-down list box, select the "test" group. In the **Toolbar Name:** edit box, type Chairs and choose the **OK** button. AutoCAD displays the new toolbar Chairs, as shown in Figure 18–12. If you cannot see the newly created toolbar on the screen, you may have to move the Toolbars dialog box out of the way to find it.

To add buttons to the empty Chairs toolbar, choose the **Customize** button in the Toolbars dialog box. The Customize Toolbars dialog box appears, as shown in Figure 18–13.

Figure 18–9 *Toolbars for a custom menu*

Figure 18–10 *Toolbars dialog box*

Figure 18–11 *New Toolbar dialog box*

Figure 18–12 *Chairs toolbar*

Figure 18–13 *Customize Toolbars dialog box*

AutoCAD lists various types of button groups in the **Categories:** drop-down list box. Select any button from the button groups and, while holding down your left mouse button, drag the outline of the button to the empty Chairs toolbar and release the mouse button. A copy of this button together with its associated code is attached to the toolbar.

Place the cursor on the button located in the Chairs toolbar, and click the right button on your pointing device. AutoCAD displays the Button Properties dialog box shown in the Figure 18–14.

Figure 18–14 *Button Properties dialog box*

Type Custom Chair in the **Name:** edit box. This is the tooltip that will be displayed when the mouse pointer passes over the button. In the **Help:** edit box, type "Custom

Office Furniture". This will be the message displayed on the status bar when the mouse pointer passes over the button. In the **Macro:** edit section, delete the existing code and replace it with the following code taken from the TEST.MNU file created earlier for Custom Chair:

```
^C^C^CATTDIA;0;-LAYER;M;FR_OFF;COLOR;3;;;-INSERT;CHAIR;\;;\\\\
    -LAYER;S;0;;ATTDIA;1
```

Do not add "+" in the macro. Although this will work for the POP menus, it will not work in the Button Property dialog box. Remember that spaces and semicolons are considered the ENTER key by AutoCAD. You may now wish to edit the Button Icon by choosing the **Edit...** button in the Button Properties dialog box. AutoCAD displays the Button Editor dialog box shown Figure 18–15.

Figure 18–15 *Button Editor dialog box*

Make the necessary changes to the icon by using the tools provided in Button Editor dialog box. Select the **Save** or **Save As...** button to save the changes. This will allow you to save the icon as a bitmap (files with the *.BMP extension). Choose the **Close** button to return to the Button Properties dialog box. For the changes to take effect, select the **Apply** button in the Button Properties dialog box. Once again AutoCAD will save the changes to the TEST.MNS file.

Repeat the procedure to add another button to the Chairs toolbar, adding a chair as standard furniture. Make sure to copy the appropriate code from the TEST.MNU file. In addition, create two more toolbars: one called Tables and the other called Computers, as shown in Figure 18–16.

Let's look at an example of creating a flyout toolbar called Office Furniture, as shown in Figure 18–17. In order to create a flyout toolbar, we need to reference existing toolbars.

Figure 18–16 *Table and Computer toolbars*

Figure 18–17 *Flyout toolbar*

The Office Furniture toolbar will include three flyout icons, namely, Computers, Chairs, and Tables. The Computers flyout will include macros for Standard Computer and Custom Computer. Similarly, the Chairs and Tables flyouts will include Standard Chair and Custom Chair and Standard Table and Custom Table, respectively.

To create a new toolbar, select Toolbars from the View menu. AutoCAD displays the Toolbars dialog box. From the Toolbars dialog box select the **New...** button. AutoCAD displays the New Toolbar dialog box.

From the **Menu Group:** drop-down list box select the "test" group. In the **Toolbar Name:** edit box, type Office Furniture and select the **OK** button. AutoCAD displays the new Office Furniture toolbar. If you cannot see the newly created toolbar on the screen, you may have to move the Toolbar dialog box out of the way to find it. To add buttons to the empty Office Furniture toolbar, click the **Customize** button in the Toolbars dialog box. AutoCAD displays the Customize Toolbars dialog box. From the **Categories:** drop-down list box, select the **Custom** option. AutoCAD provides a choice of two types of buttons; a standard button and a flyout button, indicated by the small block triangle in the lower right corner of the button, as shown in Figure 18–18.

Select the flyout button and drag it to the Office Furniture toolbar. Place two additional flyout buttons in the Office Furniture toolbar. Move your mouse pointer to the Office Furniture toolbar and right-click on the left button. The Flyout Properties dialog box appears, as shown in Figure 18–19.

Figure 18–18 *Flyout button box*

Figure 18–19 *Flyout Properties dialog box*

In the **Name:** edit box, type "Computers". In the **Help:** edit box, type "Office Computers". From the **Associated Toolbar:** drop-down list select test_Office Furniture and then select the **Apply** button. Repeat this process for the Tables and Chairs. Make sure the checkbox for **Show this Button's Icon** is turned OFF in the Flyout Properties dialog box. If it is set to ON, AutoCAD will not allow swapping of the icons to reflect the last toolbar icon selected. Once all the flyout buttons are defined, close the Toolbars dialog box.

Choose Customize Menus from the Tools menu, unload the TEST menu group, and then reload the updated test menu group. On the Menu Bars page, add the Furniture menu bar item before the AutoCAD Help item. Various toolbars will be displayed on the screen. Select the various buttons from the newly created toolbars and verify that the associated macros work.

When you create the toolbars, AutoCAD adds all descriptions of the toolbars to the TEST.MNS file under a major section called ***TOOLBARS. AutoCAD creates and maintains the *.MNS files. You can add the ***TOOLBARS section to your *.MNU template file. From the Program Manager Accessories Group, start the NOTEPAD editor program and open the TEST.MNS file. From the TEST.MNS file, select the ***TOOLBARS major section and paste it into the TEST.MNU file. You can have multiple copies of the NOTEPAD editor open at the same time. Following is the line-by-line listing of the code of the TEST.MNU file after button bars are added.

```
//This is a test menu file and it demonstrates the basic new
//functionality of the new menu name tag syntax.
***MENUGROUP=test
***POP1
**Furniture
ID_Furn [&Furniture]
ID_Furns [->&Standard Furniture]
ID_Comps    [&Computer]^C^C^CATTREQ;0;-
    LAYER;M;EL_OFFEQ;COLOR;1;;;+
INSERT;COMP;\;;\ATTREQ;1;-LAYER;SET;0;;
ID_Tables   [&Table]^C^C^CATTREQ;0;-LAYER;M;FR_OFF;COLOR;3;;;+
INSERT;TABLE1;\;;\ATTREQ;1;-LAYER;SET;0;;
ID_Chairs   [<-C&hair]^C^C^CATTREQ;0;-LAYER;M;FR_OFF;COLOR;3;;;+
INSERT;CHAIR;\;;\ATTREQ;1;-LAYER;SET;0;;
ID_Furnc [->Custo&m Furniture]
ID_Compc    [Com&uter]^C^C^CATTDIA;0;-LAYER;M;EL_OFFEQ;COLOR;1;;;+
INSERT;COMP;\;;\\\\-LAYER;S;0;;ATTDIA;1
ID_Tablec   [Tab&le]^C^C^CATTDIA;0;-LAYER;M;FR_OFF;COLOR;3;;;+
INSERT;TABLE1;\;;\\\\-LAYER;S;0;;ATTDIA;1
ID_Chairc   [<-Chai&r]^C^C^CATTDIA;0;-LAYER;M;FR_OFF;COLOR;3;;;+
INSERT;CHAIR;\;;\\\\-LAYER;S;0;;ATTDIA;1
[--]
[Layer EL_OFFEQ]^C^C^C-LAYER;M;EL_OFFEQ;COLOR;1;;;
[Layer FR_OFF]^C^C^C-LAYER;M;FR_OFF;COLOR;3;;;
[--]
ID_Lyr0 [Layer 0]^C^C^C-LAYER;SET;0;;
***HELPSTRINGS
ID_Furn [Office Furniture]
```

```
ID_Furns [Standard Office Furniture]
ID_Comps [Standard Computer]
ID_Tables [Standard Table]
ID_Chairs [Standard Chair]
ID_Furnc [Custom Office Furniture]
ID_Compc [Custom Computer]
ID_Tablec [Custom Table]
ID_Chairc [Custom Chair]
***TOOLBARS
**COMPUTER
ID_Computer [_Toolbar("Computer", _Floating, _Show, 200, 100,0)]
ID_Layers      [_Button("Standard Computer", ICON2580.bmp,
          ICON_32_-LAYERS)]^C^C^CATTREQ;0;-
     LAYER;M;EL_OFFEQ;COLOR;1;;;INSERT;COMP;\;;
          \ATTREQ;1;-LAYER;SET;0;;
ID_Layers_0    [_Button("Custom Computer", ICON7216.bmp,
     ICON_32_-LAYERS)]^C^C^CATTDIA;0;-LAYER;M;EL_OFFEQ;COLOR;1;;;INSERT;COMP;\;;
          \\\\-LAYER;S;0;;ATTDIA;1
**TABLES
ID_Tables_0   [_Toolbar("Tables", _Floating, _Show, 200, 200, 0)]
ID_StandardComputer [_Button("Standard Table", ICON1629.bmp,
     ICON499.bmp)]^C^C^CATTREQ;0;-LAYER;M;FR_OFF;COLOR;3;;;INSERT;TABLE1;\;;
          \ATTREQ;1;-LAYER;SET;0;;
ID_CustomComputer [_Button("Custom Table", ICON5498.bmp,
     ICON8877.bmp)]^C^C^CATTDIA;0;-LAYER;M;FR_OFF;COLOR;3;;;INSERT;TABLE1;\;;
          \\\\-LAYER;S;0;;ATTDIA;1
**CHAIRS
ID_Chairs      [_Toolbar("Chairs", _Floating, _Show, 200, 300, 0)]
ID_StandardTable [_Button("Standard Chair", ICON2854.bmp,
     ICON5060.bmp)]^C^C^CATTREQ;0;-LAYER;M;FR_OFF;COLOR;3;;;INSERT;CHAIR;\;;
          \ATTREQ;1;-LAYER;SET;0;;
ID_CustomTable [_Button("Custom Chair", ICON665.bmp,
     ICON4474.bmp)]^C^C^CATTDIA;0;-LAYER;M;FR_OFF;COLOR;3;;;INSERT;CHAIR;\;;
          \\\\-LAYER;S;0;;ATTDIA;1
**OFFICE_FURNITURE
**TB_OFFICE_FURNITURE
               [_Toolbar("Office Furniture", _Floating, _Show,
     200, 400, 0)]
ID_           [_Flyout("Computer", ICON8617.bmp, ICON_32_BLANK,
     _OtherIcon, test.COMPUTER)]
ID__0         [_Flyout("Tables", ICON3090.bmp, ICON_32_BLANK,
     _OtherIcon, test.TABLES)]
ID__1         [_Flyout("Chairs", ICON9328.bmp, ICON_32_BLANK,
     _OtherIcon, test.CHAIRS)]
```

Make sure you save the TEST.MNU file. If you deleted the TEST.MNS file without copying the ***TOOLBARS section to the TEST.MNU file, you will lose the newly created toolbars. During the process of adding new toolbars, it is always recommended to copy the ***TOOLBARS sections to the *.MNU file. Remember, if there is no *.MNS file, AutoCAD creates it for you based on the *.MNU template file.

You can manually insert the ***TOOLBARS section of code into your TEST.MNU file, but let AutoCAD do the job and then copy it into the *.MNU template file. Let's examine, line by line, some of the unique features of the code in the ***TOOLBARS section of the TEST.MNU file.

```
***TOOLBARS
```

Defines the major section for toolbars.

```
**COMPUTER
```

Menu subsection.

```
ID_Computer [_Toolbar("Computer", _Floating, _Show, 200, 100, 0)]
```

```
ID_Computer
```

Tag name by which the item on the menu is referenced.

```
_Toolbar
```

Signifies that the menu is dealing with a toolbar definition.

```
"Computer"
```

Name that appears above the toolbar.

```
_Floating
```

Indicates that the toolbar is floating and not docked. Possible values are _Top, _Bottom, _Left, _Right, or _Floating. The underscore character in front of the keywords is not required; however, it is used for the international versions of AutoCAD. The keywords are not case-sensitive.

```
_Show
```

Visibility keyword, available options include _Show or _Hide.

```
200, 100, 0
```

The first number is the distance in pixels from the left edge of the screen. The second number is the distance in pixels from the top edge. The final number is the number of rows in the toolbar.

```
ID_Layers    [_Button("Standard
             Computer",ICON2580.bmp,ICON_32_-LAYERS)]macro
```

```
ID_Layers
```

Tag name by which this item on the menu is referenced.

```
_Button
```

Signifies that the menu is dealing with a button definition.

```
"Standard Computer"
```

This is the tooltip text that is displayed when the mouse pointer passes over the button.

```
ICON2580.bmp
```

ID of the small-icon (16 x 16) bitmap.

```
ICON_32_-LAYERS
```

ID of the large-icon (32 x 32) bitmap.

```
macro
```

Command sequence required to complete the macro.

Flyout Section of the Menu

```
**TB_OFFICE_FURNITURE
        [_Toolbar("Office Furniture", _Floating, _Show, 200, 400,
    0)]
ID_    [_Flyout("Computer", ICON8617.bmp, ICON_32_BLANK,
        _OtherIcon, test.COMPUTER)]
ID__0 [_Flyout("Tables", ICON3090.bmp, ICON_32_BLANK, _OtherIcon,
        test.TABLES)]
ID__1 [_Flyout("Chairs", ICON9328.bmp, ICON_32_BLANK, _OtherIcon,
        test.CHAIRS)]
[_Toolbar("Office Furniture", _Floating, _Show, 200, 400, 0)]
```

The toolbar menu item is the same as for the toolbar description explained earlier.

```
ID_    [_Flyout("Computer", ICON8617.bmp, ICON_32_BLANK,
        _OtherIcon, test.COMPUTER)]
```

```
ID_
```

Tag name by which the item on the menu is referenced.

```
Flyout
```

Signifies that the menu is dealing with a flyout definition.

```
"Computer"
```

Tooltip text that is displayed when the mouse pointer passes over the button.

```
ICON8617.bmp
```

ID of the small-icon (16 x 16) bitmap.

```
ICON_32_BLANK
```

ID of the large-icon (32 x 32) bitmap.

```
_OtherIcon
```

Has one of two possible values: _OtherIcon or _OwnIcon. The OtherIcon value allows the button to switch icons and displays the last icon selected. The OwnIcon value will not permit this switching.

ACCELERATOR KEYS

AutoCAD supports user-defined accelerator keys. Add the following major section (***ACCELERATORS) to the TEST.MNU file.

```
***ACCELERATORS
ID_Comps    [SHIFT+"F1"]
ID_Tables   [SHIFT+"F2"]
ID_Chairs   [SHIFT+"F3"]
ID_Lyr0     [SHIFT+"L"]
```

The first item is the name tag followed, in brackets, by the accelerator key combination. Type the keyboard combination (for example: SHIFT + F1), and the menu item identified by ID_Comps will be executed. The accelerator key combination will not work if the menu is partially loaded as part of the ACAD.MNU file. Only when the menu is loaded in place of the AutoCAD menu will these key combinations work.

Apart from including macros, you can also include AutoLISP code in the menus (refer to Chapter 19) or have the menu macros load and run AutoLISP applications.

DIESEL

As mentioned earlier, a macro is a string of commands that AutoCAD executes when the menu item is selected. You can also add AutoLISP in a macro and call an AutoLISP routine from a macro. AutoCAD provides a macro language alternative to AutoLISP, called DIESEL. DIESEL is an acronym for Direct Interpretively Evaluated String Expression Language. The subject of DIESEL is beyond the scope of this book, but let's go through an example to see the application of DIESEL.

DIESEL allows you to customize the AutoCAD status line through the use of the MODEMACRO setvar variable, and it also allows modification of the appearance of pull-down menus. DIESEL, though a macro language and similar in style to AutoLISP, does not have the power and flexibility of AutoLISP.

The simplest use of DIESEL is through the MODEMACRO setvar variable. At the AutoCAD "Command": prompt, enter MODEMACRO and press ENTER or SPACEBAR. AutoCAD prompts:

New value for MODEMACRO, or . for none <"">: **Captain CAD**

Type in a text string as shown in this example and press ENTER. AutoCAD displays the text string in the status bar, as shown in Figure 18–20.

```
Command:    <Coords off>
Command:
```
Captain CAD 16.6111; -7.4879 , 0.0000   SNAP GRID ORTHO POLAR OSNAP OTRACK LWT MODEL

Figure 18–20 *Display of the text string in the status bar*

To return to the default value, type MODEMACRO at the "Command:" prompt and press ENTER or SPACEBAR. Type in a single period to the prompt and press ENTER or SPACEBAR. The status bar returns to the default settings.

As another example, a string with arguments is provided in response to the MODEMACRO prompt:

New value for MODEMACRO, or . for none <"">:-LAYER = $(getvar,clayer)

AutoCAD gets the current layer name and displays it in the status bar, as shown in Figure 18–21. DIESEL expressions can also be placed as part of the menus.

```
Command: '_layer
Command:
```
layer=object -8.9916, 10.7860, 0.0000   SNAP GRID ORTHO POLAR OSNAP OTRACK LWT MODEL

Figure 18–21 *Display of the current layer name in the status bar*

CUSTOMIZING TABLET MENUS

The previous descriptions and examples of menu customization can be applied to customizing the tablet menu area. One problem not encountered in a tablet menu is that of the display. The brackets that enclose non-active items will not be visible except to someone who is reading the filename.MNU file from which the filename.MNX file was compiled.

The main concern in customizing the tablet part of the menu is in placing the programming lines in the right order and in the right area so they will correspond to the preprinted overlay that will be configured on the tablet for use with the menu.

A menu may have up to four tablet areas. They will have the headings ***TABLET1, ***TABLET2, ***TABLET3, and ***TABLET4. The first program line following a heading will correspond to the configured overlay's upper-left column/row rectangle. Subsequent program lines will correspond the rectangle to the right in the same row as its predecessor until the end of the row is reached. Then the next program line will be on the extreme left rectangle of the next row. For example, a tablet menu area with 12 program lines might be in any one of the 6 following arrangements:

1	2	3	4	5	6	7	8	9	10	11	12

1	2	3	4	5	6
7	8	9	10	11	12

1	2	3	4
5	6	7	8
9	10	11	12

1	2	3
4	5	6
7	8	9
10	11	12

1	2
3	4
5	6
7	8
9	10
11	12

1
2
3
4
5
6
7
8
9
10
11
12

If there are more rectangles specified in the tablet configuration than there are program lines, the extras will be non-active when picked. If there is an excess of program lines in the menu, they will, of course, not be accessible.

DIALOG CONTROL LANGUAGE

AutoCAD Release 12 introduced programmable dialog boxes, and this feature is continued in AutoCAD 2000. Programming dialog boxes requires a thorough knowledge of AutoLISP and/or C/C++. The description of the dialog box is a text file known as a DCL (Dialog Control Language) file. The DCL file is a description of the various parts of the dialog and of the elements the dialog contains. The following is an example of a DCL file.

```
cesdoor : dialog
{
    label = "CESCO Doors";
    : row
    {
        : radio_cluster
        {
            key = "thick";
            : boxed_radio_column
            {
                label = "Thickness";
                : radio_button
                {
                    key = "t1";
                    label = "1\"";
                    value = "1";
                }
                : radio_button
                {
                    key = "t2";
                    label = "2\"";
                }
                : radio_button
                {
                    key = "t4";
                    label = "4\"";
                }
            }//boxed radio column - Thickness
        }//radio cluster - Thick
        : radio_cluster
        {
```

```
            key = "hinge";
            : boxed_radio_column
            {
                label = "Hinge";
                spacer;
                : radio_button
                {
                    key = "rt";
                    label = "Right";
                }
                : radio_button
                {
                    key = "lt";
                    label = "Left";
                    value = "1";
                }
                spacer;
            }//boxed radio column - Hinge location
        }//radio_cluster - Hinge
    }// row - top
    : row
    {
        : radio_cluster
        {
            key = "opng";
            : boxed_radio_column
            {
                label = "Open";
                width = 9;
                : radio_button
                {
                    key = "in";
                    label = "In";
                }
                : radio_button
                {
                    key = "out";
                    label = "Out";
                    value = "1";
                }
            }//boxed radio column - Opening
        }//radio_cluster - opng
        : radio_cluster
```

```
        {
            key = "hand";
            : boxed_radio_column
            {
                label = "Handles";
                : radio_button
                {
                    key = "hndl2";
                    label = "2";
                    value = "1";
                }
                : radio_button
                {
                    key = "hndl3";
                    label = "3";
                }
            }//boxed radio column - Number of handles
        }//radio_cluster - hand
    }//row - 2nd row
    spacer_1;
    ok_cancel;
}
```

When AutoCAD is loaded into memory, it automatically loads two files: ACAD.DCL and BASE.DCL. These files contain various attributes on which you can build your dialogs. Each element (:radio_button, for example) contains a key (key = "hndl3"). These keys are the names to which AutoLISP and C/C++ refer in order to retrieve values from them and act upon them. Figure 18–22 shows a sample dialog box.

Figure 18–22 *Sample dialog box*

Selecting the OK button causes AutoLISP or C/C++ to read the various keys to determine which values the user selected and pass the results to the calling program. When writing programs that utilize dialog boxes, a certain portion of the code is dedicated to reading and responding to the dialog box (apart from the DCL file describing the dialog box itself). This provides the user with a more detailed view of all the inputs and can be used to limit the user inputs to the program. Dialog boxes can utilize radio buttons, toggle boxes, horizontal and vertical sliders, list boxes, edit boxes, buttons, image tiles, and more to enhance the front end to any program requiring user input.

A complete discussion of programming dialog boxes is beyond the scope of this book. Refer to AutoCAD's Customization Guide for more details.

EXERCISES

1. What extension must a file have when you are writing a custom menu?

2. How many menus may be active simultaneously?

3. What AutoCAD command sequence will activate a menu written as a file named *XYZ.MNU?*

4. Under the ***BUTTONS menu, which program line corresponds to the second button on the mouse/puck?

5. Placing three blank spaces in a menu program line is equivalent to what action from the keyboard?

6. What is the purpose of the caret (^) in a menu program line?

7. What is the purpose of the "^c" at the beginning of a menu program line?

8. Create a menu file to include a program line that will draw a rectangle whose opposite corners are points with coordinates 1,1 and 5,2.

9. Create a menu file to include a program line that will draw a circle of diameter 2 units and whose center point has the coordinates 3,3.

10. Create a menu file to include a program line that will array the circle in Exercise 9 in a polar array every 30 degrees, with the center of the array at a point whose coordinates are 4,3.

11. Create a menu file to include a program line that will permit a line-arc-line continuation starting at point 1,1, with user picks for the second, third, and fourth points.

12. Write lines that set UNITS as follows:

    a. Decimal to four-place display

    b. Architectural to 1/8" display

    c. Each of the above to include a mechanism to return to the graph screen

CUSTOMIZING HATCH PATTERNS

Certain concepts about hatch patterns should be understood before learning to create one.

1. Hatch patterns are made up of lines or line segment/space combinations. There are no circles or arcs available in hatch patterns like there are in shapes and fonts (which will be covered next).

2. A hatch pattern may be one or more series of repeated parallel lines or repeating dot or line segment/space combinations. That is, each line in one so-called family is like every other line in that same family. And each line has the same offset and stagger (if it is a segment/space combination), relative to its adjacent sibling, as every other line.

3. One hatch pattern can contain multiple families of lines. One family of lines may or may not be parallel to other families. With properly specified basepoints, offsets, staggers, segment/space combinations, lengths, and relative angles, you can create a hatch pattern from multiple families of segment/space combinations that will display repeated closed polygons.

4. Each family of lines is drawn with offsets and staggers based on its own specified basepoint and angle.

5. All families of lines in a particular hatch pattern will be located (basepoint), rotated, and scaled as a group. These factors (location, angle of rotation, and scale factor) are determined when the hatch pattern is loaded by the HATCH command and used to fill a closed polygon in a drawing.

6. The pattern usually can be achieved by different ways of specifying parameters.

Hatch patterns are created by including their definition in a file whose extension is .PAT. This can be done by using a text editor such as NOTEPAD or a word processor in the nondocument (or programmer) mode, which will save the text in ASCII format. Your hatch pattern definition can also be added to the ACAD.PAT file. You can also create a new file specifically for a pattern.

Each pattern definition has one header line giving the pattern name/description and a separate specification line describing each family of lines in the pattern.

The header line has the following format:

```
*pattern-name[,description]
```

The pattern name will be the name for which you will be prompted when using the HATCH command. The description is optional and is there so that someone reading the .PAT file can identify the pattern. The description has no effect, nor will it be displayed while using the HATCH command. The leading asterisk denotes the beginning of a hatch pattern.

The format for a line family is as follows:

```
angle, x-origin, y-origin, delta-x, delta-y [,dash-1, dash-2...]
```

The brackets "[]" denote optional segment/space specifications used for noncontinuous-line families. Note also that any text following a semicolon (;) is for comment only and will be ignored. In all definitions, the angle, origins, and deltas are mandatory (even if their values are zero).

An example of continuous lines that are rotated at 30 degrees and separated by 0.25 units is shown in Figure 18–23:

```
*P30, 30 degree continuous
30, 0,0, 0,.25
```

The 30 specifies the angle.

The first and second zero specify the coordinates of the origin.

The third zero, though required, is meaningless for continuous lines.

The 0.25 specifies the distance between lines.

A pattern of continuous lines crossing at 60 degrees to each other could be written as follows (see Figure 18–24):

```
*PX60,x-ing @ 60
30, 0,0, 0,.25
330, 0,0, 0,.25
```

Figure 18–23 *A hatch pattern with continuous lines rotated at 30 degrees and separated by 0.25 units*

Figure 18–24 *A hatch pattern with continuous lines crossing at 60 degrees to each other and separated by 0.25 units*

A pattern of lines crossing at 90 degrees but having different offsets is as follows (see Figure 18–25):

```
*PX90, x-ing @ 90 w/ 2:1 rectangles
0, 0,0, 0,.25
90, 0,0, 0,.5
```

Note the effect of the delta-Y. It is the amount of offset between lines in one family. Hatch patterns with continuous lines do not require a value (other than zero) for delta-*X*. Orthogonal continuous lines also do not require values for the *X* origin unless they are used in a pattern that includes broken lines.

To illustrate the use of a value for the *Y* origin, two parallel families of lines can be written to define a hatch pattern for steel as follows (see Figure 18–26):

```
*steel
45, 0,0, 0,1
45, 0,.25, 0,1
```

Figure 18–25 *A hatch pattern with lines crossing at 90 degrees and having different offsets*

Figure 18–26 *Defining a hatch pattern for steel*

Three concepts are worthy of note in this example.

- If the families were not parallel, then specifying origins other than zero would serve no purpose.

- Parallel families of lines should have the same delta-Y offsets. Different offsets would serve little purpose.

- Most important, the delta-Y is at a right angle to the angle of rotation, but the Y origin is in the Y direction of the coordinate system. The steel pattern as written in the example would fill a polygon, as shown in Figure 18–27. Note the dimensions when used with no changes to the scale factor of 1.0 or a rotation angle of zero.

Figure 18–27 *The steel hatch pattern with the delta-Y at a right angle to the angle of rotation and the Y origin in the Y direction of the coordinate system*

CUSTOM HATCH PATTERNS AND TRIGONOMETRY

The dimensions in the hatch pattern resulting from a 0.25 value for the delta-Y of the second line-family definition may not be what you expected; see Figure 18–28. If you wished to have a 0.25 separation between the two line families (see Figure 18–29), then you must either know enough trigonometry/geometry to predict accurate results or else put an additional burden on the user to reply to prompts with the correct responses to achieve those results. For example, you could write the definition as follows:

```
*steel
0, 0,0, 0,1
0, 0,.25, 0,1
```

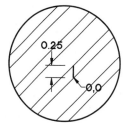

Figure 18–28 *Hatch pattern dimensions resulting from a 0.25 value for the delta Y of the second line-family definition*

Figure 18–29 *Steel hatch pattern defined with a 0.25 separation between the two line families*

In order to use this pattern as shown, the user will have to specify a 45-degree rotation when using it. This will maintain the ratio of 1 to .25 between the offset (delta-Y) and the spacing between families (Y origin). However, if you wish to avoid this inconvenience to the user, but still wish to have the families separated by .25, you can write the definition as follows:

```
*steel
45, 0,0, 0,1
45, 0,.353553391, 0,1
```

The value for the Y origin of .353553391 was obtained by dividing .25 by the sine (or cosine) of 45 degrees, which is .70710678. The X origin and Y origin specify the coordinates of a point. Therefore, setting the origins of any family of continuous lines merely tells AutoCAD that the line must pass through that point. See Figure 18–30 for the trigonometry used.

For families of lines that have segment/space distances, the point determined by the origins can tell AutoCAD not only that the line passes through that point, but that one of the segments will begin at that point. A dashed pattern can be written as follows (see Figure 18–31):

```
*dashed
0, 0,0, 0,.25, .25,-.25
```

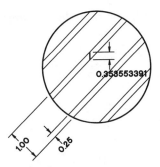

Figure 18–30 *Steel hatch pattern defined with a 45 degree rotation to maintain a 1:25 offset ratio*

Note that the value of the *X* origin is zero, thus causing the dashes of one line to line up with the dashes of other lines. Staggers can be produced by giving a value to the *X* origin as follows (see Figure 18–32):

```
*dashstagger
0, 0,0, .25,.25, .25,-.25
```

In a manner similar to defining linetypes, you can cause lines in a family to have several lengths of segments and spaces (see Figure 18–33).

Figure 18–31 *Writing a dashed pattern*

Figure 18–32 *Writing a staggered dashed pattern*

Figure 18–33 *A pattern with several lengths of segments and spaces*

```
*simple
0, 0,0, 0,.5
90, 0,0, 0,1, .5,-.5
```

A similar, but more complex hatch pattern could be written as follows (see Figure 18–34):

```
*complex
45, 0,0, 0,.5
-45, 0,0, 0,1.414213562, 0,1.41421356
```

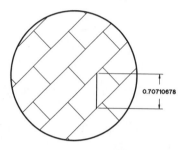

Figure 18–34 *A pattern with more complex hatch patterns*

REPEATING CLOSED POLYGONS

Creating hatch patterns with closed polygons requires planning. For example, a pattern of 45/90/45-degree triangles, as shown in Figure 18–35, should be started by first extending the lines, as shown in part a of Figure 18–35. Extend the construction lines through points of the object parallel to other lines of the object. Note the grid that emerges when you use the lines and distances obtained to determine the pattern.

It is also helpful to sketch construction lines that are perpendicular to the object lines. This will assist you in specifying segment/space values. In the example, two of the lines are perpendicular to one another, thus making this easier. Figures 18–36 through 18–39 illustrate potential patterns of triangles. Once the pattern is selected, the grid, and some knowledge of trigonometry, will assist you in specifying all of the values in the definition for each line family.

For pattern PA the horizontal line families can be written as follows (see Figure 18–40):

```
0, 0,0, 0,1, 1,-1
```

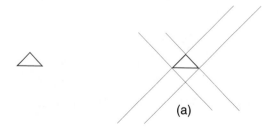

(a)

Figure 18–35 *Closed polygons*

Figure 18–36 *Creating triangular hatch patterns—method #1*

Figure 18–37 *Creating triangular hatch patterns—method #2*

Figure 18–38 *Creating triangular hatch patterns—method #3*

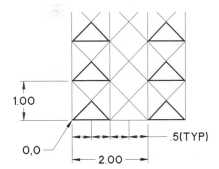

Figure 18–39 *Creating triangular hatch patterns—method #4*

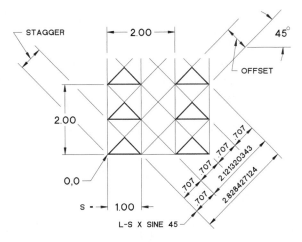

Figure 18–40 *Triangular patterns*

The specifications for the 45-degree family of lines can be determined by using the following trigonometry:

```
sin 45 degrees = 0.70710678
S = 1
L = S times sin 45 degrees
L = 1 times sin 45 degrees = 0.70710678
```

Note that the trigonometry function is applied to the hypotenuse of the right triangle. In the example, the hypotenuse is 1 unit. A different value would simply produce a proportional result; i.e., a hypotenuse of .5 would produce L = S x 0.70710678 = 0.353553391. The specifications for the 45-degree family of lines could be written as follows:

```
45, 0,0 0.70710678,0.70710678, 0.70710678,-2.121320343
angle,origin,offset,   stagger,            segment,      space
```

For the 135-degree family of lines, the offset, stagger, segment, and space have the same values (absolute) as for the 45-degree family. Only the angle, the X origin, and the sign (+ or −) of the offset or stagger may need to be changed.

The 135-degree family of lines could be written as follows:

```
135, 1,0,  -0.70710678,-0.70710678, 0.70710678,-2.121320343
```

Putting the three families of lines together under a header could be written as follows, and as shown in Figure 18–41.

```
*PA,45/90/45 triangles stacked
0, 0,0, 0,1, 1,-1
45, 0,0 0, 0.70710678,0.70710678, 0.70710678,-2.121320343
135, 1,0,  -0.70710678,0.70710678, 0.70710678,-2.121320343
```

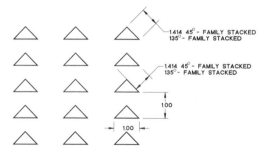

Figure 18–41 *Families of triangular patterns*

In the preceding statement, "could be written" tells you that there may be other ways to write the definitions. As an exercise, write the descriptions using 225 degrees instead of 45 degrees and 315 instead of 135 for the second and third families, respectively. As a hint, you determine the origin values of each family of lines from the standard coordinate system. But to visualize the offset and stagger, orient the layout grid so that the rotation angle coincides with the zero angle of the coordinate system. Then the signs and the values of delta-*X* and delta-*Y* will be easier to establish along the standard plus for right/up and negatives for left/down directions. Examples of two hatch patterns, PB and HONEYCOMB, follow.

The PB pattern can be written as follows, and as shown in Figure 18–42.

```
*PB, 45/90/45 triangle staggered
0, 0,0, 1,1, 1,-1
45, 0,0, 0,1.414213562, 0.70710678,-0.70710678
135, 1,0, 0,1.414213562, 0.70710678,-0.70710678
```

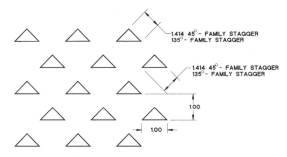

Figure 18–42 *Example of the pattern*

Note that this alignment simplifies the definitions of the second and third families of lines over the PA pattern.

The Honeycomb pattern can be written as follows, and as shown in Figure 18–43.

```
*HONEYCOMB
90, 0,0, 0.866025399,0.5, 0.577350264,-1.154700538
330, 0,0, 0.866025399,0.5, 0.577350264,-1.154700538
30, 0.5, -0.288675135, 0.866025399, 0.5, 0.577350264,-1.154700538
```

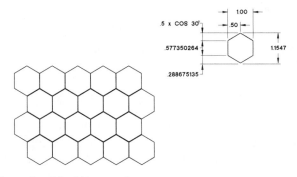

Figure 18–43 *Example of the Honeycomb pattern*

CUSTOMIZING SHAPES AND TEXT FONTS

Shapes and fonts are written in the same manner, and are both stored in files with the .SHP file extension. The .SHP files must be compiled into .SHX files. This section covers how to create and save .SHP files and how to compile .SHP files into .SHX files. To compile an .SHP file into an .SHX file for shapes or fonts, enter:

Command: **compile**

From the Select Shape File dialog box, select the file to be compiled. If the file has errors they will be reported; otherwise, you will be prompted:

Compilation successful
Output file name .SHX contains nnn bytes

The main difference between shapes and fonts is in the commands used to place them in a drawing. Shapes are drawn by using the SHAPE command, and fonts are drawn using commands that insert text, such as TEXT or DIM. Whether or not an object in an .SHP/.SHX file can be used with the SHAPE command or as a font character is partly determined by whether its shape name is written in uppercase or lowercase (explained below).

Each shape or character in a font in an .SHP or .SHX file is made up of simplified objects. These objects are simplified lines, arcs, and circles. The reason they are referred to as simplified is because in specifying their directions and distances, you cannot use decimals or architectural units. You must use only integers or integer fractions. For example, if the line distance needs to be equal to 1 divided by the square root of 2 (or .7071068), the fraction 70 divided by 99 (which equals .707070707) is as close as you can get. Rather than call the simplified lines and arcs "objects," we will refer to them as "primitives."

Individual shapes (and font characters) are written and stored in ASCII format. .SHP/.SHX files may contain up to 255 SHAPE-CHARACTERS. Each SHAPE-CHARACTER definition has a header line, as follows:

```
*shape number, defbytes, shapename
```

The codes that describe the SHAPE-CHARACTER may take up one or more lines following the header. Most of the simple shapes can be written on one or two lines. The meaning of each item in the header is as follows:

The shape number may be from 1 to 255 with no duplications within one file.

Defbytes is the number of bytes used to define the individual SHAPE-CHARACTER, including the required zero that signals the end of a definition. The maximum allowable bytes in a SHAPE-CHARACTER definition is 2,000. Defbytes (the bitcodes) in the definition are separated by commas. You may enclose pairs of bitcodes within parentheses for clarity of intent, but this does not affect the definition.

The *shapename* should be in uppercase if it is to be used by the SHAPE command. Like a block name is used in the BLOCK command, you enter the shapename when prompted to do so during the SHAPE command. If the shape is a character in a font file, you may make any or all of the shapename characters lowercase, thereby causing the name to be ignored when compiled and stored in memory. It will serve for reference only in the .SHP file for someone reading that file.

PEN MOVEMENT DISTANCES AND DIRECTIONS

The specifications for pen movement distances and directions (whether the pen is up or down) for drawing the primitives that will make up a SHAPE-CHARACTER are written in bitcodes. Each bitcode is considered one defbyte. Codes 0 through 16 are not DISTANCE-DIRECTION codes, but special instructions-to-AutoCAD codes (to be explained shortly, after the DISTANCE-DIRECTION codes discussion).

DISTANCE-DIRECTION codes have three characters. They begin with a zero. The second character specifies distance. More specifically, it specifies vector length, which may be affected by a scale factor. Vector length and scale factor combine to determine actual distances. The third character specifies direction. There are 16 standard directions available through use of the DISTANCE-DIRECTION bitcode (or defbyte). Vectors 1 unit in length are shown in the 16 standard directions in Figure 18–44.

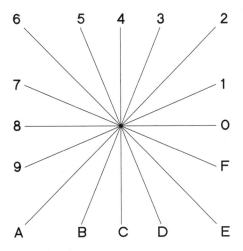

Figure 18–44 *DISTANCE-DIRECTION bitcodes*

Directions 0, 4, 8, and C are equivalent to the default 0, 90, 180, and 270 degrees, respectively. Directions 2, 6, A, and E are 45, 135, 225, and 315 degrees, respectively. But the odd-numbered direction codes are not increments of 22.5 degrees, as you might think. They are directions that coincide with a line whose delta-X and delta-Y ratio are 1 unit to 2 units. For example, the direction specified by code 1 is equivalent to drawing a line from 0,0 to 1,.5. This equates to approximately 26.56505118 degrees (or the arctangent of 0.5). The direction specified by code 3 equates to 63.434494882 degrees (or the arctangent of 2) and is the same as drawing a line from 0,0 to .5,1.

Distances specified will be measured on the nearest horizontal or vertical axis. For example, 1 unit in the 1 direction specifies a vector that will project 1 unit on the horizontal axis. Three units in the D direction will project 3 units on the vertical axis (downward). So the vector specified as 1 unit in the 1 direction will actually be 1.118033989 units long at an angle of 26.65606118 degrees, and the vector specified as 3 units in the D direction will be 3.354101967 units long at an angle of 296.5650512 degrees. See Figure 18–45 for examples of specifying direction.

To illustrate the codes specifying the DISTANCE-DIRECTION vector, the following example is a definition for a shape called "oddity" that will draw the shape shown in Figure 18–46.

Figure 18–45 *Specified directions*

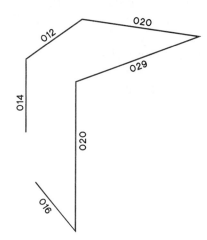

Figure 18–46 *Distance-direction vector specifying code*

```
*200,7,ODDITY
014,012,020,029,02C,016,0
```

To draw the shape named "oddity", you would first load the shape file that contains the definition and then use the SHAPE command as follows:

Command: **shape**
Name (or ?): **oddity**
Starting point: *(specify a point)*
Height <default>: *(specify a scale factor)*
Rotation angle <default>: *(specify a rotation angle)*

 Note: An alternative to the standard DISTANCE-DIRECTION codes is to use codes 8 and 9 to move the pen by paired (delta-X,delta-Y) ordinate displacements. This is explained in the following section on "Special Codes."

SPECIAL CODES

Special codes can be written in decimal or hexadecimal. You can specify a special code as 0 through 16 or as 000 through 00E. A three-character defbyte with two leading zeros will be interpreted as a hexadecimal special code. A code 10 is a special code in decimal. However, 010 is equivalent to decimal 16. But more important, it will be interpreted by AutoCAD as a DISTANCE-DIRECTION code with a vector length of 1 and a direction of 0. The hexadecimal equivalent to 10 is 00A. The code functions are as follows.

Code 0: End of Shape_The end of each separate shape definition must be marked with the code 0.

Codes 1 and 2: Pen Up and Down_The "PEN DOWN" (or DRAW) mode is on at the beginning of each shape. Code 2 turns the DRAW mode off or lifts the pen. This permits moving the pen without drawing. Code 1 turns the DRAW mode on.

Note the relationship between the insertion point specified during the SHAPE command and where you wish the object and its primitives to be located. If you wish for AutoCAD to begin drawing a primitive in the shape at a point remote from the insertion point, then you must lift the pen with a code 2 and move the pen (with the proper codes) and then lower the pen with a code 1. Movement of the pen (directed by other codes) after a "PEN DOWN" code 1 is what causes AutoCAD to draw primitives in a shape.

Codes 3 and 4: Scale Factors_Individual (and groups of) primitives within a shape can be increased or decreased in size by integer factors as follows: Code 3 tells AutoCAD to divide the subsequent vectors by the number that immediately follows the code 3. Code 4 tells AutoCAD to multiply the subsequent vectors by the number that immediately follows the code 4.

Scale factors are cumulative. The advantage of this is that you can specify a scale factor that is the quotient of two integers. A two-thirds scale factor can be achieved by a code 4 followed by a factor of 2 followed by a code 3 followed by a factor of 3. But the effects of scale factor codes must be reversed when they are no longer needed. They do not go away by themselves. Therefore, at the end of the definition (or when you wish to return to normal or other scaling within the definition), the scale factor must be countered. For example, when you wish to return to the normal scale from a two-thirds scale, you must use code 3 followed by a factor of 3 followed by a code 4 followed by a factor of 2. There is no law that states you must always return to normal from a scaled mode. You can, with codes 3 and 4 and the correct factors, change from a two-thirds scale to a one-third scale for drawing additional primitives within the shape. You should *always*, however, return to the normal scale at the end of the definition. A scale factor in effect at the end of one shape will carry over to the next shape.

Codes 5 and 6: Saving and Recalling Locations_Each location in a SHAPE definition is specified relative to a previous location. However, once the pen is at a particular location, you can store that location for later use within that SHAPE definition before moving on. This is handy when an object has several primitives starting or ending at the same location. For example, a wheel with spokes would be easier to define by using code 5 to store the center location, drawing a spoke, and then using code 6 to return to the center.

Storing and recalling locations are known as *pushing* and *popping* them, respectively, in a stack. The stack storage is limited to four locations at any one time. The order in which they are popped is the reverse of the order in which they were pushed. Every location pushed must be popped.

More pushes than pops will result in the following error message:

    Position stack overflow in shape nnn

More pops than pushes will result in the following error message:

    Position stack underflow in shape nnn

Code 7: Subshape_One shape in an .SHP/.SHX file can be included in the definition of another shape in the same file by using the code 7 followed by the inserted shape's number.

Codes 8 and 9: *X-Y* Displacements_Normal vector lengths range from 1 to 15 and can be drawn in one of the 16 standard directions unless you use a code 8 or code 9 to specify *X-Y* displacements. A code 8 tells AutoCAD to use the next two bytes as the X and Y displacements, respectively. For example, 8, (7,–8) tells AutoCAD to move the pen a distance that is 7 in the

X direction and *8* in the Y direction. The parentheses are optional, for viewing effects only. After the displacement bytes, specifications revert to normal.

Code 9_Code 9 tells AutoCAD to use all following pairs of bytes as *X-Y* displacements until terminated by a pair of zeros. For example; 9,(7,–8),(14,9),(–17,3),(0,0) tells AutoCAD to use the three pairs of values for displacements for the current mode and then revert to normal after the (0,0) pair.

Code 00A: Octant Arc_Code 00A (or 10) tells AutoCAD to use the next two bytes to define an arc. It is referred to as an octant (an increment of 45 degrees) arc. Octant arcs start and end on octant boundaries. Figure 18–47 shows the code numbers for the octants. The specification is written in the following format:

```
10, radius. (-)OSC
```

The radius may range from 1 to 255. The second byte begins with zero and specifies the direction by its sign (clockwise if negative, counterclockwise otherwise), the starting octant (S), and the number of octants it spans (C), which may be written as 0 to 7, with 0 being 8 (a full circle). Figure 18–48 shows an arc drawn with the following codes:

```
10,(2,-043)
```

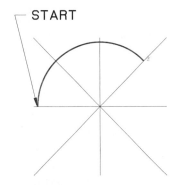

Figure 18–47 *Code numbers for octants*

Figure 18–48 *An arc drawn with code 10,(2,–043)*

The arc has a radius of 2, begins at octant arc 4, and turns 135 degrees (3 octants) clockwise.

Code 00B: Fractional Arc_Code 00B (11) can be used to specify an arc that begins and ends at points other than the octants. The definition is written as follows:

```
11,start-offset, end-offset, high-radius, low-radius, (-)OSC
```

Start and end offsets specify how far from an octant the arc starts and ends. The high-radius, if not zero, specifies a radius greater than 255. The low-radius is specified in the same manner as the radius in a code 10 arc, as are the starting octant and octants covering specifications in the last byte. The presence of the negative also signifies a clockwise direction.

The units of offset from an octant are a fraction of 1 degree times 45 divided by 256, or approximately .17578125 degrees. For example, if you wish to specify the starting value near 60 degrees, the equation would be:

```
offset = (60-45)*(256/45) = 85.333333
```

So the specification value would be 85.

To end the arc at 102 degrees, the equation would be:

```
offset = (102-90)*(256/45) = 68.2666667
```

So the specification value would be 68.

To draw an arc with a radius of 2 that starts near 60 degrees and ends near 102 degrees, the specifications would be as follows:

```
11,(85,68,0,2,012)
```

The last byte (012) specifies the starting octant to be 1 (45 degrees) and the ending octant to be 2 (90 degrees).

Codes 00C and 00D: Bulge-Specified Arc_Codes 00C and 00D (12 and 13) are used to specify arcs in a different manner from octant codes. Codes 00C and 00D call out bulge factors to be applied to a vector displacement. The effect of using code 00C or 00D involves specifying the endpoints of a flexible line by the *X-Y* displacement method and then specifying the bulge. The bulge determines the distance from the straight line between the endpoints and the extreme point on the arc. The bulge can range from −127 to 127. The maximum/minimum values (127 or −127) define a 180-degree arc (half circle). Smaller values define proportionately smaller-degree arcs. That is, an arc specified some value, say x, will be x times 180 divided by 127 degrees. A bulge value of zero will define a straight line.

Code 00C_Precedes a single-bulge-defined arcs; 00D precedes multiple arcs. This is similar to the way codes 008 and 009 work on *X-Y* displacement lines. Code 00D, like 009, must be terminated by a 0,0 byte pair. You can specify a series of bulge arcs and lines without exiting the code 00D by using the zero bulge value for the lines.

Code 00E: Flag Vertical Text Command_Code 00E (14) is only for dual-orientation text font descriptions, where a font might be used in either horizontal or vertical orientations. When code 00E is encountered in the SHAPE definition, the next code will be ignored if the text is horizontal.

TEXT FONTS

Text fonts are special SHAPE files written for use with AutoCAD TEXT drawing commands. The shape numbers should correspond to ASCII codes for characters. Table 18–1 shows the ASCII codes. Codes 1 through 31 are reserved for special control characters. Only code 10 (line feed) is used in AutoCAD. In order to be used as a font, the file must include a special shape number, 0, to describe the font. Its format is as follows:

```
*0,4,fontname
above, below, modes, 0
```

Table 18–1 ASCII Codes for Text Fonts

Code	Character	Code	Character	Code	Character
32	space	64	@	96	' left apostrophe
33	!	65	A	97	a
34	" double quote	66	B	98	b
35	#	67	C	99	c
36	$	68	D	100	d
37	%	69	E	101	e
38	&	70	F	102	f
39	' apostrophe	71	G	103	g
40	(	72	H	104	h
41	)	73	I	105	i
42	*	74	J	106	j
43	+	75	K	107	k
44	, comma	76	L	108	l
45	- hyphen	77	M	109	m
46	. period	78	N	110	n
47	/	79	O	111	o
48	0	80	P	112	p
49	1	81	Q	113	q

Code	Character	Code	Character	Code	Character
50	2	82	R	114	r
51	3	83	S	115	s
52	4	84	T	116	t
53	5	85	U	117	u
54	6	86	V	118	v
55	7	87	W	119	w
56	8	88	X	120	x
57	9	89	Y	121	y
58	: colon	90	Z	122	z
59	; semicolon	91	[	123	{
60	<	92	\ backslash	124	| vertical bar
61	=	93	]	125	}
62	>	94	^ caret	126	~ tilde
63	?	95	_ underscore		

 Note: Codes 1 to 31 are for control characters, only one of which is used in AutoCAD text fonts.

"Above" specifies the number of vector lengths that uppercase letters extend above the baseline, and "below" specifies the number of vector lengths that lowercase letters extend below the baseline. A modes byte value of zero (0) defines a horizontal (normal) mode, and a value of two (2) defines dual-orientation (horizontal or vertical). A value of 2 must be present in order for the special code 00E (14) to operate.

Standard AutoCAD fonts include special shape numbers 127, 128, and 129 for the degrees symbol, plus/minus symbol, and diameter dimensioning symbol, respectively.

The definition of a character from the TXT.SHP file is as follows:

```
*65,21,uca
2,14,8,(-2,-6),1,024,043,04D,02C,2,047,1,040,2,02E,14,8,(-4,
    -3),0
```

Note that the number 65 corresponds to the ASCII character that is an uppercase "A." The name "uca" (for uppercase a) is in lowercase to avoid taking up memory. As an exercise, you can follow the defbytes to see how the character is drawn. The given character definition starts by lifting the pen. A font containing the alphanumeric characters must take into consideration the spaces between characters. This is done by having similar starting and stopping points based on each character's particular width.

CUSTOM LINETYPES

Linetype definitions are stored in files with an .LIN extension. Approximately 40 standard linetype definitions are stored for use in the acad.lin file. The definitions are in ASCII format and can be edited, or you can add new ones of your own by using either a text editor in the non-document mode or the Create option of the LINETYPE command. Or you can save new or existing linetype definitions in another filename.lin file.

Simple linetypes consist of series of dashes, dots, and spaces. Their definitions are considered the in-line pen-up/pen-down type. Complex linetypes have repeating "out-of-line" objects, such as text and shapes, along with the optional in-line dashes, dots, and spaces. These are used in mapping/surveying drawings for such things as topography lines, fences, utilities, and many other descriptive lines. Instrumentation/control drawings also use many lines with repeating shapes to indicate graphically the purpose of each line.

Each linetype definition in a file comprises two lines. The first line must begin with an asterisk, followed by the linetype name and an optional description, in the following format:

```
*ltname,description
```

The second line gives the alignment and description by using proper codes and symbols, in the following format:

```
alignment,patdesc-1,patdesc-2,...
```

A simple linetype definition for two dashes and a dot, called DDD, could be written as follows:

```
*DDD,___  ___  .  ___  ___  .  ___  ___  .
A,.75,-.5,.75,-.5,0,-.5
```

The linetype name is DDD. A graphic description of underscores, spaces, and periods follow. The dashes are given as .75 in length (positive for pen down) separated by spaces −.5 in length (negative for pen up), with the 0 specifying a dot. No character other than the A should be entered for the alignment; it is the only one applicable at this time. This type of alignment causes the lines to begin and end with dashes (except for linetypes with dots only).

The complex linetype definitions include a descriptor (enclosed in square brackets) in addition to the alignment and dash/dot/space specification. A shape descriptor will include the shape name, shape file, and optional transform specification, as follows:

```
[shapename,filename,transform]
```

A text descriptor will include the actual text string (in quotes), the text style, and optional transform specification, as follows:

```
["string",textstyle,transform]
```

Transform specifications (if included) can be one or more of the following:

```
A=##    absolute rotation
R=##    relative rotation
S=##    scale
X=##    X offset
Y=##    Y offset
```

The ## for rotation is in decimal degrees (plus or minus); for scale and offset it is in decimal units.

The following example of an embedded shape in a line for an instrument air line (with repeating circles) could be written as follows:

```
*INSTRAIR, ___  [CIRC] ___   [CIRC] ___
A,2.0,-.5,[CIRC,ctrls.shx],-.5
```

If the ctrls.shx file contains a proper shape description of the desired circle, it will be repeated in the broken line (with spaces on each side) when applied as the INSTRAIR linetype. If the scale of the circle needed to be doubled in order to have the proper appearance, it could be written as follows:

```
*INSTRAIR, ___  [CIRC] ___   [CIRC] ___
A,2.0,-.5,[CIRC,ctrls.shx,S=2],-.5
```

The following example of an embedded text string in a line for a storm sewer (with repeating SS's) could be written as follows:

```
*STRMSWR,____  SS ___  SS ___
A,3.0,-1.0,["SS",simplex,S=1,R=0,X=0,Y=-0.125],-1.0
```

CUSTOMIZING AND PROGRAMMING LANGUAGE

There are various other topics with respect to the customization of AutoCAD, some of which are beyond the scope of this book. AutoCAD provides a programming language called AutoLISP (refer to Chapter 19). AutoLISP is a structured programming language similar in a number of ways to other programming languages. However, it is an interpreted language. Because it is an interpreted language, you can type an AutoLISP statement at the "Command:" prompt in AutoCAD and AutoCAD will execute it. AutoLISP code is created with a text editor. A compiled language (AutoCAD supports a number of these, too) is first converted into object code (this is called compiling) or machine language and then linked with various other compiled object code modules (this is called linking) to form an executable file. The executable file is then loaded into memory and executed. C/C++ and ARx (AutoCAD Runtime Extension) are examples of compiled programming languages, and they require additional software to create their executable files.

REVIEW QUESTIONS

1. The purpose of the ACAD.PGP file is to:

   a. allow other programs to be accessed while editing a drawing

   b. enable shape files to be complied

   c. store system configurations

   d. serve as a "file manager" for system variables

   e. none of the above

2. The standard AutoCAD screen menu:

   a. is stored in a file named ACAD.MNU

   b. can be viewed using the DOS TYPEcommand

   c. contains the screen menu items found in the AutoCAD screen menus

   d. all of the above

3. In an AutoCAD menu file, a semicolon contained in a menu item will tell the computer to:

   a. prompt the user for input

   b. press ENTER

   c. press CTRL

   d. ignore the rest of the line as a comment

   e. none of the above

4. AutoCAD menu files are stored with what type of file extension?

   a. DWG

   b. DXF

   c. MNU

   d. MEN

   e. none of the above

5. When developing screen menus, the information you would like to see displayed in the screen menus should be:

   a. typed in uppercase letters only

   b. enclosed with square brackets "[]"

   c. longer than four characters, but shorter than ten characters

   d. all of the above

6. How many characters between brackets will display in a screen menu?

   a. 2

   b. 4

   c. 6

   d. 8

   e. 10

7. What symbology signifies the heading of a menu device such as a digitizer or table area?

    a. ****   c. **S

    b. ***   d. none of the above

8. What is the purpose of the backslash "\" in a menu line?

    a. to pause for user input   c. to press E

    b. to terminate a command   d. none of the above

9. What file defines external commands and their parameters?

    a. ACAD.DWK   d. ACAD.LSP

    b. ACAD.EXT   e. ACAD.CMD

    c. ACAD.PGP

10. The system variable used in conjunction with DIESEL to modify the contents of the status line is:

    a. STATUSLINE   c. MODEMACRO

    b. MACRO   d. MODESTATUS

11. AutoCAD allows you to create your own icons for use in toolbar menus.

    a. True   b. False

12. To specify a cascade menu on a pull-down menu, you should use:

    a. \   c. ->

    b. >   d. #>

13. Command aliases are stored in what file?

    a. ACAD.INI   d. ACAD.MNU

    b. ACAD.ALS   e. none of the above

    c. ACAD.PGP

14. If you define a help string for a menu entry, it will display:

    a. on the status line   d. at the command prompt

    b. as a small label by the icon e. none of the above

    c. if you invoke the help command

15. When creating a partial menu, the one entry which will be included is:

    a. PARTIAL   d. MENU

    b. MENUGROUP   e. ENTRYNUMBER

    c. SECTION

16. Which of the following would not be a valid line in a definition of a custom crosshatching pattern?

a. 0,0,1,0,1

b. 90,1,0,1,0

c. 180,0,0,1,1,–1,1

d. 270,1,2,0,1,1,–1,0,–1

e. 360,1,2,1,–1

17. The first line of a custom crosshatching definition always begins with:

a. *

b. **

c. ***

d. !

e. !!

18. To define a custom linetype with three elements—a 1 unit dash, a 0.5 unit dash, and a dot, all separated by 0.25 unit spaces—what would the definition look like?

a. A,1,0.25,0.5,0.25,0

b. A,1,.25,0.5,0.25,0,0.25

c. A,1,–0.25,0.5,–0.25,0

d. A,1,–0.25,0.5,–0.25,0,–0.25

e. A,1,0.5,0,–0.25

19. The bit code which specifies a direction of 12 o'clock in a shape file is:

a. 0

b. 4

c. 8

d. 12

e. C

20. When specifying the name of a shape character, in order to conserve memory you should:

a. use upper-case

b. use lower-case

c. preface the name with an *

d. use a short name

e. it does not matter, all shapes require the same amount of memory regardless of their names

21. A macro permits the user to save and invoke a series of keystrokes with a combination of only one or two key strokes?

a. True

b. False

22. Which of the following can be customized in AutoCAD 2000?

a. aliases

b. linetypes

c. shapes

d. hatch patterns

e. all of the above

23. Command aliases can include command options?

   a. True   b. False

24. Which of the following is the correct format to define an alias using *L* for the LINE command?

   a. Line = L   c. L, *line

   b. L = <line>   d. L = *line

25. Which of the following is not a valid AutoCAD 2000 menu type?

   a. Pull-down   d. Tool tip

   b. Screen   e. all of the above

   c. Tablet

26. When writing a script file command aliases cannot be used.

   a. True   b. False

27. Which of the following is not a valid AutoCAD 2000 menu file extension?

   a. .mnu   d. .mnc

   b. .mno   e. none of the above

   c. .mns

28. How many pull-down menus does AutoCAD 2000 have available to use?

   a. 12   d. 32

   b. 16   e. unlimited

   c. 24

29. Which of the following characters, when included in a macro will repeat the command until the user cancels?

   a. :   d. *

   b. ;   e. >

   c. +

30. Which of the following characters, when placed in a macro will underscore the letter that follows and permits menu selection from the keyboard by combining the ALT key and a specified key?

   a. ~   d. @

   b. &   e. none of the above

   c. #

CHAPTER 19

Visual LISP

INTRODUCTION

This chapter will cover the fundamental concepts of the Visual LISP programming system, including writing, storing, and loading .LSP files; variables and expressions; lists; custom functions; and file handling.

After completing this chapter, you will be able to do the following:

- Grasp fundamental concepts of the Visual LISP programming system
- Decipher program files written by others
- Establish a basis for more advanced programming

BACKGROUND

In order to support Alonzo Church's lambda calculus operations, John McCarthy invented Lisp in the late 1950s. Internal data structures are ideally represented as lists, which can contain additional nested lists. The symbolic expressions for the lists are sandwiched with balanced parentheses. List processing is the intrinsic functionality of Lisp.

John Walker, the principal Autodesk co-founder, recognized the synergistic potential of Lisp for AutoCAD customization at a time when competitive programs were either offering very simple macro operations or demanding extensive investment in higher-level programming languages for similar operations. The implementation of Lisp for use within AutoCAD with some key constraints is referred to as AutoLISP™ and was introduced with AutoCAD Release 2.1 (later referenced as Release 6).

VLISP

As Autodesk placed emphasis on its ARX™ technology for the development of high-performance custom applications in recent years, developments in AutoLISP remained largely static. The introduction of Visual LISP has brought new life to legacy applications and also allowed users to maintain their investment in AutoLISP code. AutoCAD Visual LISP is referred to as VLISP™ in the remaining sections of this chapter.

INTERFACES

VLISP provides a very rich set of interfaces to the underlying AutoCAD object model. It relies on AutoCAD-specific (namely, *Proteus*) programmable dialog boxes for a graphical user interface when applications are run. (These topics are beyond the scope of this chapter.)

INTEGRATED DEVELOPMENT ENVIRONMENT

VLISP has a project window, color-coded text editor, debugger, resource windows, and many other features that simplify the development of applications that customize AutoCAD operations. The primary benefit of these features is that the time required from analysis to implementation and design is dramatically shortened. Consequently, a user is generally more productive with VLISP than previously was possible. This set of features is commonly referred to as an Integrated Development Environment (IDE).

A TEST DRIVE OF VISUAL LISP

Let us take a test drive to understand the key components of VLISP. In the remaining sections of this chapter, we will explore some simple projects to illustrate the basic components of VLISP.

SIMPLE EXAMPLE

It is customary to build a very simple program to test the installation of an integrated development environment. In this chapter we will begin with such a program that is commonly referred to as a *Hello World* example. We will then develop an additional exercise to demonstrate other key features of VLISP.

LAUNCH VISUAL LISP

VLISP operates under AutoCAD. You can edit programs without running AutoCAD, but you can only launch VLISP when AutoCAD is running.

Invoke the Visual LISP integrated development environment:

Tools menu	Choose AutoLISP > Visual LISP Editor
Command: prompt	**vlisp** (ENTER) or **vlide** (ENTER)

The Visual LISP window is displayed as a separate window with two basic child windows:

- VLISP console
- Trace (by default it's minimized)

The main areas within the VLISP window are:

- Console window
- Status bar
- Menu bar
- Toolbars
- Trace window

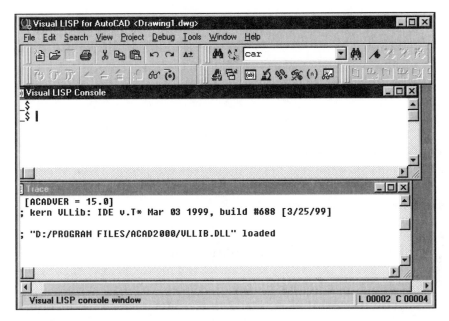

Figure 19–1 *Visual LISP initial window*

Console Window

This child window allows you to enter VLISP commands and is scrollable. You can review your previous commands or operations in this window. Most of the selections in the menu bar and toolbars have equivalent commands that you can enter in this window.

Status Bar

The current conditions related to your operations are displayed in the bottom left bar of the VLISP window.

Menu Bar

The menu bar operates like the conventional Windows menu bar. The status bar displays a short description corresponding to the current selection.

Trace Window

This window is minimized during startup. It contains information regarding the VLISP release number and other startup messages. You will also use this window during debug operations.

TOOLBARS

Toolbars allow you to issue commands using the point-and-click method. Instead of navigating the menu bar or memorizing the command syntax to be entered in the console window you can select a button to activate the command. There are five principal toolbar groups:

- Debug
- Edit
- Find
- Inspect
- Run

DEVELOPING AND APPLICATION

Our first sample program will simply display the text string "Hello World" at the AutoCAD command prompt. Let us assume that you have launched VLISP using the instructions detailed earlier.

TEXT EDITOR

The VLISP text editor allows you to write AutoLISP code. Since we have not written any code until now for the sample program we will invoke the text editor with a new file. Invoke the VLISP text editor:

File menu	Choose New
Standard toolbar	Choose New (Figure 19–2)

Figure 19–2 *Invoke New file from the Standard toolbar (VLISP)*

On completion of this command, a new child window is displayed with the title <Untitled-0> and the status bar has the message "Edit –no file- (Visual LISP)." You should proceed to enter the following code in this new window:

```
(defun c:hello ( )                     ; function definition
  (prompt "Hello World!")   ; displays the string at the command
      prompt
(princ)                                ;
)
```

Color Codes

As you enter the above code you will notice that certain elements are color-coded to assist you in recognizing the syntax. The color scheme depends on the type of file that is being edited. This file type is implicit from the file type extension in use. Some of the standard color schemes are:

Style	Description
None	No color coding
AutoLISP	LISP syntax color coding for file type extension LSP
C++	C syntax color coding for file type extensions C, CPP, and H
DCL	AutoCAD programmable dialog boxes for file type extension DCL
SQL	SQL procedure files for file type extension SQL

The code is also automatically indented for clarity. You can change the color codes yourself to suit your needs more appropriately, but these issues are beyond the scope of this chapter.

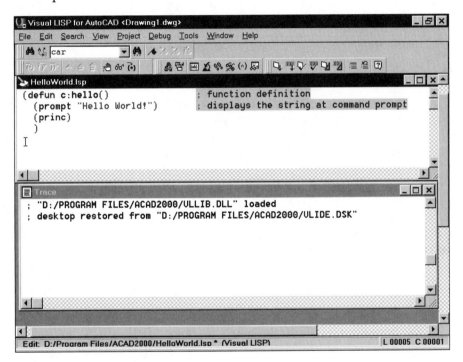

Figure 19–3 *Hello World program*

SAVE YOUR PROGRAM

After you have completed entering the code you should save your code.

Invoke the save command:

File menu	Choose Save As
Standard toolbar	Choose Save File

The Save As dialog box is displayed. For this sample program, enter the name **hello** in the **Filename** text box. Before you select the Save button in the dialog box you should ensure that the **Save As Type** text box is set to Lisp Source File and the desired path (drive:\folder(s)) is set up. AutoCAD can find source files such as your hello.lsp file easiest in the Acad2000\Support folder.

Once you have successfully saved the source program file; its text window and the status bar will reflect the file name you entered in the text box control field.

LOAD YOUR PROGRAM

Your program must be loaded before it can be run. Make your program window active by selecting anywhere within it.

Invoke the LOAD TEXT IN EDITOR command:

Tools menu	Choose Load Text in Editor
Tools toolbar	Choose Load Active Edit Window

On completion of the load operation the VLISP console window will display a confirmation message. If errors are reported, you must carefully review what you have entered and correct the mistake appropriately. Two very common sources of error messages are unbalanced parentheses and apostrophes (i.e., for strings).

RUN YOUR PROGRAM

You must switch to AutoCAD to run this short program. You can test the execution from the console window. Because this program specifically writes a string to the AutoCAD command prompt window you will not see any output in the console window except the completion code, which is *nil* for this program.

Switch to the AutoCAD window:

Window menu	Choose Activate AutoCAD
View toolbar	Choose Activate AutoCAD

 Note: If the AutoCAD window is minimized, this operation will not restore it. You must restore the AutoCAD window to run the program.

Before you can use an AutoLISP application, it must first be loaded into memory. You can use the APPLOAD command or the AutoLISP "load" function to load an application. Loading an AutoLISP application loads the AutoLISP code from the .LSP file into your system's memory. Then, you can invoke the lisp routine from the Command: prompt.

Invoke the APPLOAD Command:

Tools menu	Choose AutoLISP > Load
Command: prompt	**appload** (ENTER)

AutoCAD displays Load/Unload Applications dialog box. Select the lisp file from the appropriate folder, and click the **Load** button. AutoCAD loads the selected AutoLISP file into memory and is ready to be used for the current session. AutoCAD lists the loaded applications in the Loaded Applications list box. If necessary, you can unload the application by selecting the AutoLISP file name from the loaded applications list box and click the **Unload** button. Choose the **Close** button to close the dialog box.

You can also drag lisp files from the files list box, or from any application with dragging capabilities such as Windows Explorer, into the Startup Suite area to add them to the Startup Suite located in the Load/Unload Applications dialog box. AutoCAD automatically loads each time you start AutoCAD program the contents of the Startup Suite.

In addition to loading AutoLISP applications into memory by APPLOAD command, you can also load with AutoLISP function called "load" as shown in the following example:

Command: (load "hello")

Notes: The use of the parentheses distinguishes AutoLISP functions and routines. This is especially important when using a function, such as "load," for which there is an AutoCAD command of the same name.

Do **not** include the .LSP extension. AutoCAD appends it automatically.

You may also specify a path if necessary. For example, if the hello.LSP file is on the A: drive in a directory called LISP, the following response can be used:

Command: (load "a:/lisp/hello")

Notice the use of the nonstandard forward slashes to specify the directory path.

After loading the AutoLISP routing into memory, at the AutoCAD command prompt, enter the command **hello**. Immediately following your command you will see the string *Hello World!* If there is only one line in the command history window you may have to increase it to two lines or press F2 to see the results of your program in the AutoCAD text window. You have built and run a simple AutoLISP program using the VLISP IDE.

VLISP FUNDAMENTALS

While developing simple AutoLISP programs, the entire source code is often retained in a single file. If the programs become more complicated or more than one person is involved during development, then it is customary to divide the program into modular sections. VLISP uses a project file to allow autonomous modular development.

PROJECT FILE

To specify a project file:

Project menu	Choose New Project

VLISP responds by displaying a New Project dialog box to specify the folder and name of the project file.

You should select the correct folder where you would like to keep the project file. Enter the file name in the corresponding File name text box and ensure that the file type is set to VL project file.

After the project file name is processed, the Project Properties dialog box (as shown in Figure 19–4) is displayed to allow the selection of the files associated with the project.

Figure 19–4 *Project Properties dialog box*

PROJECT WINDOW

After you have selected the files using the appropriate buttons to move the files from one list to the other you should click the OK button. The dialog box is dismissed, and a child window listing the project files is displayed, as shown Figure 19–5.

Figure 19–5 *Project window*

The source code for the three files is listed below.

Balloon.lsp

```
; Name      : balloon
;
; Purpose : to place a balloon with leader in a 2D drawing file
;
; Last update:      12/29/89   Rev: 1.0     initial implementation
;    05/19/92        2.0       customer simplification
;    04/01/99        3.0       2000 update
;
; Usage    : balloon
;
; Prompts :
;    Scale/<Insertion point>:
;    Scale factor:
;    End point:
;
;    The author makes no warranty, either expressed or implied,
;    including but not limited to any implied warranties of
;    merchantability or fitness for a particular purpose,
;    regarding these materials and makes such materials
;    available solely on an "as-is" basis.
;
; Auxiliary functions:
;    MODER MODES
```

```
;
;                                      save entry state variables
;
; a list of variables whose entry states have to be saved
; cv          current variable
; MLST             save list
;
; Input variables:
;                  none
;
; Output variables:
;                  none
;
; Local variables:
;    atd    INSERT mode for attributes (restored upon exit)
;    bm     blip mode (restored upon exit)
;    cn     current balloon number
;    itm    current item number
;    la     current layer name (restored upon exit)
;    pt1    insertion point for block
;    pt2    terminating point for leader line
;    qty    quantity (attribute) for block (=1 default)
;    rad    radius of bubble circle
;    sx     scale factor (=1 for prototype)
;
(defun C:balloon ( / cn cw er oe la ln mors pp pt1 pt2 sx zz)
;                                  establish local error handler routine
;                                  get entry state variables
;
; Internal error handler defined locally
;
  (defun er (s)              ; If an error (such as CTRL-C) occurs
                                ; while this command is active...
    (if (/= s "Function cancelled")
      (if (= s "quit / exit abort")
        (princ)
        (princ (strcat "\nballoon: ERROR! " s))))
    (moder)
    ); defun
  (if *error*                          ; Set our new error handler
    (setq oe *error* *error* er)
    (setq *error* er)
```

```
      ); if
  (modes '("ATTDIA" 0 "ATTREQ" 0 "CMDECHO" 0 "COORDS" 1 "DRAGMODE"
    2 "GRIDMODE" 0
    "ORTHOMODE" 0 "OSMODE" 0 "OSNAPCOORD" 1 "SNAPMODE" 0))
  (setq la (getvar "CLAYER") ln "bomsym")
;                                 establish the correct layer
  (command "layer" "m" ln "")
;                                 check the scale factors
  (setq cn (getvar "USER2") cw 786 zz (getvar "USERI1"))
  (if (/= zz cw)
    (progn
      (setvar "USERI1" cw)
      (setvar "USERR1" 12)
    ); progn
  ); if
  (if (not cn)
    (setq cn 0)
    )
  (setvar "USERI2" (setq cn (1+ cn)))
;                        collect data for curr invocation of proc
  (setq pp T)
  (while pp
    (initget (+ 1 2 4 32) "Scale")
;    (graphscr)
    (setq pt1 (getpoint "\nScale/<Insertion point>: "))
    (cond ((eq pt1 "Scale") (progn
        (initget (+ 1 2 4))
        (setvar "USERR1" (getreal "Scale factor: "))
      ); progn
    )
    (T (setq pp nil))
    ); cond
  ); while
  (setq pt2 (getpoint pt1 "End point: ")
        sx (getvar "USERR1"))
  (if (or (<= 0.0 sx) (not sx))
    (setq sx (setvar "USERR1" 1.0))
    )
  (setq rad (* 0.25 sx));setq
;                             place block in dwg file
  (command "INSERT" "balloon" pt1 sx sx 0 )
;                             place leader from balloon to part geometry
```

```
                    (setq pt1 (polar pt1 (angle pt1 pt2) rad))
                    (command "LINE" pt1 pt2 "" "LAYER" "S" la "")
;                            restore entry state variables
                    (moder)
;                            restore old error handler
              (if oe
                    (setq *error* oe)
                    ); if
              (princ)
              );defun
```

Modes.lsp

```
              (defun modes ( a / cv)
              (setq saveList '())
              (repeat (/ (length a) 2)
                    (setq    cv (car a)
                       saveList (append saveList (list (list cv (getvar cv))))
                       a (cdr a))
                    (setvar cv (car a))
                    (setq    a (cdr a)))
              ); defun
```

Moder.lsp

```
              (defun moder ()
              (repeat (length saveList)
                    (setvar (caar saveList)(cadar saveList))
                    (setq saveList (cdr saveList)))
              ); defun
```

COMPILING VLISP APPLICATION (OR, SO LONG KELVINATOR)

A significant change introduced by VLISP is the ability to compile an AutoLISP application into a stand-alone executable file that is assigned a file type extension of VLX. You can load VLX files just like you would load an ARX, EXE, or EXP file. A desirable result of this technique is that your investment in your source code is protected to a reasonable extent from casual prying.

MAKING AN APPLICATION

The Make Application wizard will guide you through the steps of creating an application. For a new application the wizard offers two paths for this process—simple or expert, as shown in Figure 19–6.

Figure 19–6 *Wizard mode selection*

SIMPLE WIZARD

If you select the **Simple** radio button and click the **Next** button, the following window is displayed for you to specify the home folder and name for the application.

Figure 19–7 *Application directory data*

You can use the **Browse** button to navigate to the appropriate folder, and the corresponding folder name will be entered in Application Location text box. You must enter an application name in the corresponding text box before you click the **Next** button.

PROJECT FILE SELECTIONS

The three types of files that you can select for inclusion in your project are:

- Lisp source
- Compile lisp files
- VLISP project files

After selecting the appropriate files, select the **Next** button to proceed to the final phase of this wizard.

Figure 19–8 *Project file selections*

REVIEW SELECTIONS

If you are ready to build the applications, you should select the corresponding checkbox. Or, if you need to make changes to the code or add additional files, there is no need to build the application. Select the **Finish** button to complete the processing by the wizard.

Figure 19–9 *Build the Application wizard*

EXPERT WIZARD

If you choose the **Expert** radio button in the wizard, then you will be presented with some additional dialog boxes and options. For example, as shown in Figure 19–10, in addition to the standard Lisp and project files, you can also choose AutoCAD Visual Basic for Applications modules (DVB file type extension) and dialog declaration files (DCL file type extension).

Figure 19–10 *Additional file selections*

You can select different compile and link options, as shown below, particularly to improve the performance of the application.

Figure 19–11 *Compilation and link options*

PROJECT DEFINITION

If you are not ready to build or make your application, but still want to proceed along the development path, then you can create a project file. When you are ready to build or make the application you can simply include the project file as a source. In the meantime, you can coordinate your development as a team effort using the project file as the control source.

PROJECT FILE

In response to selecting New Project, the New Project dialog box is displayed. You should navigate to the correct folder before you enter the file name. Note that the file type extension is VLJ for VLISP project file. After you have entered the file name, click the **Save** button to proceed to the next step.

PROJECT PROPERTIES

The Project Properties dialog box has two tabs. The Project Files tab is used to collect the names of the project files. Once again, you have to be careful that you navigate to the correct folder to select the files.

Figure 19–12 *Project Files tab*

The second property tab allows you to specify the build options for the project. You can always revise these selections later as appropriate.

Figure 19–13 *Build Options tab*

VIEW TOOLBAR

As you work in the VLISP IDE you will become familiar with the toolbar buttons.

Figure 19–14 *View toolbar*

The View toolbar contains the most common navigation buttons:

- Activate AutoCAD
- Select Window
- VLISP Console
- Inspect
- Trace
- Symbol Service
- Apropos
- Watch Window

ACTIVATE AUTOCAD

This button allows you to activate the AutoCAD window. During program development and testing you may have to switch between the VLISP IDE and the AutoCAD window.

SELECT WINDOW

This button allows you set focus to an existing child window in the VLISP IDE. In addition to the VLISP console and trace windows, other windows include any source code windows.

VLISP CONSOLE

This button places the cursor at the end of the last record in the console window. You can use this command button to switch to the console command line quickly.

INSPECT

The Inspect windows allow you to browse, examine, and modify objects associated with AutoCAD or your program.. The VLISP Inspect feature can help you figure out how to process the information stored in the association list. You can scroll through complex objects to see their simpler components.

TRACE

The Trace window lists the calling sequence of the programs or procedures that lead to the current stopping point.

APROPOS

With this command, VLISP will offer to complete partially entered words from the entries in its symbol table. This feature reduces typing errors.

WATCH WINDOW

If you wish to keep track of the values of selected variables, you can place these variables in the watch window. The name and current value of the selected variable is always on display for your inspection.

> **Note:** VLISP provides an interface with the embedded AutoLISP programming language that makes it easier to program and more like other "visual" programming languages, for example Visual Basic and Visual C++. But the structure, syntax, functions, expressions, and symbols that are the body of the program are still unique and specific to AutoLISP. The following sections cover that specific structure and syntax and those functions, expressions, and symbols that must be learned before you can successfully apply AutoLISP through the VLISP interface. Once the basics of AutoLISP are mastered, there is a section on debugging at the end of the chapter that is part of the VLISP interface.

AUTOLISP BASICS

In Version 2.1 of AutoCAD (Release 6, May 1985), Autodesk first introduced AutoLISP, its embedded programming language. It provided on-board computational power for the operator while in AutoCAD. It also permitted true programming routines to be used by way of menu devices, including interactive functions to receive input from the operator from keyboard entries and screen picks for use in the routine. Version 2.18 (January 1986) included a full implementation of user-defined functions and custom commands, adding a whole new world of open architecture to AutoCAD (meaning you can customize the program to suit your needs).

EXPRESSIONS AND VARIABLES

Expressions in AutoLISP should be understood before getting into variables. The simplest application of AutoLISP is to evaluate an equation just by entering it in and pressing ENTER. Of course, you must enter the equation in the proper format, which is somewhat different from ordinary algebraic notation. It involves a format that is unique among those used in other, more popular computer programming languages. For example, if you wish to add 5 and 3, simply enter the following expression:

    Command: (+ 5 3)

The integer 8 will be displayed in the prompt area. That is, AutoLISP evaluates the expression and returns the integer 8. Throughout this chapter, the word *return* will be used to describe the result of an evaluation. The following expression returns the integer 108:

    Command: (+ 5 3 1 99)

Note: When AutoCAD sees an open parenthesis (unless responding to a prompt to enter text), it knows that it is entering an AutoLISP expression to be evaluated. The AutoLISP evaluator remains in effect until it encounters the closing parenthesis that is the mate of the first open parenthesis.

AutoLISP uses prefix notation, which means that expressions begin with the operator (after the opening parenthesis, of course). In the preceding examples, the plus sign (+) is the arithmetic operator. The *operator* tells AutoLISP what operation to perform on the items that follow. The items that follow the operator are known as the *arguments*.

As you can see by the second example, the plus operator can have more than the usual two arguments to which an algebraic plus sign is restricted.

Elements in an AutoLISP expression are separated by one or more spaces. Multiple adjoining spaces (unlike spaces in a menu line) are considered one space in an AutoLISP routine.

Variables as Symbols and Symbols that Should Not Vary

Just as algebra uses letter names for the unknown values in an equation, AutoLISP uses symbols as variables, whose name you may select during the writing of the program. In algebra, for example, a sequence might be written as follows:

Algebra	AutoLISP
a = 3	(setq a 3)
b = 7	(setq b 7)
therefore	
a + b = 10	(+ a b) returns 10
and	
ab = 21	(* a b) returns 21

Assigning Names

There are three situations for assigning names:

Expressions: For example, pi is the name for 3.14159...

Variables: A variable is an expression whose value is not known when originally written into the program. Variables will take on some value after the program has been called into use. The value of the variable is usually determined by some operation on some other value that the user has been prompted to enter while the program is in progress.

Custom-Defined Functions: AutoCAD permits users to create and name a customized function and then use it in AutoCAD in a manner similar to a standard AutoCAD command.

TERMINOLOGY AND FUNDAMENTAL CONCEPTS

Lists, operators, arguments, types, parentheses, the exclamation point, and the concept of the function list comprise the basis of AutoLISP.

Lists: Practically everything in AutoLISP is a list of one sort or another. Functions are usually represented as a list of expressions enclosed in parentheses. For example, (+ 1 2.0) is an AutoLISP function with three elements in it, the "+," the "1," and the "2.0." Other lists can be established by applying a function called list or by applying the single quote, as in '(1 2 3.0 a "b").

Operators and Arguments:: Arguments are those items in a function list on which the operator operates. For example, in the function list (+ 1 2), the operator is (+) and the arguments are "1" and "2."

Types: Arguments are classified by their type. The arguments in the example (+ 1 2) are of the type called integer. In the function list (+ 1.0 2.0), the arguments are of the type called real, signifying a real number.

Parentheses: The primary mechanisms for entering and leaving AutoLISP and entering expressions within AutoLISP are the open and close parentheses.

Note: For every open parenthesis [(] there must be a close parenthesis [)]. If AutoCAD encounters a condition wherein the close parentheses are one fewer than the open ones, the prompt will display the following:

    1>

Most often, this is caused by the need for one close parenthesis. A display of 3> could indicate the need for three closing parentheses. The problem is usually remedied by just entering the specified number of close parentheses. This type of error message can also be caused by having an odd number of double quotation marks inside of an AutoLISP expression, in which case one double quotation mark must be entered, followed by the required number of close parentheses, in order to eliminate the error message in the prompt area. Further examination will usually reveal that the program line needs to be corrected to prevent the message from recurring.

Exclamation Point: A leading exclamation point in response to a prompt is another mechanism for entering AutoLISP. This tells AutoCAD that the symbol following the exclamation point is an AutoLISP variable that has been set equal to some value and that AutoCAD should use the value of the variable as a response to the prompt.

The terms and concepts just described are explained in detail, with examples, in the sections that follow.

Functions: The first function list to be introduced involves three expressions: the operator SETQ, the variable that we have arbitrarily named x, and the real number 2.5. Within this function SETQ will perform a special operation on the variable x and the real number 2.5. SETQ is probably the most common AutoLISP function. It means "set equal." Using SETQ as the first of three expressions will set the second expression equal to the third. We are accustomed to performing this operation by the conventional expression x = 2.5. But as you will see in the next example, in order to write x = 2.5 in AutoLISP you must write in a special format: SET x EQUAL 2.5. It is done by writing three expressions in parentheses, with the first being the operator SETQ, the second being x, and the third being the value to which the variable named x is to be equal, for example:

Command: **(setq x 2.5)**

Having x = 2.5, or, more properly, to have SET x EQUAL 2.5, is necessary only if you wish to use the value of the real number 2.5 later by just entering its name, x. If, earlier, you have entered (setq x 2.5) at the "Command:" prompt, then any time during that current editing session you could re-enter AutoLISP right from the keyboard by using ! before the name of the value in response to a prompt for some real number. If the prompt is asking for a name of something (like a layer name, which requires responding with something called a string variable), trying to use a variable that has been set equal to a real number will cause an error message. The use of ! is another way to enter AutoLISP directly from the keyboard or in a string of custom menu commands and responses.

For example, if you create a unit block named UB and wish to insert it with a scale factor of 2.5, the sequence of prompts and responses would be as follows:

Command: **insert**
Block name (or ?): **ub**
Insertion point: *(pick an insertion point)*
X scale factor <1> / Corner / XYZ: **!x**

Note at this point that the use of the name x and the fact that it is being applied to the X scale is purely coincidental.

By using the ! in front of the x, you will have the value of x (or 2.5 in this case) used as a response to the prompt asking for the X scale.

This example is not very efficient. Entering !x appears to take only two keystrokes, whereas entering 2.5 takes three. True, this would save a keystroke. But because entering the exclamation point requires the SHIFT key, it is a double stroke requiring both hands. Along with the x it is actually less convenient than entering 2.5, which has three one-hand strokes. Additionally, the characters "2," ".," and "5" are all accessible on the AutoCAD tablet overlay, and the ! is not. Entering 2.5 will, however, give the same accuracy as entering 2.50000000000000.

Note: If x has not previously been set equal to a value, then entering !x will return "nil." Believe it or not, advanced programming does make use of the nil. Normally, variables and expressions retain their names and values only during the current editing session. The names are lost when a drawing is ended. At the beginning of each session they must be reestablished. But even this problem of saving variable values from one session to another can be addressed by a custom program.

Naming and SETQing the Value of an Expression: Suppose the value you wish to enter is 1 divided by the square root of 2. This could be written as 0.7071068 depending on the accuracy desired. Now compare three keystrokes using AutoLISP versus nine or ten from the keyboard or tablet. Not only will a variable having a short name (and having been set equal to that value) save time in entering but it will decrease the probability of errors in both reading and keying in a long string of characters, for example:

Command: **(setq x 0.7071068)**

If the value 0.7071068 is needed in response to a prompt, simply enter !x. Programming begins to appear both expedient and practical.

Note: You may respond with .7071068 outside AutoLISP, but while in AutoLISP you must use some value (even if zero) on both sides of the decimal point. Expressions beginning or ending with a dot (.) have another special meaning.

Before leaving this example, additional power of AutoLISP can be seen in a demonstration of using an expression within an expression within an expression. Two other AutoLISP operators will be introduced at this time. They are the SQRT and the "/" functions. SQRT returns the square root of the argument following (it must be only one argument and must be a nonnegative real number). The "/" function requires two arguments and returns the quotient of the first divided by the second, for example, the following returns 1.414213562373...

Command: **(setq y (sqrt 2))**

You may now use this value as follows:

Command: **(sqrt 2)**

This does two things. It sets y equal to the expression that follows, which is the square root of 2, and returns 1.414213562.

Having entered this, you may operate on y as follows:

Command: **(/ 1.0 y)**

This expression will return the quotient of 1 divided by the value of y, which was previously set equal to the square root of 2. Therefore, the expression evaluates to 0.7071068.

Using an AutoLISP function in this fashion neither sets values nor gives them names. It is of little use later, but does return the result of the prescribed computation for immediate viewing. Sometimes this is even handier than picking up a hand-held calculator.

A two-step use of AutoLISP could be written as follows:

> Command: **(setq y (sqrt 2))**
> Command: **(setq x (/ 1 y))**

The first step sets y equal to the square root of 2. The second step sets x equal to the inner expression, which uses the / operator (for division) and returns the value of 1 divided by the value of y. This two-step sequence involves four operators (SQRT, /, and SETQ twice). It names y and sets it equal to the square root of 2. It names x and sets it equal to the reciprocal of y.

Instead of this two-step routine, a simpler one-step routine is as follows:

> Command: **(setq x (/ 1 (sqrt 2)))**

So, in one step you can name an expression x and set it equal to the reciprocal of the square root of 2. It allows you then to respond to any prompt for a real number and apply this value by simply entering !x. This saves time and ensures accuracy to 14 decimal places, which is what CAD is all about in the first place—speed and accuracy.

EXAMPLE

At the "Command:" prompt enter the following expression:

> Command: **(setq a 1 b 2 c 99 d pi e 2.5)**

This expression is a function list using the operator SETQ. Note the opening and closing parentheses that distinguish it as an individual expression. Unlike the (+ . . .) function that performs a single operation of as many arguments as you wish to furnish, the (setq . . .) must have pairs of arguments. SETQ operates on each pair of arguments individually. After entering the preceding, you can have AutoLISP evaluate equations by entering them as follows:

> Command: **(setq x (+ a b))**   *(returns 3)*
> Command: **(setq y (- a c))**   *(returns -98)*
> Command: **(setq z (* b d))**   *(returns 6.283185308)*
> Command: **(setq q (/ c e))**   *(returns 39.6)*

Perform the following exercise:

> Name and give values to variables corresponding to the following algebraic equations:
>
> $a = 4$
> $b = 7$
> $c = a \times b$
> $d = b^2$ *(with 2 as exponent)*
> $e = \ddot{O}a^2 + b^2$ *(square root of a-squared plus b-squared)*

EXERCISES FROM THE KEYBOARD

Line: After first drawing a random line (with the Snap mode off), establish variables for the points p1 and p2 as follows (see Figure 19–15). Enter **(setq p1 (getpoint))**. Use the Osnap mode Endpoint to pick one end of the line. Enter **(setq p2 (getpoint))**, and use the Osnap mode Endpoint to pick the other endpoint of the line. Do not be alarmed that the prompt area is blank. When the points have been returned, record the *X* and *Y* coordinates on paper for use in the following exercises. For example, if (1.2345 6.7890 0.0000) is returned for p1, enter the following:

Command: **(setq X1 1.2345)**   *(returns 1.2345)*
Command: **(setq Y1 6.7890)**   *(returns 6.7890)*

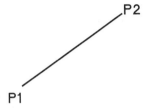

Figure 19–15 *The line for writing a routine*

Note: Enter the numbers you obtain, not the 1.2345 and 6.789 from the example!

Write in from the keyboard routines to perform the following:

1. Set *dx* equal to the horizontal distance between p1 and p2.
2. Set *dy* equal to the vertical distance between p1 and p2.
3. Set *d* equal to the distance between p1 and p2.
4. Set *a* equal to the angle between p1 and p2 in radians.

Circle: After first drawing a random circle (with the Snap mode off), set variables to the center c of the circle and to one point p on the circle (see Figure 19–16). Write routines to perform the following:

1. Set *r* equal to the radius of the circle.
2. Set *d* equal to the diameter of the circle.
3. Set *p* equal to the perimeter of the circle.
4. Set *a* equal to the area of the circle.

Arc: Establish endpoints p1 and p2 and the center c of the arc (see Figure 19–17) as in the exercises. Write routines to perform the following:

1. Set *r* equal to the radius of the arc.
2. Set *lc* equal to the chord length of the arc.
3. *(A doozie)* Set *la* equal to the arc length of the arc.

Computation: Write routines to perform the following (see Figure 19–18):

1. Given $a = 1$ and $b = 2$, find c.
2. Given $b = 2$ and $c = 3$, find a.
3. Find the area of the triangle for question 1.
4. Find the area of the triangle for question 2.

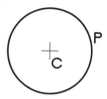

Figure 19–16 *The circle for writing a routine*

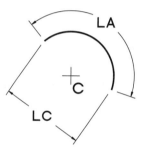

Figure 19–17 *The arc for writing a routine*

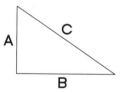

Figure 19–18 *The triangle for writing a routine*

According to Ben Shneiderman in his preface to Dan Friedman's *The Little LISPer*, "The fundamental structure of the LISP programming language was derived from the abstract notions of lambda calculus and recursive function theory by John McCarthy. His goal was to produce a programming language with a powerful notation for defining and transforming functions. Instead of operating on numeric quantities, LISP was designed to manipulate abstract symbols, called atoms, and combinations of symbols, called lists. The expressive power was recognized by a small number of researchers who were primarily concerned with difficult symbolic manipulation problems in artificial intelligence."

Translating an algebraic word problem into a true algebraic expression demands a certain symbol classification as well as list processing. Many algebra students who learn how to solve an algebraic expression (or at least memorize the procedures) once it is in the proper structure reach an impasse when asked to "interpret" the word problem into that proper structure.

Words in a sentence or paragraph have characteristics similar to AutoLISP functions or expressions. Verbs act on subjects like subroutines act on variables and expressions. The conjunction *and* joins phrases like the "+" joins real numbers. A phrase like *out of here* has the subphrase of *here* in it.

LISTS

TYPES OF ARGUMENTS

Although an argument is defined, in part, by *The American Heritage Dictionary* as "a quarrel; contention," it also is defined under the heading of math as "the independent variable of a function." *The AutoCAD Programmer's Reference* consistently describes the variables on whose value the function value depends as arguments.

The List in AutoLISP Is an Argument: When is a list not a list? As mentioned earlier, almost every function in AutoLISP is some form of list. The AutoLISP interpreter operates on lists as they occur in a routine or program. The AutoLISP expression (setq x (/ 1 (sqrt 2))) is a list. To the AutoCAD LISP interpreter it is a list to be evaluated according to the operator SETQ. Likewise, (/ 1 (sqrt 2)) and (sqrt 2) are a sublist and subsublist, respectively, to be evaluated. But none of these would be considered as the type of defined list (known as list to the type function) that some of the AutoLISP function lists, such as (CAR <list>), normally require as their argument. The lists that are acceptable arguments to these certain functions might be looked on as special lists within lists. They must take on a special format in order to satisfy the main function list as the required argument(s).

AutoLISP function lists that require lists as one of their arguments include the CAR, CADR, LAST, LENGTH, NTH, and REVERSE functions.

Whenever an AutoLISP function list requires a list as one of its arguments, you must use one of the following procedures:

1. You may use the function-list that has as its operator the function called list. This function list is written in the format (list <expr>...) as described in the *AutoCAD Programmer's Reference*. Entering in **(list 1.0 2.0)** will return (1.000000 2.000000). But more importantly, it will satisfy the requirements of this special list when needed for data input. Other function lists (such as the one with the operator SQRT) most likely will *not* satisfy the special requirement of an argument that must be of the type called list.

2. If you must respond within AutoLISP with the special form of list required, writing **(list 1.0 2.0)** every time becomes laborious (like writing 3.14159 and

so forth when the use of *pi* will do). Therefore, it is convenient to use the SETQ function and write:

(setq p (list 1.0 2.0))

The name p is strictly arbitrary in this case. Any unique variable name you want to use will suffice. Then whenever your routine requires you to respond with a point whose X and Y coordinates are 1.0 and 2.0, respectively, you can respond with p within AutoLISP or !p directly from the keyboard while in AutoCAD.

Try this exercise for an illustration of the preceding: The default limits of 0,0 and 12,9 as in the ACAD.DWG file will be helpful. Enter at the "Command:" prompt the following:

Command: **(setq p (list 1.0 2.0))** *((1.000000 2.000000) will be displayed)*
Command: **line**
Specify first point: **(list 6.0 7.0)**
Specify next point or [Undo]: **!p**

CREATING A LIST

As explained earlier in this chapter, the (list <expr>...) function list is often used under restrictive conditions. For example, using it as a response to certain commands requiring a 2D point, the "<expr>..." part of the overall expression must be "<expr> <expr>," where both <expr>s are real numbers.

Remember that an integer can be entered in AutoLISP and that AutoCAD will translate it into a real number under certain conditions. For a 3D point, the form of the response must be "<expr> <expr> <expr>." Using "<real> <real>," "<real> <int>," or "<int> <int>" may be considered proper responses to a prompt for the starting point of a 2D line. This coincides with the type of entry required for a point even when not in AutoLISP.

For example, the following are considered appropriate while in AutoCAD:

Command: **line**
Specify first point: **1.5, .7081068**
 or
Specify first point: **3.1415, 5**
 or
Specify first point: **1, 1**

A similar mixture of reals and integers is evident in the following example from within AutoLISP:

Command: **line**
Specify first point: **(list 1.5 0.7071068)**
 or
Specify first point: **(list pi 5)**

or
Specify first point: **(list 1 1)**

Note two subtle differences (not the parentheses, because they are not subtle): (1) From within AutoLISP, any real number must begin with an integer, not a decimal point. (2) The two expressions comprising the list within each function are separated by a space, as in the last examples (within AutoLISP), instead of a comma, as in the first examples (not in AutoLISP).

A List That Doesn't Look Like a List: Once a symbol (name of your choice) has been SETQed (set equal) to a list by the (setq <sym> (list <expr> <expr>...)) function, that symbol becomes a defined list. It will evaluate to that (<expr> <expr>...) for use as a response within AutoLISP or out of AutoLISP by using the ! prefix. For example, entering (setq p1 (list 2.0 3.0)) returns (2.0000 3.0000). Using the TYPE command, as in entering (type p1), will return "list." It will work for the following response to the LINE command:

Specify first point: **!p1**

But if (setq p1 (list 2.0 "z")) has been entered as the prior routine instead of (setq p1 (list 2.0 3.0)), the preceding sequence will not accept !p1 as a valid response, even though (type p1) still returns "list." Its argument, (<expr> <expr>..), as a list is not (<real> <real>..), as points require.

Before you even considered programming in AutoLISP, you were using lists. If, in response to the "From point:" prompt, you entered 2,3.75 from the keyboard (or any form of "real,real," "real,integer," or "integer,integer"), then you used a list. Even if you had entered x1,2.5 or 3,y, you would have been using a list. But you would have found out that even though it was in a similar format, the elements of the entry were not of the proper classification. Once you understood that AutoCAD rejected anything except certain classifications of elements, you learned to work using that knowledge (or went back to the drawing board, literally!).

A disguised use of the list input is when you specify a point on the screen. AutoCAD defines that specification as a list in the proper format (*X* coordinate, *Y* coordinate), and enters it for you. It is a list in the acceptable format.

Association List The following is a list, but it is a very special list:

((a 097) (b 098) (c 099) (d 100)...(x 120) (y 121) (z 122))

Each expression in the list similarly has two expressions enclosed in parentheses (making each of them a list), with the first being a lowercase letter and the second an integer.

What makes the last list special is that it satisfies the requirements of an association list. It may be used (like other qualified association lists) as an argument in the AutoLISP function ASSOC. It takes the form (assoc <item> <alist>).

EXERCISES

Determine if the following evaluation results below satisfy the requirements of the expression of the type list. If not, explain why.

a. 1 , 2

b. 1 2

c. (12)

d. (1 2)

e. (1.0 2)

f. (1.0 2.0)

g. ((1 2))

h. ((1 2)

i. (1 .2)

j. (1.0 0.2)

k. (1.0 2.0 3.0 x)

l. (1.0 2.0x3.0)

m. (* 2 4)

n. (-2 -4)

PAUSING FOR USER INPUT

There are very few AutoCAD commands that begin and end just by entering the command name. Except for END, REGEN, REDRAW, OOPS, UNDO, and the like, most commands pause for user input before they are completed. You can include this interactive capability in your AutoLISP routines through several input functions. The exercises in this section will introduce two such functions, (getpoint) and (getreal). The (getpoint) function pauses for input of a point (either from the keyboard or on screen) and returns that point in the form of a list of two reals. The (getreal) function pauses for user input of a real number and returns that real number.

AUTOLISP FUNCTIONS—THE COMMAND FUNCTION

The function list (command <args>...) provides a method of invoking an AutoCAD command from within AutoLISP. It is usually the culmination of the routine. The <args> are written in the same sequence as if entered from the keyboard from AutoCAD. The AutoCAD command name that follows the (command ...) function must be enclosed in quotation marks, as in the following example:

```
(setq p1 (list 1.0 1.0))
(setq p2 (list 2.0 2.75))
(command "line" p1 p2 "")
```

When included in an AutoLISP routine, this sequence will draw a line from 1,1 to 2,2.75.

Note that the double-double quotation mark is the equivalent of pressing ENTER while in AutoLISP. It is referred to as the null string (""). It simulates pressing the space bar.

In the (command ...) function list, the arguments that are AutoCAD entries are enclosed within quotation marks, whereas AutoLISP symbols and expressions are not.

One of the notable differences is illustrated in responding to a prompt for an angle. Remember, AutoLISP requires radians, and AutoCAD normally uses degrees (unless the units are set for angle input to be in radians). For this illustration, certain symbols were named and SETQed to values, as shown here:

```
(setq ublkname "UB1")
(setq inspt (list 2.0 2.0))
(setq xscal 2.5)
(setq yscal 2.5)
 setq ang 90)
```

Then a routine could be written in two different ways with the same results, as follows:

```
(command "insert" ublkname inspt xscal yscal ang)
```

or

```
(command "insert" "UB1" "2,2" "2.5" "" "90")
```

Although angular responses to normal AutoLISP functions must be in radians, this is an angular response while temporarily back in the AutoCAD screen and therefore must be 90, for degrees.

The first program line uses AutoLISP symbols as responses to the INSERT command that the (command ...) function invoked. The second line uses AutoCAD equivalents of the same responses. In the latter, the responses must be enclosed within quotation marks. This use of the quotation marks is different from their use to mark characters as a string type of argument in function lists other than the (command ...) function.

Two subtle lessons can be gleaned from the foregoing:

- The use of quotation marks returns you to AutoCAD types of responses.
- The double-double quotation marks (null string) is used to cause the Y scale factor to default to the X scale factor. This would not be appropriate if the Y scale needed to be different, of course. Using the null string in this manner is equivalent to returning to the AutoCAD screen, striking the space bar, and then returning to AutoLISP.

PAUSE SYMBOL

During a (command...) function you can cause a pause if you wish to allow the user input during the particular AutoCAD command that you have called up. This is done by using the pause symbol in lieu of a variable or of a fixed value where a particular response is required. For example, in the preceding program line you could have allowed the user to input a name and an angle:

```
                    (command "insert" pause inspt xscal yscal pause)
or
                    (command "insert" pause "2,2" "2.5" "" pause)
```

SYMBOLOGY USED IN THIS CHAPTER TO DESCRIBE FUNCTIONS

The function descriptions in this section include the operator (+ or -, for example), and the elements (arguments) that must (or may) follow the operator. If the arguments following the operator in the description are enclosed with angle (< >) brackets only, then those arguments must follow the operator and must be of the type specified. An argument in square ([ ]) brackets following the operator (not necessarily immediately) is optional. When an argument is followed by an ellipsis (...) then the operator will accept multiple arguments of the type specified.

ELEMENTARY FUNCTIONS

(+ <number> <number>...) This function returns the sum of the <number>s. There may be any quantity of <number>s.

(- <number> <number>...) This function returns the difference of the <number>s. There may be any quantity of <number>s.

(* <number> <number>...) This function returns the product of any quantity of <number>s.

(\ <number> <number>...) This function returns the quotient of the first <number> divided by the second <number>. If there are more than two arguments, the quotient of the first and second will be divided by the third, and so on.

RULES OF PROMOTION OF AN INTEGER TO A REAL

As noted in the *Reference Manual*, integers may range from -32,768 to 32,767, depending on the platform. Adding or multiplying integers whose sum or product exceeds 32,767 (or is less than -32,768) will not provide an acceptable result.

1. If any argument in one of the functions is entered as an integer and that integer is outside the integer limits, then the result is not usable.

2. If all of the number arguments entered are integers, then the result will be an integer, and if that result is outside the integer limits, the result will be subject to error.

3. If any of the arguments is entered as a real (and none is an integer exceeding the integer limits), then the result will be a real without limits.

MORE ELEMENTARY FUNCTIONS

(1+ <number>) and (1- <number>) These functions are just different methods of writing (+ <number> 1) and (- <number> 1), respectively. For example:

```
(1+7)  returns 8
(1-7)  returns 6
```

(abs <number>) This function returns the absolute value of the <number>. For example:

```
(abs (- 4 7))  returns 3
```

(ascii <string>) This function returns the ASCII value of the first character of the <string>. For example:

```
(ascii "All")  returns 65
(ascii "a")    returns 97
(ascii "B")    returns 66
```

(chr <number>) This function returns the character (as a one-character string) whose ASCII code is <number>. For example:

```
(chr 65)   returns "A"
(chr 97)   returns "a"
(chr 100)  returns "d"
```

(eval <expr>) This function returns the result of evaluating <expr>, where <expr> is any LISP expression. For example:

```
(setq z 6)
(setq q 'z)
(eval z)  returns 6
(eval q)  returns 6
```

(exp <number>) This function returns e raised to the <number> power (natural antilog). It returns a real. For example:

```
(exp 1.0)   returns 2.718282
(exp -0.2)  returns 0.818730753
```

(expt <base> <power>) This function returns <base> raised to the specified <power>. If both arguments are integers, the result is an integer; otherwise, the result is a real. For example:

```
(expt 3 4)  returns 81
```

(log <number>) This function returns the natural log of <number> as a real. For example:

```
(log 3.74)   returns 1.32175584
(log 1.025)  returns 0.024692613
```

(sqrt <number>) This function returns the square root of <number> as a real. For example:

```
(sqrt 16)   returns 4.000000
(sqrt 2.0)  returns 1.414213562
```

(type <*item*>) This function returns the type of <*item*>, where type is one of the following (as an atom):

REAL	floating point numbers
FILE	file descriptors
STR	strings
INT	integer
SYM	symbol
list	lists (and user functions)
SUBR	internal AutoLISP functions
PICKSET	AutoCAD selection sets
ENAME	AutoCAD entity names
PAGETB	function paging table

TRIGONOMETRY FUNCTIONS

(sin <angle>) This function returns the sine of <angle>, where <angle> is expressed in radians.

(cos <angle>) This function returns the cosine of <angle>, where <angle> is expressed in radians. For example:

```
(cos 1)        returns .540302306
(sin (/ pi 2)) returns 1.00000
```

(atan <num1> [<num2>]) If only <num1> is present, then (atan ...) returns the angle (in radians) whose tangent is <num1>. If <num1> and <num2> are present, then (atan ...) returns the angle whose tangent is the dividend of <num1> divided by <num2>. For example:

```
(atan 0.75)    returns 0.643501109
(atan 1.0 2.0) returns 0.463647609
```

LIST HANDLING FUNCTIONS

(list <expr>...) This function has expression(s) as its arguments. It is included in this section because it is the function that creates the lists that the other functions in this section require as arguments.

(list ...) This function takes any number of expressions and makes them into a list. For example:

```
(list 1 1)      returns (1 1)
```

Remember: A created list is enclosed in parentheses.

A specific application of the (list ...) function is to combine two or three reals in the format required by a function whose argument is <pt>, which is a 2D or 3D point. The point is a special form of a list. For example:

```
(setq p1 (list 1.0 1.0))
(setq p2 (list 2.0 2.0))
```

allows the following:

```
(setq a (angle p1 p2))
```

which returns 0.785398163.

But

```
(setq p1 (list "you" 1.0))
(setq p2 (list 2.0 2.0))
(setq a (angle p1 p2))
```

returns

error: bad point value

(angle <pt1> <pt2>) This function returns the angle (in radians) between the baseline of angle zero and the line from pt1 to pt2.

(distance <pt1> <pt2>) This function returns the distance in decimal units from pt1 to pt2. If the units are set to architectural and the distance between two points is 6'3", the (distance ...) function will return 75.000000.

(polar <pt> <angle> <distance>) This function returns a point. It can be one of the most useful tools in the AutoLISP arsenal. The arguments must be of the proper type. The <pt> must evaluate to a list of two reals. The <angle> and <distance> are each a real. The value of the <angle> is in radians.

(osnap <pt> <mode-string>) This function returns a point. It allows object snapping to a point while in AutoLISP. Like the (polar ...) function, it returns a point in the form of a list of two or three reals. For example, if a circle has been drawn using p1 and p2 in the 2P method as follows:

```
(setq p1 (list 1.0 3.0))
(setq p2 (list 4.0 3.0))
(command "circle" "2P" p1 p2)
```

then

```
(command "line" "0,0" (setq c (osnap p1 "center")) "")
```

returns (2.5 3.0), or the list or two reals representing the center of the circle as the endpoint of the line.

(inters <pt1> <pt2> <pt3> <pt4> [<onseg>]) This function returns a point. It can be used for the following:

1. To determine if two nonparallel lines intersect.
2. If they intersect, the location of that point.
3. If they do not intersect, where they would intersect if one or both were extended until they intersected.

If the optional <onseg> argument is present and is nil, the lines will be considered infinite in length and the function will return a list of three reals designating that intersection point.

If the optional <onseg> argument is not present or is not nil, then their intersection must be on both segments in order for a point to be returned; otherwise, the function will return nil.

CAR, CDR, AND COMBINATIONS

CAR and CDR are the primary functions that select and return element(s) of a list. Unlike the (angle ...) and (distance ...) functions, these functions will operate on a list comprised of elements of any type. The elements can be atoms or lists within the list. The atoms can be reals, integers, strings, or symbols. The argument to the CAR and CDR functions can even be a list of mixed types of elements. These were the foundation functions designed to analyze a list of symbols (which is what language is). In AutoLISP, these functions break down points into coordinates. For example:

```
(setq p1 '(1.0 2.0)) returns (1.0 2.0)
```

Then

```
(car p1) returns 1.0, a real
```

But

```
(cdr p1) returns (2.0), a list
```

Note the parentheses enclosing 2.0. If

```
(setq L1 (list 1.0 2.0)) and (setq L2 (list L1 3.0))
```

then

```
(car L2) returns (1.0 2.0), a list
```

Here is a list within a list. Or in the function list (setq L2 (list L1 3.0)), L1 is a list within a list within a function list.

The preceding expression is not used very often in AutoLISP, but its capability is noteworthy.

> (car <list>) returns the first element of <list>.

> (cdr <list>) returns <list> without the first element.

The type of the return of the (car <list>) function depends on the type of the first element.

If

```
(setq L (list 1.0 2.0))
```

then

```
(car L)  returns 1.0, a real
```

If

```
(setq L (list 1 2.0))
```

then

```
(car L)  returns 1, an integer
```

If

```
(setq L (list "1" 2))
```

then

```
(car L)  returns "1," a string
```

If the symbol "a" evaluates to nil and

```
(setq L (list a 2))
```

then

```
(car L)  returns  the symbol "a"
```

But if

```
(setq a 1.0)  and  (setq L (list a 2.0))
```

then

```
(car L)  returns  1.0, a real
```

Whereas (car <list>) can return any type of expression, (cdr <list>) always returns a list. The return may not look like a list, but its type will be a list (even if nil). For example:

```
(setq L1 (1.0 2.0))
(cdr L1)
```

returns (2.0), a list. Remember, the parentheses designate the list and (car L1) returns 1.0, a real. So how can the list (2.0) be used as a real? By using the following combination:

```
(car (cdr L1))
```

which is the same as

```
(car (2.0))
```

which returns 2.0, a real. Note how (car <list>) breaks the first expression out of the <list> and returns it evaluated to its type.

Although (car <list>) and (cdr <list>) operate on the same list, (cdr <list>) always returns a list. Consider the following:

```
(setq L1 (list 1.0))  returns (1.0), a list
(car L1)  returns 1.0, a real
(cdr L1)  returns nil
```

This is the null list, or the same as (list ( )).

If

```
(setq L1 (list 1.0 2.0))
```

then

```
(car L1)  returns 1.0
(cdr L1)  returns (2.0)
(car (cdr L1))  returns 2.0
```

A shortcut for this is: (cadr <list>), or

```
(cadr L1)
```

which returns 2.0, a real. CADR is one of several short forms of combinations of CAR and CDR.

 Note: (car <list>) may return any type. (cdr <list>) is a list.

The short forms of CAR and CDR begin with C and end with R. The characters between will be either an A or a D and will determine the sequence of combined CAR(s) and CDR(s). Here are some short forms:

CAAR	(car (car <list>))
CDDR	(cdr (cdr <list>))
CADR	(car (cdr <list>))
CDAR	(cdr (car <list>))

Any form that begins with CA will return an expression.

Any form that begins with CD will return a list.

All CARs and CDRs represented by A's and D's in such forms as CADAR and CDDAR, however deep, must have lists as their individual arguments.

For example: If

```
(setq L1 (list (1.0 2.0) 3.0)) returns ((1.0 2.0) 3.0)
(car L1)  returns (1.0 2.0), a list
(cdr L1)  returns (3.0), also a list
```

(list (1.0 2.0) 3.0) makes a list out of the list (1.0 2.0) and the atom 3.0. That is why (car L1) returned a list, even though 3.0 is a real in the list "L1;" (cdr <list>) always returns a list. The first element of L1 is (1.0 2.0), when it is removed from ((1.0 2.0) 3.0), (3.0) remains. And that is the function of (cdr L1), to return a list with its first element removed.

So

```
(caar L1) is the same as (car (car L1))
```

and returns 1.0, a real. But

```
(cdar L1), which is the same as (cdr (car L1))
```

returns (2.0), a list. Because 2.0 and 3.0 are not first elements, in order to return their values as reals, they must first be returned as the first elements of a list, as follows:

```
(cadar L1) means (car (cdr (car L1)))
```

and returns 2.0, which is the same as

```
(car (cdr (1.0 2.0)))
```

and the same as

```
(car (2.0))
```

And

```
(cadr L1)  means  (car (cdr L1))
```

and returns 3.0, or

```
(car (3.0))
```

CAR AND CADR MAINLY FOR GRAPHICS

(car ...) and (cadr ...) are the primary functions for accessing the X and Y coordinates of a point in AutoCAD. Remember, if (setq L1 (list 1.0 2.0)) is entered, (cdr L1) returns the not-so-useful list (2.0). But by using (cadr L1), which means (car (cdr L1)), in the following manner (cadr L1) returns 2.0, which is no longer a list but a real.

REVIEW

All forms of the CAR-CDR combination that begin with CA will return the first expression. All forms that begin with CD will return a list.

For graphics applications, lists represent the X and Y coordinates of a 2D point and the X, Y, and Z coordinates of a 3D point. Therefore, the more useful CAR-CDR combination forms are as follows:

In a 2D point:

```
(setq p2d (list 1.0 2.0))
(car p2d)  returns  1.0, the X coordinate
(cadr p2d)  returns  2.0, the Y coordinate
(caddr p2d)  returns  nil
```

In a 3D point:

```
(setq p3d (list 1.0 2.0 3.0))
(car p3d) returns 1.0, the X coordinate
(cadr p3d) returns 2.0, the Y coordinate
(caddr p3d) returns 3.0, the Z coordinate
(cdr p2d) returns (2.0), a list
(cdr p3d) returns (2.0 3.0), a list
```

Neither of the last two statements is very useful unless the programmer wishes to project all of the 3D points on the Y-Z plane.

(last <list>) This function returns the last expression in the <list>. Although (last <list>) can be used to return the Y coordinate of a 2D point and the Z coordinate of a 3D point, it is not recommended for that purpose. Because there might be an erroneous return, it is recommended that (cadr <list>) be used for the Y coordinate and (caddr <list>) for the Z coordinate.

(cons \<new first element> \<list>) This function returns the \<first new element> and \<list> combined into a new list, as in the following:

```
(setq a 2.0)
(setq b 3.0)
(setq L1 (list a b))
(cons 1.0 L1)  returns (1.0 2.0 3.0)
```

(cons ...) This function will also return what is known as a "dotted pair" when there is an atom in place of the \<list> argument, as in the following:

```
(cons 1.0 2.0) returns (1.000000 . 2.000000)
```

This special form of a list requires less memory than ordinary lists.

(length \<list>) This function returns the number of elements in \<list> as in the following:

```
(setq L1 (list "you" (1.0 2.0) 3.0))
returns ("you" (1.0 2.0) 3.0)
(length L1)  returns 3
(length (car L1))  returns nil, because (car L1) is an atom
(length (cdr L1))  returns 2
(length (cadr L1))  returns 2
(length (caddr L1))  returns nil
```

If

```
(setq L1 (list "you" (1.0) 2.0))
(length L1)  returns 3
(length (cadr L1))  returns 1
```

(nth \<n> \<list>) This function returns the nth element of \<list>. (Zero is the first element.) For example:

```
(setq L1 (list "you" (1.0 2.0) 3))
(nth 0 L1)  returns "you," a string
(nth 1 L1)  returns (1.0 2.0), a list
(nth 2 L1)  returns 3, an integer
(nth 0 (cadr L1))  returns 1.0, a real
(nth 1 (cadr L1))  returns 2.0, a real
(nth 3 L1)  returns nil
```

(reverse \<list>) This function returns the \<list> with the elements in reverse order, as in the following:

```
(setq L1 (list (1.0 2.0) 3.0))
(reverse L1)  returns (3.0 (1.0 2.0))
(reverse (car L1))  returns (2.0 1.0)
```

TYPE-CHANGING FUNCTIONS

In order for the AutoCAD operator, AutoCAD, and AutoLISP to communicate properly between (and within) themselves, data is constantly exchanged. Of the different data types (as classified by AutoLISP), there are three types that can be stored as one type but need to be communicated as another type: integer, the real, and the string.

Some AutoLISP functions are designed to take data of one type as their argument and return that data as another type. The basic outline below, Figure 19–19, shows the functions and the types that they are designed to translate from and to.

		FROM		
		INTEGER	REAL	STRING
TO	INTEGER	(FIX)	FIX	ATOI
	REAL	FLOAT	(FLOAT)	ATOF
	STRING	ITOA	ANGTOS	
			RTOS	

Figure 19–19 *The basic translation for communication between AutoCAD and AutoLISP*

(angtos <angle> [<mode> [<precision>]]) This function takes <angle> input as a real in radians and returns a string in the format determined by <mode>. The values of <mode> and their corresponding format are as follows:

Angtos Mode	Format
0	degrees
1	degrees/minutes/seconds
2	grads
3	radians
4	surveyor's units

For example: If

```
(setq p1 (1.0 1.0))
(setq p2 (2.0 2.0))
(setq a (angle p1 p2))  returns 0.78539816
```

then

```
(angtos a 0) returns "45"
(angtos a 1) returns "45.000000"
(angtos a 2) returns "45d0'0.0000""
(angtos a 3) returns "0.78539816r"
(angtos a 4)  returns "N 45d0'0.0000 W""
```

The optional <precision> determines the decimal places to be displayed.

(**atof** <**string**>) This function takes a <string> and returns a real. For example:

```
(atof "3.75")  returns 3.750000
(atof "4")  returns 4.000000
```

(**atoi** <**string**>) This function takes a <string> and returns an integer. For example:

```
(atoi "3.75")  returns 3
(atoi "4")  returns 4
```

(**itoa** <**int**>) This function takes an <integer> and returns a string. For example:

```
(itoa 33)  returns "33"
(itoa -4)  returns "-4"
```

(**rtos** <**number**> [<**mode**> [<**precision**>]]) This function takes a real input and re-turns a string in the format determined by <mode>. The values of <mode> and their corresponding format is as follows:

Rtos Mode	Format
1	scientific
2	decimal
3	engineering
4	architectural
5	arbitrary Fractional Units

The optional <precision> determines the decimal places to be displayed.

(**fix** <**number**>) This function takes a real or integer and returns an integer. For example:

```
(fix 4)  returns 4
(fix 4.25)  returns 4
```

(**float** <**number**>) This function takes a real or integer and returns a real. For example:

```
(float 4)  returns 4.000000
(float 4.25)  returns 4.250000
```

INPUT FUNCTIONS

The input functions cause a program to pause for user input of a particular type and return data in the format of a specified type.

The optional [<prompt>] in all (get ...) functions allows the programmer to display the <prompt> message in the prompt area on the screen. It will be demonstrated in the first (get ...) function description.

Note: The (getvar ...) function is not a function for user input.

Caution: For all (get ...) functions, the user input cannot be in the form of an AutoLISP function.

(getangle [<pt>] [<prompt>]) This function will return an angle, in radians, between two points, the first of which may be the optional [<pt>] in the function. Note the option of either selecting two points or inputting the first point into the AutoLISP function and selecting the second point. This method is used in the (getdist ...) and the (getorient ...) functions and will be referred to in their descriptions. For example:

```
(setq a (getangle "PICK TWO POINTS: "))
```

will pause for the user to input two points, either of which may be entered in at the keyboard (as in 1'2,3'6-1/2 if in the architectural units mode) or picked on the screen. If the response to the preceding (getangle ...) function were 1,1 and 2,2, then the function would return 0.785398. Or if

```
(setq p1 (list 14.0 42.5))
```

then

```
(setq a (getangle p1 "PICK SECOND PT: "))
```

will use p1 for the first point and pause for the user to input the second point from the keyboard or on the screen.

If

```
(setq p1 (list 1 1)) and (setq a (getangle p1 "PICK SECOND PT: "))
```

and 2,2 were entered, it would return 0.785398.

Caution: Unlike the (angle ...) function, the angle returned by the (getangle ...) function is affected by a change in the ANGBASE system variable. If ANGBASE were changed from 0 degrees to 45 degrees, the preceding (getangle p1 ...) function with a response of 2,2 would return 0.000000. (See the upcoming (getorient ...) function description.)

(getcorner <pt> [<prompt>]) This function returns a point selected during the pause. As the user places the cursor for selection, a rectangle is displayed with the <pt> as one corner of the rectangle and the cursor location as the diagonally opposite corner.

(getdist [<pt>] [<prompt>]) This function returns the distance between two points in the same manner and with the same options as the (getangle ...) function returns an angle. The return will be a real. If the units are set to architectural a length of 3'-6 1/2" would be returned as 42.500000.

(getint [<prompt>]) This function pauses for an integer input and returns that integer.

(getkword [<prompt>]) This function pauses for user input of a keyword that must correspond to a word on a list set up by the (initget ...) function prior to using the (getkword ...) function. If the response is not appropriate, AutoCAD will retry. This function prevents a program from terminating prematurely due to the wrong type of data being input by mistake and gives the user another chance. It also permits returning a string by just inputting initial letters. For example:

```
(initget 1 "SET Make New")
(setq g (getkword "LAYER CHOICES? (SET, M, or N): "))
```

AutoCAD will reject any response that is not made up of the initial uppercase characters of the options in the list set by (initget ...). The response may also include any and all of the lowercase characters in the string, but nothing in addition to the characters of any of the options. The responses that are valid to the example here are: SET, M, Ma, Mak, Make, N, Ne, and New, with any of the preceding in uppercase.

```
(initget 1 "SET MaKe New")
(getkword)
```

will accept m because the K was preceded by a lowercase a.

```
(initget 1 "SET MAke New")
(getkword)
```

will not accept m, but will accept ma, mak, or make, but not makeup.

(getorient [<pt>] [<prompt>]) This function will pause for the user to input two points and will return an angle, in radians, between two points.

The point selection options are the same as for the (getangle ...) function. Unlike the (getangle ...) function, the (getorient ...) function is not affected by a change in the ANGBASE system variable. It will return an angle determined by the line connecting the two input points. The angle will be measured between that line and the zero East baseline regardless of the ANGBASE setting.

(getpoint [<pt>] [<prompt>]) This function pauses for input of a point (either from the keyboard or a pick on the screen), and returns that point in the form of a list of two reals.

The optional <pt>, if used, will cause a rubber-band line from <pt> to the placement of the cursor until a pick is made.

(getreal [<prompt>]) This function pauses for user input of a real number and returns that real number.

(getstring [<cr>] [<prompt>]) This function pauses for keyboard characters to be entered and returns them as a string. It is not necessary to enclose the input in quotation marks. AutoLISP will do that automatically. The optional <cr>, if present and not nil, will permit the string to have blank spaces. The string must be terminated by striking ENTER. Otherwise, if <cr> is present and nil, striking the space bar will terminate the entry.

(initget [<bits>] [<string>]) This function offers the programmer a one-time control of the user's response to the next (get ...) function, and that (get ...) function only.

This means that anytime control is needed for a (get ...) function, a new (initget ...) function must precede it.

The type of control that is offered by the (initget ...) function is as follows:

The program can be set up to reject responses of a certain unwanted type or value (without terminating the program) and offer the user a second chance to enter an acceptable response.

The program can be made to accept points outside of the limits even when LIMCHECK is on.

The program can be made to return 3D points rather than 2D points.

Dashed lines can be used when drawing a rubber band or a box.

The program can be made to accept a string when the (get ...) function normally requires a specific type, such as point or real.

The controls offered by using the <bits> option are shown in Figure 19–20. The <bits> may be a sum of whichever values in Figure 19–20 correspond to the controls desired for the next (get ...) function. For example:

```
(setq p1 (list 0 0))
(initget 9)
(setq d (getdist p1 "SECOND POINT: "))
```

The bits in the next to last line are a sum of 1 (rejects null input) and 8 (allows input outside limits). This will allow the second point to be outside the limits even if the LIMITS mode is on. It also will not accept a null return.

BITS VALUE	MEANING
1	REJECTS NULL INPUT
2	REJECTS ZERO VALUES
4	REJECTS NEGATIVE VALUES
8	ALLOWS INPUT OUTSIDE LIMITS
16	RETURNS 3D POINTS RATHER THAN 2D
32	USES DASHED LINES FOR RUBBERBAND/BOX

Figure 19–20 *The controls offered by the <bits> option*

CONDITIONAL AND LOGIC FUNCTIONS

AutoLISP conditional and logic functions allow the user to have a program test certain conditions and proceed according to the result of those tests. Or, by using the WHILE function, a programmer's LOOP situation will allow iteration of a changing variable between the extents of a specified range.

The symbol T is used when an expression is needed that will never evaluate to nil.

(if <testexpr> <thenexpr> [<elseexpr>]) This function evaluates <testexpr>. If <textexpr> does not return nil, the function returns the evaluation of <thenexpr>. If the optional <elseexpr> is present and <testexpr> evaluates to nil, the function returns the evaluation of <elseexpr>. Otherwise, the function returns nil. For example:

```
(setq q (getint "ENTER QUANTITY FROM 1-99: " ))
(if (< q 10)
    (setq c q)
    (setq c (fix (/ q 10))))
)
```

This program will take a number (from user input) and test to see if it is less than 10. If so, it will SETQ the symbol c to that number. If it is 10 or greater, it will divide the number by 10 and SETQ c to the whole number of the result; for instance, 37 becomes 3.7 becomes 3.

(cond (<test1> <result1> ...) ...) This function accepts any number of arguments. The first item in each list is evaluated, and when one returns not nil, the following expressions in that list are evaluated and the function returns the value of the last expression. For example, a routine could be written to return the angle of a line to be only in the first or fourth quadrants, regardless of how it was originally selected. Note the four possibilities shown in Figure 19–21.

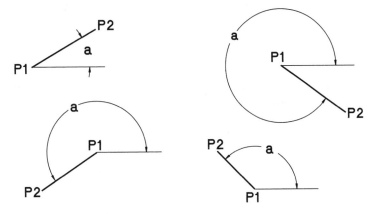

Figure 19–21 *Conditional and logic function return possibilities*

We will not consider the four ortho directions, N, E, S, or W, in this example. If an angle is returned that was determined by the function (setq a (angle p1 p2)), then it could be in one of four quadrants. Then to ensure that no matter which angle was set by p1-p2, the function would return an angle in the first or fourth quadrant. For example:

```
(setq a (angle p1 p2))
(cond
    ((and (> a pi) (< a (* 3 (/ pi 2))))) (setq a (- a pi))
    ((and (> a (/ pi 2)) (< a pi)) (setq a (+ a pi))
)
```

(while <testexpr> <expr>...) This function evaluates <textexpr>, and, if it is not nil, evaluates the following expressions and then repeats the procedure again with <testexpr>.

This repetition continues until <textexpr> evaluates to nil; the function then returns the evaluation of <lastexpr>.

TEST EXPRESSIONS

(if ...), (cond ...), and (while ...) These functions normally use one of the logic functions as <textexpr>. For example, the following program lines were previously entered:

```
(setq a 10.0)
(if (null a) (setq a 6.0))
```

This sequence will not change the value of the symbol a because the evaluation of the expression (null a) is nil. Therefore, the following expression will not be evaluated. Had a not been previously SETQed, the expression (null a) would evaluate to T (for TRUE), and the expression following would be evaluated and would have SETQed the symbol a to 6.0.

(= <atom> <atom> ...) This expression returns T if all of the <atom>s evaluate to the same thing.

(/= <atom1> <atom2>) This expression returns T if <atom1> is "not equal to" <atom2>. It is nil otherwise.

(< <atom> <atom> ...) This expression returns T if each <atom> is "less than" the <atom> to its right. It is nil otherwise.

(<= <atom> <atom> ...) This expression returns T if each <atom> is "less than or equal to" the <atom> to its right. It is nil otherwise.

(> <atom> <atom> ...) This expression returns T if each <atom> is "greater than" the <atom> to its right. It is nil otherwise.

(>= <atom> <atom> ...) This expression returns T if each <atom> is "greater than or equal to" the <atom> to its right. It is nil otherwise.

(**and <expr>...**) This expression returns T if all <expr>s return T. It is nil otherwise.

(**boundp <atom>**) This expression returns T if <atom> has a value bound to it. It is nil otherwise.

(**eq <expr1> <expr2>**) This expression returns T if <expr1> and <expr2> are identical and are bound to the same object. It is nil otherwise.

(**equal <expr1> <expr2>**) This expression returns T if <expr1> and <expr2> evaluate to the same thing. It is nil otherwise.

(**not <expr>**) This expression returns T if <expr> is nil. Otherwise, the function returns nil.

(**null <item>**) This expression returns T if <item> is bound to nil. The function returns nil otherwise.

(**or <expr>...**) This expression returns T if any of the <expr>s evaluate to something that is not nil. Otherwise the function returns nil.

EXERCISES

1. Identify the atoms in the following expressions:

```
(if (not (null dfr)) (setq dfr (rtos rad)) (setq dfr "0"))
```

Write the AutoLISP expression equivalent to each of the following:

2. *a* = 1
3. *b* = 2.0
4. *c* = a
5. *d* = *l* + 2
6. *e* = Ö2
7. *f* = 1 + *b*²
8. *g* = 7 + 9 + 3 + 7
9. *h* = (3 + 5 + 6) + 7
10. *i* = (7 + 3) + (4 - (6 + 3))

11. *j* = 3 (7 + 6)
12. *k* = 3 + (7 × 6)
13. *l* = 5 - (7 + 2)
14. *m* = sin 0.75
15. *n* = cos 1.75
16. *o* = sin (pi , 2)
17. *p* = absolute value of a
18. *q* = *x*³
19. *r* = tan ⁻¹((3 pi) + 4)

In the rectangle shown in Figure Ex19–1, write the expressions that evaluate to the following:

20. the vertical distance between p1 and p3
21. the distance between p1 and p3
22. the horizontal distance between p2 and p4
23. the distance between p1 and p4
24. p2 in terms of p1 and p3
25. p3 in terms of p2 and p4
26. the sum of the four sides (perimeter)
27. the area enclosed by the four sides
28. the area of a triangle whose vertices are p1, p2, and p3
29. the angle between lines from p1 to p3 and from p1 to p4

Figure Ex19–1 *The rectangle for writing expressions*

Answer the following:

30. In (setq L1 (list 1.0 2.0)) what is (car L1)?

31. In (setq L2 (list 3.0 4.0)) what is (cdr L2)?

32. In (setq L3 (list 5.0 6.0)) what is (cadr L3)?

For the following expression, write the combination of CAR and CDR that will return the given values from list L5:

```
(setq L5 (list 1.0 '(2.0 (3.0 4.0)) 5.0))
```

33. 1.0   35. 3.0   37. 5.0

34. 2.0   36. 4.0

Give the evaluations of the following expressions:

40. (+ 5 6.0)   67. (/ 2 40000)
41. (- 7 6)   68. (abs (+ 4 2))
42. (- 8 9)   69. (abs (- 4 2))
43. (+ 1 20 300)   70. (ascii "ABC")
44. (- 1 20 300)   71. (ascii "aBC")
45. (- 300 20 1)   72. (chr 66)
46. (- 30000 1.0)   73. (chr 98)
47. (+ 30000 30000)   74. (expt 3 2)
48. (* 1 2)   75. (expt 2 3)
49. (* 1.0 2.0)   76. (sqrt 9)
50. (* 3 4)   77. (sqrt (* 2 8))
51. (* 5.0 6)   78. (type 1)
52. (* 1 2 3)   79. (type pi)
53. (* 4 4 6.0)   80. (type 3.0)
54. (* 7 8000)   81. (type T)
55. (* 8.0 9000)   82. (sin 0)
56. (* 3 pi)   83. (cos (/ pi 2))
57. (* pi pi)   84. (atan 1)
58. (/ 2.0 1.0)   85. (angle (list 1.0 1.0) (list 2.0 2.0))
59. (/ 2.0 1)   86. (angle (list 2.0 1.0) (list 0.0 1.0))
60. (/ 2 1)   87. (distance (list 1.0 1.0) (list 3.5 1.0))
61. (/ 1 2)   88. (distance (list 6.0 5.0) (list 2.0 2.0))
62. (/ 3 2)   89. (polar (list 1.0 1.0) 0 2.0)
63. (/ 3.0 2)   90. (polar (list 2.0 3.0) pi 1)
64. (/ 3 2.0)   91. (inters (list 0.0 0.0) (list 3.0 3.0)
65. (/ 1.0 2)   92. (list 3.0 0.0) (list 0.0 3.0))
66. (/ 40000 2)

CUSTOM COMMANDS AND FUNCTIONS

(**defun ...**) and (**defun C: ...**) This is an AutoLISP function called DEFUN that will create a custom-designed function. Its format is:

    (defun <sym> <argument list> <expr>...)

The DEFUN function operates in a manner similar to the SETQ function. SETQ names a variable and sets it equal to a value. DEFUN names a new function so that when its name is entered in a routine, it will evaluate to the subsequent expression(s). In addition, it allows variables to be operated on by the function by entering the name of that defined function, followed by a variable or group of variables.

The customizing power of this feature cannot be emphasized enough. What this means to the programmer is that when a routine has been worked out in AutoLISP, the entire routine can be given a name. Then, when the routine needs to be used again, the programmer simply enters the name of the routine. This makes it a user-defined or custom-defined function.

The name of a custom-defined function can also be entered while in AutoCAD, or it can be included in a menu string by entering its name within parentheses.

In addition to having a one-word entry perform the task of many lines of programming, there is an optional feature that makes the custom-defined function even more effective. The user can define the custom function to perform the prescribed task on a dummy variable in the definition and then enter the defined function followed by some real variable later and have the task performed on the real variable. The dummy variables are listed in the <argument list>.

ARGUMENTS AND LOCAL SYMBOLS

Every custom-defined function and command must include the parentheses used to enclose the arguments. Even if there are no arguments or local symbols (which means the opening and closing parentheses would be empty), those opening and closing parentheses must follow the user-designated function name. Remember, however, that arguments (the dummy variables) are not permitted in custom commands using the "C:FNAME" format. The following forms are examples of how function and command definitions may be written:

(defun fname ( )...def exprs...)	No arguments or local symbols
(defun C:FNAME ( )...def exprs...)	No arguments or local symbols
(defun fname (/ x y z)...def exprs...)	Three local symbols only
(defun C:FNAME (/ x y z)...def exprs...)	Three local symbols only
(defun fname (a b)...def exprs...)	Two arguments only
(defun fname (a b / x y z)...def exprs...)	Two arguments and three local symbols

Caution: If the arguments' enclosing parentheses are inadvertently omitted, AutoLISP will take whatever is in the first set of parentheses of the definition expressions and try to use them as arguments and/or local symbols. Naturally, this will not operate as expected and will result in an error message. Therefore, include parentheses as shown, even if empty.

Also, within the <argument list>, optional local symbols may be listed (after the slash) to name symbols to be used within the function definition only, without having any effect on the same symbol name outside the defined function. This form of creating a defined function is for use within AutoLISP and cannot be entered without the enclosing parentheses. An added feature, to be discussed later, will allow the user-defined function to be invoked in the same manner as AutoCAD commands. That is, with the added feature, the user-defined functions can be entered from the keyboard or in the menu without parentheses and then recalled immediately afterward by simply pressing ENTER.

DEFINED FUNCTIONS AND COMMANDS

When one starts to delve into custom programs in many graphic applications, one of the first geometric-trigonometric stumbling blocks to overcome is the use of radians to measure angles. AutoCAD is no exception. Even if the use of radians is second nature to someone, using degrees seems to be first nature with almost all designers/drafters. The main problem is in converting one to the other. This is often necessary because AutoCAD uses degrees for the screen and radians in AutoLISP. Therefore, one of the first AutoLISP routines introduced in articles and books on AutoLISP is one that converts the value of an angle in degrees to its value in radians. The second routine is usually one that converts from radians back to degrees. The usual name for the degrees-to-radians function is the abbreviation dtr. The dtr function is written as follows:

```
(defun dtr (a)
    (* pi (/ a 180.0))
)
```

In the format (defun <sym> <argument list> <expr>...), dtr is the symbol (<sym>), a is the only argument in the argument list, and (* pi (/ a 180.0)) is the expression (<expr>).

Though (/ a 180.0) is an expression, it resides within the expression (* pi (/ a 180.0)) and is not considered one of the separate <expr> expressions in the format: (defun <sym> <argument list> <expr>...). The <expr>... indicates there is no limit to the number of expressions possible in a defined function. Although this is a rather minor point, it may help in understanding what an expression is and how AutoLISP looks at them.

WRITING A DEFINED FUNCTION

The dtr function may be entered in from the keyboard while in the AutoCAD screen, as follows:

Command: **(defun dtr (a) (* pi (/ a 180.0)))**

This function may be used and applied to any real number in the following manner:

Command: **(dtr 90.0)** (returns 1.570796327)

or

Command: **(dtr 180.0)** (returns 3.141592654)

or

Command: **(dtr 7.5)** (returns 0.130899694)

Although entering this function from the keyboard will define the dtr function for use at any time during the current editing session, the definition will be lost when the drawing is ended. Although writing this routine is fairly simple, it is still not very economical. There is a better way.

WRITING AND STORING A DEFINED FUNCTION

Usually, any function that needs to be defined in one drawing will be useful enough to warrant saving the definition for instant reuse later in any drawing without having to enter it in repeatedly. Therefore, defined functions may be saved for loading into a drawing. By using a text editor or word processor in the programmer mode, the custom programmer can create a file having the file extension of .LSP and then enter in the routine. The following example is similar to the first example, only with a slight variation in format:

File name: ANGLE.LSP

```
(defun dtr (a)              ;degrees to radians
    (* pi (/ a 180.0))      ;pi times "a" divided by 180
)                           ;leave function definition
```

The indentation is permissible in writing a defined function in a FILENAME.LSP file (unlike the program lines in a FILENAME.MNU file). It is even desirable in order to be able to identify easily the distinct elements and subelements of the function definition. The value of this procedure will become more evident with more elaborate definitions. The closing parenthesis, on a line by itself, is lined up with the opening parenthesis. This is an example of how indentation coincides with the depth of nesting of expressions.

In this last example, remarks have been written to the right, following semicolons. AutoLISP ignores anything on a line following a semicolon.

Function Names The name of the file may or may not be the same as any of the function names it includes. It may even be the same as a function in another file, although that would not be logical. Uniqueness of names is critical within a group of named items, such as files, functions, variables, and drawings in one directory. But in most cases it is not a problem to duplicate names across groups, such as having a FILENM1.LSP and a FILENM1.DWG, and a FILENM1.SHP at the same time along with defining a custom function as follows:

```
(defun filenm1 (...
```

Duplication in this case is not only acceptable but sometimes desirable in coordinating a group of specially named custom files and functions that are all in a single enhancement program being used for a singular purpose.

Loading a Defined Function As mentioned earlier in this chapter, custom-defined functions must be written into a FILENAME.LSP file in order to be saved and usable later. Once a custom-defined function is stored in that file, the user must load the file in order to use it. For example, the file named ANGLE.LSP has been created to store the custom-defined function named dtr as follows:

File name: ANGLE.LSP

```
(defun dtr (a)             ;degrees to radians
    (* pi (/ a 180.0))     ;pi times a divided by 180
)                          ;leave function definition
```

To make dtr usable, the user must invoke the AutoLISP function named LOAD. It has the format of:

(load <filename>)

with <filename> being a string without the extension of .LSP. Therefore, the file name must be enclosed within quotations, as follows:

(load "angle")

 Caution: Be aware of the distinction between the AutoLISP load function and the AutoCAD load command. The AutoCAD command named LOAD is used to load FILENAME.SHP (shape) files and is entered without the parentheses, in the following manner:

Command: **load**

The AutoLISP function named LOAD" is used to load FILENAME.LSP (AutoLISP) files and is entered with the parentheses, in the following manner:

Command: **(load "angle")**

The preceding sequence will make the custom-defined function named dtr usable in the current editing session. The user must be careful when trying to load .LSP files from some other drive or directory. If the file named ANGLE.LSP is stored on the directory named LISPFILE, the format would be as follows:

Command: **(load "/lispfile/angle")**

Special note should be made of the forward slash versus the backslash. The file name is a string, and a leading backslash in a string is itself a special control character used in conjunction with other code characters for specific operations on the string. Thus, in order to have a backslash read as a backslash in a string, there must be a double backslash. Therefore, the forward slash is recommended in designating a directory path in PC-DOS/MS-DOS.

Drive specifications can also be included in the AutoLISP LOAD function format as follows:

Command: **(load "a:angle")**

or

Command: **(load "a:/lispfile/angle")**

The last entry may be used if the ANGLE.LSP file is in the LISPFILE directory on the A: drive.

If the custom programmer has created custom-defined functions in a file called ACAD.LSP and that file is accessible when a drawing is edited, AutoCAD will "load" that file automatically. Only functions that might be used in all drawings should be included in that file if it is created.

SIN and COS, but not TAN! Sometimes designers/drafters wish to determine some distances in the all-powerful right triangle (from which trigonometry is derived) by using the opposite and adjacent sides instead of the hypotenuse. So why doesn't AutoLISP have the TAN (tangent) function in addition to the SIN and COS? First of all, an angle is defined in *Webster's New Collegiate Dictionary* as "a measure of the amount of turning necessary to bring one line or plane into coincidence with or parallel to another." Two special cases exist in computing the tangent of an angle, one of which causes a problem that the sine and cosine do not cause. When two lines exist, it is assumed that they must have length and therefore have a nonzero value. When considering the sine or cosine of an angle, the hypotenuse is always one of the existing nonzero lines. In the sine and cosine of any angle, the hypotenuse is always the divisor. So, even if the opposite side (in the case of the sine) or the adjacent side (in the case of the cosine) turn out to be zero, the worst that can happen is that the function will result in zero. In the case of the tangent, where the divisor is the adjacent side, if it is zero the result approaches infinity and is therefore not valid for use within the program. This occurs, of course, at 90 degrees.

The other special case is where the angle is zero and the tangent is zero. Zero may be a valid entry, whereas infinity cannot be. This problem does not occur in the sine and cosine functions because the results of either range from 1 to 0 to -1 and back to 1. All results within this range are valid for use within programs.

An approach to arriving at the tangent is possible by using existing AutoLISP functions named SIN and COS. The tangent can be expressed as the quotient of the sine divided by the cosine. But here again is the possibility of the divisor being zero because the cosine of 90 degrees is just that. If used for specific purposes where it is known that the value of the adjacent side involved is nonzero, then a custom-defined function that will return the tangent of an angle (along with a radians-to-degrees function) might be included in the ANGLE.LSP file, as follows:

File name: ANGLE.LSP

```
(defun dtr (a)              ;degrees to radians
    (* pi (/ a 180.0))      ;pi times a divided by 180
  )                         ;leave function definition
(defun rtd (b)              ;radians to degrees
    (* 180.0 (/ b pi))      ;180 times b divided by pi
  )                         ;leave function definition
(defun tan (c)              ;tangent
    (/ (sin c) (cos c))     ;sine divided by cosine
  )                         ;leave function definition
```

Defined AutoLISP Commands AutoCAD programming features allow the creation of custom-defined AutoLISP commands with two characteristics similar to regular AutoCAD commands.

- AutoLISP commands can be entered from the keyboard or within a menu string by just entering the name of the custom command without having to enclose it in parentheses.

- If a particular AutoLISP command was the last command used, then typing « will recall that same command for immediate use.

 Note: AutoLISP commands cannot operate on external variables (arguments) of an <argument list>. However, local symbols are allowed.

Writing an AutoLISP Function The format used in creating an AutoLISP command is as follows:

(defun C:CMDNAME </ local symbols> <expr>...)

where CMDNAME can be a name of your choice and can duplicate the name of an existing AutoCAD command or function, if your purpose is to override that command with a definition of your own.

WHAT IS THE DATABASE?

The file that stores information about the drawing has tables and Block sections that include data associated with linetype, layer, style (for text), view, UCS, viewport, and blocks. These are accessible through the (tblnext...) and (tblsearch...) functions.

The Entities section contains data associated individually with each entity in the drawing. There are functions that point to the entity data and that can use or manipulate that data.

ENTITY NAMES, ENTITY DATA, AND SELECTION SETS

Information is continually being updated during an editing session. An ongoing record is being kept for each new, modified, or deleted entity. The information or data concerning each entity is stored at some particular location in the drawing file. (This location is not to be confused with its X, Y, and Z coordinates.)

AutoLISP gains access to an entity's data through its entity name, which points to the location of the entity's data within the drawing file. The entity name is the address of that entity's data. You must grasp the concepts of entity name and entity data and be able to distinguish between them.

ENTITY NAME FUNCTIONS

Entity name functions evaluate to an item whose data type is called AutoCAD entity name. An entity name can be used as a response when an AutoCAD command prompts you to "Select object:."

The (entnext [<ename>]) function, if performed with no arguments, returns the entity name of the first nondeleted entity in the database. If (entnext...) is performed with an entity name (which we will call en1 for illustration purposes) as the argument, it returns the entity name of the first nondeleted entity that follows en1 in the database.

WALKING THROUGH THE DATA BASE

Certain functions can search through the database. The (entlast) function returns the entity name of the last nondeleted main entity in the database. It may be used to return the entity name of a new entity that has just been added by a previous (command...) function.

The (entsel [<prompt>]) function returns a list with the entity name as the first element and the point by which the entity was selected as the second element. Note that the second element (a point) is a list itself. For example:

```
Command: line
From point: 1,1
to point: 4,4
```

to point: ENTER
Command: **(setq es (entsel "Select an Entity: "))**
Select an Entity: **2.5,2.5** *(returns (<Entity name: 60000018> (2.5 2.5 0.0)))*

While the evaluation of (entsel...) is a list and not an AutoCAD entity name, its first element is an entity name, obtainable by using the following:

Command: **(car es)** *(returns <Entity name: 60000018>, an entity name)*

Note that the address of this entity is displayed in hexadecimal form, 60000018, and is different for each different entity name. It will also differ from one editing session to the next for the same entity.

Point data is also obtainable by:

Command: **(cadr es)** *(returns (2.5 2.5 0.0),a list)*

And the *X* coordinate is obtainable by:

Command: **(caadr es)** *(returns 2.5, a real)*
Note that

Command: (cdr es)

returns ((2.5 2.5 0.0)), which is a list whose only element is also the list (2.5 2.5 0.0).

The (handent <handle>) function returns the entity name associated with the specified handle. This more advanced concept will not be covered within this section. It is noteworthy, however, that the (handent...) function assists in the problem of entities changing their names from one editing session to the next.

ENTITY DATA FUNCTIONS
Entity data functions have entity names as their arguments.

ENTGET Function The primary entity data function is (entget <ename>). This function returns a list of entity data describing <ename> in that special format called an association list.

Data Types for Database Functions Before continuing with detailed descriptions of the entity data functions and even before introducing selection sets, it would be convenient to discuss certain data types that AutoLISP has set aside especially for use when accessing the database. Some of the concepts used in this discussion of data types will not be described in detail until later in this section. Therefore, the reader will probably wish to refer back here later for review after studying those concepts.

In a manner similar to integers, reals, strings, and lists, the special data types called AutoCAD selection sets and AutoCAD entity names can be used as arguments required by certain AutoLISP functions. It should be noted that the special functions designed to operate on these data types can operate on them only. For example; having performed the following sequence:

```
(setq en1 (entnext))
(setq ss1 (ssget "w" '(0 0) '(12 9)))
(setq od1 (list 1.0 2.0))
(setq od2 "HELLO")
```

the following are valid entries:

```
(setq ed1 (entget en1)) and (setq en2 (ssname ss1 0))
```

But the following are not valid:

```
(setq ed2 (entget od1)) and (setq ed3 (entget od2))
```

The first two are valid entries because en1 is an AutoCAD entity name, which is the required data type for an argument to the (entget...) function. The second two are not valid entries because od1 and od2 are other data types and so are not valid as arguments. Od1 is a list and od2 is a string.

Similarly, ss1 is a special data type called an AutoCAD selection set and is valid as an argument to the (ssname...) function.

Further study will show that the (ssname...) function returns a value in the form of that special data type called an AutoCAD entity name. Therefore, if the preceding entries had been performed, then the following is a valid entry:

```
(setq ed4 (entget (ssname ss1 0)))
```

returning the data (as an association list) about the first entity in the drawing file that is included in a window whose corners are 0,0 and 12,9. Even if no entities had been found, and the operation had returned nil, the operation would still have been proper, having had the required data type as an argument to the (entget...) function.

Back to Entget For an example let's say a newly created drawing had the following as the first entries:

```
Command: line
Specify first point: 1,1
Specify next point or [Undo]: 4,4
Specify next point or [Undo]: (ENTER)

Command: line
Specify first point: 1,4
Specify next point or [Undo]: 4,1
Specify next point or [Undo]: (ENTER)
```

then

```
(setq L1 (entget (entnext)))
```

returns

((-1 . <Entity name: 60000018>) (0 . "LINE") (8 . "0")
      (10 1.0 1.0 0.0) (11 4.0 4.0 0.0) (210 0.0 0.0 1.0))

and

```
(setq L2 (entget (entlast)) or (setq L2 (entget (entnext L1))
```

returns

((-1 . <Entity name: 60000030>) (0 . "LINE") (8 . "0")
      (10 1.0 4.0 0.0) (11 4.0 1.0 0.0) (210 0.0 0.0 1.0))

Also note the syntax in the following example:

Having entered

```
(setq en1 (entnext))
(setq en2 (entlast))
```

then

```
(setq L1 (entget en1))
```

returns the same as

```
(setq L1 (entget (entnext)))
```

and

```
(setq L2 (entget en2))
```

returns the same as

```
(setq L2 (entget (entlast))) or (setq L2 (entget (entnext L1)))
```

This is to emphasize that the entity names en1 and en2 are only addresses (or pointers) to where the data associated with the entities are located within the drawing file. The data list itself is gotten through the (entget...) function, which must have that address or entity name as its argument.

THE ASSOCIATION LIST

Once the concepts of <ename>s and <elist>s are understood and the throes of creating those first routines to extract them are survived, the custom programmer must now deal with how to make use of the results.

The (assoc <item> <alist>) function is the primary mechanism used to extract specific data associated with a selected entity. To store data in an organized, efficient, and economical fashion, each common group of data is assigned what is called a group code. That group code is simply an integer. For example, the starting point for a line is a group code 10 and its layer is a group code 8.

Using the first entity in the previous example, you may then enter the following sequence:

Command: **(setq ed1 (entget (entnext)))**

This returns the association list we saw earlier, which we will display as follows:

```
((-1 . <Entity name:60000018>)
  (0 . "LINE")
  (8 . "0")
  (10 1.0 1.0 0.0)
  (11 4.0 4.0 0.0)
  (210 0.0 0.0 1.0)
)
```

Dotted Pairs The first and last parentheses enclose the association list. Each of the other matched pairs of parentheses enclose lists, with each list comprising a specific group code integer as the first element followed by its associated value. In the case of the group code 8, its value is the string "0," which is the name of the layer that this entity is on. Likewise, the value of the group code 0 is the entity type, which is the string "LINE." Note that these two sublists have a period between the group code and its associated value (separated by spaces). This is a special list, called a dotted pair, that requires less memory in storage. For economy, AutoCAD uses these dotted pairs where feasible. Some group codes, like the starting point, may be in the more common form of the list, such as (10 1.0 1.0 0.0) with the group code as the first element followed by the X, Y, and Z coordinates, respectively.

Group codes are broken down as shown in Tables 19–1 and 19–2.

Table 19–1 Group Codes and Their Respective Value

GROUP CODE RANGE	FOLLOWING VALUE
0–9	String
10–59	Floating-point
60–79	Integer
210–239	Floating-point
999	Comment (String)

Table 19–2 Group Codes and Their Respective Value Type

GROUP CODE RANGE	VALUE TYPE
0	Identifies the start of an entity, table entry, or file separator.
1	The primary text value for an entity
2	A name, attribute tag, block name, etc.
3–4	Other textual or name values
5	Entity handle expressed as a hexadecimal string
6	Linetype name (fixed)
7	Text style name (fixed)
8	Layer name (fixed)
9	Variable name identifier (used only in the Header section of the DXF file)
10	Primary point (start point of a line or text entity, center of a circle, etc.)
11–18	Other points
39	This entity's thickness if nonzero (fixed)
40–48	Floating-point values (text height, scale, etc.)
49	Repeated value—multiple 49 groups may appear in one entity for variable-length tables (such as the dash lengths in the Ltype table)
50–58	Angles
62	Color number (fixed)
66	"Entities follow" flag (fixed)
70-78	Integer values, such as repeat counts, modes.
210, 220, 230	*X, Y,* and *Z* components of extrusion direction.
999	Comments

EXTRACTING DATA FROM A LIST

```
((-1 . <Entity name:60000018>)
  (0 . "LINE")
  (8 . "0")
  (10 1.0 1.0 0.0)
  (11 4.0 4.0 0.0)
  (210 0.0 0.0 1.0)
)
```

If the preceding list had been SETQed to the variable named ed1, examples of the (assoc...) function would be as follows:

Command: **(assoc 0 ed1)** *(returns (0 . "LINE"))*

Command: **(assoc 8 ed1)** *(returns (8 . "0"))*

Command: **(assoc 10 ed1)** *(returns (10 1.0 1.0 0.0))*

These results are lists. The first two are dotted pairs. To extract data from these, the following may be entered:

Command: **(cdr (assoc 0 ed1))** *(returns "LINE," a string)*

Command: **(cdr (assoc 8 ed1))** *(returns "0," also a string)*

But the following returns (1.0 1.0 0.0),a list:

Command: **(cdr (assoc 10 ed1))**

Therefore, in order to extract the *X* coordinate, enter:

Command: **(cadr (assoc 10 ed1))**

which returns 1.0, a real. Note how CDR returns the second element of a dotted pair as an atom, whereas it requires CADR to return the second element of an ordinary list as an atom. CDR applied to an ordinary list returns a list (unless it is applied to a single-element list).

OTHER ENTITY DATA FUNCTIONS

The **(entdel <ename>)** function deletes the entity specified by <ename> from the drawing or restores the entity previously deleted during the current editing session.

The **(entmod <elist>)** updates the database information for the entity specified. Using this function in conjunction with the AutoLISP (subst...) function is a convenient way to make specific changes to selected entities.

The **(entupd <ename>)** function is used for more advanced handling of block and polyline subentities. The reader is referred to the *AutoLISP Programmer's Reference* for use of this function.

SELECTION SETS

A selection set is a collection of entity names. It is the programmer's equivalent to a group of entities selected by one or more of the optional methods of selecting objects from the screen; except that, by using the selection set functions, certain entities not visible on the screen may even be included in the group. It should be emphasized that a selection set comprises the entity names of the entities in the group. The selection set can be used in response to any AutoCAD command where selection by "Last" is valid. It also is a valid argument to selection set functions that supply entity names to the (entget) function, which then returns entity data in the form of an association list.

The **(ssget [<mode>] [<pt1> [<pt2>]])** function returns the selection set and prompts something like <Selection set: 1>. This indicates that the selection set contains one or more entities. If no objects meet the selection method, then nil will be returned.

The <mode>, if included, determines by what method the selection process is made. The "W," "L," "C," and "P" correspond to the window, last, crossing, and previous selection modes, respectively. Examples are as follows:

(ssget)	Asks the user for entity selection with one or more standard options
(ssget "W" '(2 2) '(8 9))	Selects the entities inside the window from 2,2 to 8,9
(ssget "L")	Selects the last entity added to the database
(ssget "C" '(0 0) '(5 3))	Selects the entities crossing the box from 0,0 to 5,3
(ssget "P")	Selects the most recently selected objects
(ssget '(1.0 1.0))	Selects the entity passing through the point 1,1
(ssget "X" <filter-list>)	Selects the entities matching the "filter-list"

SSGET Filters   The **(ssget "X" <filter-list>)** function provides a method of scanning the entire drawing file and selecting the entities having certain values associated with specified group codes. This function is used in conjunction with the (cons...) function. Note that while the (ssget...) function returns a group of entity names, it scans a group of association lists (the drawing file, that is). So the filter will be constructed in a manner that can be tested against entity data. Examples are as follows:

`(ssget "X"(list(cons 0 "CIRCLE")))`	Returns a selection set consisting of all the circles in the drawing
`(ssget "X"(list(cons 8 "0")))`	Returns all entities on layer "0"
`(ssget "X"(list(cons 0 "CIRCLE")` `(cons 8 "0")))`	Returns all circles on layer "0"

Note that the (ssget "X"...) function selects only from main entities. Special methods must be used to gain data about subentities, such as attributes and polyline vertices.

The **(sslength <ss>)** function returns an integer containing the number entities in selection set <ss>. For example, the following returns 1:

```
(setq sset (ssget "L"))
(sslength sset)
```

The **(ssname <ss> <index>)** function returns the entity name of the <index>th element of selection set <ss>. The first element begins with number 0. For example, let's say that (setq sset (ssget)) results in five items:

```
(setq en1 (ssname sset 0))  (returns the first entity)
(setq en3 (ssname sset 2))  (returns the third entity)
```

The **(ssadd [<ename> [<ss>]])** function without arguments constructs a selection set with no members. If performed with a single-entity-name argument, it constructs a new selection set containing that single entity name. If performed with an entity name and a selection set, it adds the named entity to the selection set.

The **(ssdel <ename> <ss>)** function deletes entity name <ename> from selection set <ss>.

The **(ssmemb <ename> <ss>)** function tests whether entity name <ename> is a member of selection set <ss>.

ADVANCED ASSOCIATION LIST AND (SCANNING) FUNCTIONS

Examples of **(assoc <item> <alist>)** have been presented previously in this section. It will now be used in our custom-designed command.

The **(subst <newitem> <olditem> <list>)** function searches <list> for <olditem>, and returns a copy of <list> with <newitem> substituted in every place where <olditem> occurred.

Two sample routines are listed next. The TSAVE routine redefines the REDRAW command and then uses a time-checking part to cause the newly defined REDRAW command to invoke a SAVE command after a predetermined time has passed. The PARAB routine draws a parabola.

Saved in the file named TSAVE.LSP:

```
;-Autosave-
(defun S::STARTUP ( )
          (command "undefine" "redraw")
          (setq savetime 0.25)
)

(defun C:TSAVE ( )
          (setq temptime (getreal "ENTER INTERVAL OF TIME IN MINUTES
```

```
                    FOR SAVING: "))
                (setq savetime (/ temptime 60.0))
)


(defun C:REDRAW ( )
        (if (null cdate1)
           (setq cdate1 (decihr (getvar "cdate"))))
        (setq cdate2 (decihr (getvar "cdate")))
        (if (or (> (- cdate2 cdate1) savetime)
           (> cdate1 cdate2)
           )
           (progn
                    (setq tempex (getvar "expert"))
                    (setvar "expert" 2)
                        (command "save" "c:backup")
                    (setq cdate1 cdate2)
                    (setvar "expert" tempex)
              )
        )
           (command ".redraw")
)
(defun decihr (dt / hms dh dm ds hd)
        (setq hms (* (- dt (fix dt)) 10000.0))
        (setq dh (/ (fix hms) 100.0))
        (setq dm (/ (- dh (fix dh)) 0.6))
        (setq ds (/ (- hms (fix hms)) 36.0))
        (setq hd (+ (fix dh) dm ds))
)
```

Saved in the file named PARAB.LSP:

——————————— PARABOLA ———————————
```
(defun c:parab ( )
        (setq point1 (getpoint "ENTER POINT: " ))
        (setq number (getint "ENTER ITERATIONS: "))
        (setq i (getreal "ENTER INCREMENTS: "))
        (setq p1 point1 p (car point1) counter 0)
        (while (< counter number)
          (setq p2
            (list
                    (+ (car point1) i)
                    (+ (cadr point1) (* 2.0 (sqrt (* p i))))
            )
          )
          (command "line" p1 p2 "")
          (setq p1 p2 counter (1+ counter) i (+ i i))
          )
)
```

DEBUG YOUR PROGRAM

As you develop an integrated suite of procedures for your applications, you will undoubtedly be confronted with the situation in which the applications are not producing the intended results. You may no longer find it practical to visually inspect your source code to locate the cause. You may prefer to use the debugging techniques in VLISP. Using the sample application developed earlier, this section illustrates the basic debug operations.

LOAD PROJECT

If the project needs to be loaded you may do so by selecting from the Visual LISP for AutoCAD window:

Project menu	Choose Open Project

Or from the AutoCAD window select:

Tools menu	Choose Macro > Load Project

You may use the BROWSE command button to navigate to the folder when the project file is located and then select it. If the selection is successful, then a project window will pop up and list the names of the files associated with the project.

These files can be loaded in a single step by selecting:

Project menu	Choose Load Project Source Files

The VLISP console window will display the results of this selection as shown below.

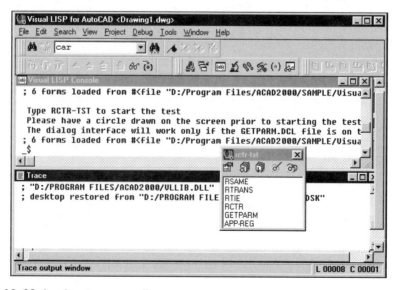

Figure 19–22 *Load project source files*

SET BREAK POINT

In order to debug the source code you will need to step through the statements as the program executes. Since it may be impractical, especially for the larger or more iterative programs, to patiently step through each statement, you can insert break points in your source code. If the program reaches a break point statement during a run, execution is halted. In this state or condition you can examine the values of program variables, list the call sequence, and perform other debug operations.

To set a break point, place the cursor at the position where you would like to halt the execution. Then select:

Debug menu	Choose Toggle Breakpoint

The break point is identified with an inverse video attribute (*viz.*, default is red foreground, white background). If you attempt to set the break point again at the same position, you will notice that the command operates as a toggle switch. In other words, selecting it again disables the break point. If at least one break point is still set then the Break on Error option in the Debug pop-up menu is set with a check mark to indicate the current status.

DEBUG TOOLBAR

All the buttons in the Debug toolbar are active when a break point is reached. The key buttons are:

- Step into
- Step over
- Step out
- Continue
- Quit
- Reset
- Toggle breakpoint
- Add Watch
- Last Break

The View toolbar contains additional buttons that may be useful during debugging operations

STEP INTO

When you click this button, the next expression in the code is evaluated and the program halts once again. If the expression is nested then the first expression at next level of the nest is evaluated.

STEP OVER

You should use this button to proceed to the closing parenthesis of the current expression. You can avoid having to step through nested expressions with this button

STEP OUT

This button will allow you to proceed to end of the current function before halting. If you do not want to step through the remainder of the code in the current function, you can use this button.

CONTINUE

This button allows you to resume the execution of the program. The program will subsequently halt at the next break point, if any, or any other program stop condition.

QUIT

This button terminates the current break loop. The execution of the program is returned to the next higher loop level.

RESET

This button terminates all the loops. The execution of the program is terminated and control is returned to the console window.

ADD WATCH

This button allows you to examine the values of the selected variables in dynamic mode. As the values of the variables change during the execution of the program the current values are displayed in the watch window.

AUTOCAD DEVELOPMENT SYSTEM (ADS)

The AutoCAD Development System (ADS) is a programming interface that, like AutoLISP, permits you to write or use applications for AutoCAD in high-level languages, such as C. The applications are loaded in a similar manner to AutoLISP. AutoLISP is more appropriate for smaller applications, whereas ADS makes use of an extensive, powerful, and more complex library of programming functions. ADS, because of the large library, demands more of your system. ADS is not stand-alone, but is tied to AutoLISP; therefore, it is not a good substitute for simpler tasks that can be so easily implemented with AutoLISP.

REVIEW QUESTIONS

1. To load the LISP file SETUP into the current AutoCAD drawing, enter the following at the "Command:" prompt:

   a. SETUP

   b. LOAD SETUP

   c. (LOAD SETUP)

   d. (LOAD "SETUP")

   e. none of the above

2. The command used to easily select and load AutoLISP and ADS routines is:

   a. APPLOAD

   b. LOAD

   c. DBLIST

   d. ADS

   e. none of the above

3. What does the (CADR L) function do?

   a. returns the first element of list L

   b. returns the second element of list L

   c. returns the third element of list L

   d. none of the above

4. What does the (atan N) function do?

   a. returns the angle, in degrees, whose tangent is N

   b. draws a line tangent to the last object drawn

   c. returns the tangent of the angle N, where N is in radians

   d. none of the above

5. What does the function (/ A B) do?

   a. returns the quotient of A divided by B

   b. returns the quotient of B divided by A

   c. returns the remainder of A divided by B

   d. returns the remainder of B divided by A

6. In (+ a 7) the + is called the:

   a. operand

   b. operator

   c. symbol

   d. none of the above

7. The following are acceptable AutoLISP data types, **except**:

   a. real

   b. integer

   c. string

   d. storage

   e. symbol

8. If (setq a 7), what does (list 4 a) return?

   a. (4 7)   d. an error message

   b. (7 4)   e. none of the above

   c. (11)

9. What does (/ 3 7) return?

   a. 10   d. 4

   b. 0   e. none of the above

   c. 0.4285714285714

10. The expression to add 7 to 3 is:

   a. (7 + 3)   c. (7 3 +)

   b. (+ 7 3)   d. none of the above

11. The expression (* 10 .5) returns:

   a. 5   c. an error

   b. 20   d. (5)

12. The statement (setq pi 3) will:

   a. redefine the symbol PI

   b. give an error message

   c. be ignored by AutoCAD because PI is a predefined symbol

   d. none of the above

13. To associate a LISP file with a menu file, so that it will be automatically loaded when the menu file is loaded you must use the same file name for both, with only the extensions varying. The proper extension for the LISP file is:

   a. MLP   d. LSM

   b. MNL   e. LLP

   c. LSP

14. When working in AutoCAD and you get the prompt "1>", it means:

   a. you have one minute until the network shuts down

   b. you need to specify one more argument for the previous function

   c. you need to enter one more close parenthesis ")"

Hardware Requirements

INTRODUCTION

The configuration of your CAD system is a combination of the hardware and software you have assembled to create your system. There are countless PC configurations available on the market. The goal for a new computer user should be to assemble a PC workstation that does not block future software and hardware upgrades. This section lists the essential hardware required to run AutoCAD 2000.

When you install AutoCAD 2000 it automatically configures itself to the various drivers already set up for Windows 95/Windows NT. The Options... dialog box can be used to custom configure your installation, as explained in Chapter 13.

RECOMMENDED CONFIGURATION

The minimum configuration required by Autodesk for a personal computer CAD workstation includes the following:

1. Intel Pentium® 133 or better, or compatible processor.

2. Operating System: Windows NT® version 4.0, Windows® 95, or Windows® 98.

3. RAM and Hard-Disk Space: 32MB of RAM (64MB of RAM recommended), 130 MB of hard disk space (minimum), 50MB of free hard-disk space in your system folder, 64MB of disk swap space (minimum).

4. Video 1024 x 768 VGA video display recommended, with Windows-supported display adapter.

5. Peripherals: CD ROM drive required for initial installation only; mouse/pointing device.
   Optional Peripherals: Printer or plotter, digitizer.

WINDOWS: THE OPERATING ENVIRONMENT

Windows 95, Windows 98, or Windows NT is the name of the operating environment. The operating system provides the graphical user interface and device independence, and lets you run more than one program at a time.

The graphical user interface (or GUI, for short) consists of the windows, icons, scroll bars, menu bar, button bar, and other graphical elements you see on the Windows screen. The GUI lets you use the mouse to control most aspects of a software program, including selecting commands, resizing windows, clicking on icons, and drawing.

Device independence means that all software applications running under Windows use the same device drivers, such as for the display, the mouse, and the printer. AutoCAD uses the Windows 95, Windows 98, or Windows NT device driver.

The multitasking and task switching feature of Windows 95, Windows 98, or Windows NT lets you be more productive. Multitasking means that Windows can run more than one program at a time, unlike DOS which runs just one program at a time. Task switching means that you can quickly switch from task to task (or from one program to another).

INPUT DEVICES

AutoCAD supports several input device configurations. Data may be entered via the keyboard, the mouse, or a digitizing tablet.

KEYBOARD

The keyboard is one of the primary input methods. It can be used to enter commands and responses.

MOUSE

The mouse is used with the keyboard as a tracking device to move the crosshairs on the screen. This method is fast and far surpasses the keyboard for positioning the screen crosshairs.

The mouse is equipped with two or more buttons. The left button is the pick button; the other buttons can be programmed to perform any of the AutoCAD commands.

DIGITIZER

The digitizer is another means of input supported by AutoCAD, but is not well supported by Windows. Hence, Autodesk provides a "mole" mode that lets the mouse and digitizer work at the same time in Windows while AutoCAD is running.

The digitizing tablet is a flat, sensitized electromechanical device that can recognize the location of the tablet cursor. The tablet cursor moves over the surface of the digitizing tablet, causing a corresponding movement of the screen cursor.

Another use of the digitizing tablet is to overlay a tablet menu, configured to enter commands when they are selected by placing the crosshairs of the puck at the corresponding digitizer coordinates. This process automates any AutoCAD command options that are represented on the menu, and eliminates the use of the keyboard except to enter an occasional command value.

The tablet cursor (puck) may have from 4 to 16 buttons. Except for the first button, the current menu can be programmed to cause the remaining buttons to perform any of the regular AutoCAD commands. The first button is always the pick button.

PLOTTERS AND PRINTERS

AutoCAD supports several types of output devices for producing hard copies of drawings. The most common devices are the thermal plotter, pen plotter, inkjet printer, and laser printer.

THERMAL PLOTTERS

The thermal plotter heats up the drawing paper, changing the color to black to create a hard copy of the drawing. The resolution of the output, which is measured in dots per inch (dpi), may range from a low-quality (100 dpi) product to medium-quality (400 dpi) product.

PEN PLOTTERS

The pen plotter draws continuous lines on the paper. The drawing pen is driven by vector commands that correspond to the X and Y coordinates on the drawing paper. Drawing media vary between mylar, vellum, and bond. The pens used in such plotters also vary in size and quality, thus affecting the drawing resolution. Resolution is typically between the equivalent of 1000 and 2000 dpi.

Pen plotters come in all shapes and sizes, and over the years pen plotters have improved in the areas of speed, accuracy, and price. The pen plotter is an established favorite for those who demand clear line and shape definition. Although prices vary on the different types of pen plotters, it is possible to get a good desktop pen plotter in the area of $2,000. Contact your dealer to find out if a particular model is supported.

INKJET PRINTERS

AutoCAD supports the ability to generate a hard copy of your drawing with an inkjet printer, either in black-and-white or in color. Inkjet printers form characters and graphics by squirting ink onto standard paper. Inkjet printers provide the most inexpensive means for color printing. They are generally used to provide rough draft hard copies for graphics and text data. They are an excellent and cost-effective way to produce check plots in A-, B-, and C-size.

LASER PRINTERS

Laser printers are the newest printing device on the CAD market. They deliver a low-cost, medium- to high-resolution hard copy that can be driven from a personal computer. Laser printers boast resolutions from 300 dpi to 1200 dpi and printing speeds of about 4–12 pages per minute (ppm) for text and 1–2 pages per minute for graphics.

As with any raster device, laser printers require a vector-to-raster conversion to generate most graphics. AutoCAD can plot design files on laser printers that support the HP LaserJet PCL and PostScript graphic output languages. The cost of laser printers ranges from under $1,000 to over $12,000, yet the increase in resolution makes the price a good investment for many serious CAD users.

Listing of AutoCAD Commands and Dimensioning Variables

Command Alliases	Explanation	Options		Toolbar	Pull-Down Menu
'About	Displays a dialog box with the AutoCAD version and serial numbers, a scrolling window with the text of the *acad.msg* file, and other information				Help
Acisin	Imports an ACIS file				Insert
Acisout	Exports AutoCAD solid objects to an ACIS file				Export
Adcclose	Closes AutoCAD DesignCenter				TYPE IN
Adcenter	Manages content			Standard	
Adcnavigate	Directs the Desktop in AutoCAD DesignCenter to the file name, directory location, or network path you specify				TYPE IN
Align	Moves and rotates objects to align with other objects				Modify
Ameconvert	Converts AME solid models to AutoCAD solid objects				TYPE IN
Aperture	Regulates the size of the object snap target box	Select (1–50 pixels) to increase or reduce size of box			TYPE IN
Appload	Loads AuotLISP, ADS, and ARX applications				Tools
Arc (A)	Draws an arc of any size	A C D E L R S ENTER	Included angle Center point Starting direction Endpoint Length of chord Radius Start point Continues arc from endpoint of line or arc	Draw	Draw
Area	Calculates the area of a polygon, pline, or circle	A S O	Sets add mode Sets subtract mode Calculates area of the circles, ellipses, splines, polylines, polygons, regions, and solids	Object Properties	Tools

Command Alliases	Explanation	Options		Toolbar	Pull-Down Menu
Array	Copies selected objects in circular or rectangular pattern	P	Polar (circular) arrays about a center point	Modify I	Modify
		R	Rectangular arrays objects in horizontal rows and vertical columns		
Arx	Loads, unloads, acquaints you with ARX applications				
Asexxx	Related to Structured Query Launguage applications				
Attdef	Creates an attribute definition that assigns (tags) textual information to a block	I	Regulates visibility		TYPE IN
		C	Regulates constant/variable mode		
		V	Regulates verify mode		
		P	Regulates preset mode		
Attdisp	Regulates the visibility of attributes in the drawing	ON	Makes all attribute tags visible		View
		OFF	Makes all attributes invisible		
		N	Normal visibility set individually		
Attedit	Permits the editing of attributes				Modify
Attext	Extracts attribute information from a drawing	C	CDF comma-delimited format		TYPE IN
		D	DXF format		
		S	SDF format		
		E	Select objects		
Attredef	Redefines a block and updates associated attributes				TYPE IN
Audit	Performs drawing integrity check while in AutoCAD	Y	Fixes errors encountered		File
		N	Reports, but does not fix, errors		
Background	Sets up the background for your scene	Solid	Selects one color background	Render	
		Gradient	Specifies a two- or three-color gradient background		
		Image	Uses a bitmap file for the background		
		Merge	Uses the current AutoCAD® image as the background		
		Colors	Sets color for a solid or gradient background		
		Preview	Displays a preview of the current Background settings		
		Image	Specifies the image file name including BMP, JPG, PCX, TGA, & TIFF		
		Environment	Defines an environment for creating reflection and refraction effects on		

Command Alliases	Explanation		Options	Toolbar	Pull-Down Menu
			objects with reflective, raytraced materials		
		Horizon	Represents the percentage of unrotated height		
		Height	Represents a percentage of the second color in a three-color gradient. Use the box or scroll bar to set the value		
		Rotation	Sets an angle at which you can rotate a gradient background		
Base	Defines the origin point for insertion of one drawing into another				Draw
Bhatch	Fills an automatically defined boundary with a hatch pattern through the use of dialog boxes; also allows previewing and repeated adjustments without starting over each time			Draw	Draw
Blipmode	Turns blip markers on and off				TYPE IN
Block	Makes a compound object from one or more objects	?	Lists names of defined blocks a group of objects	Draw	Draw
Blockicon	Generates preview images for blocks created with Release 14 or earlier				TYPE IN
Bmpout	Exports selected objects to bitmap format				
Boundary	Creates a polyline of a closed boundary				Draw
Box	Creates a 3D solid box			Solids	Draw
Break	Breaks out (erases) part of an object or splits it into parts	F	Allows you to reselect the first point again	Modify I	Modify
Browser	Launches the Web browser				
Cal	Evaluates mathematical and geometric expressions				TYPE IN
Camera	Sets a different camera and target location			View	
Chamfer	Makes a chamfer at the intersection of two lines	D	Sets chamfer distance	Modify I	Modify
		P	Chamfers all intersections of a pline figure		
		A	Sets the chamfer distances using a chamfer distance for the first line and an angle for the second		
		T	Controls whether AutoCAD trims the selected edges to the chamfer line endpoints		
		M	Controls whether AutoCAD uses two distances or a distance and an angle to create the chamfer		

Command Alliases	Explanation	Options		Toolbar	Pull-Down Menu
Change	Makes changes in the location, size, orientation, and other properties of selected objects; is very helpful for editing text	P E C LA LT T LW PL	Changes properties of objects Elevation Color Layer Linetype Thickness Lineweight Plotstyle		TYPE IN
Chprop	Makes changes in the properties of selected objects	C LA LT T LW PL	Color Layer Linetype Thickness Lineweight Plotstyle		TYPE IN
Circle (C)	Draws a circle of any size; default is center point and radius	2P 3P D TTR R	Two endpoints on diameter Three points on circle. Enters circle diameter Tangent, Tangent, Radius Enter radius	Draw	Draw
Close	Closes the current drawing				File
Color	Sets color for objects by name or number; also sets color to be by block or layer BYBLOCK BYLAYER	number name Retains color of block Uses color of layer 1. Red 3. Green 5. Blue 7. White	Sets color by number Sets a color by name 2. Yellow 4. Cyan 6. Magenta		TYPE IN
Compile	Compiles shape and font files				TYPE IN
Cone	Creates a 3D solid cone			Solids	Solids
Convert	Converts associative hatches and 2D polylines to optimized format				
Copy (CP)	Makes one or more copies of selected objects	M	Makes more than one copy of the selected object	Modify I	Modify
Copybase	Copies objects with a specified base point				Edit
Copyclip	Copies selected objects to the Windows clipboard				
Copyhist	Text in command line history is copied to the clipboard				
Copylink	Current view is copied to the clipboard for OLE applications				
Cutclip	Cuts and copies selected objects from the drawing to the clipboard				
Cylinder	Creates a 3D solid cylinder			Solids	Solids
Dbcclose	Closes the dbConnect Manager			Standard	

Command Aliases	Explanation	Options		Toolbar	Pull-Down Menu
Dbconnect	Provides an AutoCAD interface to external database tables			Standard	
Dblist	Makes a listing of every object in the drawing database			Inquiry	Tools
Ddedit	Edits text and attribute definitions			Modify II	
Ddptype	Specifies the display mode and size of point objects				Format
'Ddptype	Specifies point object display mode and sizes				
Ddvpoint	Sets the 3D viewing direction				View
Delay	Sets the timing for a sequence of command used in a script file				TYPE IN
Dim	Accesses the dimensioning mode				TYPE IN
Dist	Determines the distance between two points			Inquiry	Tool
Divide	Places markers along selected objects, dividing them into a specified number of parts	B	Sets a specified block as a marker		Draw
Doughnut (Donut)	Draws a solid circle or a ring with a specified inside and outside diameter			Draw	Draw
Dragmode	Allows control of the dynamic specification (dragging) feature for all appropriate commands	ON OFF A	Honors drag requests when applicable Ignores drag requests Sets Auto mode: drags whenever possible		TYPE IN
Draworder	Changes the display order of images and objects				
Dsettings	Specifies settings for Snap mode, grid, and polar and object snap tracking				Tools
Dsviewer	Opens the Aerial View window			View (Aerial View)	
Dview (DV)	Defines parallel or visual perspective views dynamically	CA CL D H OFF PA PO TA TW U X Z	Selects the camera angle relative to the target Sets front and back clipping planes Sets camera-to-target distance, turns on perspective Removes hidden lines on the selection set Turns perspective off Pans the drawing across the screen Specifies the camera and target points Rotates the target point about the camera Twists the view around your line of sight Undoes a Dview subcommand Exits the Dview command Zooms in/out, or sets lens length		View

Command Alliases	Explanation	Options		Toolbar	Pull-Down Menu
Dwgprops	Sets and displays the properties of the current drawing			File	
Dxbin	Inserts specially coded binary files into a drawing				File
Edge	Changes the visibility of 3D face edges			Surfaces	Draw
Edgesurf	Constructs a 3D polygon mesh approximating a Coons surface patch (a bicubic surface interpolated between four adjoining edges)			Surfaces	Draw
Elev	Sets the elevation and extrusion thickness for entities to be drawn in 3D drawings				TYPE IN
Ellipse	Draws ellipses using any of several methods	C	Selects center point	Draw	Draw
		R	Selects rotation rather than second axis		
		I	Draws isometric circle in current isoplane		
		A	Creates an elliptical arc		
Erase (E)	Deletes objects from the drawing			Modify I	Modify
Explode	Changes a block or polyline back into its original objects			Modify I	Modify
Export	Saves objects to other file formats				
Expresstools	Activates the installed AutoCAD Express Tools if currently unavailable				TYPE IN
Extend	Extends a line, arc, or polyline to meet another object	U	Undoes last extension	Modify I	Modify
		P	Specifies the Projection mode AutoCAD uses when extending objects		
		E	Extends the object to another object's implied edge, or only to an object that actually intersects it in 3D space		
Extrude	Creates unique solid primitives by extruding existing 2D objects			Solids	Draw
Fill	Determines if solids, traces, and wide polylines are automatically filled	ON	Solids, traces, and wide polylines filled		TYPE IN
		OFF	Solids, traces, and wide polylines outlined		
Fillet	Constructs an arc of specified radius between two lines, arcs, or circles	P	Fillets an entire polyline; sets fillet radius	Modify I	Modify
		R	Sets fillet radius		
		T	Controls whether AutoCAD trims the selected edges to the fillet arc endpoints		
Filter	Creates lists to select objects based on properties				TYPE IN

Command Alliases	Explanation	Options		Toolbar	Pull-Down Menu
Find	Finds, replaces, selects, or zooms to specified text			Stanadard	
Fog	Provides visual cues for the apparent distance of objects			Render	
Graphscr F2	Flips to the graphics display on single-screen systems; used in command scripts and menus				TYPE IN
Grid F7 On/Off toggle	Displays a grid of dots, at desired spacing, on the screen	ON	Turns grid on	Status bar	
		OFF	Turns grid off		
		S	Locks grid spacing to snap resolution		
		A	Sets grid aspect (differing *X–Y* spacings)		
		number	Sets grid spacing (0 = use snap spacing)		
		number X	Sets spacing to multiple of snap spacing		
Group	Creates a named selection set of objects			Standard	Tools
Hatch	Creates crosshatching and patternfilling	name	Uses hatch pattern name from library file		TYPE IN
		U	Uses simple user-defined hatch pattern		
		?	Lists selected names of available hatch patterns		
		NAME and U can be followed by a comma and a hatch style from the following list:			
		I	Ignores internal structure		
		N	Normal style: turns hatch lines off and on when internal structure is encountered		
		O	Hatches outermost portion only		
Hatchedit	Modifies an existing associative hatch block			Modify II	Modify
'Help or '?	Displays a list of valid commands and data entry options or obtains help for a specific command or prompt	To get a set of Help modes, use ESC and F2 for flip screen		Standard	Help
Hide	Regenerates a 3D visualization with hidden lines removed			Render	View
Hyperlink	Attaches a hyperlink to a graphical object or modifies an existing hyperlink			Standard	
Hyperlinkoptions	Controls the visibility of the hyperlink cursor and the display of hyperlink tooltips				TYPE IN
Id	Displays the coordinates of a point selected on the drawing				Tools

Command Alliases	Explanation	Options		Toolbar	Pull-Down Menu
Imagexxx	Commands used for modifying and displaying images to the clipboard				
Import	Imports various file formats into AutoCAD				
Insert	Inserts a copy of a block or Wblock complete drawing into the current drawing	fname	Loads fname as block	Draw	Draw
		fname=f	Creates block fname from file f		
		*name	Retains individual part objects		
		C	(as reply to X scale prompt) Specifies scale via two points (Corner specification of scale)		
		XYZ	(as reply to X scale prompt) Readies Insert for X, Y, and Z scales		
		~	Displays a File dialog box		
		?	Lists names of defined blocks		
Insertobj	Inserts embedded or linked objects				
Interfere	Finds the interference of two or more solids and creates a composite solid from their common volume			Solids	Solids
Intersect	Creates composite solids or regions from the intersection of two or more solids or regions			Modify II	Modify
Isoplane CTRL + E	Changes the location of the isometric crosshairs to left, right, and top plane	L	Left plane		TYPE IN
		R	Right plane		
		T	Top plane		
		ENTER	Toggle to next plane		
Layer (LA)	Allows for the creation of drawing layers and the assigning of color and linetype properties	C	Sets layers to color selected	Object Properties	Format
		F	Freezes layers		
		LT	Sets specified layers to linetype		
		M	Makes a layer the current layer, creating it if necessary		
		N	Creates new layers		
		ON	Turns on layers		
		OFF	Turns off layers		
		S	Sets current layer to existing layer		
		T	Thaws layers		
		?	Lists specified layers and their associated colors, linetypes, and visibility		
		L	Lock		
		U	Unlock		
		LW	Changes the lineweight associated with a layer		
		PS	Sets the plot style assigned to a layer		
Layout	Creates a new layout and renames, copies, saves, or deletes an existing layout	Copy	Copies a layout.	Layout	
		Delete	Deletes a layout		
		New	Creates a new layout tab		
		Template	Creates a new template based on an existing		

Command Aliases	Explanation	Options		Toolbar	Pull-Down Menu
			layout in a template (DWT) or drawing (DWG) file		
		Rename	Renames a layout		
		Save	Saves a layout		
		Set	Makes a layout current		
		?-List Layouts	Lists all the layouts defined in the drawing		
Layoutwizard	Starts the Layout wizard, in which you can designate page and plot settings for a new layout				TYPE IN
Leader	Draws a line from an object to, and including an annotation				
Lengthen	Lengthens an object			Modify I	Modify
Light	Manages lights and lighting effects			Render	View
Limits	Sets up the drawing size	2 points	Sets lower left/upper right drawing limits		Format
		ON	Enables limits checking		
		OFF	Disables limits checking		
Line (L)	Draws straight lines of any length	ENTER	(as reply to "From point:") Starts at end of previous line or arc	Draw	Draw
		C	(as reply to "To point:") Closes polygon		
		U	(as reply to "To point:") Undoes segment		
Linetype	Defines, loads, and sets the linetype	?	Lists a linetype library		TYPE IN
		C	Creates a linetype definition		
		L	Loads a linetype definition		
		S	Sets current object linetype; *set suboptions:*		
		name	Sets object linetype name		
		BYBLOCK	Sets floating object linetype		
		BYLAYER	Uses layer's linetype for objects		
		?	Lists specified loaded linetypes		
List	Provides database information for objects that are selected			Inquiry	Tools
Load	Loads a file of user-defined shapes to be used with the SHAPE command	?	Lists the names of loaded shape files		TYPE IN
Logfileoff	Closes the log file opened by LOGFILEON				TYPE IN
Logfileon	Writes the text window contents to a file				TYPE IN
Lsedit	Edits a landscape object			Render	
Lslib	Maintains libraries of landscape objects			Render	

Command Alliases	Explanation	Options		Toolbar	Pull-Down Menu
Lsnew	Adds realistic landscape items, such as trees and bushes, to your drawings			Render	
Ltscale	Regulates the scale factor to be applied to all linetypes within the drawing				TYPE IN
Lweight	Sets the current lineweight, lineweight display options, and lineweight units				Format
Massprop	Calculates and displays the mass properties of regions or solids			Inquiry	Tools
Matchprop	Causes properties of one object to be assigned to selected objects				
Matlib	Imports and exports materials to and from a library of materials				Tools
Measure	Inserts markers at measured distances along a selected object	B	Uses specified block as marker	Inquiry	Tools
Menu	Loads a menu into the menu areas (screen, pull-down, tablet, and button)				Tools
Menuload	Loads partial menu files				TYPE IN
Menuunload	Unloads partial menu files				TYPE IN
Minsert	Inserts multiple copies of a block in a rectangular array	fname	Loads fname and forms a rectangular array of the resulting block	Draw	Draw
		fname=f	Creates block fname from file f and forms a rectangular array		
		?	Lists names of defined blocks		
		C	(as reply to X scale prompt) Specifies scale via two points (Corner specification of scale)		
		XYZ	(as reply to X scale prompt) Readies Multiple Insert for $X, Y,$ and Z scales		
		~	Displays a File dialog box		
		S	Sets the scale factor for the $X, Y,$ and Z axes		
		PD	Sets the scale factor for the $X, Y,$ and Z axes to control the display of the block as it is dragged into position		
		PR	Sets the rotation angle of the block as it is dragged into position		
Mirror	Reflects selected objects about a user-specified axis, vertical, horizontal, or inclined			Modify I	Modify
Mirror3D	Creates a mirror image copy of objects about a plane				Modify
Mledit	Edits multiple parallel lines			Modify II	Modify
Mline	Draws multiple parallel lines				

Command Alliases	Explanation	Options		Toolbar	Pull-Down Menu
Mlstyle	Defines a style for multiple parallel lines				Format
Model	Switches from a layout tab to the Model tab and makes it current				TYPE IN
Move (M)	Moves selected objects to another location in the drawing			Modify I	Modify
Mslide	Creates a slide of what is displayed on the screen				TYPE IN
Mspace (MS)	Switches to model space from paper space			Status bar	View
Mtext	Creates paragraph text			Draw	Draw
Multiple	Allows the next command to repeat until canceled				TYPE IN
Mview	Sets up and controls viewports	ON	Turns selected viewport(s) on; causes model to be regenerated in the selected viewport(s)		View
		OFF	Turns selected viewport(s) off; causes model not to be displayed in the selected viewport(s)		
		Hideplot	Causes hidden lines to be removed in selected viewport(s) during paper space plotting		
		Fit	Creates a single viewport to fit the current paper space view		
		2	Creates two viewports in specified area or to fit the current paper space view		
		4	Creates four equal viewports in specified area or to fit the current paper space view		
		Restore	Translates viewport configurations saved with the VPORTS command into individual viewport objects in paper space		
		<point>	Creates a new viewport within the area specified by two points		
		O	Specifies a closed polyline, ellipse, spline, region, or circle to convert into a viewport		
		P	Creates an irregularly shaped viewport using specified points		
		3	Creates three viewports in specified area or to fit the current paper space view		

Command Alliases	Explanation	Options		Toolbar	Pull-Down Menu
Mvsetup	Sets up the specifications of a drawing				TYPE IN
New	Creates a new drawing			Standard	File
Offset	Reproduces curves or lines parallel to the one selected	number T	Specifies offset distance Through: allows specification of a point through which the offset curve is to pass	Modify I	Modify
Olescale	Displays the OLE Properties dialog box				TYPE IN
Olelinks	Cancels, changes, updates OLE links				
Oops	Recalls last set of objects previously erased				TYPE IN
Open	Opens an existing drawing			Standard	File
Options	Customizes the AutoCAD settings				Tools
Ortho F8	Restricts the cursor to vertical or horizontal use	ON OFF	Forces cursor to horizontal or vertical use Does not constrain cursor movement	Status bar	
Osnap	Allows for selection of precise points on existing objects	CEN END INS INT MID NEA NOD NON PER QUA QUI TAN EXT APP PAR	Center of arc or circle Closest endpoint of arc or line Insertion point of text/block/shape Intersection of line/arc/circle Midpoint of arc or line Nearest point of arc/circle/line/point Node (point) None (off) Perpendicular to arc/line/circle Quadrant point of arc or circle Quick mode (first find, not closest) Tangent to arc or circle Snaps to the extension point of an object Apparent Intersection includes two separate snap modes: Snaps to an extension in parallel with an object	Object Snap	Tools
Pagesetup	Specifies the layout page, plotting device, paper size, and settings for each new layout			Layout	
'Pan (P)	Moves the display window			Standard	View
Partiaload	Loads additional geometry into a partially opened drawing				File

Command Alliases	Explanation	Options		Toolbar	Pull-Down Menu
Partialopen	Loads geometry from a selected view or layer into a drawing				TYPE IN
Pasteblock	Pastes a copied block into a new drawing				Edit
Pasteclip	Inserts clipboard data				
Pasteorig	Pastes a copied object in a new drawing using the coordinates from the original drawing				Edit
Pastespec	Specifies fomat of data imported from the clipboard				
Pcinwizard	Displays a wizard to import PCP and PC2 configuration file plot settings into the Model tab or current layout				Tools
Pedit (2D)	Permits editing of 2D polylines	C	Closes to start point	Modify II	Modify
		D	Decurves, or returns a spline curve to its control frame		
		F	Fits curve to polyline		
		J	Joins to polyline		
		O	Opens a closed polyline		
		S	Uses the polyline vertices as the frame for a spline curve (type set by SPLINETYPE)		
		U	Undoes one editing operation		
		W	Sets uniform width for polyline		
		X	Exits PEDIT command during vertex editing		
		E	Edits vertices during vertex editing		
		B	Sets first vertex for Break		
		G	Go (performs Break or Straighten operation)		
		I	Inserts new vertex after current one		
		M	Moves current vertex		
		N	Makes next vertex current		
		P	Makes previous vertex current		
		R	Regenerates the polyline		
		S	Sets first vertex for Straighten		
		T	Sets tangent direction for current vertex		
		W	Sets new width for following segment		
		X	Exits vertex editing, or cancels Break/Straighten		
		L	Generates the linetype in a continuous pattern through the vertices of the polyline		

Command Alliases	Explanation	Options		Toolbar	Pull-Down Menu
Pedit (3D)	Allows editing of 3D polylines	C	Closes to start point.	Modify II	Modify
		D	Decurves, or returns a spline curve to its control frame		
		O	Opens a closed polyline		
		S	Uses the polyline vertices as the frame for a spline curve (type set by SPLINETYPE)		
		U	Undoes one editing operation		
		X	Exits PEDIT command		
		E	Edits vertices		
		During vertex editing:			
		B	Sets first vertex for Break		
		G	Go (performs Break or Straighten operation)		
		I	Inserts new vertex after current one		
		M	Moves current vertex		
		N	Makes next vertex current		
		P	Makes previous vertex current		
		R	Regenerates the polyline		
		S	Sets first vertex for Straighten		
		X	Exits vertex editing, or cancels Break/Straighten		
Pedit (mesh)	Allows editing of 3D polygon meshes	D	Desmoothes-restores original mesh	Modify II	Modify
		M	Opens (or closes) the mesh in the *M* direction		
		N	Opens (or closes) the mesh in the *N* direction		
		S	Fits a smooth surface as defined by SURFTYPE		
		U	Undoes one editing operation		
		X	Exits PEDIT command		
		E	Edits mesh vertices during vertex editing		
		D	Moves down to previous vertex in *M* direction		
		L	Moves left to previous vertex in *N* direction		
		M	Repositions the marked vertex		
		N	Moves to next vertex		
		P	Moves to previous vertex		
		R	Moves right to next vertex in *N* direction		
		RE	Redisplays the polygon mesh		
		U	Moves up to next vertex in *M* direction		
		X	Exits vertex editing during vertex editing		
Pface	Creates a 3D mesh of arbitrary complexity and surface characteristics				TYPE IN

Command Aliases	Explanation	Options		Toolbar	Pull-Down Menu
Plan	Puts the display in plan view (Vpoint 0,0,1) relative to either the current UCS, a specified UCS, or the WCS	C	Establishes a plan view of the current UCS		View
		U	Establishes a plan view of the specified UCS		
		W	Establishes a plan view of the WCS		
Pline (PL)	Draws 2D polylines	H	Sets new half-width	Draw	Draw
		U	Undoes previous segment		
		W	Sets new line width		
		ENTER	Exits PLINE command		
		C	Closes with straight segment		
		L	Segment length (continues previous segment)		
		A	Switches to arc mode		
		In arc mode:			
		A	Included angle		
		CE	Center point		
		CL	Closes with arc segment		
		D	Starting direction		
		L	Chord length, or switches to line mode		
		R	Radius		
		S	Second point of three-point arc		
Plot (Print)	Plots a drawing to a plotting device or a file			Standard	File
Plotstyle	Sets the current plot style for new objects, or the assigned plot style for selected objects				TYPE IN
Plottermanager	Displays the Plotter Manager, where you can launch the Add-a-Plotter wizard and the Plotter configuration editor				File
Point	Draws a single point on the drawing			Draw	Draw
Polygon	Creates regular polygons with the specified number of sides indicated	E	Specifies polygon by showing one edge	Draw	Draw
		C	Circumscribes around circle		
		I	Inscribes within circle		
Preview	Displays plotted view of drawing				
Properties	Controls properties of existing objects			Standard	
Propertiesclose	Closes the Properties window				TYPE IN
Psdrag	Controls the appearance of an imported PostScript image that is being dragged (that is, positioned and scaled) into place by the PSIN command	0	Only the image's bounding box is displayed as you drag it into place		File
		1	The rendered PostScript image is displayed as you drag it into place		
Psetupin	Imports a user-defined page setup into a new drawing layout				TYPE IN

Command Alliases	Explanation	Options		Toolbar	Pull-Down Menu
Psfill	Fills 2D polyline outlines with PostScript fill patterns defined in the AutoCAD PostScript support file *(acad.psf)*				TYPE IN
Psin	Imports encapsulated PostScript (EPS) files				File
Psout	Exports the current view of your drawing to an encapsulated Postscript (EPS) file				File
Pspace (PS)	Switches to paper space			Status bar	View
Purge	Removes unused Blocks, text styles, layers, linetypes, and dimension styles from the drawing	A B D LA SH ST LT	Purges all unused named objects Purges unused blocks Purges unused dimstyles Purges unused layers Purges unused shape files Purges unused text styles Purges linetypes		File
Qdim	Quickly creates a dimension	Continuous Staggered Baseline Ordinate Radius Diameter Datum Point Edit	Creates a series of continued dimensions Creates a series of staggered dimensions Creates a series of baseline dimensions Creates a series of ordinate dimensions Creates a series of radius dimensions Creates a series of diameter dimensions Sets a new datum point for baseline and ordinate dimensions Edits a series of dimensions	Dimension	
Qleader	Quickly creates a leader and leader annotation				Dimension
Qsave	Saves the drawing without requesting a file name			Standard	File
Qselect	Quickly creates selection sets based on filtering criteria				Tools
Qtext	Enables text objects to be identified without drawing the test detail	ON OFF	Quick text mode on. Quick text mode off		TYPE IN
Quit	Exit AutoCAD				File
Ray	Creates a semi-infinite line			Draw	
Recover	Attempts to recover damaged or corrupted drawings				File
Rectang	Draws a rectangular polyline			Draw	Draw
Redefine	Restores a built-in command deleted by UNDEFINE				TYPE IN

Command Alliases	Explanation	Options		Toolbar	Pull-Down Menu
Redo	Reverses the previous command if it was U or Undo			Standard	Edit
'Redraw (R)	Refreshes or cleans up the current viewport			Standard	View
'Redrawall	Redraws all viewports			Standard	View
Refclose	Saves back or discards changes made during in-place editing of a reference (an xref or a block)				Modify
Refedit	Selects a reference for editing			Refedit	
Refset	Adds or removes objects from a working set during in-place editing of a reference (an xref or a block)	Add Remove	Adds objects to the working set Removes objects from the working set		Modify
Regen	Regenerates the current viewport				TYPE IN
Regenall	Regenerates all viewports				TYPE IN
Regenauto	Controls automatic regeneration performed by other commands	ON OFF	Allows automatic regenerations Prevents automatic regenerations		TYPE IN
Region	Creates a region object from a selection set of existing objects			Draw	Draw
Reinit	Allows the I/O ports, digitizer, display, plotter, and *PGP* file to be reinitialized				TYPE IN
Rename	Changes the names associated with text styles, layers, linetypes, blocks, views, UCSs, viewport configurations, and dimension styles	B D LA LT S U VI VP	Renames block. Renames dimension style Renames layer Renames linetype Renames text style Renames UCS Renames view Renames viewport configuration		TYPE IN
Render	Creates a realistically shaded image of a 3D wireframe or solid model			Render	View
Rendscr	Redisplays the last rendering created with the RENDER command				TYPE IN
Replay	Displays a GIF, TGA, or TIFF image				Tools
'Resume	Resumes an interrupted command script				TYPE IN
Revolve	Creates a solid by revolving a 2D object about an axis			Solids	Draw
Revsurf	Creates a 3D polygon mesh approximating a surface of revolution, by rotating a curve around a selected axis			Surfaces	Draw
Rmat	Manages rendering materials			Render	View

Command Alliases	Explanation		Options	Toolbar	Pull-Down Menu
Rotate	Rotates existing objects to the angle selected	R	Rotates with respect to reference angles	Modify I	Modify
Rotate3D	Moves objects about a 3D axis				
Rpref	Sets rendering preferences			Render	View
Rscript	Restarts a command script from the beginning				TYPE IN
Rulesurf	Creates a 3D polygon mesh approximating a ruled surface between two curves			Surfaces	Draw
Save	Updates the current drawing file without exiting the Drawing Editor			Standard	File
Saveas	Same as SAVE, but also renames the current drawing				File
Saveimg	Saves a rendered image to a file				Tools
Scale	Changes the size of existing objects to the selected scale factor	R	Resizes with respect to reference size	Modify I	Modify
Scene	Manages scenes in model space			Render	View
Script	Executes a command script				Tools
Section	Uses the intersection of a plane and solids to create a region			Section	Modify
Select	Groups objects into a selection set for use in subsequent commands				TYPE IN
Setuv	Maps materials onto objects			Render	
'Setvar	Allows you to display or change the value of system variables	?	Lists specified system variables		TYPE IN
Shademode	Shades the objects in the current viewport	2D Wireframe	Displays the objects using lines and curves to represent the boundaries		View
		3D Wireframe	Displays the objects using lines and curves to represent the boundaries		
		Hidden	Displays the objects using 3D wireframe representation and hides lines representing back faces		
		Flat Shaded	Shades the objects between the polygon faces		
		Gouraud Shaded	Shades the objects and smooths the edges between polygon faces		
		Flat Shaded, Edges On	Combines the Flat Shaded and Wireframe options		

Command Alliases	Explanation	Options		Toolbar	Pull-Down Menu
		Gouraud Shaded, Edges On	Combines the Gouraud Shaded and Wireframe options		
Shape	Draws predefined shapes	?	Lists available shape names		TYPE IN
Shell	Allows access to other programs while running AutoCAD				TYPE IN
Showmat	Lists material type and method of attachment for the selected object				
Sketch	Allows freehand sketching	C	Connect: restarts sketch at endpoint	Draw	Draw
		E	Erases (backs up over) temporary lines		
		P	Raises/lowers sketching pen		
		Q	Discards temporary lines, remains in SKETCH		
		R	Records temporary lines, remains in SKETCH		
		X	Records temporary lines, exits SKETCH; draws line to current point		
Slice	Slices a set of solids with a plane			Solids	Draw
Snap F9	Allows for precision alignment of points	number	Sets snap resolution	Status bar	
		ON	Aligns designated points		
		OFF	Does not align designated points		
		A	Sets aspect (differing X–Y spacing)		
		R	Rotates snap grid		
		S	Selects style, standard or isometric		
Soldraw	Generates sections and profiles in viewports created with SOLVIEW				
Solid	Creates filled-in polygons			Draw	Draw
Solidedit	Edits faces and edges of 3D solid objects				Modify
Solprof	Creates a profile image of a 3D solid				
Solview	Creates floating viewports for 3D solid objects				
Spell	Checks the spelling in a drawing			Standard	Tools
Sphere	Creates a 3D solid sphere			Solids	Draw
Spline	Creates a quadratic or cubic spline (NURBS) curve			Draw	Draw
Splinedit	Edits a spline object			Modify II	Modify
Stats	Displays rendering statistics			Render	View
Status	Displays drawing setup				Tools
Stlout	Stores a solid in ASCII or binary file				File

Command Alliases	Explanation	Options		Toolbar	Pull-Down Menu
Stretch	Allows you to move a portion of a drawing while retaining connections to other parts of the drawing			Modify I	Modify
Style	Sets up named text styles, with various combinations of font, mirroring, obliquing, and horizontal scaling	?	Lists specified currently defined text style		Format
Stylesmanager	Displays the Plot Style Manager				File
Subtract	Creates a composite region or solid by subtracting the area of one set of regions from another and subtracting the volume of one set of solids from another			Modify II	Modify
Syswindows	Arranges windows				
Tablet	Allows for configuration of a tablet menu or digitizing of an existing drawing	ON	Turns tablet mode on		Tools
		OFF	Turns tablet mode off		
		CAL	Calibrates tablet for use in the current space		
Tabsurf	Creates a polygon mesh approximating a general tabulated suface defined by a path and a direction vector			Surfaces	Draw
Text	Enters text on the drawing	J	Prompts for justification options	Draw	Draw
		S	Lists or selects text style		
		A	Aligns text between two points, with style-specified width factor; AutoCAD computes appropriate height		
		C	Centers text horizontally		
		F	Fits text between two points, with specified height; AutoCAD computes an appropriate width factor		
		M	centers text horizontally and vertically		
		R	Right-justifies text		
		BL	Bottom left		
		BC	Bottom center		
		BR	Bottom right		
		ML	Middle left		
		MC	Middle center		
		MR	Middle right		
		TL	Top left		
		TC	Top center		
		TR	Top right		
'Textscr F2	Flips to the text display on singlescreen systems; used in command scripts and menus				TYPE IN
Time	Indicates total elapsed time for each drawing	D	Displays current times		Tools
		ON	Starts user elapsed timer		
		OFF	Stops user elapsed timer		
		R	Resets user elapsed timer		

Command Alliases	Explanation	Options		Toolbar	Pull-Down Menu
Tolerance	Creates geometric tolerances			Dimension	Draw
Toolbar	Customizes, hides, and displays toolbars				
Torus	Creates a donut-shaped solid			Solids	Draw
Trace	Creates solid lines of specified width			Draw	Draw
Transparency	Determines transparency of opacity of bitonal image background				
Treestat	Displays information on the drawing's current spatial index, such as the number and depth of nodes in the drawing's database; use this information with the TREEDEPTH system variable setting to fine-tune performances for large drawings				TYPE IN
Trim	Deletes portions of selected entities that cross a selected boundary edge	U	Undoes last trim operation	Modify I	Modify
U	Reverses the effect of the previous command			Standard	Edit
UCS	Defines or modifies the current User Coordinate System	D	Deletes one or more saved coordinate systems	Standard	View
		E	Sets a UCS with the same extrusion direction as that of the selected object		
		O	Shifts the origin of the current coordinate system		
		P	Restores the previous UCS		
		R	Restores a previously saved UCS		
		S	Saves the current UCS		
		V	Establishes a new UCS whose Z Axis is parallel to the current viewing direction		
		W	Sets the current UCS equal to the WCS		
		X	Rotates the current UCS around the *X* axis		
		Y	Rotates the current UCS around the *Y* axis		
		Z	Rotates the current UCS around the *Z* axis		
		ZA	Defines a UCS using an origin point and a point on the positive portion of the *Z* axis		
		3	Defines a UCS using an origin point, a point on the positive portion of the *X* axis, and a point on the positive *Y* portion of the *X* plane		
		?	Lists specified saved coordinate systems		

Command Alliases	Explanation	Options		Toolbar	Pull-Down Menu
Ucsicon	Controls visibility and placement of the UCS icon, which indicates the origin and orientation of the current UCS; the options normally affect only the current viewport	A	Changes settings in all active viewports		Tools
		N	Displays the icon at the lower-left corner of the viewport		
		OR	Displays the icon at the origin of the current UCS if possible		
		ON	Enables the coordinate system icon		
Ucsman	Manages defined user coordinate systems			UCS	
Undefine	Deletes the definition of a built-in AutoCAD command				TYPE IN
Undo	Reverses the effect of multiple commands, and provides control over the Undo facility	number	Undoes the number most recent commands	Standard	Edit
		A	Auto: controls treatment of menu items as Undo groups		
		B	Back: undoes back to previous Undo mark		
		C	Control: enables/disables the Undo mark		
		E	End: terminates an Undo group		
		G	Group: begins sequence to be treated as one command		
		M	Mark: places marker in Undo file (for back)		
Union	Creates a composite region or solid			Modify II	Modify
Units	Selects coordinate and angle display formats and precision				Format
Vbaide	Displays the Visual Basic Editor				Tools
Vbaload	Loads a global VBA project into the current AutoCAD session				Tools
Vbaman	Loads, unloads, saves, creates, embeds, and extracts VBA projects				Tools
Vbarun	Runs a VBA macro				Tools
Vbastmt	Executes a VBA statement on the AutoCAD command line				TYPE IN
Vbaunload	Unloads a global VBA project				TYPE IN
'View	Saves the current graphics display and space as a named view, or restores a saved view and space to the display	D	Deletes named view		View
		R	Restores named view to screen		
		S	Saves current display as named view		
		W	Saves specified window as named view		
		?	Lists specified named views		
Viewports or Vports	Divides the AutoCAD graphics display into multiple viewports, each of which can contain a different view of the current drawing	D	Deletes a saved viewport configuration		View
		J	Joins (merges) two viewports		
		R	Restores a saved viewport configuration		

Command Alliases	Explanation	Options		Toolbar	Pull-Down Menu
		S	Saves the current viewport configuration		
		S1	Displays a single viewport filling the entire graphics area		
		2	Divides the current viewport into viewports		
		3	Divides the current viewport into three viewports		
		4	Divides the current viewport into four viewports		
		?	Lists the current and saved viewport configurations		
Viewres	Adjusts the precision and speed of circle and arc drawing on the monitor				Format
Vlisp	Displays the Visual LISP interactive development environment (IDE)				Tools
Vpclip	Clips viewport objects	Object	Specifies an object to act as a clipping boundary		TYPE IN
		Polygonal	Draws a clipping boundary		
		Delete	Deletes the clipping boundary of a selected viewport		
Vplayer	Sets viewport visibility for new and existing layers	?	Lists layers frozen in a selected viewport		Type In
		Freeze	Freezes specified layers in selected viewport(s)		
		Thaw	Thaws specified layers in selected viewport(s)		
		Reset	Resets specified layers to their default visibility		
		Newfz	Creates new layers that are frozen in all viewports		
		Vpvisdfit	Sets the default viewport visibility for existing layers		
Vpoint	Selects the viewpoint for a 3D visualization	R	Selects viewpoint via two rotation angles		View
		ENTER	Selects viewpoint via compass and axes tripod		
		x,y,z	Specifies viewpoint		
Vslide	Displays a previously created slide file	file	Views slide		Tools
		*file	Preloads next Vslide you will view		
Wblock	Creates a block as a separate drawing	name	Writes specified block definition		File
		=	Block name same as file name		
		*	Writes entire drawing		
		ENTER	Writes selected objects		

Command Alliases	Explanation	Options		Toolbar	Pull-Down Menu
Wedge	Creates a 3D solid with a tapered sloping face			Solids	Draw
Whohas	Displays ownership information for opened drawing files				TYPE IN
Wmxxx	Controls windows metafiles				
Xattach	Attaches an external reference				
Xbind	Permanently adds a selected subset of an external reference's dependent symbols to your drawing	Block	Adds a Block.	Reference	Modify
		Dimstyle	Adds a dimstyle		
		Layer	Adds a layer		
		Ltype	Adds a linetype		
		Style	Adds a style		
Xclip	Defines and external reference				
Xline	Creates an infinite line			Draw	Draw
Xplode	Breaks a compound object into its component objects				TYPE IN
Xref	Allows you to work with other AutoCAD drawings without adding them permanently to your drawing and without altering their contents	Attach	Attaches a new Xref or a copy of an Xref that you have already attached	Reference	Insert
		Bind	Makes an Xref a permanent part of your drawing		
		Detach	Removes an Xref from your drawing		
		Path	Allows you to view and edit the file name AutoCAD uses when loading a particular Xref		
		Reload	Updates one or more contents Xrefs at any time, without leaving and re-entering the Drawing Editor		
		?	Lists Xrefs in your drawing and the drawing associated with each one		
'Zoom (Z)	Enlarges or reduces the display area of a drawing	number	Multiplier from original scale	Standard	View
		numberX	Multiplier from current scale		
		number XP	Scale relative to paper space		
		A	All		
		C	Center		
		D	Dynamic Pan Zoom		
		E	Extents ("drawing uses")		
		L	Lower left corner		
		P	Previous		
		V	Virtual screen maximum		
		W	Window		
3D	Creates three-dimensional polygon mesh objects				Draw

Command Alliases	Explanation	Options		Toolbar	Pull-Down Menu
3Darray	Creates a three-dimensional array				Modify
3Dclip	Invokes the interactive 3D view and opens the Adjust Clipping Planes window				
3Dcorbit	Invokes the interactive 3D view and enables you to set the objects in the 3D view into continuous motion			Orbit	
3Ddistance	Invokes the interactive 3D view and makes objects appear closer or farther away			Orbit	
3Dface	Draws 3D plane sections	I	Makes the following edge invisible	Surfaces	Draw
3Dmesh	Defines a 3D polygon mesh (by specifying its size in terms of *M* and *N*) and the location of each vertex in the mesh			Surfaces	Draw
3Dorbit	Controls the interactive viewing of objects in 3D			Standard	
3Dpan	Invokes the interactive 3D view and enables you to drag the view horizontally and vertically			Orbit	
3Dpoly	Creates a 3D polyline	C	Closes the Polyline back to the first point	Draw	Draw
		U	Undoes (deletes) the last segment entered		
		ENTER	Exits 3Dpoly command		
3Dsin	Imports a 3D Studio (3DS) file				Insert
3Dsout	Exports to a 3D Studio (3DS) file				TYPE IN
3Dswivel	Invokes the interactive 3D view and simulates the effect of turning the camera			3D Orbit	
3Dzoom	Invokes the interactive 3D view so you can zoom in and out on the view			3D Orbit	

DIMENSIONING COMMANDS

Command	Explanation
Dimaligned	Aligns dimension parallel with objects
Dimangular	Draws an arc to show the angle between two nonparallel lines or three specified points
Dimbaseline	Continues a linear dimension from the baseline (first extension line) of the previous or selected dimension
Dimcenter	Draws a circle/arc center mark or centerlines
Dimcontinue	Continues a linear dimension from the second extension line of the previous dimension
Dimdiameter	Dimensions the diameter of a circle or arc
Dimedit	Edits dimensions
Dimlinear	Creates linear dimensions
Dimordinate	Creates ordinate point associative dimensions
Dimoverride	Overrides a subset of the dimension variable settings associated with selected dimension objects
Dimradius	Dimensions the radius of a circle or arc, with an optional center mark or centerlines
Dimtedit	Allows repositioning and rotation of text items in an associative dimension without affecting other dimension subentities
Leader	Draws a line with an arrowhead placement of dimension text

DIMENSIONING VARIABLES

Name	Description	Type	Default
DIMADEC	Controls number of places of precision displayed for angular dimension text		
DIMALT	Alternate units	Switch	Off
DIMALTD	Alternate units decimal places	Integer	2
DIMALTF	Alternate units scale factor	Scale	25.4
DIMALTRND	Alternate units rounding value		0.0000
DIMALTTD	Alternate units tolerance value	Integer	2
DIMALTTZ	Toggles suppression of zeros for tolerance values	Integer	0
DIMALTU	Sets unit format for alternate units	Integer	2
DIMALTZ	Toggles suppression of zeros for alternate values	Integer	0
DIMAPOST	Alternate units text suffix	String	None
DIMASZ	Arrow size.	Distance	0.18
DIMATFIT	Arrow and text fit		3
DIMAUNIT	Angle format	Integer	0
DIMBLK	Arrow block	String	None
DIMBLK1	Separate arrow block 1	String	None
DIMBLK2	Separate arrow block 2	String	None
DIMCEN	Center mark size	Distance	0.09
DIMCLRD	Dimension line color	Color number	BYBLOCK
DIMCLRE	Extension line color	Color number	BYBLOCK
DIMCLRT	Dimension text color	Color number	BYBLOCK
DIMDEC	Decimal place for tolerance values	Integer	4
DIMDLE	Dimension Line extension	Distance	0.0
DIMDLI	Dimension line increment	Distance	0.38

Name	Description	Type	Default
DIMDSEP	Decimal separator		.
DIMEXE	Extension line extension	Distance	0.18
DIMEXO	Extension line offset	Distance	0.0625
DIMFRAC	Fraction format		0
DIMGAP	Dimension line gap	Distance	0.09
DIMJUST	Controls horizontal text position	Imteger	0
DIMLDRBLK	Leader arrowhead block name		ClosedFilled
DIMLFAC	Length factor	Scale	1.0
DIMLIM	Limits dimensioning	Switch	Off
DIMLUNIT	Linear unit format		2
DIMLWD	Dimension line and leader lineweight		-2
DIMLWE	Extension line lineweight		-2
DIMPOST	Dimension text suffix	String	None
DIMRND	Rounding value	Scaled distance	0.0
DIMSAH	Separate arrow blocks	Switch	Off
DIMSCALE	Dimension feature scale factor	Switch	1.0
DIMSD1	Suppresses first dimension line	Switch	Off
DIMSD2	Suppresses second dimension line	Switch	Off
DIMSE1	Suppresses extension line 1	Switch	Off
DIMSE2	Suppresses extension line 2	Switch	Off
DIMSOXD	Suppresses outside dimension lines	Switch	Off
DIMTAD	Text above dimension line	Switch	Off
DIMTDEC	Tolerance values	Integer	4
DIMTFAC	Tolerance text scale factor	Scale	1.0
DIMTIH	Text inside horizontal	Switch	On
DIMTIX	Text inside extension lines	Switch	Off
DIMTM	Minus tolerance value	Scaled distance	0.0

Name	Description	Type	Default
DIMTMOVE	Text movement		0
DIMTOFL	Text outside, force line inside	Switch	Off
DIMTOH	Text outside horizontal	Switch	On
DIMTOL	Tolerance dimensioning	Switch	Off
DIMTOLJ	Tolerance dimensioning justification	Integer	1
DIMTP	Plus tolerance value	Scaled distance	0.0
DIMTSZ	Tick size	Distance	0.0
DIMTVP	Text vertical position	Scale	0.0
DIMTXSTY	Dimension text style	String	Standard
DIMTXT	Text size	Distance	0.18
DIMTZIN	Controls suppression of zeros in tolerance values		
DIMUNIT	Sets unit format	Integer	2
DIMUPT	User-positioned text	Switch	Off
DIMZIN	Zero suppression	Integer	0

OBJECT SELECTION

Object Selection Option	Meaning
point	Selects one object that crosses the small pick box. If no object crosses the pick box and Auto mode has been selected, this designated point is taken as the first corner of a Crossing or Window box
Multiple	Allows selection of multiple objects using a single search of the drawing. The search is not performed until you give a null response to the "Select objects:" prompt
Window	Selects all objects that lie entirely within a window
WPolygon	Selects objects that lie entirely within a polygon shaped selection area
Crossing	Selects all objects that lie within *or cross* a window
CPolygon	Selects all objects that lie within and crossing a polygon-shaped selection area
Fence	Selects all objects that cross a selection fence line
BOX	Prompts for two points. If the second point is to the right of the first point, selects all objects inside the box (like "Window"); otherwise, selects all objects within or crossing the box (like "Crossing")
AUto	Accepts a point, which can select an object using the small pick box; if the point you pick is in an empty area, it is taken as the first corner of a BOX (see above)
ALL	Selects all entities in the drawing except entities on frozen or locked layers
Last	Selects the most recently drawn object that is currently visible
Previous	Selects the previous selection set
Add	Establishes Add mode to add following objects to the selection set
Remove	Sets Remove mode to remove following objects from the selection set
SIngle	Sets single selection mode; as soon as one object (or one group of objects via Window/Crossing box) is selected, the selection set is considered complete and the editing command uses it without further user interaction
Undo	Undoes (removes objects last added)

APPENDIX
C

AutoCAD Toolbars

Standard

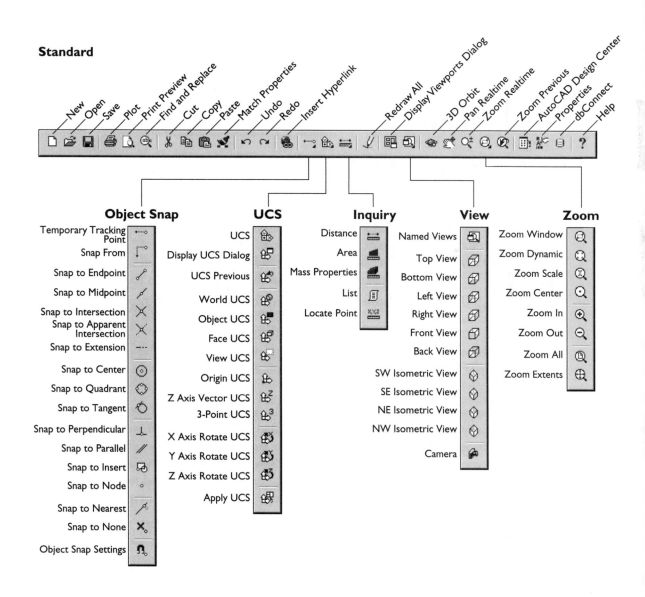

Object Properties

Make Object's
Layer Current
Layers: On/Off
Layers: Freeze/Thaw in all viewports
Layers: Lock/Unlock
Make layer plottable or nonplottable
Layers: Color

Layers

| | ByLayer | ByLayer | ByLayer | ByLayer |

Layer Control   Color Control   Linetype Control   Lineweight Control   Plot Styles Control

Modify

Erase	
Copy Object	
Mirror	
Offset	
Array	
Move	
Rotate	
Scale	
Stretch	
Lengthen	
Trim	
Extend	
Break	
Chamfer	
Fillet	
Explode	

Draw

Line	
Construction Line	
Multiline	
Polyline	
Polygon	
Rectangle	
Arc	
Circle	
Spline	
Ellipse	
Insert Block	
Make Block	
Point	
Hatch	
Region	
Multiline Text	

Layouts

New Layout	
Layout From Template	
Paper Setup	
Display Viewports Dialog	

Insert

Insert Block	
External Reference	
Image	
Import	
OLE Object	

Dimension

Linear Dimension	
Aligned Dimension	
Ordinate Dimension	
Radius Dimension	
Diameter Dimension	
Angular Dimension	
Quick Dimension	
Baseline Dimension	
Continue Dimension	
Quick Leader	
Tolerance	
Center Mark	
Dimension Edit	
Dimension Text Edit	
Dimension Update	
Dimension Style	

Modify II

Draworder	
Edit Hatch	
Edit Polyline	
Edit Spline	
Edit Multiline	
Edit Attribute	
Edit Text	

3D Orbit

3D Pan	
3D Zoom	
3D Orbit	
3D Continuous Orbit	
3D Swivel	
3D Adjust Distance	
3D Adjust Clip Planes	
Front Clip On/Off	
Back Clip On/Off	

Viewports

Display Viewports Dialog	
Single Viewport	
Polygonal Viewport	
Convert Object to Viewport	
Clip Existing Viewport	

Shade

2D Wireframe	
3D Wireframe	
Hidden	
Flat Shaded	
Gouraud Shaded	
Flat Shaded, Edges On	
Gouraud Shaded, Edges On	

Solids Editing

Union	
Subtract	
Intersect	
Extrude Faces	
Move Faces	
Offset Faces	
Delete Faces	
Rotate Faces	
Taper Faces	
Copy Faces	
Color Faces	
Copy Edges	
Color Edges	
Imprint	
Clean	
Separate	
Shell	
Check	

Surfaces

2D Solid
3D Face
Box
Wedge
Pyramid
Cone
Sphere
Dome
Dish
Torus
Edge
3D Mesh
Revolved Surface
Tabulated Surface
Ruled Surface
Edge Surface

Reference

External Reference
External Reference Attach
External Reference Clip
External Reference Bind
External Reference Clip Frame
Image
Image Attach
Image Clip
Image Adjust
Image Quality
Image Transparency
Image Frame

Solids

Box
Sphere
Cylinder
Cone
Wedge
Torus
Extrude
Revolve
Slice
Section
Interfere
Set Up Drawing
Set Up View
Set Up Profile

Render

Hide
Render
Scenes
Lights
Materials
Materials Library
Mapping
Background
Fog
Landscape New
Landscape Edit
Landscape Library
Render Preferences
Statistics

Refedit

Edit Block or Xref
Add Objects to Working Set
Remove Objects from Working Set
Discard Changes to Reference
Save Back Changes to Reference

UCS II

Display UCS Dialog
Move UCS Origin

Web

Go Back
Go Froward
Stop Navigation
Browse the Web

1298

APPENDIX d

System Variables

This is a complete listing of AutoCAD system variables. Each variable has an associated type: integer, real, point, or text string. These variables can be examined and changed (unless read-only) by means of the SETVAR command and AutoLISP (getvar and setvar) functions. Many of the system variables are saved across editing sessions; as indicated in the table, some are saved in the drawing itself, while others are saved in the AutoCAD general configuration file, ACAD.CFG

Variable	Default Setting	Type	Saved In	Explanation
ACADLSPASDOC	0	Integer	Registry	Controls whether AutoCAD loads the acad.lsp file into every drawing or just the first drawing opened in an AutoCAD session
ACADPREFIX	" "	String	Read-only	The directory path, if any specified by the ACAD environment variable, with path separators appended if necessary (read-only)
ACADVER		String		This is the AutoCAD version number, which can have values only like "15.0" (read-only). Note that this differs from the DXF file $ACADVER header variable, which contains the drawing database level number.
ACISOUTVER	40	Integer		Controls the ACIS version of SAT files created using the ACISOUTcommand
AFLAGS	0	Integer		Attribute flags bit-code for ATTDEF command (sum of the following): 0 = No Attribute Mode Selected 1 = Invisible 2 = Constant 4 = Verify 8 = Preset
ANGBASE	0	Real	Drawing	Angle 0 direction (with respect to the current UCS)
ANGDIR	0	Integer	Drawing	1= clockwise angles, 0 = counterclockwise (with respect to the current UCS)
APERTURE	10	Integer	Config	Object Snap target height, in pixels (default value = 10)
APBOX				Sets AutoSnap aperture box to ON or OFF
APERTURE	10	Integer	Registry	Sets object snap target height, in pixels

Reprinted with permission of Autodesk Inc.

Variable	Default Setting	Type	Saved In	Explanation
AREA		Real		True area computed by Area, List, or Dblist (read-only)
ATTDIA	0	Integer	Registry	1 causes the INSERT command to use a dialog box for entry of attribute values; 0 to issue prompts
ATTMODE	1	Integer	Drawing	Attribute display mode (0 = OFF, 1 = normal, 2 = ON)
ATTREQ	1	Integer	Registry	0 assumes defaults for the values of all attributes during insert of blocks; 1 enables prompts (or dialog box) for attribute values, as selected by ATTDIA
AUDITCTL	0	Integer	Config	Controls whether an .adt log file (audit report file) is created 0 = Disables (or prevents) the writing of adt log files 1 = Enables the writing of .adt log files by the AUDIT command
AUNITS	0	Integer	Drawing	Angular units mode (0 = decimal degrees, 1 = degrees/minutes/seconds, 2 = grads, 3 = radians, 4 = surveyor's units)
AUPREC	0	Integer	Drawing	Angular units decimal places
AUTOSNAP	63	Integer	Registry	Controls the display of the AutoSnap marker and SnapTips and sets the AutoSnap magnet to ON or OFF
BACKZ	0.0000	Real	Drawing	Back clipping plane offset for the current viewport, in drawing units. Meaningful only if the back clipping bit in VIEWMODE is on. The distance of the back clipping plane from the camera point can be found by subtracting BACKZ from the camera-to-target distance (read-only)
BINDTYPE	0	Integer		Controls how xref names are handled when binding xrefs or editing xrefs in-place
BLIPMODE	0	Integer	Registry	Marker blips ON if 1, OFF if 0
CDATE		Real		Calendar date/time (read-only)
CECOLOR	"BY-LAYER"	String	Drawing	Current object color (read-only)
CELTSCALE	1	Real	Drawing	Sets the current global linetype scale for objects
CELTYPE	"BY-LAYER"	String	Drawing	Current object linetype (read-only)
CELWEIGHT	-1	Integer		Sets the lineweight of new objects
CHAMFERA	0.5000	Real	Drawing	First chamfer distance
CHAMFERB	1.0000	Real	Drawing	Second chamfer distance
CHAMFERC	1.0000	Real	Drawing	Sets the chamfer length
CHAMFERD	0.0000	Real	Drawing	Sets the chamfer angle
CHAMMODE	0	Integer		Sets the input method by which AutoCAD creates chamfers 0 = Requires two chamfer distances 1 = Requires one chamfer length and an angle
CIRCLERAD	0.0000	Real		Sets the default circle radius; to specify no default, enter 0 (zero)

Variable	Default Setting	Type	Saved In	Explanation
CLAYER	"0"	String	Drawing	Sets the current layer (read-only)
CMDACTIVE		Integer		Bitcode that indicates whether an ordinary command, transparent command, script, or dialog box is active (read-only). It is the sum of the following: 1 = Ordinary command is active 2 = Ordinary command and a transparent command are active 4 = Script is active 8 = Dialog box is active 16= Autolist active (only visible to an object arcs-defined)
CMDDIA	1	Integer	Config	1 = Use dialog boxes for PLOT commands 0 = Don't use dialog boxes for PLOT command
CMDECHO	1	Integer		When the AutoLISP (command) function is used, prompts and input are echoed if this variable is 1 but not if it is 0
CMDNAMES		String		Displays in English the name of the command (and transparent command) that is currently active, for example; LINE'ZOOM indicates that the ZOOM command is being used transparently during the LINE command
CMLJUST	0	Integer	Drawing	Specifies multiline justification 0 = Top 1 = Middle 2 = Bottom
CMSCALE	1.0000	Real	Drawing	Controls the overall width of a multiline
CMSTYLE	"Standard"	String	Drawing	Sets the name of the multiline style that AutoCAD uses to draw the multiline
COMPASS	0	Integer		Controls whether the 3D compass is on or off in the current viewport
COORDS	1	Integer	Registry	If 0, coordinate display is updated on point picks only; if 1, display of absolute coordinates is continuously updated; if 2, distance and angle from last point are displayed when a distance or angle is requested
CPLOTSTYLE	"ByLayer"	Drawing		Controls the current plot style for new objects
CPROFILE		String	Registry	Stores the name of the current profile (read-only)
CTAB		String	Drawing	(Read-Only) Returns the name of the current (model or layout) tab in the drawing. Provides a means for the user to determine which tab is active
CURSORSIZE				Sets the size of the crosshairs as a percentage of screen size
CVPORT	2	Integer	Drawing	The identification number of the current viewport
DATE		Real		Julian date/time (read-only)
DBMOD		Integer		Bitcode that indicates the drawing modification status (read-only); it is the sum of the following: 1 = Entity database modified 2 = Symbol table modified 4 = Database variable modified 8 = Window modified 16 = View modified

Variable	Default Setting	Type	Saved In	Explanation
DCTCUST		String	Registry	Displays the current custom spelling dictionary path and file name
DCTMAIN		String	Registry	Displays the current main spelling dictionary file name
DEFLPLSTYLE	""	String	Registry	Specifies the default plot style for new layers
DEFPLSTYLE	"ByLayer"		Registry	Specifies the default plot style for new objects
DELOBJ	1	Integer	Registry	Controls whether objects used to create other objects are retained or deleted from the drawing database 0 = Objects are retained 1 = Objects are deleted
DEMANDLOAD	3	Integer	Registry	Determines demand loading of a third-party application if a drawing contains custom objects created in that application
DIASTAT		Integer		Dialog box exit status: if 0, the most recent dialog box was exited via CANCEL; if 1, the most recent dialog box was exited by pressing OK (read-only)
DIMxxx		Assorted	Drawing	All the dimensioning variables are also accessible as system variables (see Dimensioning Variables, Appendix B)
DISPSILH	0	Integer	Drawing	Sets display of silhouette curves of body objects in wireframe mode
DISTANCE		Real		Distance computed by DIST command (read-only)
DONUTID	0.5000	Real		Default donut inside diameter; can be zero
DONUTOD	1.0000	Real		Default donut outside diameter; must be nonzero. If DONUTID is larger than DONUTOD, the two values are swapped by the next command
DRAGMODE	2	Integer	Registry	0 = no dragging, 1 = on if requested, 2 = auto
DRAGP1	10	Integer	Registry	Regeneration-drag input sampling rate
DRAGP2	25	Integer	Registry	Fast-drag input sampling rate
DWGCHECK	0	Integer	Registry	Determines whether a drawing was last edited by a product other than AutoCAD
DWGCODEPAGE		String	Drawing	Drawing code page: This variable is set to the system code page when a new drawing is created, but otherwise AutoCAD doesn't maintain it. It should reflect the code page of the drawing and you can set it to any of the values used by the SYSCODEPAGE system variable or to "undefined." It is saved in the header DWGNAME String Drawing name as entered by the user; if the user specified a drive/directory prefix, it is included as well (read-only)
DWGPREFIX		String		Drive/directory prefix for drawing (read-only)
DWGTITLED		Integer		Bitcode that indicates whether the current drawing has been named (read-only) 0 = The drawing hasn't been named 1 = The drawing has been named

Variable	Default Setting	Type	Saved In	Explanation
EDGEMODE	0	Integer	Registry	Sets how TRIM and EXTEND determine cutting and boundary edges
ELEVATION	0.0000	Real	Drawing	Current 3D elevation, relative to the current UCS for the current space
EXPERT	0	Integer		Controls the issuance of certain "are you sure?" prompts: 0 = Issues all prompts normally 1 = Suppresses "About to regen, proceed?" and "Really want to turn the current layer off?" 2 = Suppresses the preceding prompts and Block's "Block already defined. Redefine it?" and Save/Wblock's "A drawing with this name already exists. Overwrite it?" 3 = Suppresses the preceding prompts and those issued by linetype if you try to load a linetype that is already loaded or create a new linetype in a file that already defines it 4 = Suppresses the preceding prompts and those issued by "Ucs Save" and "Vports Save" if the name you supply already exists 5 = Suppresses the preceding prompts and those issued by "Dim Save" and "Dim Override" if the dimension style name you supply already exists (the entries are redefined) When a prompt is suppressed, EXPERT, the operation in question, is performed as though you had responded Y to the prompt. In the future, values greater than 5 may be used to suppress additional safety prompts. The setting of EXPERT can affect scripts, menu macros, AutoLISP, and the command functions. The default value is 0.
EXPLMODE	0	Integer		Determines whether EXPLODE supports nonuniformly scaled (NUS) blocks
EXTMAX		3D point	Drawing	Upper right drawing uses extents. Expands outward as new objects are drawn; shrinks only by ZOOM All or ZOOM Extents. Reported in World coordinates for the current space (read-only)
EXTMIN		3D point	Drawing	Lower left drawing uses extents. Expands outward as new objects are drawn; shrinks only by ZOOM All or ZOOM Extents. Reported in World coordinates for the current space (read-only)
EXTNAMES	1	Integer	Drawing	Sets the parameters for named object names (such as linetypes and layers) stored in symbol tables
FACETRATIO	0	Integer		Controls the aspect ratio of faceting for cylindrical and conic ACIS solids
FACETRES	0.5	Real	Drawing	Adjusts smoothness of shaded and hidden-line-removed objects
FFLIMIT	0	Integer	Config	Limits numbers of PostScript and TrueType fonts in memory

Variable	Default Setting	Type	Saved In	Explanation
FILEDIA	1	Integer	Registry	1 = Use file dialog boxes if possible; 0 = do not use File dialog boxes unless requested via ~ (tilde)
FILLETRAD	0.5000	Real	Drawing	Fillet radius
FILLMODE	1	Integer	Drawing	Fill mode ON if 1, OFF if 0
FONTALT	"Simplex.shp"	String	Registry	Specifies alternate font
FONTMAP	"acad.fmp"	String	Registry	Specifies font mapping file
FRONTZ	0.0000	Real	Drawing	Front clipping plane offset for the current viewport, in drawing units. Meaningful only if the front clipping bit in VIEWMODE is ON and the "Front clip not at eye" bit is also ON. The distance of the front clipping bit from the camera point can be found by subtracting FRONTZ from the camera-to-target distance (read-only)
FULLOPEN		Integer		Indicates whether the current drawing is partially open (read-only)
GRIDMODE	0	Integer	Drawing	1 = Grid on for current viewport, X and Y
GRIDUNIT	0.5000,0.5000	2D point	Drawing	Grid spacing for current viewport, X and Y
GRIPBLOCK	0	Integer	Registry	Controls the assignment of grips in blocks 0 = Assigns grip only to the insertion point of the block 1 = Assigns grips to entities within the block
GRIPCOLOR	5	Integer (1–255)	Registry	Color of nonselected grips; drawn as a box outline
GRIPHOT	1	Integer (1–255)	Registry	Color of selected grips; drawn as a filled box
GRIPS	1	Integer	Registry	Allows the use of selection set grips for the Stretch, Move, Rotate, Scale, and Mirror modes 0 = Disables grips 1 = Enables grips
GRIPSIZE	3	Integer (1–255)	Registry	The size in pixels of the box drawn to display the grip
HANDLES	1	Integer	Drawing	If 0, entity handles are disabled; if 1, handles are on (read-only)
HIDEPRECISION	0	Integer		Controls the accuracy of hides and shades
HIGHLIGHT	1	Integer		Object selection highlighting ON if 1, OFF if 0
HPANG	0.0000	Real		Default hatch pattern angle
HPBOUND	1			Controls BHATCH and BOUNDARY object types
HPDOUBLE	0	Integer		Default hatch pattern doubling for "U" user defined patterns 0 = Disables doubling 1 = Enables doubling
HPNAME	"ANSI31"	String		Default hatch pattern name. Up to 34 characters no spaces allowed. Returns " " if there is no default. Enter . (period) to set no default
HPSCALE	1.0000	Real		Default hatch pattern scale factor; must be nonzero

Variable	Default Setting	Type	Saved In	Explanation
HPSPACE	1.0000	Real		Default hatch pattern line spacing for "U" user defined simple patterns; must be nonzero
HYPERLINKBASE	""	String	Drawing	Specifies the path used for all relative hyperlinks in the drawing
IMAGEHLT	0	Integer	Registry	Controls whether the entire raster image or only the raster image frame is highlighted
INDEXCTL	0	Integer	Drawing	Controls whether layer and spatial indexes are created and saved in drawing files
INETLOCATION	www.acad.com/ acaduser			Saves the Browser location used by the Internet
INSBASE	0.0000, 0.0000, 0.0000	3D point	Drawing	Insertion basepoint (set by BASE command) expressed in UCS coordinates for the current space
INSNAME	" "	String		Default block name for DDINSERT or INSERT. The name must conform to symbol-naming conventions. Returns " " if there is no default. Enter . (period) to set no default.
INSUNITS	0	Integer	Registry	When you drag a block from AutoCAD DesignCenter, specifies a drawing units value
INSUNITSDEFSOURCE	0	Integer	Registry	Sets source content units value
INSUNITSDEFTARGET	0	Integer	Registry	Sets target drawing units value
ISAVEBAK	1	Integer	Registry	Optimizes the speed of periodic saves, especially for large drawings in Windows
ISAVEPERCENT	50	Integer	Registry	Specifies the amount of wasted space tolerated in a drawing file
ISOLINES	4	Integer	Drawing	Specifies the number of iso lines first surface on the object
LASTANGLE	0	Real		The end angle of the last arc entered, relative to the XY plane of the current UCS for the current space (read-only)
LASTPOINT	0.0000, 0.0000, 0.0000,	3D point		The last point entered, expressed in UCS coordinates for the current space; referenced by @ during keyboard entry
LASTPROMPT	" "	String		Saves the last string echoed to the command line (read-only)
LENSLENGTH		Real	Drawing	Length of the lens (in millimeters) used in perspective viewing, for current viewport (read-only)
LIMCHECK	0	Integer	Drawing	Limits checking for the current space: ON if 1, OFF if 0
LIMMAX	12,000, 9,000	2D point	Drawing	Upper right drawing limits for the current space, expressed in World coordinates
LIMMIN	0.0000, 0.0000,	2D point	Drawing	Lower left drawing limits for the current space, expressed in World coordinates
LISPRINT				Detrmines whether names and values of AutoLISP-defined functions and variables are preserved when you open a new drawing
LOCALE	"enu"	String		Displays the ISO language code of the current AutoCAD version

Variable	Default Setting	Type	Saved In	Explanation
LOGFILEMODE	0	Integer	Registry	Determines whether the contents of the text window are written to a log file
LOGFILENAME				Determines the path for the log file (read-only)
LOGFILEPATH		String	Registry	Specifies the path for the log files for all drawings in a session
LOGINNAME		String		Displays the user's name as configured or input when AutoCAD is loaded (read-only)
LTSCALE	1.000	Real	Drawing	Sets global linetype scale factor
LUNITS	2	Integer	Drawing	Linear units units mode (1 = scientific, 2 = decimal, 3 = engineering, 4 = architectural, 5 = fractional)
LUPREC	4	Integer	Drawing	Linear units decimal places
LWDEFAULT	25	Enum	Registry	Sets the value for the default linweight
LWDISPLAY	0	Integer	Drawing	Controls whether the lineweight is displayed in the Model or Layout tab
LWUNITS	1	Integer	Registry	Controls whether lineweight units are displayed in inches or millimeters
MAXACTVP	64	Integer	Drawing	Maximum number of viewports to regenerate at one time (read-only)
MAXSORT	200	Integer	Config	Maximum number of symbol/file names to be sorted by listing commands; if the total number of items exceeds this number, then none of the items are sorted (default value is 200)
MBUTTONPAN	1	Integer	Registry	Controls the behavior of the third button or wheel on the pointing device
MEASUREMENT	0	Integer	Drawing	Sets drawing units as English or metric
MENUCTL	1	Integer	Config	Controls the page switching of the screen menu 0 = Screen menu doesn't switch pages in response to keyboard command entry 1 = Screen menu switches pages in response to keyboard command entry
MENUECHO	0	Integer		Menu echo/prompt control bits (sum of the following): 1 = Suppresses echo of menu items (^P in a menu item toggles echoing) 2 = Suppresses printing of system prompts during menu 4 = Disables ^P toggle of menu echoing 8 = Displays input/output strings The default value is 0 (all menu items and style prompts are displayed)
MENUNAME	"Acad"	Integer	Drawing	The name of the currently loaded menu file; includes a drive/path prefix if you entered it (read-only)
MIRRTEXT	1	Integer	Drawing	Mirror reflects text if nonzero, retains text direction if 0

Variable	Default Setting	Type	Saved In	Explanation
MODEMACRO		String		Allows you to display a text string in the status line, such as the name of the current drawing, time/date stamp, or special modes. You can use MODEMACRO to display a simple string of text, or use special text strings written in the DIESEL macro language to have AutoCAD evaluate the macro from time to time and base the status line on user-selected conditions
MTEXTED	"Internal"	String	Config	Sets the name of the program to use for editing mtext objects
NOMUTT	0	Short		Suppresses the message display (muttering) when it wouldn't normally be suppressed
OFFSETDIST	1.0000	Real		Sets the default offset distance; if you enter a negative value, it defaults to Through mode
OFFSETGAPTYPE	0	Integer	Registry	Controls how to offset polylines when a gap is created as a result of offsetting the individual polyline segments
OLEHIDE	0	Integer	Registry	Controls the display of OLE objects in AutoCAD
OLEQUALITY	1	Integer	Registry	Controls the default quality level for embedded OLE objects
OLESTARTUP	0	Integer	Drawing	Controls whether the source application of an embedded OLE object loads when plotting
ORTHOMODE	0	Integer	Drawing	Ortho mode ON if 1, OFF if 0
OSMODE	0	Integer	Registry	Object Snap modes bitcode (sum of the following): 1 = Endpoint 128 = Perpendicular 2 = Midpoint 256 = Tangent 4 = Center 512 = Nearest 8 = Node 1024 = Quick 16 = Quadrant 2048 = Appint 32 = Intersection 4096 = Extension 64 = Insertion 8192 = Parallel
OSNAPCOORD	2	Integer	Registry	Controls whether coordinates entered on the command line override running object snaps
PAPERUPDATE	0	Integer	Registry	Controls the display of a warning dialog when attempting to print a layout with a paper size different from the paper size specified by the default for the plotter configuration file
PDMODE	0	Integer	Drawing	Point entity display mode
PDSIZE	0.0000	Real	Drawing	Point entity display size
PELLIPSE	0	Integer	Drawing	Controls the ellipse type created with ELLIPSE 0 = Creates a true ellipse object 1 = Creates a polyline representation of an ellipse
PERIMETER		Real		Perimeter computed by Area, List, or Dblist (read-only)
PFACEMAX	4	Integer		Maximum number of vertices per face (read-only)

Variable	Default Setting	Type	Saved In	Explanation
PICKADD	1	Integer	Config	Controls additive selection of objects 0 = Disables PICKADD. The most recently selected objects, either by an individual pick or windowing, become the selection set. Previously selected objects are removed from the selection set. You can add more objects to the selection set, however, by holding down SHIFT while selecting 1 = Enables PICKADD. Each object you select, either individually or by windowing, is added to the current selection set. To remove objects from the selection set, hold down SHIFT while selecting
PICKAUTO	1	Integer	Config	Controls automatic windowing when the "Select objects:" prompt appears 0 = Disables PICKAUTO 1 = Allows you to draw a selection window (both window and crossing window) automatically at the "Select objects:" prompt
PICKBOX	3	Integer	Config	Object selection target height, in pixels
PICKDRAG	0	Integer	Config	Controls the method of drawing a selection window 0 = You draw the selection window by clicking the mouse at one corner and then at the other corner 1 = You draw the selection window by clicking at one corner, holding down the mouse button, dragging, and releasing the mouse button at the other corner
PICKFIRST	1	Integer	Config	Controls the method of object selection so that you can select objects first and then use an edit/inquiry command 0 = Disables PICKFIRST 1 = Enables PICKFIRST
PICKSTYLE	1	Integer	Registry	Controls group selection and associative hatch selection
PLATFORM		String		Read-only message that indicates which version of AutoCAD is in use
PLINEGEN	1	Integer	Drawing	Sets the linetype pattern generation around the vertices of a 2D polyline. When set to 1, PLINEGEN causes the netype to be generated in a continuous pattern around the vertices of the polyline. When set to 0, polylines are generated with the linetype to start and end with a dash at each vertex. PLINEGEN doesn't apply to polylines with tapered segments
PLINETYPE	2	Integer	Registry	Determines whether AutoCAD uses optimized 2D polylines
PLINEWID	0.0000	Real	Drawing	Default polyline width; it can be zero
PLOTID		String	Config	Changes the default plotter, based on its assigned description
PLOTROTMODE	1	Integer	Registry	Controls orientation of plots

Variable	Default Setting	Type	Saved In	Explanation
PLOTTER		Integer	Config	Changes the default plotter, based on its assigned integer (0–maximum configured); you can create up to 29 configurations
PLQUIET	0	Integer	Registry	Controls the display of optional dialog boxes and nonfatal errors for batch plotting and scripts
POLARADDANG	null	String	Registry	Contains user-defined polar angles
POLARANG	90.0000	Real	Registry	Sets the polar angle increment
POLARDIST	0.000	Real	Registry	Sets the snap increment when the SNAPSTYL system variable is set to 1 (polar snap)
POLARMODE	1	Integer	Registry	Controls settings for polar and object snap tracking
POLYSIDES	4	Integer		Default number of sides for the POLYGON command; the range is 3–1024
POPUPS	1	Integer		1 if the currently configured display driver supports dialog boxes, the menu bar, pull-down menus, and icon menus; 0 if these advanced user interface features are not available (read-only)
PRODUCT	"AutoCAD"			Returns the product name (read-only)
PROGRAM	"acad"			Returns the program name (read-only)
PROJECTNAME	""	String	Drawing	Saves the current project name
PROJMODE	1	Integer	Config	Sets the current Projection mode for Trim or Extend operations
PROXYGRAPHICS	1	Integer	Drawing	Determines whether images of proxy objects are saved in the drawing
PROXYNOTICE	1	Integer	Registry	Displays a notice when you open a drawing containing custom objects created by an application that is not present
PROXYSHOW	1	Integer	Registry	Controls the display of proxy objects in a drawing
PSLTSCALE	1	Integer	Drawing	Controls paper space linetype scaling 0 = No special linetype scaling 1 = Viewport scaling governs linetype scaling
PSPROLOG	""	String	Config	Assigns a name for a prolog section to be read from the acad.psf file when using the PSOUT command

PSQUALITY	75	Integer	Config	Controls the rendering quality of PostScript images and whether they are drawn as filled objects or as outlines. A zero setting disables PostScript image generation and a nonzero setting enables PostScript generation
				Positive setting:
				    Sets the number of pixels per AutoCAD drawing unit for the PostScript resolution
				Negative setting:
				    Still sets the number of pixels per drawing unit, but uses the absolute value; causes AutoCAD to show the PostScript paths as outlines, and doesn't fill them

PSTYLEMODE	0	Integer	Drawing	Indicates whether the current drawing is in a Color-Dependent or Named Plot Style mode (read-only)

Variable	Default Setting	Type	Saved In	Explanation
PSTYLEPOLICY	1	Integer	Registry	Controls whether an object's color property is associated with its plot style
PSVPSCALE	0	Real		Sets the view scale factor for all newly created viewports
PUCSBASE	""	String	Drawing	Stores the name of the UCS that defines the origin and orientation of orthographic UCS settings in paper space only
QTEXTMODE	0	Integer	Drawing	Quick text mode ON if 1, OFF if 0
RASTERPREVIEW	1	Integer	Drawing	Controls whether drawing preview images are saved with the drawing
REFEDITNAME	""	String		Indicates whether a drawing is in a reference-editing state and stores the reference file name (read-only)
REGENMODE	1	Integer	Drawing	Regenauto ON if 1, OFF if 0
RE-INIT		Integer		Reinitializes the I/O ports, digitizer, display, plotter, and *acad.pgp* file using the following bit codes. To specify more than one reinitialization, enter the sum of their values, for example, 3 to specify both digitizer port (1) and plotter port (2) reinitialization 1 = Digitizer port reinitialization 2 = Plotter port reinitialization 4 = Digitizer reinitialization 8 = Display reinitialization 16 = PGP file reinitialization (reload)
RTDISPLAY	1	Integer	Registry	Controls the display of raster images during Realtime ZOOM
SAVEFILE	""	String	Config	Current auto-save file name (read-only)
SAVEFILEPATH	"C\TEMP\"	String	Registry	Specifies the path to the directory for all automatic save files for the AutoCAD session.
SAVENAME		String		The file name you save the drawing to (read-only)
SAVETIME	120	Integer	Config	Automatic save interval, in minutes (or 0 to disable automatic saves). The SAVETIME timer starts as soon as you make a change to a drawing, and is reset and restarts by a manual SAVE, SAVEAS, or QSAVE. The current drawing is saved to auto.sv$
SCREENBOXES		Integer		The number of boxes in the screen menu area of the graphics area. If the screen menu is disabled configured off), SCREENBOXES is zero. On platforms that permit the AutoCAD graphics window to be resized or the screen menu to be econfigured during an editing session, the value of this variable might change during the editing session (read-only)
SCREENMODE		Integer		A (read-only) bit code indicating the graphics/text state of the AutoCAD display. It is the sum of the following bit values: 0 = Text screen is displayed 1 = Graphics mode is displayed 2 = Dual-screen display configuration

Variable	Default Setting	Type	Saved In	Explanation
SCREENSIZE		2D point		Current viewpoint size in pixels, X and Y (read-only)
SDI	0	Integer	Registry	Controls whether AutoCAD runs in single- or multiple-document interface
SHADEDGE	3	Integer	Drawing	0 = Faces shaded, edges not highlighted 1 = Faces shaded, edges drawn in background color 2 = Faces not filled, edges in object color 3 = Faces in entity color, edges in background color
SHADEDIF	70	Integer	Drawing	Ratio of ambient to diffuse light (in percentage of ambient light)
SHORTCUTMENU	11	Integer	Registry	Controls whether Default, Edit, and Command mode shortcut menus are available in the drawing area
SHPNAME	" "	String		Default shape name; must conform to symbol-naming conventions. If no default is set, it returns a " ". Enter . (period) to set no default.
SKETCHINC	0.1000	Real	Drawing	Sketch record increment
SKPOLY	0	Integer	Drawing	Sketch generates lines if 0, polylines if 1
SNAPANG	0	Real	Drawing	Snap/Grid rotation angle (UCS-relative) for the current viewport
SNAPBASE	0.0000, 0.0000,	2D point	Drawing	Snap/Grid origin point for the current viewport (in UCS XY coordinates)
SNAPISOPAIR	0	Integer	Drawing	Current isometric plane (0 = left, 1 = top, 2 = right) for the current viewport
SNAPMODE	0	Integer	Drawing	1 = Snap on for current viewport; 0 = Snap off
SNAPSTYL	0	Integer	Drawing	Snap style for current viewport (0 = standard, 1 = isometric)
SNAPTYPE	0	Integer	Registry	Sets the snap style for the current viewport
SNAPUNIT	0.5000, 0.5000,	2D point	Drawing	Snap spacing for current viewport, X and Y
SOLIDCHECK	1	Integer		Turns the solid validation on and off for the current AutoCAD session
SORTENTS	96	Integer	Drawing	Controls the display of objects sort order perations using the following codes. To select more than one, enter the sum of their codes; for example, enter 3 to specify codes 1 and 2 The default, 96, specifies sort operations for plotting and PostScript output 0 = Disables SORTENTS 1 = Sort for object selection 2 = Sort for object snap 4 = Sort for redraws 8 = Sort for MSLIDE slide creation 16 = Sort for regenerations 32 = Sort for plotting 64 = Sort for PostScript output

Variable	Default Setting	Type	Saved In	Explanation
SPLFRAME	0	Integer	Drawing	If = 1: − the control polygon for spline-fit polylines is to be displayed − only the defining mesh of a surface-fit polygon mesh is displayed (the fit surface is not displayed) − invisible edges of 3D faces are displayed If = 0: − does not display the control polygon for spline-fit polylines − displays the fit surface of a polygon mesh, not the defining mesh − does not display the invisible edges of 3D faces
SPLINESEGS	8	Integer	Drawing	The number of line segments to be generated for each spline patch
SPLINETYPE	6	Integer	Drawing	Type of spline curve to be generated by PEDIT Spline. The valid values are: 5 = Quadratic B-spline 6 = Cubic B-spline
SURFTAB1	6	Integer	Drawing	Number of tabulations to be generated for Rulesurf and Tabsurf; also mesh density in the M direction for Resurf and Edgesurf
SURFTAB2	6	Integer	Drawing	Mesh density in the N direction for Revsurf and Edgesurf
SURFTYPE	6	Integer	Drawing	Type of surface fitting to be performed by PEDIT Smooth. The valid values are: 5 = Quadratic B-spline surface 6 = Cubic B-spline surface 8 = Bezier surface
SURFU	6	Integer	Drawing	Surface density in the M direction
SURFV	6	Integer	Drawing	Surface density in the N direction
SYSCODEPAGE		String	Drawing	Indicates the system code page specified in acad.xmf (read-only)
TABMODE	0	Integer		Controls the use of tablet mode 0 = Disables tablet mode 1 = Enables tablet mode
TARGET	0.0000, 0.0000, 0.0000,	3D point	Drawing	Location (in UCS coordinates) of the target (look-at) point for the current viewport (read-only)
TDCREATE		Real	Drawing	Time and date of drawing creation (read-only)
TDINDWG		Real	Drawing	Total editing time (read-only)
TDUCREATE		Real	Drawing	Stores the universal time and date the drawing was created
TDUPDATE		Real	Drawing	Time and date of last update/save (read-only)
TDSURTIMER		Real	Drawing	User elapsed timer (read-only)
TDUUPDATE		Real	Drawing	Stores the universal time and date of the last update/save

Variable	Default Setting	Type	Saved In	Explanation
TEMPPREFIX	" "	String		Contains the directory name (if any) configured for placement of temporary files, with a path separator appended if necessary (read-only)
TEXTEVAL	0	Integer		If = 0, all responses to prompts for text strings and attribute values are taken literally. If = 1, text starting with "(" or "!" is evaluated as an AutoLISP expression, as for nontextual input. Note: The DTEXT command takes all input literally, regardless of the setting of TEXTEVAL
TEXTFILL	1	Integer	Registry	Controls the filling of Bitstream, TrueType, and Adobe Type 1 fonts
TEXTQLTY	50	Integer		Sets the resolution of Bitstream, TrueType, and Adobe Type 1 fonts
TEXTSIZE	0.2000	Real	Drawing	The default height for new text objects drawn with the current text style (meaningless if the style has a fixed height)
TEXTSTYLE	"STANDARD"	String	Drawing	Contains the name of the current text style (read-only)
THICKNESS	0.0000	Real	Drawing	Current 3D thickness
TILEMODE	1	Integer	Drawing	1 = Release 10 compatibility mode (uses Vports) 0 = Enables paper space and viewport entities (uses MVIEW)
TOOLTIPS	1	Integer	Registry	Controls the display of Tool Tips
TRACEWID	0.0500	Real	Drawing	Default trace width
TRACKPATH	0	Integer	Registry	Controls the display of polar and object snap tracking alignment paths
TREEDEPTH	3020	Integer	Drawing	A 4-digit (maximum) code that specifies the number of times the tree-structured spatial index may divide into branches, hence affecting the speed in which AutoCAD searches the database before completing an action. The first two digits refer to the depth of the model space nodes, and the second two digits refer to the depth of paper space nodes. Use a positive setting for 3D drawings and a negative setting for 2D drawings
TREEMAX	10000000	Integer	Registry	Limits memory consumption during drawing regeneration
TRIMMODE	1	Integer	Registry	Controls whether AutoCAD trims selected edges for chamfers and fillets
TSPACEFAC	1	Real		Controls the multiline text line spacing distance measured as a factor of text height
TSPACETYPE	1	Integer		Controls the type of line spacing used in multiline text
TSTACKALIGN	1	Integer	Drawing	Controls the vertical alignment of stacked text
TSTACKSIZE	70	Integer	Drawing	Controls the percentage of stacked text fraction height relative to selected text's current height

Variable	Default Setting	Type	Saved In	Explanation
UCSAXISANG	90	Integer	Registry	Stores the default angle when rotating the UCS around one of its axes using the X, Y, or Z options of the UCS command
UCSBASE	"World"	String	Drawing	Stores the name of the UCS that defines the origin and orientation of orthographic UCS settings
UCSFOLLOW	0	Integer	Drawing	The setting is maintained separately for both spaces and can be accessed in either space, but the setting is ignored while in paper space (it is always treated as if set to 0)
UCSICON	3	Integer	Drawing	The coordinate system icon bitcode for the current viewport (sum of the following): 1 = On — icon display enabled 2 = Origin — if icon display is enabled, the icon floats to the UCS origin if possible 3 = On and displayed at origin
UCSNAME	" "	String	Drawing	Name of the current coordinate system for the current space; returns a null string if the current UCS is unnamed (read-only)
UCSORG	0.0000, 0.0000, 0.0000,	3D point	Drawing	The origin point of the current coordinate system for the current space; this value is always returned in World coordinates (read-only)
UCSORTHO	1	Integer	Registry	Determines whether the related orthographicUCS setting is restored automatically when an orthographic view is restored
UCSVIEW	1	Integer	Registry	Determines whether the current UCS is saved with a named view
UCSVP	1	Integer	Drawing	Determines whether the UCS in active viewports remains fixed or changes to reflect the UCS of the currently active viewport
UCSXDIR	1.0000, 0.0000, 0.0000,	3D point	Drawing	The X direction of the current UCS for the current space (read-only)
UCSYDIR	0.0000, 1.0000, 0.0000,	3D point	Drawing	The Y direction of the current UCS for the current space (read-only)
UNDOCTL	1	Integer		A (read-only) code indicating the state of the UNDO feature; it is the sum of the following values: 0 = undo is turned off 1 = Set if UNDO is enabled 2 = Set if only one command can be undone 4 = Set if Auto-group mode is enabled 8 = Set if a group is currently active
UNDOMARKS		Integer		The (read-only) number of marks that have been placed in the UNDO control stream by the UNDO command Mark option. The Mark and Back options are unavailable if a group is currently active

Variable	Default Setting	Type	Saved In	Explanation
UNITMODE	0	Integer	Drawing	0 = Displays fractional, feet and inches, and surveyor's angles as previously 1 = Displays fractional, feet and inches, and surveyor's angles in input format
USERI1-5	0	Integer	Drawing	Saves and recalls integer values
USERR1-5	0.0000	Real	Drawing	Saves and recalls real numbers
USERS1-5	" "	String		Saves and recalls text string data
VIEWCTR		3D point	Drawing	Center of view in current viewport, expressed in UCS coordinates (read-only)
VIEWDIR	3D Vector	3D point	Drawing	The current viewport's viewing direction expressed in World coordinates; this describes the camera point as a 3D offset from the TARGET point (read-only)
VIEWMODE	0	Integer	Drawing	Viewing mode bitcode for the current viewport (read-only); the value is the sum of the following: 0 = turned off 1 = Perspective view active 2 = Front clipping on 4 = Back clipping on 8 = UCS follow mode 16 = Front clip not at eye. If On, the front clip distance (FRONTZ) determines the front clipping plane. If Off, FRONZ is ignored and the front clipping is set to pass through the camera point (i.e., vectors behind the camera are not displayed). This flag is ignored if the front clipping bit (2) is off
VIEWSIZE		Real	Drawing	Height of view in current viewport, expressed in drawing units (read-only)
VIEWTWIST	0	Real	Drawing	View twist angle for the current viewport (read-only)
VISRETAIN	1	Integer	Drawing	If = 0, the current drawing's On/Off, Freeze/Thaw, color, and linetype settings for Xref-dependent layers take precedence over the Xref's layer definition; if = 1, these settings don't take precedence
VSMAX		3D point		The upper right corner of the current viewport's virtual screen, expressed in UCS coordinates (read-only)
VSMIN		3D point		The lower left corner of the current viewport's virtual screen, expressed in UCS coordinates (read-only)
WHIPARC	0	Integer	Registry	Controls whether the display of circles and arcs is smooth
WMFBKGND	1	Integer		Controls the background of the output Windows metafile resulting from the WMFOUT command, and for the metafile format of objects placed on the Clipboard or dragged and dropped into other applications
WORLDUCS	1	Integer		If = 1, the current UCS is the same as the WCS; if = 0, it is not (read-only)

Variable	Default Setting	Type	Saved In	Explanation
WORLDVIEW	1	Integer	Drawing	Dview and Vpoint command input is relative to the current UCS. If this variable is set to 1, the current UCS is changed to the WCS for the duration of a DVIEW or VPOINT command. Default value = 1
WRITESTAT	1	Integer	Read-only	Indicates whether a drawing file is read-only or can be written to, for developers who need to determine write status through AutoLISP
XCLIPFRAME	0	Integer	Drawing	Sets visibilty of Xref clipping boundaries
XEDIT	1	Integer	Drawing	Controls whether the current drawing can be edited in-place when being referenced by another drawing
XFADECTL	50	Integer	Registry	Controls the fading intensity for references being edited in-place
XLOADCTL	1	Integer	Registry	Turns XREF demand loading on/off
XLOADPATH				Creates a path for storing temporary copies of demand-loaded Xref files
XREFCTL	0	Integer	Config	Controls whether *.xlg* files (external reference log files) are written 0 = Xref log (*.xlg*) files not written 1 = Xref log (*.xlg*) files written
ZOOMFACTOR	10	Integer	Registry	Controls the incremental change in zoom with each IntelliMouse wheel action, whether forward or backward

APPENDIX e

Hatch and Fill Patterns

INTRODUCTION

AutoCAD supports two types of hatch patterns: vector patterns and PostScript fill patterns. Vector patterns are made of straight lines and dots; they are defined in the Acad.Pat pattern file. You can create custom hatch patterns or purchase patterns created by third-party vendors. You place a hatch pattern with the HATCH and BHATCH commands.

PostScript fill patterns are made via PostScript PDL (page description language); they are defined in the Acad.Psf file. To create a custom fill pattern, you need to know PostScript programming. You place a fill pattern with the PSFILL command.

Over 60 hatch patterns and 12 PostScript fills are supplied with the AutoCAD package. Some are shown on the following pages.

HATCH PATTERNS

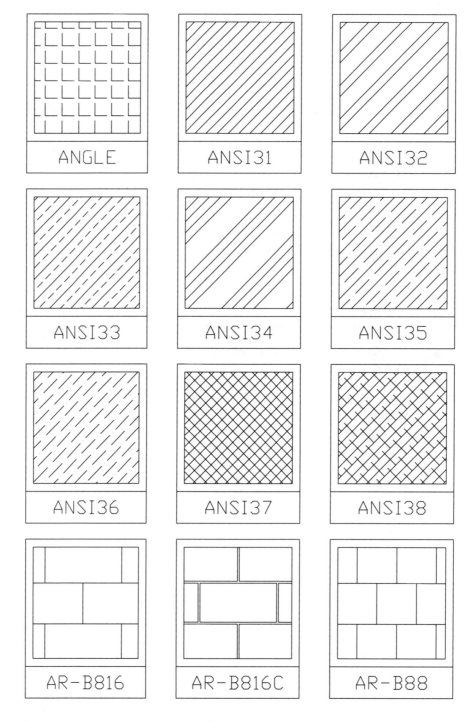

ANGLE   ANSI31   ANSI32

ANSI33   ANSI34   ANSI35

ANSI36   ANSI37   ANSI38

AR-B816   AR-B816C   AR-B88

1320

CLAY

CORK

CROSS

DASH

DOLMIT

DOTS

EARTH

ESCHER

FLEX

GRASS

GRATE

HEX

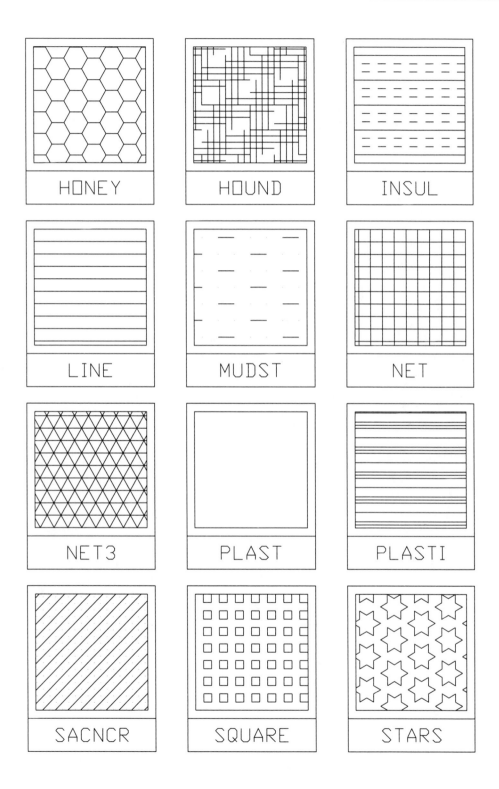

HONEY	HOUND	INSUL
LINE	MUDST	NET
NET3	PLAST	PLASTI
SACNCR	SQUARE	STARS

STEEL

SWAMP

TRANS

TRIANG

ZIGZAG

POSTSCRIPT FILL PATERNS

Square	Waffle	Zigzag
Square	Waffle	Zigzag
Stars	Brick	Specks
Stars	Brick	Specks

Fonts

INTRODUCTION

AutoCAD works with two types of text fonts: the original SHX-format font files and PFB PostScript font files. The AutoCAD package includes 17 SHX text fonts, five SHX symbol fonts, and 16 PFB text fonts. Some of the available fonts are shown on the following pages.

USING SHX AND PFB FILES

Other SHX font files are available from third-party developers. In addition, AutoCAD can use PostScript fonts from any source, many of which are included free with other software packages. Postscript fonts are usually stored in the \Psfonts folder.

AutoCAD does not store text fonts in the drawing file. Instead, the DWG file references SHX font definition files stored elsewhere on the hard drive. Thus, if you receive a drawing from another AutoCAD system, you might have to tell AutoCAD where to find the font files on your system.

PostScript fonts placed in an AutoCAD drawing have two anomilies: the fonts are unfilled and they are drawn 30% too small. To compensate for the reduced size, specify a text height 50% larger.

STANDARD TEXT FONTS

FAST FONTS

TXT   ABCDEFGHIJKLMNOPQRSTUVWXYZ 1234567890

MONOTXT   ABCDEFGHIJKLMNOPQRSTUVWXYZ 1234567890

SIMPLEX FONTS

ROMANS   ABCDEFGHIJKLMNOPQRSTUVWXYZ 1234567890

SCRIPTS   *ABCDEFGHIJKLMNOPQRSTUVWXYZ* 1234567890

GREEKS   ABΧΔΕΦΓΗΙϑΚΛΜΝΟΠΘΡΣΤΥΩΞΨΖ 1234567890

DUPLEX FONTS

ROMAND   **ABCDEFGHIJKLMNOPQRSTUVWXYZ 1234567890**

COMPLEX FONTS

ROMANC   ABCDEFGHIJKLMNOPQRSTUVWXYZ 1234567890

ITALICC   *ABCDEFGHIJKLMNOPQRSTUVWXYZ 1234567890*

SCRIPTC   *ABCDEFGHIJKLMNOPQRSTUVWXYZ 1234567890*

GREEKC   ABΧΔΕΦΓΗΙϑΚΛΜΝΟΠΘΡΣΤΥΩΞΨΖ 1234567890

TRIPLEX FONTS

ROMANT   **ABCDEFGHIJKLMNOPQRSTUVWXYZ 1234567890**

ITALICT   ***ABCDEFGHIJKLMNOPQRSTUVWXYZ 1234567890***

STANDARD TEXT FONTS

GOTHIC FONTS

GOTHICE   ABCDEFGHIJKLMNOPQRSTUVWXYZ 1234567890

GOTHICG   ABCDEFGHIJKLMNOPQRSTUVWXYZ 1234567890

GOTHICI   ABCDEFGHIJKLMNOPQRSTUVWXYZ 1234567890

SYMBOL FONTS

A B C D E F G H I J K L M N O P Q R S T U V W X Y Z
a b c d e f g h i j k l m n o p q r s t u v w x y z

SYASTRO

SYMAP

SYMATH

SYMETEO

SYMUSIC

STANDARD TEXT FONTS

POSTSCRIPT FONTS

CIBT.PFB ABCDEFGHIJKLMNOPQRSTUVWXYZ 1234567890

COBT.PFB ABCDEFGHIJKLMNOPQRSTUVWXYZ 1234567890

EUR.PFB ABCDEFGHIJKLMNOPQRSTUVWXYZ 1234567890

EURO.PFB ABCDEFGHIJKLMNOPQRSTUVWXYZ 1234567890

PAR.PFB ABCDEFGHIJKLMNOPQRSTUVWXYZ 1234567890

ROM.PFB ABCDEFGHIJKLMNOPQRSTUVWXYZ 1234567890

ROMB.PFB ABCDEFGHIJKLMNOPQRSTUVWXYZ 1234567890

ROMI.PFB ABCDEFGHIJKLMNOPQRSTUVWXYZ 1234567890

SAS.PFB ABCDEFGHIJKLMNOPQRSTUVWXYZ 1234567890

SASB.PFB ABCDEFGHIJKLMNOPQRSTUVWXYZ 1234567890

SASBO.PFB ABCDEFGHIJKLMNOPQRSTUVWXYZ 1234567890

SASO.PFB ABCDEFGHIJKLMNOPQRSTUVWXYZ 1234567890

SUF.PFB ABCDEFGHIJKLMNOPQRSTUVWXYZ 1234567890

TE.PFB ABCDEFGHIJKLMNOPQRSTUVWXYZ 1234567890

TEB.PFB ABCDEFGHIJKLMNOPQRSTUVWXYZ 1234567890

CYRILLIC FONTS

CYRILLIC АБВГДЕЖЗИЙКЛМНОПРСТУФХЦЧШЩ 1234567890

CYRILTLC АБЧДЕФГХИЩКЛМНОПЦРСТУВШЖЙЗ 1234567890

Linetypes and Lineweights

INTRODUCTION

Linetypes are defined by the ACAD.LIN file. In addition to the continuous linetype, the AutoCAD program comes with the 45 linetypes. Some of them are listed on the next page. You can add custom linetypes to the ACAD.LIN file.

Before you can use a linetype in a drawing, it must be loaded by means of the LINETYPE command. Set the linetype scale with the LTSCALE command; set independent linetype scaling in paper space with the PSLTSCALE system variable; control the generation of linetype along a polyline with the PLINEGEN system variable.

AutoCAD Standard Lineweights

mm	inch		ISO
0.00			
0.05	.002		
0.09	.003		
0.13	.005		
0.15	.006		
0.18	.007		X
0.20	.008		
0.25	.010		X
0.30	.012		
0.35	.014		X
0.40	.016		
0.50	.020		X
0.53	.021		
0.60	.024		
0.70	.028		X
0.80	.031		
0.90	.035		
1.00	.039		X
1.06	.042		
1.20	.047		
1.40	.056		X
1.58	.062		
2.00	.078		X
2.11	.083		

STANDARD LINETYPES

BORDER	
BORDER2	
BORDERX2	
CENTER	
CENTER2	
CENTERX2	
DASHDOT	
DASHDOT2	
DASHDOTX2	
DASHED	
DASHED2	
DASHEDX2	
DIVIDE	
DIVIDE2	
DIVIDEX2	
DOT	
DOT2	
DOTX2	
HIDDEN	
HIDDEN2	
HIDDENX2	
PHANTOM	
PHANTOM2	
PHANTOMX2	

Command Aliases

ALIAS NAME	COMMAND NAME
3A	3DARRAY
3DO	3DORBIT
3F	3DFACE
3P	3DPOLY
A	ARC
ADC	ADCENTER
AA	AREA
AL	ALIGN
AP	APPLOAD
AR	ARRAY
ATT	ATTDEF
-ATT	-ATTDEF
ATE	ATTEDIT
-ATE	-ATTEDIT
ATTE	-ATTEDIT
B	BLOCK
-B	-BLOCK
BH	BHATCH
BO	BOUNDARY
-BO	-BOUNDARY
BR	BREAK
C	CIRCLE
CH	PROPERTIES

ALIAS NAME	COMMAND NAME
-CH	CHANGE
CHA	CHAMFER
COL	COLOR
COLOUR	COLOR
CO	COPY
D	DIMSTYLE
DAL	DIMALIGNED
DAN	DIMANGULAR
DBA	DIMBASELINE
DBC	DBCONNECT
DCE	DIMCENTER
DCO	DIMCONTINUE
DDI	DIMDIAMETER
DED	DIMEDIT
DI	DIST
DIV	DIVIDE
DLI	DIMLINEAR
DO	DONUT
DOR	DIMORDINATE
DOV	DIMOVERRIDE
DR	DRAWORDER
DRA	DIMRADIUS
DS	DSETTINGS
DST	DIMSTYLE
DT	DTEXT
DV	DVIEW
E	ERASE
ED	DDEDIT
EL	ELLIPSE
EX	EXTEND
EXIT	QUIT

ALIAS NAME	COMMAND NAME
EXP	EXPORT
EXT	EXTRUDE
F	FILLET
FI	FILTER
G	GROUP
-G	-GROUP
GR	DDGRIPS
H	BHATCH
-H	HATCH
HE	HATCHEDIT
HI	HIDE
I	INSERT
-I	-INSERT
IAD	IMAGEADJUST
IAT	IMAGEATTACH
ICL	IMAGECLIP
IM	IMAGE
-IM	-IMAGE
IMP	IMPORT
IN	INTERSECT
INF	INTERFERE
IO	INSERTOBJ
L	LINE
LA	LAYER
-LA	-LAYER
LE	QLEADER
LEN	LENGTHEN
LI	LIST
LINEWEIGHT	LWEIGHT
LO	-LAYOUT
LS	LIST

ALIAS NAME	COMMAND NAME
LT	LINETYPE
-LT	-LINETYPE
LTYPE	LINETYPE
-LTYPE	-LINETYPE
LTS	LTSCALE
LW	LWEIGHT
M	MOVE
MA	MATCHPROP
ME	MEASURE
MI	MIRROR
ML	MLINE
MO	PROPERTIES
MS	MSPACE
MT	MTEXT
MV	MVIEW
O	OFFSET
OP	OPTIONS
ORBIT	3DORBIT
OS	OSNAP
-OS	-OSNAP
P	PAN
-P	-PAN
PA	PASTESPEC
PARTIALOPEN	-PARTIALOPEN
PE	PEDIT
PL	PLINE
PO	POINT
POL	POLYGON
PR	OPTIONS
PRCLOSE	PROPERTIESCLOSE
PROPS	PROPERTIES

ALIAS NAME	COMMAND NAME
PRE	PREVIEW
PRINT	PLOT
PS	PSPACE
PU	PURGE
R	REDRAW
RA	REDRAWALL
RE	REGEN
REA	REGENALL
REC	RECTANGLE
REG	REGION
REN	RENAME
-REN	-RENAME
REV	REVOLVE
RM	DDRMODES
RO	ROTATE
RPR	RPREF
RR	RENDER
S	STRETCH
SC	SCALE
SCR	SCRIPT
SE	DSETTINGS
SEC	SECTION
SET	SETVAR
SHA	SHADE
SL	SLICE
SN	SNAP
SO	SOLID
SP	SPELL
SPL	SPLINE
SPE	SPLINEDIT
ST	STYLE

ALIAS NAME	COMMAND NAME
SU	SUBTRACT
T	MTEXT
-T	-MTEXT
TA	TABLET
TH	THICKNESS
TI	TILEMODE
TO	TOOLBAR
TOL	TOLERANCE
TOR	TORUS
TR	TRIM
UC	DDUCS
UCP	DDUCSP
UN	UNITS
-UN	-UNITS
UNI	UNION
V	VIEW
-V	-VIEW
VP	DDVPOINT
-VP	VPOINT
W	WBLOCK
-W	-WBLOCK
WE	WEDGE
X	EXPLODE
XA	XATTACH
XB	XBIND
-XB	-XBIND
XC	XCLIP
XL	XLINE
XR	XREF
-XR	-XREF
Z	ZOOM

INDEX